BIOTECHNOLOGY FUNDAMENTALS

BIOTECHNOLOGY FUNDAMENTALS

FIRDOS ALAM KHAN

CRC Press is an imprint of the
Taylor & Francis Group, an **informa** business

CRC Press
Taylor & Francis Group
6000 Broken Sound Parkway NW, Suite 300
Boca Raton, FL 33487-2742

Printed and bound in India by Replika Press Pvt. Ltd.

International Standard Book Number: 978-1-4398-2009-4 (Hardback)

Library of Congress Cataloging-in-Publication Data

Khan, Firdos Alam.
Biotechnology fundamentals / by Firdos Alam Khan.
p. ; cm.
Includes bibliographical references and index.
ISBN 978-1-4398-2009-4 (hardcover)
1. Biotechnology. I. Title.
[DNLM: 1. Biotechnology. TP 248.2]

TP248.2.K43 2012
660.6--dc22 2011011801

Visit the Taylor & Francis Web site at
http://www.taylorandfrancis.com

and the CRC Press Web site at
http://www.crcpress.com

Contents

Preface

This book has been written to share the information related to all aspects of the field of biotechnology in a concise manner without sacrificing the content. Being a teacher of biotechnology, I had been searching for a book that covers all aspects of biotechnology from academic to industry-related issues. I could never find the book that I was looking for, and therefore, I decided to write one. In addition, there is a genuine requirement for a book that teaches all the aspects of biotechnology at the undergraduate level. Many students decide to follow a career path that is related to the field of biotechnology, such as microbial or industrial biotechnology, agricultural biotechnology, medical biotechnology, and animal biotechnology. Students also want to know how biotechnology products are produced, how scientific research conducted in universities is different from that of industrial research, and which area of biotechnology brings challenging career opportunities. At the same time, they want to learn the laboratory techniques employed in the field of biotechnology, which will enable them to shape their career intelligently. There are a great number of textbooks available that deal with these issues independently. In this book we have discussed all aspects of biotechnology in great detail with excellent illustrations. The idea for *Biotechnology Fundamentals* arose initially from the discussions between various teachers and undergraduate students. It was felt that such a textbook would assist students who want to pursue a challenging career in various aspects of biotechnology. To begin this, I first approached T. Michael Slaughter, executive editor, CRC Press/Taylor & Francis Group, for his opinion. He immediately wrote back to me saying that "the concept of this project is one that I would definitely be interested in." I thus started working on the proposal. Initially, I thought of some other title for this book but Michael suggested the current title, which I found very appropriate and unique.

The field of biotechnology is advancing at a rapid pace and it becomes very important to keep track of all new information pertaining to it. In this book, I have made an attempt to include all the topics that are directly or indirectly related to various fields of biotechnology. The book discusses both conventional and modern aspects of biotechnology and provides suitable examples, giving the impression that the field of biotechnology has existed for ages under different names. In fact, plant breeding, cheese making, *in vitro* fertilization, and alcohol fermentation are all fruits of biotechnology.

The main objective of this book is to help students learn both the classical and modern forms of biotechnology and to expose them to a range of topics from basic information to complex technicalities. There are a total of fifteen chapters in this textbook covering topics ranging from introduction to biotechnology, genes and genomics, proteins and proteomics, recombinant DNA technology, microbial biotechnology, agricultural biotechnology, animal biotechnology, environmental biotechnology, medical biotechnology, nanobiotechnology, product development in biotechnology, industrial biotechnology, ethics in biotechnology, careers in biotechnology, and laboratory tutorials. All chapters begin with a brief summary followed by text with suitable examples. Each chapter is illustrated by simple line diagrams, pictures, and tables and concludes with a question section, an assignment, and field trip information. Brief answers to all questions are provided in a solution manual that is available to instructors. I have included laboratory tutorials as a separate chapter to expose students to various laboratory techniques and protocols. This practical information provides be an added advantage to students while they learn the theoretical aspects of biotechnology.

No one walks alone on the journey of life, and when one is finishing one's walk, it is time to thank those that joined you, walked beside you, and helped you along the way. First, I am grateful to Almighty God who gave me the strength to write this book on stipulated time and helped me remain focused for the entire duration of the writing. I am thankful to all those who helped me in producing this book, especially executive editor T. Michael Slaughter, who encouraged me, provided invaluable suggestions, and gave me an opportunity to write this book. I am also thankful to the entire team of CRC Press/Taylor & Francis Group for making this book a reality, especially Jennifer Ahringer, project coordinator, editorial project development, and Richard Tressider, project editor, for their constant support and understanding. I would like to thank to Dr. Alexandra Gorgevska for her critical review of the manuscript and for her valuable suggestions to improve the quality of this book.

I would like to thank the entire management team, faculty, and staff of Manipal University, India and Dubai, for their constant support, especially Dr. Vinod Bhatt, pro-vice-chancellor, Manipal University, India; Dr. B. Ramjee, director, Manipal University Dubai Campus; and Professor K. Sathyamoorthy, director, Manipal Life Sciences Center, India, for their constant support.

I am thankful to all my teachers and mentors, especially to Professor Nishikant Subhedar, PhD, who introduced me to scientific research, and to Professor Obaid Siddiqi, PhD, FRS, for making me a true scientist. I am also thankful to all my friends, well-wishers, and colleagues for their support and cooperation.

I am grateful to all my family members, especially my parents, my brothers, my sisters, my wife (Samina), and my sons (Zuhayr, Zaid, and Zahid) and daughter (Azraa). All of them, in their own ways, inspired and supported me to complete this project by sacrificing many weekends and holidays.

I dedicate this book to all students who want to pursue a challenging career in the field of biotechnology. I welcome your comments and suggestions to make this book error-free and more interesting in the future. You may send your comments or recommendations to the address below.

Enjoy reading!

Firdos Alam Khan, PhD
Department of Biotechnology
Manipal University Dubai Campus
Dubai, United Arab Emirates
E-mail: firdoskhan@manipaldubai.com

Author

Firdos Alam Khan is a professor and chairperson in the Department of Biotechnology, Manipal University Dubai Campus, United Arab Emirates. He received his doctoral degree from Nagpur University, India, in 1997. He has more than 13 years of research and teaching experience in various domains of biotechnology. He did his first postdoctoral research at the National Centre for Biological Sciences in Bangalore, India, where he worked on a World Health Organization–sponsored research project entitled "Olfactory learning in *Drosophila melanogaster*." In 1998, Dr. Khan moved to the United States and joined the Department of Brain and Cognitive Sciences, Massachusetts Institute of Technology, where he worked on a research project entitled "Axonal nerve regeneration in adult Syrian hamster." In 2001, he returned to India and joined Reliance Life Sciences, a Mumbai-based biotechnology company, where he was associated with adult and embryonic stem cell research projects. Dr. Khan showed that both adult and embryonic stem cells have the ability to differentiate into neuronal cells. He also developed novel protocols to derive neuronal cells from both adult and embryonic stem cells. Over the past five years, he has been associated with the Department of Biotechnology, Manipal University Dubai, as chairperson, and he has been teaching the business of biotechnology course to graduate students. His area of specialty in biotechnology includes stem cell technology, pharmacology, and neuroscience. He has written numerous articles in various national and international journals in the areas of neuroscience, neuropharmacology, and stem cell biology. He has filed a number of patents in the field of stem cell technology in India, Europe, and the United States. He has also been associated with various international scientific organizations like the International Brain Research Organization, France, and the Society for Neuroscience, United States. He has presented at more than twenty different national and international conferences in India, China, Singapore, Thailand, UAE, and the United States.

Author

1 Introduction to Biotechnology

LEARNING OBJECTIVES

- Define biotechnology
- Discuss the historical perspectives of biotechnology
- Explain the classifications of biotechnology based on its applications
- Explain how biotechnology has revolutionized the healthcare, agricultural, and environmental sectors
- Explain how biotechnology became the science of integration of diverse fields
- Explain the rule of ethical issues on biotechnology

1.1 WHAT IS BIOTECHNOLOGY?

The fruits of biotechnology are so evident in everyday life, but sometimes, we do not realize that we are actually benefitting from it such as when we eat yogurt or when we receive a vaccine. Everyone may not be aware of the formal definition of biotechnology, but one thing is certain, we all have benefitted from the products of biotechnology such as cheese, detergents, biodegradable plastics, and antibiotics. It is important to know how these useful products are developed and passed on to us for our own benefits. One may argue that cheese making is no big deal, as cheese can be found in almost every city in the world. In addition, what is the relationship between cheese making and biotechnology? Yes, there is a relationship if you know different ingredients that are required for making cheese. Let us first learn how to make cheese at home, which will give us a fair idea about cheese making, and then we can learn how to make cheese at the industrial level.

Cheese making at home:

- Place one cup of milk in a saucepan; bring the milk slowly to a boil while stirring constantly. It is very important to constantly stir the milk or it will burn.
- Turn the burner off once the milk is boiling, but leave the saucepan on the element or gas grate.
- Add vinegar to the boiling milk, at which point the milk should turn into curd and whey.
- Stir well with spoon and let it sit on the element for 5–10 min.
- Pass the curd and whey through cheesecloth or a handkerchief to separate the curd from the whey.
- Press the cheese using the cloth to get as much of the moisture out.
- Open the cloth and add a pinch of salt if desired.
- Mix the cheese and salt and then press again to remove any extra moisture.
- Put the cheese in a mold or just leave it in the form of a ball.
- Refrigerate for a while before eating.

It is not so easy to make cheese at home. Initially, cheese making in industries is done by acidifying (souring) the milk and adding an acid-like vinegar in a few cases, but sometimes bacteria are also used. These starter bacteria convert milk sugars into lactic acid. The same bacteria and the enzymes they produce also play a large role in the eventual flavor of aged cheeses. Most cheeses are made

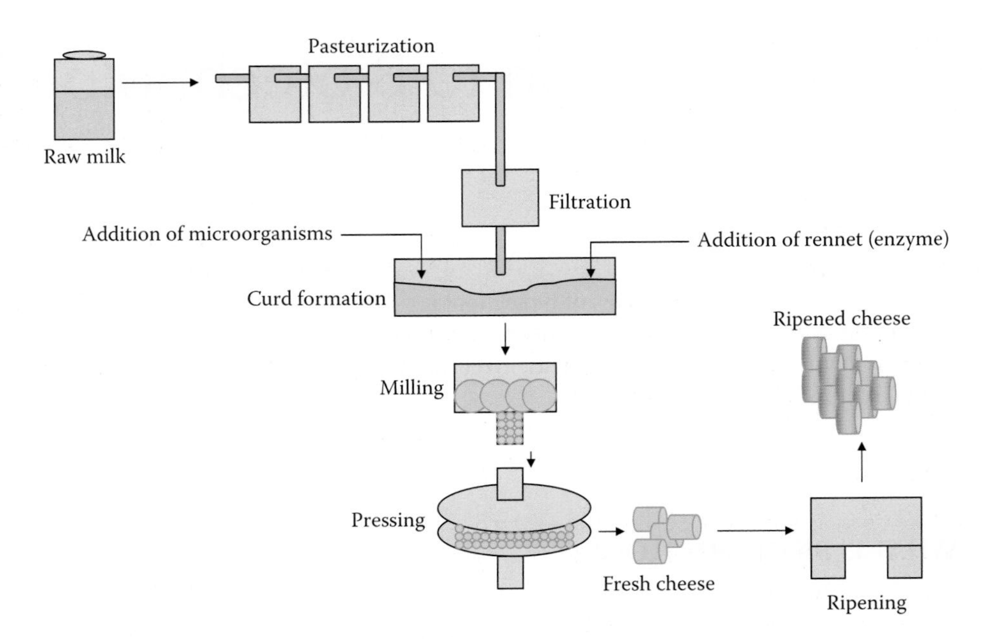

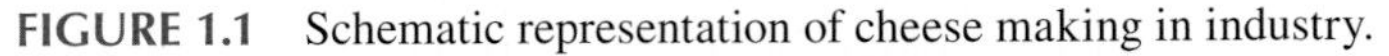

FIGURE 1.1 Schematic representation of cheese making in industry.

with starter bacteria from the *Lactococci*, *Lactobacilli*, or *Streptococci* families. Swiss starter cultures also include *Propionibacterium shermanii*, which produces carbon dioxide gas bubbles during aging, giving Swiss cheese (or Emmental) its holes (Figure 1.1).

You may now fairly know the application of microorganisms in industrial level production of cheese, but the role of microorganisms is not limited to cheese making. It has multiple applications in other products as well such as curd and antibiotic productions. It is now easy to define biotechnology as "the application of microorganisms in industrial level cheese making."

1.1.1 Definitions of Biotechnology

In general, *biotechnology* is a field that involves the use of biological systems or living organisms to manufacture products or develop processes that ultimately benefit humans. The following are some of the most commonly used definitions of biotechnology:

- The use of living organisms (especially microorganisms) in industrial, agricultural, medical, and other technological applications.
- The application of the principles and practices of engineering and technology to the life sciences.
- The use of biological processes to make products.
- The production of genetically modified organisms or the manufacture of products from genetically modified organisms.
- The use of living organisms or their products to make or modify a substance. Biotechnology includes recombinant DNA (deoxyribonucleic acid) techniques (genetic engineering) and hybridoma technology.
- A set of biological techniques developed through basic research and applied to research and product development.
- The use of cellular and biomolecular processes to solve problems or make useful products.
- An industrial process that involves the use of biological systems to make monoclonal antibodies and genetically engineered recombinant proteins.

We should not debate on which of the given definitions is true because all of them are true in their respective senses. For example, if you ask a farmer about what biotechnology is, he or she may say, "Biotechnology is to produce high yield or pest-resistant crops." If you pose the same question to a doctor, he or she may say, "Biotechnology is about making new vaccines and antibiotics." If you ask the question to an engineer, he or she may say, "Biotechnology is about designing new diagnostic tools for better understanding of human diseases," and if you ask the question to a patient suffering from Parkinson's disease, he or she may say "Biotechnology is about stem-cell-based therapy and has tremendous capability to cure Parkinson's disease." All these different definitions of biotechnology suggest that biotechnology has immensely impacted our daily life with arrays of products. As the field of biotechnology keeps expanding, efforts are being made to subclassify this field into various types. The field of biotechnology may be broadly subclassified into animal, agricultural, medical, industrial, and environmental biotechnology.

1.2 ANIMAL BIOTECHNOLOGY

Animal biotechnology is the application of scientific and engineering principles to the processing or production of materials by animals or aquatic species, to provide research models, and to make health products. Some examples of animal biotechnology are generation of transgenic animals (animals with one or more genes introduced by human intervention), use of gene knockout technology to generate animals with a specified gene-inactivated production of nearly identical animals by somatic cell nuclear transfer (also referred to as *clones*), and production of infertile aquatic species. Since the early 1980s, methods have been developed and refined to generate transgenic animals. For example, transgenic livestock and transgenic aquatic species have been generated with increased growth rates, enhanced lean muscle mass, enhanced resistance to disease, or improved use of dietary phosphorous to lessen the environmental impacts of animal manure. Transgenic poultry, swine, goats, and cattle have also been produced to generate large quantities of human proteins in eggs, milk, blood, or urine, with the goal of using these products as human pharmaceuticals. Some examples of human pharmaceutical proteins are enzymes, clotting factors, albumin, and antibodies. The major factor limiting the widespread use of transgenic animals in agricultural production systems is the relatively inefficient rate (success rate <10%) of production of transgenic animals (Figure 1.2).

Animal biotechnology can also knock out or inactivate a specific gene. Knockout technology creates a possible source of replacement organs for humans. The process of transplanting cells, tissues, or organs from one species to another is referred to as *xenotransplantation*. Currently, pigs are the best candidates as xenotransplant donors to humans. Unfortunately, pig cells and human cells are not immunologically compatible. Pig cells express a carbohydrate epitope (alpha1, 3 galactose) on their surface that is not normally found on human cells. Humans will generate antibodies to this epitope, which will result in acute rejection of the xenograft. Genetic engineering is used to knock out or inactivate the pig gene (alpha1, 3 galactosyl transferase) that attaches this carbohydrate epitope on pig cells. Another example of knockout technology in animals is the inactivation of the prion-related peptide gene that may produce animals that are resistant to diseases associated with prions such as bovine spongiform encephalopathy, Creutzfeldt–Jakob disease, etc.

Another application of animal biotechnology is the use of somatic cell nuclear transfer to produce genetically identical copies of an organism. This process has been referred to as *cloning*. To date, somatic cell nuclear transfer has been used to clone cattle, sheep, pigs, goats, horses, mules, cats, rats, and mice. The technique involves culturing somatic cells from an appropriate tissue (fibroblasts) from the animal to be cloned. Nuclei from the cultured somatic cells are then microinjected into an enucleated oocyte obtained from another individual of the same or a closely related species. Through a process that is not yet understood, the nucleus from the somatic cell is reprogrammed to a pattern of gene expression suitable for directing normal development of the embryo. The embryo is further cultured in *in vitro* environment, and then it is transferred to a recipient female for normal fetal development.

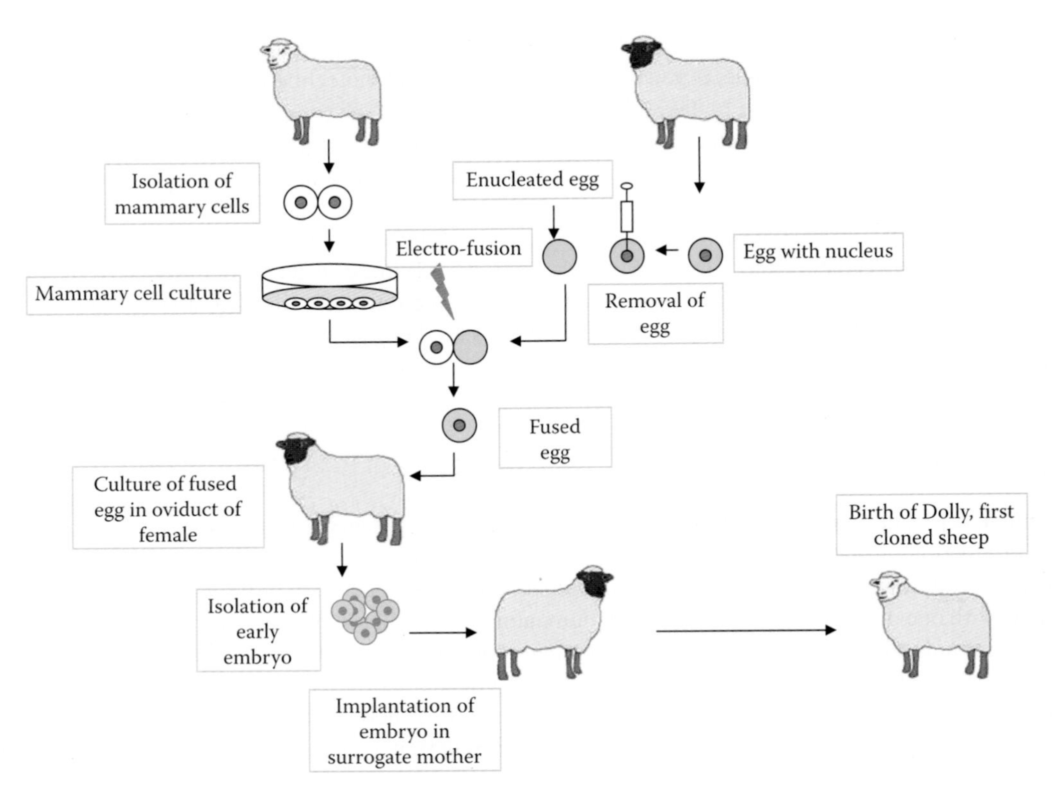

FIGURE 1.2 Animal cloning.

1.3 AGRICULTURAL BIOTECHNOLOGY

Agricultural biotechnology, which is also known as *green biotechnology*, involves the use of environment-friendly solutions as an alternative to traditional industrial agriculture, horticulture, and animal breeding processes. The following are some examples of green biotechnology:

- Use of bacteria to facilitate the growth of plants
- Development of pest-resistant grains
- Engineering of plants to express pesticides
- Accelerated evolution of disease-resistant animals
- Use of bacteria to assure better crop yields (instead of using pesticides and herbicides)
- Production of superior plants by stimulating the early development of their root systems
- Use of plants to remove heavy metals such as lead, nickel, or silver, which can then be extracted (or "mined") from the plants
- Genetic manipulation to allow plant strains to be frost-resistant
- Use of genes from soil bacteria to genetically alter plants to promote tolerance to fungal pathogens
- Use of bacteria to have the plants grow faster, resist frost, and ripen earlier (Figure 1.3)

1.4 MEDICAL BIOTECHNOLOGY

Medical biotechnology deals with the development of therapy using cells or microorganisms by employing molecular engineering techniques. It includes the designing of organisms to manufacture pharmaceutical products like therapeutic proteins, antibiotics, vaccines, regenerative medicine, and gene therapy. Medical biotechnology is also used in forensics through DNA profiling (Figure 1.4).

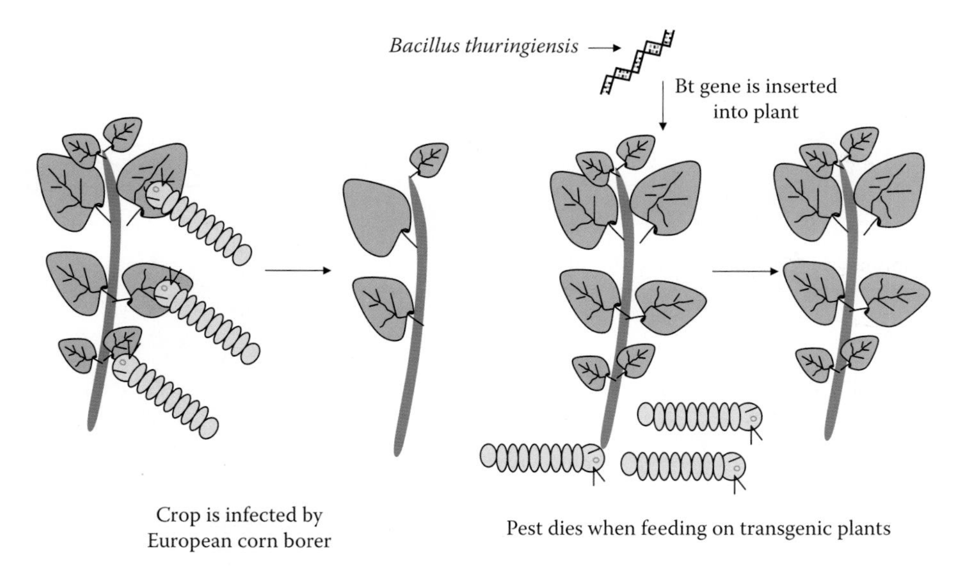

FIGURE 1.3 Engineering-resistant corn. Following the insertion of a gene from the bacteria *Bacillus thuringiensis*, corn becomes resistant to corn borer infection. This allows farmers to use less insecticides.

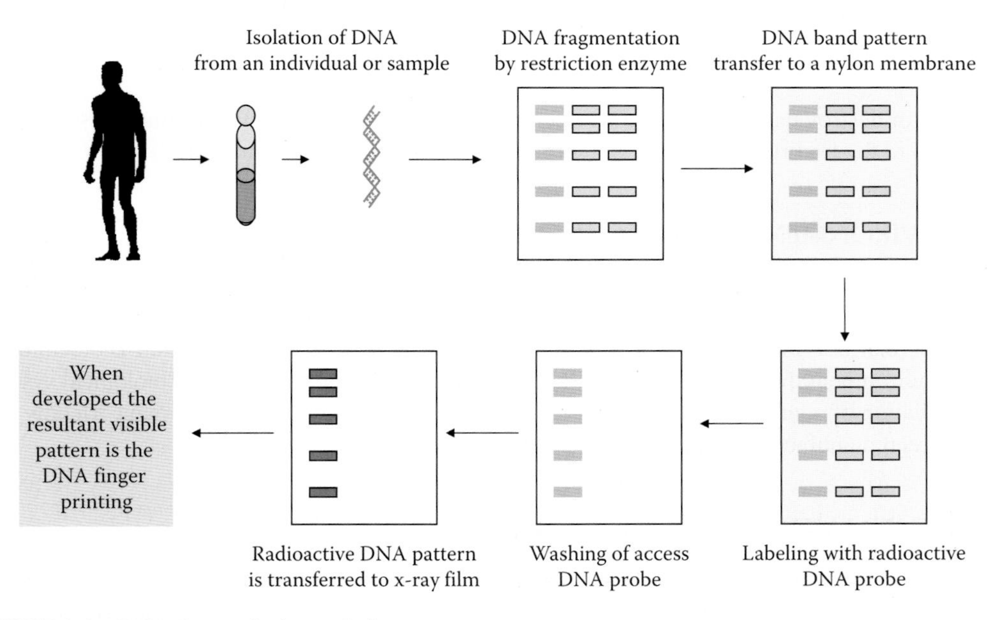

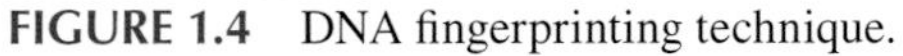
FIGURE 1.4 DNA fingerprinting technique.

1.5 INDUSTRIAL BIOTECHNOLOGY

Industrial biotechnology, which is known mainly in Europe as *white biotechnology*, is the application of biotechnology for industrial purposes such as manufacturing of biomolecules, enzymes or chemicals, and biomaterials. It includes the practice of using cells or components of cells like enzymes to generate industrially useful products. It uses living cells from yeast, moulds, bacteria, plants, and enzymes to synthesize products that are easily degradable, require less energy, and create less waste during their production. Some examples include the designing of an organism to produce a useful chemical and the use of enzymes as industrial catalysts to either produce valuable

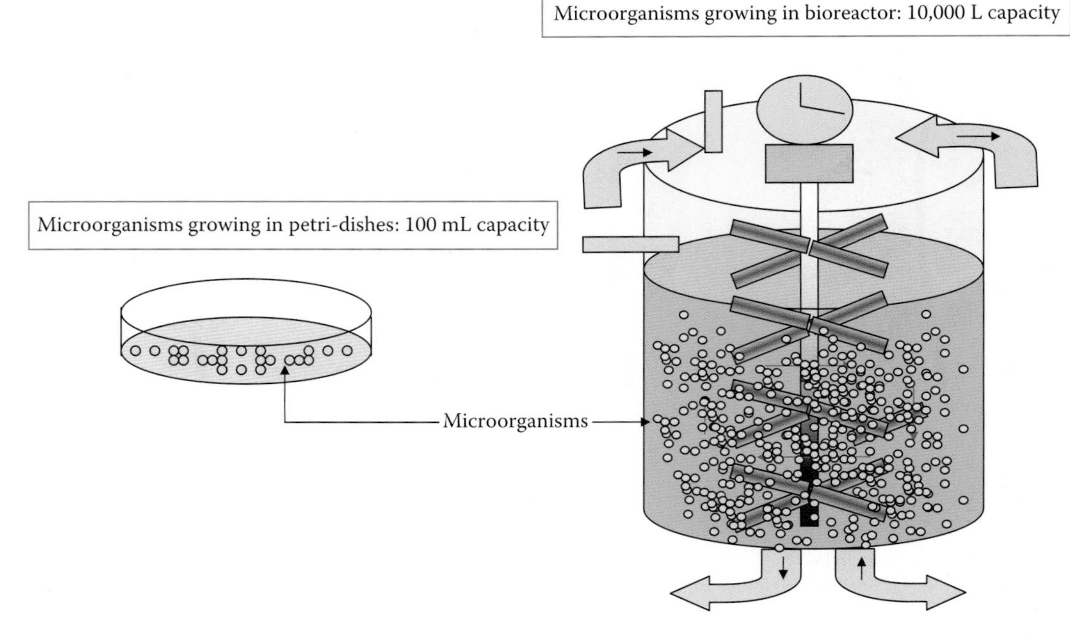

FIGURE 1.5 Industrial biotechnology: Difference between laboratory level production and industrial level production of microorganisms.

chemicals or destroy hazardous/polluting chemicals. White biotechnology consumes less resources (compared to the traditional processes) to produce industrial goods (Figure 1.5).

1.6 ENVIRONMENTAL BIOTECHNOLOGY

Environmental biotechnology is when biotechnology is applied to and used to study the natural environment. Environmental biotechnology could also imply that one try to harness biological process for commercial use and exploitation. The International Society for Environmental Biotechnology defines environmental biotechnology as "the development, use and regulation of biological systems for remediation of contaminated environments (land, air, water), and for environment-friendly processes (green manufacturing technologies and sustainable development)."

1.7 OTHER EMERGING FIELDS OF BIOTECHNOLOGY

In addition to the major fields of biotechnology, there are other fields that are directly or indirectly associated with biotechnology. In this section, we will describe them briefly in order to understand their importance.

1.7.1 Nanobiotechnology

Nanobiotechnology is the branch of nanotechnology with biological and biochemical applications or uses. It also studies existing elements in nature to develop new devices. The term *bionanotechnology* is often used interchangeably with *nanobiotechnology*, though a distinction is sometimes drawn between the two. If the two are distinguished, nanobiotechnology usually refers to the use of nanotechnology to further the goals of biotechnology, while bionanotechnology might refer to any overlap between biology and nanotechnology, including the use of biomolecules as part of or as an inspiration for nanotechnological devices (Figure 1.6).

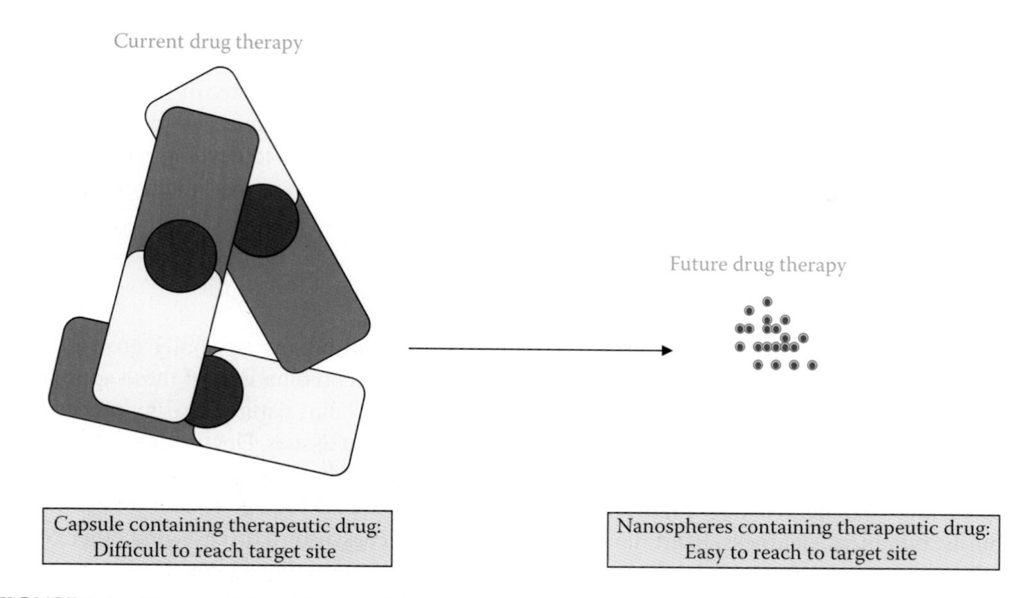

FIGURE 1.6 Nanomedicine: Nanoparticles constructed to carry a therapeutic payload.

1.7.2 Bioinformatics

Bioinformatics is an interdisciplinary field that addresses biological problems by using computational techniques, and makes the rapid organization and analysis of biological data possible. The field may also be referred to as computational biology and can be defined as, "conceptualizing biology in terms of molecules and then applying informatics techniques to understand and organize the information associated with these molecules, on a large scale." Bioinformatics plays a key role in various areas such as functional genomics, structural genomics, and proteomics. It also forms a key component in the biotechnology and drug discovery industries (Figure 1.7).

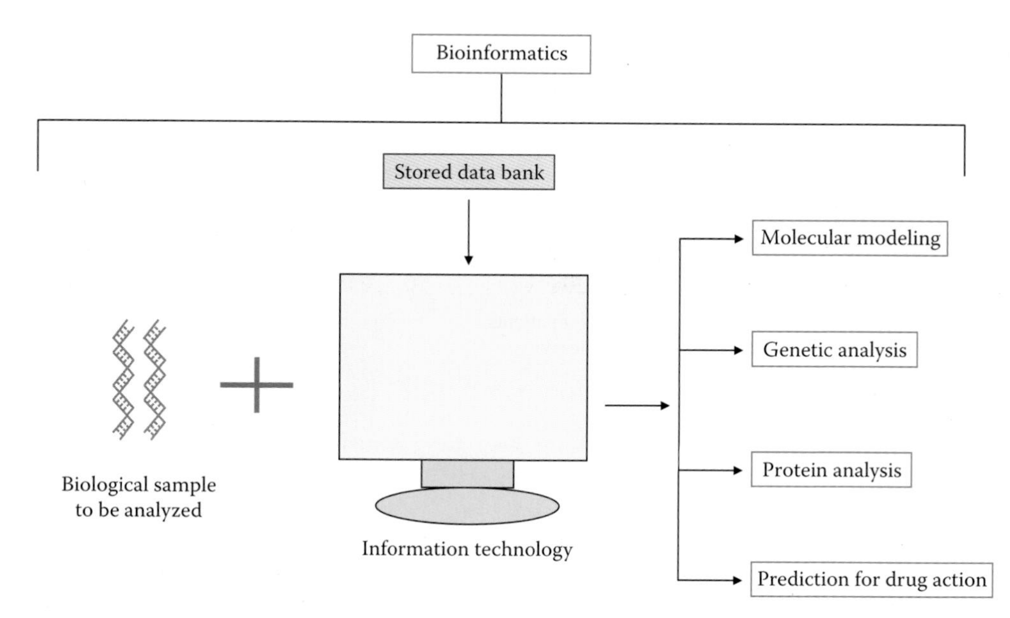

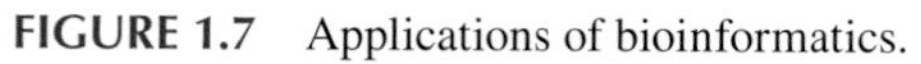

FIGURE 1.7 Applications of bioinformatics.

1.7.3 Pharmacogenomics

Pharmacogenomics is the branch of pharmacology that deals with the influence of genetic variation on drug response in patients by correlating gene expression or single-nucleotide polymorphisms with a drug's efficacy or toxicity. By doing so, pharmacogenomics aims to develop rational means to optimize drug therapy with respect to the patients' genotype to ensure maximum efficacy with minimal adverse effects.

1.7.4 Regenerative Medicine

Regenerative medicine is the study and development of artificial organs, specially grown tissues and cells including stem cells, laboratory-made compounds, and combinations of these approaches for the treatment of injuries and diseases. Regenerative medicine helps natural healing processes to work faster, or uses special materials to regrow missing or damaged tissues. Doctors use regenerative medicine to speed up healing and to help heal injuries that cannot heal on their own. Regenerative therapies have been demonstrated in trials or laboratory experiments to heal or treat broken bones, bad burns, blindness, deafness, heart damage, nerve damage, Parkinson's disease, and other conditions. Recently, stem cell research has generated quite a lot of interest among patients suffering from Parkinson's disease because stem-cell-based therapy may provide cure for these patients. Although it may be too early to comment on the effectiveness of such a novel therapy, trials on animals have proved to be quite interesting (Figure 1.8).

1.7.5 Therapeutic Proteins

Therapeutic proteins are proteins that are either manufactured using genetic engineering by extracting protein from human cells or engineered in the laboratory for pharmaceutical use. The majority of therapeutic proteins are recombinant human proteins manufactured using nonhuman mammalian cell lines that are engineered to express certain human genetic sequences to produce specific proteins. Recombinant proteins are an important class of therapeutics used to replace deficiencies in critical blood borne growth factors and to strengthen the immune system to fight cancer and

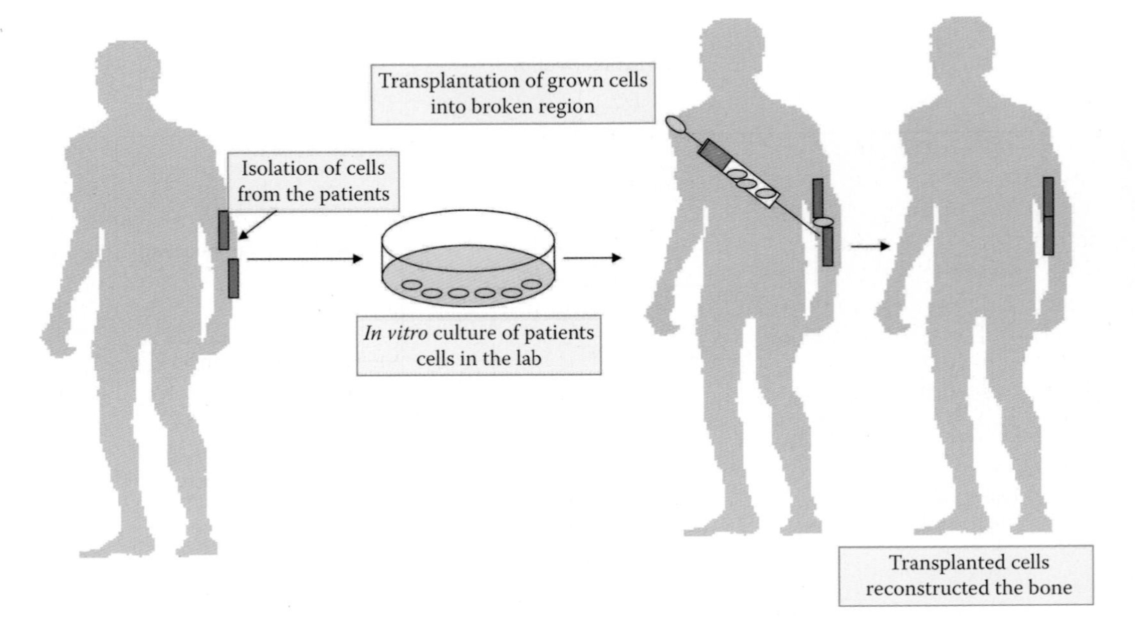

FIGURE 1.8 Regenerative medicine: Reconstruction of bone by using patient's own cells.

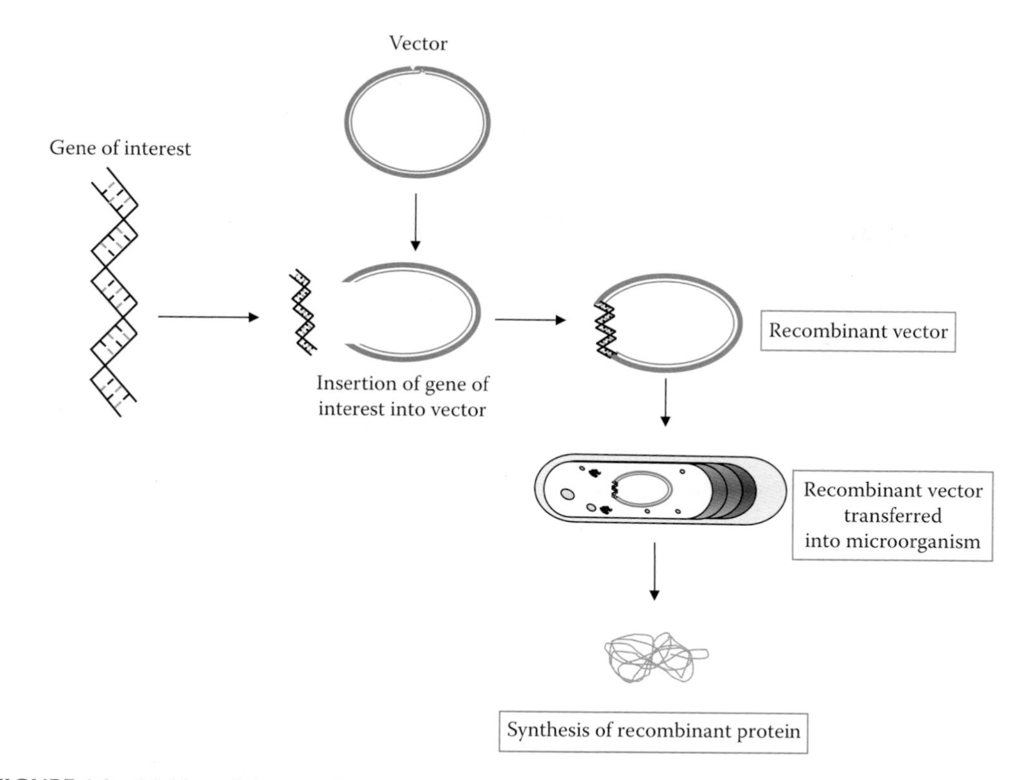

FIGURE 1.9 Making of therapeutic or recombinant protein.

infectious disease. Therapeutic proteins are also used to relieve patients' suffering from many conditions such as various cancers (treated by monoclonal antibodies and interferon), heart attacks, strokes, cystic fibrosis and Gaucher's disease (treated by enzymes and blood factors), diabetes (treated by insulin), anemia (treated by erythropoietin), and hemophilia (treated by blood-clotting factors) (Figure 1.9).

1.8 HISTORY OF BIOTECHNOLOGY

After learning various definitions and applications of biotechnology, we will now go through the historical aspects of biotechnology. Most people still think that biotechnology is a relatively new discipline that is only recently getting a lot of attention. It may surprise you to know that, in many ways, this technology involves several ancient practices and methodologies. In fact, our ancestors have been using biotechnology for their benefit in many processes for many centuries. Although the word "biotechnology" was not yet in use at that time, the application itself already existed. One of the early applications of biotechnology was the use of microorganisms in making bread, cheese, yogurt, and alcoholic beverages.

We can view the development of biotechnology through the traditional as well as the modern window. One of the common links between traditional and modern biotechnology is that in both periods, man has been exploiting organisms to generate products. In modern biotechnology, human manipulations of the genes of organisms and their insertions into other organisms are used to acquire desired traits, whereas in traditional biotechnology, microorganisms were used in fermentation.

1.8.1 Ancient Biotechnology

An old saying quotes, "Necessity is the mother of all inventions." Over the centuries, our ancestors have been using breeding techniques based on the phenotype characteristics to create animals and plants of desirable traits (high milk-producing cows) and benefits (high-yield crops). During this period, the best animals and plants have been bred together, and each successive generation has been more likely to carry the desirable traits of the parent animal or plant. A hundred years ago, an organism's DNA would have been scanned first to look for desirable traits, and then organisms with those traits would have been bred. Today this is no longer necessary since we can now genetically engineer animals.

Another form of biotechnology that has been around for thousands of years is the use of microorganisms in food. Microorganisms are used to turn milk into cheese and yogurt, and to ferment alcohol. Yeast is used in bread to make it rise. These are considered as biotechnology because they utilize microorganisms.

Biotechnology is as old as the ancient cultures of the Indians, Chinese, Greeks, Romans, Egyptians, Sumerians, and other ancient communities of the world. Some examples of ancient biotechnology are the use of microorganisms for fermentation, domesticating animals for livestock, alcohol in the form of wine and beer, herbal remedies, and plant balms for treatment of wounds and ailments. The contribution of other scientific fields have greatly helped the development of biotechnology as it utilizes the sciences of biology, chemistry, physics, engineering, computers, and information technology to develop tools and products that hold great promise and hope for thousands of patients who are suffering from various incurable diseases such as cancer, diabetes, and Parkinson's disease.

1.8.2 Modern Biotechnology

Modern biotechnology deals more with the treatment of ailments and alteration of organisms to better human life. Most breakthroughs in biotechnology have been relatively recent, with the earliest advancement about 170 years ago with the discovery of microbes. Proteins were discovered only in 1830, with the isolation of the first enzyme following closely 3 years later. In 1859, Charles Darwin published his revolutionary book *On the Origin of Species*. Six years later, Gregor Mendel, who is considered the father of modern genetics, discovered the laws of heredity and laid the groundwork for genetic research. Near the turn of the century, Louis Pasteur and Robert Koch provided the basis for research in microbiology. These numerous advancements allowed modern biotechnology to rise. In early twentieth century, the modern biotechnology movement started, particularly in immunology and genetics. Penicillin, computers, discovery of DNA as the genetic basis, use of bacteria to treat raw sewage (bioremediation project) are significant developments in this direction. Revolution in forensics and biomedical science took place with the new laboratory methods such as DNA sequencing, protein analysis, and polymerase chain reaction (PCR). The millennium ended with the introduction of the first cloned sheep (named Dolly), which started the debate over the ethical issues relating to biotechnology, stem cell research, genetic testing, and genetically modified organisms. Modern biotechnology received a big boost when Watson and Crick discovered the double helix of DNA structure, which allowed researchers to study the genetic code of life in great detail and opened an era of genetic engineering, genetic mapping, or genetic manipulation. The twenty-first century started with the development of the rough draft of the human genome, or the map of human life. The milestones in the field of biotechnology are listed in Table 1.1.

1.9 HUMAN GENOME PROJECT

In the quest to chart the innermost reaches of the human cell, scientists have set out biology's most important mapping expedition, the Human Genome Project (HGP). Its mission is to identify the full set of genetic instructions contained inside the human cells and to read the complete text written in the language of the hereditary chemical DNA. The project began with the culmination of several years of work supported and subsequently initiated by the United States Department of Energy. A 1987 report stated boldly, "The ultimate goal of this initiative was to understand the human

TABLE 1.1
Milestones in Biotechnology

1830	Proteins were discovered
1855	*Escherichia coli* (*E. coli*) bacterium was discovered (later became a major tool in biotechnology)
1882	Chromosomes were discovered
1940	Avery demonstrated that DNA is the "transforming factor" and material of genes
1944	Scientists demonstrated that DNA, not the protein, is a hereditary material
1953	Double helix structure of DNA was first described by Watson and Crick
1971	Cetus, the world's first biotech company, was founded in Emeryville, California
1973	Cohen and Boyer developed genetic engineering techniques to "cut and paste" DNA and reproduce the new DNA in bacteria
1976	The first working synthetic gene was developed
1977	Genentech scientists and their collaborators produced the first human protein (somatostatin) in a bacterium (*E. coli*)
1978	Genentech scientists and their collaborators produced recombinant human insulin
1979	Genentech scientists produced recombinant human growth hormone.
1982	Eli Lilly and Company marketed Genentech-licensed recombinant human insulin—the first such product on the market
1983	PCR technique conceived (which will become a major means of copying genes and gene fragments)
1985	Genentech received FDA approval for protropin for growth hormone deficiency in children—the first biotech drug manufactured and marketed by a biotech company
1988	The "Harvard Mouse" became the first mammal patented in the United States
1990	HGP, an international effort to map all the genes in the human body, was launched
1994	BRCA1, the first breast cancer susceptibility gene, was discovered
1995	The first full gene sequence of a living organism other than a virus was completed for the bacterium *Haemophilus influenzae*
2000	The first draft of human genome sequence was completed by the HGP and Celera Genomics
2002	The first cervical cancer vaccine was developed
2003	The SARS (Severe Acute Respiratory Syndrome) virus was sequenced 3 weeks after its discovery
2004	The first cloned pet, a kitten, was delivered to its owner
2005	Review of stem cell and cloning legislation
2006	Recombinant vaccine against human papillomavirus (HPV) received FDA approval
2007	U.S. FDA concluded that food and food products derived from cloned animals or their offspring are as safe to eat as that from non-cloned animals
2008	Japanese researchers successfully developed the world's first genetically modified blue rose

genome" and "knowledge of the human is necessary to the continuing progress of medicine and other health sciences as knowledge of human anatomy has been for the present state of medicine." Candidate technologies were already being considered for the proposed undertaking at least as early as 1985. James D. Watson was the head of the National Center for Human Genome Research at the National Institutes of Health (NIH) in the United States starting from 1988. Largely due to his disagreement over the issue of patenting genes, Watson was replaced by Francis Collins in April 1993, and the name of the center was changed to the National Human Genome Research Institute (NHGRI) in 1997.

The $3-billion project was founded in 1990 by the United States Department of Energy and the U.S. NIH, and took almost 15 years to complete. In addition to the United States, the international consortium comprised of geneticists from the United Kingdom, France, Germany, Japan, China, and India. Due to widespread international cooperation and advances in the field of genomics, the study of genomes of organisms, as well as major advances in computing technology, a "rough draft" of the genome was finished in 2000. Ongoing sequencing led to the announcement of the essentially complete genome in April 2003, 2 years earlier than planned. In May 2006, another milestone was passed on the way to completion of the project, when the sequence of the last chromosome was published in the journal *Nature*.

1.9.1 Need for Human Genome Project

Inherited diseases are rare but there are more than 3000 disorders which are known to be caused due to single altered genes. Current knowledge of genetic tools is not sufficient to provide a successful cure for most of these disorders. However, having a gene in hand allows scientists to study its structure and characterize the molecular alterations, or mutations that result in disease. Progress in understanding the causes of cancer, for example, has taken a leap forward by the recent discovery of cancer genes. The goal of the HGP was to provide scientists with powerful new tools to help them clear the research hurdles that now keep them from understanding the molecular essence of other tragic and devastating illnesses, such as schizophrenia, alcoholism, Alzheimer's disease, and manic depression.

Gene mutations probably play a role in many of today's most common diseases such as heart disease, diabetes, immune system disorders, and birth defects. These diseases are believed to result from complex interactions between genes and environmental factors. When genes for diseases have been identified, scientists can study how specific environmental factors, such as food, drugs, or pollutants, interact with those genes. Once a gene is located on a chromosome and its DNA sequence worked out, scientists can then determine which protein in the gene is responsible for the disease and then find out its function in the body. This is the first step in understanding the mechanism of a genetic disease and eventually conquering it. One day, it may be possible to treat genetic diseases by correcting errors in the gene itself, replacing its abnormal protein with a normal one, or by switching the faulty gene off. Finally, the HGP research will help solve one of the greatest mysteries of life: How does one fertilized egg "know" how to generate so many different specialized cells, such as those making up muscles, brain, heart, eyes, skin, blood, and so on? For a human being or any organism to develop normally, a specific gene or sets of genes must be switched on in the right place in the body at exactly the right moment in development. Information generated by the HGP will shed light on how this intimate dance of gene activity is choreographed into the wide variety of organs and tissues that make up a human being.

1.10 MAJOR SCIENTIFIC DISCOVERIES IN BIOTECHNOLOGY

In order to understand the significance of biotechnology, it is very important to know the various constituents of science and technology that greatly helped the field of biotechnology. In this section, we will go through various milestones that greatly impacted biotechnology. Among these discoveries, the application of selective breeding in plants and animals has produced great impact in improving the crop and livestock productions for human consumption. In selective breeding, organisms with desired features are purposely mated to produce offsprings with the same desirable characteristics. For example, mating plants that produce the largest, sweetest, and most tender ears of corn is a good way for farmers to maximize their land to produce the most desirable crops. Selective breeding in plants has been done primarily based on the phenotypic information of plants and recent advancement in the field of genetic engineering. This made it possible to create a unique individual plant, called a *transgenic plant*. The selective breeding technique was not confined to plants alone. It had also been extensively used in animals such as cows, chickens, goats, and pigs to improve the population and quality of farm animals. Like plants, selective breeding in animals has been done primarily based on the phenotypic information of animals and recent advancement in the field of genetic engineering. It also made it possible to create transgenic animals by manipulating specific gene of interest.

Another discovery which revolutionized the treatment procedure for microbial infections in humans is the use of antibiotics in treating microbial infections. In 1918, Sir Alexander Fleming discovered that the mold *Penicillium* inhibited the growth of human skin disease-causing bacteria called *Staphylococcus aureus*. He was then able to successfully isolate and purify the antibiotic substance of the mold to use it for human purpose. *Antibiotics* are substances produced by microorganisms that normally inhibit the growth of other microorganisms. In 1940s, penicillin became a widely available drug for treating microbial infections in human beings. Currently, a wide variety of microorganisms have been used to generate thousands of liters of antibiotic drugs by using advanced biotechnology tools.

Between 1950 and 1960, a series of discoveries unfolded about the human genetic code. The whole thing started in 1953 with the publication of a research article by James Watson and Francis Crick who claimed to have discovered the structure of the human DNA. Nine years later, in 1962, they shared the Nobel Prize in Physiology or Medicine with Maurice Wilkins, for solving one of the most important biological riddles. This discovery has led to the birth of genetic manipulation and genetic engineering. With the help of genetic engineering, the genes of interest can then be identified and manipulated to develop the desired product. This process of genetic manipulation is called *recombinant DNA technology*. The recombinant DNA technique was first proposed by Peter Lobban, a graduate student with A. Dale Kaiser at the Stanford University Department of Biochemistry. The technique was then realized by a group of researchers and the finding was published in reputed international journals. Recombinant DNA technology was made possible by the discovery, isolation, and application of restriction endonucleases by Werner Arber, Daniel Nathans, and Hamilton Smith, for which they received the 1978 Nobel Prize in Medicine. Cohen and Boyer applied for a patent on the process for producing biologically functional molecular chimeras which could not exist in nature in 1974.

In recent years, recombinant DNA technology has been extensively employed to generate therapeutic products for treating human diseases. With a rapid increase in diabetic patients, there is an urgent need for a drug which can solve this problem. Surprisingly, with the efforts of scientists, they were able to successfully synthesize insulin molecules by using recombinant DNA technology. The application of recombinant DNA technology produced arrays of therapeutic products and disease-resistant crops, thus producing greater harvest. Scientists were also able to generate better quality of rice such as the golden rice. Recombinant DNA technology has also been applied to create engineered bacteria which are capable of degrading environmental pollutants.

1.11 BIOTECHNOLOGY AS THE SCIENCE OF INTEGRATION

So far, we have learned about various applications of biotechnology to create diverse products, and that this could not have been achieved without the contributions of various fields of science and technology. In this section, we will discuss the fundamentals of biotechnology.

The foundation of biotechnology is based on biology and chemistry supported by mathematics, physics, engineering, and computer and information technology. The integration of various disciplines in biotechnology is well described in Figure 1.10. We know that biotechnology is initially known for microbial related products. So, to make a microbial product, microbiology is integrated with *bioreactor technology*, a machine used to manufacture microbes in great number, to develop a new technology known as microbial biotechnology.

The discovery of DNA double helix and its recombinant capabilities has given a completely new dimension to the field of biotechnology by integrating molecular biology, genetics, human pathology, biochemistry, and microbiology, and gives a recombinant insulin product. The integration is not limited only to the fields of biology and engineering. We also find integration in the field of information technology. One such example is bioinformatics which studies and analyzes genetic data from a large databank using specialized tools and software. The integration of the science of pharmacology, genetics, and information technology had given birth to a new technology called pharmacogenomics. Pharmacogenomics is an emerging field of biotechnology that makes genetically tailored medicines. All these examples suggest that biotechnology is a highly diversified field

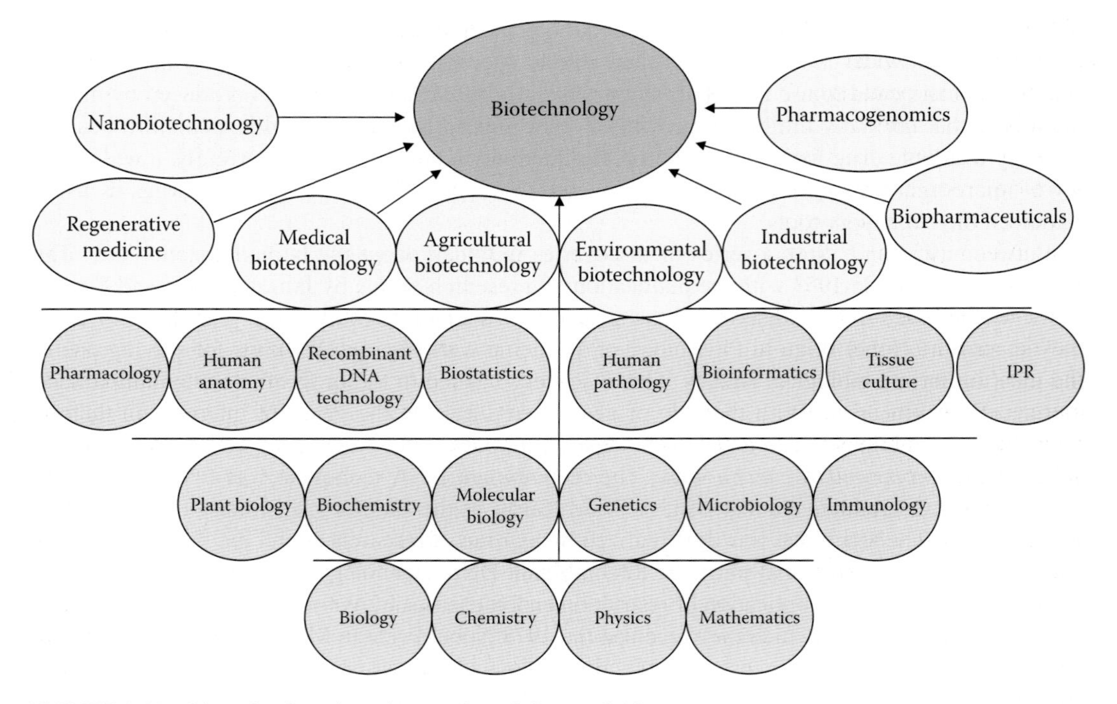

FIGURE 1.10 Biotechnology is an integration of diverse fields.

with enormous applications. Some new integration is also happening in other fields such as bionanotechnology, biometrics, and biomedical equipment. The field of biotechnology keeps expanding with the emergence of new technology and knowledge, so we expect to see more integration in the future as well.

1.12 BIO-REVOLUTION

The bio-revolution resulting from advances in molecular biosciences and biotechnology had already outstripped the advances of the "Green Revolution." In the early 1960s, the pioneering studies of Nobel Prize winner Norman Borlaug, using crossbreeding techniques based on classical genetics, offered for the first time a weapon against hunger in the countries of Latin America, Asia, and Africa. As a direct result of the comprehensive studies of Borlaug and his contemporaries, new wheat hybrids began to transform the harvests of India and China, although they had a relatively minor influence on agriculture in more temperate climates. There is little doubt that genetic manipulation will open more new doors in this field, and will dramatically alter farming worldwide. It does not require a crystal ball to imagine the potential of the immediate biotechnological future. From the advances in recent years, it is possible to extrapolate the number of likely developments based on the researches which are now in progress.

In the plant world, the 1978 development of the "pomato," a laboratory-generated combination of two members of the Solanaceae family (the potato and the tomato), was very significant. The Flavr Savr tomato was reviewed by the Food and Drug Administration (FDA) in the spring of 1994 and found to be as safe as conventionally produced tomatoes. This is the first time the FDA had evaluated a whole food produced by biotechnology. Exciting prospects are likely to result from industrial-scale plant tissue culture. This may soon obviate the need for rearing whole plants in order to generate valuable commodities such as dyes, flavorings, drugs, and chemicals. Cloning techniques could prove to be the way to tackle some of the acute problems of reforesting in semidesert areas. Seedlings grown from the cells of mature trees could greatly speed up the process. In the summer of 1987, a Belgian team introduced into crop plants a group of genes encoding for insect resistance and resistance to widely used herbicides. This combination of advantageous genes brought about a new era in plant protection. The crop can be treated safely with more effective doses of weed killer, and it is also engineered to be less susceptible to insect damage.

Dairy farming is also benefiting from advances in biotechnology. Bovine somatotropin (growth hormone) enhances milk yields, with no increase in feed costs. Embryo duplication methods mean that cows will bear more calves than in the past, and embryo transfer techniques are enabling cattle of indifferent quality to rear good quality stock, a potentially important development for nations with less advanced agriculture. Genetic manipulation of other stock, such as sheep and pigs, appears to be feasible, and work is in progress on new growth factors for poultry.

The outcome of this intense activity will be improvements in the texture, quality, variety, and availability of traditional farm products, as well as the emergence of newly engineered food sources. Such bioengineered super-foods will be welcomed, and will offer new varieties, and hence find new markets in the quality-conscious advanced countries. Despite the enormous potential gains, the economic consequences of possible overproduction in certain areas must also be faced. It will be essential for those concerned with making agricultural policies to keep abreast of the pace of modern biotechnology. Short-term benefits to the consumer of lower agricultural prices must be weighed against a long-term assessment of the impact of new discoveries on the farming industry.

In the medical field, considerable efforts will be devoted to the development of vaccines for killer diseases such as AIDS. Monoclonal antibodies will be used to boost the body's defenses and guide anticancer drugs to their target sites. This technology may also help to rid the human and animal world of a range of parasitic diseases by producing specific antibodies to particular parasites. Synthesis of drugs, hormones, and animal health products, together with drug-delivery mechanisms, are all advancing rapidly. Enzyme replacement and gene replacement therapy are other areas where progress is anticipated. The next decade will see significant advances in medicine, agriculture, and animal health directly attributable to biotechnology. The impact of new technology will not, however, be confined to bio-based industries. Genetically engineered microbes may become more widely used to extract oil from the ground and valuable metals from factory wastes. Although most biotechnology companies are primarily based in North America and Europe, a tremendous growth has been witnessed in biotechnology-related activities in Singapore, Australia, China, and India.

1.13 ETHICAL AND REGULATORY ISSUES IN BIOTECHNOLOGY

One of the greatest challenges of modern biotechnology is to face various ethical, social, and regulatory issues. These issues thoroughly scrutinize the technology before it reaches to human. To the extent possible, we should consider these ethical issues as critical, scientific, and technological. Recall that there was once a huge debate in the United States over the use of human embryo in making embryonic stem cell lines. The notion then was that killing a human embryo is the same as killing a human life, and so the president banned all embryonic stem cell research which uses fresh embryos. Because of the ban, scientific communities were very upset. Still, they worked hard and tried to find out alternative ways of generating embryonic stem cells from non-embryonic sources. New research findings suggest that this is possible. In Chapter 13, we will discuss in great detail various ethical and regulatory issues related to biotechnology products and research.

1.14 FUTURE OF BIOTECHNOLOGY

The recent emphasis on environmental awareness has challenged scientists to find solutions for better and safer living conditions. The added threat of deadly diseases such as AIDS and resistant strains of tuberculosis, gonorrhea, bird flu, and swine flu have forced scientists to look for new therapies within the field of biotechnology. The structure of DNA was deciphered by James Watson, a geneticist, and Francis Crick, a physicist, thus marking the beginning of molecular biology in the twentieth century. Their determination of the physical structure of the DNA molecule became the foundation for modern biotechnology, enabling scientists to develop new tools to improve the future of mankind. The HGP is a major biotechnological endeavor, the aim of which is to make a detailed map of the human DNA. The hereditary instructions inscribed in the DNA guide the development of the human being from the fertilized egg cell to his death. In this project, which took 15 years to complete, chromosome maps are developed in various laboratories worldwide through a coordinated effort guided by the National Institute of Health. The genetic markers for over 4000 diseases caused by single mutant genes have been mapped. To get an idea of the magnitude of this project, imagine a stack of 25,000 books. If each book is 2 cm thick, the stack would measure 500 m, the height of a 15-story building. Consider locating a particular word within one of the books in the stack. For a molecular biologist, this would be analogous to finding one gene in the human genome. Up to this point, molecular biologists have mapped only a tiny fraction of the genome. The 23 pairs of human chromosomes are estimated to contain between 50,000 and 100,000 genes, of which apparently only about 5% have so far been transcribed.

Recombinant DNA biotechnology has aroused public interest and concern and has influenced medicine, industry, agriculture, and environmental problem solving in the past 20 years since its inception. In medicine, faster and more efficient diagnosis and treatment of diseases such as cystic fibrosis, cancer, sickle cell anemia, and diabetes are soon to be developed. Recombinant organisms will be used in industry to produce new vaccines, solvents, and chemicals of all kinds. Biotechnology also has applications in both plant and animal breeding. Scientists are developing disease- and herbicide-resistant crops, disease-resistant animals, seedless fruits and rapidly growing chickens. Microbes are also being engineered to digest compounds that are currently polluting our environment.

Some of the more exciting frontiers of biotechnology include protein-based "biochips" which may replace silicon chips. It is believed that biochips would be faster and more energy efficient. Biochip implants in the body could deliver precise amounts of drugs to affect heart rate and hormone secretion or to control artificial limbs. Biosensors are monitors that use enzymes, monoclonal antibodies, or other proteins to test air and water quality, to detect hazardous substances, and to monitor blood components *in vivo*.

Gene therapy involves correction of defects in genetic material. In this process, a normal gene is introduced to replace a malfunctioning one. Gene therapy will be the "expression" of the medical research branch of biotechnology. It may, in time, form the basis of its own industry or join the traditional pharmaceutical industry. New delivery systems, called *liposomes*, are being developed to get cytotoxic drugs to tumor sites with minimal damage to surrounding healthy tissues. New monoclonal antibodies will be isolated for use in cancer treatment, diagnostic testing, bone marrow transplantation, and other applications.

Progress in biotechnology is currently working on environmental-friendly biodegradation processes for a cleaner and healthier planet, experimenting with until-now untapped energy sources, and devising useful consumer chemicals such as adhesives, detergents, dyes, flavors, perfumes, and plastics. With the progress made thus far in the fight against deadly diseases such as polio and smallpox, it is not unreasonable to expect biotechnology to hold the promise for effective treatments or even cures for, say, cancer and AIDS. Gene therapy may well become the method whereby we correct congenital disease caused by faulty genes. Stem cell research may prove the panacea for Parkinson's disease, multiple sclerosis, and muscular dystrophy. Moreover, given the genetic improvements made with crop yield and nutritive value, world hunger and malnutrition may witness their end with the continual advancement of biotechnology. Whatever the future of these particular ventures, it seems molecular biology and biotechnology will be important sciences of the coming century.

PROBLEMS

Section A: Descriptive Type

Q1. Cite and explain some definitions of biotechnology.
Q2. Differentiate ancient biotechnology from modern biotechnology.
Q3. Explain the significance of the HGP.
Q4. Explain the status of career opportunities in the field of biotechnology.
Q5. How does academic research differ from industrial research?

Section B: Multiple Choice

Q1. Most cheeses are made with starter bacteria from the *Lactococci, Lactobacilli*, or *Streptococci* families. True/False

Q2. Which of the following is not a product of recombinant DNA technology?
a. Human insulin
b. Human growth hormone
c. Antibiotic
d. Antibody

Q3. Which is the branch of pharmacology that deals with the influence of genetic variation on drug response?
a. Gene therapy
b. Bioinformatics
c. Pharmacogenomics
d. Medicine

Q4. Recombinant human proteins are manufactured using nonhuman mammalian cells that are engineered to express certain human genetic sequences to produce specific proteins. True/False

Q5. Which of the following is not a product of ancient biotechnology?
a. Cheese
b. Alcohol
c. Human insulin
d. Animal breeding based on traits

Q6. When was the HGP founded?
a. 1989
b. 1990
c. 1991
d. 1995

Q7. Sir Alexander Fleming discovered that the mold Penicillium promoted the growth of human skin disease-causing bacteria called *Staphylococcus aureus*. True/False

Q8. Recombinant DNA technology was made possible by the discovery, isolation, and application of …
a. Exonucleases
b. Restriction endonucleases
c. RNAases
d. Telemerases

Q9. Flavr Savr tomato is the first genetically engineered fruit approved by U.S. FDA. True/False

Q10. Which of the following is the first genetically engineered rice?
a. Silver rice
b. Golden rice
c. Brown rice
d. Basmati rice

Section C: Critical Thinking

Q1. Why is biotechnology a field of diverse sciences? Explain.
Q2. More than 3000 diseases are known to result from genetic mutations, so how can one correct these genetic mutations to cure the disorders?
Q3. Do you believe that ethical issues are hindrances in the development and progress of the field of biotechnology? Why?

ASSIGNMENT

Prepare a poster based on various applications of biotechnology in medicine, agriculture, environment, and industry. Describe the application in each field using suitable examples. Posters can be displayed in the classroom or in the lobby to enhance general awareness.

ONLINE RESOURCES

Current information pertaining to biotechnology may be directly accessed from biotechnology journals, discussion panels, and societies in addition to pubmed (www.pubmed.com).

REFERENCES AND FURTHER READING

Bains, W.E. *Biotechnology from A to Z.* Oxford University Press, Oxford, U.K., 1993.

Barnhart, B.J. DOE human genome program. *Hum. Genome Quart.* 1: 1, 1989. http://www.ornl.gov/sci/techresources/Human_Genome/publicat /hgn/v1n1/01doehgp. html (retrieved February 3, 2005).

Benton, D. Bioinformatics-principles and potentials of a new multidisciplinary tool. *Trends Biotechnol.* 14: 261–272, 1996.

Cavalieri, D., McGovern, P.E., Hart, D.L., Mortimer, R., and Polsinelli, M. Evidence for *S. cerevisiae* fermentation in ancient wine. *J. Mol. Evol.* 57: S226–S232, 2003.

DeLisi, C. Genomes: 15 years later a perspective by Charles DeLisi, HGP pioneer. *Hum. Genome News* 11: 3–4, 2001. http://genome.gsc.riken.go.jp/ hgmis/publicat/hgn/v11n3/05delisi.html (retrieved February 3, 2005).

Dirar, H. *The Indigenous Fermented Foods of the Sudan: A Study in African Food and Nutrition. CAB International.* Cambridge, U.K., 1993.

Fermented fruits and vegetables. A global perspective. *FAO Agricultural Services Bulletins*, 134. January 19, 2007. http://www.fao.org/docrep/x0560e/x0560e05.htm (retrieved on January 28, 2007).

Kreuzer, H. and Massay, A. *Recombinant DNA and Biotechnology: A Guide for Students.* ASM Press, Washington, DC, 2001.

Pederson, R.A. Embryonic stem cells for medicine. *Sci. Am.* 280: 68–73, 1999.

Steinkraus, K.H. *Handbook of Indigenous Fermented Foods.* Marcel Dekker, Inc., New York, 1995. http://www.phppo.cdc.gov/phtn/botulism/alaska/alaska.asp (retrieved on January 28, 2007).

Sugihara, T.F. Microbiology of breadmaking, in: *Microbiology of Fermented Foods*, B.J.B. Wood (ed.). Elsevier Applied Science Publishers, London, U.K., 1985.

2 Genes and Genomics

LEARNING OBJECTIVES

- Define cell as the building block of human body
- Discuss intracellular and extracellular organization of cell
- Discuss about nucleus and its constituents
- Explain structure and function of DNA
- Explain cell division by meiosis and mitosis
- Explain DNA replication in cell division
- Discuss about molecular and genetic tools of biotechnology

2.1 INTRODUCTION

In Chapter 1, we learned various attributes of biotechnology and how the fields of biotechnology have revolutionized agricultural, environmental, and healthcare sectors. This incredible phase of biotechnology is called modern biotechnology, where modern tools are employed in creating arrays of products which include antibiotics, vaccines, monoclonal antibodies, and recombinant insulin. In this chapter, we will learn about modern molecular biology tools and how these tools influence the development of new biotechnology products. We know the fact that the molecule *deoxyribonucleic acid* (DNA) plays a critical role in the development and functioning of all known living organisms and some viruses. The main role of DNA molecules is the long-term storage of information. It is often compared to a set of blueprints, a recipe, or a code since it contains the instructions needed to construct other components of cells, such as proteins and *ribonucleic acid* (RNA) molecules. It would be interesting to know the structure and function of DNA in various species and to understand how DNA controls the physiological functions in the human body, plants, animals, and microbes. In this chapter, we will compare the difference between animal cells and plant cells, or between plant cells and microbial cells. Let us first learn about the organization of cells in animals, plants, and microorganisms which contain genetic material.

2.2 CELL AS THE BUILDING BLOCK OF LIFE

The cell is the structural and functional unit of all known living organisms. It is the smallest unit of a living organism, and is often called the building block of life. Humans have approximately 100 trillion or 10^{14} cells, an example of a *multicellular* organism. On the other hand, a single-celled bacterium is called *unicellular*. A typical cell size is 10 μm while a typical cell mass is 1 ng. All animals and plants are made of cells and this concept was originally coined by Aristotle (384–322 BC). In 1665, Robert Hooke observed for the first time the structure of a cell under a very primitive microscope. In 1674, Antonie van Leeuwenhoek discovered cells with structural organization within the cell.

Earlier in the nineteenth century, scientists proposed theories about the origin of a cell and how a cell becomes a living organism. In 1824, H.J. Dutrochet, a French scientist, gave the idea of the cell theory, while German botanist M.I. Schleiden and German zoologist T. Schwann formulated and outlined the basic features of the theory in 1839. In 1858, R. Virchow extended the cell theory and suggested that all living cells arise from preexisting living cells. To prove Virchow's hypothesis,

Louis Pasteur performed some experiments and concluded that living things are composed of cells, and all living cells arise from preexisting cells. One exception that does not fit into the cell theory are viruses which may be defined as an infectious subcellular and ultramicroscopic organism. Viruses are simple as they lack the internal organization which is the main characteristic of a living cell. Due to this unique characteristic, viruses, mycoplasma, viroids, and prions do not easily fit in the definition of the cell theory. In addition, there are other organisms such as protozoa and algae which also do not fit in the definition of the cell theory.

2.3 CLASSIFICATION OF CELLS

There are two different types of cells, the *eukaryotic* and *prokaryotic* cells. Prokaryotic cells are usually independent, while eukaryotic cells are often found in multicellular organisms. Let us first learn to understand the basic similarities and differences between them.

2.3.1 Prokaryotic Cell

The prokaryote cell is simpler than the eukaryote cell, lacking a nucleus and most of the other organelles of eukaryotes. There are two kinds of prokaryotes, *bacteria* and *archaea*. These prokaryotes share a similar overall structure. A prokaryotic cell has three architectural regions:

1. On the outside, *flagella* and *pili* project from the cell's surface. These are structures (not present in all prokaryotes) made of proteins that facilitate movement and communication between cells.
2. Enclosing the cell is the *cell envelope* generally consisting of a cell wall covering a plasma membrane though some bacteria also have a further covering layer called a *capsule*. The envelope gives rigidity to the cell and separates the interior of the cell from its environment, serving as a protective filter. Though most prokaryotes have a cell wall, there are exceptions such as *Mycoplasma* (bacteria) and *Thermoplasma* (archaea). The cell wall consists of *peptidoglycan* in bacteria, and acts as an additional barrier against exterior forces. It also prevents the cell from expanding and finally bursting, called *cytolysis*, from osmotic pressure against a hypotonic environment. Some eukaryote cells (plant cells and fungi cells) also have a cell wall.
3. Inside the cell is the cytoplasmic region that contains the cell genome (DNA) and ribosomes and various sorts of inclusions. A *prokaryotic chromosome* is usually circular in shape. An exception of this is the bacterium *Borrelia burgdorferi*, which causes Lyme disease, whose DNA is linear in shape.

One of the distinct features of a prokaryote is the absence of nucleus in the cytoplasm and DNA is usually condensed in a *nucleoid*. Prokaryotes can carry extra-chromosomal DNA elements called *plasmids*, which are usually circular. Plasmids enable additional functions, such as antibiotic resistance. The presence of plasmid DNA in bacteria not only makes bacteria distinct from animals and plants but also makes it an important genetic engineering tool (Figure 2.1).

2.3.2 Eukaryotic Cell

Eukaryotic cells are about 10 times the size of a typical prokaryote and can be as much as 1000 times greater in volume. The major difference between prokaryotes and eukaryotes is that eukaryotic cells contain membrane-bound compartments in which specific metabolic activities take place. Most important among these is the presence of a cell nucleus, a membrane-delineated compartment that houses the eukaryotic cell's DNA. It is this nucleus that gives the eukaryote its name, which

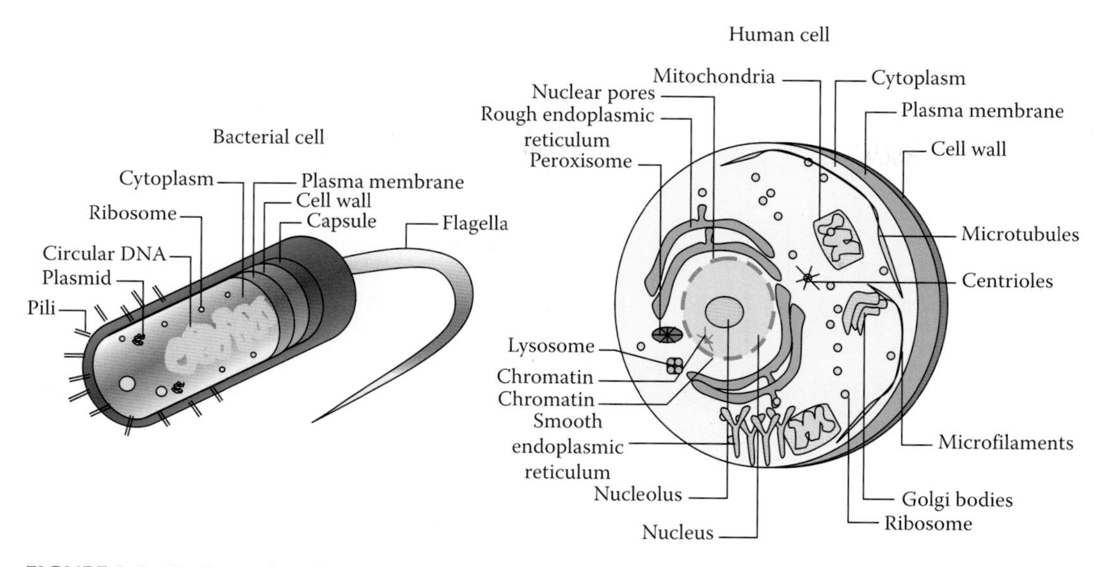

FIGURE 2.1 Prokaryotic cell and eukaryotic cell: Diagrammatic illustration.

means "true nucleus." The plasma membrane resembles that of prokaryotes in function, with minor differences in the setup. The eukaryotic DNA is organized in one or more linear molecules, called *chromosomes*, which are associated with histone proteins. All chromosomal DNA is stored in the cell nucleus, separated from the cytoplasm by a membrane. Some eukaryotic organelles such as mitochondria also contain some DNA. Eukaryotes can move using *cilia* or *flagella*. The flagella of a eukaryote are more complex than those of prokaryotes.

All cells, whether prokaryotic or eukaryotic, have a membrane that envelops the cell, separates its interior from its environment, regulates what moves in and out (selectively permeable), and maintains the electric potential of the cell. Inside the membrane, a salty cytoplasm takes up most of the cell volume. All cells possess DNA and RNA to build various proteins such as enzymes, the cell's primary machinery. There are also other kinds of biomolecules in cells. In the next section, we will look into all the constituents present in bacterial, animal or plant cell and try to understand their function (Figure 2.1).

2.4 EXTRACELLULAR ORGANIZATION

The cell consist of various components which play critical role in protecting cell from outside invasion, to make cell–cell contact and to help the cell to perform regulated physiological functions.

2.4.1 Cell Membrane

The cytoplasm of a cell is surrounded by a *cell membrane* or *plasma membrane*. The plasma membrane in plants and prokaryotes is usually covered by a cell wall. This membrane serves to separate and protect a cell from its surrounding environment and is made mostly from a double layer of lipids (hydrophobic fat-like molecules) and hydrophilic phosphorus molecules. Hence, the layer is called a *phospholipids bilayer*. It may also be called a *fluid mosaic membrane*. Embedded within this membrane is a variety of protein molecules that act as channels and pumps that move different molecules into and out of the cell. The membrane is said to be "semipermeable," in that it can either let a substance (molecule or ion) pass through freely, pass through to a limited extent, or not pass through at all. Cell surface membranes also contain receptor proteins that allow cells to detect external signaling molecules such as hormones (Figure 2.2).

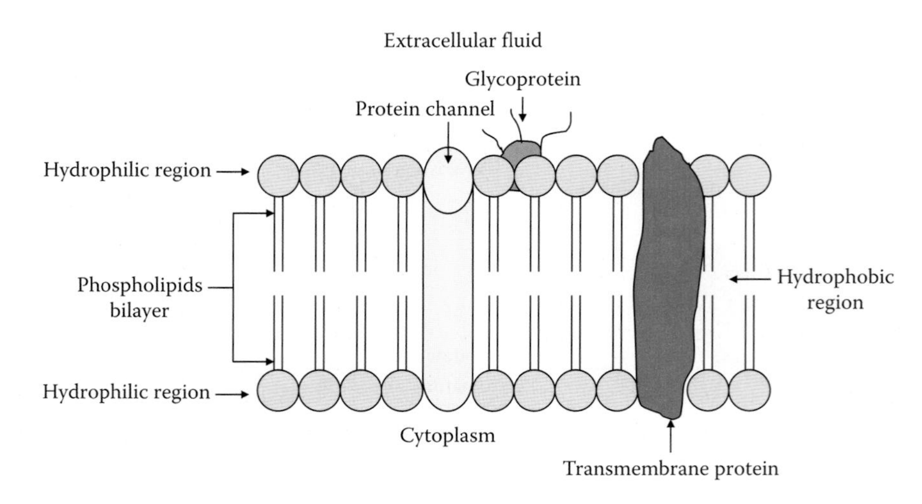

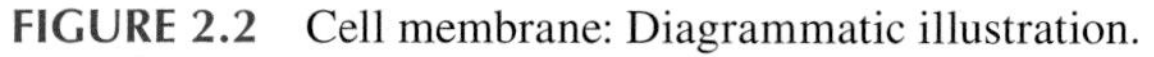
FIGURE 2.2 Cell membrane: Diagrammatic illustration.

2.4.2 Cell Capsule

The *cell capsule* is a very large organelle of some prokaryotic cells, such as bacterial cells. It is a layer that lies outside the cell wall of bacteria. It is a well-organized layer, not easily washed off, and it can be the cause of various diseases. It is usually composed of polysaccharides, but could be composed of other materials (e.g., polypeptide in *B. anthracis*). Because most capsules are water soluble, they are difficult to stain using standard stains because most stains do not adhere to the capsule. For examination under the microscope, the bacteria and their background are stained darker than the capsule, which does not stain. Since the capsule protects bacteria against phagocytosis, it is considered a virulence factor. A capsule-specific antibody may be required for phagocytosis to occur. Capsules also contain water which protects bacteria against desiccation. They also exclude bacterial viruses and most hydrophobic toxic materials such as detergents. Further than that, bacterial capsules allow bacteria to adhere to surfaces and other cells.

2.4.3 Flagella

A *flagellum* is a tail-like structure that projects from the cell body of certain prokaryotic and eukaryotic cells, and functions in locomotion. There are some notable differences between prokaryotic and eukaryotic flagella, such as protein composition, structure, and mechanism of propulsion. An example of a flagellated bacterium is the ulcer-causing *Helicobacter pylori*, which uses multiple flagella to propel itself through the mucus lining to reach the stomach epithelium. An example of a eukaryotic flagellated cell is the sperm cell, which uses its flagellum to propel itself through the female reproductive tract. Eukaryotic flagella are structurally identical to eukaryotic cilia, although distinctions are sometimes made according to function and/or length.

2.5 INTRACELLULAR ORGANIZATION

2.5.1 Cytoplasmic Constituents

Like the human body which contains many different organs performing different functions, cells, too, have a set of "little organs," called *organelles* that are specialized for carrying out one or more vital functions. There are several types of organelles within an animal cell. Some such as nucleus and Golgi apparatus are typically solitary, while others such as mitochondria, peroxisomes, and lysosomes can be numerous (hundreds to thousands). The *cytosol* is the gelatinous fluid that fills the cell and surrounds the organelles.

2.5.1.1 Mitochondria and Chloroplasts

Mitochondria are self-replicating organelles that occur in various numbers, shapes, and sizes in the cytoplasm of all eukaryotic cells. Mitochondria play a critical role in generating energy in the eukaryotic cell. Mitochondria generate the cell's energy by the process of oxidative phosphorylation, utilizing oxygen to release energy stored in cellular nutrients (typically pertaining to glucose) to generate adenosine triphosphate (ATP). Mitochondria multiply by splitting in two. Organelles that are modified chloroplasts are broadly called *plastids*, and are involved in energy storage through the process of photosynthesis, which utilizes solar energy to generate carbohydrates and oxygen from carbon dioxide and water. Mitochondria and chloroplasts each contain their own genome, which is separate and distinct from the nuclear genome of a cell. Both of these organelles contain this DNA in circular plasmids, much like prokaryotic cells, strongly supporting the evolutionary theory of endosymbiosis. Since these organelles contain their own genomes and have other similarities to prokaryotes, they are thought to have developed through a symbiotic relationship after being engulfed by a primitive cell.

2.5.1.2 Ribosomes

The *ribosome* is a large complex of RNA and protein molecules. This is where proteins are produced. Ribosomes can be found either floating freely or bound to a membrane (the rough endoplasmatic reticulum in eukaryotes, or the cell membrane in prokaryotes).

2.5.1.3 Endoplasmic Reticulum

The *endoplasmic reticulum* (ER) is only present in eukaryote cells and is the transport network for molecules targeted for certain modifications and specific destinations, as compared to molecules that will float freely in the cytoplasm. The ER has two forms: the rough ER, which has ribosome on its surface and secretes proteins into the cytoplasm, and the smooth ER, which lacks them. Smooth ER plays a role in calcium sequestration and release.

2.5.1.4 Golgi Apparatus

The primary function of the Golgi apparatus is to process and package the macromolecules such as proteins and lipids that are synthesized by the cell. It is particularly important in the processing of proteins for secretion. The Golgi apparatus forms a part of the endomembrane system of eukaryotic cells. Vesicles that enter the Golgi apparatus are processed in a *cis* to *trans* direction, meaning they coalesce on the *cis* side of the apparatus, and after processing pinch off on the opposite (*trans*) side to form a new vesicle in the animal cell.

2.5.1.5 Lysosomes and Peroxisomes

Lysosomes contain digestive enzymes (acid hydrolases). They digest excess or worn-out organelles, food particles, and engulfed viruses or bacteria. *Peroxisomes* have enzymes that rid the cell of toxic peroxides. The cell could not house these destructive enzymes if they were not contained in a membrane-bound system. These organelles are often called a "suicide bag" because of their ability to detonate and destroy the cell.

2.5.1.6 Centrosome

The *centrosome* produces the microtubules of a cell—a key component of the cytoskeleton. It directs the transport through the ER and the Golgi apparatus. Centrosomes are composed of two centrioles, which separate during cell division and help in the formation of the mitotic spindle. A single centrosome is present in the animal cells. They are also found in some fungi and algae cells.

2.5.1.7 Vacuoles

Vacuoles store food and waste. Some vacuoles store extra water. They are often described as a liquid-filled space and are surrounded by a membrane. Some cells, such as *Amoeba*, have contractile vacuoles, which can pump excess water out of the cell.

2.5.1.8 Cytoskeleton

The *cytoskeleton* organizes and maintains the shape of a cell, anchors organelles in place, helps during *endocytosis* (the uptake of external materials by a cell) and *cytokinesis* (the separation of daughter cells after cell division), and moves parts of the cell in processes of growth and mobility. The eukaryotic cytoskeleton is composed of microfilaments, intermediate filaments, and microtubules. There are a great number of proteins associated with them, each controlling a cell's structure by directing, bundling, and aligning filaments. The prokaryotic cytoskeleton is less studied but it plays an essential role in the maintenance of a cell's shape, polarity, and cytokinesis.

2.5.2 Nuclear Constituents

2.5.2.1 Deoxyribonucleic Acid

The cell nucleus is the most important component found in a eukaryotic cell and it contains the genetic information, where almost all DNA replications and RNA syntheses occur. The nucleus is spherical in shape and separated from the cytoplasm by a double membrane called the nuclear envelope. The nuclear envelope isolates and protects a cell's DNA from various molecules that could accidentally damage its structure or interfere with its processing. In most living organisms (except for viruses), genetic information is stored in the DNA, which resides in the nucleus of living cells. It gets its name from the sugar molecule contained in its backbone (deoxyribose). However, it gets its significance from its unique structure. Four different nucleotide bases occur in DNA: adenine (A), cytosine (C), guanine (G), and thymine (T). The versatility of DNA comes from the fact that the molecule is actually double-stranded. The nucleotide bases of the DNA molecule form complementary pairs. The nucleotides hydrogen bond to another nucleotide base in a strand of DNA opposite to the original. This bonding is specific, and adenine always bonds to thymine (and vice versa) and guanine always bonds to cytosine (and vice versa). This bonding occurs across the molecule, leading to a double-stranded system as shown in Figure 2.3. The human DNA consists of about 3 billion bases, and more than 99% of those bases are the same in all people. DNA bases pair up with each other, A with T and C with G, to form units called *base pairs*. Each base is also attached to a sugar molecule and a phosphate molecule together making up a nucleotide. Nucleotides are arranged in two long strands forming a *double helix*. The structure of the double helix is somewhat like a ladder,

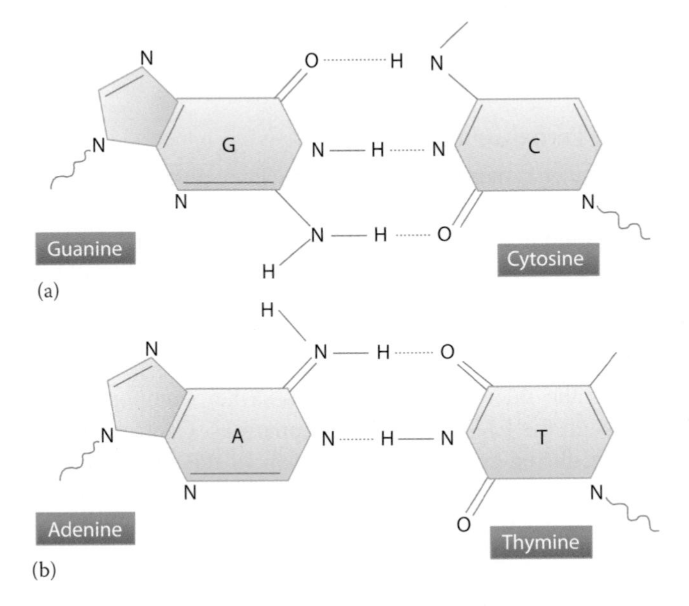

FIGURE 2.3 (a) A GC base pair with three hydrogen bonds. (b) An AT base pair with two hydrogen bonds. Non-covalent hydrogen bonds between the pairs are shown as dashed lines.

with the base pairs forming the ladder's rungs and the sugar and phosphate molecules forming the vertical sidepieces of the ladder. An important property of DNA is that it can replicate, or make copies of itself. Each strand of DNA in the double helix can serve as a pattern for duplicating the sequence of bases. This is critical when cells divide because each new cell needs to have an exact copy of the DNA present in the old cell.

Recall that the DNA consists of nucleotides, with backbones made of sugars and phosphate groups joined by ester bonds. The DNA chain is 22–26 Å wide (or 2.2–2.6 nm), and one nucleotide unit is 3.3 Å (or 0.33 nm) long. Although each individual repeating unit is very small, DNA polymers can be very large molecules containing millions of nucleotides. For instance, the largest human chromosome, chromosome number 1, is approximately 220 million base pairs long. These two strands run in opposite directions to each other and are therefore anti-parallel. Attached to each sugar is one of four types of molecules called bases. It is the sequence of these four bases along the backbone that encodes information. This information is read using the genetic code, which specifies the sequence of the amino acids within proteins.

In all living organisms, DNA usually exists as a pair of molecules that are held tightly together in the shape of a double helix. These two long strands entwine like vines, in the shape of a double helix. The nucleotide repeats contain both the segment of the backbone of the molecule, which holds the chain together, and a base, which interacts with the other DNA strand in the helix. In general, a base linked to a sugar is called a *nucleoside,* and a base linked to a sugar and one or more phosphate groups is called a *nucleotide*. If multiple nucleotides are linked together, as in DNA, this polymer is called a *polynucleotide*. The backbone of the DNA strand is made from alternating phosphate and sugar residues. The sugar in DNA is 2-deoxyribose, which is a pentose (five-carbon) sugar. The sugars are joined together by phosphate groups that form phosphodiester bonds between the third and fifth carbon atoms of adjacent sugar rings. These asymmetric bonds mean a strand of DNA has a direction. In a double helix, the direction of the nucleotides in one strand is opposite to their direction in the other strand. This arrangement of DNA strands is called *antiparallel*. The asymmetric ends of DNA strands are referred to as the 5′ (*five prime*) and 3′ (*three prime*) ends, with the 5′ end being that with a terminal phosphate group and the 3′ end that with a terminal hydroxyl group. One of the major differences between DNA and RNA is the sugar, with 2-deoxyribose being replaced by the alternative pentose sugar ribose in RNA.

One of the interesting characteristics of DNA is its ability to form a coil-like structure. This process of coiling is called as *DNA supercoiling*. With DNA in its "relaxed" state, a strand usually circles the axis of the double helix once every 10.4 base pairs, but if the DNA is twisted, the strands become tighter. When the DNA is twisted in the direction of the helix, it is called *positive supercoiling*, and the bases are held more tightly together. When they are twisted in the opposite direction, it is called *negative supercoiling*, and the bases come apart more easily. In nature, most DNA has slight negative supercoiling that is introduced by enzymes called *topoisomerases*. These enzymes are also needed to relieve the twisting stresses introduced into DNA strands during transcription and DNA replication (Figure 2.4).

2.5.2.2 Ribonucleic Acid

Ribonucleic acid (RNA) is a biologically important type of molecule that consists of a long chain of nucleotide units. RNA and DNA are both nucleic acids, but differ in three main ways. First, unlike DNA which is double-stranded, RNA is a single-stranded molecule in most of its biological roles and has a much shorter chain of nucleotides. Second, while DNA contains *deoxyribose*, RNA contains *ribose*, (there is no hydroxyl group attached to the pentose ring in the 2′ position in DNA). These hydroxyl groups make RNA less stable than DNA because it is more prone to hydrolysis. Third, the complementary base to adenine is not thymine, as it is in DNA, but rather uracil, which is an unmethylated form of thymine. Like DNA, most biologically active RNAs, including mRNA, tRNA, rRNA, snRNAs and other non-coding RNAs, contain self-complementary sequences that allow parts of the RNA to fold and pair with itself to form double helices. Structural analysis

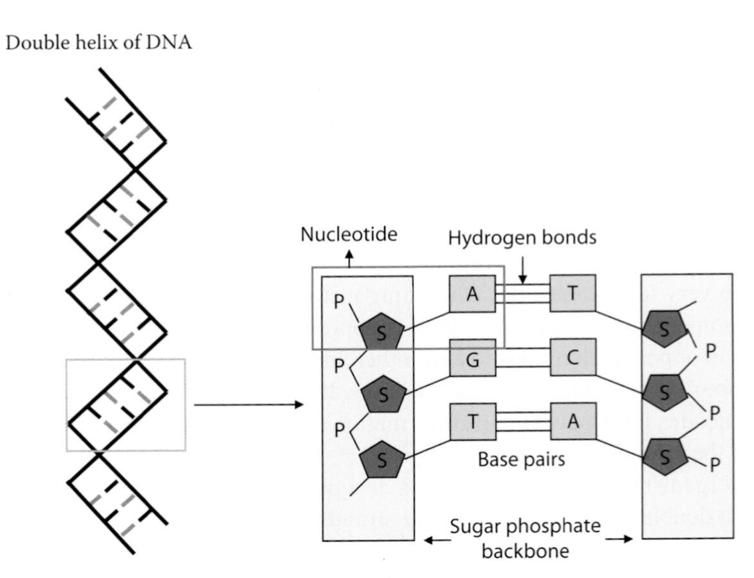

FIGURE 2.4 The structure of DNA: pieces of DNA are pairs of molecules, which entwine like vines to form a double helix. DNA strands are composed of four nucleotide subunits. These are adenine (A), thymine (T), cytosine (C), and guanine (G). Each base forms hydrogen bonds readily to only one other—A to T and C to G. The entire nucleotide sequence of each strand is complementary to that of the other.

of these RNAs has revealed that they are highly structured. Unlike DNA, their structures do not consist of long double helices but rather collections of short helices packed together into structures akin to proteins. In this fashion, RNAs can achieve chemical catalysis, like enzymes. For instance, determination of the structure of the ribosome, an enzyme that catalyzes peptide bond formation, revealed that its active site is composed entirely of RNA.

2.5.2.3 Messenger Ribonucleic Acid

Messenger ribonucleic acid (mRNA) is involved in protein synthesis by carrying coded information to the sites of protein synthesis: the ribosomes. Here, the nucleic acid polymer is translated into a polymer of amino acids: a protein. In mRNA, as in DNA, genetic information is encoded in the sequence of nucleotides arranged into codons consisting of three bases each. Each codon encodes for a specific amino acid, except the stop codons that terminate protein synthesis. This process requires two other types of RNA: *transfer RNA* (tRNA) which mediates recognition of the codon and provides the corresponding amino acid, and *ribosomal RNA* (rRNA) which is the central component of the ribosome's protein manufacturing machinery.

2.5.2.4 Transfer RNA

The tRNA is a small RNA molecule (usually about 74–95 nucleotides) that transfers a specific active amino acid to a growing polypeptide chain at the ribosomal site of protein synthesis during translation. It has a 3′ terminal site for amino acid attachment. This covalent linkage is catalyzed by an aminoacyl tRNA synthetase. It also contains a three-base region called the *anticodon* that can base pair to the corresponding three-base codon regions on mRNA. Each type of tRNA molecule can be attached to only one type of amino acid, but because the genetic code contains multiple codons that specify the same amino acid, tRNA molecules bearing different anticodons may also carry the same amino acid (Figure 2.5).

2.5.2.5 Ribosomal RNA

The rRNA is the central component of the ribosome, the protein manufacturing machinery of all living cells. The function of the rRNA is to provide a mechanism for decoding mRNA into

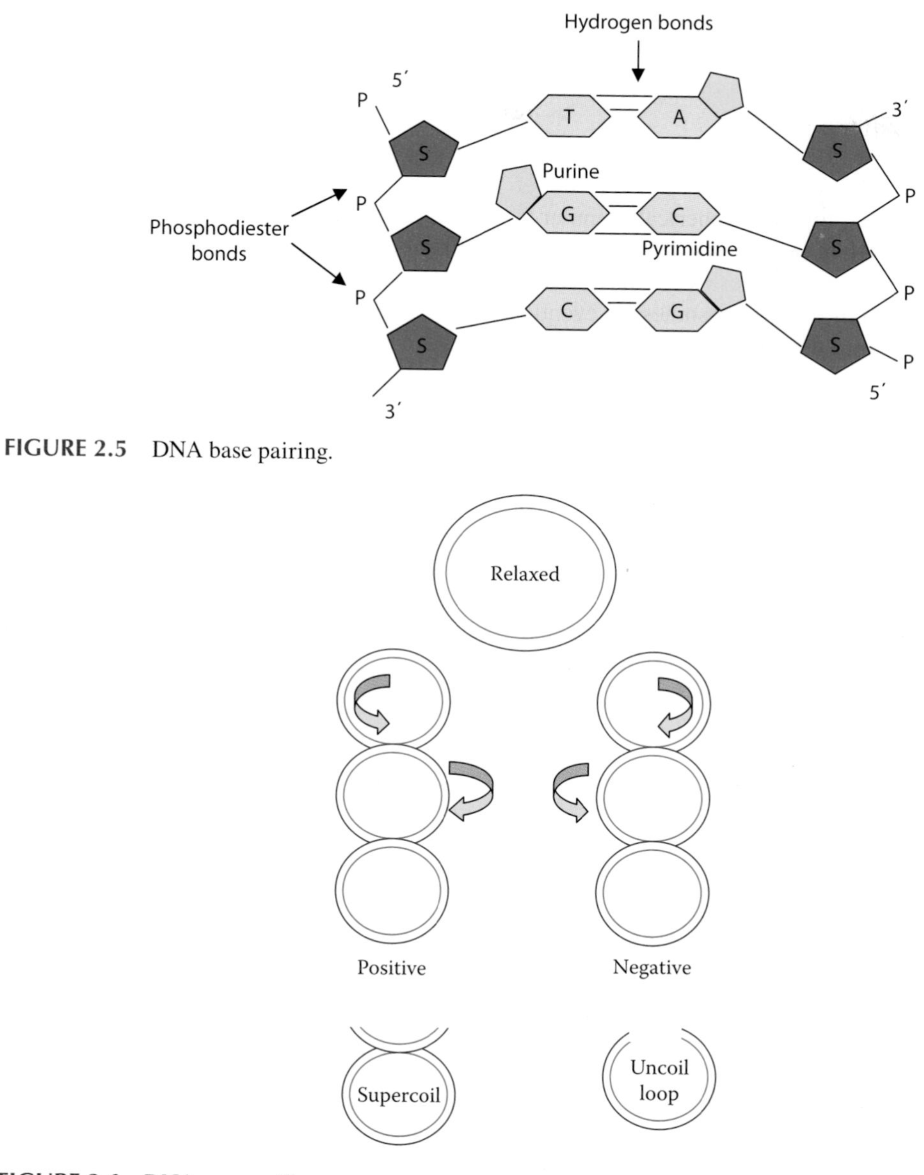

FIGURE 2.5 DNA base pairing.

FIGURE 2.6 DNA supercoiling.

amino acids and to interact with the tRNAs during translation by providing peptidyl transferase activity. The tRNA then brings the necessary amino acids corresponding to the appropriate mRNA codon (Figure 2.6).

2.5.2.6 Small Nuclear RNA

The *small nuclear RNA* (snRNA) is a class of small RNA molecules that are found within the nucleus of eukaryotic cells. They are transcribed by RNA polymerase II or RNA polymerase III and are involved in a variety of important processes such as *RNA splicing* (removal of introns from hnRNA), regulation of transcription factors (7SK RNA) or RNA polymerase II (B2 RNA), and maintaining the telomeres. They are always associated with specific proteins, and the complexes are referred to as *small nuclear ribonucleoproteins* (snRNP) or sometimes as *snurps*. These elements are rich in uridine content. A large group of snRNAs is known as *small nucleolar RNAs* (snoRNAs). These are small RNA molecules that play an essential role in RNA biogenesis and guide chemical modifications of rRNAs and other RNA genes (tRNA and snRNAs). They are located in the nucleolus and the cajal bodies of eukaryotic cells (the major sites of RNA synthesis).

2.5.2.7 Nucleoli

Nucleoli are a small, typically spherical granular body located in the nucleus of a eukaryotic cell. It is composed largely of protein and RNA. When the cell is not undergoing division, loops of DNA from one or more chromosomes extend into the nucleolus and direct the synthesis of rRNA and the formation of ribosomes. The ribosomes are eventually transferred out of the nucleus via pores in the nuclear envelope into the cytoplasm.

2.5.2.8 Chromatin

Chromatin is the complex combination of DNA, RNA, and protein that makes up chromosomes. It is found inside the nuclei of eukaryotic cells, and within the nucleoid in prokaryotic cells. It is divided between heterochromatin (condensed) and euchromatin (extended) forms. The major components of chromatin are DNA and histone proteins, although many other chromosomal proteins have prominent roles, too. The functions of chromatin are to package DNA into a smaller volume to fit in the cell, to strengthen the DNA to allow mitosis and meiosis, and to serve as a mechanism to control expression and DNA replication. Chromatin contains genetic material-instructions to direct cell functions. Changes in chromatin structure are affected by chemical modifications of histone proteins such as methylation (DNA and proteins) and acetylation (proteins), and by non-histone DNA-binding proteins.

2.6 MACROMOLECULES

Although living cells are primarily made up of water, a number of other molecules are also abundant within a cell such as *macromolecules*. Macromolecules provide structural support, store fuel, store and retrieve genetic information, and speed up biochemical reactions. There are four major types of macromolecules that play important functions in the life of a cell: proteins, carbohydrates, nucleic acid, and lipids.

2.6.1 Proteins

Like other biological macromolecules, such as polysaccharides and nucleic acids, proteins are essential parts of organisms and participate in virtually every process within cells. Many proteins are enzymes that catalyze biochemical reactions and are vital to metabolism. Proteins also have structural or mechanical functions, such as actin and myosin in muscle and the proteins in the cytoskeleton, which form a system of scaffolding that maintains cell shape. Other proteins are important in cell signaling, immune responses, cell adhesion, and the cell cycle. Proteins are also necessary in animals' diets, since animals cannot synthesize all the amino acids they need and must obtain essential amino acids from food. Through the process of digestion, animals break down ingested proteins into free amino acids that are then used in metabolism.

Proteins were first described and named by the Swedish chemist Jöns Jakob Berzelius in 1838. However, the central role of proteins in living organisms was not fully appreciated until 1926, when James B. Sumner showed that the enzyme urease was a protein. The first protein to be sequenced was insulin, by Frederick Sanger, who won the Nobel Prize for this achievement in 1958. The first protein structures to be solved were hemoglobin and myoglobin, by Max Perutz and Sir John Cowdery Kendrew, respectively, in 1958. The three-dimensional structures of both proteins were first determined by x-ray diffraction analysis. Perutz and Kendrew shared the 1962 Nobel Prize in Chemistry for these discoveries. Proteins may be purified from other cellular components using a variety of techniques such as ultracentrifugation, precipitation, electrophoresis, and chromatography. The advent of genetic engineering has made possible a number of methods to facilitate purification. Methods commonly used to study protein structure and function include immunohistochemistry, site-directed mutagenesis, and mass spectrometry. Proteins also known as *polypeptides* are organic compounds made of amino acids arranged in a linear chain and folded into a globular form. The amino acids in a polymer chain are joined together by the peptide bonds between the carboxyl and amino groups of adjacent amino acid residues. The sequence of amino acids in a protein is defined by the sequence of a gene, which is encoded in the genetic code. In general, the genetic code specifies 20 standard amino acids. However, in certain organisms, the genetic code can include selenocysteine and in certain archaea pyrrolysine. Shortly after or even during synthesis, the residues in a protein are often chemically modified by post-translational modification, which alters the physical and chemical properties—folding, stability, activity, and ultimately, the function of the proteins. Proteins can also work together to achieve a particular function, and they often associate to form stable complexes.

2.6.2 Carbohydrates

Carbohydrates are the main energy source for the human body. Chemically, carbohydrates are organic molecules in which carbon, hydrogen, and oxygen bond together in the ratio: $C_x\ (H_2O)_y$, where x and y are whole numbers that differ depending on the specific carbohydrate to which we are referring. Both animals and humans break down carbohydrates during the process of metabolism to release energy. Animals obtain carbohydrates by eating foods that contain them such as potatoes, rice, and bread. These carbohydrates are manufactured by plants during the process of photosynthesis. Plants harvest energy from sunlight to run the reaction. A potato, for example, is primarily a chemical storage system containing glucose molecules manufactured during photosynthesis. In a potato, however, those glucose molecules are bound together in a long chain. As it turns out, there are two types of carbohydrates, the *simple sugars* and those carbohydrates that are made of long chains of sugars—the *complex carbohydrates.*

2.6.2.1 Simple Sugars

All carbohydrates are made up of units of sugar. Carbohydrates that contain only one sugar unit (monosaccharides) or two sugar units (disaccharides) are called *simple sugars.* Simple sugars are sweet in taste and are broken down quickly in the body to release energy. Two of the most common monosaccharides are glucose and fructose. *Glucose* is the primary form of sugar stored in the human body for energy. *Fructose* is the main sugar found in most fruits. Both glucose and fructose have the same chemical formula ($C_6H_{12}O_6$). However, they have different structures (Figure 2.7).

2.6.2.2 Complex Carbohydrates

Complex carbohydrates are polymers of the simple sugars. In other words, the complex carbohydrates are long chains of simple sugar units bonded together (for this reason, the complex carbohydrates are often referred to as *polysaccharides*). The potato we discussed earlier actually contains the complex carbohydrate *starch.* Starch is a polymer of the monosaccharide glucose. It is the principal polysaccharide used by plants to store glucose for later use as energy. Plants often store starch in seeds or other specialized organs. Some sources of starch are rice, beans, wheat, corn, and potatoes. When humans eat starch, an enzyme called *amylase* breaks the bonds between the repeating glucose units, thus allowing the sugar to be absorbed into the bloodstream. Once absorbed into the bloodstream, the human body distributes glucose to the different organs and stores it in the form of glycogen. *Glycogen*, a polymer of glucose, is the polysaccharide used by animals to store energy. Excess glucose is bonded together to form glycogen molecules, which the animal stores in the liver and muscle tissue as an "instant" source of energy. Both starch

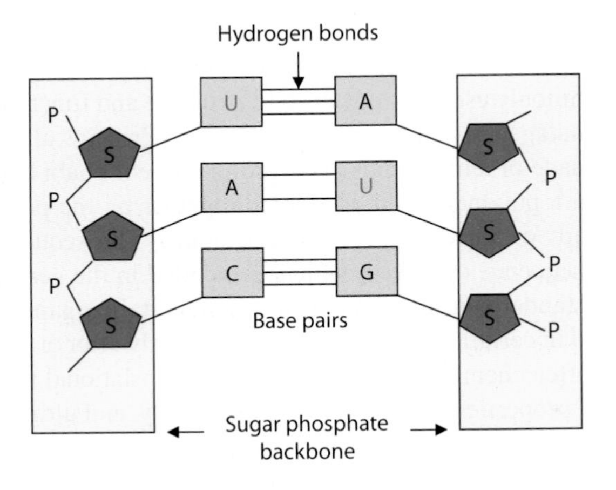

FIGURE 2.7 RNA structure.

and glycogen are polymers of glucose. However, starch is a long, straight chain of glucose units, whereas glycogen is a branched chain of glucose units.

Another important polysaccharide is *cellulose*. Cellulose is yet a third polymer of the monosaccharide glucose. Cellulose differs from starch and glycogen because the glucose units form a two-dimensional structure, with hydrogen bonds holding together nearby polymers, thus giving the molecule added stability. Cellulose, also known as *plant fiber*, cannot be digested by humans; therefore cellulose passes through the digestive tract without being absorbed into the body. Some animals, such as cows and termites, contain bacteria in their digestive tract that help them to digest cellulose. Cellulose is a relatively stiff material. In plants, cellulose is used as a structural molecule to add support to the leaves, stem, and other plant parts. Despite the fact that it cannot be used as an energy source in most animals, cellulose fiber is essential in the diet because it helps exercise the digestive track and keep it clean and healthy.

2.6.3 Lipids

Lipids are a broad group of naturally-occurring molecules which includes fats, waxes, sterols, fat-soluble vitamins (such as vitamins A, D, E, and K), monoglycerides, diglycerides, phospholipids, and others. The main functions of lipids are as energy storage, as structural components of cell membranes, and as important signaling molecules. *Glycerophospholipids* are the main structural component of biological membranes such as the cellular plasma membrane and the intracellular membranes of organelles. In animal cells, the plasma membrane physically separates the intracellular components from the extracellular environment. *Triacylglycerols*, stored in adipose tissue, are a major form of energy storage in animals. In recent years, evidence showed that lipid signaling is a vital part of the cell signaling. Lipid signaling may occur via activation of G protein-coupled or nuclear receptors, and members of several different lipid categories have been identified as signaling molecules and cellular messengers. Lipids may be broadly defined as *hydrophobic* or *amphiphilic* small molecules. The amphiphilic nature of some lipids allows them to form structures such as vesicles, liposomes, or membranes in an aqueous environment. Biological lipids originate entirely or in part from two distinct types of biochemical subunits or "building blocks": *ketoacyl* and *isoprene* groups. Using this approach, lipids may be divided into eight categories: fatty acyls, glycerolipids, glycerophospholipids, sphingolipids, saccharolipids, polyketides (derived from condensation of ketoacyl subunits), sterol lipids, and prenol lipids (derived from condensation of isoprene subunits). Although the term *lipid* is sometimes used as a synonym for *fats*, fats are a subgroup of lipids called *triglycerides*. Lipids also encompass molecules such as fatty acids and their derivatives (including tri-, di-, and monoglycerides and phospholipids), as well as other sterol-containing metabolites such as cholesterol. Although humans and other mammals use various biosynthetic pathways to both break down and synthesize lipids, some essential lipids cannot be made this way and must be obtained from the diet.

2.6.3.1 Nucleic Acid

Living organisms are complex systems. Hundreds of thousands of proteins exist inside each one of us to help carry out our daily functions. These proteins are produced locally and assembled piece-by-piece to exact specifications. An enormous amount of information is required to manage this complex system correctly. This information, detailing the specific structure of the proteins inside of our bodies, is stored in a set of molecules called *nucleic acids.*

The nucleic acids are very large molecules that have two main parts. The backbone of a nucleic acid is made of alternating sugar and phosphate molecules bonded together in a long chain. Though only four different nucleotide bases can occur in a nucleic acid, each nucleic acid contains millions of bases bonded to it. The order in which these nucleotide bases appear in the nucleic acid is the coding for the information carried in the molecule. In other words, the nucleotide bases serve as a sort of genetic alphabet on which the structure of each protein in our bodies is encoded.

2.7 GENES AND GENETICS

Recall that DNA contains the genetic information that allows all living organisms to function and reproduce. However, it is unclear how long in the 4-billion-year history of life DNA has performed this function, as it has been proposed that the earliest forms of life may have used RNA as their genetic material. RNA may have acted as the central part of early cell metabolism as it can both transmit genetic information and carry out catalysis as part of ribozymes. This ancient RNA world where nucleic acid would have been used for both catalysis and genetics may have influenced the evolution of the current genetic code based on four nucleotide bases. This would occur since the number of unique bases in such an organism is a trade-off between a small number of bases increasing replication accuracy and a large number of bases increasing the catalytic efficiency of ribozymes. Unfortunately, there is no direct evidence of ancient genetic systems, as recovery of DNA from most fossils is impossible. This is because DNA can survive in the environment for less than 1 million years and slowly degrades into short fragments in solution. Claims for older DNA have been made, most notably a report of the isolation of a viable bacterium from a salt crystal 250-million years old, but these claims are controversial.

2.7.1 Mendelian Genetics

For thousands of years, farmers and herders have been selectively breeding their plants and animals to produce more useful hybrids. It was somewhat of a hit or miss process since the actual mechanisms governing inheritance were unknown. Knowledge of these genetic mechanisms finally came as a result of careful laboratory breeding experiments carried out over the last century and a half. By the 1890s, the invention of better microscopes allowed biologists to discover the basic facts of cell division and sexual reproduction. The focus of genetics research then shifted to understanding what really happens in the transmission of hereditary traits from parents to children. A number of hypotheses were suggested to explain heredity, but Gregor Mendel, a little known Central European monk, was the only one who got it more or less right. His ideas published in 1866 but largely went unrecognized until 1900, which was long after his death. While Mendel's research was with plants, the basic underlying principles of heredity that he discovered also apply to people and other animals because the mechanisms of heredity are essentially the same for all complex life forms. Through the selective crossbreeding of common pea plants (*Pisum sativum*) over many generations, Mendel discovered that certain traits show up in offspring without any blending of parent characteristics. For instance, the pea flowers are either purple or

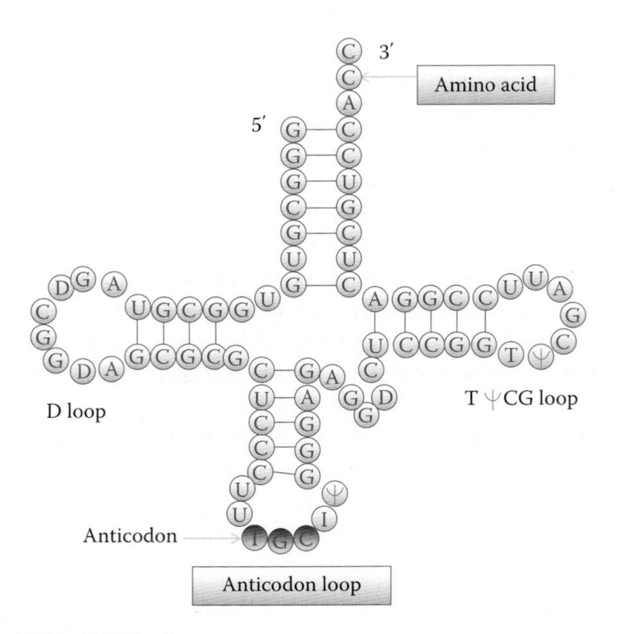

FIGURE 2.8 Transfer RNA (tRNA) structure.

white—intermediate colors do not appear in the offspring of cross-pollinated pea plants. Mendel observed seven traits that were easily recognized as shown in Figure 2.8.

The observation that these traits do not show up in offspring plants with intermediate forms was critically important because the leading theory in biology at the time was that inherited traits blend from generation to generation. Most of the leading scientists in the nineteenth century accepted this "blending theory." Charles Darwin proposed another equally wrong theory known as "pangenesis." This held that hereditary "particles" in our bodies are affected by the things we do during our lifetime. These modified particles were thought to migrate via blood to the reproductive cells and subsequently could be inherited by the next generation. This was essentially a variation of Lamarck's incorrect idea of the "inheritance of acquired characteristics." Mendel picked common garden pea plants for the focus of his research because they can be grown easily in large numbers and their reproduction can be manipulated. Pea plants have both male and female reproductive organs. As a result, they can either self-pollinate themselves or cross-pollinate with another plant. In his experiments, Mendel was able to selectively cross-pollinate purebred plants with particular traits and observe the outcome over many generations. This was the basis for his conclusions about the nature of genetic inheritance. The following lists Mendel's conclusions from the result of his experiments:

1. The inheritance of each trait is determined by genes that are passed on to descendants unchanged.
2. An individual inherits one such gene from each parent for each trait.
3. A trait may not show up in an individual but can still be passed on to the next generation.

It is important to realize that, in this experiment, the starting parent plants were homozygous for pea seed color. That is to say, they each had two identical forms (or alleles) of the gene for this trait—2 yellows (Y) or 2 greens (G). The plants in the first offspring (F1) generation were all heterozygous. In other words, they each had inherited two different alleles—one from each parent plant. It becomes clearer when we look at the actual genetic makeup, or genotype, of the pea plants instead of only the phenotype, or observable physical characteristics. Note that each of the F1 generation plants inherited a Y allele from one parent and a G allele from the other. When the F1 plants breed, each has an equal chance of passing on either Y or G alleles to each offspring. With all of the seven pea plant traits that Mendel examined, one form appeared dominant over the other, which means it masked the presence of the other allele. For example, when the genotype for pea seed color is YG (heterozygous), the phenotype is yellow. However, the dominant yellow allele does not alter the recessive green one in any way. Both alleles can be passed on to the next generation unchanged.

Mendel's observations from his experiments can be summarized in two principles: the principle of segregation and the principle of independent assortment. According to the principle of segregation, for any particular trait, the pair of alleles of each parent separate and only one allele passes on from each parent to an offspring. Which allele in a parent's pair of alleles is inherited is a matter of chance. This segregation of alleles occurs during meiosis. According to the principle of independent assortment, different pairs of alleles are passed on to offspring independently of each other. The result is that, new combinations of genes present in neither parent are possible. For example, a pea plant's inheritance of the ability to produce purple flowers instead of white ones does not make it more likely that it will also inherit the ability to produce yellow pea seeds in contrast to green ones. Likewise, the principle of independent assortment explains why the human inheritance of a particular eye color does not increase or decrease the likelihood of having six fingers on each hand. Today, we know this is due to the fact that the genes for independently assorted traits are located in different chromosomes. These two principles of inheritance, along with the understanding of unit inheritance and dominance, were the beginnings of our modern science of genetics.

2.7.2 Modern Genetics

It was not until the late 1940s and early 1950s that most biologists accepted the evidence showing that DNA must be the chromosomal component that carries hereditary information. One of the most convincing experiments was that of Alfred Hershey and Martha Chase who, in 1952, used radioactive labeling to reach this conclusion. This team of biologists grew a particular type of phage, known as T2, in the presence of two different radioactive labels so that the phage DNA incorporated radioactive phosphorus (32P), while the protein incorporated radioactive sulfur (35S). They then allowed the labeled phage particles to infect non-radioactive bacteria trying to find the label associated with the infected cell. Their analysis showed that most of the 32P-label was found inside of the cell, while most of the 35S was found outside. This suggested that the proteins of the T2 phage remained outside of the newly infected bacterium while the phage-derived DNA was injected into the cell. They then showed that the phage-derived DNA caused the infected cells to produce new phage particles. This elegant work showed, conclusively, that DNA is the molecule which holds genetic information. Meanwhile, much of the scientific world was asking questions about the physical structure of the DNA molecule, and the relationship of that structure to its complex functioning. In 1951, the then 23-year-old biologist James Watson traveled from the United States to work with Francis Crick, an English physicist at the University of Cambridge. Crick was already using the process of x-ray crystallography to study the structure of protein molecules. Together, Watson and Crick used x-ray crystallography data, produced by Rosalind Franklin and Maurice Wilkins at King's College in London, to decipher DNA's structure. The discovery of double helix structure of DNA marks the beginning of modern biotechnology.

2.8 CELL DIVISION

We have learned in the previous section about DNA and its structural and functional attributes especially during meiosis and mitosis when a cell undergoes division. When cells are divided to yield two daughter cells, the genetic material must be divided equally so that each daughter cell contains identical DNA copies. In the next section, we will learn how DNA helps the cells to divide by meiotic and mitotic pathways.

2.8.1 Meiosis

Meiosis was discovered and described for the first time in sea urchin eggs in 1876, by noted German biologist Oscar Hertwig (1849–1922). It was described again in 1883, at the level of chromosomes in *Ascaris* worm's eggs, by Belgian zoologist Edouard Van Beneden (1846–1910). The significance of meiosis for reproduction and inheritance, however, was described only in 1890 by German biologist August Weismann (1834–1914), who noted that two cell divisions were necessary to transform one diploid cell into four haploid cells if the number of chromosomes had to be maintained. In 1911, the American geneticist Thomas Hunt Morgan (1866–1945) observed crossover in *Drosophila melanogaster* meiosis and provided the first genetic evidence that genes are transmitted on chromosomes. Meiosis is a process of reduction division in which the number of chromosomes per cell is reduced to half. In animals, meiosis always results in the formation of gametes, while in other organisms it can give rise to spores. As with mitosis, before meiosis begins, the DNA in the original cell is replicated during the S-phase of the cell cycle. Two cell divisions separate the replicated chromosomes into four haploid gametes or spores.

Meiosis is essential for sexual reproduction and therefore occurs in all eukaryotes (including single-celled organisms) that reproduce sexually. A few eukaryotes, notably the Bdelloid rotifers, have lost the ability to carry out meiosis and have acquired the ability to reproduce by parthenogenesis. Meiosis does not occur in archaea or bacteria, which reproduce via asexual processes such as binary fission. During meiosis, the genome of a diploid germ cell, which is composed of long segments of DNA packaged into chromosomes, undergoes DNA replication followed by two rounds of division, resulting in four haploid cells. Each of these cells contains one complete set of chromosomes, or half of the genetic content of the original cell. If meiosis produces gametes, these cells must fuse during fertilization to create a new diploid cell, or zygote before any new growth can occur. Thus, the division mechanism of meiosis is a reciprocal process to the joining of two genomes that occurs at fertilization. Because the chromosomes of each parent undergo genetic recombination during meiosis, each gamete, and thus each zygote, will have a unique genetic *blueprint* encoded in its DNA. Together, meiosis and fertilization constitute sexuality in the eukaryotes, and generate genetically distinct individuals in populations. In all plants, and in many protists, meiosis results in the formation of haploid cells that can divide vegetatively without undergoing fertilization, referred to as spores. In these groups, gametes are produced by mitosis.

Meiosis uses many of the same biochemical mechanisms employed during mitosis to accomplish the redistribution of chromosomes. There are several features unique to meiosis, most importantly the pairing and genetic recombination between homologous chromosomes (Figure 2.9).

2.8.2 Mitosis

Mitosis is the process in which a eukaryotic cell separates the chromosomes in its cell nucleus into two identical sets in two daughter nuclei. It is generally followed immediately by *cytokinesis*, which divides the nuclei, cytoplasm, organelles, and cell membrane into two daughter cells containing roughly equal shares of these cellular components. Mitosis and cytokinesis together define the *mitotic (M) phase* of the cell cycle—the division of the mother cell into two daughter cells, genetically identical to each other and to their parent cell. Mitosis divides the chromosomes in a cell nucleus. Mitosis occurs exclusively in eukaryotic cells, but occurs in different ways in different species. For example, animals undergo an "open" mitosis, where the nuclear envelope breaks down before the chromosomes separate, while fungi such as *Aspergillus nidulans* and *Saccharomyces cerevisiae* (yeast) undergo a "closed" mitosis, where chromosomes divide within an intact cell nucleus. Prokaryotic cells, which lack a nucleus, divide using binary fission. The process of mitosis is complex and highly regulated. The sequence of events is divided into phases, corresponding to the completion of one set of activities and the start of the next. These stages are prophase, prometaphase, metaphase, anaphase, and telophase. During the process of mitosis, the pairs of chromosomes condense and attach to fibers that pull the sister chromatids to opposite

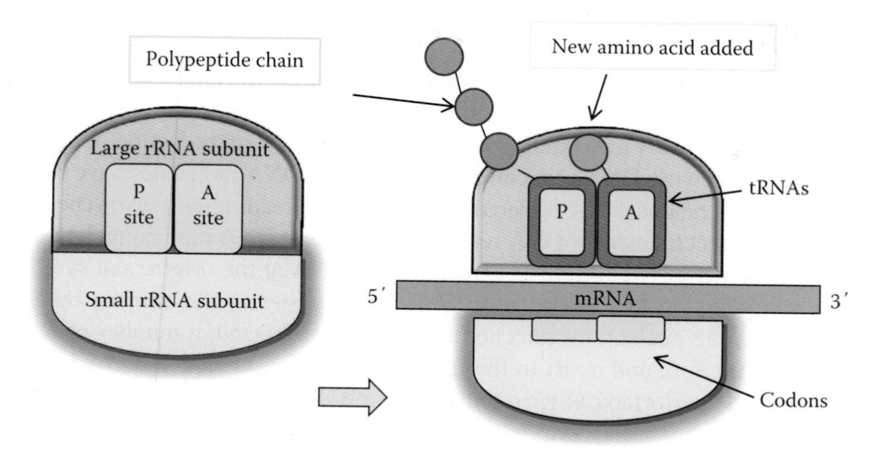

FIGURE 2.9 Ribosomal RNA (rRNA).

sides of the cell. The cell then divides in cytokinesis, to produce two identical daughter cells. Because cytokinesis usually occurs in conjunction with mitosis, "mitosis" is often used interchangeably with "mitotic phase." However, there are many cells where mitosis and cytokinesis occur separately, forming single cells with multiple nuclei. This occurs mostly among the fungi and slime moulds, but also found in various different groups. Even in animals, cytokinesis and mitosis may occur independently such as during certain stages of fruit fly embryonic development. Errors in mitosis can either kill a cell through apoptosis or cause mutations that may lead to cancer (Figure 2.10).

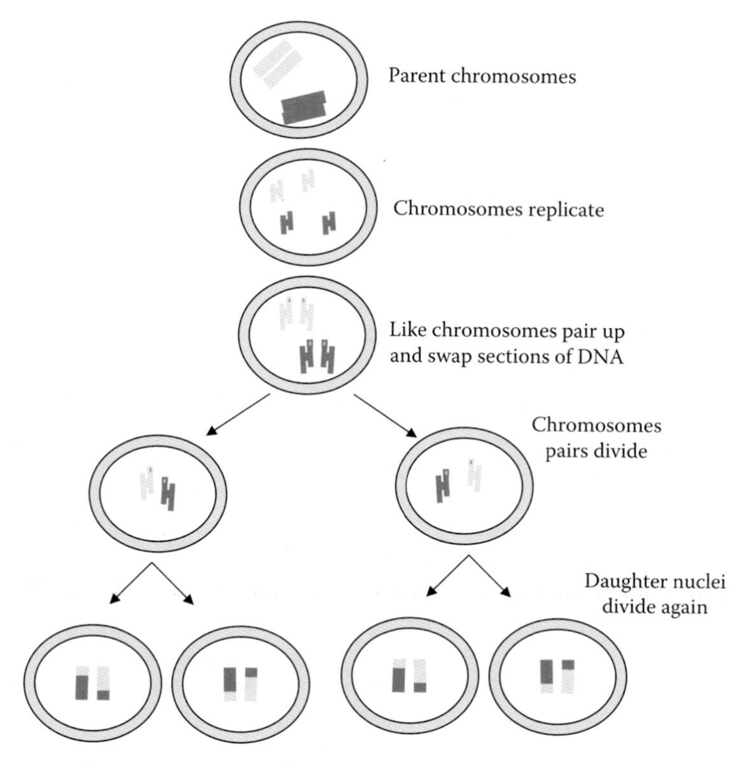

FIGURE 2.10 Stages of meiosis.

2.9 DNA REPLICATION

DNA *replication*, the basis for biological inheritance, is a fundamental process occurring in all living organisms to copy their DNA. This process is "semi-conservative" in that each strand of the original double-stranded DNA molecule serves as template for the reproduction of the complementary strand. Hence, following DNA replication, two identical DNA molecules have been produced from a single double-stranded DNA molecule. Cellular proofreading and error-checking mechanisms ensure near perfect fidelity for DNA replication. In a cell, DNA replication begins at specific locations in the genome, called "origins." Unwinding of DNA at the origin, and synthesis of new strands, forms a replication fork. In addition to DNA *polymerase*, the enzyme that synthesizes the new DNA by adding nucleotides matched to the template strand, a number of other proteins are associated with the fork and assist in the initiation and continuation of DNA synthesis. DNA replication can also be performed *in vitro* (outside a cell). DNA polymerases, isolated from cells, and artificial DNA primers are used to initiate DNA synthesis at known sequences in a template molecule. The *polymerase chain reaction* (PCR), a common laboratory technique, employs such artificial synthesis in a cyclic manner to amplify a specific target DNA fragment from a pool of DNA (Figure 2.11).

2.9.1 Role of DNA Polymerase in Replication

DNA polymerases are a family of enzymes that carry out all forms of DNA replication. A DNA polymerase can only extend an existing DNA strand paired with a template strand; it cannot begin the synthesis of a new strand. To begin synthesis of a new strand, a short fragment of DNA or RNA,

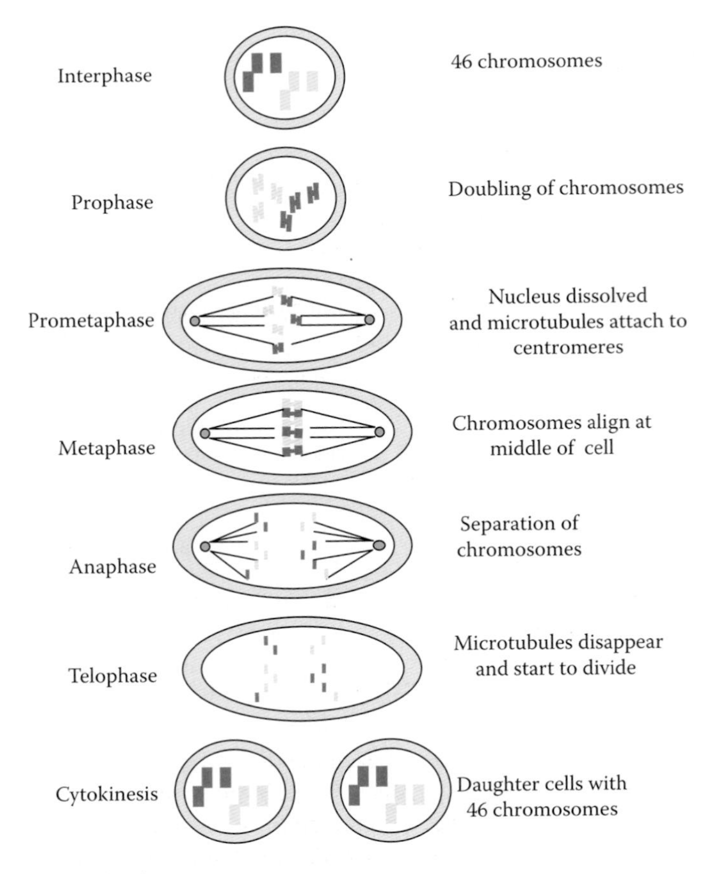

FIGURE 2.11 Stages of mitosis.

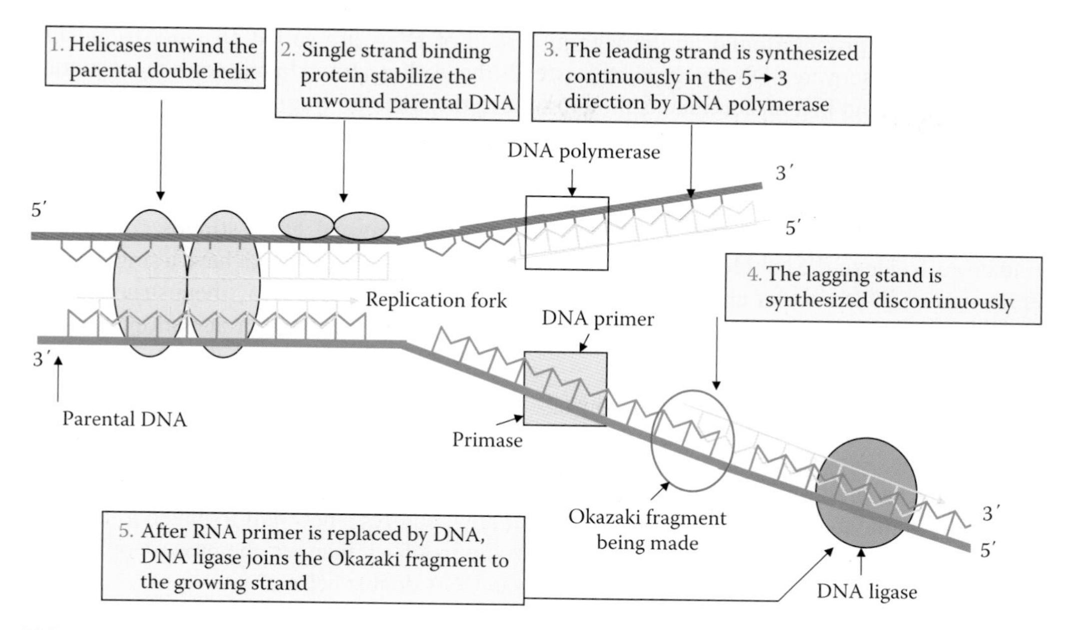

FIGURE 2.12 DNA replication process.

called a *primer* must be created, and paired with the DNA template. Once a primer pairs with DNA to be replicated, DNA polymerase synthesizes a new strand of DNA by extending the 3′ end of an existing nucleotide chain, adding new nucleotides matched to the template strand one at a time via the creation of phosphodiester bonds. The energy for this process of DNA polymerization comes from two of the three total phosphates attached to each unincorporated base. Free bases with their attached phosphate groups are called *nucleoside triphosphates*. When a nucleotide is being added to a growing DNA strand, two of the phosphates are removed and the energy produced creates a phosphodiester (chemical) bond that attaches the remaining phosphate to the growing chain. The energetics of this process also help explain the directionality of synthesis; if DNA were synthesized in the 3′–5′ direction, the energy for the process would come from the 5′ end of the growing strand rather than from free nucleotides. DNA polymerases are generally extremely accurate, making less than one error for every 10^7 nucleotides added. Even so, some DNA polymerases also have proofreading ability; they can remove nucleotides from the end of a strand in order to correct mismatched bases. If the 5′ nucleotide needs to be removed during proofreading, the triphosphate end is lost. Hence, the energy source that usually provides energy to add a new nucleotide is also lost (Figure 2.12).

2.9.2 DNA Replication within the Cell

For a cell to divide, it must first replicate its DNA. This process is initiated at particular points within the DNA, known as "origins," which are targeted by proteins that separate the two strands and initiate DNA synthesis. Origins contain DNA sequences recognized by replication initiator proteins. These initiator proteins recruit other proteins to separate the two strands and initiate replication forks. Initiator proteins recruit other proteins to separate the DNA strands at the origin, forming a bubble. Origins tend to be "AT-rich" (rich in adenine and thymine bases) to assist this process, because A-T base pairs have two hydrogen bonds (rather than the three formed in a C-G pair). Strands rich in these nucleotides are generally easier to separate due to the few flexibility/many durability relationships found in hydrogen bonding. Once strands are separated, RNA primers are created on the template strands. More specifically, the leading strand receives one RNA primer per active origin of replication while the lagging strand receives several. These several fragments of RNA primers found on the lagging strand of DNA are called *Okazaki fragments*, named after their discoverer. DNA polymerase extends the leading strand in one continuous motion and the lagging strand in a discontinuous motion (due to the Okazaki fragments). RNAse removes the RNA fragments used to initiate replication by DNA polymerase, and another DNA polymerase enters to fill the gaps. When this is complete, a single nick on the leading strand and several nicks on the lagging strand can be found. Ligase works to fill in these nicks, thus completing the newly replicated DNA molecule. As DNA synthesis continues, the original DNA strands continue to unwind on each side of the bubble, forming replication forks. In bacteria, which have a single origin of replication on their circular chromosome, this process eventually creates a "theta structure." In contrast, eukaryotes have longer linear chromosomes and initiate replication at multiple origins within these.

2.9.2.1 Replication Fork

When replicating, the original DNA splits in two, forming two "prongs" which resemble a fork, hence the name *replication fork*. DNA has a ladder-like structure. Now, imagine a ladder broken in half vertically, along the steps. Each half of the ladder now requires a new half to match it. Because DNA polymerase can only synthesize a new DNA strand in a 5′–3′ manner, the process of replication goes differently for the two strands comprising the DNA double helix.

2.9.2.2 Leading Strand

The *leading strand* is that strand of the DNA double helix that is orientated in a 5′–3′ manner. On the leading strand, a polymerase "reads" the DNA and continuously adds nucleotides to it. This polymerase is DNA polymerase III (DNA Pol III) in prokaryotes and presumably Pol ε in eukaryotes.

2.9.2.3 Lagging Strand

The *lagging strand* is that strand of the DNA that is orientated in a 3′–5′ manner. Since its orientation is opposite to the working orientation of DNA polymerase III, which is in a 5′–3′ manner, replication of the lagging strand is more complicated than that of the leading strand. On the lagging strand, primase "reads" the DNA and adds RNA to it in short, separated segments. In eukaryotes, primase is intrinsic to Pol αDN. A polymerase III or Pol δ lengthens the primed segments, forming Okazaki fragments. Primer removal in eukaryotes is also performed by Pol δ in prokaryotes, DNA polymerase I "reads" the fragments, removes the RNA using its flap endonuclease domain, and replaces the RNA nucleotides with DNA nucleotides. This is necessary because RNA and DNA use slightly different kinds of nucleotides. DNA ligase joins the fragments together.

2.9.3 Regulation of DNA Replication

In eukaryotes, DNA replication is controlled within the context of the cell cycle. As the cell grows and divides, it progresses through stages in the cell cycle. DNA replication occurs during the S phase (Synthesis phase). The progress of the eukaryotic cell through the cycle is controlled by cell cycle checkpoints. Progression through checkpoints is controlled through complex interactions between various proteins, including cyclins and cyclin-dependent kinases. The G1/S checkpoint (or restriction checkpoint) regulates whether eukaryotic cells enter the process of DNA replication and subsequent division. Cells which do not proceed through this checkpoint are quiescent in the "G0" stage and do not replicate their DNA. Replication of chloroplast and mitochondrial genomes occur independent of the cell cycle, through the process of D-loop replication.

Most bacteria do not go through a well-defined cell cycle and instead continuously copy their DNA. During rapid growth, this can result in multiple rounds of replication occurring concurrently. Within *E. coli*, the well-characterized bacteria, regulation of DNA replication can be achieved through several mechanisms, including the hemi-methylation and sequestering of the origin sequence, the ratio of ATP to ADP, and the levels of protein DnaA. All these control the process of initiator proteins binding to the origin sequences. Because *E. coli* methylates GATC DNA sequences, DNA synthesis results in hemimethylated sequences. This hemimethylated DNA is recognized by a protein (SeqA) which binds and sequesters the origin sequence. In addition, DnaA (required for initiation of replication) binds less well to hemimethylated DNA. As a result, newly replicated origins are prevented from immediately initiating another round of DNA replication. ATP builds up when the cell is in a rich medium, triggering DNA replication once the cell has reached a specific size. ATP competes with ADP to bind with DnaA, and the DnaA-ATP complex is able to initiate replication. A certain number of DnaA proteins are also required for DNA replication. Each time the origin is copied, the number of binding sites for DnaA doubles, requiring the synthesis of more DnaA to enable another initiation of replication.

2.9.4 Termination of Replication

The chromosomes in bacteria are circular, so termination of replication occurs when the two replication forks meet each other on the opposite end of the parental chromosome. *E. coli* regulates this process through the use of termination sequences which, when bound by the Tus protein, enable only one direction of replication fork to pass through. As a result, the replication forks are constrained to always meet within the termination region of the chromosome. Eukaryotes initiate DNA replication at multiple points in the chromosome, so replication forks meet and terminate at many points in the chromosome. These are not known to be regulated in any particular manner. Because eukaryotes have linear chromosomes, DNA replication often fails to synthesize to the very end of the chromosomes (telomeres), resulting in telomere shortening. This is a normal process in somatic cells. Here, cells are only able to divide a certain number of times before the DNA loss prevents further division. This is known as the *Hayflick limit*. Within the germ cell line, which passes DNA to the next generation, the enzyme telomerase extends the repetitive sequences of the telomere region to prevent degradation. Telomerase can become mistakenly active in somatic cells, sometimes leading to cancer formation.

2.10 DNA INTERACTIONS WITH PROTEINS

Besides DNA's role in protein synthesis, the function of DNA depends on interactions with proteins and these protein interactions can be nonspecific, or the protein can bind specifically to a single DNA sequence. Enzymes can also bind to DNA and of these, the polymerases that copy the DNA base sequence in transcription and DNA replication are particularly important. Structural proteins that bind DNA are well-understood examples of nonspecific DNA–protein interactions. Within chromosomes, DNA is held in complexes with structural proteins. These proteins organize the DNA into chromatin. In eukaryotes, this structure involves DNA binding to histones, while in prokaryotes multiple types of proteins are involved. The histones form a disk-shaped complex called a *nucleosome*, which contains two complete turns of double-stranded DNA wrapped around its surface. These nonspecific interactions are formed through basic residues in the histones making ionic bonds to the acidic sugar-phosphate backbone of the DNA, and are therefore largely independent of the base sequence. Chemical modifications of these basic amino acid residues include methylation, phosphorylation, and acetylation. These chemical changes alter the strength of the interaction between the DNA and the histones, making the DNA more or less accessible to transcription factors and changing the rate of transcription. Other nonspecific DNA-binding proteins in chromatin include the high-mobility group proteins, which bind to bent or distorted DNA. These proteins are important in bending arrays of nucleosomes and arranging them into the larger structures that make up chromosomes.

A distinct group of DNA-binding proteins is one that specifically binds single-stranded DNA. In humans, replication protein A is the best-understood member of this group and is used in processes where the double helix is separated, including DNA replication, recombination, and DNA repair.

These binding proteins seem to stabilize single-stranded DNA and protect it from forming stem-loops or being degraded by nucleases. In contrast, other proteins have evolved to bind to particular DNA sequences. The most intensively studied of these are the various transcription factors, which are proteins that regulate transcription. Each transcription factor binds to one particular set of DNA sequences and activates or inhibits the transcription of genes that have these sequences close to their promoters. The transcription factors can do this in two ways. They can bind the RNA polymerase responsible for transcription directly or through other mediator proteins. This locates the polymerase at the promoter and allows it to begin transcription. Alternatively, transcription factors can bind enzymes that modify the histones at the promoter. This will change the accessibility of the DNA template to the polymerase. As these DNA targets can occur throughout an organism's genome, changes in the activity of one type of transcription factor can affect thousands of genes. Consequently, these proteins are often the targets of the signal transduction processes that control responses to environmental changes or cellular differentiation and development. The specificity of these transcription factor interactions with DNA come from the proteins making multiple contacts to the edges of the DNA bases, allowing them to read the DNA sequence. Most of these base-interactions are made in the major groove, where the bases are most accessible.

2.11 DNA-MODIFYING ENZYMES

Nucleases are enzymes that cut DNA strands by catalyzing the hydrolysis of the phosphodiester bonds. Nucleases that hydrolyse nucleotides from the ends of DNA strands are called *exonucleases*, while *endonucleases* cut within strands. The most frequently used nucleases in molecular biology are the *restriction endonucleases*, which cut DNA at specific sequences. For instance, the EcoRV enzyme recognizes the 6-base sequence 5′-GAT|ATC-3′ and makes a cut at the vertical line. In nature, these enzymes protect bacteria against phage infection by digesting the phage DNA when it enters the bacterial cell, acting as part of the restriction modification system. In technology, these sequence-specific nucleases are used in molecular cloning and DNA fingerprinting. Enzymes called DNA ligases can rejoin cut or broken DNA strands. Ligases are particularly important in lagging strand DNA replication, as they join together the short segments of DNA produced at the replication fork into a complete copy of the DNA template. They are also used in DNA repair and genetic recombination.

2.11.1 Topoisomerases and Helicases

Topoisomerases are enzymes with both nuclease and ligase activity. These proteins change the amount of supercoiling in DNA and some of these enzymes work by cutting the DNA helix and allowing one section to rotate, thereby reducing its level of supercoiling; the enzymes then seal the DNA break. Other types of enzymes are capable of cutting one DNA helix and then passing a second strand of DNA through this break, before rejoining the helix. Topoisomerases are required for many processes involving DNA, such as DNA replication and transcription. *Helicases* are proteins that are a type of molecular motor. They use the chemical energy in nucleoside triphosphates, predominantly ATP, to break hydrogen bonds between bases and unwind the DNA double helix into single strands. These enzymes are essential for most processes where enzymes need to access the DNA bases.

2.12 DNA METHYLATION

DNA methylation is one such post-synthesis modification. DNA methylation has been proven by research to be manifested in a number of biological processes such as regulation of imprinted genes, X chromosome inactivation, and tumor suppressor gene silencing in cancerous cells. DNA methylation also acts as a protection mechanism bacterial pathogen DNA against the endonuclease activity that destroys any foreign DNA. The expression of genes is influenced by how the DNA is packaged in chromosomes through chromatin. Base modifications can be involved in packaging, with

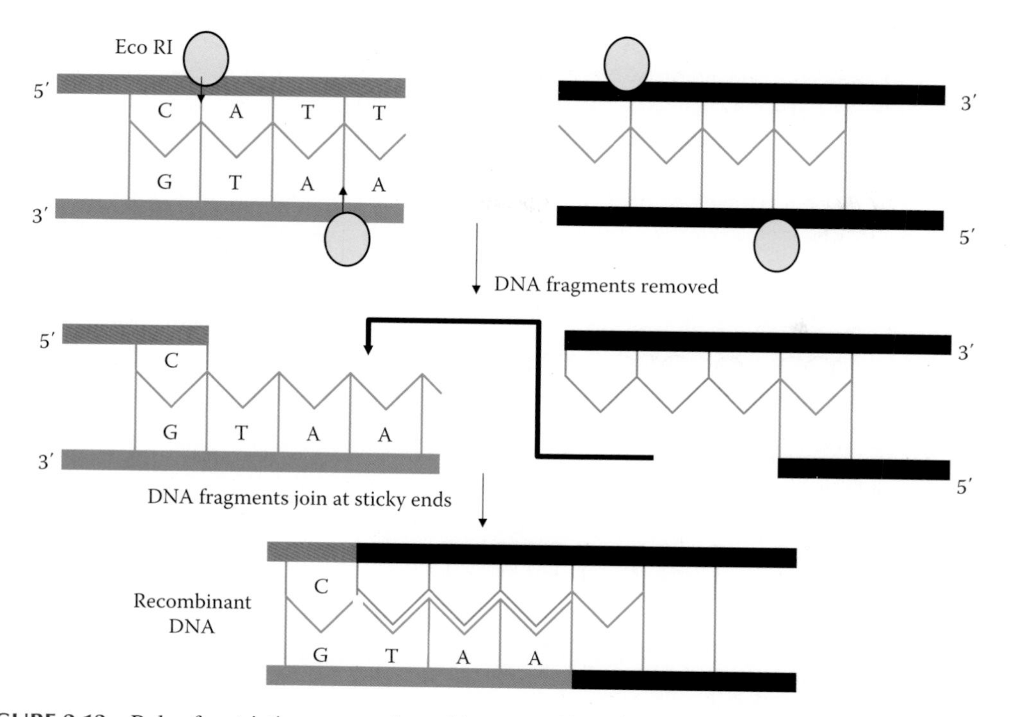

FIGURE 2.13 Role of restriction enzyme in making recombinant DNAATTCCC5.

regions that have low or no gene expression, usually containing high levels of methylation of cytosine bases. For example, cytosine methylation produces 5-methylcytosine, which is important for X-chromosome inactivation. The average level of methylation varies among organisms—the worm *Caenorhabditis elegans* lacks cytosine methylation, while vertebrates have higher levels, with up to 1% of their DNA containing 5-methylcytosine. Despite the importance of 5-methylcytosine, it can deaminate to leave a thymine base; methylated cytosines are therefore particularly prone to mutations. Other base modifications include adenine methylation in bacteria and the glycosylation of uracil to produce the "J-base" in kinetoplastids (Figure 2.13).

2.13 DNA MUTATION

DNA mutation is a change in the sequence of DNA. It can be caused by copying errors in the genetic material during cell division, by exposure to ultraviolet/ionizing radiation, chemical mutagens, viruses, or by cellular processes such as hyper-mutation. It can also be induced by the organism itself. In multicellular organisms with dedicated reproductive cells, mutations can be subdivided into *germ line mutations*, which can be passed on to descendants through the reproductive cells, and *somatic mutations*, which involve cells outside the dedicated reproductive group and which are not usually transmitted to descendants. If an organism can reproduce asexually through mechanisms such as budding, the distinction can become blurred. For example, plants can sometimes transmit somatic mutations to their descendants asexually or sexually where flower buds develop in somatically mutated parts of plants. A new mutation that was not inherited from either parent is called a *de novo* mutation. The source of the mutation is unrelated to the consequence, although the consequences are related to which cells were mutated. DNA can be damaged by many different sorts of mutagens, which change the DNA sequence. Mutagens include oxidizing agents, alkylating agents, and also high-energy electromagnetic radiation such as ultraviolet light and x-rays. The type of DNA damage produced depends on the type of mutagen. For example, UV light can damage DNA by producing thymine dimers, which

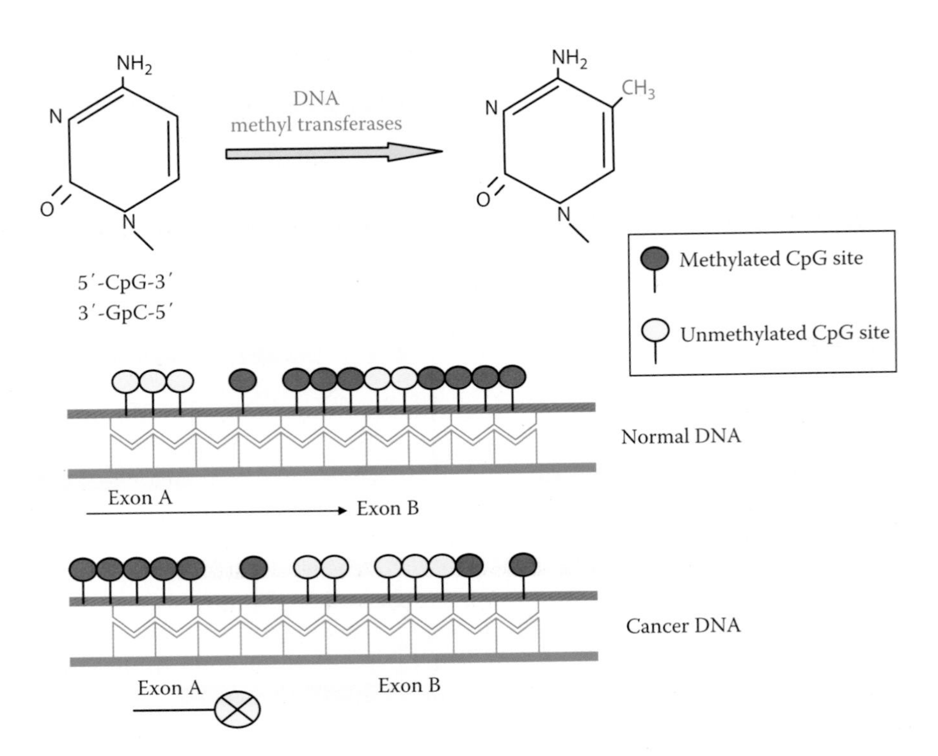

FIGURE 2.14 DNA methylation.

are cross-links between pyrimidine bases. On the other hand, oxidants such as free radicals or hydrogen peroxide produce multiple forms of damage, including base modifications, particularly of guanosine and double-strand breaks.

A typical human cell contains about 150,000 bases that have suffered oxidative damage. Of these oxidative lesions, the most dangerous are double-strand breaks since these are difficult to repair and can produce point mutations, insertions/deletions from the DNA sequence, and chromosomal translocations. Many mutagens fit into the space between two adjacent base pairs in a process called *intercalation*. Most intercalators are aromatic and planar molecules such as ethidium bromide, daunomycin, and doxorubicin. For an intercalator to fit between base pairs, the bases must separate, distorting the DNA strands by unwinding of the double helix. This inhibits both transcription and DNA replication, causing toxicity and mutations. As a result, DNA intercalators are often carcinogens. Some well-known examples of DNA intercalators are Benzo[*a*]pyrene diol epoxide, acridines, aflatoxin and ethidium bromide. Nevertheless, due to their ability to inhibit DNA transcription and replication, other similar toxins are also used in chemotherapy to inhibit rapidly growing cancer cells (Figure 2.14).

2.14 TOOLS OF BIOTECHNOLOGY

With the advancement of molecular biology, various tools have been produced to study DNA and RNA with utmost accuracy and precision. In this section, we will discuss some of the major tools which have greatly impacted the field of biotechnology.

2.14.1 Polymerase Chain Reaction

PCR is a technique to amplify a single or a few copies of a piece of DNA across several orders of magnitude, generating thousands to millions of copies of a particular DNA sequence. PCR was developed in 1983 by Kary Mullis. Now, it is a common and most indispensable technique used in medical and biological research laboratories for a variety of applications such as DNA cloning for sequencing, DNA-based phylogeny, functional analysis of genes, diagnosis of hereditary diseases, identification of genetic fingerprints, and detection and diagnosis of infectious diseases. The method relies on thermal cycling, consisting of cycles of repeated heating and cooling of the reaction for DNA melting and enzymatic replication of the DNA. Primers which are basically short DNA fragments containing sequences complementary to the target region along with a DNA polymerase are key components to enable selective and repeated amplification. As PCR progresses, the DNA generated is itself used as a template for replication, setting in motion a chain reaction in which the DNA template is exponentially amplified. PCR can also be extensively modified to perform a wide array of genetic manipulations.

Almost all PCR applications employ a heat-stable DNA polymerase, such as *Taq polymerase*, an enzyme originally isolated from the bacterium *Thermus aquaticus*. This DNA polymerase enzymatically assembles a new DNA strand from nucleotides by using single-stranded DNA as a template and DNA oligonucleotides (also called DNA primers) which are required for initiation of DNA synthesis. The vast majority of PCR methods use thermal cycling, that is, alternately heating and cooling the PCR sample in a defined series of temperature steps. These thermal cycling steps are necessary to physically separate the two strands in a DNA double helix at a high temperature in a process called *DNA melting*. At a lower temperature, each strand is then used as the template in DNA synthesis by the DNA polymerase to selectively amplify the target DNA. The selectivity of PCR results from the use of primers that are complementary to the DNA region targeted for amplification under specific thermal cycling conditions (Figure 2.15).

Most PCR methods typically amplify DNA fragments of up to ~10 kb pairs, although some techniques allow for amplification of fragments up to 40 kb in size.

The PCR is commonly carried out in a reaction volume of 10–200 μL in small reaction tubes (0.2–0.5 mL volumes) in a thermal cycler. The thermal cycler heats and cools the reaction tubes to achieve the temperatures required at each step of the reaction. Many modern thermal cyclers make use of the *Peltier effect* which permits both heating and cooling of the block holding the PCR tubes simply by reversing the electric current. Thin-walled reaction tubes permit favorable thermal conductivity to allow for rapid thermal equilibration. Most thermal cyclers have heated lids to prevent condensation at the top of the reaction tube. Older thermocyclers lacking a heated lid require a layer of oil on top of the reaction mixture or a ball of wax inside the tube.

2.14.1.1 Principle and Practice of PCR

The PCR usually consists of a series of 20–40 repeated temperature changes called *cycles*. Each cycle typically consists of 2–3 discrete temperature steps. Most commonly, PCR is carried out with

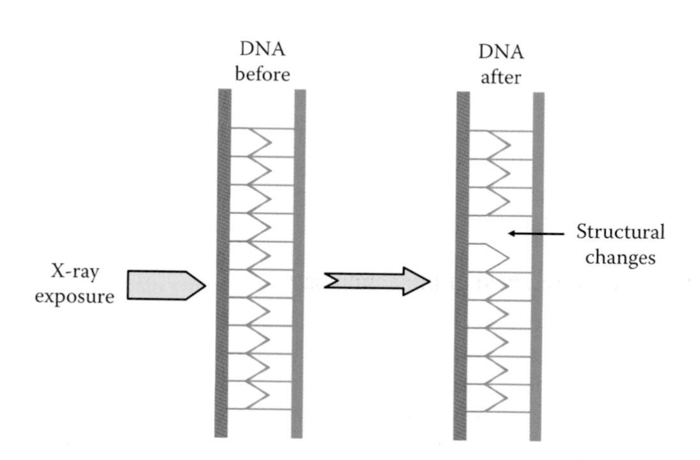

FIGURE 2.15 DNA mutation.

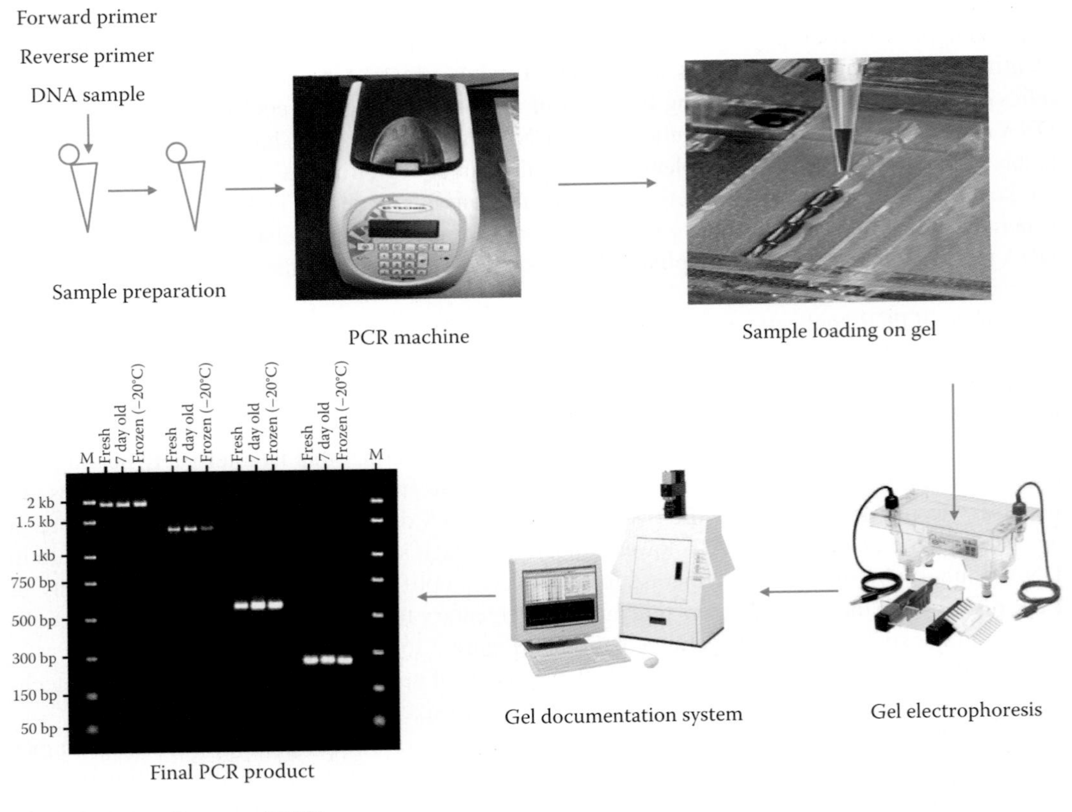

FIGURE 2.16 Process of PCR.

cycles that have three temperature steps. The cycling is often preceded by a single temperature step (called *hold*) at a high temperature (>90°C), and followed by one hold at the end for a final product extension or brief storage. The temperatures used and the length of time they are applied in each cycle depend on a variety of parameters. These include the enzyme used for DNA synthesis, the concentration of divalent ions and dNTPs in the reaction, and the melting temperature (Tm) of the primers (Figure 2.16).

- *Initialization step*: This step consists of heating the reaction to a temperature of 94°C–96°C (or 98°C if extremely thermostable polymerases are used), which is held for 1–9 min. It is only required for DNA polymerases that require heat activation by hot-start PCR.
- *Denaturation step*: This step is the first regular cycling event and consists of heating the reaction to 94°C–98°C for 20–30 s. It causes DNA melting of the DNA template by disrupting the hydrogen bonds between complementary bases, yielding single strands of DNA.
- *Annealing step*: The reaction temperature is lowered to 50°C–65°C for 20–40 s allowing annealing of the primers to the single-stranded DNA template. Typically the annealing temperature is about 3°C–5°C below the Tm of the primers used. Stable DNA–DNA hydrogen bonds are only formed when the primer sequence very closely matches the template sequence. The polymerase binds to the primer-template hybrid and begins DNA synthesis.
- *Elongation step*: The temperature at this step depends on the DNA polymerase used. Taq polymerase has its optimum activity temperature at 75°C–80°C, and commonly a temperature of 72°C is used with this enzyme. At this step, the DNA polymerase synthesizes a new DNA strand complementary to the DNA template strand by adding dNTPs that are complementary to the template in 5′–3′ direction, condensing the 5′-phosphate group of

the dNTPs with the 3′-hydroxyl group at the end of the nascent (extending) DNA strand. The extension time depends both on the DNA polymerase used and on the length of the DNA fragment to be amplified. As a rule-of-thumb, at its optimum temperature, the DNA polymerase will polymerize a thousand bases per minute. Under optimum conditions, that is, if there are no limitations due to limiting substrates or reagents, at each extension step, the amount of DNA target is doubled, leading to exponential (geometric) amplification of the specific DNA fragment.

- *Final elongation*: This single step is occasionally performed at a temperature of 70°C–74°C for 5–15 m after the last PCR cycle to ensure that any remaining single-stranded DNA is fully extended.
- *Final hold*: This step at 4°C–15°C for an indefinite time may be employed for short-term storage of the reaction. The size of PCR products is determined by comparison with a DNA ladder which is basically a molecular weight marker, containing DNA fragments of known size, run on the gel alongside the PCR products.

2.14.1.2 PCR Optimization

In order to get correct amplifications, PCR reactions must be standardized by trial- and-error steps. PCR can fail for various reasons, in part due to its sensitivity to contamination causing amplification of spurious DNA products. Because of this, a number of techniques and procedures have been developed for optimizing PCR conditions. Contamination with extraneous DNA is addressed with laboratory protocols and procedures that separate pre-PCR mixtures from potential DNA contaminants. This usually involves spatial separation of PCR-setup areas from areas for analysis or purification of PCR products, use of disposable plastic wares, and thoroughly cleaning the work surface between reaction setups. Primer-design techniques are important in improving PCR product yield and in avoiding the formation of spurious products, and the usage of alternate buffer components or polymerase enzymes can help with amplification of long or otherwise problematic regions of DNA.

2.14.2 Types of PCR Reactions

2.14.2.1 Allele-Specific PCR

Allele-specific PCR is a diagnostic or cloning technique which is based on single-nucleotide polymorphisms (SNPs), or single-base differences in DNA. Selective PCR amplification of one of the alleles has been used to detect SNP. Selective amplification is usually achieved by designing a primer such that the primer will match or mismatch one of the alleles at the 3′-end of the primer.

2.14.2.2 Polymerase Cycling Assembly

Polymerase cycling assembly (PCA) is a method for the assembly of large DNA oligonucleotides from shorter fragments. This process uses the same technology as PCR, but takes advantage of DNA hybridization and annealing as well as DNA polymerase to amplify a complete sequence of DNA in a precise order based on the single stranded oligonucleotides used in the process. It thus allows for the production of synthetic genes and even entire synthetic genomes.

2.14.2.3 Asymmetric PCR

Asymmetric PCR is used to preferentially amplify one DNA strand in a double-stranded DNA template. It is used in some types of sequencing and hybridization probing where amplification of only one of two complementary strands is ideal. PCR is carried out as usual, but with a great excess of the primers for the chosen strand. Due to the slow (arithmetic) amplification later in the reaction after the limiting primer has been used up, extra cycles of PCR are required.

2.14.2.4 Helicase-Dependent Amplification

Helicase-dependent amplification (HDA) is a method for *in vitro* DNA amplification like PCR, but that works at constant temperature. PCR is the most widely used method for *in vitro* DNA amplification for purposes of molecular biology and biomedical research. This process involves the separation of the double-stranded DNA in high heat into single strands (the denaturation step, typically achieved at 95°C–97°C), annealing of the primers to the single stranded DNA (the annealing step), and copying the single strands to create new double-stranded DNA (the extension step that requires the DNA polymerase). It also requires the reaction to be done in a thermal cycler. These bench-top machines are large, expensive, and costly to run and maintain, limiting the potential applications of DNA amplification in situations outside the laboratory (e.g., in the identification of potentially hazardous microorganisms at the scene of investigation, or at the point of care of a patient). *In vivo*, DNA is replicated by DNA polymerases with various accessory proteins, including a DNA helicase that acts to separate the DNA by unwinding the DNA double helix. HDA was developed from this concept, using a helicase (an enzyme) to denature the DNA.

2.14.2.5 Hot-Start PCR

Hot-start PCR is a modification of the conventional PCR that reduces nonspecific product amplification. In this procedure, amplification cannot occur until the reaction temperature is above that where nonspecific annealing of primers to targets occurs. This block in amplification is usually accomplished by using a DNA polymerase that is inactive until higher temperatures are reached.

2.14.2.6 AFLP-PCR

Amplified fragment length polymorphism (AFLP)-PCR was originally described by Zabeau and Vos in 1993. The procedure of this technique is divided into three steps: (1) Digestion of total cellular DNA with one or more restriction enzymes and ligation of restriction half-site specific adaptors to all restriction fragments; (2) Selective amplification of some of these fragments with two PCR primers that have corresponding adaptor and restriction site specific sequences; and (3) Electrophoretic separation and amplicons on a gel matrix, followed by visualization of the band pattern.

2.14.2.7 Alu PCR

Alu PCR is a PCR which uses a primer that anneals to Alu repeats to amplify DNA located between two oppositely oriented Alu sequences. It is also used as a method to obtain a fingerprint of bands from an uncharacterized human DNA.

2.14.2.8 Colony PCR

The *Colony PCR* is the screening of bacterial (*E. coli*) or yeast clones for correct ligation or plasmid products. Selected colonies of bacteria or yeast are picked with a sterile toothpick or pipette tip from a growth (agarose) plate. This is then inserted into the PCR master mix or pre-inserted into autoclaved water. PCR is then conducted to determine if the colony contains the DNA fragment or plasmid of interest.

2.14.2.9 Inverse PCR

Inverse PCR is a variant of the polymerase chain reaction that is used to amplify DNA with only one known sepuence. One limitation of conventional PCR is that it requires primers complementary to both termini of the target DNA, but this method allows PCR to be carried out even if only one sequence is avilable from which primers may be designed. Inverse PCR is especially useful for the determination of insert locations. For example, various retroviruses and transposons randomly integrate into genomic DNA. To identify the sites where they have entered, the known, "internal" viral or transposon sequences can be used to design primers that will amplify a small portion of the flanking, "external" genomic DNA. The amplified product can then be sequenced and compared with DNA databases to locate the sequence which has been disrupted.

2.14.2.10 Ligation-Mediated PCR

Ligation-mediated PCR uses small DNA linkers ligated to the DNA of interest and multiple primers annealing to the DNA linkers. It has been used for DNA sequencing, genome walking, and DNA footprinting.

2.14.2.11 Methylation-Specific PCR

Methylation-specific PCR (MSP) is developed by Stephen Baylin and Jim Herman at the Johns Hopkins School of Medicine. It is used to detect methylation of CpG islands in genomic DNA. DNA is first treated with sodium bisulfate, which converts un-methylated cytosine bases to uracil, which is recognized by PCR primers as thymine. Two PCRs are then carried out on the modified DNA using primer sets identical except at any CpG islands within the primer sequences. At these points, one primer set recognizes DNA with cytosines to amplify methylated DNA, and one set recognizes DNA with uracil or thymine to amplify unmethylated DNA. MSP using qPCR can also be performed to obtain quantitative rather than qualitative information about methylation.

2.14.2.12 Miniprimer PCR

Miniprimer PCR uses a thermostable polymerase (S-Tbr) that can extend from short primers ("smalligos") as short as 9 or 10 nucleotides. This method permits PCR to target smaller primer binding regions, and is used to amplify conserved DNA sequences, such as the 16S (or eukaryotic 18S) rRNA gene.

2.14.2.13 Multiplex Ligation-Dependent Probe Amplification

Multiplex ligation-dependent probe amplification (MLPA) permits multiple targets to be amplified with only a single primer pair, thus avoiding the resolution limitations of multiplex PCR.

2.14.2.14 Multiplex PCR

Multiplex PCR consists of multiple primer sets within a single PCR mixture to produce amplicons of varying sizes that are specific to different DNA sequences. By targeting multiple genes at once, additional information may be gained from a single test run that otherwise would require several times the reagents and more time to perform. Annealing temperatures for each of the primer sets must be optimized to work correctly within a single reaction, and amplicon sizes (i.e., their base pair length) should be different enough to form distinct bands when visualized by gel electrophoresis.

2.14.2.15 Nested PCR

Nested PCR is normally used to increase the specificity of DNA amplification (Figure 2.17), by reducing background due to nonspecific amplification of DNA. Two sets of primers are used in two successive PCRs. In the first reaction, one pair of primers is used to generate DNA products, which besides the intended target, may still consist of nonspecifically amplified DNA fragments. The product(s) are then used in a second PCR with a set of primers whose binding sites are completely or partially different from and located at 3′ of each of the primers used in the first reaction. Nested PCR is often more successful in specifically amplifying long DNA fragments than conventional PCR, but it requires more detailed knowledge of the target sequences.

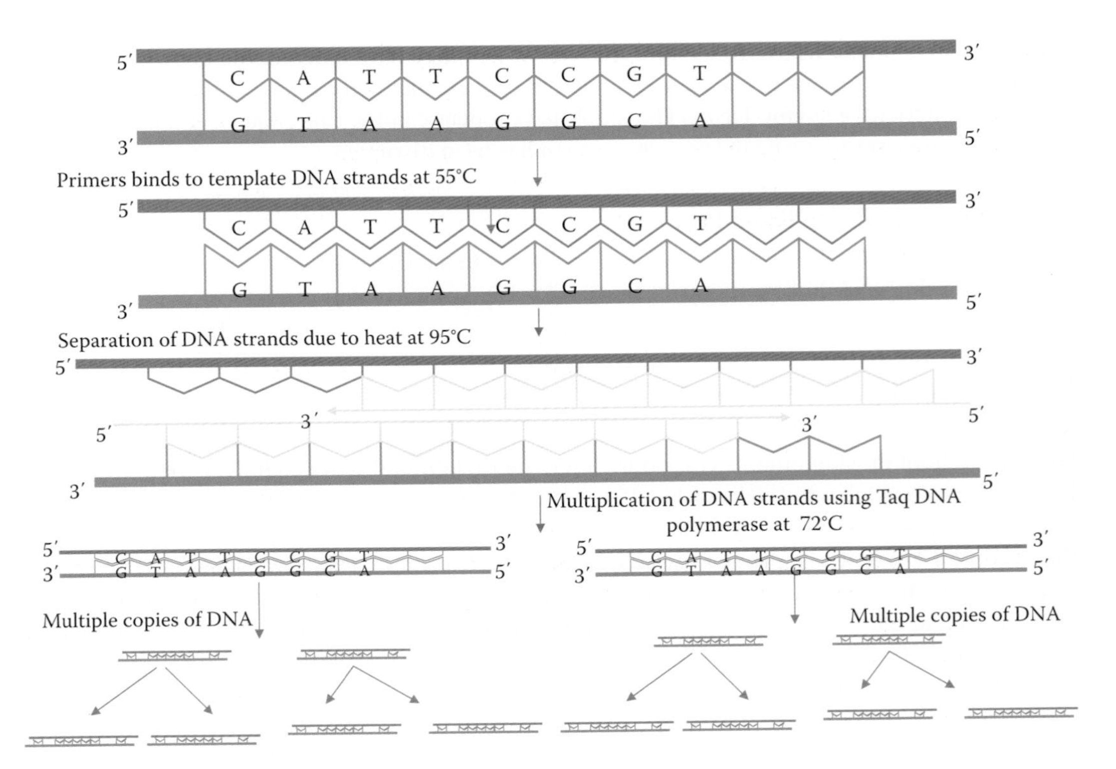

FIGURE 2.17 Amplifications of DNA fragment.

2.14.2.16 Overlap-Extension PCR

Overlap-extension PCR is a genetic engineering technique which allows the construction of a DNA sequence with an alteration inserted beyond the limit of the longest practical primer length.

2.14.2.17 Quantitative PCR

Quantitative PCR (Q-PCR) is normally used to measure the quantity of a PCR product, commonly in real-time. It quantitatively measures starting amounts of DNA, cDNA, or RNA. Q-PCR is also commonly used to determine whether a DNA sequence is present in a sample and the number of its copies in the sample. *Quantitative real-time PCR* (QRT-PCR) has a very high degree of precision. QRT-PCR methods use fluorescent dyes, such as Sybr Green, EvaGreen, or fluorophore-containing DNA probes, such as TaqMan, to measure the amount of amplified product in real time. It is also sometimes abbreviated to RT-PCR (*R*eal *T*ime PCR) or RQ-PCR. QRT-PCR or RTQ-PCR are more appropriate contractions, since RT-PCR commonly refers to reverse transcription PCR (see the next section), often used in conjunction with Q-PCR.

2.14.2.18 Reverse Transcription PCR

Reverse transcription PCR (RT-PCR) is used to amplify DNA from RNA where reverse transcriptase reversed transcribes RNA into cDNA, which is then amplified by PCR. RT-PCR is widely used in expression profiling, to determine the expression of a gene or to identify the sequence of an RNA transcript, including transcription start and termination sites. If the genomic DNA sequence of a gene is known, RT-PCR can be used to map the location of exons and introns in the gene. The 5′ end of a gene (corresponding to the transcription start site) is typically identified by RACE-PCR (Rapid Amplification of cDNA Ends).

2.14.2.19 Solid Phase PCR

Solid phase PCR has multiple applications such as (a) polony amplification where PCR colonies are derived in a gel matrix, for example; (b) bridge PCR where primers are covalently linked to a solid-support surface; (c) conventional solid phase PCR where an asymmetric PCR is applied in the presence of a solid support bearing primer with sequence matching one of the aqueous primers; and (d) enhanced solid phase PCR where a conventional solid phase PCR can be improved by employing high T_m and a nested solid support primer with optional application of a thermal step to favor solid support priming.

2.14.2.20 Thermal Asymmetric Interlaced PCR

Thermal asymmetric interlaced PCR (TAIL-PCR) is used for isolation of an unknown sequence flanking a known sequence. Within the known sequence, TAIL-PCR uses a nested pair of primers with differing annealing temperatures; a degenerate primer is used to amplify in the other direction from the unknown sequence.

2.14.2.21 Touchdown PCR (Step-Down PCR)

Touchdown PCR is a variant of PCR that aims to reduce nonspecific background by gradually lowering the annealing temperature as PCR cycling progresses. The annealing temperature at the initial cycles is usually a few degrees (3°C–5°C) above the T_m of the primers used, while at the later cycles, it is a few degrees (3°C–5°C) below the primer T_m. The higher temperatures give greater specificity for primer binding, and the lower temperatures permit more efficient amplification from the specific products formed during the initial cycles.

2.14.2.22 Universal Fast Walking

This method is used for genome walking and genetic fingerprinting using a more specific "two-sided" PCR than conventional "one-sided" approaches using only one gene-specific primer and one general primer, which can lead to artefactual "noise" by virtue of a mechanism involving lariat structure formation. Streamlined derivatives of universal fast walking (UFW) are LaNe RAGE (lariat-dependent nested PCR for rapid amplification of genomic DNA ends), 5′ rapid amplification of cDNA ends (RACE) by lariat-dependent nested PCR 5 (RACE LaNe), and 3′ RACE LaNe.

2.14.3 Applications of PCR

2.14.3.1 Diagnostic Assay

For the past few years, there has been a tremendous demand for genetic-based diagnostic tests for almost all diseases. The reason for this, obviously, is the rapidness and accuracy of the data generated by the PCR technique. PCR assays can be performed directly on genomic DNA samples to detect translocation-specific malignant cells at a sensitivity which is at least 10,000 fold higher than other methods. PCR also permits identification of noncultivatable or slow-growing microorganisms such as mycobacteria, anaerobic bacteria, or viruses from tissue culture assays and animal models. The basis for PCR diagnostic applications in microbiology is the detection of infectious agents and the discrimination of nonpathogenic from pathogenic strains by virtue of specific genes. Viral DNA can likewise be detected by PCR. The primers used need to be specific to the targeted sequences in the DNA of a virus, and the PCR can be used for diagnostic analyses or DNA sequencing of the viral genome. The high sensitivity of the PCR permits virus detection soon after infection and even before the onset of disease. Such early detection may give physicians a significant lead in treatment. The amount of virus (or viral load) in a patient can also be quantified by PCR-based DNA quantitation techniques.

2.14.3.2 Genetic Engineering

Methods have been developed to purify DNA from organisms, such as phenol-chloroform extraction, and to manipulate it in the laboratory, such as restriction digests and PCR. Modern biology and biochemistry make intensive use of these techniques in recombinant DNA technology. *Recombinant DNA* is a man-made DNA sequence that has been assembled from other DNA sequences. They can be transformed into organisms in the form of plasmids or in the appropriate format, by using a viral vector. The genetically modified organisms produced can be used to cultivate in agriculture, or to produce products such as recombinant proteins which are used in medical research.

2.14.3.3 Forensic DNA Profiling

Forensic scientists can use the DNA in blood, semen, skin, saliva, or hair found at a crime scene to identify a matching DNA of an individual, such as a perpetrator. This process is called *genetic fingerprinting*, or more accurately, *DNA profiling*. In DNA profiling, the lengths of variable sections of repetitive DNA, such as short tandem repeats and minisatellites, are compared between people. This method is usually an extremely reliable technique for identifying a matching DNA. However, identification can be complicated if the scene is contaminated with DNA from several people. DNA profiling was developed in 1984 by British geneticist Sir Alec Jeffreys, and first used in forensic science to convict Colin Pitchfork in the 1988 Enderby murder case. People convicted of certain types of crimes may be required to provide a sample of DNA for a database. This has helped investigators solve old cases where only a DNA sample was obtained from the scene. DNA profiling can also be used to identify victims of mass casualty incidents. On the other hand, many convicted people have been released from prison on the basis of DNA techniques, which were not available when a crime had originally been committed.

PROBLEMS

Section A: Descriptive Type

Q1. What is the cell theory?
Q2. Describe the characteristics of a prokaryotic cell.
Q3. Explain Mendelian genetics.
Q4. Explain supercoiling in a DNA molecule.
Q5. Describe the role of DNA polymerase in replication.
Q6. What are topoisomerases and helicases?
Q7. How does DNA methylation occur?
Q8. What is a PCR?
Q9. How is forensic DNA profiling done with a PCR tool?

Section B: Multiple Choices

Q1. Humans have an estimated number of ________ cells.
   a. 100 billion
   b. 1000 billion
   c. 100 trillion
   d. 5 trillion

Q2. Who was the first to study the internal structure of a cell?
   a. Robert Hooke
   b. Leeuwenhoek
   c. Dutrochet
   d. Charles Darwin

Q3. Do viruses fit in the cell theory concept?
 a. Yes
 b. No
Q4. Prokaryotes carry extra-chromosomal DNA molecules which are called …
 a. Nucleus
 b. Plasmids
 c. Mitochondria
 d. Ribosome
Q5 In eukaryotes, non-nuclear DNA is located in …
 a. Endoplasmic reticulum
 b. Mitochondria
 c. Golgi bodies
 d. Chromatin
Q6. Cell surface membranes contain receptor proteins that allow cells to detect external signaling molecules such as hormones. True/False
Q7. What is common to mitochondria and chloroplast?
 a. Both do not contain their own genome.
 b. Both contain their own genome.
 c. Both are present in prokaryotes.
Q8. Except for ________, all living organisms have genetic information stored in their DNA.
 a. Retroviruses
 b. Bacteria
 c. Fungi
Q9. Nuclear DNA is linear whereas mitochondria DNA is circular. True/False
Q10. Glycogen is a polysaccharide used by animals to store energy. True/False
Q11. Mendel observed that organisms inherit traits called …
 a. DNA
 b. Genes
 c. Proteins
Q12. One of the major differences between DNA and RNA is …
 a. Protein
 b. Hormones
 c. Sugar
 d. Phosphate
Q13. When DNA is twisted in the direction of helix, this is called ________ supercoiling.
 a. Positive
 b. Negative
 c. Linear
Q14. Messenger RNA encodes for …
 a. Gene expression
 b. Protein synthesis
 c. Both protein and gene expression
Q15. What is the protein manufacturing machine of all living cells?
 a. Ribosome
 b. Golgi bodies
 c. Endoplasmic reticulum
Q16. Meiosis is a process of reduction of division in which the number of chromosomes increases to double. False/True
Q17. Replication in DNA is done by the enzyme …
 a. Polymerase
 b. Endonuclease

c. Exonuclease
d. Telomerase

Q18. The structural change in the DNA sequence is called …
a. DNA methylation
b. DNA mutation
c. DNA replication

Q19. Who invented the PCR?
a. James Watson
b. Kary Mullis
c. Ian Wilmut

Q20. Nested PCR is used to increase the specificity of ________ amplification.
a. RNA
b. DNA
c. Both DNA and RNA
d. None of them

SECTION C: CRITICAL THINKING

Q1. In order to identify the real culprit among a group of crime suspects, what technique can be used to establish the identity of the culprit? Explain with suitable examples.

Q2. Is it possible to study the genetic information of an individual by working with mRNA only? Explain.

Q3. What would be the status of gene expression in case mRNA is not available?

Q4. What will happen if nuclear DNA is circular in shape and mitochondrial DNA is linear in shape?

ASSIGNMENT

With the help of your course instructor, organize special lectures on the origin of DNA and submit your analysis in the form of a written assignment.

REFERENCES AND FURTHER READING

Alberts, B., Johnson, A., Lewis, J. et al. *Molecular Biology of the Cell*, 4th edn. Garland Science, New York, 2002.

Aldaye, F.A., Palmer, A.L., and Sleiman, H.F. Assembling materials with DNA as the guide. *Science* 321: 1795–1799, 2008.

Ananthakrishnan, R. and Ehrlicher, A. The forces behind cell movement. *Int. J. Biol. Sci.* 3: 303–317, 2007.

Baianu, I.C. X-ray scattering by partially disordered membrane systems. *Acta Cryst.* A34(5): 751–753, 1978.

Baianu, I.C. Structural order and partial disorder in biological systems. *Bull. Math. Biol.* 42(4): 137–141, 1980.

Bickle, T. and Kruger, D. Biology of DNA restriction. *Microbiol. Rev.* 57: 434–450, 1993.

Bird, A. DNA methylation patterns and epigenetic memory. *Genes Dev.* 16: 6–21, 2002.

Burt, D.W. Origin and evolution of avian microchromosomes. *Cytogenet. Genome. Res.* 96: 97–112, 2002.

Campbell, N.A., Williamson, B., and Heyden, R.J. *Biology: Exploring Life*. Pearson Prentice Hall, Boston, MA, 2006.

Leslie, A.G., Arnott, S., Chandrasekaran, R., and Ratliff, R.L. Polymorphism of DNA double helices. *J. Mol. Biol.* 143: 49–72, 1980.

Li, Z., Van Calcar, S., Qu, C., Cavenee, W., Zhang, M., and Ren, B. A global transcriptional regulatory role for c-Myc in Burkitt's lymphoma cells. *Proc. Natl. Acad. Sci. USA* 100: 8164–8169, 2003.

Lindahl, T. Instability and decay of the primary structure of DNA. *Nature* 362(6422): 709–715, 1993.

Luger, K., Mäder, A., Richmond, R., Sargent, D., and Richmond, T. Crystal structure of the nucleosome core particle at 2.8 A resolution. *Nature* 389: 251–260, 1997.

Maddox, B. The double helix and the wronged heroine [PDF]. *Nature* 421: 407–408, 2003.

Makalowska, I., Lin, C., and Makalowski, W. Overlapping genes in vertebrate genomes. *Comput. Biol. Chem.* 29: 1–12, 2005.

Martinez, E. Multi-protein complexes in eukaryotic gene transcription. *Plant. Mol. Biol.* 50: 925–947, 2002.

Mandelkern, M., Elias, J., Eden, D., and Crothers, D. The dimensions of DNA in solution. *J. Mol. Biol.* 152: 153–161, 1981.

Mendell, J.E., Clements, K.D., Choat, J.H., and Angert, E.R. Extreme polyploidy in a large bacterium. *Proc. Natl. Acad. Sci. USA* 105: 6730–6734, 2008.

Ménétret, J.F., Schaletzky, J., Clemons, W.M. et al. Ribosome binding of a single copy of the SecY complex: Implications for protein translocation. *Mol. Cell.* 28: 1083–1092, 2007.

Michie, K. and Löwe, J. Dynamic filaments of the bacterial cytoskeleton. *Annu. Rev. Biochem.* 75: 467–492, 2006.

Nakabachi, A., Yamashita, A., Toh, H., Ishikawa, H., Dunbar, H., Moran, N., and Hattori, M. The 160-kb genome of the bacterial endosymbiont Carsonella. *Science* 314: 267, 2006.

Neale, M.J. and Keeney, S. Clarifying the mechanics of DNA strand exchange in meiotic recombination. *Nature* 442: 153–158, 2006.

Nickle, D., Learn, G., Rain, M., Mullins, J., and Mittler, J. Curiously modern DNA for a 250 million-year-old bacterium. *J. Mol. Evol.* 54: 134–137, 2002.

Painter, T.S. The spermatogenesis of man. *Anat. Res.* 23: 129, 1922.

Thanbichler, M. and Shapiro, L. Chromosome organization and segregation in bacteria. *J. Struct. Biol.* 156(2): 292–303, 2006.

Thanbichler, M., Wang, S.C., and Shapiro, L. 2005. The bacterial nucleoid: A highly organized and dynamic structure. *J. Cell. Biochem.* 96(3): 506–521, 2005.

The ENCODE Project Consortium. Identification and analysis of functional elements in 1% of the human genome by the ENCODE pilot project. *Nature* 447(7146): 799–816, 2007.

Thomas, J. HMG1 and 2: Architectural DNA-binding proteins. *Biochem. Soc. Trans.* 29(Pt 4): 395–401, 2001.

Tjio, J.H. and Levan, A. The chromosome number of man. *Hereditas* 42: 1–6, 1956.

Tuteja, N. and Tuteja, R. Unraveling DNA helicases. Motif, structure, mechanism and function. *Eur. J. Biochem.* 271(10): 1849–1863, 2004.

Valerie, K. and Povirk, L. Regulation and mechanisms of mammalian double-strand break repair. *Oncogene* 22(37): 5792–5812, 2003.

Watson, J.D. and Crick, F.H.C. A structure for deoxyribose nucleic acid (PDF). *Nature* 171: 737–738, 1953.

3 Proteins and Proteomics

LEARNING OBJECTIVES

- Define protein and explain its significance
- Discuss the role of protein in cell signaling
- Discuss the structural and functional attributes of protein
- Explain the process of protein biosynthesis
- Explain different methods of protein prediction models
- Explain the significance of protein folding and protein modification
- Discuss about protein transport and degradation
- Discuss genetic regulation of protein synthesis
- Discuss about tools and techniques used to analyze proteins

3.1 INTRODUCTION

After genomics, proteomics is often considered the next step in the study of biological systems. It is much more complicated than genomics mostly because while an organism's genome is more or less constant, the proteome differs from cell to cell and from time to time. This is because distinct genes are expressed in distinct cell types. This means that even the basic set of proteins that are produced in a cell needs to be determined. Nowadays, we keep hearing of high-protein or low-protein diets. But why do we need to keep our protein level under control? This is because proteins are very important for body functions, and any deficiency or malfunction in protein may lead to serious ailments.

Proteins are the primary components of numerous body tissues such as muscle tissues. They help to increase strength, improve athletic performance, and develop muscles. Proteins also make up the outer layers of hair, nails, and skin. The most important function of protein is to build up, maintain, and replace the tissues in the body. Muscles, organs, and some hormones (insulin) are made up mostly of protein. Proteins also make antibodies and hemoglobin (responsible for delivering oxygen to the blood cells). They make up half the dry weight of an *Escherichia coli* cell. On the other hand, they make up 3% of a DNA molecule and 20% of an RNA molecule. The set of proteins expressed in a particular cell or cell type is known as its *proteome*.

The main characteristic of proteins that allows their diverse set of functions is their ability to bind with other molecules specifically and tightly. The region of the protein responsible for binding with another molecule is called the *binding site*. It is often a depression or "pocket" on the molecular surface of a protein. This binding ability is mediated by the tertiary structure of the protein, which defines the binding site pocket, and by the chemical properties of the surrounding amino acids' side chains. Protein binding can be extraordinarily tight and specific. For example, the ribonuclease inhibitor protein binds to human angiogenin with a sub-femtomolar dissociation constant ($<10^{-15}$ M) but does not bind at all to its amphibian homolog onconase (>1 M). Extremely minor chemical changes such as the addition of a single methyl group to a binding partner can sometimes suffice to nearly eliminate binding. For example, the aminoacyl tRNA synthetase specific to the amino acid valine discriminates against the very similar side chain of the amino acid isoleucine. Proteins can bind to other proteins as well as to small-molecule substrates. When proteins bind specifically to other copies of the same molecule, they can oligomerize to form fibrils. This process occurs often in structural proteins that consist of globular monomers that self-associate to form rigid fibers. Protein-to-protein interactions also regulate enzymatic activity, control progression through the cell cycle, and allow the assembly of large protein complexes that carry out many closely related reactions with a common biological function. The ability of binding partners to induce conformational changes in proteins allows the construction of enormously complex signaling networks. Importantly, as interactions between proteins are reversible, and depend heavily on the availability of different groups of partner proteins to form aggregates that are capable of carrying out discrete sets of functions, study of the interactions between specific proteins is a key to understanding important aspects of cellular functions, and ultimately the properties that distinguish particular cell types.

3.2 SIGNIFICANCE OF PROTEINS

Protein plays a key role in food intake regulation through satiety related to diet-induced thermogenesis. Protein also plays a key role in body weight regulation through its effect on thermogenesis and body composition. In this section we have analyzed various applications of proteins in great detail.

3.2.1 Proteins for Body Functions

There are 20 different identified types of amino acids which are necessary for normal body functions. Of these amino acids, 14 are produced by the body while 6 are ingested through the foods we eat. Those amino acids that the body can produce are called nonessential amino acids while those that the body cannot produce are called essential amino acids.

3.2.1.1 Nonessential Amino Acids

The following lists the nonessential amino acids with some of their functions, benefits and side effects:

- Alanine: It removes toxic substances released from breakdown of muscle protein during intensive exercise. Excessive alanine level in the body is associated with chronic fatigue.
- Cysteine: It is a component of protein type abundant in nails, skin, and hair. It also acts as an antioxidant (free radical scavenger), and has synergetic effect when taken along with other antioxidants such as vitamin E and selenium.
- Cystine: The same as cysteine, it aids in removal of toxins and formation of skin.
- Glutamine: It promotes healthy brain function. It is also necessary for the synthesis of RNA and DNA molecules.
- Glutathione: It is an antioxidant and has an antiaging effect. It is useful in removal of toxins.
- Glycine: It is a component of skin and is beneficial for wound healing. It also acts as neurotransmitter. High level of glycine in the body may cause fatigue.
- Histidine: It is important in the synthesis of red and white blood cells and a precursor for histamine which is good for sexual arousal. It also improves blood flow. High dosage of histidine may cause stress and anxiety.
- Serine: It is a constituent of brain proteins. It aids in the synthesis of immune system proteins and helps improve muscle development.
- Taurine: It is necessary for proper brain functioning and synthesis of amino acids. It is also important in the assimilation of mineral nutrients such as magnesium, calcium, and potassium.
- Threonine: It balances the protein level in the body and promotes the immune system. It is also beneficial for the synthesis of tooth enamel and collagen.

- Asparagine: It helps promote equilibrium in the central nervous system, thus balancing the state of emotion.
- Aspartic acid: It enhances stamina, aids in removal of toxins and ammonia from the body, and beneficial in the synthesis of proteins involved in the immune system.
- Proline: It plays a role in intracellular signaling.
- L-arginine: It plays a role in blood vessel relaxation and removal of excess ammonia from the body.

3.2.1.2 Essential Amino Acids

The following lists the eight amino acids which are generally essential in humans.

- *Phenylalanine*: Phenylalanine (abbreviated as Phe or F) is an α-amino acid with the formula $HO_2CCH(NH_2)CH_2C_6H_5$. This essential amino acid is classified as nonpolar because of the hydrophobic nature of the benzyl side chain. The codons for L-phenylalanine are UUU and UUC. L-Phenylalanine (LPA) is an electrically neutral amino acid, one of the 20 common amino acids used to biochemically form proteins, coded for by DNA. Phenylalanine is structurally closely related to dopamine, epinephrine (adrenaline), and tyrosine. Phenylalanine is found naturally in the breast milk of mammals. It is manufactured for food and drink products and is also sold as nutritional supplements for its reputed analgesic and antidepressant effects. It is a direct precursor to the neuromodulator phenylethylamine, a commonly used dietary supplement.
- *Valine*: Valine (abbreviated as Val or V) is an α-amino acid with the chemical formula $HO_2CCH(NH_2)CH(CH_3)_2$. L-Valine is one of 20 proteinogenic amino acids. Its codons are GUU, GUC, GUA, and GUG. This essential amino acid is classified as nonpolar. Human dietary sources include cottage cheese, fish, poultry, peanuts, sesame seeds, and lentils. Along with leucine and isoleucine, valine is a branched-chain amino acid. It is named after the plant valerian. In sickle-cell disease, valine substitutes for the hydrophilic amino acid glutamic acid in hemoglobin. Because valine is hydrophobic, the hemoglobin does not fold correctly.
- *Threonin*: Threonine is an α-amino acid with the chemical formula $HO_2CCH(NH_2)CH(OH)CH_3$. Its codons are ACU, ACA, ACC, and ACG. This essential amino acid is classified as polar. Together with serine and tyrosine, threonine is one of three proteinogenic amino acids bearing an alcohol group. The threonine residue is susceptible to numerous post-translational modifications (PTMs). The hydroxyl side chain can undergo O-linked glycosylation. In addition, threonine residues undergo phosphorylation through the action of a threonine kinase. In its phosphorylated form, it is referred to as phosphothreonine.
- *Tryptophan*: Tryptophan (abbreviated as Trp or W and sold as Tryptan) is one of the 20 standard amino acids, as well as an essential amino acid in the human diet. It is encoded in the standard genetic code as the codon UGG. Only the L-stereoisomer of tryptophan is used in structural or enzyme proteins, but the D-stereoisomer is occasionally found in naturally produced peptides (such as the marine venom peptide contryphan). The distinguishing structural characteristic of tryptophan is that it contains an indole functional group.
- *Isoleucine*: Isoleucine (abbreviated as Ile or I) is an α-amino acid with the chemical formula $HO_2CCH(NH_2)CH(CH_3)CH_2CH_3$. The codons of these essential amino acids are AUU, AUC, and AUA. With a hydrocarbon side chain, isoleucine is classified as a hydrophobic amino acid. Together with threonine, isoleucine is one of two common amino acids that have a chiral side chain. Four stereoisomers of isoleucine are possible, including two possible diastereomers of L-isoleucine. However, isoleucine present in nature exists in an enantiomeric form, (2S, 3S)-2-amino-3-methylpentanoic acid.
- *Methionine*: Methionine is one of only two amino acids encoded by a single codon (AUG) in the standard genetic code (tryptophan, encoded by UGG, is the other). The codon AUG

is also the "Start" message for a ribosome that signals the initiation of protein translation from mRNA. As a consequence, methionine is incorporated into the N-terminal position of all proteins in eukaryotes and archaea during translation, although it is usually removed by PTM.

- *Leucine*: Leucine (abbreviated as Leu or L) is an α-amino acid with the chemical formula $HO_2CCH(NH_2)CH_2CH(CH_3)_2$. The codons of this essential amino acid are UUA, UUG, CUU, CUC, CUA, and CUG. With a hydrocarbon side chain, leucine is classified as a hydrophobic amino acid. It has an isobutyl R group. Leucine is a major component of the subunits in ferritin, astacin, and other "buffer" proteins.
- *Lysine*: Lysine is an α-amino acid with the chemical formula $HO_2CCH(NH_2)(CH_2)4NH_2$. The codons of this essential amino acid are AAA and AAG. Lysine is a base, as are arginine and histidine. The ε-amino group often participates in hydrogen bonding and as a general base in catalysis. Common PTMs include methylation of the ε-amino group, giving methyl-, dimethyl-, and trimethyllysine. The latter occurs in calmodulin. Other PTMs at lysine residues include acetylation and ubiquitination. Collagen contains hydroxylysine which is derived from lysine by lysyl hydroxylase. O-Glycosylation of lysine residues in the endoplasmic reticulum or Golgi apparatus is used to mark certain proteins for secretion from the cell.

3.2.1.3 Other Amino Acids

Essential amino acids are called essential not because they are more important than the others, but because the body does not synthesize them, making it essential to include them in one's diet in order to obtain them. On the other hand, the amino acids arginine, cysteine, glycine, glutamine, histidine, proline, serine, and tyrosine are considered conditionally essential, meaning they are not normally required in the diet, but must be supplied exogenously to specific populations who do not synthesize it in adequate amounts. For example, individuals living with phenylketonuria (PKU) disease must keep their intake of phenylalanine extremely low to prevent mental retardation and other metabolic complications. However, phenylalanine is the precursor for tyrosine synthesis. Without phenylalanine, tyrosine cannot be made and so tyrosine becomes essential in the diet of PKU patients.

3.2.2 Proteins as Enzymes

The best-known role of proteins in the cell is as enzymes, which catalyze chemical reactions. Enzymes are usually highly specific and accelerate only one or a few chemical reactions. Enzymes carry out most of the reactions involved in metabolism, as well as DNA manipulation in processes such as DNA replication, DNA repair, and transcription. Some enzymes act on other proteins to add or remove chemical groups in a process known as PTM. About 4000 reactions are known to be catalyzed by enzymes. The rate of acceleration conferred by enzymatic catalysis is often enormous, as much as a 10^{17}-fold increase in rate over the uncatalyzed reaction rate in the case of orotate decarboxylase (78 million years without the enzyme, 18 ms with the enzyme). The molecules bound and acted upon by enzymes are called *substrates*. Although enzymes can consist of hundreds of amino acids, it is usually only a small fraction of the residues that come in contact with the substrate, and an even smaller fraction, 3–4 residues on the average, are directly involved in catalysis. The region of the enzyme that binds the substrate and contains the catalytic residues is known as the *active site*.

3.2.3 Proteins in Cell Signaling and Ligand Binding

Many proteins are involved in the process of cell signaling and signal transduction. Some proteins such as insulin are extracellular proteins that transmit a signal from the cell in which they were synthesized to other cells in distant tissues. Others are membrane proteins that act as

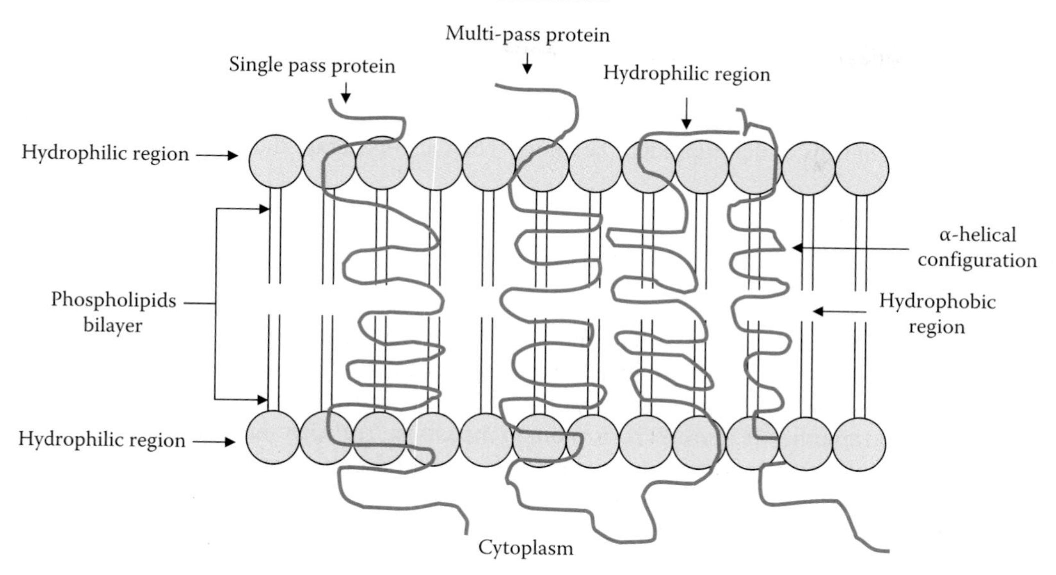

FIGURE 3.1 Cell signaling through proteins.

receptors whose main function is to bind a signaling molecule and induce a biochemical response in the cell. Many receptors have a binding site exposed on the cell surface and an effectors domain within the cell, which may have enzymatic activity or may undergo a conformational change detected by other proteins within the cell. *Antibodies* are protein components of adaptive immune system whose main function is to bind antigens, or foreign substances in the body, and target them for destruction. Antibodies can be secreted into the extracellular environment or anchored in the membranes of specialized B cells known as *plasma cells*. Whereas enzymes are limited in their binding affinity for their substrates by the necessity of conducting their reaction, antibodies have no such constraints. An antibody's binding affinity to its target is extraordinarily high (Figure 3.1).

Many ligand transport proteins bind particular small biomolecules and transport them to other locations in the body of a multicellular organism. These proteins must have a high binding affinity when their ligand is present in high concentrations, but must also release the ligand when it is present at low concentrations in the target tissues. The canonical example of a ligand-binding protein is hemoglobin, which transports oxygen from the lungs to other organs and tissues in all vertebrates and has close homologs in every biological kingdom. *Lectins* are sugar-binding proteins which are highly specific for their sugar moieties. Lectins typically play a role in biological recognition phenomena involving cells and proteins. Receptors and hormones are highly specific binding proteins. Transmembrane proteins can also serve as ligand transport proteins that alter the permeability of the cell membrane to small molecules and ions. The membrane alone has a hydrophobic core through which polar or charged molecules cannot diffuse. Membrane proteins contain internal channels that allow such molecules to enter and exit the cell. Many ion channel proteins are specialized to select for only a particular ion. For example, potassium and sodium channels often discriminate for only one of the two ions.

3.2.4 Structural Proteins

Structural proteins confer stiffness and rigidity to otherwise-fluid biological components. Most structural proteins are fibrous proteins. For example, actin and tubulin are globular and soluble as monomers, but polymerize to form long, stiff fibers that comprise the cytoskeleton, which allows the cell to maintain its shape and size. Collagen and elastin are critical components of connective tissue such as cartilage, while keratin is found in hard or filamentous structures such as hair, nails, feathers, hooves, and some animal shells. Other proteins that serve structural functions are motor proteins such as myosin, kinesin, and dynein, which are capable of generating mechanical forces. These proteins are crucial for cellular motility of single-celled organisms and the sperm of many multicellular organisms which reproduce sexually. They also generate the forces exerted by contracting muscles.

3.3 PROTEIN BIOSYNTHESIS

After learning various attributes of proteins, the next question that arises is how our body makes these proteins. In this section, we will learn how our body makes proteins in a step-by-step manner. Recall that a single gene expression results in the formation of protein, and protein synthesis is a multistep process which not only involves DNA but also mRNA, tRNA, and rRNA in a well-coordinated manner (Figure 3.2). Protein synthesis starts with the first gene expression (or *transcription phase*), leading to translation, and to final protein formation. Also, recall that gene expression is used by all known life, eukaryotes (including multicellular organisms), prokaryotes (bacteria and archea), and viruses, to generate the macromolecular machinery for life. Several steps in the gene expression process may be modulated, including the transcription, RNA splicing, translation, and PTM of a protein.

3.3.1 TRANSCRIPTION STAGE

Transcription, also called RNA synthesis, is the process of creating an equivalent RNA copy of a sequence of DNA. Both RNA and DNA are nucleic acids, which use base pairs of nucleotides as a complementary language that can be converted back and forth from DNA to RNA in the presence of the correct enzymes. During transcription, a DNA sequence is read by an RNA polymerase, which produces a complementary and antiparallel RNA strand. As opposed

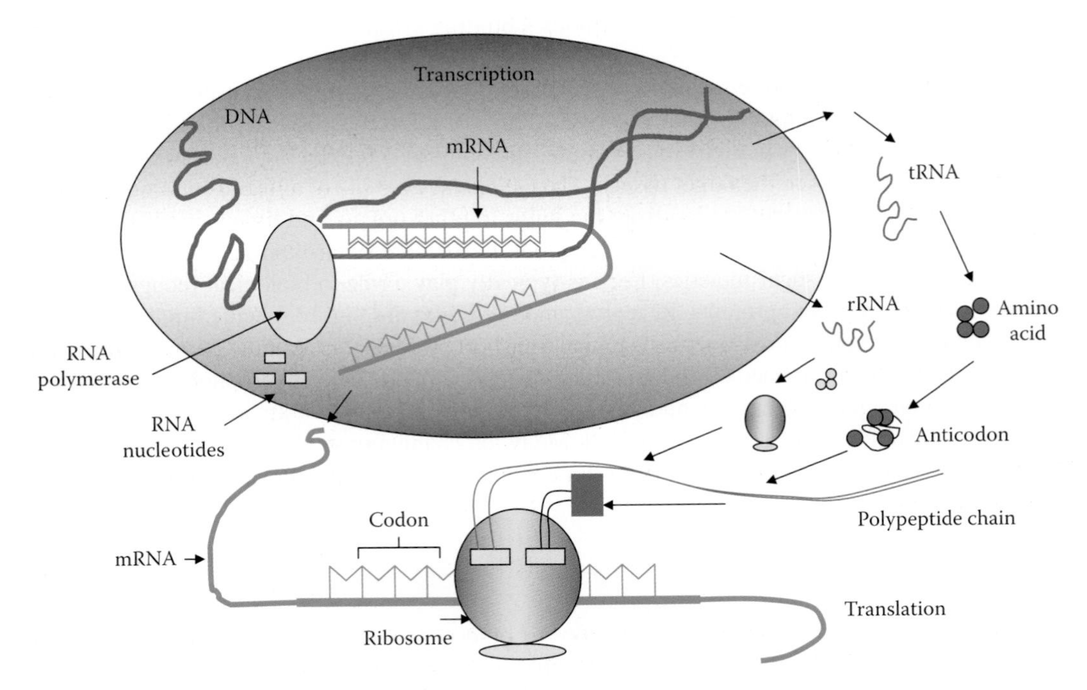

FIGURE 3.2 Process of protein biosynthesis.

to DNA replication, transcription results in an RNA compliment that includes uracil (U) in all instances where thymine (T) would have occurred in a DNA compliment. Transcription is the first step leading to gene expression. The stretch of DNA transcribed into an RNA molecule is called a *transcription unit*, which encodes at least one gene. If the gene transcribed encodes for a protein, the result of the transcription is a messenger RNA (mRNA), which will then be used to create that protein via the process of translation. In addition to this, the transcribed ribosomal RNA (rRNA), other components of the protein-assembly process, or other ribozymes.

A DNA transcription unit encoding for a protein contains not only the sequence that will eventually be directly translated into the protein but also regulatory sequences that direct and regulate the synthesis of that protein. The regulatory sequence before the coding sequence is called the *five prime untranslated region* (5′UTR) and is also known as *upstream process*. The sequence following the coding sequence is called the *three prime untranslated regions* (3′UTR) and is also known as *downstream process*. Transcription has some proofreading mechanisms, but they are fewer and less effective than the controls for copying DNA. Therefore, transcription has a lower copying fidelity than DNA replication. As in DNA replication, DNA is read from 3′–5′ during transcription. Meanwhile, the complimentary RNA is created from the 5′–3′ direction. Although DNA is arranged as two antiparallel strands in a double helix, only one of the two DNA strands, called the *template strand*, is used for transcription. This is because RNA is only single-stranded, as opposed to double-stranded DNA. The other DNA strand is called the *coding strand*, because its sequence is the same as the newly created RNA transcript except for the substitution of uracil by thymine. The use of only the 3′–5′ strand eliminates the need for the Okazaki fragments seen in DNA replication.

3.3.2 Transcription in Prokaryotes and Eukaryotes

One major difference between prokaryotes and eukaryotes is the existence of membrane-bound structures within eukaryotes, including a cell nucleus with a nuclear membrane that encapsulates cellular DNA. As a result, transcription varies between the two, with prokaryotic transcription occurring in the cytoplasm alongside translation and eukaryotic transcription occurring only in the nucleus, where it is separated from the cytoplasm by the nuclear membrane. Following transcription, the resulting RNA is transported into the cytoplasm, where translation then occurs. Another important difference is that eukaryotic DNA not currently in use is stored as heterochromatin around histones to form nucleosomes and must be unwound as euchromatin to be transcribed. Chromatin has a strong influence on the accessibility of the DNA to transcription factors (TFs) and the transcriptional machinery, including RNA polymerase. Finally, in prokaryotes, mRNA usually remains unmodified while eukaryotic mRNA is heavily processed through RNA splicing, 5′ end capping (5′ cap), and the addition of a polyA tail (Figure 3.3).

3.3.3 Stages of Transcription

Transcription is divided into five stages: pre-initiation, initiation, promoter clearance, elongation, and termination. We will describe each step to understand the beginning of protein synthesis (Figure 3.4).

3.3.3.1 Pre-Initiation

In eukaryotes, RNA polymerase, and therefore the initiation of transcription, requires the presence of a core promoter sequence in the DNA. Promoters are regions of DNA that promote transcription and are found around 10–35 base pairs upstream from the start site of transcription. Core promoters are sequences within the promoter which are essential for transcription initiation. RNA polymerase is able to bind to core promoters in the presence of various specific TFs. The most common type of core promoter in eukaryotes is a short DNA sequence known as a *TATA box*. The TATA box, as a core promoter, is the binding site for a TF known as *TATA binding protein* (TBP), which is itself a subunit of another TF, called *transcription factor II D* (TFIID). After TFIID binds to the

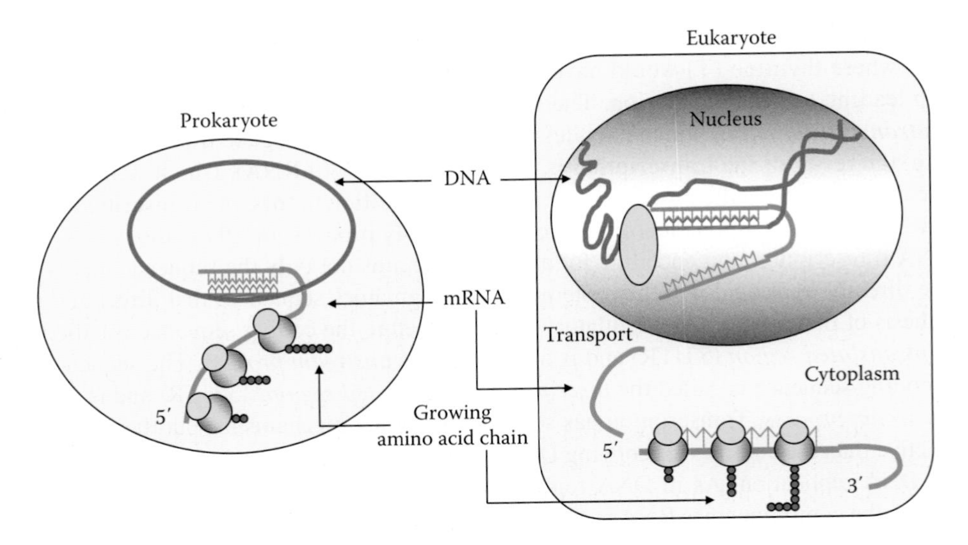

FIGURE 3.3 Transcription in prokaryote and eukaryote.

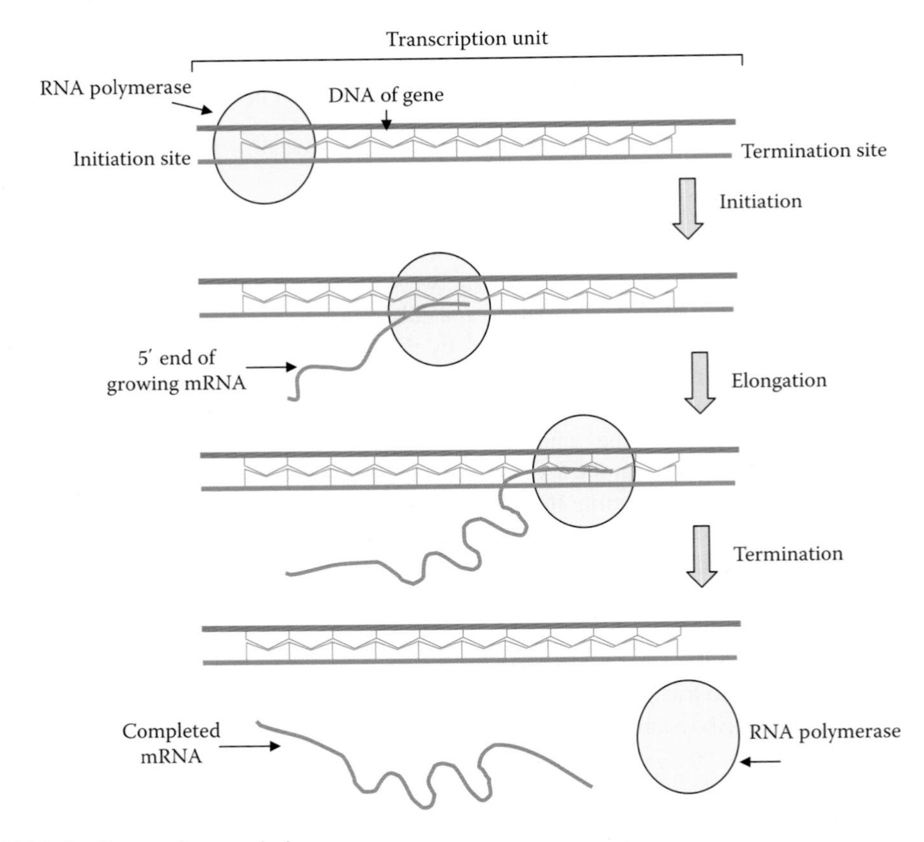

FIGURE 3.4 Stages of transcription.

TATA box via the TBP, five more TFs, and RNA polymerase combine around the TATA box in a series of stages to form a pre-initiation complex. One TF, DNA helicase, has helicase activity and so is involved in the separating of opposing strands of double-stranded DNA to provide access to a single-stranded DNA template. However, only a low or basal rate of transcription is driven by the pre-initiation complex alone. Other proteins known as activators and repressors, along with any associated coactivators or corepressors, are responsible for modulating transcription rate. The transcription pre-initiation in archaea, formerly a domain of prokaryote, is essentially homologous to that of eukaryotes, but is much less complex. The archaeal pre-initiation complex assembles at a TATA-box binding site. However, in archaea, this complex is composed of only RNA polymerase II, TBP, and TFB (the archaeal homologue of eukaryotic TF II B (TFIIB)).

3.3.3.2 Initiation

In bacteria, transcription begins with the binding of RNA polymerase to the promoter in DNA. RNA polymerase is a core enzyme consisting of five subunits: 2 α subunits, 1 β subunit, 1 β' subunit, and 1 ω subunit. At the start of initiation, the core enzyme is associated with a sigma factor (number 70) that aids in finding the appropriate –35 and –10 base pairs downstream of promoter sequences. Transcription initiation is more complex in eukaryotes. Eukaryotic RNA polymerase does not directly recognize the core promoter sequences. Instead, a collection of proteins called TFs mediates the binding of RNA polymerase and the initiation of transcription. Only after certain TFs are attached to the promoter does the RNA polymerase bind to it. The completed assembly of TFs and RNA polymerase bind to the promoter, forming a transcription initiation complex. Transcription in the archaea domain is similar to transcription in eukaryotes.

3.3.3.3 Promoter Clearance

After the first bond is synthesized, the RNA polymerase must clear the promoter. During this time, there is a tendency to release the RNA transcript and produce truncated transcripts. This is called *abortive initiation* and is common for both eukaryotes and prokaryotes. Abortive initiation continues to occur until the σ factor rearranges, resulting in the transcription elongation complex (which gives a 35 base pairs moving footprint). The σ factor is released before 80 nucleotides of mRNA are synthesized. Once the transcript reaches approximately 23 nucleotides, it no longer slips and elongation can occur. This, like most of the remainder of transcription, is an energy-dependent process, consuming adenosine triphosphate (ATP). Promoter clearance coincides with phosphorylation of serine 5 on the carboxyl terminal domain of RNA Pol in prokaryotes, which is phosphorylated by transcription factor II H (TFIIH).

3.3.3.4 Elongation

One strand of DNA, the template strand (or noncoding strand), is used as a template for RNA synthesis. As transcription proceeds, RNA polymerase traverses the template strand and uses base pairing complementarily with the DNA template to create an RNA copy. Although RNA polymerase traverses the template strand from $3' \rightarrow 5'$, the coding (non-template) strand and newly-formed RNA can also be used as reference points, so transcription can be described as occurring $5' \rightarrow 3'$. This produces an RNA molecule from $5' \rightarrow 3'$, an exact copy of the coding strand except that thymines are replaced with uracils and the nucleotides are composed of a ribose (5-carbon) sugar where DNA has deoxyribose in its sugar-phosphate backbone. Unlike DNA replication, mRNA transcription can involve multiple RNA polymerases on a single DNA template and multiple rounds of transcription (amplification of particular mRNA), so many mRNA molecules can be rapidly produced from a single copy of a gene. Elongation also involves a proofreading mechanism that can replace incorrectly incorporated bases. In eukaryotes, this may correspond with short pauses during transcription that allow appropriate RNA editing factors to bind. These pauses may be intrinsic to the RNA polymerase or due to chromatin structure.

3.3.3.5 Termination Stage

Bacteria use two different strategies for transcription termination: Rho-independent and Rho-dependent transcription termination. In Rho-independent transcription termination, RNA transcription stops when the newly synthesized RNA molecule forms a G-C rich hairpin loop followed by a run of Us, which makes it detach from the DNA template. In the Rho-dependent type of termination, a protein factor called “Rho” destabilizes the interaction between the template and the mRNA, thus releasing the newly synthesized mRNA from the elongation complex. Transcription termination in eukaryotes is less understood but involves cleavage of the new transcript followed by template-independent addition of *A*s at its new 3′ end, in a process called *polyadenylation*.

3.3.3.6 Reverse Transcription

Some viruses (such as HIV, the cause of AIDS) have the ability to transcribe RNA into DNA. HIV has an RNA genome that is duplicated into DNA. The resulting DNA can be merged with the DNA genome of the host cell. The main enzyme responsible for synthesis of DNA from an RNA template is called *reverse transcriptase*. In the case of HIV, reverse transcriptase is responsible for synthesizing a complementary DNA strand (cDNA) to the viral RNA genome. An associated enzyme, *ribonuclease H*, digests the RNA strand, and reverse transcriptase synthesizes a complementary strand of DNA to form a double helix DNA structure. This cDNA is integrated into the host cell’s genome via another enzyme (integrase) causing the host cell to generate viral proteins which reassemble into new viral particles. Subsequently, the host cell undergoes programmed cell death, *apoptosis*. Some eukaryotic cells contain an enzyme with reverse transcription activity called *telomerase*. Telomerase is a reverse transcriptase that lengthens the ends of linear chromosomes. Telomerase carries an RNA template from which it synthesizes DNA repeating sequence, or “junk” DNA. This repeated sequence of DNA is important because every time a linear chromosome is duplicated, it is shortened in length. With “junk” DNA at the ends of chromosomes, the shortening eliminates some of the nonessential, repeated sequence rather than the protein-encoding DNA sequence farther away from the chromosome end. Telomerase is often activated in cancer cells to enable cancer cells to duplicate their genomes indefinitely without losing important protein-coding DNA sequence. Activation of telomerase could be part of the process that allows cancer cells to become technically immortal. However, the true *in vivo* significance of telomerase has still not been empirically proven.

3.3.3.7 RNA Splicing

In molecular biology, *splicing* is a modification of RNA after transcription, in which introns are removed and exons are joined. This is needed for the typical eukaryotic messenger RNA before it can be used to produce a correct protein through translation. For many eukaryotic introns, splicing is done in a series of reactions which are catalyzed by the *spliceosome*, a complex of small nuclear ribonucleoproteins (snRNPs), but there are also self-splicing introns. Several methods of RNA splicing occur in nature. The type of splicing depends on the structure of the spliced intron and the catalysts required for splicing to occur.

3.3.3.8 Spliceosomal Introns

Spliceosomal introns often reside in eukaryotic protein-coding genes. Within the intron, a 3′ splice site, 5′ splice site, and branch site are required for splicing. The 5′ splice site or splice donor site includes an almost invariant sequence GU at the 5′ end of the intron, within a larger, less highly conserved consensus region. The 3′ splice site or splice acceptor site terminates the intron with an almost invariant AG sequence. Upstream (5′-ward) from the AG, there is a region high in pyrimidines (C and U), or polypyrimidine tract. Upstream from the polypyrimidine tract is the branch point, which includes an adenine nucleotide.

3.3.3.9 Spliceosome Formation and Activity

Splicing is catalyzed by the spliceosome which is a large RNA-protein complex composed of five snRNPs also pronounced as ‘snurps.’ The RNA components of snRNPs interact with the intron and may be involved in catalysis. Two types of spliceosomes have been identified (the major and minor) which contain different snRNPs. *Trans-splicing* is a form of splicing that joins two exons that are not within the same RNA transcript.

3.3.3.10 Self-Splicing

Self-splicing occurs for rare introns that form a ribozyme, performing the functions of the spliceosome by RNA alone. There are three kinds of self-splicing introns: Group I, Group II, and Group III. Groups I and II introns perform splicing similar to the spliceosome without requiring any protein. This similarity suggests that Groups I and II introns may be evolutionarily related to the spliceosome. Self-splicing may also be very ancient, and may have existed in an RNA world present before protein. The two splicing mechanisms, do not require any proteins to occur, they use 5 additional RNA molecules and over 50 proteins to hydrolyze many ATP molecules. The splicing mechanisms use ATP in order to accurately splice mRNA's. If the cell were to not use any ATPs, the process would be highly inaccurate and many mistakes would occur.

3.3.3.11 tRNA Splicing

The tRNA splicing is another rare form of splicing that usually occurs in tRNA. The splicing reaction involves a different biochemistry than the spliceosomal and self-splicing pathways. Ribonucleases cleave the RNA and ligases join the exons together.

3.3.3.12 RNA Export

The translation of mRNA can also be controlled by a number of mechanisms, mostly at the level of initiation. Recruitment of the small ribosomal subunit can indeed be modulated by mRNA secondary structure, anti-sense RNA binding or protein binding. In both prokaryotes and eukaryotes, a large number of RNA binding proteins exist which often are directed to their target sequence by the secondary structure of the transcript, which may change depending on certain conditions such as temperature or presence of a ligand (aptamer). Some transcripts act as ribosome and self-regulate their expression.

3.3.4 Translation

Translation is the first stage of protein biosynthesis and this is a part of the overall process of gene expression. It is the production of proteins by decoding mRNA produced in transcription It also occurs in the cytoplasm where the ribosomes are located. Ribosomes are made of a small and large subunit which surrounds the mRNA. In translation, messenger RNA is decoded to produce a specific polypeptide according to the rules specified by the genetic code. This uses an mRNA sequence as a template to guide the synthesis of a chain of amino acids that form a protein. Many types of transcribed RNA, such as transfer RNA, ribosomal RNA, and small nuclear RNA are not necessarily translated into an amino acid sequence (Figure 3.5).

Translation proceeds in four phases: activation, initiation, elongation, and termination, and all describing the growth of the amino acid chain, or polypeptide that is the product of translation. Amino acids are brought to ribosomes and assembled into proteins. In activation, the correct amino acid is covalently bonded to the correct tRNA. While this is not technically a step in translation, it is required for translation to proceed. The amino acid is joined by its carboxyl group to the 3′ OH of the tRNA by an ester bond. When the tRNA has an amino acid linked to it, it is termed "charged." Initiation involves the small subunit of the ribosome binding to 5′ end of mRNA with the help of initiation factors (IF). Termination of the polypeptide happens when the A site of the ribosome faces a stop codon (UAA, UAG, or UGA). When this happens, no tRNA can recognize it, but a releasing factor can recognize nonsense codons and causes the release of the polypeptide chain. The 5′ end of the mRNA gives rise to the protein's N-terminus, and the direction of translation can

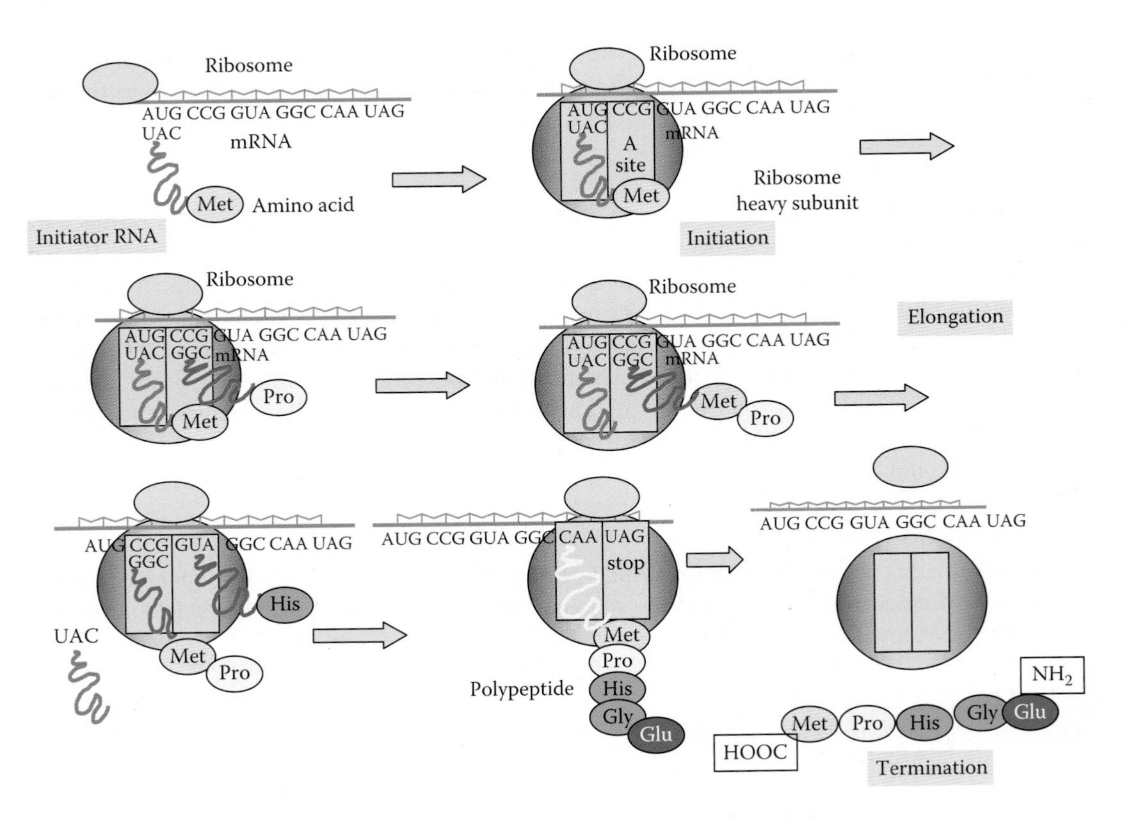

FIGURE 3.5 The process of protein biosynthesis and various stages of translation.

therefore be stated as N->C. A number of antibiotics act by inhibiting translation. Some of these are anisomycin, cycloheximide, chloramphenicol, tetracycline, streptomycin, erythromycin, and puromycin. Prokaryotic ribosomes have a different structure from that of eukaryotic ribosomes, and thus antibiotics can specifically target bacterial infections without any detriment to a eukaryotic host's cells. The mRNA carries genetic information encoded as a ribonucleotide sequence from the chromosomes to the ribosomes. The ribonucleotides are "read" by translational machinery in a sequence of nucleotide triplets called codons. Each of those triplets codes for a specific amino acid. The ribosome and tRNA molecules translate this code to a specific sequence of amino acids. The ribosome is a multisubunit structure containing rRNA and proteins. It is the "factory" where amino acids are assembled into proteins. tRNAs are small noncoding RNA chains (74–93 nucleotides) that transport amino acids to the ribosome. tRNAs have a site for amino acid attachment called an anticodon. The anticodon is an RNA triplet complementary to the mRNA triplet that codes for their cargo amino acid.

Aminoacyl tRNA synthetase catalyzes the bonding between specific tRNAs and the amino acids that their anticodons sequences call for. The product of this reaction is an aminoacyl-tRNA molecule. This aminoacyl-tRNA travels inside the ribosome, where mRNA codons are matched through complementary base pairing to specific tRNA anticodons. The amino acids that the tRNAs carry are then used to assemble a protein. The rate of translation varies. It is significantly higher in prokaryotic cells, up to 17–21 amino acid residues per second than in eukaryotic cells, up to 6–7 amino acid residues per second.

3.3.4.1 Post-Translational Modification

PTM is the chemical modification of a protein after its translation. It is one of the later steps in protein biosynthesis. After translation, the PTM of amino acids extends the range of functions of the protein by attaching to it other biochemical functional groups such as acetate, phosphate, various lipids, and carbohydrates, by changing the chemical nature of an amino acid (or by making structural changes, like the formation of disulfide bridges). In addition, enzymes may remove amino acids from the amino end of the protein, or cut the peptide chain in the middle. For instance, the peptide hormone insulin is cut twice after disulfide bonds are formed, and a propeptide is removed from the middle of the chain. The resulting protein consists of two polypeptide chains connected by disulfide bonds. In addition, most nascent polypeptides start with the amino acid methionine because the "start" codon on mRNA also codes for this amino acid. This amino acid is usually taken off during PTM. Other modifications, like phosphorylation, are part of common mechanisms for controlling the behavior of a protein such as activating or inactivating an enzyme. PTM of proteins is detected by *mass spectrometry* (MS) or *eastern blotting*.

3.4 PROTEIN STRUCTURE

There are four distinct types of protein structures:

- *Primary structure*: The amino acid sequence of the peptide chains. The wide variety of three-dimensional protein structures corresponds to the diversity of functions proteins fulfill. Proteins fold in three dimensions. Protein structure is organized hierarchically from so-called primary structure to quaternary structure. Higher-level structures are motifs and domains. Above all, the wide variety of conformations is due to the huge amount of different sequences of amino acid residues. The primary structure is the sequence of residues in the polypedptide chain.
- *Secondary structure*: Highly regular sub-structures (alpha helix and strands of β-sheet) which are locally defined which means that there can be many different secondary motifs present in one single protein molecule. Secondary structure is a local regularly occurring structure in proteins and is mainly formed through hydrogen bonds between backbone atoms. So-called random coils, loops, or turns do not have a stable secondary structure. There are two types of stable secondary structures, α-helices and β-sheets which are usually located at the core of the protein. On the other hand, loops are usually located in outer regions.
- *Tertiary structure*: Three-dimensional structure of a single protein molecule; a spatial arrangement of the secondary structures. It also describes the completely folded and compacted polypeptide chain. Tertiary structure describes the packing of α-helices, β-sheets, and random coils with respect to each other on the level of one whole polypeptide chain.
- *Quaternary structure*: Complex of several protein molecules or polypeptide chains, usually called *protein subunits* in this context, which function as part of the larger assembly or protein complex. Quaternary structure only exists if there is more than one polypeptide chain present in a complex protein. Only then will quaternary structure describe the spatial organization of the chains (Figure 3.6).

In addition to these levels of structure, a protein may shift between several similar structures in performing its biological function. This process is also reversible. In the context of these functional rearrangements, these tertiary or quaternary structures are usually referred to as *chemical conformation*, and transitions between them are called *conformational changes*.

The primary structure is held together by covalent or peptide bonds, which are made during the process of protein biosynthesis or translation. These peptide bonds provide rigidity to the protein. The two ends of the amino acid chain are referred to as the *C-terminal end* or *carboxyl terminus* (C-terminus) and the *N-terminal end* or *amino terminus* (N-terminus) based on the nature of the free group on each extremity. The various types of secondary structures are defined by their patterns of hydrogen bonds between the main-chain peptide groups. However, these hydrogen bonds

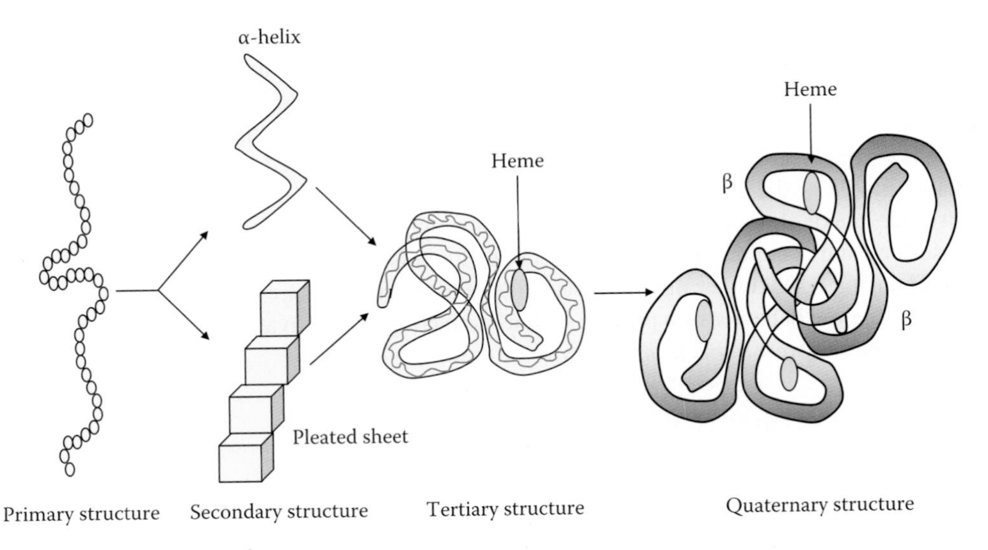

FIGURE 3.6 Different protein structures.

are generally not stable by themselves, since the water-amide hydrogen bond is generally more favorable than the amide-amide hydrogen bond. Thus, the secondary structure is stable only when the local concentration of water is sufficiently low, for example, in the molten globule or fully folded states. Similarly, the formation of molten globules and tertiary structure is driven mainly by structurally nonspecific interactions such as the rough propensities of the amino acids and hydrophobic interactions. However, the tertiary structure is fixed only when the parts of a protein domain are locked into place by structurally specific interactions which include ionic interactions (salt bridges), hydrogen bonds, and the tight packing of side chains. The tertiary structure of extracellular proteins can also be stabilized by disulfide bonds which reduce the entropy of the unfolded state. Disulfide bonds are extremely rare in cytosolic proteins since the cytosol is generally a reducing environment.

The subunits of a protein are amino acids or, to be precise, amino acid residues. An amino acid consists of a central carbon atom (the alpha carbon, C_α), an amino group (NH_2), a hydrogen atom (H), a carboxyl group (COOH), and a side chain (R) which is bound to the C_α. Different side chains (R_i) make up different amino acids with different physicochemical properties. A peptide bond is formed via covalent binding of the carbon atom of the carboxy group of one amino acid to the nitrogen atom of the amino group of another amino acid by dehydration.

3.4.1 Protein Prediction Methods

Knowing a protein's three-dimensional structure helps us to understand its functionality and provides means for planning experiments and drug design. Experimental methods given by x-ray crystallography and nuclear magnetic resonance (NMR) spectroscopy to determine protein structure are essential. The Brookhaven Protein Data Bank (PDB) is the repository for those structures. Files including atom coordinates which are suited for visualization by graphical molecule viewers like rasmol can be obtained at this site. PDB is also searchable with a sequence as a query, for example, with the BLAST service located at NCBI with a polypeptide as a query. Nevertheless, experimental methods are technically very difficult and expensive and the gap in the number between sequenced proteins and known structures increases. Thus, model building of proteins is of great importance. When a protein first is unfolded *in vitro* and then released again, it folds back to the same three-dimensional structure it had before. Thus, various prediction methods are based on the assumption that the three-dimensional protein structure is determined by its primary structure. Structure prediction methods are coarsely divided into three categories as described in the following section.

3.5 PROTEIN FOLDING

Protein folding is the physical process by which a polypeptide folds into its characteristic and functional three-dimensional structure from random coil (Figure 3.7). Each protein exists as an unfolded polypeptide or random coil when translated from a sequence of mRNA to a linear chain of amino acids. This polypeptide lacks any developed three-dimensional structure. However, amino acids interact with each other to produce a well-defined three-dimensional structure, the folded protein, known as the *native state*. The resulting three-dimensional structure is determined by the amino acid sequence. For many proteins, the correct three-dimensional structure is essential for normal function and failure to fold into the intended shape usually produces inactive proteins with different properties including toxic prions. Several neurodegenerative and other diseases are believed to result from the accumulation of incorrectly folded proteins. Aggregated proteins are associated with prion-related illnesses such as Creutzfeldt-Jakob disease and bovine spongiform encephalopathy (mad cow disease), and amyloid-related illnesses such as Alzheimer's disease and familial amyloid cardiomyopathy or polyneuropathy, as well as intracytoplasmic aggregation diseases such as Huntington's and Parkinson's diseases. These age onset degenerative diseases are associated with the multi-merization of misfolded proteins into insoluble, extracellular aggregates, and/or intracellular inclusions including cross-β-sheet amyloid fibrils. It is not clear whether the aggregates are the cause or merely a reflection of the loss of protein homeostasis, or the balance between synthesis, folding, aggregation, and protein turnover. The excessive misfolding and degradation of protein leads to a number of proteopathy diseases such as antitrypsin-associated emphysema, cystic fibrosis, and the lysosomal storage diseases, where loss of function is the origin of the disorder. You might be wondering how to treat the problems associated with protein dysfunction, while protein replacement therapy has historically been used to correct the defect. An emerging approach now is to use pharmaceutical chaperones to fold mutated proteins to render them functional. One of the basic problems in protein folding is how to study their structures. With the help of current advancement in scientific techniques, it can now be studied by using a number of methods as described in the following section.

3.5.1 Tools of Protein Folding

More than a half century ago, evidence began to accumulate that a major part of most proteins' folded structure consists of two regular, highly periodic arrangements, designated alpha and beta.

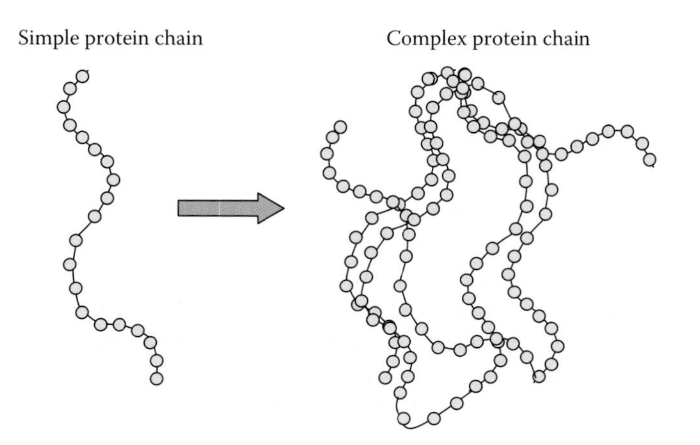

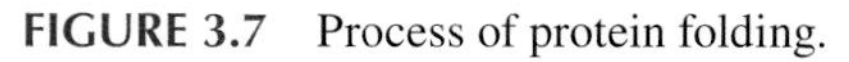
FIGURE 3.7 Process of protein folding.

In 1951 researchers worked out the precise nature of these arrangements. In this section, we have discussed about various tools and techniques used to study the protein folding.

3.5.1.1 Circular Dichroism

Circular dichroism is one of the most general and basic tools to study protein folding. *Circular dichroism spectroscopy* measures the absorption of circularly polarized light. In proteins, structures such as alpha helices and β-sheets are chiral, and thus absorb such light. The absorption of this light acts as a marker of the degree of foldedness of the protein ensemble. This technique can be used to measure equilibrium unfolding of the protein by measuring the change in this absorption as a function of denaturant concentration or temperature. A denaturant melt measures the free energy of unfolding as well as the protein's m value, or denaturant dependence. A temperature melt measures the melting temperature (T_m) of the protein. This type of spectroscopy can also be combined with fast-mixing devices, such as stopped flow, to measure protein folding kinetics and to generate chevron plots.

3.5.1.2 Dual Polarization Interferometry

Dual polarization interferometry is a relatively new bench top technique for measuring the overall change in protein size and fold density during interactions or other stimulus. The technique captures a layer of protein on a glass slide and, using two polarizations of light, measures the conformation and conformational changes with a time resolution of circa 10 Hz at a dimensional resolution of 0.01 nm. The method is quantitative and can be compared directly to what one would expect of crystallography data.

3.5.1.3 Vibration Circular Dichroism of Proteins

The more recent developments of vibration circular dichroism (VCD) techniques for proteins, currently involving Fourier transform (FFT) instruments, provide powerful means for determining protein conformations in solution even for very large protein molecules. Such VCD studies of proteins are often combined with x-ray diffraction of protein crystals, FT-IR data for protein solutions in heavy water (D_2O), or ab initio quantum computations to provide unambiguous structural assignments that are unobtainable from CD.

3.5.1.4 Protein Folding with High Time Resolution

The study of protein folding has been greatly advanced in recent years by the development of fast, time-resolved techniques. These are experimental methods for rapidly triggering the folding of a sample of unfolded protein, and then observing the resulting dynamics. Fast techniques in widespread use include neutron scattering, ultrafast mixing of solutions, photochemical methods, and laser temperature jump spectroscopy.

3.5.1.5 Energy Landscape Theory

The protein folding phenomenon was largely an experimental endeavor until the formulation of energy landscape theory by Joseph Bryngelson and Peter Wolynes in the late 1980s and early 1990s. This approach introduced the principle of minimal frustration, which asserts that evolution has selected the amino acid sequences of natural proteins so that interactions between side chains largely favor the molecule's acquisition of the folded state. Interactions that do not favor folding are selected against, although some residual frustration is expected to exist. This "folding funnel" landscape allows the protein to fold to the native state through any of a large number of pathways and intermediates, rather than being restricted to a single mechanism. The theory is supported by both computational simulations of model proteins and numerous experimental studies, and it has been used to improve methods for protein structure prediction and design.

3.6 PROTEIN MODIFICATION

When protein synthesis is over, proteins undergo structural alteration which affects various physiological functions of our body. These structural alterations, also called *modification*, are caused by phosphorylation, ubiquitination, and other modifications which we will describe briefly in the following subsections.

3.6.1 Protein Modification by Phosphorylation

During cell signaling, many enzymes and structural proteins undergo phosphorylation. The addition of a phosphate to particular amino acids, most commonly serine and threonine mediated by serine or threonine kinases, or more rarely tyrosine mediated by tyrosine kinases, causes a protein to become a target for binding or interacting with a distinct set of other proteins that recognize the phosphorylated domain. Because protein phosphorylation is one of the most-studied protein modifications, many "proteomic" efforts are geared to determining the set of phosphorylated proteins in a particular cell or tissue-type under particular circumstances.

3.6.2 Protein Modification by Ubiquitination

Ubiquitin is a small protein that can be affixed to certain protein substrates by enzymes called *E3 ubiquitin ligases*. Determining which proteins are poly-ubiquitinated can be helpful in understanding how protein pathways are regulated. This is therefore an additional legitimate "proteomic" study. Similarly, once it is determined what substrates are ubiquitinated by each ligase, determining the set of ligases expressed in a particular cell type will be helpful.

3.6.3 Additional Modifications

Listing all the protein modifications that might be studied in a "proteomics" project would require a discussion of most of biochemistry. Therefore, a short list will serve here to illustrate the complexity of the problem. In addition to phosphorylation and ubiquitination, proteins can be subjected to methylation, acetylation, glycosylation, oxidation, nitrosylation, etc., Some proteins undergo all of these modifications, which nicely illustrate the potential complexity one has to deal with when studying protein structure and function.

3.7 PROTEIN TRANSPORT

Upon successful synthesis of proteins, the next obvious step will be to transport the proteins for the required or assigned function. For that reason, proteins must be transported to an organ or tissue. Many proteins are destined for other parts of the cell than the cytosol and a wide range of signaling sequences are used to direct proteins to where they are supposed to be. In prokaryotes, this is normally a simple process due to limited compartmentalization of the cell. However, in eukaryotes, there is a great variety of different targeting processes to ensure the protein arrives at the correct organelle. Not all proteins remain within the cell and many are exported such as digestive enzymes, hormones, and extracellular matrix proteins. In eukaryotes, the export pathway is well developed and the main mechanism for the export of these proteins is translocation to the endoplasmatic reticulum, followed by transport via the Golgi apparatus (Figure 3.8).

3.8 PROTEIN DYSFUNCTION AND DEGRADATION

Just like the rest of our body, our brain changes as we age. Most of us notice some slowed thinking and occasional problems with remembering certain things. However, serious memory loss, confusion, and other major changes in the way our mind works are not a normal part of aging. They may

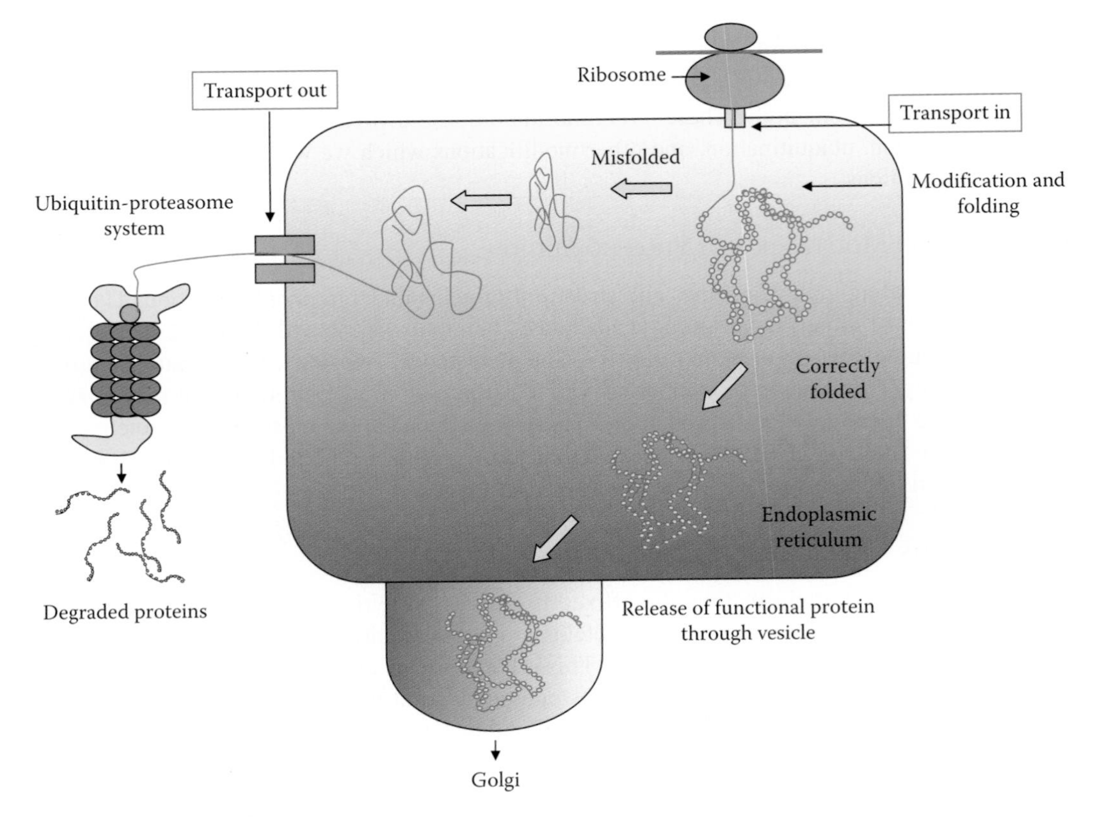

FIGURE 3.8 Mechanism of protein transport.

be a sign that brain cells are failing. The brain has 100 billion nerve cells (neurons). Each nerve cell communicates with many others to form networks. Nerve cell networks have special jobs. Some are involved in thinking, learning, and remembering. Others help us see, hear, and smell. Still others tell our muscles when to move. In Alzheimer's disease, as in other types of dementia, increasing numbers of brain cells deteriorate and die. Two abnormal structures called *plaques* and *tangles* are prime suspects in damaging and killing nerve cells. Plaques and tangles were among the abnormalities that Dr. Alois Alzheimer saw in the brain of Auguste D., although he called them different names. Plaques build up between nerve cells. They contain deposits of a protein fragment called *beta-amyloid*. Tangles are twisted fibers of another protein called *tau*. Tangles form inside dying cells. Though most people develop some plaques and tangles as they age, those with Alzheimer's tend to develop far more. The plaques and tangles tend to form in a predictable pattern, beginning in areas important in learning and memory and then spreading to other regions. Scientists are not absolutely sure what role plaques and tangles play in Alzheimer's disease. Most experts believe they somehow block communication among nerve cells and disrupt activities that cells need to survive (Figure 3.9).

Protein molecules are continuously synthesized and degraded in all living organisms. The concentration of individual cellular proteins is determined by a balance between the rates of synthesis and degradation, which in turn are controlled by a series of regulated biochemical mechanisms. Differences in the rates of protein synthesis and breakdown result in cellular and tissue atrophy (loss of proteins from cells) and hypertrophy (increase in protein content of cells). The degradation rates of proteins are important in determining their cellular concentrations. Protein degradation exhibits first-order kinetics unlike protein synthesis, which is zero-order. Protein degradation is

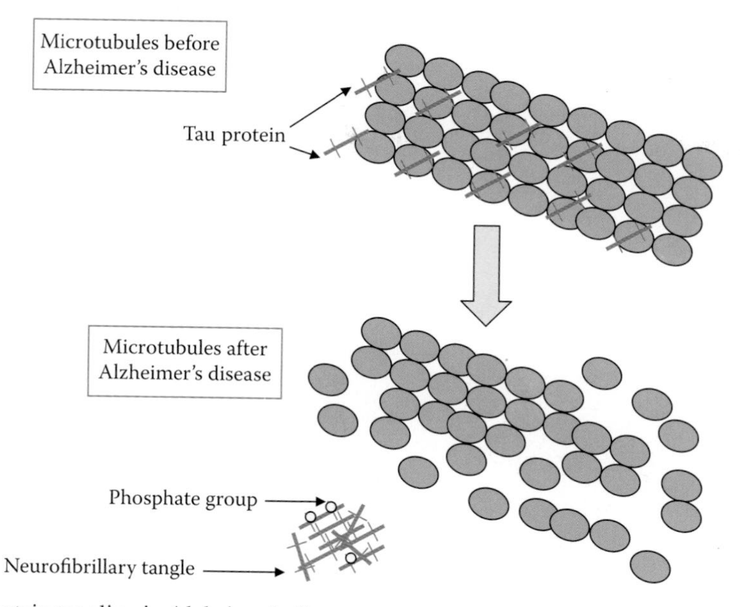

FIGURE 3.9 Protein tangling in Alzheimer's disease.

energy dependent, requiring ATP, and is limited by the concentration of the reactants, whereas protein synthesis cannot be completed in the absence of any one of the necessary reactants.

Proteins breakdown at rates, ranging from 100% per hour to less than 10% per hour and their half-lives (time taken for loss of half the protein molecules) vary between 24 and 72h. Regulatory enzymes and proteins have much shorter half-lives of the order of 5–120min. Protein breakdown can take place in the mitochondria, chloroplasts, lumen of the endoplasmic reticulum, and endosomes, but occurs most commonly in one of two major sites of intracellular proteolysis, lysosomes and the cytosol. The individual degradation rates of proteins vary within a single organelle or cell compartment and also from compartment to compartment, due to either differing sensitivity to local proteases or differing rates of transfer to the cytosol or lysosomes. The range of protein degradation rates within a single organelle is limited, suggesting that the proteins may be treated as groups or families (Figure 3.10).

Most nonselective protein degradation takes place in the lysosomes, where changes in the supply of nutrients and growth factors can influence the rates of protein breakdown. Proteins enter lysosomes by *macroautophagy*, which is the enclosure of a volume of the cytoplasm by an intracellular membrane. The rates of lysosomal degradation can vary greatly with cell type and conditions, ranging from less than 1%/h of total cell protein to 5%–10%/h. The lysosomal degradation of some cytosolic proteins increases in cells deprived of nutrients. It is assumed that the proteins undergoing enhanced degradation are of limited importance for cell viability, and can be sacrificed to support the continuing synthesis of key proteins. Short-lived regulatory proteins are degraded in the cytosol by local proteolytic mechanisms. All short-lived proteins are thought to contain recognition signals that mark them for early degradation. One commonly employed method is the selective labeling of targeted proteins by ubiquitin molecules. Ubiquitin, a protein of 76 amino acids, binds covalently to available lysine residues on target proteins, which are then recognized by proteases. A number of molecular recognition signals for intracellular protein degradation have been identified, and there are likely to be others as yet undiscovered. Additional degradative mechanisms exist for the identification and rapid degradation of proteins that contain translational or post-translational errors, or have been damaged in some way. The degradation of red blood cells is unusual in that it is age dependent. As a consequence, the hemoglobin also exhibits age-linked degradation.

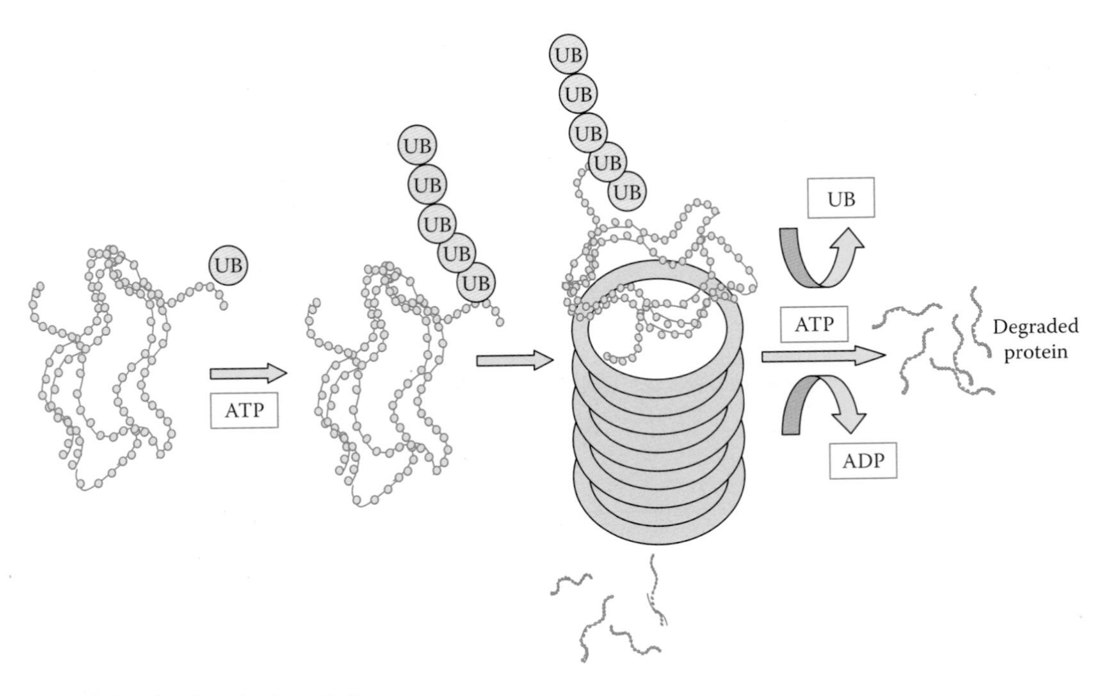

FIGURE 3.10 Protein degradation.

3.9 REGULATION OF PROTEIN SYNTHESIS

In order to control the process of protein synthesis in the body, the function of genes must be regulated with great accuracy and precision. The total amount of DNA present in a cell may contain from a few to thousands of genes. Although the different types of cells in the body of a multi cellular organism differ in structure and function, their genes are identical, since all the cells are ultimately derived from the zygote. The problem therefore is how do cells with identical genetic complements differ so much in structure and function? The answer is that not all genes are active at one time. As development proceeds certain genes become active while others become inactive, that is, the genes are "switched on" and "switched off" at different times. This process is called differential gene action. When genes are active they direct the formation of enzymes which affect certain traits. The metabolic products formed may repress synthesis of enzymes (feedback or end product inhibition). Thus enzyme synthesis is induced and repressed at different times. Although a cell has the genes to produce hundreds of enzymes, only the enzymes required at a particular time are produced. This control mechanism ensures that the cell is not flooded with unnecessary enzymes. A hypothesis to explain induction and repression of enzyme synthesis was first put forward by Francois Jacob and Jacques Monod (1961) of the Institute Pasteur in Paris. For this and some other major contributions in biochemistry Jacob and Monod were awarded the Nobel Prize in Medicine in 1965. The scheme proposed by these workers is called the operon model, and has been considered to be the leading biological discovery of the present century, along with the elucidation of the structure of DNA by Watson and Crick (1953).

Furthermore, gene regulation drives the processes of cellular differentiation and morphogenesis, leading to the creation of different cell types in multicellular organisms where the different types of cells may possess different gene expression profiles though they all possess the same genome sequence.

3.9.1 Stages of Gene Expression

The gene expression may be regulated from the DNA–RNA transcription step to PTM, RNA transport, translation, mRNA degradation, or PTMs.

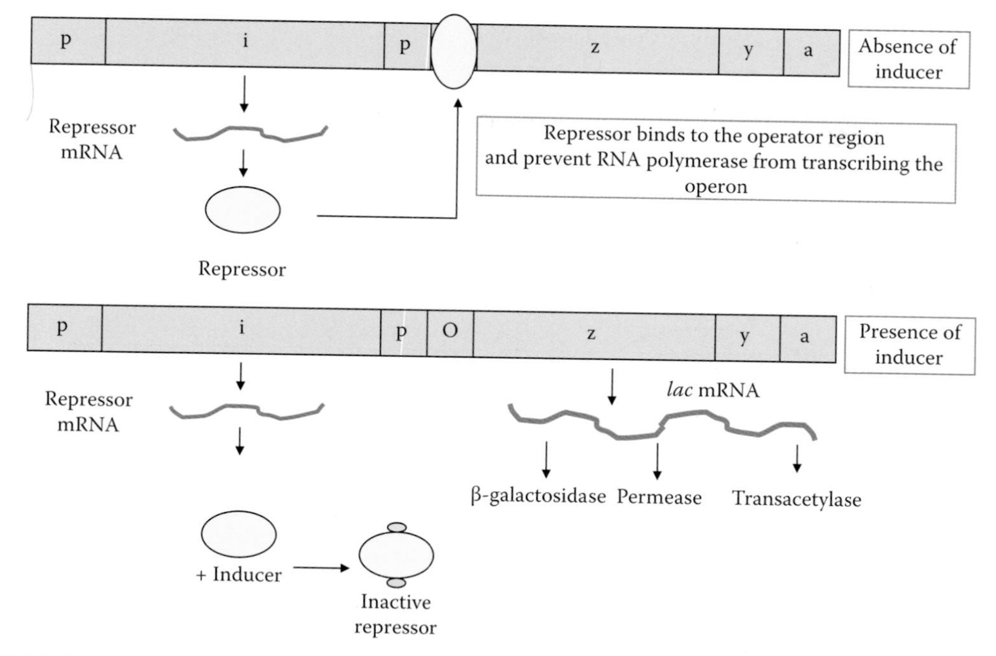

FIGURE 3.11 The *lac* operon model 1.

3.9.1.1 Operon Model for Gene Regulation

The operon consists of the following components: regulator gene, promoter gene (a relatively recent concept), operator gene, structural genes, repressor, corepressor, and inducer. For example, to synthesize β-galactosidase in *E. coli*, Jacob and Monod in 1961 proposed a model based on inducible system which is also known as *operon model*. An *operon* consists of an operator gene which controls the activity of protein synthesis, and a number of structural genes which take part in the synthesis of proteins. In brief, the structural genes will synthesize mRNA under the operational control of an operator gene which in turn is under the control of a repressor molecule synthesized by a regulator gene which is not a part of the operon (Figure 3.11).

3.9.1.2 General Transcription Factors

These TFs position RNA polymerase at the start of a protein-coding sequence and then release the polymerase to transcribe the mRNA. *General transcription factors* (GTF's) or *basal TFs* are protein TFs that have been shown to be important in the transcription of class II genes to mRNA templates. Many of them are involved in the formation of a pre-initiation complex, which, together with RNA polymerase II, bind to and read the single-stranded DNA gene template (Figure 3.12).

3.9.1.3 Enhancers

Enhancers are sites on the DNA helix that are bound to by activators in order to loop the DNA bringing a specific promoter to the initiation complex.

3.9.1.4 Induction and Repression

Escherichia coli synthesis of β-galactosidase has been extensively studied where lactose is converted into glucose and galactose. In order to understand the role of *E. coli* in the synthesis of β-galactosidase, experiments were performed and it has been observed that if β-galactosides are not supplied to *E. coli* cells, the presence of β-galactosidase is hardly detectable but as soon as lactose is added, production of enzyme β-galactosidase increases as much as 10,000 times. The enzyme quantity again falls down as quickly as the substrate (lactose) is removed. Such enzymes whose synthesis can be induced by adding a substrate are known as inducible enzymes and the genetic

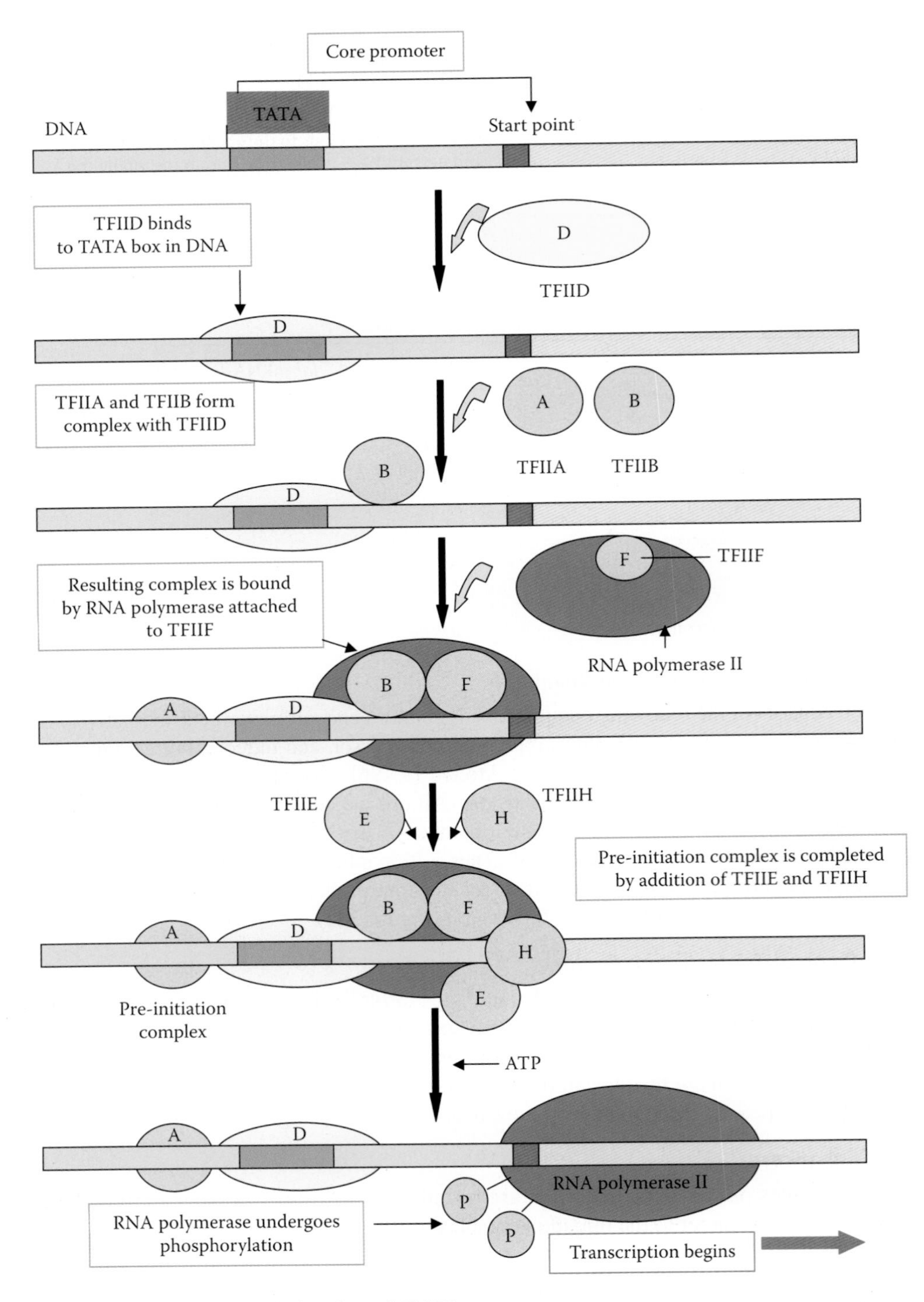

FIGURE 3.12 Protein transcription through TATA.

systems responsible for the synthesis of such enzymes are known as inducible systems. In another situation where no amino acids are supplied from outside, *E. coli* cells can synthesize all the enzymes needed for the synthesis of different amino acids. However, if a particular amino acid like histidine is added, the production of histidine synthesized enzymes declines. In such a scenario, the addition of an end product of a biosynthetic pathway will check synthesis of the enzymes needed for its biosynthesis. The enzymes whose synthesis can be repressed by adding an end product are known as repressible enzymes and their genetic systems would be known as repressible systems. The substrate whose addition induced the synthesis of an enzyme (as lactose in case of synthesis of β-galactosidase) is called inducer. In the same way, the end product whose addition repressed the synthesis of biosynthetic enzymes is called co-repressor. Note that in the absence of lactose, no β-galactosidase is synthesized. This would mean that in the absence of an inducer, the gene or genes responsible for the synthesis of β-galactosidase do not function.

3.10 REGULATORY PROTEIN

Regulatory protein is a term used in genetics to describe a protein involved in regulating gene expression. It is usually bound to a DNA binding site which is sometimes located near the promoter, although this is not always the case. Regulatory proteins are often needed to be bound to a regulatory binding site to switch a gene on (activator) or to shut off a gene (repressor). Generally, as the organism grows more sophisticated, its cellular protein regulation becomes more complicated and indeed some human genes can be controlled by many activators and repressors working together.

In prokaryotes, regulation of transcription is needed for the cell to quickly adapt to the ever-changing outer environment. The presence or the quantity and type of nutrients determine which genes are expressed. In order to do that, genes must be regulated in some fashion. In prokaryotes, repressors bind to regions called *operators* that are generally located downstream from and near the promoter (normally part of the transcript). Activators bind to the upstream portion of the promoter, such as the CAP region (completely upstream from the transcript). A combination of activators, repressors, and rarely enhancers (in prokaryotes) determines whether a gene is transcribed. In eukaryotes, transcriptional regulation tends to involve combinatorial interactions between several TFs, which allow for a sophisticated response to multiple conditions in the environment. This permits spatial and temporal differences in gene expression. Eukaryotes also make use of enhancers. A major difference between eukaryotes and prokaryotes is the fact the eukaryotes have a nuclear envelope, which prevents simultaneous transcription and translation. RNA interference also regulates gene expression in most eukaryotes, both by epigenetic modification of promoters and by breaking down mRNA.

3.11 METHODS FOR PROTEIN ANALYSIS

In order to understand the role of protein in biological functions, various methods have been developed to accurately identify the nature of proteins and as well as to quantify them precisely. In this section, we will describe principles of some of the major methods used in protein identification and quantification.

3.11.1 Protein Identification and Quantification

Like genetic mapping, it is also important to know the methods and procedures to identify and quantify proteins. One of the interesting features of proteomics is that it gives a much better understanding of an organism than genomics. First, the level of transcription of a gene gives only a rough estimate of its level of expression into a protein. An mRNA produced in abundance may be degraded rapidly or translated inefficiently, resulting in a small amount of protein. Second, as mentioned earlier, many proteins experience PTMs that profoundly affect their activities. For example, some proteins are not active until they become phosphorylated. Methods such as phosphoproteomics and glycoproteomics are used to study PTMs. Third, many transcripts give rise to more than one protein, through alternative splicing or alternative PTMs. Fourth, many proteins form complexes with other proteins or RNA molecules, and only function in the presence of these other molecules. Finally, protein degradation rate plays an important role in protein content. In this section, we will learn various techniques and tools used in protein identification and analysis. We will specifically deal with tools and techniques which are being used to analyze the protein structurally, morphologically, and functionally. Ideally, measurement of expression is done by detecting the final gene product (for many genes, this is the protein). However, it is often easier to detect one of the precursors, typically mRNA, and infer gene expression level.

3.11.1.1 Native Gel

Native poly acrylamide gel electrophoresis is an electrophoretic separation method typically used in proteomics and metallomics. Native PAGE (poly acrylamide gel electrophoresis) separations are run in non-denaturing conditions. Detergents are used only to the extent that they are necessary to lyse lipid membranes in the cell. Complexes remain—for the most part—associated and folded as they would be in the cell. One downside, however, is that complexes may not separate cleanly or predictably, since they cannot move through the poly acrylamide gel as quickly as individual, denatured proteins. Take care not to confuse native PAGE with SDS-PAGE (sodium dodecyl sulfate poly acrylamide gel electrophoresis). SDS-PAGE uses a detergent, sodium dodecyl sulfate, to denature the proteins and provide a negative charge that is proportional to the protein mass, allowing electrophoretic separation. There are three popular methods of native PAGE, *blue native* (BN-PAGE), *clear native* (CN-PAGE), and *quantitative preparative native continuous* (QPNC-PAGE) (Figure 3.13).

3.11.1.2 SDS-PAGE

SDS-PAGE is a technique widely used in biochemistry, forensics, genetics and molecular biology to separate proteins according to their electrophoretic mobility (a function of length of polypeptide chain or molecular weight as well as higher order protein folding, PTMs, and other factors). The SDS gel electrophoresis of samples having identical charge to mass ratios results in fractionation by size and is probably the world's most widely used biochemical method. The solution of proteins to be analyzed is first mixed with SDS, an anionic detergent which denatures secondary and non-disulfide-linked tertiary structures, and applies a negative charge to each protein in proportion to its mass. Without SDS, different proteins with similar molecular weights would migrate differently due to differences in mass charge ratio, as each protein has an isoelectric point and molecular weight particular to its primary structure. This is known as native PAGE. Adding SDS solves this problem, as it binds to and unfolds the protein, giving a near uniform negative charge along the length of the

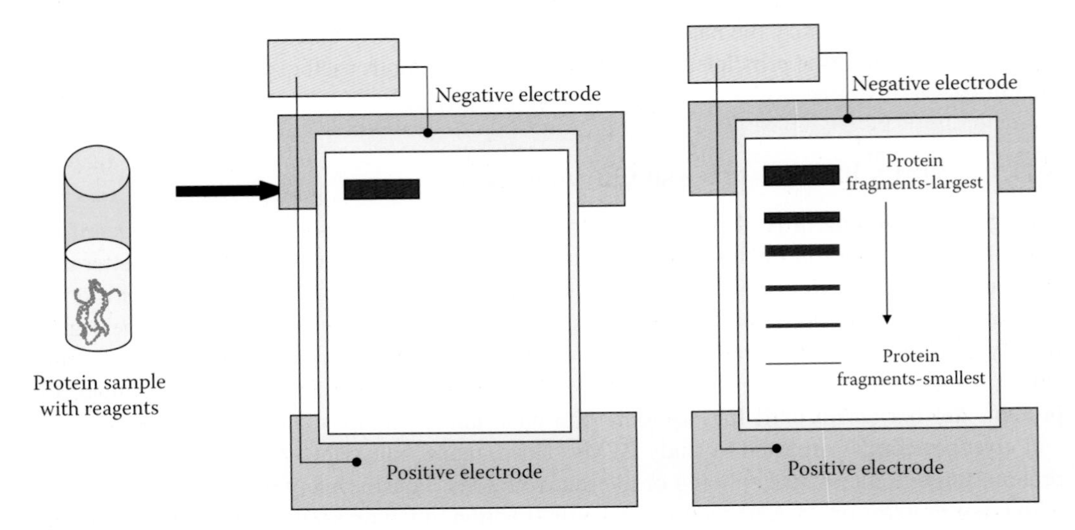

FIGURE 3.13 Gel electrophoresis of protein.

polypeptide. SDS binds in a ratio of approximately 1.4 g SDS per 1.0 g protein (although binding ratios can vary from 1.1–2.2 g SDS/g protein), giving an approximately uniform mass: charge ratio for most proteins, so that the distance of migration through the gel can be assumed to be directly related to only the size of the protein. A tracking dye may be added to the protein solution (of a size smaller than protein) to allow the experimenter to track the progress of the protein solution through the gel during the electrophoretic run.

3.11.1.3 QPNC-PAGE

QPNC-PAGE or quantitative preparative native continuous poly acrylamide gel electrophoresis, is a high-resolution technique applied in biochemistry and bioinorganic chemistry to separate proteins by isoelectric point. This variant of gel electrophoresis is used by biologists to isolate active or native metalloproteins in biological samples and to resolve properly and improperly folded metal cofactor-containing proteins in complex protein mixtures.

3.11.1.4 Protomap

Protomap is a recently developed proteomic technology for identifying changes in proteins that manifest in altered migration by one-dimensional SDS-PAGE. It is similar, conceptually, to two-dimensional gel electrophoresis. The only difference is that it enables global identification of proteins that undergo altered electrophoretic migration resulting from, for example, proteolysis or PTM. However, it is unique in that all proteins are sequenced using MS which provides information on the sequence coverage detected in each isoform of each protein thereby facilitating interpretation of proteolytic events. Protomap is performed by resolving control and experimental samples in separate lanes of a 1D SDS-PAGE gel. Each lane is cut into evenly spaced bands (usually 15–30 bands) and proteins in these bands are sequenced using shotgun proteomics. Sequence information from all of these bands are bioinformatically integrated into a visual format called a *peptograph* which plots gel-migration in the vertical dimension (high- to low-molecular weight, top to bottom) and sequence coverage in the horizontal dimension (N- to C-terminus, left to right). A peptograph is generated for each protein in the sample (thousands of peptographs are generated from a single experiment) and this data format enables rapid identification of proteins undergoing proteolytic cleavage by making evident changes in gel-migration that are accompanied by altered topography.

3.11.1.5 Western Blot

The *western blot* (alternatively, protein immunoblot) is an analytical technique used to detect specific proteins in a given sample of tissue homogenate or extract. It uses gel electrophoresis to separate native or denatured proteins by the length of the polypeptide (denaturing conditions) or by the 3-D structure of the protein (native/non-denaturing conditions). The proteins are then transferred to a membrane (typically nitrocellulose or PVDF), where they are probed (detected) using antibodies specific to the target protein. There are now many reagent companies that specialize in providing antibodies (both monoclonal and polyclonal antibodies) against tens of thousands of different proteins. Commercial antibodies can be expensive, although the unbound antibody can be reused between experiments. This method is used in the fields of molecular biology, biochemistry, immunogenetics, and other molecular biology disciplines. The method originated from the laboratory of George Stark at Stanford. The name western blot was given to the technique by W. Neal Burnette and is a play on the name Southern blot, a technique for DNA detection developed earlier by Edwin Southern. Detection of RNA is termed *northern blotting* and the detection of PTM of protein is termed *eastern blotting* (Figures 3.14 and 3.15).

3.11.1.6 Cellular Techniques

The study of proteins *in vivo* is often concerned with the synthesis and localization of the protein within the cell. Although many intracellular proteins are synthesized in the cytoplasm and

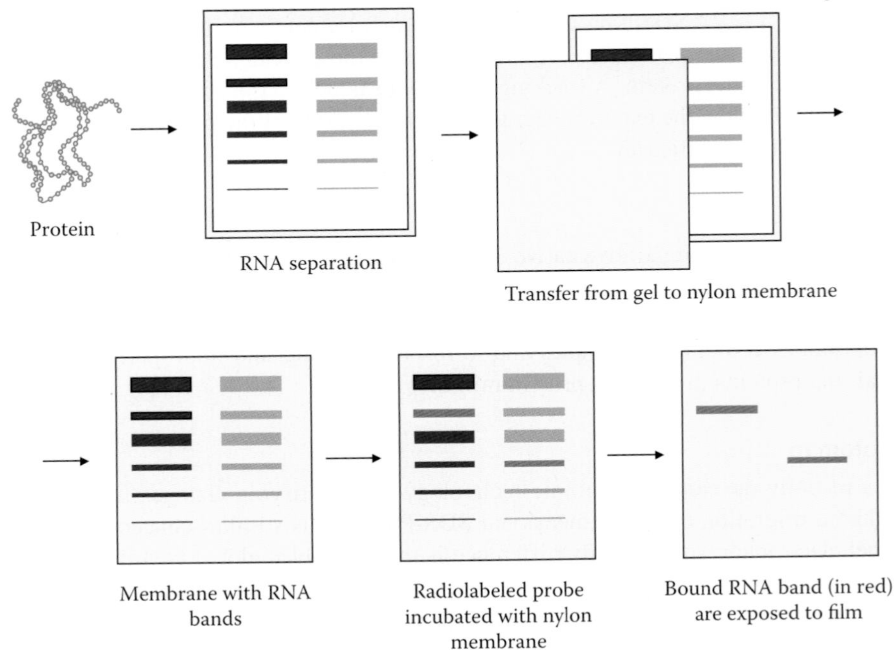

FIGURE 3.14 Western blotting technique.

membrane-bound or secreted proteins in the endoplasmic reticulum, the specifics of how proteins are targeted to specific organelles or cellular structures is often unclear. A useful technique for assessing cellular localization of antigens (proteins) in the cells by using green fluorescent protein (GFP) as a marker and proteins can be visualized by using fluorescent microscopy. Other methods for elucidating the cellular location of proteins requires the use of known compartmental markers for regions such as the ER, the Golgi, lysosomes, or vacuoles, mitochondria, chloroplasts, plasma membrane, etc. With the use of fluorescently-tagged versions of these markers or of antibodies to known markers, it becomes much simpler to identify the localization of a protein of interest. For example, indirect immunofluorescence will allow for fluorescence co-localization and demonstration of location. Fluorescent dyes are used to label cellular compartments for a similar purpose.

Other possibilities exist, as well. For example, immunohistochemistry usually utilizes an antibody to one or more proteins of interest that are conjugated to enzymes yielding either luminescent or chromogenic signals that can be compared between samples, allowing for localization information. Another applicable technique is cofractionation in sucrose (or other material) gradients using isopycnic centrifugation. While this technique does not prove co-localization of a compartment of known density and the protein of interest, it does increase the likelihood, and is more amenable to large-scale studies. Finally, the gold-standard method of cellular localization is immunoelectron microscopy. This technique also uses an antibody to the protein of interest, along with classical electron microscopy techniques. The sample is prepared for normal electron microscopic examination, and then treated with an antibody to the protein of interest that is conjugated to an extremely electro-dense material, usually gold. This allows for the localization of both ultrastructural details as well as the protein of interest. Through another genetic engineering application known as site-directed mutagenesis, researchers can alter the protein sequence and hence its structure, cellular localization, and susceptibility to regulation. This technique even allows the incorporation of unnatural amino acids into proteins, using modified tRNAs, and may allow the rational design of new proteins with novel properties.

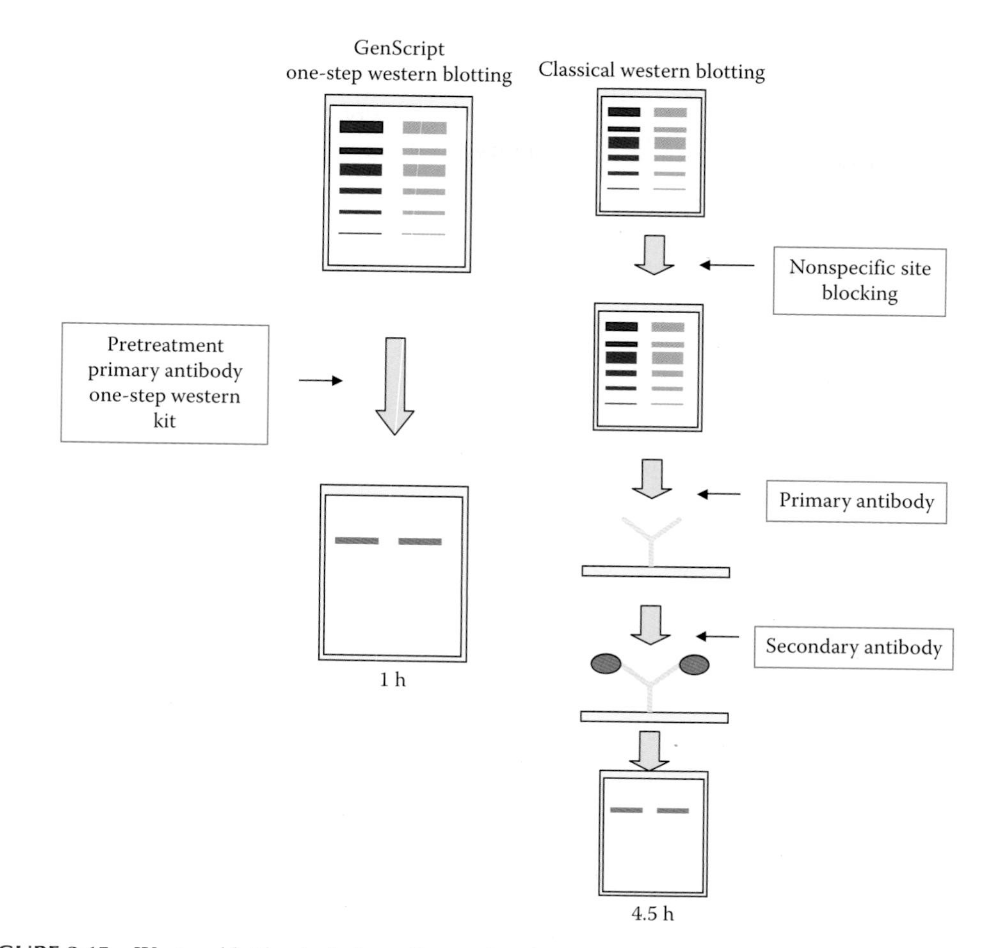

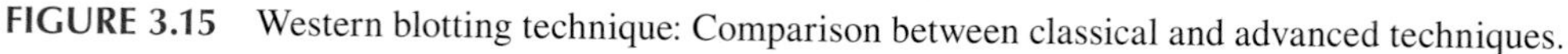

FIGURE 3.15 Western blotting technique: Comparison between classical and advanced techniques.

3.11.1.7 Enzyme-Linked Immunosorbent Assay

Enzyme-linked immunosorbent assay, also called (ELISA), or *enzyme immunoassay* (EIA), is a biochemical technique used mainly in immunology to detect the presence of an antibody or an antigen in a sample. ELISA has been used as a diagnostic tool in medicine and plant pathology, as well as a quality control check in various industries. In simple terms, in ELISA an unknown amount of antigen is affixed to a surface, and then a specific antibody is washed over the surface so that it can bind to the antigen. This antibody is linked to an enzyme, and in the final step a substance is added that the enzyme can convert to some detectable signal. Thus, in the case of fluorescence ELISA, when light of the appropriate wavelength is shown upon the sample, any antigen/antibody complexes will fluoresce so that the amount of antigen in the sample can be inferred through the magnitude of the fluorescence. Performing an ELISA involves at least one antibody with specificity for a particular antigen. The sample with an unknown amount of antigen is immobilized on a solid support (usually a polystyrene microtiter plate) either nonspecifically (via adsorption to the surface) or specifically (via capture by another antibody specific to the same antigen, in a "sandwich" ELISA). After the antigen is immobilized the detection antibody is added, forming a complex with the antigen. The detection of antibody in the sample can be covalently linked to an enzyme, or can itself be detected by a secondary antibody linked to an enzyme through bio-conjugation. Between each step, the plate is typically washed with a mild detergent solution to remove any proteins or antibodies that are not specifically bound. After the final wash step, the plate is developed by adding an enzymatic substrate to produce a visible signal, which indicates the quantity of antigen in

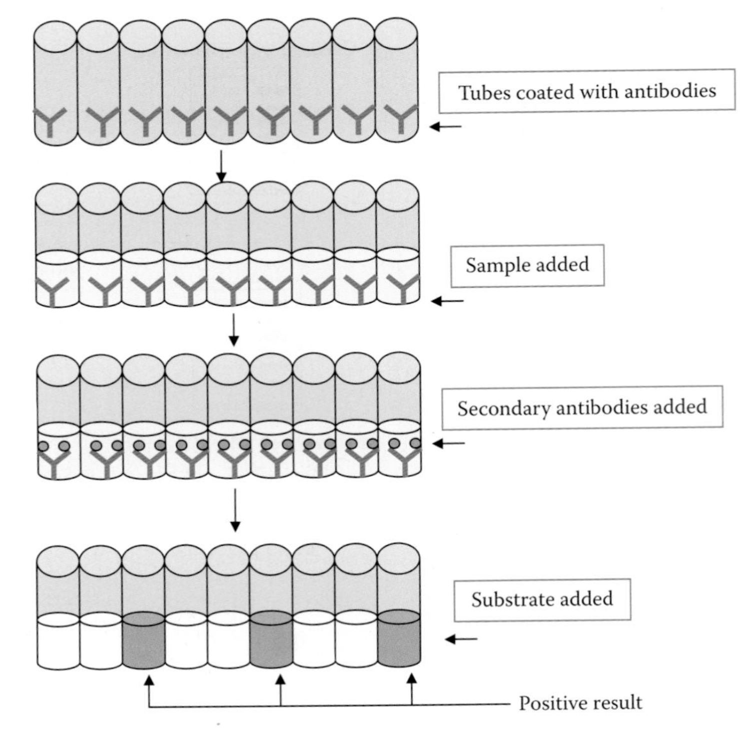

FIGURE 3.16 ELISA technique.

the sample. As technology advances, there is also improvement in analysis; for instance, traditional ELISA utilizes chromogenic substrates, but in the modern ELISA fluorescence-tag is being used for higher sensitivity (Figure 3.16).

3.12 PROTEIN PURIFICATION

In order to perform *in vitro* analysis, a protein must be purified away from other cellular components. This process usually begins with cell lysis, in which a cell's membrane is disrupted and its internal contents released into a solution known as a *crude lysate*. The resulting mixture can be purified using ultracentrifugation, which fractionates the various cellular components into fractions containing soluble proteins, membrane lipids and proteins, cellular organelles, and nucleic acids. Precipitation by a method known as *salting out* can concentrate the proteins from this lysate. Various types of chromatography are then used to isolate the protein or proteins of interest based on properties such as molecular weight, net charge, and binding affinity. The level of purification can be monitored using various types of gel electrophoresis if the desired protein's molecular weight and isoelectric point are known, by spectroscopy if the protein has distinguishable spectroscopic features, or by enzyme assays if the protein has enzymatic activity. Additionally, proteins can be isolated according their charge using electrofocusing. For natural proteins, a series of purification steps may be necessary to obtain protein sufficiently pure for laboratory applications. To simplify this process, genetic engineering is often used to add chemical features to proteins that make them easier to purify without affecting their structure or activity. Here, a "tag" consisting of a specific amino acid sequence, often a series of histidine residues (a "His-tag"), is attached to one terminus of the protein. As a result, when the lysate is passed over a chromatography column containing nickel, the histidine residues ligate the nickel and attach to the column while the untagged components of the lysate pass unimpeded. A number of different tags have been developed to help researchers purify specific proteins from complex mixtures.

3.13 TOOLS OF PROTEOMICS

The total complement of proteins present at a time in a cell or cell type is known as its *proteome*, and the study of such large-scale data sets defines the field of *proteomics*, named by analogy to the related field of genomics. The key experimental techniques in proteomics are described in the following text.

3.13.1 2D Electrophoresis

Two-dimensional gel electrophoresis or 2-D electrophoresis is a form of gel electrophoresis commonly used to separate a large number of proteins. Mixtures of proteins are separated by two properties in two dimensions on 2D gels. 2-D electrophoresis begins with 1-D electrophoresis but then separates the molecules by a second property in the direction 90° from the first. In 1-D electrophoresis, proteins are separated in one dimension, so that all the proteins will lie along a lane but separated from each other by an isoelectric point. The result is that the molecules are spread out across a 2-D gel. Because it is unlikely that two molecules will be similar in two distinct properties, molecules are more effectively separated in 2-D electrophoresis than in 1-D electrophoresis. To separate the proteins by isoelectric point is called *isoelectric focusing* (IEF). Thereby, a gradient of pH is applied to a gel and an electric potential is applied across the gel, making one end more positive than the other. At all pHs other than their isoelectric point, proteins will be charged. If they are positively charged, they will be pulled towards the more negative end of the gel and if they are negatively charged they will be pulled to the more positive end of the gel. The proteins applied in the first dimension will move along the gel and will accumulate at their isoelectric point. That is, the point at which the overall charge on the protein is 0 (a neutral charge). The result of this is a gel with proteins spread out on its surface. These proteins can then be detected by a variety of means, but the most commonly used stains are silver and coomassie staining. In this case, a silver colloid is applied to the gel. The silver binds to cysteine groups within the protein. The silver is darkened by exposure to ultraviolet light. The darkness of the silver can be related to the amount of silver and therefore the amount of protein at a given location on the gel. This measurement can only give approximate amounts, but is adequate for most purposes (Figure 3.17).

3.13.2 Mass Spectrometry

MS is an analytical technique for the determination of the elemental composition of a sample or molecule and which allows rapid high-throughput identification of proteins and sequencing of peptides most often after in-gel digestion. It is also used for elucidating the chemical structures of molecules, such as peptides and other chemical compounds. The MS principle consists of ionizing chemical compounds to generate charged molecules or molecule fragments and measurement of their mass-to-charge ratios. MS instruments consist of three modules: an *ion source*, which can convert gas phase sample molecules into ions (or, in the case of electrospray ionization (ESI), move ions that exist in solution into the gas phase); a *mass analyzer*, which sorts the ions by their masses by applying electromagnetic fields; and a *detector*, which measures the value of an indicator quantity and thus provides data for calculating the abundances of each ion present. The technique has both qualitative and quantitative uses. These include identifying unknown compounds, determining the isotopic composition of elements in a molecule, and determining the structure of a compound by observing its fragmentation. MS is now in very common use in analytical laboratories that study physical, chemical, or biological properties of a great variety of compounds. MS is an important emerging method for the characterization of proteins. The two primary methods for ionization of whole proteins are ESI and *matrix-assisted laser desorption/ionization* (MALDI). In keeping with the performance and mass range of available mass spectrometers, two approaches are used for characterizing proteins. In the first, intact proteins are ionized by either of the two techniques described

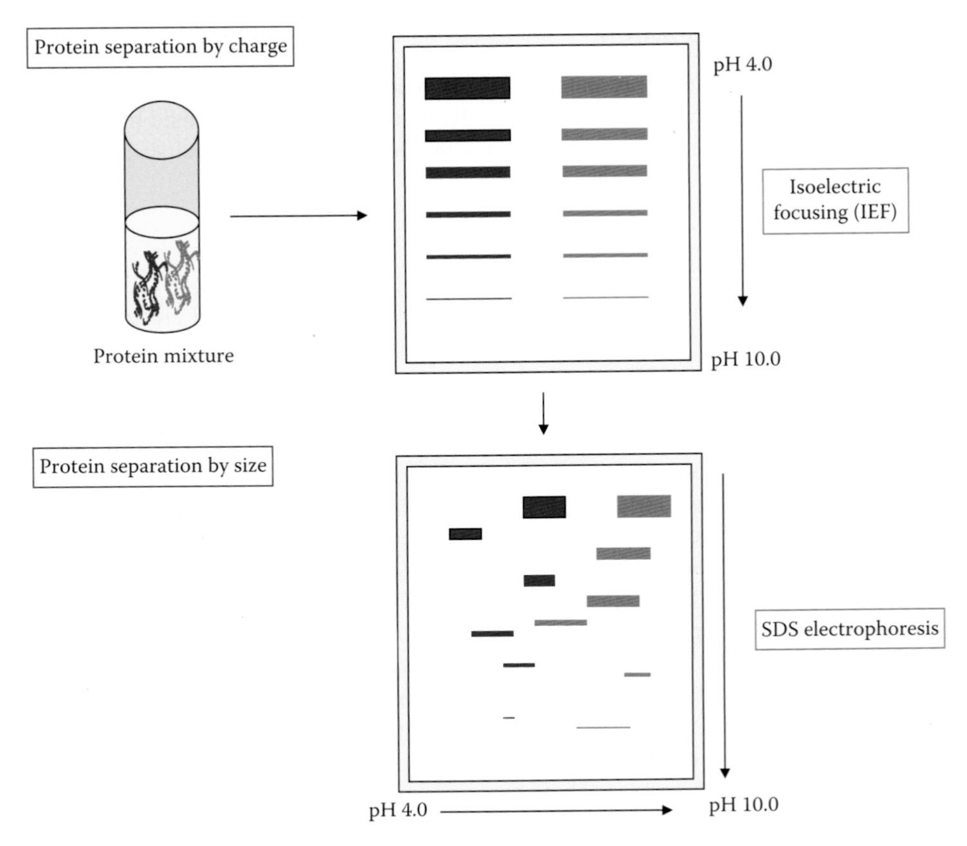

FIGURE 3.17 2D Electrophoresis.

earlier, and then introduced to a mass analyzer. This approach is referred to as "top-down" strategy of protein analysis. In the second, proteins are enzymatically digested into smaller peptides using proteases such as trypsin or pepsin, either in solution or in gel after electrophoretic separation. Other proteolytic agents are also used. The collection of peptide products are then introduced to the mass analyzer. When the characteristic pattern of peptides is used for the identification of the protein, the method is called *peptide mass fingerprinting* (PMF). If the identification is performed using the sequence data determined in tandem with MS analysis, it is called *de novo sequencing*. These procedures of protein analysis are also referred to as the "bottom-up" approach.

3.13.3 Protein Microarray

A *protein microarray*, sometimes referred to as a *protein binding microarray*, provides a multiplex approach to identify protein–protein interactions, to identify the substrates of protein kinases, to identify TF protein-activation, or to identify the targets of biologically active small molecules. The array is a piece of glass on which different molecules of protein or specific DNA binding sequences (as capture probes for the proteins) have been affixed at separate locations in an ordered manner thus forming a microscopic array. The most common protein microarray is the *antibody microarray*, where antibodies are spotted onto the protein chip and are used as *capture molecules* to detect proteins from cell lysate solutions. Protein microarrays (also biochip, proteinchip) are measurement devices used in biomedical applications to determine the presence and/or amount (referred to as relative quantitation) of proteins in biological samples, for example, blood. They have the potential to be an important tool for proteomics research. Usually different capture agents, most frequently monoclonal antibodies, are deposited on a chip surface (glass or silicon) in a miniature array. This format is often referred to as a microarray (a more general term

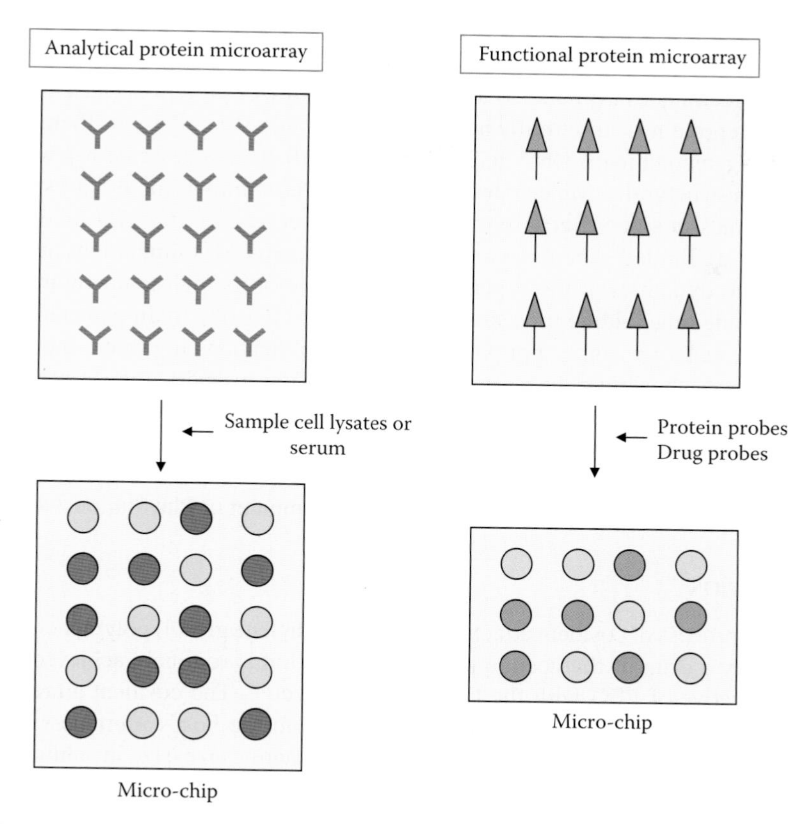

FIGURE 3.18 Protein microarray.

for chip based biological measurement devices). There are several types of protein chips, the most common being glass slide chips and nano-well arrays (Figure 3.18).

3.13.4 Two-Hybrid Screening

Two-hybrid screening allows the systematic exploration of protein–protein interactions. It is also known as yeast two-hybrid system or Y2H, a molecular biology technique used to discover protein–protein interactions and protein–DNA interactions by testing for physical interactions such as binding between two proteins or a single protein and a DNA molecule, respectively. The premise behind the test is the activation of downstream reporter gene(s) by the binding of a TF onto an upstream activating sequence (UAS). For two-hybrid screening, the TF is split into two separate fragments, called the *binding domain* (BD) and *activating domain* (AD). The BD is the domain responsible for binding to the UAS, while AD is the domain responsible for the activation of transcription. One limitation of classic yeast two-hybrid screens is that they are limited to soluble proteins. It is therefore impossible to use them to study the protein–protein interactions between insoluble integral membrane proteins. The split-ubiquitin system provides a method for overcoming this limitation. In the split-ubiquitin system, two integral membrane proteins to be studied are fused to two different ubiquitin moieties: a C-terminal ubiquitin moiety ("Cub," residues 35–76) and an N-terminal ubiquitin moiety ("Nub," residues 1–34). These fused proteins are called the *bait* and *fish*, respectively. In addition to being fused to an integral membrane protein, the Cub moiety is also fused to a TF that can be cleaved off by ubiquitin specific proteases. Upon bait-fish interaction, Nub- and Cub-moieties assemble, reconstituting the split-ubiquitin. The reconstituted split-ubiquitin molecule is recognized by ubiquitin specific proteases which cleave off the reporter protein, allowing it to induce the transcription of reporter genes.

3.13.5 Protein Structure Prediction

There are various techniques available to analyze the structure of protein because it is very important to know which protein is structurally normal or abnormal especially in the therapeutic proteins developed by using recombinant DNA technology. We will briefly describe a few techniques to analyze the protein structurally. Molecular dynamics (MD) is an important tool for studying protein folding and dynamics in silico. Because of computational cost, *ab initio* MD folding simulations with explicit water are limited to peptides and very small proteins. MD simulations of larger proteins remain restricted to dynamics of the experimental structure or its high-temperature unfolding. In order to simulate long time folding processes (beyond about 1 μs), like folding of small-size proteins (about 50 residues) or larger, some approximations or simplifications in protein models need to be introduced. An approach using reduced protein representation (pseudo-atoms representing groups of atoms are defined) and statistical potential is not only useful in protein structure prediction, but is also capable of reproducing the folding pathways. There are distributed computing projects which use idle CPU time of personal computers to solve problems such as protein folding or prediction of protein structure. People can run these programs on their computer or PlayStation 3 to support them.

3.13.6 PEGylation

PEGylation is the process of covalent attachment of poly (ethylene glycol) polymer chains to another molecule, normally a drug or therapeutic protein. PEGylation is routinely achieved by incubation of a reactive derivative of PEG with the target macromolecule. The covalent attachment of PEG to a drug or therapeutic protein can "mask" the agent from the host's immune system (reduced immunogenicity and antigenicity) and increase the hydrodynamic size (size in solution) of the agent which prolongs its circulatory time by reducing renal clearance. PEGylation can also provide water solubility to hydrophobic drugs and proteins.

3.13.7 High-Performance Liquid Chromatography

High-performance liquid chromatography (or high pressure liquid chromatography, HPLC) is a form of column chromatography used frequently in biochemistry and analytical chemistry to separate, identify, and quantify compounds including proteins. HPLC utilizes a column that holds chromatographic packing material (stationary phase), a pump that moves the mobile phase(s) through the column, and a detector that shows the retention times of the molecules. Retention time varies depending on the interactions between the stationary phase, the molecules being analyzed, and the solvent(s) used.

3.13.8 Shotgun Proteomics

Shotgun proteomics is a method of identifying proteins in complex mixtures using a combination of high-performance liquid chromatography combined with MS. The name is derived from shotgun sequencing of DNA which is itself named by analogy with the rapidly-expanding, quasi-random firing pattern of a shotgun. In shotgun proteomics, the proteins in the mixture are digested and the resulting peptides are separated by liquid chromatography. Tandem mass spectrometry is then used to identify the peptides.

3.13.9 Top-Down Proteomics

Top-down proteomics is a method of protein identification that uses an ion trapping mass spectrometer to store an isolated protein ion for mass measurement and tandem mass spectrometry analysis. The name is derived from the similar approach to DNA sequencing. Proteins are typically ionized by ESI and trapped in a Fourier transform ion cyclotron resonance (Penning trap) or quadruple ion trap (Paul trap) mass spectrometer. Fragmentation for tandem mass spectrometry is accomplished by electron capture dissociation or electron transfer dissociation.

PROBLEMS

Section A: Descriptive Type

Q1. Explain the function of proteins as enzymes.
Q2. Explain the role of proteins in cell signaling.
Q3. What are the different kinds of nonessential amino acids?
Q4. What is a protein biosynthesis?
Q5. Discuss transcription in prokaryotes and eukaryotes.
Q6. Describe the tools for studying the structure of proteins.
Q7. What is protein folding?
Q8. Describe the regulation of gene expression for protein synthesis.
Q9. What is an operon model for gene regulation?

Section B: Multiple Choice

Q1. About how many reactions are known to be catalyzed by enzymes?
  a. 3000
  b. 4500
  c. 4000
  d. 5000
Q2. Cystine and aspartic acids are the nonessential amino acids. True/False
Q3. Protein synthesis starts with translation of proteins. True/False
Q4. Make correct pair
  a. Upstream process 3′UTR
  b. Down stream process 5′UTR
Q5. What is the most common type of core promoter in eukaryotes?
  a. GATA box
  b. TATA box
  c. BATA box
Q6. In bacteria, transcription begins with the binding of RNA polymerase to the promoter in DNA. True/False
Q7. Which of the following has the ability to transcribe RNA into DNA?
  a. Bacteria
  b. Fungi
  c. Viruses
Q8. How many stages are there in the translation phase of protein synthesis?
  a. 3
  b. 4
  c. 5
Q9. Among the following diseases, which one is not caused by incorrect protein folding?
  a. Creutzfeldt-Jakob disease
  b. Alzheimer's disease
  c. Epilepsy
  d. Amyloid cardiomyopathy
Q10. Protein molecules are continuously synthesized and degraded in all living organisms. True/False
Q11. Where does protein breakdown take place?
  a. Nucleus
  b. Microtubules
  c. Mitochondria
  d. Golgi bodies

Q12. During protein synthesis, operon model is used for …
   a. Protein transport
   b. Gene regulation
   c. Amino acid regulation
Q13. Which of the following is not native gel?
   a. BN-PAGE
   b. CN-PAGE
   c. SDS-PAGE
   d. QPNC-PAGE
Q14. What does ELISA mean?
   a. Enzyme assay
   b. Enzyme-linked immuno assay
   c. Enzyme-linked immunosorbent assay
   d. Enzyme-linked sorbent assay
Q15. What does HPLC mean?
   a. High protein liquid chromatography
   b. High-performance liquid chromatography
   c. High purified liquid chromatography

Section C: Critical Thinking

Q1. How can one distinguish normal protein from abnormal protein using proteomics tools?
Q2. Why do proteins undergo folding and refolding phases? Explain using examples.
Q3. What method would you employ to identify the structure of a newly synthesized protein?

ASSIGNMENT

Make a chart of the essential and nonessential amino acids showing their beneficial attributes. Discuss your chart with your classmates.

REFERENCES AND FURTHER READING

Anfinsen, C. The formation and stabilization of protein structure. *Biochem. J.* 128: 737–749, 1972.

Alberts, B., Johnson, A., Lewis, J., Raff, M., Roberts, K., and Walters, P. The shape and structure of proteins, in: *Molecular Biology of the Cell*, 4th edn. Garland Science, New York, 2002.

Alexander, P.A.Y., He, Y., Chen, J., Orban, P.N., and Bryan, P.N. The design and characterization of two proteins with 88% sequence identity but different structure and function. *Proc Natl. Acad. Sci. USA* 104(29): 11963–11968, 2007.

Belay, E. Transmissible spongiform encephalopathies in humans. *Annu. Rev. Microbiol.* 53: 283–314, 1999.

Bu, Z., Cook, J., and Callaway, D.J.E. Dynamic regimes and correlated structural dynamics in native and denatured alpha-lactalbumin. *J. Mol. Biol.* 312: 865–873, 2001.

Büeler, H.A., Aguzzi, A., Sailer, R., Greiner, P., Autenried, M., Aguet, C., and Weissmann, C. Mice devoid of PrP are resistant to scrapie. *Cell* 73: 1339–1347, 1993.

Collinge, J. Prion diseases of humans and animals: Their causes and molecular basis. *Annu. Rev. Neurosci.* 24: 519–550, 2001.

Deechongkit, S., Nguyen, H., Dawson, P.E., Gruebele, M., and Kelly, J.W. Context dependent contributions of backbone h-bonding to β-sheet folding energetics. *Nature* 403: 101–105, 2004.

Fürst, P. and Stehle, P. What are the essential elements needed for the determination of amino acid requirements in humans? *J. Nutr.* 134: 1558–1565, 2004.

Gilch, S. et al. Intracellular re-routing of prion protein prevents propagation of PrPSc and delays onset of prion disease. *The EMBO J.* 20: 3957–3966, 2001.

Ironside, J.W. Variant Creutzfeldt-Jakob disease: Risk of transmission by blood transfusion and blood therapies. *Hemophilia* 12: 8–15, 2006.

Imura, K. and Okada, A. Amino acid metabolism in pediatric patients. *Nutrition* 14: 143–148, 1998.

Kim, P.S. and Baldwin, R.L. Intermediates in the folding reactions of small proteins. *Annu. Rev. Biochem.* 59: 631–660, 1990.

Kmiecik, S. and Kolinski, A. Characterization of protein-folding pathways by reduced-space modeling. *Proc. Natl. Acad. Sci. USA* 104: 12330–12335, 2007.

Kubelka, J., Hofrichter, J., and Eaton, W.A. The protein folding speed limit. *Curr. Opin. Struct. Biol.* 14: 76–88, 2004.

Lee, S. and Tsai, F. Molecular chaperones in protein quality control. *J. Biochem. Mol. Biol.* 38(3): 259–265, 2005.

Levinthal, C. Are there pathways for protein folding?. *J. Chim. Phys.* 65: 44–45, 1968.

Pace, C., Shirley, B., McNutt, M., and Gajiwala, K. Forces contributing to the conformational stability of proteins. *Faseb. J.* 10: 75–83, 1996.

Reeds, P.J. Dispensable and indispensable amino acids for humans. *J. Nutr.* 130: 1835–1840, 2000.

Rose, G., Fleming, P., Banavar, J., and Maritan, A. A backbone-based theory of protein folding. *Proc. Natl. Acad. Sci. USA* 103: 16623–16633, 2006.

Shortle, D. The denatured state (the other half of the folding equation) and its role in protein stability. *Faseb J.* 10: 27–34, 1996.

Telling, G., Scott, M., Mastrianni, J., Gabizon, R., Torchia, M., Cohen, F., DeArmond, S., and Prusiner, S. Prion propagation in mice expressing human and chimeric PrP transgenes implicates the interaction of cellular PrP with another protein. *Cell* 83: 93, 1995.

Van den Berg, B., Wain, R., Dobson, C.M., and Ellis, R.J. Macromolecular crowding perturbs protein refolding kinetics. Implications for folding inside the cell. *Embo J.* 19(15): 3870–3935, 2000.

Young, V.R. Adult amino acid requirements: The case for a major revision in current recommendations. *J. Nutr.* 124: 1517S–1523S, 1994.

4 Recombinant DNA Technology

LEARNING OBJECTIVES

- Define recombinant DNA (rDNA) technology
- Discuss the significance of rDNA technology
- Explain the steps involved in making rDNA products
- Discuss the role of restriction enzymes in rDNA technology
- Explain the significance of vectors and their characteristics
- Discuss applications of rDNA technology
- Explain DNA sequence and its methods
- Explain DNA microarray, DNA chips, and complementary DNA (cDNA) library topics

4.1 INTRODUCTION

In the previous chapters, we learned about the structure and function of genes and proteins. The next question that comes to mind is how we can manipulate genes to make customized proteins. The use of technology to manipulate genes is called *genetic engineering* or *recombinant DNA technology* (rDNA). rDNA technology is a field of molecular biology in which scientists "edit" DNA to form new synthetic molecules, called *chimeras*. The practice of cutting, pasting, and copying DNA dates back to Arthur Kornberg's successful replication of viral DNA in a breakthrough that served as a proof-of-concept for cloning. This was followed by the Swiss biochemist Werner Arber's discovery of restriction enzymes in bacteria that degrade foreign viral DNA molecules while sparing their own DNA. Arber effectively showed geneticists how to "cut" DNA molecules. Soon to follow was the understanding that ligase could be used to "glue" them together. These two achievements launched rDNA technology research. The most common recombinant process involves combining the DNA of two different organisms. The rDNA technique was first proposed by Peter Lobban, a graduate student with A. Dale Kaiser at Stanford University, Department of Biochemistry. The technique was then realized by Lobban and Kaiser; Jackson, Symons, and Berg; and Stanley Norman Cohen, Chang, Boyer, and Helling in 1972–1974. rDNA technology was made possible by the discovery, isolation, and application of restriction endonucleases by Werner Arber, Daniel Nathans, and Hamilton Smith, for which they received the 1978 Nobel Prize in Medicine (Figure 4.1).

4.2 MAKING OF RECOMBINANT DNA

The making of rDNA is multistep process, which includes isolation and insertion of the gene of interest in a specific vector. The steps are described briefly in the following.

4.2.1 Steps in Making a Recombinant DNA Product

The following are the steps involved in making an rDNA product:

Step 1: rDNA technology begins with the isolation of the gene of interest (foreign DNA). The gene is then inserted into a vector and cloned. A *vector* is a piece of DNA that is capable of independent growth. The commonly used vectors are bacterial plasmids and viral phages. The gene of interest is integrated into the plasmid or phage, which is referred to as *rDNA*.

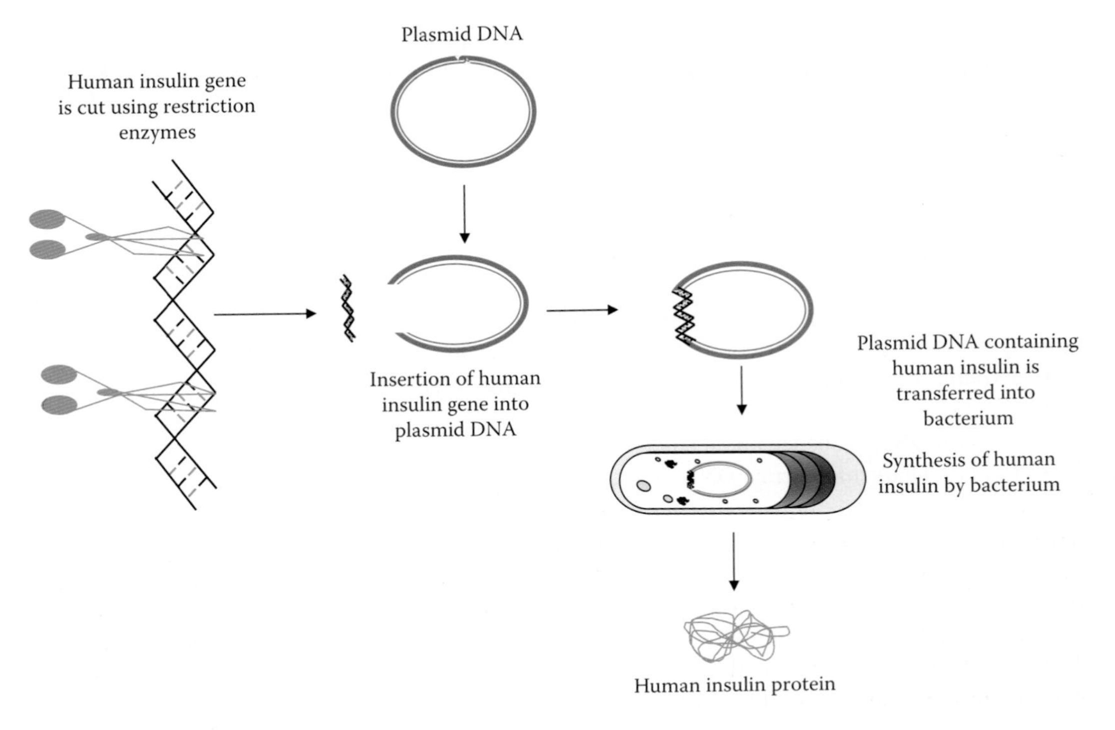

FIGURE 4.1 rDNA technology.

Step 2: Before introducing the vector containing the foreign DNA into host cells to express the protein, it must be cloned. Cloning is necessary to produce numerous copies of the DNA since the initial supply is inadequate to insert into host cells.

Step 3: Once the vector is isolated in large quantities, it can be introduced into the desired host cells such as mammalian, yeast, or special bacterial cells. The host cells will then synthesize the foreign protein from the rDNA. When the cells are grown in vast quantities, the foreign or recombinant protein can be isolated and purified in large amounts.

4.2.2 Methods Involved in Making Recombinant DNA Product

There are three methods involved in making rDNA: *transformation*, *phage introduction*, and *non-bacterial transformation*.

4.2.2.1 Transformation

The first step in transformation is to select a piece of DNA to be inserted into a vector. The next step is to cut that DNA with a restriction enzyme and then ligate the DNA insert into the vector with DNA ligase. The insert contains a selectable marker that allows for identification of recombinant molecules. An antibiotic marker is often used so that a host cell without a vector dies when exposed to a certain antibiotic, and the host with the vector will live because it is resistant. The vector is inserted into a host cell by a process called *transformation*. One example of a possible host cell is *E. coli*. The host cells must be specially prepared to take up the foreign DNA. Selectable markers can be for antibiotic resistance, color changes, or any other characteristic that can distinguish transformed hosts from untransformed hosts. Different vectors have different properties to make them suitable for different applications. Some properties can include symmetrical cloning sites, size, and high copy number.

4.2.2.2 Nonbacterial Transformation

Nonbacterial transformation is a process very similar to transformation. The only difference is that nonbacterial transformation does not use bacteria such as *E. coli* for the host. In microinjection, the DNA is injected directly into the nucleus of the cell being transformed. In biolistics, the host cells are bombarded with high-velocity microprojectiles, such as particles of gold or tungsten that have been coated with DNA.

4.2.2.3 Phage Introduction

Phage introduction is a process of transfection that is very similar to transformation, except a phage is used instead of a bacterium. *In vitro* packaging of a vector is also used. This uses lambda or MI3 phages to produce phage plaques that contain recombinants. The recombinants that are created can be identified by differences in the recombinants and nonrecombinants by using various selection methods.

4.3 SIGNIFICANCE OF RECOMBINANT DNA TECHNOLOGY

rDNA has been gaining importance over the past few years, especially now that genetic diseases have become more prevalent and the agricultural area is reduced. rDNA technology has a great impact on growing better crops (drought- and heat-resistant crops), making of recombinant vaccines (such as for Hepatitis B), prevention and cure of sickle cell anemia and cystic fibrosis, production of clotting factors, insulin and recombinant pharmaceuticals, plants that produce their own insecticides, and germ line and somatic gene therapy. rDNA works when the host cell expresses protein from the recombinant genes. The host will only produce significant amounts of recombinant protein if expression factors are added. Protein expression depends upon the gene being surrounded by a collection of signals that provide instructions for the transcription and translation of the gene by the cell. These signals include the promoter, the ribosome binding site, and the terminator. Expression vectors in which the foreign DNA is inserted contain these signals. Signals are species specific. In the case of *E. coli*, these signals must be *E. coli* signals, as *E. coli* is unlikely to understand the signals of human promoters and terminators. Problems are encountered if the gene contains introns or contains signals that act as terminators to a bacterial host. This results in premature termination, and the recombinant protein may not be processed correctly, may be folded incorrectly, or may even be degraded. Production of recombinant proteins in eukaryotic systems generally takes place in yeast and filamentous fungi. The use of animal cells is difficult, because many need a solid support surface, unlike bacteria, and have complex growth needs. However, some proteins are too complex to be produced in bacterium, so eukaryotic cells must be used.

4.4 ROLE OF RESTRICTION ENZYMES IN rDNA TECHNOLOGY

Restriction enzyme (or *restriction endonuclease*) is an enzyme that cuts double-stranded or single-stranded DNA at specific recognition nucleotide sequences, known as *restriction sites*. These enzymes are very critical components of rDNA technology. In fact, without restriction enzymes, it is not possible to make rDNA products. Restriction enzymes, which are found in bacteria and archaea, are thought to have evolved to provide a defense mechanism against invading viruses. Inside a bacterial host, the restriction enzymes selectively cut up foreign DNA in a process called *restriction*. Host DNA is then methylated by a modification enzyme (a methylase) to protect it from the restriction enzyme's activity. Collectively, these two processes form the restriction modification system. To cut the DNA, a restriction enzyme makes two incisions, once through each sugar-phosphate backbone (i.e., each strand) of the DNA double helix.

The discovery of restriction enzymes led to the development of rDNA technology. Over 3000 restriction enzymes have been studied in detail; more than 600 of these are available commercially and are routinely used for DNA modification and manipulation in laboratories. Since their discovery in the 1970s, more than 100 different restriction enzymes have been identified in different bacteria.

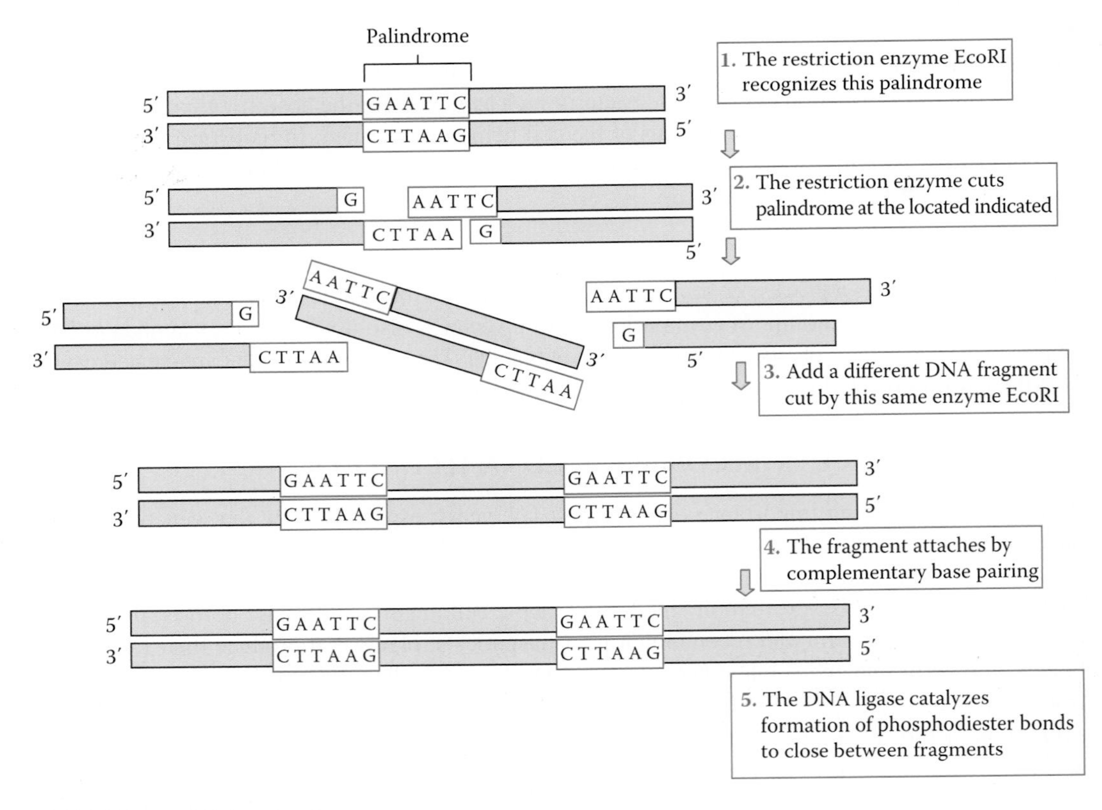

FIGURE 4.2 rDNA technology.

Each enzyme is named after the bacterium from which it was isolated using a naming system based on bacterial genus, species, and strain (Figure 4.2).

4.4.1 Types of Restriction Enzymes

Restriction enzymes are categorized into three general groups (types I, II, and III) based on their composition and enzyme cofactor requirements, the nature of their target sequence, and the position of their DNA cleavage site relative to the target sequence.

4.4.1.1 Type I Restriction Enzymes

Type I restriction enzymes were the first to be identified, and are characteristic of two different strains of *E. coli* (K-12 and B). These enzymes cut at a site that differs and is some distance (at least 1000 bp) away from their recognition site. The recognition site is asymmetrical and is composed of two portions: one containing 3–4 nucleotides, and another containing 4–5 nucleotides separated by a spacer of approximately 6–8 nucleotides. Several enzyme cofactors—including *S*-adenosylmethionine (AdoMet), hydrolyzed adenosine triphosphate (ATP), and magnesium (Mg^{2+}) ions—are required for their activity. Type I restriction enzymes possess three subunits: HsdR, HsdM, and HsdS. The subunit HsdR is required for restriction, HsdM for adding methyl groups to host DNA (methyltransferase activity), and HsdS for specificity of cut site recognition in addition to its methyltransferase activity (Figure 4.3).

4.4.1.2 Type II Restriction Enzymes

Type II restriction enzymes are the most commonly available and widely used restriction enzymes. They differ from type I restriction enzymes in several ways: they are composed of only one subunit; their recognition sites are usually undivided and palindromic and measure 4–8 nucleotides in length;

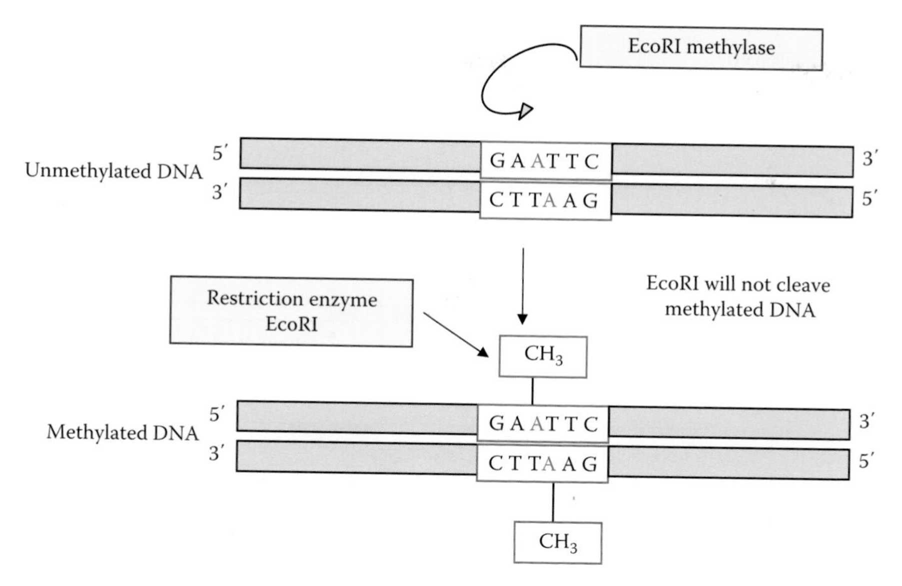

FIGURE 4.3 Restriction enzyme type I.

they recognize and cleave DNA at the same site; and they do not use ATP or AdoMet for their activity, since they usually require only Mg^{2+} as a cofactor. In the 1990s and early 2000s, new enzymes from this family were discovered that did not follow all the classical criteria of this enzyme class; consequently, new subfamily nomenclature was developed to divide this large family into subcategories based on deviations from typical characteristics of type II enzymes. These subgroups are defined using a letter suffix.

Type IIB restriction enzymes (e.g., BcgI and BplI) are multimers, containing more than one subunit. They cleave DNA on both sides of their recognition to cut out the recognition site. They require both AdoMet and Mg^{2+} cofactors. *Type IIE restriction enzymes* (e.g., NaeI) cleave DNA, following interaction with two copies of their recognition sequence. One recognition site acts as the target for cleavage, while the other acts as an allosteric effector that speeds up or improves the efficiency of enzyme cleavage. Similar to type IIE enzymes, *type IIF restriction enzymes* (such as NgoMIV), interact with two copies of their recognition sequence, but cleave both sequences at the same time. *Type IIG restriction enzymes* (such as Eco57I) do have a single subunit, like classical Type II restriction enzymes, but require the cofactor AdoMet to be active. *Type IIM restriction enzymes* (such as DpnI) are able to recognize and cut methylated DNA. *Type IIS restriction enzymes* (such as FokI) cleave DNA at a defined distance from their nonpalindromic asymmetric recognition sites. These enzymes may function as dimers. Similarly, *Type IIT restriction enzymes* (such as Bpu10I and BslI) are composed of two different subunits. Some recognize palindromic sequences, while others have asymmetric recognition sites (Figure 4.4).

4.4.1.3 Type III Restriction Enzymes

Type III restriction enzymes (such as EcoP15) recognize two separate nonpalindromic sequences that are inversely oriented. They cut DNA approximately 20–30 bp after the recognition site. These enzymes contain more than one subunit, and require AdoMet and ATP cofactors for their roles in DNA methylation and restriction, respectively.

4.4.2 Nomenclature of Restriction Enzymes

The nomenclature of restriction enzymes follows a general pattern. The first letter of the name of genus in which a given enzyme is first discovered is capitalized. This is followed by the first two letters of species name of the organism. These three letters are generally written in italics, such as Eco

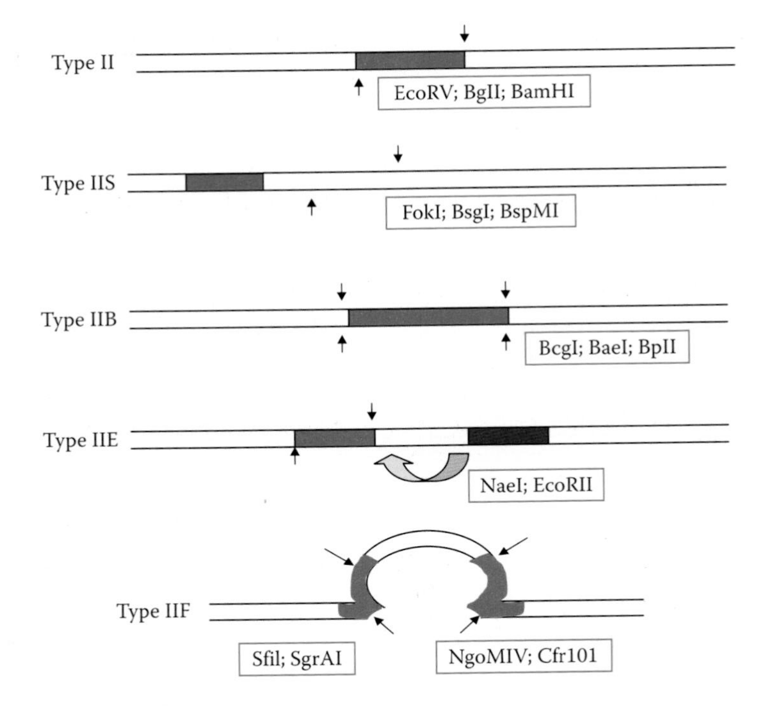

FIGURE 4.4 Restriction enzyme type II.

from *Escherichia coli*, Hin from *Haemophilus influenzae*, etc. The fourth letter identifies in Roman the particular strain or type, such as Ecok. When an organism produces more than one enzyme, these are identified by sequential Roman numerals. For example, the different enzymes produced by *H. influenzae* strain Rd are named Hind II, Hind III, etc.

4.4.3 Recognition Sequences for Type II Restriction Enzymes

The recognition sequences for Type II restriction enzymes form palindromes with rotational symmetry. In a palindrome, the base sequence in the second half of a DNA strand is the mirror image of the sequence in its first half. Consequently, the complementary DNA (cDNA) strand of a double helix also shows the same situation. However, in a palindrome with rotational symmetry, the base sequence in the first half of one strand of a DNA double helix is the mirror image of the second half of its complementary strand. Thus, in such palindromes, the base sequence in both the strands of a DNA duplex reads the same when read from the same end (either 5′ or 3′) of both the strands. Most of the type II restriction enzymes have recognition sites of 4, 5, or 6 bp, which are predominantly GC-rich. Longer palindromic target sequences are also known, and so are nonpalindromic ones (specific for some enzymes). Some restriction enzymes, such as EcoRII, have ambiguities in their recognition sites, so that they may recognize up to four different target sequences.

4.4.4 Cleavage Pattern of Type II Restriction Enzymes

Most type II restriction enzymes cleave the DNA molecules within their specific recognition sequences, but some (such as NlaIII and Sau3A) produce cuts immediately outside the target sequence. These cuts are either staggered or even, depending on the enzyme. Most enzymes produce staggered cuts in which the two strands of a DNA double helix are cleaved at different locations. This generates protruding (3′- or 5′-) ends. For example, one strand of the double helix extends some bases beyond the other. Due to the palindromic (symmetrical) nature of the target sites, the two protruding ends generated by such a cleavage by a given enzyme have a complementary base sequence.

As a result, they readily pair with each other, and such ends are called *cohesive* or *sticky ends*. An important consequence of this fact is that when fragments generated by a single restriction enzyme from different DNAs are mixed, they join together due to their sticky ends. Therefore, this property of the restriction enzymes is of great value for the construction of rDNAs. Some restriction enzymes, on the other hand, cut both the strands of a DNA molecule at the same site, so that the resulting termini or ends have blunt or flush ends in which the two strands end at the same point. The blunt cut ends also can be effectively utilized for construction of rDNAs following one of several strategies.

4.4.5 Modification of Cut Ends

The 3′-ends of DNA strands always carry a free hydroxyl (–OH) group, while their 5′-ends always bear a phosphate group. Often, the ends produced by restriction enzymes have to be modified for further manipulation of the fragments. Some of the modifications are summarized in the following:

- Removal of the 5′-phosphate group of vector DNA by alkaline phosphatase treatment in order to prevent vector circularization during DNA insert integration.
- Addition of a phosphate group to a free 5′-hydroxyl group by T4 polynucleotide kinase.
- Removal of the protruding ends by digestion with, for instance, S1 nuclease. This enzyme digests both 3′- and 5′-protruding ends.
- Filling in of the protruding ends by extending the recessed (shorter) strand with, for instance, Klenow fragment of *E. coli* DNA polymerase I. (Both the third and fourth strategies generate blunt ends that can be ligated by T4 polynucleotide ligase.)
- Synthesis of single-stranded tails (protruding ends) at the 3′-ends of blunt-ended fragments by the enzyme terminal deoxynucleotidyl transferase. This is called *tailing*. This reaction can be used to generate protruding ends of defined sequence, such as poly-A tails on the 3′-ends of the DNA insert and poly-T tails on the 3′-ends of the vector. The protruding ends of the DNA insert and the vector will, therefore, base pair under annealing conditions.
- Linker and/or adaptor molecules can be joined to the cut ends. Linkers are short, chemically synthesized, self-complementary, double-stranded oligonucleotides, which contain within them one or more restriction enzyme sites. For example, linker 5′-CCGAA TTCGG (only one strand of the linker is shown here) contains one EcoRI site. Linkers are joined with blunt-ended DNA fragments. Cleavage of the linker with the appropriate restriction enzyme creates suitable cohesive protruding ends. Linkers create cohesive ends on blunt-ended DNA fragments, and on fragments having unmatched or undefined sequences in their protruding ends. In the latter situation, the DNA fragments are first made blunt-ended, following which the selected linkers are ligated to them by T4 ligase.
- Adaptors are short, chemically synthesized DNA double strands, which can be used to link the ends of two DNA molecules that have different sequences at their ends. There are different kinds of adaptors suited for different purposes. For example, a *conversion adaptor* is used to join a DNA fragment or insert cut with one restriction enzyme (for instance, EcoRI), with a vector opened with another enzyme such as BamHI. These adaptors have the recognition sequences of different endonucleases at their ends.

4.5 STEPS IN GENE CLONING

The entire procedure of cloning or rDNA technology may be classified into the following five steps for convenience in description and on the basis of the chief activity performed (Figure 4.5):

- Identification and isolation of the desired gene or DNA fragment to be cloned
- Insertion of the isolated gene in a suitable vector

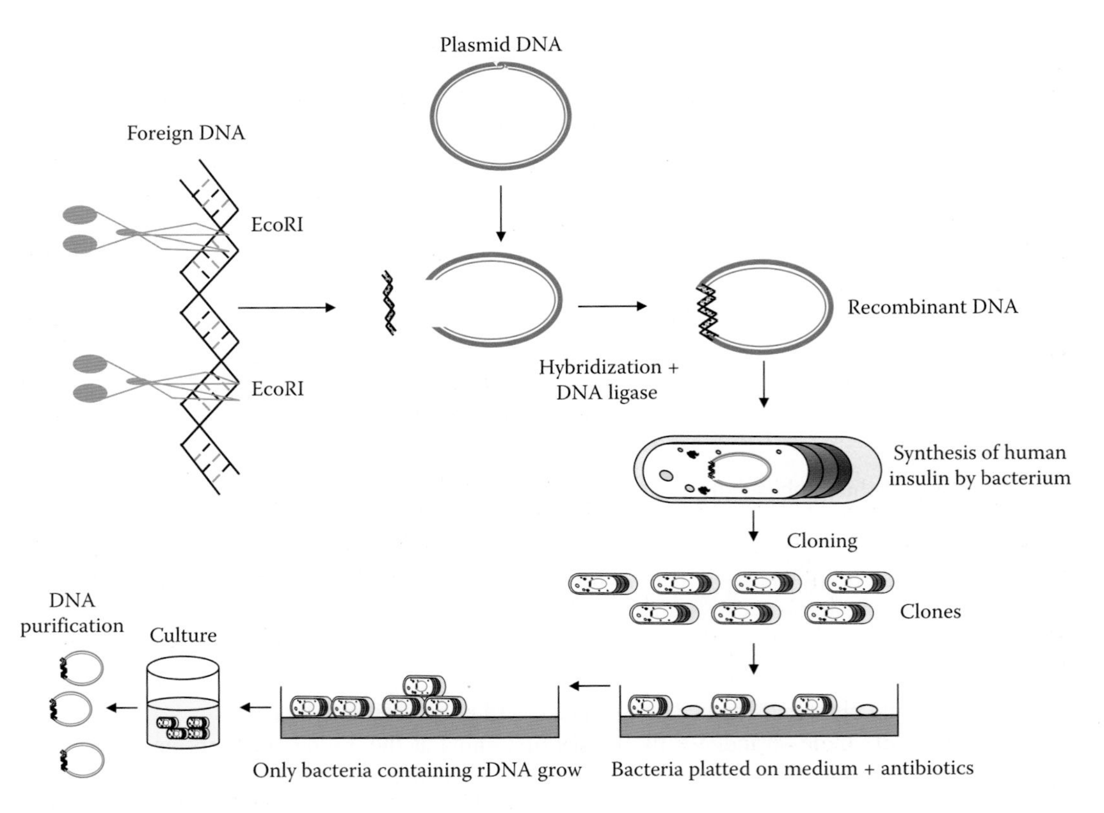

FIGURE 4.5 Gene cloning.

- Introduction of this vector into a suitable organism/cell, called the host (transformation)
- Selection of the transformed host cells
- Multiplication/expression/integration, followed by expression of the introduced gene in the host

4.6 SYNTHESIS OF COMPLETE GENE

It is not possible to synthesize the entire gene directly, but oligonucleotides of up to approximately 50 nucleotides are rapidly synthesized. One of the strategies for synthesizing the complete gene is to synthesize short stretches of the two DNA strands in such a way that each segment of one strand overlaps two segments of the other strands. Therefore, the segments of the two strands become properly aligned due to complementary base pairing among the various segments of the two strands. For example, the somatostatin gene was synthesized as eight overlapping segments of the two strands. Similarly, the 514-bp-long interferon gene was assembled from 66 short fragments representing the two strands. However, it is now far more convenient, cheap, and rapid to isolate the concerned gene from the genome or from total messenger RNA (mRNA) using polymerase chain reaction (PCR). Chemical synthesis is only used to produce the two primers specific to the 3′-ends of the gene. However, synthesis of the complete gene sometimes becomes necessary even when the original gene is available. This need arises because of the desirability of introducing extensive and specific changes in the base sequence of the gene for one reason or another—e.g., to increase codon usage in the new genetic background (new host or organism). An important example of such gene synthesis concerns the cry genes of *Bacillus thuringiensis*, which encodes the famous insecticidal crystal (Cry) proteins.

4.7 POLYMERASE CHAIN REACTION AND GENE CLONING

PCR has many exciting and varied applications. It can be used to amplify a specific gene present in different individuals of a species and even in different somatic cells or gametes, such as human sperm. These copies can be used for cloning. Alternatively, they can be sequenced to obtain information on the mutational changes in the genes of different individuals, cells, or gametes. Such data can be used in disease diagnosis, population genetics, estimation of recombination frequencies, etc. PCR has been used to study DNA polymorphism in the genome using known sequences as primers. Synthetic nucleotides of any sequence can be used as random primers to amplify polymorphic DNAs having sequences specific to the primers used. Such an application of PCR generates random amplified polymorphic DNA (RAPD, pronounced "rapid"), which is detected as bands after electrophoresis. RAPD bands of different strains or species can be compared. They can be used to construct RAPD maps, similar to restriction fragment length polymorphism (RFLP) maps. PCR can be used to detect the presence of a gene transferred into an organism (transgene) by using the end sequences of the transgene for amplification of DNA from the putative transgenic organism. Amplification will occur only when the transgene is present in the organism. The amplified DNA is detected as a band on the electrophoretic gel. Microdissected segments of chromosomes, such as salivary gland chromosomes of Drosophila, can be used for PCR amplification to determine the physical location of specific genes in chromosomes. PCR can be used to determine the sex of embryos. Thus, the sex of *in vitro* fertilized cattle embryos could be determined using Y-chromosome-specific primers before their implantation in the uterus.

4.7.1 Comparison of PCR versus Gene Cloning

PCR is a revolutionary technology and is more efficient than gene cloning, as it needs much less amount of the desired DNA (a single copy is enough). It is not difficult to store and does not require costly restriction enzymes, ligase, and vector DNA, thus reducing experimental cost drastically. It needs far less work, time, and skill, and has many more applications than gene cloning. Typically, gene cloning experiments take 2–4 days, while PCR takes only up to 4–5 h. In addition, PCR is fully automated, while gene cloning is not. Nevertheless, for PCR, one does need sequence information for construction of primers, and a thermal cycler or PCR machine. It is expected that PCR will eventually take over most of the applications of gene cloning and will find many novel applications as well.

4.8 SIGNIFICANCE OF VECTORS IN rDNA TECHNOLOGY

Recall that in molecular biology, a *vector* is a DNA molecule used as a vehicle to transfer foreign genetic material into another cell. Viral vectors are tools commonly used by molecular biologists to deliver genetic material into cells. This process can be performed inside a living organism (*in vivo*) or in cell culture (*in vitro*). Viruses have evolved specialized molecular mechanisms to efficiently transport their genomes inside the cells they infect. Delivery of genes by a virus is termed *transduction* and the infected cells are called *transduced*. Molecular biologists first harnessed this machinery in the 1970s. Paul Berg used a modified SV40 virus containing DNA from the bacteriophage lambda to infect monkey kidney cells maintained in culture. The four major types of vectors are plasmids, bacteriophages and other viruses, cosmids, and artificial chromosomes. Common to all engineered vectors are an origin of replication, a multi-cloning site, and a selectable marker. The vector itself is generally a DNA sequence that consists of an insert (transgene) and a larger sequence that serves as the "backbone" of the vector. The purpose of a vector, which transfers genetic information to another cell, is typically to isolate, multiply, or express the insert in the target cell. Vectors called *expression vectors* (or *expression constructs*) are specifically for the expression of the transgene in the target cell, and generally have a promoter sequence that drives expression of the transgene.

Simpler vectors called *transcription vectors* are only capable of being transcribed but not translated. They can be replicated in a target cell but not expressed, unlike expression vectors. Transcription vectors are used to amplify their insert. Insertion of a vector into the target cell is generally called *transfection*.

4.8.1 Properties of Good Vectors

Vectors should be able to replicate autonomously. When the objective of cloning is to obtain a large number of copies of the DNA insert, the vector replication must be under relaxed control so that it can generate multiple copies of itself in a single host cell. It should also be easy to isolate, purify, and introduce into the host cells. Transformation of the host with the vector should be easy. The vector should have suitable marker genes that allow easy detection and/or selection of the transformed host cells. When the objective is gene transfer, it should have the ability to integrate either itself or the DNA insert it carries into the genome of the host cell. The cells transformed with the vector containing the DNA insert (rDNA) should be easily identifiable and selectable from those transformed by the unaltered vector. A vector should contain unique target sites for as many restriction enzymes as possible into which the DNA insert can be integrated. When the expression of the DNA insert is desired, the vector should at least contain suitable control elements such as promoter, operator, and ribosome binding sites. It should be kept in mind that the DNA molecules used as vectors have coevolved with their specific natural host species, and hence are adapted to function well in them and in their closely related species. Therefore, the choice of vector largely depends on the host species into which the DNA insert of gene is to be cloned. In addition, most naturally occurring vectors do not have all the required functions. Therefore, useful vectors have been created by joining together segments performing specific functions (called *modules*) from two or more natural entities.

4.8.2 Cloning and Expression Vectors

All vectors used for propagation of DNA inserts in a suitable host are called *cloning vectors*. However, when a vector is designed for the expression or production of the protein specified by the DNA insert, it is called an *expression vector*. As a rule, such vectors contain at least the regulatory sequences such as promoters, operators, and ribosomal binding sites having optimum function in the chosen host. When a eukaryotic gene is to be expressed in a prokaryote, the eukaryotic coding sequence has to be placed after prokaryotic promoter and ribosome building site since the regulatory sequences of eukaryotic are not recognized in prokaryotes. In addition, eukaryotes genes, as a rule, contain introns (noncoding regions) present within their coding regions. These introns must be removed from the DNA insert to enable the proper expression of eukaryotic genes, since prokaryotes lack the machinery needed for their removal from the RNA transcripts. When eukaryotic genes are isolated as cDNA, they are intron-free and therefore suitable for expression in prokaryotes. Expression vectors can be constructed by allowing the synthesis of fusion proteins—which comprises of amino acids encoded by a sequence—in the vector and those encoded by the DNA insert (translational fusion). Another way to construct expression vectors is by permitting the synthesis of pure proteins encoded exclusively by the DNA inserts (transcriptional fusion). Some examples of the first strategy, which produces fusion proteins, are the expression of rat insulin, rat growth hormone, structural protein VP1 of foot and mouth disease virus, and human growth hormone. On the other hand, some examples of the second strategy, which produces unique proteins, are the rabbit β-globin, small t-antigen of SV40, human fibroblast interferon, and human IGF-I protein. It may be pointed out that in the case of translational fusion the undesired amino acids encoded by the vector sequence must be removed from the fusion proteins by a suitable chemical cleavage. Several other problems arise when eukaryotic genes are expressed in a prokaryotic system such as removal of signal sequences from precursor proteins to obtain active mature protein molecules. Various strategies are being rapidly devised to effectively overcome these problems.

4.8.3 Applications of Viral Vectors

Vectors have been extensively used in molecular biology-based researches due to their advantage as the most desirable transfection vehicle. Compared to traditional methods such as calcium phosphate precipitation, transduction can ensure that nearly 100% of cells are infected without severely affecting cell viability. Furthermore, some viruses integrate into the cell genome, facilitating stable expression. However, transfection is still the method of choice for many applications, as construction of a viral vector is a much more laborious process. Protein coding genes can be expressed using viral vectors, commonly to study the function of the particular protein. Viral vectors, especially retroviruses, stably expressing marker genes such as green fluorescent protein (GFP) are widely used to permanently label cells to track them and their progeny, such as in xenotransplantation experiments when infected cells are implanted into a host animal.

One of the major hurdles in treating cancer patients is to deliver genes, and success for gene therapy is primarily based on how effective genes can be delivered to the target site. In the future, gene therapy may provide a way to cure genetic disorders such as severe combined immunodeficiency, cystic fibrosis, or even *Haemophilia A*. Because these diseases result from mutations in the DNA sequence for specific genes, gene therapy trials have used viruses to deliver unmutated copies of these genes to the cells of the patient's body. There have been a huge number of laboratory successes with gene therapy. However, several problems of viral gene therapy must be overcome before it gains widespread use. Immune response to viruses not only impedes the delivery of genes to target cells, but can also cause severe complications for the patient. One of the early gene therapy trials in 1999 led to the death of Jesse Gelsinger, who was treated using an adenoviral vector. Some viral vectors, such as lentiviruses, insert their genomes at a seemingly random location on one of the host chromosomes, which can disturb the function of cellular genes and lead to cancer. In a severe combined immunodeficiency retroviral gene therapy trial conducted in 2002, two of the patients developed leukemia as a consequence of the treatment. Adeno-associated virus (AAV)-based vectors are much safer in this respect, as they always integrate at the same site in the human genome.

One of the main applications of viral vectors is to develop vaccines to protect humans from various pathogens. Viruses expressing pathogen proteins are currently being developed as vaccines against these pathogens, based on the same rationale as DNA vaccines. T-lymphocytes recognize cells infected with intracellular parasites based on the foreign proteins produced within the cell. T-cell immunity is crucial for protection against viral infections and diseases such as malaria. A viral vaccine induces expression of pathogen proteins within host cells similar to the Sabin polio vaccine and other attenuated vaccines. However, since viral vaccines contain only a small fraction of pathogen genes, they are much safer, and sporadic infection by the pathogen is impossible. Adenoviruses are being actively developed as vaccines.

4.9 CLASSIFICATION OF VECTORS

There are two different classifications of vectors that have been extensively used in rDNA technology. In this section, we will describe these vector classifications in detail.

4.9.1 Bacterial Vectors

Bacterial vectors are broadly classified into *E. coli* vectors and plasmid vectors as described below.

4.9.1.1 *E. coli* Vectors

Bacteria are the hosts of choice for DNA cloning. Among them, *E. coli* occupies a prominent position since cloning and isolating DNA inserts for structural analysis is easiest when using this host. Therefore, the initial cloning experiments are generally carried out using *E. coli*. The *E. coli* strain K12 is the most commonly used strain. It has several substrains such as C600, RRI, and HB101, each of which has some specific features important in cloning. For example, the substrain RRI has, in addition to certain other features, the mutation HsdR, which inactivates the restriction enzyme endogenous to *E. coli* K12. This minimizes the degradation of rDNA introduced into it.

4.9.1.2 Plasmid Vectors

Plasmid vectors are shortened linear A. genomes containing DNA replication and lytic functions plus the cohesive ends of the phage. Their middle nonessential segment is replaced by a linearized plasmid with an intact replication module. In practice, a plasmid vector contains several tandem copies of the plasmid to make it longer than 38 kb (kilobase pairs), the minimum size needed for packaging in A. particles. During construction of the rDNA, one or more copies of the plasmid are deleted from the vector and the DNA insert is integrated into it, but generally one copy of the plasmid is retained in the rDNA. Plasmids, both recombinant and unaltered, are packaged in A. particles *in vitro* and used for infection of appropriate *E. coli* cells. If a plasmid lacks the A. gene el, which produces the lysis repressor, it multiplies like a phage and produces plaques on a bacterial lawn. However, if el gene is present, the plasmid replicates like a plasmid. Furthermore, a plasmid may contain a mutant cI gene, which produces a temperature-sensitive CI protein (inactive at higher temperatures). Such vectors replicate as plasmids at lower temperatures, but behave like phage at higher temperatures. This feature is quite useful in some experiments.

It is possible for plasmids of different types to coexist in a single cell. Several different plasmids have been found in *E. coli*. However, related plasmids are often incompatible, in the sense that only one of them survives in the cell line, due to the regulation of vital plasmid functions. Therefore, plasmids can be assigned into compatibility groups.

Another way to classify plasmids is by function. There are five main groups of plasmids classified according to function:

- *Fertility-F-plasmids*, which contain transfer operon (tra genes), are capable of conjugation (transfer of genetic material between bacteria which are touching).
- *Resistance-(R) plasmids* contain genes that can build a resistance against antibiotics or poisons and help bacteria to produce pili. Before the nature of plasmids was understood, resistance-(R) plasmids were historically known as *R-factors*.
- *Col-plasmids* contain genes that code for bacteriocins or proteins that can kill other bacteria.
- *Degradative plasmids* enable the digestion of unusual substances such as toluene and salicylic acid.
- *Virulence plasmids* turn a bacterium into a pathogen (one that causes disease).

Plasmids can belong to more than one of these functional groups. Plasmids that exist only as one or a few copies in each bacterium are in danger of being lost in one of the segregating bacteria during cell division. Such single-copy plasmids have systems that attempt to actively distribute a copy to both daughter cells. Some plasmids include an *addiction system* or "post-segregation killing system (PSK)," such as the host killing/suppressor of killing (hok/sok) system of plasmid R1 in *E. coli*. They produce both a long-lived poison and a short-lived antidote. Daughter cells that retain a copy of the plasmid survive, while a daughter cell that fails to inherit the plasmid dies or suffers a reduced growth-rate because of the lingering poison from the parent cell.

4.9.2 Viral Vectors

In Section 4.8, we learned how viral vectors are important in rDNA technology. We will now study th unique characteristics that are tailored to their specific applications. Although viral vectors

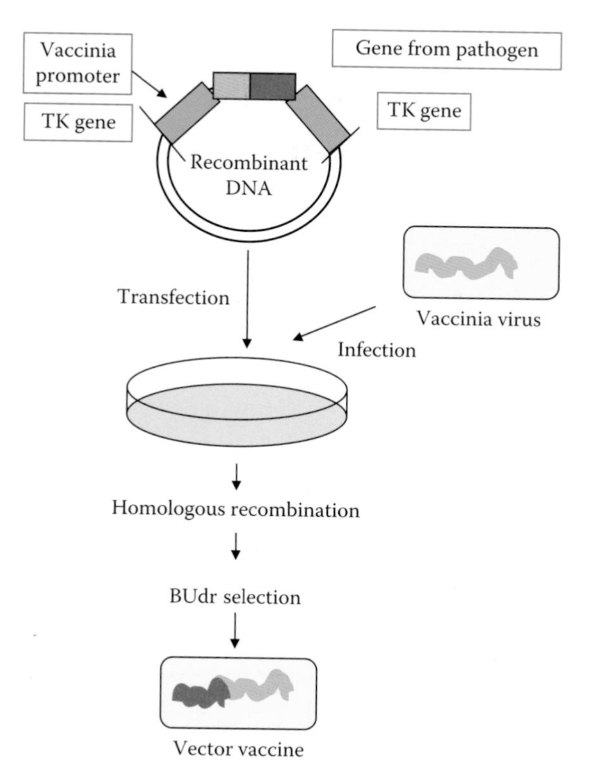

FIGURE 4.6 Viral vector in vaccine development.

are occasionally created from pathogenic viruses, they are modified so as to minimize the risk of handling them. This usually involves the deletion of a part of the viral genome critical for viral replication. Such a virus can efficiently infect cells but, once the infection has taken place, requires a helper virus to provide the missing proteins for production of new virions. The viral vector should have a minimal effect on the physiology of the cell it infects. Another issue related to viral vectors is that some viruses are genetically unstable and can rapidly rearrange their genomes. Most viral vectors are engineered to infect as wide a range of cell types as possible. However, sometimes the opposite is preferred. The viral receptor can be modified to target the virus to a specific kind of cell (Figure 4.6).

4.9.2.1 Retroviruses

Retroviruses are one of the mainstays of current gene therapy approaches. The recombinant retroviruses such as the Moloney murine leukemia virus have the ability to integrate into the host genome in a stable fashion. They contain a reverse transcriptase that allows integration into the host genome. They have been used in a number of FDA-approved clinical trials, such as the SCID-X1 trial. Retroviral vectors can be either replication-competent or replication-defective. Replication-defective vectors are the most common choice in studies, because the viruses have had the coding regions for the genes necessary for additional rounds of virion replication and packaging replaced with other genes or deleted. These viruses are capable of infecting their target cells and delivering their viral payload, but then fail to continue typical lytic pathway, which would typically result in cell lysis and death. Conversely, replication-competent viral vectors contain all the necessary genes for virion synthesis, and will continue to propagate themselves once infection occurs. Because the viral genome for these vectors is much lengthier, the length of the actual inserted gene of interest is limited compared to the possible length of the insert for replication-defective vectors (Figure 4.7).

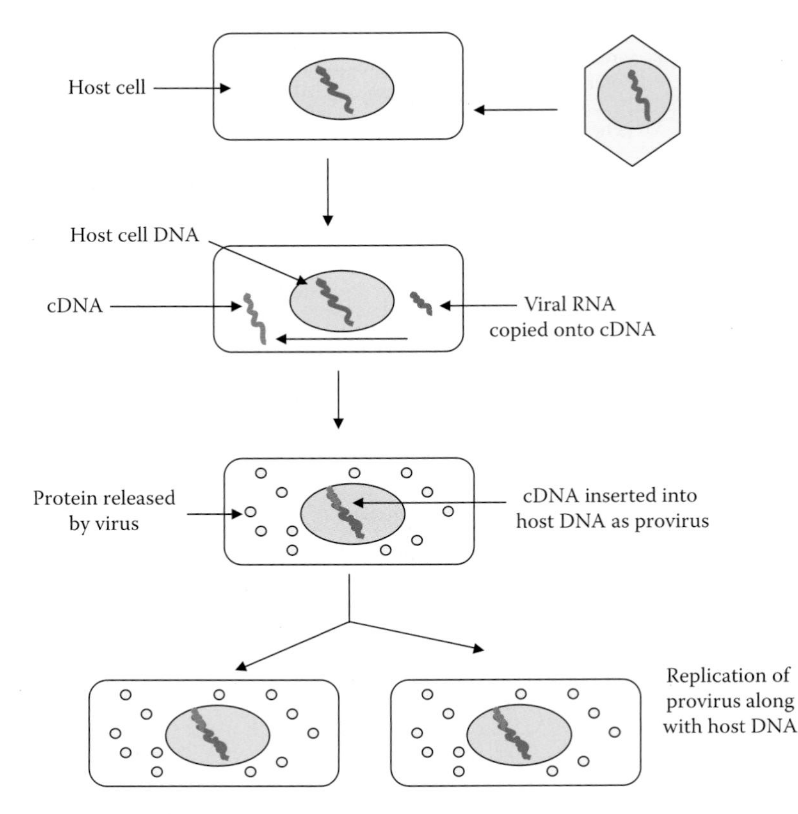

FIGURE 4.7 Replication of retrovirus.

Depending on the viral vector, the typical maximum length of an allowable DNA insert in a replication-defective viral vector is usually about 8–10 kb. While this limits the introduction of many genomic sequences, most cDNA sequences can still be accommodated. The primary drawback to the use of retroviruses such as the Moloney retrovirus involves the requirement for cells to be actively dividing for transduction. As a result, cells such as neurons are very resistant to infection and transduction by retroviruses. There is a concern for insertional mutagenesis due to the integration into the host genome, which can lead to cancer or leukemia.

4.9.2.2 Lentiviruses

Lentiviruses are basically a subclass of retroviruses. They have recently been adapted as vectors because of their ability to integrate into the genome of nondividing cells; this is a unique feature of lentiviruses, as other retroviruses can infect only dividing cells. The viral genome in the form of RNA is reverse-transcribed when the virus enters the cell to produce DNA, which is then inserted into the genome at a random position by the viral integrase enzyme. The vector, now called a *provirus*, remains in the genome and is passed on to the progeny of the cell when it divides. The site of integration is unpredictable, which can pose a problem. The provirus can disturb the function of cellular genes and lead to activation of oncogenes that promote the development of cancer, which raises concerns for possible applications of lentiviruses in gene therapy. However, studies have shown that lentivirus vectors have a lower tendency than gamma-retroviral vectors to integrate in places that potentially cause cancer. More specifically, one study found that lentiviral vectors did not cause either an increase in tumor incidence or an earlier onset of tumors in a mouse strain with a much higher incidence of tumors. Moreover, clinical trials that utilized lentiviral vectors to deliver gene therapy for the treatment of HIV experienced no increase in mutagenic or oncogenic activities in the patients. For safety reasons, lentiviral vectors never carry the genes required for their replication.

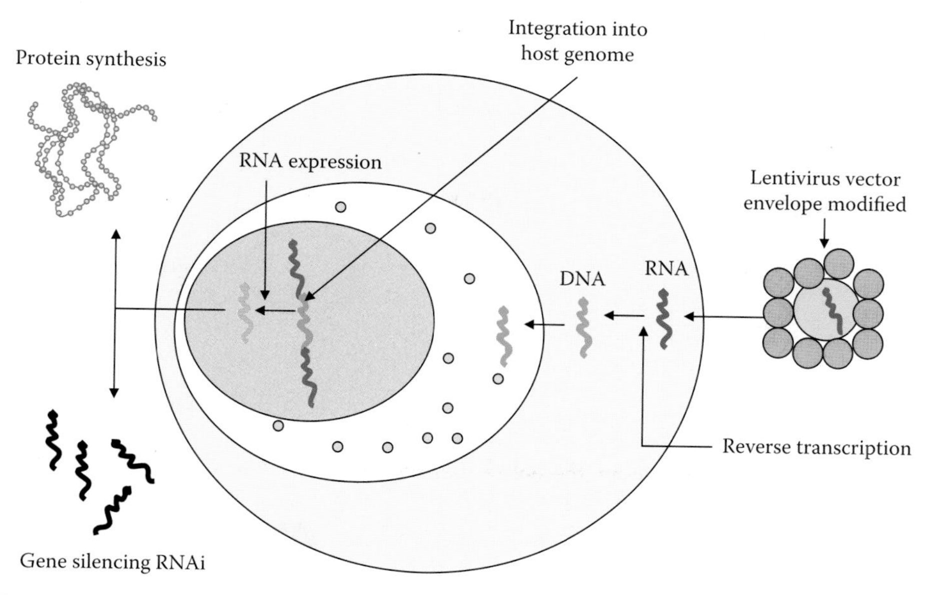

FIGURE 4.8 Lentivirus vector.

To produce a lentivirus, several plasmids are transfected into a so-called *packaging cell line*, commonly HEK 293. One or more plasmids, generally referred to as *packaging plasmids*, encode the virion proteins, such as the capsid and the reverse transcriptase. Another plasmid contains the genetic material to be delivered by the vector. It is transcribed to produce the single-stranded RNA viral genome and is marked by the presence of the ψ (psi) sequence. This sequence is used to package the genome into the virion (Figure 4.8).

4.9.2.3 Adenoviruses

Unlike lentiviruses, adenoviruses do not integrate into the genome and do not replicate during cell division. This limits their use in basic research, although adenoviral vectors are occasionally used in *in vitro* experiments. Their primary applications are in gene therapy and vaccination. Since humans commonly come in contact with adenoviruses, which cause respiratory, gastrointestinal and eye infections, they trigger a rapid immune response with potentially dangerous consequences. To overcome this problem, scientists are currently investigating adenoviruses to which humans do not have immunity (Figure 4.9).

4.9.2.4 Adeno-Associated Viruses

AAV is a small virus that infects humans and some other primate species. AAV is not currently known to cause disease, and consequently the virus causes a very mild immune response. AAV can infect both dividing and nondividing cells and may incorporate its genome into that of the host cell. These features make AAV a very attractive candidate for creating viral vectors for gene therapy.

4.9.3 ARS Vectors

In yeast chromosomes, origin of replication is specified by about 100 bp sequences called *autonomously replicating sequences* (ARS). All ARS sequences have an 11 bp consensus sequence, which is essential for their function. Other functional but variable sequences are also present in ARS. Any DNA double helix containing an ARS can serve as a yeast vector. Such a vector will be maintained in yeast cells only if it is essential for their survival—for example, if is the only source of an essential gene (such as, TRP1) in yeast cells mutant for that gene (in this case, Trp1-yeast cells). In the absence of such a selection pressure, ARS vectors are rapidly lost.

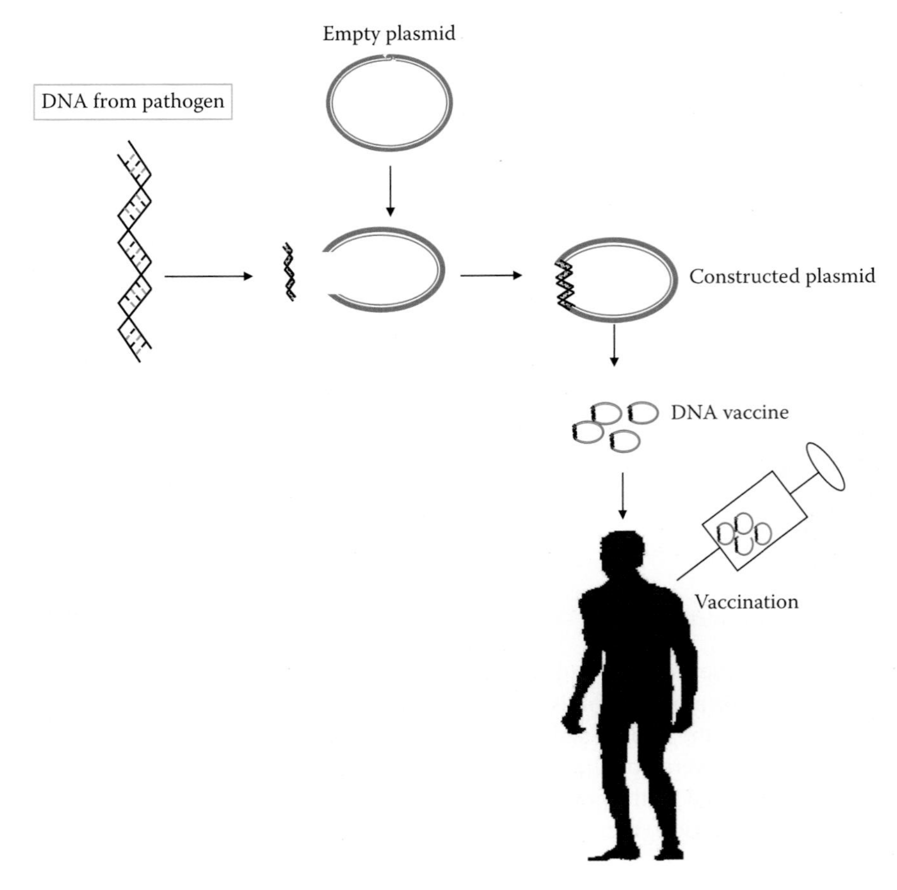

FIGURE 4.9 Adenovirus vector for vaccine development.

4.9.4 Minichromosome Vectors

These shuttle vectors behave like very small chromosomes, in that they replicate only once during each cell division and are distributed to daughter cells like true chromosomes. A typical minichromosome vector contains the following functions:

From yeast

- An ARS sequence (provides replication origin).
- A CEN sequence (regulates replication and distribution during cell division similar to those of chromosomes).
- A functional gene, such as LEU2, that serves as a selectable marker in appropriate yeast strains (in this case LEU2-cells).

From a bacterial plasmid

- The ori or replication.
- A selectable marker (e.g., ampr). CEN sequences are of about 500 bp and come from the centromeric regions of yeast chromosomes. They are responsible for centromeric functions during cell division.

4.9.5 Yeast Artificial Chromosome Vectors

Linear vectors that behave like a yeast chromosome are called *yeast artificial chromosomes* (YACs). A typical YAC such as pYAC3 contains the following functional elements from yeast:

- An ARS sequence for replication.
- A CEN4 sequence for centromeric function.
- Telomeric sequences at the two ends for protection from exonuclease action.
- One or two selectable marker genes such as TRP J and URA3 (strategy similar to other vectors).
- SUP4, a selectable marker into which the DNA insert is integrated.
- The necessary sequences from *E. coli* plasmid for selection and propagation in *E. coli*. The telomeric sequence in yeast chromosomes is a 20–70 tandem repeat of the six base sequence 5′CCCCAA3′ (its complementary sequence, 5TIGGGG3′, occurs in the other strand). Theirs is a hairpin loop formation at the terminus, which makes the DNA duplex resistant to exonuclease action.
- Vector pYAC3 is essentially a pBR322 plasmid into which the above described yeast sequences have been integrated. Subsequently, several YAC vectors have been constructed on the basic scheme of pYAC3. The YAC vector itself is propagated in *E. coli*, but cloning is done in yeast.

For cloning, the vector is restricted with a combination of BarnHI and SnaBI. BamBI cleaves the vector at the junctions of the two TEL sequences with the fragment that is used to circularize the vector for propagation in *E. coli*; this fragment is discarded. The enzyme SnaBI recognizes the single sequence 5′T ACGT A3′ located in SUP4 and produces blunt-ended cleavage, thereby generating two arms of the YAC, each ending in a TEL sequence. The DNA insert, therefore, must have blunt ends. It is integrated within SUP4 to generate the linear YAC. The recombinant YAC is introduced into TRP 1^- URA3^- yeast cells by protoplast transformation. Transformed cells are selected by plating them onto the minimal medium. Only those cells that have correctly constructed YAC, containing one left and one right arm of each chromosome, are able to grow on this medium. Recombinant clones are identified due to the insertional inactivation of SUP4, detected by a simple color test: recombinant colonies are white, while nonrecombinant ones are red. The TEL sequence of the vector is not the complete telomeric sequence, but it contains enough of this sequence to be able to support the creation of complete telomere once the YAC is inside a yeast cell. Thus, a YAC is a shuttle vector that is propagated in circular form in *E. coli* and is used for cloning in yeast in a linear form. When a YAC is less than approximately 20 kb, the centromeric function is unable to control copy number during mitosis, so that several copies of YAC accumulate per yeast cell.

The centromeric function improves in YACs of 50 kb or more. YACs of 150 kb or more behave like regular yeast chromosomes. YACs are the predominant vector system used for cloning of very large (up to 100–1400 kb) DNA segments for mapping of complex eukaryotic chromosomes. YACs are reported to suffer from many problems, including chimerism, tedious steps in YAC library construction, and low yields of YAC insert DNA. The yeast genes present in different yeast vectors can become integrated into the host genome. This is called *permanent transformation*. It generally occurs through homologous recombination between the gene present in a vector (e.g., LEU2) and that present in the yeast chromosomes (e.g., LEU2-). Rarely, the gene may become inserted at a random chromosome site. The homologous recombination may occur by regular crossing over or it may involve gene conversion (a nonreciprocal recombination).

4.9.6 Vectors for Animals

Animal cells such as *in vitro* cultured cell lines, xenopus oocytes, and early embryo obtained from transgenic animals may be used for transfection or transformation. DNA fragments from animals are generally cloned in *E. coli* to obtain sufficient copies for structural analysis.

However, expression of the DNA inserts has to be studied in animal cells. Therefore, many of the vectors are shuttle vectors for *E. coli* and animal cells. The different vectors used in animal cells are derived from viral genomes, and are either virus-like (rDNAs produce virions, which are virus particles) or plasmid-like vectors (these replicate but do not produce virions). Some other vectors are designed to be unable to replicate. Another way of looking at animal vectors relates to their ability to replicate in animal cells. Some vectors are capable of replication (replicating vectors), while others are not (non-replicating vectors). Furthermore, any DNA fragment, or a mixture of fragments, can be used to transfect animal cells and this DNA segment becomes integrated into the cellular genome and expresses itself. In animal cells, vectors that remain and replicate in extra-chromosomal state are quickly lost in a few days, even when selection conditions favoring their retention are imposed. This is called *transient transfection*. However, some animal vectors do remain stable in the extra-chromosomal state, such as those derived from bovine papillomavirus (BPV) and those that contain the origin of DNA replication from herpes virus Epstein-Barr. However, permanent or stable transfection is most often due to the integration of transfecting DNA into the genome of animal cells. This is also known as *insertional transfection*. The DNA integration is usually by nonhomologous recombination, hence in random locations in the genome. This is in contrast to the situation in yeast, where recombination is homologous and, as a result, site-specific. The first animal vector was devised from the primate papova virus, simian virus 40 (SV40). Subsequently, vectors have been developed from many other viruses such as papillomavirus, adenoviruses, the Epstein-Barr herpes virus, vaccinia viruses (all for mammals), and baculoviruses (for insects).

4.9.7 SV40 Vectors

The SV40 vector is a spherical virus with a circular, double-stranded 5243 bp chromosome, which encodes five proteins: small-T, large-T (both early proteins), and the virion proteins (VP) VP1, VP2, and VP3. It has an origin of replication (approximately 80 bp) and is complexed with histones to form chromatin. Large-T is essential for viral replication, while VP1, VP2, and VP3 form the viral capsid. In the laboratory, it is multiplied in cultured kidney cells of the African green monkey. The SV40 genome has been mainly used to develop three types of vectors: transducing vectors, plasmid vectors, and transforming vectors.

4.9.8 Bovine Papillomavirus Vectors

BPVs belong to the papovavirus class and cause warts. They have a circular 8 kb genome organized in nucleosomes. The BPV replicates as a stable plasmid in rodent and many bovine cells, and the cells are not killed. The viral genome transforms cells that behave like tumor cells and form piled up colonies of cells instead of the typical monolayer. (Transformation describes the conversion of normal cells into timorous cells). The transformed state is due to the genes present in the "transforming region" (approximately 5500 bp) of the virus genome. The virus genome is generally used to produce shuttle vectors by using the transforming region of viral genome. Eukaryotic DNA segments are first cloned in *E. coli* to select rDNAs. Then the *E. coli* plasmid, such as pBR322, is deleted from the vector and the linear rDNA is introduced into animal cells. The rDNA now becomes circular and replicates as a plasmid. The *E. coli* neo gene may be included within the vector. This allows easy selection of transfected cells by culturing them on a medium containing the aminoglycoside G-418.

4.10 INTEGRATION OF THE DNA INSERT INTO THE VECTOR

Once the DNA fragments to be cloned are prepared and the appropriate vector is selected (the selection will be based on the purpose of cloning and the host[s] to be employed), the DNA segments will have to be integrated into the vector at an appropriate site. The vector is cut open with a restriction enzyme that has a unique (single) target site located in the sequence where the DNA insert is to be integrated. There are five possible situations with respect to the vector and the DNA insert:

- Both have completely matched or compatible cohesive ends.
- One end of each of the vector and the DNA insert are compatible and protruding, while their other ends are different from their first ends, but are protruding and compatible.
- They have different or unmatched cohesive ends.
- Both have flush or blunt ends.
- One end of both is cohesive and matched, while the other is blunt.

The strategy of integration of the DNA insert in these situations is briefly described.

4.10.1 Both Ends Cohesive and Compatible

The simplest strategy concerns the presence of compatible cohesive ends in both the vector and the DNA insert. This happens when the vector is cut open by the same restriction enzyme that was used to isolate the DNA insert. The opened up vector and the DNA insert are mixed under annealing conditions that allow pairing between the compatible cohesive ends of the vector and the DNA insert. The nicks (the broken covalent bond between the 5′-phosphate and the 3′-OH of two neighboring nucleotides within a DNA strand) remaining after annealing are sealed, joined, or ligated by DNA ligase. Initially, DNA ligase from *E. coli* was used, but T4 (an *E. coli* phage) ligase is currently preferred. DNA inserts joined in this manner can be easily and precisely isolated from the rDNA using the same restriction enzyme that was used to generate the inserts and to open the vector. In the rDNAs produced in this way, the DNA insert may be present in either of the two orientations relative to the sequences of the vector. If the two cohesive ends of the vector are marked as 1 and 2 and those of the DNA insert as 1′ and 2′, the insert may join the vector to either yield 1–1′ and 2–2′ junctions, or 1–2′ and 2–1′ junctions.

The orientation of DNA insert within the vector is not important when only copies of the insert are to be obtained. However, it is extremely important when expression of the DNA insert is desired. In addition to the formation of rDNA, the cohesive ends of the vector itself will pair together to produce unaltered vector molecules. Similarly, the two ends of the DNA insert will also join to yield a circular DNA insert molecule. A circularized insert is not a problem since it lacks an origin of replication and, as a result, is diluted out of the transformed cells. The formation of unchanged circularized vector can be prevented by treating the opened-up vector with alkaline phosphatase, which removes the 5′-phosphate present as monoester at the vector ends. When such a vector is mixed, annealed, and ligated with the DNA insert, two nicks remain in the rDNA due to a lack of phosphate at the 5′-ends of the vector. These nicks are readily repaired once the rDNA is introduced into appropriate host cells.

4.10.2 Both Ends Cohesive and Separately Matched

One protruding end of the vector may be compatible with one end of the DNA insert, while the other end of vector is compatible with the second end of the insert. This situation arises when one end of the vector as well as that of the DNA insert is generated by one restriction enzyme, while their other end is cut by a different enzyme. As earlier, the opened vector and the DNA insert are mixed under annealing conditions to allow pairing between the compatible ends of the vector and the DNA insert. T4 ligase is then used to seal the nicks to yield rDNA. In such a situation, only the rDNA is circularized (since the two ends of vector are unmatched as are the two ends of DNA insert) and the DNA insert is integrated in only one orientation or direction.

4.10.3 Both Ends Cohesive and Unmatched

Often, the DNA insert is prepared using one restriction enzyme, while the vector is opened with another enzyme. This generates cohesive ends in the vector and the DNA insert, which are unmatched. In this situation, the following approaches are available. The protruding ends are converted into blunt ends by removing the protruding ends either by digestion or by extending the recessed ends, using Klenow fragments or reverse transcriptase. The blunt ends can then be joined together by T4 ligase. The blunt ends so produced can again be changed into protruding ends by 3′-tailing. A poly-T tail may be added to the 3′-ends of the vector, while poly-A tails are added to the DNA insert. The two now have matched protruding ends, which pair together under annealing conditions to yield rDNA. Often, the tails are of different sizes. Therefore, the gaps remaining in the rDNA are filled with DNA polymerase I prior to ligation. Alternatively, linkers can be attached to the blunt ends of the vector and the DNA insert. The linkers are cleaved with the appropriate restriction enzyme to generate protruding ends that are compatible. The vector and the insert are now joined together. In addition, appropriate adaptors may be joined to the protruding ends of the vector and/or the DNA insert to generate completely matched cohesive ends.

4.10.4 Both Ends Flush/Blunt

Flush or blunt-ended vector and DNA insert can be joined together by T4 DNA ligase. However, high concentrations of both the enzyme and the vector and insert DNAs are required. The DNA insert gets ligated in either orientation, and it can be easily and precisely separated from the vector only if the two were cleaved by the same restriction enzyme (producing blunt ends). Alternatively, the blunt ends can be converted into cohesive ends, as outlined above.

4.10.5 One End Cohesive and Compatible, the Other End Blunt

If the vector and the DNA insert have one compatible and cohesive end and one flush end, their cohesive ends pair together when they are mixed under annealing conditions. The T4 ligase seals the nick at the cohesive end and joins the blunt ends as well. It should be noted that under such a situation, only rDNAs are produced, since vector and DNA insert molecules cannot circularize due to their one cohesive end and one blunt end.

4.11 INTRODUCTION OF THE RECOMBINANT DNA INTO THE SUITABLE HOST

rDNA is constructed *in vitro*. It is then generally introduced into *E. coli* to select the rDNA from the unchanged vector, to obtain many copies of the rDNA, or to express the DNA insert in *E. coli* itself. Purified rDNA may subsequently be introduced into another bacterium (such as *Bacillus subtilis* and Streptomyces), yeast, and higher plants or animals. The various approaches for introducing rDNAs into bacteria, especially *E. coli*, are briefly summarized here.

4.12 INCREASED COMPETENCE OF *E. COLI* BY $CACL_2$ TREATMENT

E. coli cells are generally poorly accessible to DNA molecules, but treatment with $CaCl_2$ makes them permeable to DNA. The rDNA is then added. Efficient transformation takes only a few minutes, and the cells are plated on a suitable medium for the selection of transformed clones. The frequency of transformed cells is 106–107/μg of plasmid DNA. This is approximately one transformation per 10,000 plasmid molecules. This frequency can be further improved by using special *E. coli* strains such as SK1590, SK1592, and X1766, and by some specific conditions during transformation. These may raise the frequency to 5 × 108 transformed cells/μg of plasmid DNA. The transformed cells are suitably diluted and spread thinly on a suitable medium so that each cell is well separated and produces a separate colony. Generally, the medium is designed so as to permit only the transformed cells to divide and produce colonies.

4.13 INFECTION BY RECOMBINANT DNAs PACKAGED AS VIRIONS

Alternatively, those rDNA that have the λ phage cos sequences such as those derived from cosmids, phasmids, and λ vectors, are generally packaged *in vitro* into specially produced empty λ phage heads and complete A particles are constituted. These phage particles are used to infect *E. coli* cells. This process is often called transfection. These rDNAs can also be used to transform *E. coli* cells directly as naked DNA, using the $CaCl_2$ technique. Generally, transfection is far more efficient than direct transformation. For example, the frequency of transfection by recombinant λ phage DNAs packaged in phage particles is up to 108 plaques/μg of DNA, while it is <103 plaques/μg DNA when the rDNA is used for transformation by the $CaCl_2$ technique. The infected/transformed bacterial cells are spread on a lawn of susceptible cells, where clear areas or plaques develop in the lawn. Plaques containing the rDNA (A vector and phasmids) are identified, and the phage particles collected from such plaques provide the purified vector/rDNA.

4.14 SELECTION OF RECOMBINANT CLONES

When rDNA is constructed and used for transformation of *E. coli*, cells following types of bacterial cells are obtained. The majority of the cells are nontransformed. A proportion of the transformed cells contain an unaltered vector, while the remainder cells have rDNA. The first objective of cloning experiments is to identify and isolate the small number of cells that contain the rDNA from among a very large number of nontransformed cells. Since the DNA inserts are generally mixtures, particularly when cDNA preparations and genomic DNA fragments are used, the various transformed clones would contain a variety of different DNA inserts. The next step, therefore, is to identify the clone having the desired DNA insert from among the large number of clones containing the rDNAs. Suitable selection strategies have been devised to achieve these two critical objectives. This is the most important step in DNA cloning.

4.15 IDENTIFICATION OF CLONES HAVING RECOMBINANT DNAs

The next step consists of identification and isolation of those clones that are transformed by the rDNAs from among those that contain the unaltered vector. This may be achieved in one of several ways listed in the following:

- In case the vector has two selectable markers such as pBR322, the DNA insert may be placed within one of these markers (for instance, the ampT gene). The other marker (in this case, tetr) is used for elimination of the nontransformed cells. The transformed clones are then replica plated on an ampicillin-containing medium. The clones containing the rDNAs will be sensitive to ampicillin due to inactivation of the gene ampT by insertion of the DNA fragment. Such clones are identified and isolated from the master plate.
- Some vectors contain a gene, or sometimes only part of a gene, which complements a function missing in their host cells, such as gene lacZα in the pUC vectors, which complements such lacZ—*E. coli* strains in which lacZα is deleted. The same combination is used for some A. vectors and M13 phage vectors. In all such cases, the DNA insert is so placed that it disrupts the expression of lacZα.
- Therefore, *E. coli* cells containing the rDNA are deficient in β-galactosidase and produce white colonies or plaques on a medium containing X-gal and IPTG. On the other hand, cells having the unchanged vector produce active β-galactosidase and give rise to blue colonies or plaques on the same medium. This allows an easy identification of the clones containing the rDNAs.

- When the DNA insert codes for a gene product which is defective in the auxotrophic host cells, a direct selection for the rDNA is possible. The host cells are grown on a medium lacking the compound needed by the auxotrophic host. Only those cells which contain the rDNA can grow and form colonies. Obviously, this approach is limited in application.
- Similarly, selection by suppression of nonsense mutations present in the host also permits a direct selection for the rDNA.
- Some A. vectors, such as λgt10, retain the lysogenic function as well. In such vectors, the DNA insert may be placed within the lysis repressor gene cI-, so that the vector becomes cI. As a result, cells transfected by the rDNA will give rise to clear plaques, whereas those infected by the unaltered vector will yield cloudy or turbid plaques. Thus, the rDNAs are readily identified and isolated.
- Some vectors such as A. replacement vectors and cosmids are much shorter than the minimum genome length needed for their packaging within virus particles. In such cases, the length of the DNA insert can be adjusted so as to allow the packaging of only the rDNA. This provides an efficient selection strategy for rDNA.

4.16 SELECTION OF CLONE CONTAINING A SPECIFIC DNA INSERT

Once we obtain a population of recombinant clones, the next step is to identify a clone that has the DNA insert of interest. The technique used for identification has to be highly precise and extremely sensitive to allow an accurate detection of a single clone from among the thousands obtained from a cloning experiment. The various strategies used for the purpose are briefly outlined in the following.

4.16.1 Colony Hybridization

The most efficient and rapid strategy for identification of a clone having the desired insert uses the technique of colony hybridization. The bacterial colonies are replica plated or phage plaques are directly lifted on nitrocellulose filters. The cells are lysed and their DNA is denatured. The filter is incubated with the specific radioactive (32p-labelled) probe under annealing conditions. After some time, the probe is washed out leaving only those probe molecules that have hybridized with the denatured DNA from bacterial cells or phage particles. The colonies/plaques with whose DNA the probe has hybridized are identified by autoradiography. These contain the desired DNA insert. These colonies/plaques are isolated from the master plate used for replica plating. A very large number of colonies or plaques (up to 10,000 plaques) can be lifted on to a single 10 cm diameter filter. However, it is essential that a specific probe for the DNA insert is available.

A *probe* is a polynucleotide molecule (DNA or RNA and usually small molecules of as few as 15 bases, but more often of 2530 bases) of a specific base sequence, which is used to detect DNA molecules having the same base sequence by complementary base pairing. Generally, the probes are labeled with 32p to enable autoradiography for easy identification of the DNA samples that base pair with the probe. It is desirable that the probes are single-stranded to avoid pairing between the two strands of the probe itself. Either DNA or RNA can be used as a probe. There are several approaches for developing specific probes.

4.16.2 Other Approaches for Developing Specific Probes

When specific probes are not available, many indirect approaches may be used for identifying clones having the desired DNA insert. These approaches or procedures are not generally convenient for screening of a large number of clones. Two such procedures, called *hybrid arrested translation* (HART) and *hybrid selection*, use *in vitro* translation systems, followed by identification of the resulting polypeptide(s). It is therefore necessary that the protein product of the DNA insert being searched should be known, at least in terms of its electrophoretic mobility.

4.16.3 Complementation

The cloned DNA insert may express itself in the bacterial cells. This is possible for prokaryotic genes, for some yeast genes, and for eukaryotic cDNAs cloned in suitable expression vectors. Eukaryotic sequences isolated from genomic DNA have to be expressed in appropriate eukaryotic hosts, such as yeast cells or animal cells. If the protein produced by the desired DNA insert is deficient in the host cells, this insert will correct the deficiency of the cells transformed by it, which means it will complement the deficiency of the host cells. This can be stated in general terms as follows. The host cells are deficient in a protein A, which means they are A-. These cells can be used to isolate the DNA fragment coding for protein A from a mixture of DNA fragments. Expression of rDNAs is prepared from the DNA fragments and A- host cells are transformed. These cells are now cultured under selective conditions that require functional A product. Only those host cells that contain the DNA insert encoding protein A will be able to multiply under the selective conditions (since the DNA insert will provide functional protein A). This strategy is limited in application by the availability of appropriate host cells.

4.16.4 Unique Gene Products

Alternatively, the protein product of the DNA insert can be identified by its unique function—that is, a function not performed by the proteins of nontransformed host cells. Such functions may relate to enzyme activities or hormone effects for which appropriate assays exist.

4.16.5 Antibodies Specific to the Protein Product

Finally, if the protein lacks a recognizable and measurable function, it can be detected by using specific antibodies. A practical approach is to divide the large number of recombinant clones into a convenient number of groups and to assay for the presence of the protein. The positive group is again divided into subgroups and assayed. In this manner, the positive groups are subdivided again and again, till a single positive clone is identified. This approach is applicable to the previous strategy as well. The identification of proteins using antibodies may be achieved by western blotting, precipitation and electrophoresis, or enzyme-linked immunosorbent assay (ELISA).

4.16.6 Colony Screening with Antibodies

An efficient and rapid screening using antibodies is as follows. The antibody specific to the concerned gene product (i.e., protein) is spread uniformly over a solid support such as a plastic or paper disc, which is placed in contact with an agar layer containing lysed bacterial colonies or phage plaques. If any clone is producing the protein in question, it will bind to the antibody molecules present on the disc. The disc is removed from the agar and is treated with a second radiolabeled (generally with L25I) antibody, which is also specific to the same protein but in a region different from that recognized by the first antibody. These antibodies will therefore also bind to the protein molecule held by the first antibody. The location of radioactivity on the disc is determined by autoradiography. The colonies/plaques producing the protein are then identified and isolated from the master plate. This technique is analogous to colony hybridization and is able to screen large numbers of clones rather rapidly. However, for this technique, we require two different antibodies that bind to two distinct domains of the desired protein. This protein must not be produced by the nontransformed host cells.

4.16.7 Fluorescence Activated Cell Sorter

In case of animal cells, an automated system, called *fluorescence activated cell sorter* (FACS), can be used for very rapid (up to 1000 cells/s) sorting of transformed cells. This is applicable to all the genes whose products become arranged on the cell surface and are available for binding of specific antibodies. Therefore, these proteins must not be produced by the nontransformed host cells. The antibody molecules are attached to a fluorescent molecule, and the transformed cells are treated with this antibody specific for the desired protein. The cells containing the protein in question on their surface will interact with the fluorescent antibodies. Cells are then passed one by one in a stream between a laser and a fluorescence detector. The cells that fluoresce are deflected into a microculture tray, while the nonfluorescing cells are drawn away by an aspirator. This approach is also applicable to the genes encoding receptor proteins present on the cell surface. In such cases, fluorescent ligands (the concerned molecule to which the receptor binds) are used in the place of fluorescent antibodies.

4.17 APPLICATIONS OF RECOMBINANT DNA TECHNOLOGY

rDNA technology has revolutionized therapeutic cloning by allowing us to create millions of DNA copies in a short span of time. The method entails clipping the desired segment out of the surrounding DNA and copying it millions of times. The success of rDNA technology, by which microbial cells can be engineered to produce foreign proteins, relies on the faithful reading of the corresponding genes by bacterial cell machinery, and has fueled most of the recent advances in modern molecular biology. During the last 20 years, studies of cloned DNA sequences have given us a detailed knowledge of gene structure and organization, and have provided clues to the regulatory pathways by which the cell controls gene expression in the multiple cell types comprising the basic vertebrate body plan. Genetic engineering, by which an organism can be modified to include new genes designed with desired characteristics, is now routine practice in basic research laboratories. It has provided the means to produce large amounts of highly purified human proteins such as follicle-stimulating hormone (FSH), insulin, and growth hormone. rDNA technology has been employed to identify mutations associated with breast cancer, retinoblastoma, and neurofibromatosis; in the diagnosis of affected and carrier states for hereditary diseases such as cystic fibrosis gene, the Huntington's disease gene, the Tay-Sachs disease gene, or the Duchenne muscular dystrophy gene; and transferring of genes from one organism to another.

Recent advances in this technology have also changed the course of medical research. Exciting new approaches are being developed to exploit the enormous potential of rDNA research in the analysis of genetic disorders. The new ability to manipulate human genetic material has opened radically new avenues for diagnosis and treatment, and has far-reaching consequences for the future of medicine. Yet the basic principles of rDNA, like the structure of DNA itself, are surprisingly simple. In addition, the use of rDNA has generated considerable interest among biotechnology and biopharmaceutical industries with regard to its benefits in cloning and construction of vectors. In this section, we will learn various diverse applications of rDNA technology in brief.

4.17.1 Genetically Modified Organisms

In this subsection, we will concentrate on how genetically modified organisms (GMOs) have revolutionized the research, development, and industrial sectors. A *genetically modified organism* or *genetically engineered organism* (GEO) is an organism whose genetic material has been altered using rDNA technology. The rDNA technology uses DNA molecules from different sources, which are combined into one molecule to create a new set of genes. This DNA is then transferred into an organism, giving it modified or novel genes. Transgenic organisms, a subset of GMOs, are organisms which have inserted DNA that originated in a different species.

4.17.1.1 Transgenic Microbes

A bacterium having a simple genetic makeup was the first organism to be modified in the laboratory. Bacteria are now used for several purposes, and are particularly important in producing large amounts of therapeutic proteins for treating various ailments and diseases, such as the genetically modified (GM) bacteria used to produce the protein insulin to treat diabetes. Similar bacteria have been used to produce clotting factors to treat hemophilia, and human growth hormone to treat various forms of dwarfism. In addition to use of bacteria in making therapeutic proteins, GM bacteria are also being used in treating dental disease. For example, tooth decay is caused due to the bacteria *Streptococcus mutans*; these bacteria consume leftover sugars in the mouth, producing lactic acid that corrodes tooth enamel and ultimately causes cavities. Scientists have recently modified *Streptococcus mutans* so that they do not produce lactic acid. These transgenic bacteria, if properly colonized in a person's mouth, could reduce the formation of cavities. In recent research, transgenic microbes have also been used to kill or hinder tumors. GM bacteria are also used in some soils to facilitate crop growth and to produce chemicals that are toxic to crop pests.

4.17.1.2 Transgenic Animals

Transgenic animals are used as experimental models to perform phenotypic tests with genes whose function is unknown. Genetic modification can also produce animals that are susceptible to certain compounds or stresses for testing in biomedical research. In biological research, transgenic fruit flies (*Drosophila melanogaster*) are model organisms used to study the effects of genetic changes on development. Fruit flies are often preferred over other animals due to their short life cycle, low maintenance requirements, and relatively simple genome compared with many vertebrates. Transgenic mice are often used to study cellular and tissue-specific responses to disease. This is possible since mice can be created with the same mutations that occur in human genetic disorders. The production of the human disease in these mice then allows treatments to be tested. In 2009, scientists in Japan announced that they had successfully transferred a gene into a primate species (marmosets) and produced a stable line of breeding transgenic primates for the first time. It is hoped that this will aid research into human diseases that cannot be studied in mice, such as Huntington's disease and strokes.

Besides mammalians and flies, rDNA technology has been used to create transgenic fish and cnidarians such as *Hydra* to study the evolution of immunity. For analytical purposes, an important technical breakthrough was the development of a transgenic procedure for generation of stably transgenic hydras by embryo microinjection. The creation of transgenic fish to produce higher levels of growth hormone has resulted in dramatic growth enhancement in several species, including salmonids, carps, and tilapias. These fish have been created for use in the aquaculture industry to increase the speed of development and to potentially reduce fishing pressure on wild stocks.

4.17.1.3 Transgenic Plants

rDNA technology has been widely used in generating plants with desirable traits, including resistance to pests, herbicides, or harsh environmental conditions; improved product shelf-life; and increased nutritional value. In 1996, transgenic plants were cultivated at the commercial level. Thereafter, there was tremendous increase in generation of transgenic plants that are not only tolerant to the herbicides glufosinate and glyphosate, but are also resistant to virus damage, as in Ringspot virus-resistant GM papaya grown in Hawaii, and to produce the Bt toxin, a potent insecticide.

4.17.1.4 Cisgenic Plants

In some GMOs, cells do not contain DNA from other species and are therefore not transgenic, but what is called *cisgenic*. GM sweet potatoes have been enhanced with protein and other nutrient values, while golden rice, developed by the International Rice Research Institute (IRRI), is a good source for Vitamin A. As Vitamin A deficiency causes deformities in children, eating cisgenic golden rice is highly recommended for children with a Vitamin A deficiency. In January 2008, scientists altered a carrot so that it would produce calcium and become a possible cure for osteoporosis. However, people would need to eat 1.5 kg of carrots/day to reach the required amount of calcium. The coexistence of GM plants with conventional and organic crops has raised significant concern in many European countries. Since there is separate legislation for GM crops and high demand from consumers for the freedom of choice between GM and non-GM foods, measures are required to separate foods and feed produced from GMO plants and from conventional and organic foods. European research programs such as Co-Extra, Transcontainer, and SIGMEA are investigating appropriate tools and rules. At the field level, biological containment methods include isolation distances and pollen barriers.

4.18 DNA SEQUENCING

Determination of the nucleotide or base sequence of a DNA molecule/fragment is known as *DNA sequencing*. At present, DNA sequencing is possible for only 700–800-bp-long DNA fragments. DNA sequencing has become feasible as a result of

- The availability of restriction endonucleases.
- The development of highly sensitive gel electrophoretic techniques that can separate DNA fragments differing by only one nucleotide.
- The development of gene cloning.
- The development of PCR techniques making available large quantities of individual DNA fragments.
- The development of two DNA sequencing procedures: the Maxam and Gilbert procedure and the enzymatic procedure (Figure 4.10).

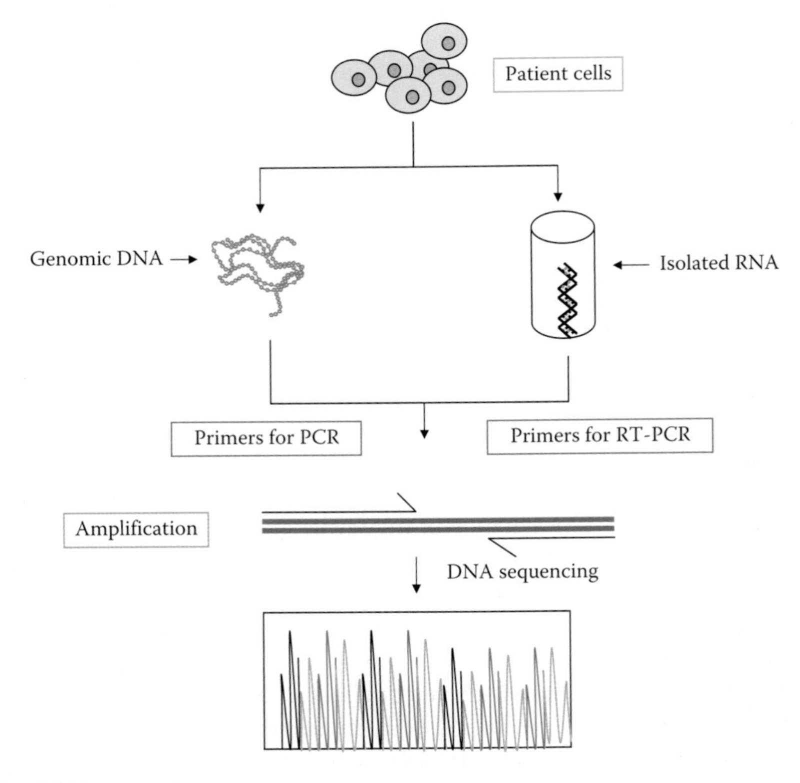

FIGURE 4.10 DNA sequencing.

4.18.1 Maxam and Gilbert Procedure

In the Maxam and Gilbert procedure, the DNA fragment to be sequenced is end-labeled by the addition of 32p-dATP, either at the 5′-ends (by the enzyme polynucleotide kinase) or at the 3′-ends (by the enzyme deoxynucleotidyl transferase) of its two strands. The end-labeled fragment is now digested with a restriction endonuclease, which cleaves it into only two fragments of unequal lengths. As a result, only one end of each of the two fragments thus produced will be labeled. The two unequal fragments are separated through gel electrophoresis and they are sequenced separately. Alternatively, the end-labeled fragment is denatured and its two complementary strands are separated through gel electrophoresis. For some reasons, the two complementary strands of a DNA molecule generally show different mobilities during gel electrophoresis. The samples of complementary strands thus separated are sequenced separately. It may be noted that each strand will be labeled only at one end (either the 5′-end or the 3′-end).

The single end-labeled double- or single-stranded DNA samples thus produced are subjected to base specific chemical cleavage so that a reaction mixture cleavage occurs only at one of the following four sites: Sites having G, C, G + A, or C + T. Each DNA sample is partially digested in four separate reaction mixtures (one for each of the specific cleavage at the sites having G, C, G + A, or C + T). In these reaction mixtures, each DNA fragment/strand is expected to be cleaved on an average of only once at any one of the sites having the particular base for which the reaction mixture is specific, each such site in the DNA fragment/strand being equally likely to be cleaved.

The base specific cleavage of DNA fragments involves the following steps: (1) modification of the concerned base, (2) removal of the modified base from the DNA strand, and (3) induction of strand break (break in the sugar phosphate backbone) in the position from which the modified base has been removed. Such a cleavage generates a mixture of end-labeled DNA fragments of variable lengths. Digests of double-stranded DNA are denatured before they are subjected to electrophoresis. The digests from the four reaction mixtures are then subjected to gel electrophoresis in separate lanes of the same gel to separate the fragments according to their lengths. The base sequence is determined by sequential reading of the bands developed in the four lanes of the gel through autoradiography.

4.18.2 Enzymatic Procedure

The enzymatic procedure, commonly referred to as the *Sanger–Coulson method*, was developed by F. Sanger and coworkers. In this technique, the DNA fragment to be sequenced is denatured and the complementary strands are separated through electrophoresis. One of the two complementary strands (or often both the strands, but in separate experiments) is used as a template for DNA replication catalyzed by the Klenow fragment (*E. coli* DNA polymerase I minus the first of its 323 amino acids—that is, the sequence having 5′ >— 3′ exonuclease activity). Single-stranded samples of DNA fragments may also be obtained by cloning DNA fragments in a single-stranded DNA virus such as M13 vector. In the reaction system for DNA replication, at least one of the four deoxyribonucleotides is radioactive in order to permit the autoradiographic development of bands after gel electrophoresis.

A small primer sequence with a free 3′-OH group must be provided with the template strand for DNA replication to proceed, since a free 3′-OH is absolutely essential for DNA polymerase I to catalyze DNA replication. Four different reaction mixtures are prepared for the replication of each DNA strand to be sequenced. In one of the reaction systems, 2′, 3′-dideoxycytidine triphosphate (ddCTP) is added in a concentration of approximately 1/100th of the ddCTP present in the system. ddCTP acts as a terminator of the polynucleotide chain being newly synthesized on the template strand. Chain termination by ddCTP is achieved because it (and the other 2′-3′-dideoxynucleotides) does not have a free 3′-OH group, as a result of which further nucleotides cannot be added to the new chain. At the concentration used here, ddCTP would terminate the newly synthesized polynucleotide chains at anyone of all the possible sites where cytosine is to be incorporated in the new chain. In each of the three other reaction mixtures using the same DNA fragment as template, 2′,3′-dideoxythymidine triphosphate (ddTTP), 2′,3′-dideoxyadenosine triphosphate (ddATP), or 2′,3′-dideoxyguanosine triphosphate (ddGTP) is used as chain terminator to terminate the polynucleotide chains at anyone of all the positions where T, A, or G, respectively, are to be incorporated in the new chain (each 2′,3′-dideoxynucleotide is used in a separate reaction mixture).

The partially synthesized DNA chains (due to chain termination) from each of the above four reaction mixtures are separated from the template strand by denaturation. The four single-stranded samples are now separately subjected to gel electrophoresis (in separate lanes of a gel) in order to separate the strands according to their size. The bands in the gels are developed onto an x-ray film through autoradiography. The fastest moving fragment will be the smallest one, and each subsequent band will be one nucleotide longer than the previous one. Therefore, by comparing the bands of the four gel lanes thus obtained, the nucleotide sequence of the DNA fragment can be determined. The position of a band in the gel from a reaction mixture will indicate the position of the base of which 2′,3′-dideoxynucleotide triphosphate was used as chain terminator in that mixture.

4.18.3 Automated DNA Sequencing

Automated DNA sequencing is based on the Sanger–Coulson method, with two notable differences from the standard procedure. The first difference concerns the labeling of the products of PCR. Automated procedures use fluorescent labels in the place of radioactive labeling used in the standard procedure. The fluorescent labels are usually attached to the four dideoxynucleotides used for chain termination. In the four-track system of automated DNA sequencing, each of the four dideoxynucleotides is used in a separate reaction, and the products are run in four adjacent lanes of the gel.

If a different fluorochrome is attached to each of the four dideoxynucleotides, all of them could be used in the same reaction in place of preparing a separate reaction for each dideoxynucleotide. This is called the *single-track system* since the reaction products are run in a single gel lane or capillary. Generally, the DNA to be sequenced is subjected to thermal cycle sequencing to generate the chain-terminated polynucleotides required for sequencing. The reaction products are subjected to polyacrylamide gel electrophoresis under denaturing conditions or are loaded into a capillary filled with a sequencing gel. The bands produced in the polyacrylamide gel/capillary are identified with the help of a fluorescence detector, which identifies the fluorescent signal emitted by each band. The fluorochromes are excited by a laser beam, and the resulting fluorescence signal is sensed by a photovoltaic cell.

The resulting data are fed into a computer, which in turn converts these signals into the base sequence of the DNA molecule. The sequence information can be printed out or stored in a data storage device for future use; this is the second major deviation from the standard Sanger–Coulson procedure. In the four-track system, the sequence can be recognized from the raw data, but it has to be interpreted using an appropriate computer program in the single-track system. This becomes necessary in order to compensate for the shifts in mobility due to the different fluorochromes. Automated DNA sequencers can read up to 96 DNA sequences in a 2-h period, which is extremely fast when compared to manual DNA sequencing. Automated DNA sequencing has several advantages over manual DNA sequencing, which include not using radioactivity, and gel processing after electrophoresis and autoradiography are not needed. In addition, the tedious manual reading of gels is not required, as data are processed in a computer. The sequence data is directly fed into and stored in a computer. Another advantage of automated sequencing is that the separation of the same reaction products can be repeated, to recheck the results in cases of doubt, and data can be stored for a long period.

4.18.4 Current Challenges in DNA Sequencing

As is true of all technology, there are common challenges for DNA sequencing that hamper the quality of sequencing to a great extent, including poor quality in the first 15–40 bases of the sequence and deteriorating quality of sequencing traces after 700–900 bases. In cases where DNA fragments are cloned before sequencing, the resulting sequence may contain parts of the cloning vector. In contrast, PCR-based cloning and emerging sequencing technologies based on pyro-sequencing often avoid using cloning vectors. Automated DNA-sequencing instruments (DNA sequencers) can sequence up to 384 DNA samples in a single day. DNA sequencers carry out capillary electrophoresis for size separation, detection, and recording of dye fluorescence, and data output as fluorescent peak trace chromatograms. Sequencing reactions by thermocycling, cleanup, and re-suspension in a buffer solution before loading onto the sequencer are performed separately. A number of commercial and noncommercial software packages can trim low-quality DNA traces automatically. These programs score the quality of each peak and remove low-quality base peaks (generally located at the ends of the sequence). The accuracy of such algorithms is below visual examination by a human operator, but sufficient for automated processing of large sequence data sets.

Current methods of sequencing can directly sequence only relatively short (300–1000-nucleotides long) DNA fragments in a single reaction. The main obstacle to sequencing DNA fragments above this size limit is insufficient power of separation for resolving large DNA fragments that differ in length by only one nucleotide. Large-scale sequencing aims at sequencing very long DNA pieces, such as whole chromosomes. Common approaches consist of cutting (with restriction enzymes) or shearing (with mechanical forces) large DNA fragments into shorter DNA fragments. The fragmented DNA is cloned into a DNA vector and amplified in *E. coli*. Short DNA fragments purified from individual bacterial colonies are individually sequenced and assembled electronically into one long, contiguous sequence. This method does not require any preexisting information about the sequence of the DNA and is referred to as *de novo* sequencing. Gaps in the assembled sequence may be filled by primer walking. The different strategies have different tradeoffs in speed and accuracy. Shotgun methods are often used for sequencing large genomes, but its assembly is complex and difficult, particularly with sequence repeats often causing gaps in genome assembly.

4.18.5 Trends in DNA Sequencing

4.18.5.1 High-Throughput Sequencing

The high demand for low-cost sequencing has driven the development of high-throughput sequencing technologies that parallelize the sequencing process, producing thousands or millions of sequences at once. High-throughput sequencing technologies are intended to lower the cost of DNA sequencing beyond what is possible with standard dye-terminator methods.

4.18.5.2 *In Vitro* Clonal Amplification

Molecular detection methods are not sensitive enough for single molecule sequencing, and most approaches therefore use an *in vitro* cloning step to amplify individual DNA molecules. Emulsion PCR isolates individual DNA molecules along with primer-coated beads in aqueous droplets within an oil phase. PCR then coats each bead with clonal copies of the DNA molecule, followed by immobilization for later sequencing. Emulsion PCR is used in the methods published by Marguilis et al. (commercialized by 454 Life Sciences), Shendure and Porreca et al. (also known as "polony sequencing"), and SOLiD sequencing, (developed by Agencourt, now Applied Biosystems). Another method for *in vitro* clonal amplification is *bridge PCR*, where fragments are amplified upon primers attached to a solid surface, used in the Illumina Genome Analyzer. The single-molecule method developed by Stephen Quake's laboratory (later commercialized by Helicos) skips this amplification step, directly fixing DNA molecules to a surface.

4.18.5.3 Parallelized Sequencing

DNA molecules are physically bound to a surface and sequenced in parallel. Sequencing by synthesis, like dye-termination electrophoretic sequencing, uses a DNA polymerase to determine the base sequence. Reversible terminator methods (used by Illumina and Helicos) use reversible versions of dye-terminators, adding one nucleotide at a time, and detect fluorescence at each position in real time by repeated removal of the blocking group to allow polymerization of another nucleotide. Pyro-sequencing (used by 454 sequencing) also uses DNA polymerization, adding one nucleotide species at a time, detecting and quantifying the number of nucleotides added to a given location through the light emitted by the release of attached pyrophosphates.

4.18.5.4 Sequencing by Ligation

Sequencing by ligation uses a DNA ligase to determine the target sequence. Used in the polony method and in the SOLiD technology, it uses a pool of all possible oligonucleotides of a fixed length, labeled according to the sequenced position. Oligonucleotides are annealed and ligated. The preferential ligation by DNA ligase for matching sequences results in a signal informative of the nucleotide at that position.

4.18.5.5 Microfluidic Sanger Sequencing

In microfluidic Sanger sequencing, the entire thermocycling amplification of DNA fragments as well as their separation by electrophoresis is done on a single glass wafer (approximately 10 cm in diameter), thus reducing the reagent usage. Research will still need to be done in order to make this use of technology effective.

4.18.5.6 Other Sequencing Technologies

Sequencing by hybridization is a nonenzymatic method that uses a DNA microarray. A single pool of DNA whose sequence is to be determined is fluorescently labeled and hybridized to an array containing known sequences. A strong hybridization signal from a given spot on the array identifies its sequence in the DNA being sequenced. Mass spectrometry may be used to determine mass differences between DNA fragments produced in chain-termination reactions. DNA sequencing methods currently under development include labeling the DNA polymerase, reading the sequence as a DNA strand transits through nanopores, and microscopy-based techniques such as AFM or electron microscopy. These techniques are used to identify the positions of individual nucleotides within long DNA fragments (>5000 bp) by nucleotide labeling with heavier elements (e.g., halogens) for visual detection and recording.

4.19 MICROARRAYS

In case of microarrays, a series of probes are immobilized on a glass slide as microdots, which are then hybridized with a mixture of test DNA sequences that are labeled with a fluorochrome. An extremely large number of probes are spotted onto the slide, each probe is a pure preparation, the test DNA is a mixture of sequences, and the results are visualized by confocal microscopy. Microarrays were first used in the case of yeast, which has 6000 genes. Every yeast gene was obtained as an individual clone and a sample of each gene was spotted onto glass sides in arrays of 80 spots × 80 spots. In order to determine the identity of genes active in yeast cells under a set of given conditions, mRNA is extracted from these cells. The mRNA is converted into cDNA, and the cDNA is fluorescently labeled. The labeled cDNA is used for hybridization with the microarray, and the identity of spots showing fluorescence (i.e., hybridization) is determined by confocal microscopy. The spots showing fluorescence represent the genes that were expressed in the cells. It may be pointed out that single-stranded DNA preparations are spotted using a laboratory robot (Figure 4.11).

4.20 DNA CHIPS

DNA chips are thin wafers of silicon glass carrying many different oligonucleotides synthesized at a very high density (300,000 to over 1 million oligonucleotides/cm^2) directly onto the wafer. The oligonucleotides are synthesized at a high spatial resolution and in precise locations.

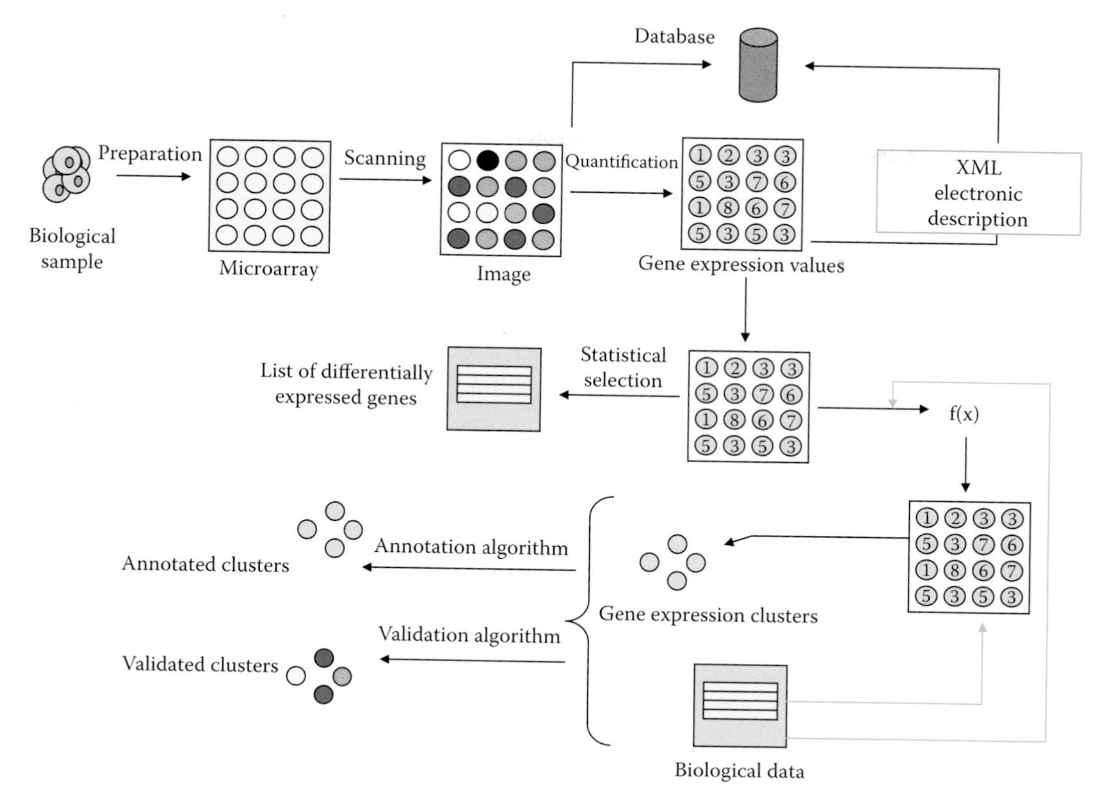

FIGURE 4.11 Microarray technique.

Each oligonucleotide has the sequence of a different gene present in the genome. Therefore, sequence information for the genes to be represented in the DNA chip must be available. The oligonucleotide synthesis is based on two techniques called *photolithography* and *solid-phase DNA synthesis*. It uses a series of building blocks that contain photochemically removable protective groups. The DNA chips are inverted and mounted in a temperature-controlled hybridization chamber into which fluorescently labeled cDNA preparation is injected and allowed to hybridize with the oligonucleotides. Laser excitation enters through the back of the glass support focused at the interface of the array surface and target solution. Fluorescent emission is collected by a lens and passed on to a sensitive detector, and a quantitative assay of hybridization intensity is obtained. DNA chips present an alternative to DNA microarrays. Study of the gene expression pattern of an organism is affected by the stage of development and/or the environment. DNA chips can be prepared for the detection of single nucleotide polymorphisms (SNPs). These chips are called *SNP chips*. DNA chips can also be used for the detection of genetic diseases, such as for detection of mutant alleles causing cystic fibrosis and mutant alleles of gene BRCA1 (gene involved in breast cancer) (Figure 4.12).

4.21 ISOLATION OF DESIRED DNA

The identification and isolation of the gene or DNA fragment to be cloned, called *DNA insert*, is a critical step in gene cloning. The desired DNA inserts can be obtained from the following: (1) cDNA libraries, (2) genomic libraries, (3) chemical (or enzymatic) synthesis, and (4) amplification through PCR.

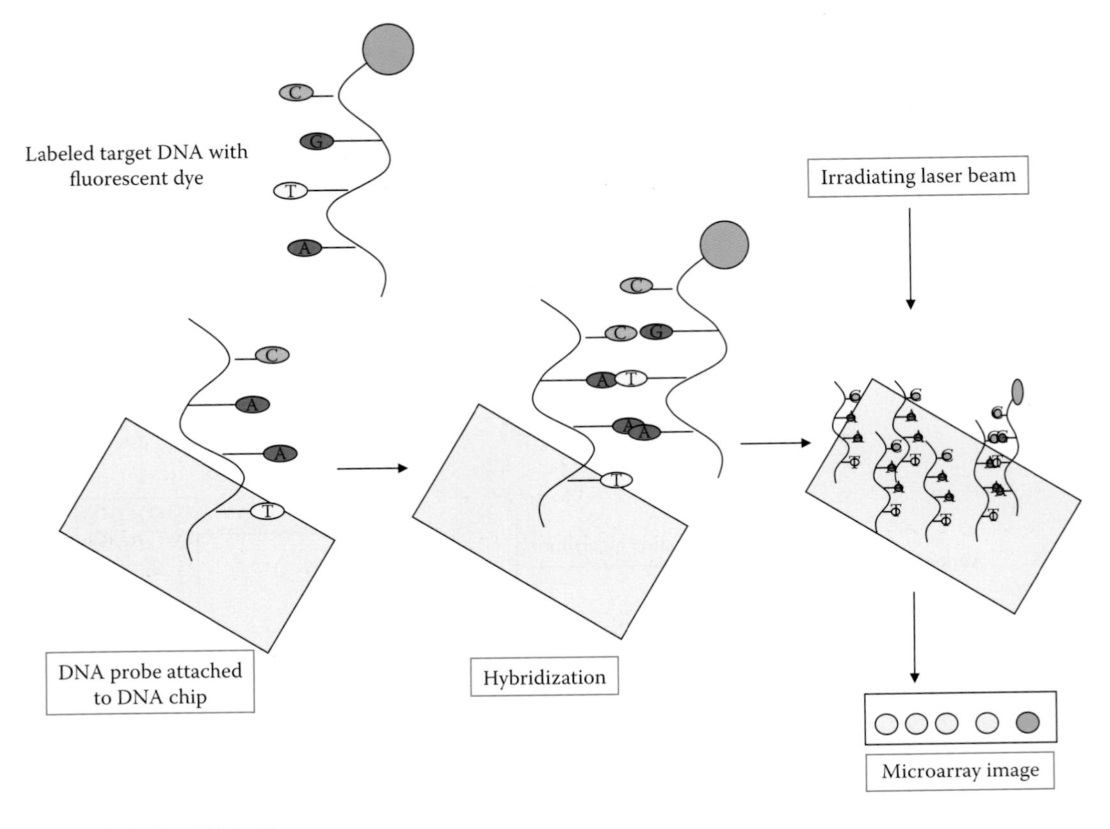

FIGURE 4.12 DNA chip.

4.22 cDNA LIBRARY

A *cDNA library* is a population of bacterial transformants or phage lysates in which each mRNA isolated from an organism or tissue is represented as its cDNA insertion in a plasmid or a phage vector. The frequency of a specific cDNA in such a library would ordinarily depend on the frequency of the concerned mRNA in the tissue/organism in question. This enzyme performs similar reactions as DNA polymerase, and has an absolute requirement for a primer with a free 3′-OH. When eukaryotic mRNA is used as a template, a poly-T oligonucleotide (more specifically, oligodeoxynucleotide) is conveniently used as the primer since these mRNAs have a poly-A tail at their 3′-ends. However, special tricks are required to utilize primers for other RNAs such as prokaryotic mRNA, rRNA, and RNA virus genomes.

For example, a poly-A tail may be added to the 3′-end of the RNA to make it analogous to eukaryotic mRNA (oligo-T is now used as primer). This reaction is catalyzed by the enzyme poly-A polymerase. The appropriate oligonucleotide primer (oligo-T for eukaryotic mRNA) is annealed with the mRNA. These primers will base-pair to the 3′-end of mRNA. Reverse transcriptase extends the 3′-end of the primer using the mRNA molecule as a template. This produces an RNA–DNA hybrid molecule, the DNA strand being the cDNA. The RNA strand is digested by either RNase H or alkaline hydrolysis. This frees the single-stranded cDNA. Curiously, the 3′-end of this cDNA serves as its own primer and provides the free 3′-OH required for the synthesis of its complementary strand. Therefore, a primer is not required for this step. The complementary strand of the cDNA single strand is synthesized by either the reverse transcriptase itself or by *E. coli* DNA polymerase. This generates a hairpin loop in the cDNA. The hairpin loop is cleaved by a single strand specific nuclease to yield a regular DNA duplex.

4.22.1 Problems in cDNA Preparation

There is always a problem in preparing a cDNA library. This is because the double-strand cDNA preparations are always a mixture of different kinds of molecules, which causes problems in the copying of the RNA, and also because even highly purified mRNAs are never absolutely pure. Physical and chemical methods are incapable of resolving these mixtures. Therefore, the cDNA mixture itself is used for cloning, and the desired cDNA is identified and isolated in pure form from the appropriate bacterial clone.

4.22.2 Isolation of mRNA

To prepare cDNA, mRNA is essential and to get mRNA, total RNA is first extracted from a suitable organism/tissue. The amount of desired mRNA in this sample is then increased by using one of several procedures. Use of cDNA is absolutely essential when the expression of a eukaryotic gene is required in a prokaryote such as a bacterium. This is because eukaryotic genes have introns, which must be removed from their transcripts to yield mature mRNA, and bacteria do not possess the enzymes necessary for removal of introns. For example, cDNAs for interferon, blood clotting factor VIIIC (both human), and several other mRNAs have been expressed in bacteria.

4.23 PREPARATION OF cDNA

cDNA is the copy or complementary DNA produced by (usually) using mRNA as a template. In fact, any RNA molecule can be used to produce cDNA. The DNA copy of an RNA molecule is produced by the enzyme reverse transcriptase (RNA-dependent DNA polymerase; discovered by Temin and Baltimore in 1970) generally obtained from avian myeloblastosis virus (AMV).

4.24 GENOMIC LIBRARY

A *genomic library* is a collection of plasmid clones or phage lysates containing rDNA molecules so that the sum total of DNA inserts in this collection ideally represents the entire genome of the concerned organism. In spite of all the care taken in the production of genomic libraries, certain DNA fragments should be expected to be underrepresented, overrepresented, or even missing. There are several possible reasons for this, and they cannot be addressed at present.

4.24.1 Construction of a Genomic Library

A genomic library can be constructed by extracting the total genomic DNA of an organism. The DNA is broken into fragments of appropriate size, either by mechanical shearing sonication or by using a suitable restriction endonuclease for partial digestion of the DNA. Complete digestion is avoided, since it generates fragments that are too heterogeneous in size. For partial digestion, restriction enzymes having four-base (tetrameric) recognition sequences are employed in preference to those having six-base (hexameric) target sites. This is because a given four-base recognition site is expected to occur every 4^4 (=256) bp in a DNA molecule, while a six-base target site would occur only after every 4^6 (=4096) bp. (It is assumed here that the arrangement of the four bases in DNA molecules is random.) Therefore, the fragments produced in partial digests with enzymes having four-base recognition sites are more likely to be of appropriate size for cloning than those generated by enzymes having six-base recognition sites.

Single or mixed digestions with the enzymes AluI, HaeIII, or Sau3A have been used for constructing genomic libraries. The use of restriction enzymes has the advantage that the same set of fragments are obtained from a DNA each time a specific enzyme is used, and many of the enzymes produce cohesive ends. The partial digests of genomic DNA are subjected to agarose gel electrophoresis or sucrose gradient centrifugation for separation from the mixture of fragments of appropriate size.

These fragments are then inserted into a suitable vector for cloning. This constitutes the shotgun approach to gene cloning. In principle, any vector can be used, but A vectors and cosmids have been the most commonly used, since DNA inserts of up to 23–25kb can be cloned in these vectors. The vectors containing the inserts are cloned in a suitable bacterial host.

4.25 DNA LIBRARIES

DNA libraries, like conventional libraries, are used to collect and store information. In DNA libraries, the information is stored as a set of DNA molecules, each of which contains biological sequences that can be used for a variety of applications. All DNA libraries are collections of DNA fragments that represent a particular biological system of interest. By analyzing the DNA from a particular organism or tissue, researchers can answer a variety of important questions. The two most common uses for these DNA collections are DNA sequencing and gene cloning. Several types of DNA libraries have been developed for specific purposes, but all share some common features. The DNA fragments that make up the library are attached to other DNA sequences that are used as "handles" to maintain the fragments. These "handles," called vectors, allow the DNA to be replicated and stored, typically within model organisms such as yeast or bacteria.

Different types of vectors can be used to store DNA fragments of different lengths. For example, plasmid vectors can store small fragments (from a few hundred bases up to ten or twenty thousand bases of sequence), while viral vectors or viral-plasmid hybrids such as cosmids can store up to fifty thousand bases, and YAC vectors can store hundreds of thousands of bases. In general, plasmid-based vectors are the easiest to manipulate, but store the smallest fragments. They are commonly used for applications that involve complex manipulations, such as cloning or gene expression, but that require only small DNA fragments such as cDNA libraries (Figure 4.13).

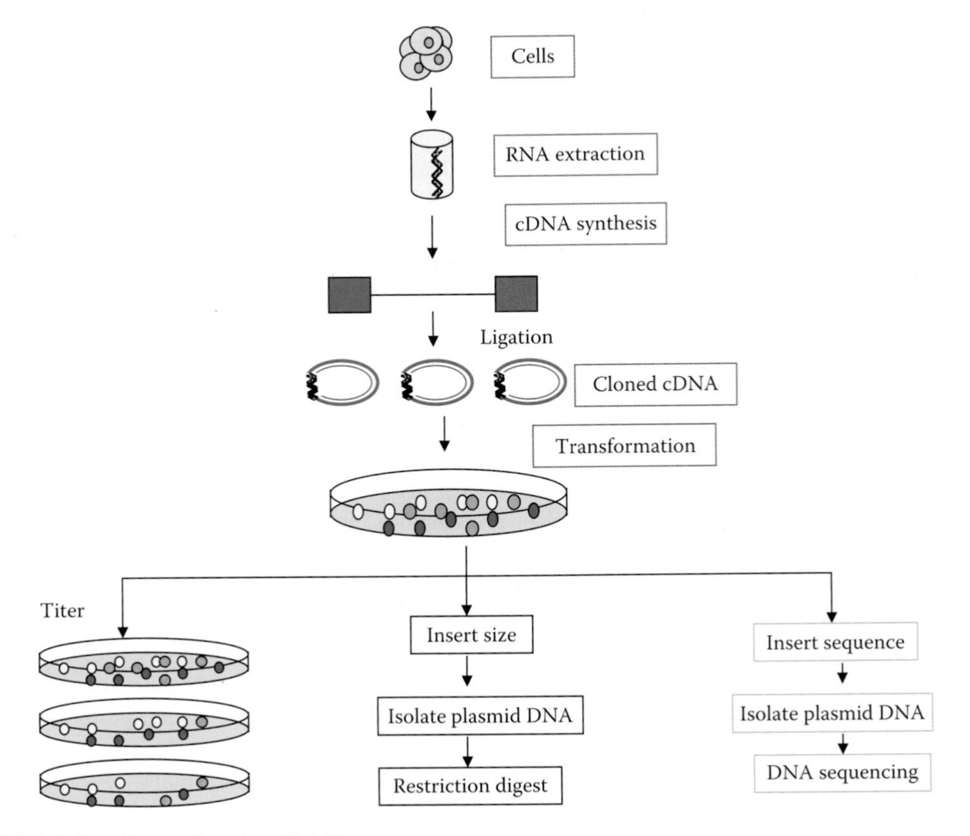

FIGURE 4.13 Generation of cDNA library.

4.26 CHEMICAL SYNTHESIS GENE

The amino acid sequence of the protein (or the base sequence of mRNA) produced by a gene enables the deduction of the base sequence of the concerned gene on the basis of the codons for the various amino acids. However, the degeneracy of the genetic code may present some problems, but a functional sequence of the gene can nonetheless be worked out. Once the base sequence of a gene is deduced, a polynucleotide of the same base sequence can be synthesized either chemically or even enzymatically. There are three distinct methods of doing this, differing mainly in the strategy of protection of OH groups of the phosphate residues: phosphodiester approach, phosphotriester or phosphate triester approach, and phosphite triester or phosphoramidite approach.

4.26.1 Phosphodiester Approach

The first significant success was achieved by this approach. Khorana and coworkers synthesized the gene for alanine (suppressor) tRNA of yeast in 1970, while Brown and coworkers synthesized the gene for tyrosine (suppressor) tRNA of *E. coli*. This approach has now been almost completely replaced by the more convenient approaches since it presents a variety of solubility problems.

4.26.2 Phosphotriester Approach

Both phosphotriester and the phosphite triester methods utilize deoxyribonucleosides as starting materials, and involve the stepwise addition of mononucleotides and oligonucleotides. The amino groups of the nucleosides deoxyadenosine and deoxcytidine are usually benzoylated, and that on deoxyguanosine is protected by an isobutyryl group, thymidine, which requires no protection. The 5′-OH group is protected by dimethoxytrityl, commonly abbreviated as DMTr or (MeO)2 Tr. The amino groups of the bases are freed by mild alkaline hydrolysis, while (MeO)2 Tr is removed by gentle acid hydrolysis. In the phosphotriester approach, the 3′-OH is coupled with a suitable phosphorylating agent such as p-chlorophenyl phosphorodichloridate. The phosphate residue has a free OH group that accepts any nucleotide or oligonucleotide with a free 5′-OH group. Therefore, such protected nucleotides have phosphodiesters and serve as the 5′-terminus residue in oligonucleotide synthesis.

Such protected and phosphorylated nucleotides are further modified to make them suitable for joining to the -OH group of the 3′-phosphate residue. The OH group of the 3′-phosphate residue of such nucleotides is blocked by a suitable agent such as β-cyanoethanol, following which the 5′-OH is freed by mild acid hydrolysis (this removes the (MeO)2Tr). This yields phosphotriester nucleotides, which have a free 5′-OH. A desired diester nucleotide (free-OH at 3′-phosphate residue) is now mixed with the desired triester nucleotide, and agents that promote their coupling are added to the mixture. Coupling is promoted by arylsulfonyl compounds such as tri-iso-propyl-benzene sulfonyl chloride (TPS). This reaction yields a fully protected dinucleotide, which can be either fully unprotected or may be selectively unprotected to be used as the starting material for construction of larger molecules. The DNA chains can be constructed either in $3' \rightarrow 5'$ or $5' \rightarrow 3'$ direction. Tedious purifications are essential after every addition to the growing chain to remove the uncoupled mononucleotides/oligonucleotides. This and some other problems are eliminated by using a solid support to which the first nucleotide is fixed. Somehow, fixing the 3′-OH is better than fixing the 5′-0R. Generally, this is done by forming an ester between the 3′-OH and a carboxyl group on a solid support such as controlled pore glass beads. This procedure has been adopted for automated stepwise synthesis; a 10–20-nucleotide-long chain is synthesized in a few days.

4.26.3 Phosphite Triester Approach

Nucleosides having protected bases and 5′-OH groups are the basic materials. The 3′-OH of the terminal nucleotide is fixed to a solid support, after which its 5′-OH is freed by gentle acid hydrolysis. The subsequent nucleotides are used as 3′-phosphoramidites, which are readily produced by coupling the nucleotides with di-isopropylammonium tetrazolide. The phosphoramidites are stable and efficient coupling agents, and are readily synthesized. The desired nucleotide 3′-phosphoramidite is now added and is activated for coupling by the addition of tetrazole. The immediate product is a phosphite, which is oxidized to phosphate by iodine (12). This phosphate remains as a triester. The 5′-OH of the second nucleotide is now freed, and the third desired nucleotide 3′-phosphoramidite is added and joined to the growing chain. In this manner, the oligo nucleotide chain is elongated. Finally, the various protective groups are removed and the chain is freed from the solid support by alkaline hydrolysis. This approach using silica-based or controlled pore glass beads solid support is used for automated synthesis of oligonucleotides. It takes less than15 min for adding one nucleotide to the chain, and chains as long as 50 nucleotides can be prepared in good yields. The automated DNA synthesizers are popularly called gene machines. They are microprocessor controlled and carry out all the operations automatically.

4.27 APPLICATIONS OF SYNTHETIC OLIGONUCLEOTIDES

Synthetic oligonucleotides are used for the sequencing of 12–20 bases and as hybridization probes when other probes are not available. They are used as primers for cDNA preparation using reverse transcriptase, amplification of DNA segments using PCR, and enzymatic DNA sequencing. Oligonucleotides are also used as linkers and adaptors in gene cloning. Finally, they are used to produce complete gene sequences by linking them in a defined order. Attempts also are being made to use antisense oligonucleotides as therapeutic agents.

PROBLEMS

Section A: Descriptive Type

Q1. What is rDNA technology?
Q2. What are the steps involved in making rDNA?
Q3. Describe nonbacterial transformation.
Q4. Why are restriction enzymes so important in rDNA technology?
Q5. How can we amplify genes using PCR?
Q6. Differentiate between PCR and gene cloning.
Q7. What is a vector? Describe its role in rDNA technology
Q8. What is colony hybridization?
Q9. Describe DNA sequencing by the Maxam and Gilbert procedure
Q10. What are the different benefits of automated DNA sequencing?
Q11. What is parallelized sequencing?
Q12. What is a DNA chip?

Section B: Multiple Choice

Q1. Which of the following methods is not used in making rDNA product?
  a. Translation
  b. Transformation
  c. Phage introduction
  d. Nonbacterial transformation
Q2. Restriction enzyme cuts double-stranded or single-stranded DNA at a specific site. True/False
Q3. Type IIM restriction endonucleases such as Dpnl are able to recognize and cut methylated DNA. True/False
Q4. All vectors used for propagation of DNA inserts in a suitable host are called cloning vectors. True/False

Q5. Which of the following is *not* an *E. coli* strain?
a. K12
b. C600
c. HB101
d. B250

Q6. Resistance plasmids contain genes that do not build resistance against antibiotics or poisons. True/False

Q7. What is the spherical virus with a circular, double-stranded 5243 bp chromosome?
a. SV40 vector
b. YAC vector
c. ARS vector
d. Adenovirus

Q8. What does FACS mean?
a. Fluorescence acquired cell sorter
b. Fluorescence activated cell sorter
c. Fluorescence auto cell sorter

Q9. What do you call GM plants that do not contain the DNA of other species?
a. Cisgenic plants
b. Transgenic plants
c. Autogenic plants
d. Heterogenic plants

Q10. At present, DNA sequencing is possible for only ________ bp long DNA fragments.
a. 600–700
b. 700–800
c. 900–1000
d. >10,000

Q11. What do you call the first DNA sequencing technique?
a. Watson and Crick procedure
b. Maxam and Gilbert procedure
c. Karry Mullis's method
d. None of above

Q12. For which organism was microarray first used?
a. Bacteria
b. Virus
c. Fungi
d. Yeast

Q13. DNA chips are thin wafers of silicon glass carrying many different oligonucleotides synthesized at a very high density. True/False

Q14. The genomic library can be constructed by extracting total genomic ________ of an organism.
a. DNA
b. cDNA
c. Mitochondrial DNA

Q15. Oligonucleotides are also used as linkers and adaptors in gene cloning. True/False

Section C: Critical Thinking

Q1. What problems may arise if, instead of microbes, mammalian cells are used as vectors for producing rDNA products?

Q2. Is it possible to cut DNA fragments without restriction enzyme? Explain why.

Q3. Explain why 30–45 cycles are usually carried out in most PCR reactions.

ASSIGNMENTS

Make a poster based on the various rDNA products such as insulin, growth hormones, or erythropoietin. Display it in the classroom and discuss it with your colleagues.

REFERENCES AND FURTHER READING

Berg, P., Baltimore, D., Brenner, S., Roblin III, R.O., and Singer, M.F. Summary statement of the Asilomar conference on recombinant DNA molecules. *Proc. Natl. Acad. Sci. USA* 72(6): 1981–1984, 1975.

Braslavsky, I., Hebert, B., Kartalov, E., and Quake, S.R. Sequence information can be obtained from single DNA molecules. *Proc. Natl. Acad. Sci. USA* 100: 3960–3964, 2003.

Church, G.M. Genomes for All. *Sci. Am.* 294: 46–54, 2006.

Cohen, S.N., Chang, A.C.Y., Boyer, H.W., and Helling, R.B. Construction of biologically functional bacterial plasmids in vitro. *Proc. Natl. Acad. Sci. USA* 70(11): 3240–3244, 1973.

Colowick, S.P. and Kapian, O.N. *Methods in Enzymology—Volume 68; Recombinant DNA*. Academic Press, Burlington, MA, 1980.

Ewing, B. and Green, P. Base-calling of automated sequencer traces using phred. II. Error probabilities. *Genome Res.* 8: 186–194, 1998.

Fiers, W., Contreras, R., Duerinck, F. et al. Complete nucleotide sequence of bacteriophage MS2 RNA: Primary and secondary structure of the replicase gene. *Nature* 260: 500–507, 1976.

Garret, R.H. and Grisham, C.M. *Biochemistry*. Saunders College Publishers, Philadelphia, PA, 2000.

Genentech 1978. The insulin synthesis is the first laboratory production DNA technology. Press release. Archived from the original on May 9, 2006. http://web.archive.org/web/20060509151511/http://www.gene.com/gene/news/press-releases/display.do?method=detail&id=4160 (retrieved January 7, 2009).

Gilbert, W. DNA sequencing and gene structure. Nobel lecture. December 8, 1980.

Gilbert, W. and Maxam, A. The nucleotide sequence of the lac operator. *Proc. Natl. Acad. Sci. USA* 70: 3581–3584, 1973.

Hall, N. Advanced sequencing technologies and their wider impact in microbiology. *J. Exp. Biol.* 210: 1518–1525, 2007.

Hanna, G.J., Johnson, V.A., Kuritzkes, D.R. et al. Comparison of sequencing by hybridization and cycle sequencing for genotyping of human immunodeficiency virus type 1 reverse transcriptase. *J. Clin. Microbiol.* 38: 2715–2721, 2000.

Inoue, N., Takeuchi, H., Ohashi, M., and Suzuki, T. The production of recombinant human erythropoietin. *Biotechnol. Ann. Rev.* 1: 297–300, 1995.

Johnston, S.A. and Tang, D.C. Gene gun transfection of animal cells and genetic immunization. *Method. Cell Biol.* 43(Pt A): 353–365, 1994.

Ju, J., Ruan, C., Fuller, C.W., Glazer, A.N., and Mathies, R.A. Fluorescence energy transfer dye-labeled primers for DNA sequencing and analysis. *Proc. Natl. Acad. Sci. USA* 92: 4347–4351, 1995.

Kruzer, H. and Massay, A. *Recombinant DNA and Biotechnology: A Guide for Student*. ASM Press, Washington, DC, 2001.

Leader, B., Baca, Q.J., and Golan, D.E. Protein therapeutics: A summary and pharmacological classification. *Nat. Rev. Drug Discov.* A guide to drug discovery 7: 21–39, 2008.

Lee, L.Y. and Gelvin, S.B. T-DNA binary vectors and systems. *Plant Physiol.* 146(2): 325–332, 2008.

Margulies, M., Egholm, M., Altman, W.E. et al. Genome sequencing in microfabricated high-density picolitre reactors. *Nature* 437: 376–380, 2005.

Maxam, A.M. and Gilbert, W. A new method for sequencing DNA. *Proc. Natl. Acad. Sci. USA* 74: 560–564, 1977.

Min Jou, W., Haegeman, G., Ysebaert, M., and Fiers, W. Nucleotide sequence of the gene coding for the bacteriophage MS2 coat protein. *Nature* 237: 82–88, 1972.

Park, F. Lentiviral vectors: Are they the future of animal transgenesis? *Physiol. Genomics* 31(2): 159–173, 2007.

Pipe, S.W. Recombinant clotting factors. *Thromb. Haemost.* 99: 840–850, 2008.

Ronaghi, M., Karamohamed, S., Pettersson, B., Uhlen, M., and Nyren, P. Real-time DNA sequencing using detection of pyrophosphate release. *Anal. Biochem.* 242: 84–89, 1996.

Sanger, F. Determination of nucleotide sequences in DNA. Nobel lecture. December 8, 1980.

Sanger, F. and Coulson, A.R. A rapid method for determining sequences in DNA by primed synthesis with DNA polymerase. *J. Mol. Biol.* 94: 441–448, 1975.

Sanger, F., Nicklen, S., and Coulson, A.R. DNA sequencing with chain-terminating inhibitors. *Proc. Natl. Acad. Sci. USA* 74: 5463–5467, 1977.

Shendure, J. Accurate multiplex polony sequencing of an evolved bacterial genome. *Science* 309: 1728–1732, 2005.

Shreeve, J. Secretes of the gene. *Natl. Geogr.* 1966: 42–75, 1999.

Smith, L.M., Fung, S., Hunkapiller, M.W., Hunkapiller, T.J., and Hood, L.E. The synthesis of oligonucleotides containing an aliphatic amino group at the 5′ terminus: Synthesis of fluorescent DNA primers for use in DNA sequence analysis. *Nucleic Acids Res.* 13: 2399–2412, 1985.

Smith, L.M., Sanders, J.Z., Kaiser, R.J. et al. Fluorescence detection in automated DNA sequence analysis. *Nature* 321: 674–679, 1986.

Walsh, G. Therapeutic insulins and their large-scale manufacture. *Appl. Microbiol. Biotechnol.* 67: 151–159, 2005.

Watson, J. and Tooze, J. *The DNA Story: A Documentary History of Gene Cloning*. W.H. Freeman & Co., San Francisco, CA, 1981.

Weiner, D. and Kenendy, R. Genetic vaccines. *Sci. Am.* 281: 50–57, 1999.

5 Microbial Biotechnology

LEARNING OBJECTIVES

- Define microbes and explain their various attributes
- Discuss structural and functional characteristics of microbes
- Explain the growth and culture of microbes
- Discuss microbial genetics and genetic transformations
- Discuss the role of microbes in food, medical, agricultural, and environmental biotechnology

5.1 INTRODUCTION

Microbes are single-celled organisms, so tiny that millions can fit into the eye of a needle. They are the oldest form of life on earth. Microbe fossils date back more than 3.5 billion years, to a time when the earth was covered with oceans that regularly reached boiling point, hundreds of millions of years before dinosaurs roamed the earth. Without microbes, we could not eat or breathe. Without us, they would probably be just fine. Understanding microbes is vital to understanding our own past and future and that of our planet. Microbes are everywhere. There are more of them on a person's hand than there are people on the entire planet! Microbes are in the air we breathe, the ground we walk on, the food we eat—they are even inside us! We could not digest food without them, and neither could animals. Without microbes, plants could not grow, garbage would not decay, and there would be a lot less oxygen to breathe in. In fact, without these invisible companions, our planet as we know it would not survive!

With a view to understanding their significance in our daily life, it is very important to have historical information pertaining to microorganisms. It is believed that the ancestors of modern bacteria were single-celled microorganisms that were the first forms of life to develop on earth, approximately 4 billion years ago. For about 3 billion years, all organisms were microscopic, and bacteria and archaea were the dominant forms of life. Although bacterial fossils such as stromatolites exist, their lack of distinctive morphology prevents them from being used to examine the history of bacterial evolution or to date the time of origin of a particular bacterial species. However, gene sequences can be used to reconstruct bacterial phylogeny, and these studies indicate that bacteria diverged first from the archaeal/eukaryotic lineage. The most recent common ancestor of bacteria and archaea was a hyperthermophile that lived approximately 2.5–3.2 billion years ago.

Bacteria were first observed by Antonie van Leeuwenhoek in 1676, using a single-lens microscope of his own design. He called them "animalcules" and published his observations in a series of letters to the Royal Society. The name *bacterium* was introduced much later, in 1838, by Christian Gottfried Ehrenberg. In 1859, Louis Pasteur demonstrated that the fermentation process is caused by the growth of microorganisms, and that this growth is not due to spontaneous generation. (Yeasts and molds that are commonly associated with fermentation are not bacteria, but rather fungi.) Along with his contemporary, Robert Koch, Pasteur was an early advocate of the germ theory of disease. Robert Koch was a pioneer in medical microbiology and worked on cholera, anthrax, and tuberculosis. In his research into tuberculosis, Koch finally proved the germ theory, for which he was awarded a Nobel Prize in 1905. In *Koch's postulates*, he set out criteria to test if an organism is the cause of a disease. These postulates are still used today. Though it was known in the nineteenth century that bacteria are the cause of many diseases, no effective antibacterial treatments were available. In 1910, Paul Ehrlich developed the first antibiotic, by changing dyes that selectively stained *Treponema pallidum*—the spirochaete that causes syphilis—into compounds that selectively killed the pathogen. In 1908, Ehrlich was awarded a Nobel Prize for his work on immunology, and pioneered the use of stains to detect and identify bacteria, with his work being the basis of the Gram stain and the Ziehl-Neelsen stain. A major step forward in the study of bacteria was the recognition, in 1977, by Carl Woese that archaea have a separate line of evolutionary descent from bacteria. This new phylogenetic taxonomy was based on the sequencing of 16S ribosomal RNA, and divided prokaryotes into two evolutionary domains as part of the three-domain system.

5.2 STRUCTURAL ORGANIZATION OF MICROBES

5.2.1 Structure

Bacteria display a wide diversity of shapes and sizes, called *morphologies*. Bacterial cells are approximately one-tenth the sizes of eukaryotic cells and are typically 0.5–5.0 μm in length. However, a few species, such as *Thiomargarita namibiensis* and *Epulopiscium fishelsoni*, are up to half a millimeter long and are visible to the unaided eye. Among the smallest bacteria are members of the genus *Mycoplasma*, which measure only 0.3 μm—as small as the largest viruses. Some bacteria may be even smaller, but these ultramicrobacteria are not well studied. Most bacterial species are either spherical, called *cocci*, or rod-shaped, called *bacilli*. Some rod-shaped bacteria, called *vibrio*, are slightly curved or comma-shaped. Others can be spiral-shaped, called *spirilla*, or tightly coiled, called *spirochaetes*. A small number of species even have tetrahedral or cuboidal shapes. More recently, bacteria that grow as long rods with a star-shaped cross section were discovered deep under the earth's crust. The large surface area to volume ratio of this morphology may give these bacteria an advantage in nutrient-poor environments. This wide variety of shapes is determined by the bacterial cell wall and cytoskeleton, and is important because it can influence the ability of the bacteria to acquire nutrients, attach to surfaces, swim through liquids, and escape predators. Many bacterial species exist simply as single cells, while others associate in characteristic patterns: *Neisseria* form diploids (pairs), *Streptococcus* form chains, and *Staphylococcus* group together in "bunch of grapes" clusters. Bacteria such as *Actinobacteria* can also be elongated to form filaments.

Filamentous bacteria are often surrounded by a sheath that contains many individual cells. Certain types, such as species of the genus *Nocardia*, even form complex, branched filaments, similar in appearance to fungal mycelia. Bacteria often attach to surfaces and form dense aggregations called *biofilms* or *bacterial mats*. These films can range from a few micrometers in thickness to up to half a meter in depth, and may contain multiple species of bacteria, protists, and Achaea. Bacteria living in biofilms display a complex arrangement of cells and extracellular components, forming secondary structures such as microcolonies through which there are networks of channels to enable better diffusion of nutrients. In natural environments, such as soil or the surfaces of plants, the majority of bacteria are bound to surfaces in biofilms. Biofilms are also important in medicine, as these structures are often present during chronic bacterial infections or in infections of implanted medical devices, and bacteria protected within biofilms are much harder to kill than individual isolated bacteria. Even more complex morphological changes are sometimes possible. For example, when starved of amino acids, *Mycobacterium* detect surrounding cells in a process known as *quorum sensing*, migrate toward each other, and aggregate to form fruiting bodies up to 500 μm long and containing approximately 100,000 bacterial cells. In these fruiting bodies, the bacteria perform separate tasks. This type of cooperation is a simple type of multicellular organization. For example, approximately 1 in 10 cells migrate to the top of these fruiting bodies and differentiate into a specialized dormant state called *myxospores*, which are more resistant to drying and other adverse environmental conditions than are ordinary cells (Figure 5.1).

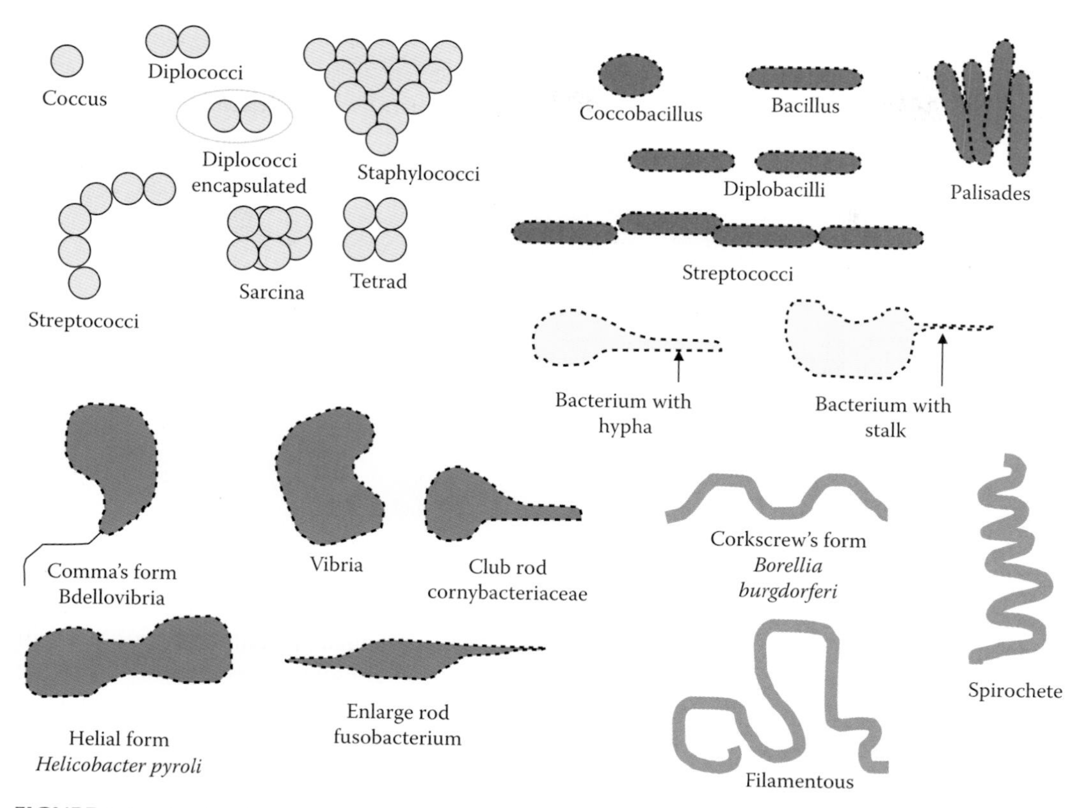

FIGURE 5.1 Family of microorganisms.

5.2.2 Intracellular Organization

The bacterial cell is surrounded by a lipid membrane, or *cell membrane*, which encloses the contents of the cell and acts as a barrier to hold nutrients, proteins, and other essential components of the cytoplasm within the cell. As they are prokaryotes, bacteria do not tend to have membrane-bound organelles in their cytoplasm, and thus contain few large intracellular structures. They consequently lack a nucleus, mitochondria, chloroplasts, and the other organelles present in eukaryotic cells, such as the Golgi apparatus and endoplasmic reticulum. Bacteria were once seen as simple bags of cytoplasm, but elements such as prokaryotic cytoskeleton and the localization of proteins to specific locations within the cytoplasm have been found to show levels of complexity. These subcellular compartments have been called *bacterial hyperstructures*. Microcompartments such as carboxysome provide a further level of organization; microcompartments are compartments within bacteria that are surrounded by polyhedral protein shells, rather than by lipid membranes. These polyhedral organelles localize and compartmentalize bacterial metabolism, a function performed by the membrane-bound organelles in eukaryotes. Many important biochemical reactions, such as energy generation, occur by concentration gradients across membranes, a potential difference also found in a battery. The general lack of internal membranes in bacteria means reactions such as electron transport occur across the cell membrane between the cytoplasm and the periplasmic space. However, in many photosynthetic bacteria the plasma membrane is highly folded and fills most of the cell with layers of light-gathering membrane. These light-gathering complexes may even form lipid-enclosed structures called *chlorosomes* in green sulfur bacteria. Other proteins import nutrients across the cell membrane or expel undesired molecules from the cytoplasm (Figure 5.2).

Bacteria do not have a membrane-bound nucleus, and their genetic material is typically a single circular chromosome located in the cytoplasm in an irregularly shaped body called the *nucleoid*.

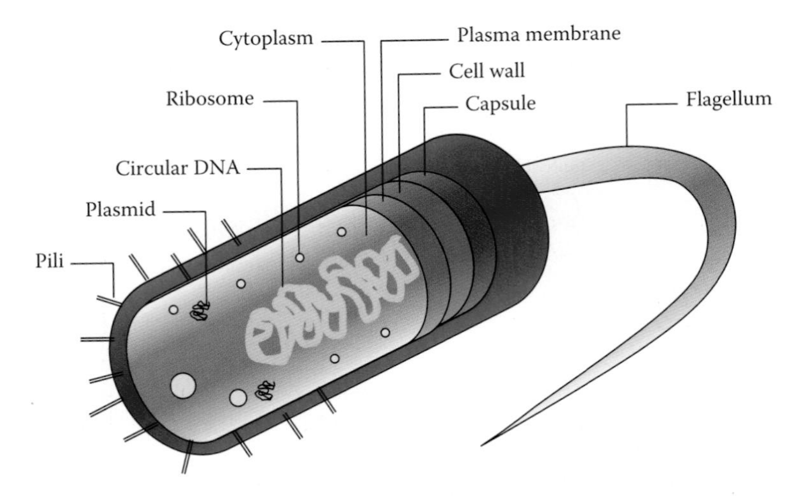

FIGURE 5.2 Internal organization of a microorganism.

The nucleoid contains the chromosome with associated proteins and RNA. The order *Planctomycetes* are an exception to the general absence of internal membranes in bacteria, because they have a membrane around their nucleoid and contain other membrane-bound cellular structures. Like all living organisms, bacteria contain ribosomes for the production of proteins, but the structure of the bacterial ribosome is different from those of eukaryotes and *Archaea*. Some bacteria produce intracellular nutrient storage granules such as glycogen, polyphosphate, or sulfur. These granules enable bacteria to store compounds for later use. Certain bacterial species, such as the photosynthetic *Cyanobacteria*, produce internal gas vesicles which they use to regulate their buoyancy, allowing them to move up or down into water layers with different light intensities and nutrient levels.

5.2.3 Extracellular Organization

Around the outside of the cell membrane is the bacterial *cell wall*. Bacterial cell walls are made of peptidoglycan, called *murein* in older sources, which is made from polysaccharide chains cross-linked by unusual peptides containing D-amino acids. Bacterial cell walls are different from the cell walls of plants and fungi, which are made of cellulose and chitin, respectively. The cell walls of bacteria are also distinct from that of *Archaea*, which do not contain peptidoglycan. The cell wall is essential to the survival of many bacteria, and the antibiotic penicillin is able to kill bacteria by inhibiting a step in the synthesis of peptidoglycan. Broadly speaking, there are two different types of cell walls in bacteria, called *Gram-positive* and *Gram-negative*. The names originate from the reaction of cells to the *Gram stain*, a test long employed for the classification of bacterial species. Gram-positive bacteria possess a thick cell wall containing many layers of peptidoglycan and teichoic acids. In contrast, Gram-negative bacteria have a relatively thin cell wall consisting of a few layers of peptidoglycan surrounded by a second lipid membrane containing lipopolysaccharides and lipoproteins. Most bacteria have the Gram-negative cell wall, and only the *Firmicutes* and *Actinobacteria* (previously known as the *low G + C Gram-positive bacteria* and *high G + C Gram-positive bacteria*, respectively) have the alternative Gram-positive arrangement. These differences in structure can produce differences in antibiotic susceptibility. For instance, vancomycin can kill only Gram-positive bacteria and is ineffective against Gram-negative pathogens such as *Haemophilus influenzae* or *Pseudomonas aeruginosa*.

In many bacteria, an S-layer of rigidly arrayed protein molecules covers the outside of the cell. This layer provides chemical and physical protection for the cell surface and can act as a macromolecular diffusion barrier. S-layers have diverse but mostly poorly understood functions, although they are known to act as virulence factors in *Campylobacter* and contain surface enzymes in

Bacillus stearothermophilus. *Flagella* are rigid protein structures, approximately 20 nm in diameter and up to 20 μm in length, that are used for motility. Flagella are driven by the energy released by the transfer of ions down an electrochemical gradient across the cell membrane. *Fimbriae* are fine filaments of protein, just 2–10 nm in diameter and up to several micrometers in length. They are distributed over the surface of the cell, and resemble fine hairs when seen under the electron microscope. Fimbriae are believed to be involved in attachment to solid surfaces or to other cells and are essential for the virulence of some bacterial pathogens. *Pili* (pilus in the singular) are cellular appendages slightly larger than fimbriae, which can transfer genetic material between bacterial cells in a process called *conjugation*.

Capsules or slime layers are produced by many bacteria to surround their cells, and vary in structural complexity, ranging from a disorganized slime layer of extra-cellular polymer to a highly structured capsule or *glycocalyx*. These structures can protect cells from engulfment by eukaryotic cells, such as macrophages. They can also act as antigens and be involved in cell recognition, as well as aiding attachment to surfaces and the formation of biofilms. The assembly of these extracellular structures is dependent on bacterial secretion systems. These transfer proteins from the cytoplasm into the periplasm or into the environment around the cell. Many types of secretion systems are known, and as these structures are often essential for the virulence of pathogens, they are intensively studied. Certain genera of Gram-positive bacteria, such as *Bacillus*, *Clostridium*, *Sporohalobacter*, *Anaerobacter*, and *Heliobacterium*, can form highly resistant, dormant structures called *endospores*. In almost all cases, one endospore is formed and this is not a reproductive process, although *Anaerobacter* can make up to seven endospores in a single cell. Endospores have a central core of cytoplasm containing DNA and ribosomes, surrounded by a cortex layer and protected by an impermeable and rigid coat.

Endospores show no detectable metabolism and can survive extreme physical and chemical stresses, such as high levels of UV light, gamma radiation, detergents, disinfectants, heat, pressure, and desiccation. In this dormant state, these organisms may remain viable for millions of years, and endospores even allow bacteria to survive exposure to the vacuum and radiation in space. Endospore-forming bacteria can also cause disease. For example, anthrax can be contracted by the inhalation of *Bacillus anthracis* endospores, and contamination of deep puncture wounds with *Clostridium tetani* endospores causes tetanus (Figures 5.3 and 5.4).

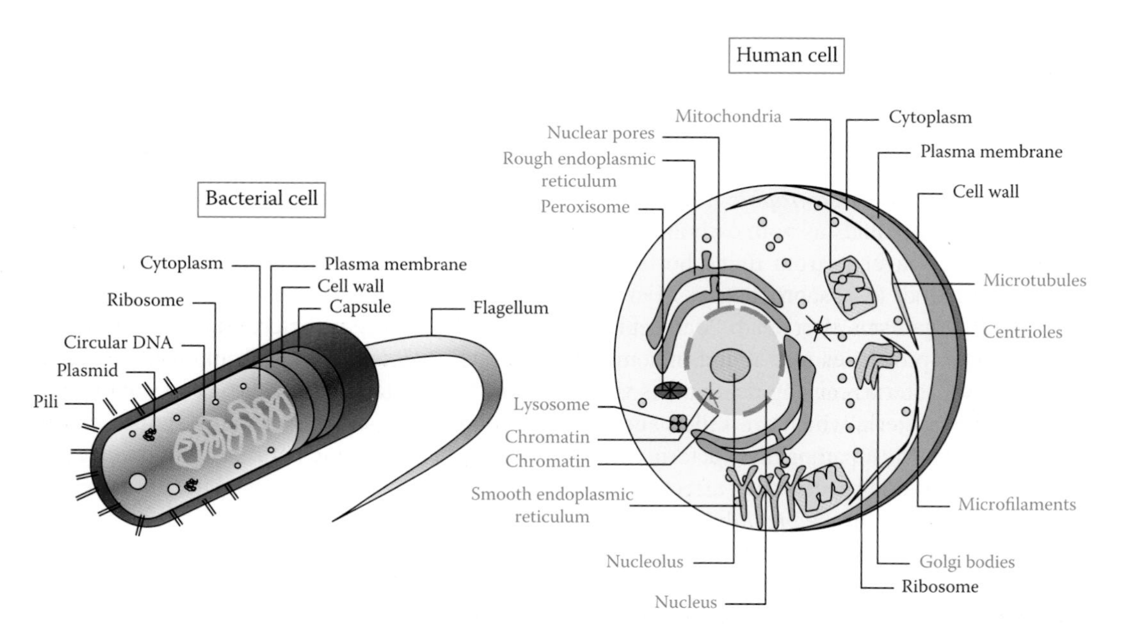

FIGURE 5.3 Similarities and differences between a bacterial cell and a human cell.

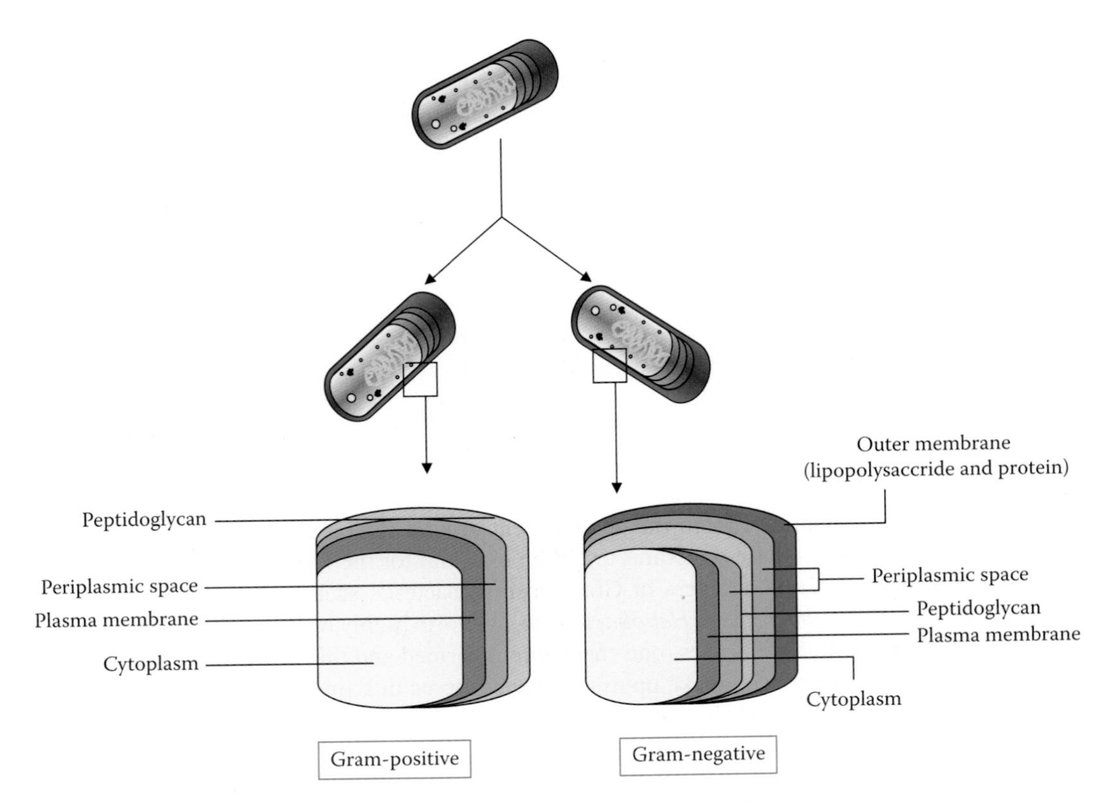

FIGURE 5.4 Differences between Gram-positive and Gram-negative microbes.

5.3 MICROBIAL METABOLISM

Microbial metabolism is the process through which microorganisms obtain the energy and nutrients in order to live and reproduce. Microbes use many different types of metabolic strategies, and microbes can often be differentiated from each other based on metabolic characteristics. The specific metabolic properties of a microbe are the major factors in determining their usefulness in industrial applications.

5.3.1 Heterotrophic Microbial Metabolism

Most microbes are *heterotrophic* (more precisely, chemoorganoheterotrophic), which means they use organic compounds as both carbon and energy sources. Heterotrophic microbes live off nutrients that they scavenge from living hosts (e.g., commensals or parasites) or find in dead organic matter of all kinds (e.g., saprophages). Microbial metabolism is the main contribution for the bodily decay of all organisms after death. Many eukaryotic microorganisms are heterotrophic by predation or parasitism, properties also found in some bacteria such as *Bdellovibrio* (an intracellular parasite of other bacteria, causing death of its victims) and Myxobacteria such as *Myxococcus* (predators of other bacteria which are killed and lysed by cooperating swarms of many single cells of Myxobacteria). Most pathogenic bacteria can be viewed as heterotrophic parasites of humans or of the other eukaryotic species they affect. Heterotrophic microbes are extremely abundant in nature and are responsible for the breakdown of large organic polymers such as cellulose, chitin, or lignin, which are generally indigestible to larger animals. Generally, the breakdown of large polymers to carbon dioxide (mineralization) requires several different organisms, with one breaking down the polymer into its constituent monomers, one able to use the monomers and excreting simpler waste compounds as by-products, and one able to use the excreted wastes. There are many variations on

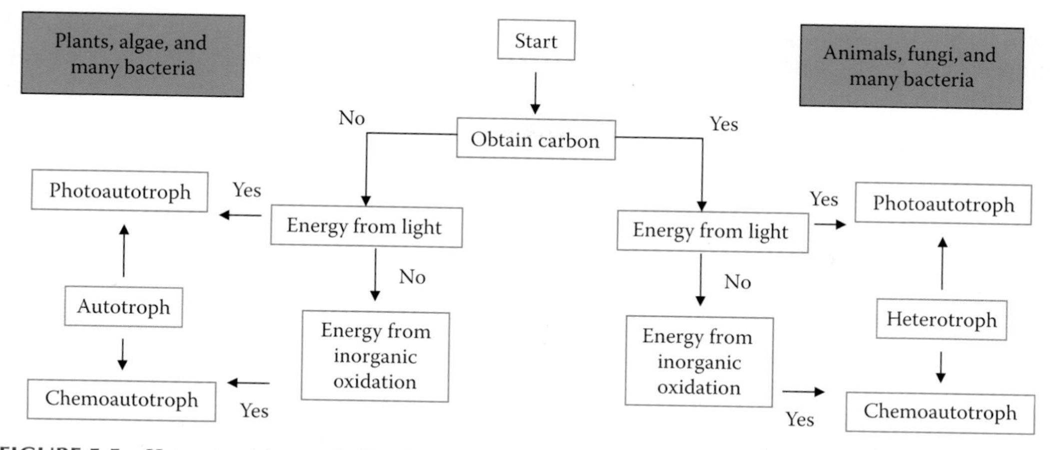

FIGURE 5.5 Heterotrophic metabolism in bacteria.

this theme, as different organisms are able to degrade different polymers and secrete different waste products. Some organisms are even able to degrade more recalcitrant compounds such as petroleum compounds or pesticides, making them useful in bioremediation (Figure 5.5).

Biochemically, prokaryotic heterotrophic metabolism is much more versatile than that of eukaryotic organisms, although many prokaryotes share the most basic metabolic models with eukaryotes, such as using glycolysis (also called EMP pathway) for sugar metabolism and the citric acid cycle to degrade acetate, producing energy in the form of ATP, and reducing power in the form of NADH or quinols. These basic pathways are well conserved, because they are also involved in biosynthesis of many conserved building blocks needed for cell growth (sometimes in reverse direction). However, many bacteria and archaea utilize alternative metabolic pathways other than glycolysis and the citric acid cycle. A well-studied example is sugar metabolism via the keto-deoxy-phosphogluconate pathway (also called *ED pathway*) in *Pseudomonas*. Moreover, there is a third alternative sugar-catabolic pathway used by some bacteria, the pentose-phosphate pathway. The metabolic diversity and ability of prokaryotes to use a large variety of organic compounds arises from the much deeper evolutionary history and diversity of prokaryotes, as compared to eukaryotes. It is also noteworthy that the mitochondrion, the small membrane-bound intracellular organelle that is the site of eukaryotic energy metabolism, arose from the endosymbiosis of a bacterium related to obligate intracellular *Rickettsia*, and also to plant-associated *Rhizobium* or *Agrobacterium*. Therefore, it is not surprising that all mitrochondriate eukaryotes share metabolic properties with these *Proteobacteria*. Most microbes respire (use an electron transport chain), although oxygen is not the only terminal electron acceptor that may be used. As discussed below, the use of terminal electron acceptors other than oxygen has important biogeochemical consequences.

5.3.2 Fermentation

Fermentation is a specific type of heterotrophic metabolism that uses organic carbon instead of oxygen as a terminal electron acceptor. This means that these organisms do not use an electron transport chain to oxidize nicotinamide adenine dinucleotide (NADH) to NAD^+, and therefore must have an alternative method of using this reducing power and maintaining a supply of NAD^+ for the proper functioning of normal metabolic pathways (e.g., glycolysis). As oxygen is not required, fermentative organisms are anaerobic. Many organisms can use fermentation under anaerobic conditions and anaerobic respiration when oxygen is not present. These organisms are facultative anaerobes. To avoid the overproduction of NADH, obligately fermentative organisms usually do not have a complete citric acid cycle. Instead of using an adenosine triphosphatase (ATPase) as in respiration, adenosine triphosphate (ATP) in fermentative organisms is produced by substrate-level phosphorylation, where a phosphate group is transferred from a high-energy organic compound to adenosine diphosphate (ADP) to form ATP. As a result of the need to produce high-energy phosphate-containing organic compounds (generally in the form of CoA-esters), fermentative organisms use NADH and other cofactors to produce many different reduced metabolic by-products, often including hydrogen gas (H_2). These reduced organic compounds are generally small organic acids and alcohols derived from pyruvate, the end product of glycolysis. Examples include ethanol, acetate, lactate, and butyrate. Fermentative organisms are very important industrially and are used to make many different types of food products. The different metabolic end products produced by each specific bacterial species are responsible for the different tastes and properties of each food.

Not all fermentative organisms use substrate-level phosphorylation. Instead, some organisms are able to couple the oxidation of low-energy organic compounds directly to the formation of a proton (or sodium) motive force, and therefore ATP synthesis. Examples of these unusual forms of fermentation include succinate fermentation by *Propionigenium modestum* and oxalate fermentation by *Oxalobacter formigenes*. These are extremely low-energy-yielding reactions. Humans and other higher animals also use fermentation to produce lactate from excess NADH, although this is not the major form of metabolism as it is in fermentative microorganisms (Figure 5.6).

5.3.3 Aerobic Respiration

While aerobic organisms use oxygen as a terminal electron acceptor during respiration, anaerobic organisms use other electron acceptors. These inorganic compounds have a lower reduction potential than oxygen, which means respiration is less efficient in these organisms and leads to slower growth rates than aerobes. Many facultative anaerobes can use either oxygen or alternative terminal electron acceptors for respiration, depending on the environmental conditions. Most respiring anaerobes are heterotrophs, although some do live autotrophically. All of the processes described below are dissimilative, which means that they are used during energy production and not to provide nutrients for the cell (assimilative). Assimilative pathways for many forms of anaerobic respiration are also known.

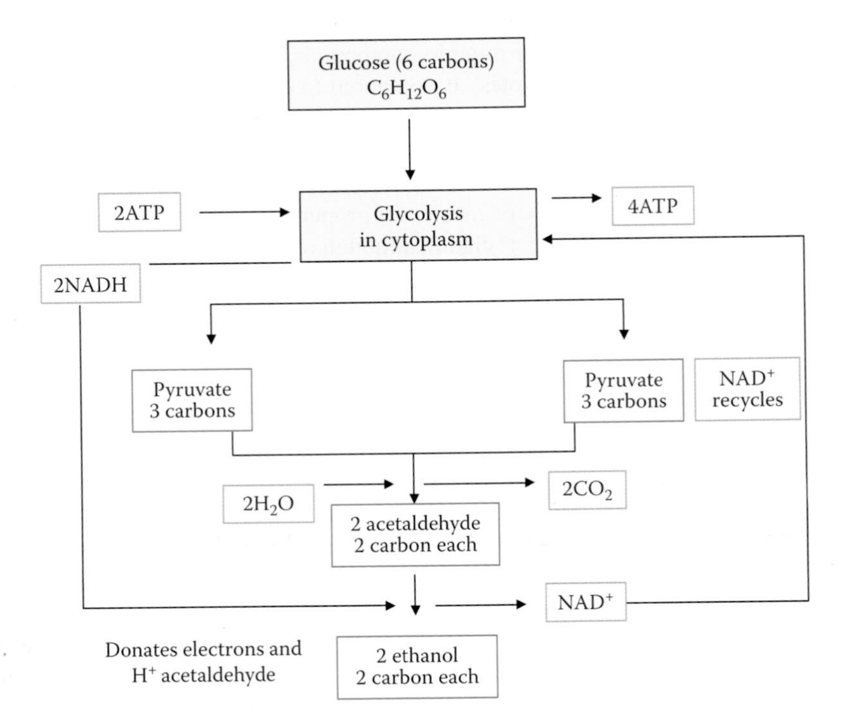

FIGURE 5.6 Bacterial metabolism through fermentation.

5.3.4 Denitrification

Denitrification is the utilization of nitrate (NO_3^-) as a terminal electron acceptor. It is a widespread process that is used by many members of *Proteobacteria*. Many facultative anaerobes use denitrification because nitrate, like oxygen, has a high reduction potential. Many denitrifying bacteria can also use ferric iron (Fe^{3+}) and some organic electron acceptors. Denitrification involves the stepwise reduction of nitrate to nitrite (NO_2^-), nitric oxide (NO), nitrous oxide (N_2O), and dinitrogen (N_2) by the enzymes nitrate reductase, nitrite reductase, nitric oxide reductase, and nitrous oxide reductase, respectively. Protons are transported across the membrane by the initial NADH reductase, quinones, and nitrous oxide reductase to produce the electrochemical gradient critical for respiration. Some organisms (such as *E. coli*) only produce nitrate reductase and therefore can accomplish only the first reduction, leading to the accumulation of nitrite. Others (such as *Paracoccus denitrificans* and *Pseudomonas stutzeri*) reduce nitrate completely. Complete denitrification is an environmentally significant process because some intermediates of denitrification (nitric oxide and nitrous oxide) are important greenhouse gases that react with sunlight and ozone to produce nitric acid, a component of acid rain. Denitrification is also important in biological wastewater treatment, where it is used to reduce the amount of nitrogen released into the environment, thereby reducing eutrophication (Figure 5.7).

5.3.5 Nitrogen Fixation

Nitrogen is an element required for growth by all biological systems. While extremely common in the atmosphere (80% by volume), dinitrogen gas (N_2) is generally biologically inaccessible due to its high activation energy. Throughout all of nature, only specialized bacteria and *Archaea* are capable of nitrogen fixation, converting dinitrogen gas into ammonia (NH_3), which is easily assimilated by all organisms. These prokaryotes are therefore very important ecologically and are often essential for the survival of entire ecosystems. This is especially true in the ocean, where nitrogen-fixing cyanobacteria are often the only sources of fixed nitrogen, and in soils, where specialized symbioses exist between legumes and their nitrogen-fixing partners to provide the nitrogen needed by these plants for growth.

Nitrogen fixation can be found distributed throughout nearly all bacterial lineages and physiological classes, but is not a universal property. The enzyme *nitrogenase* is responsible for nitrogen fixation and is very sensitive to oxygen, which will inhibit it irreversibly; consequently,

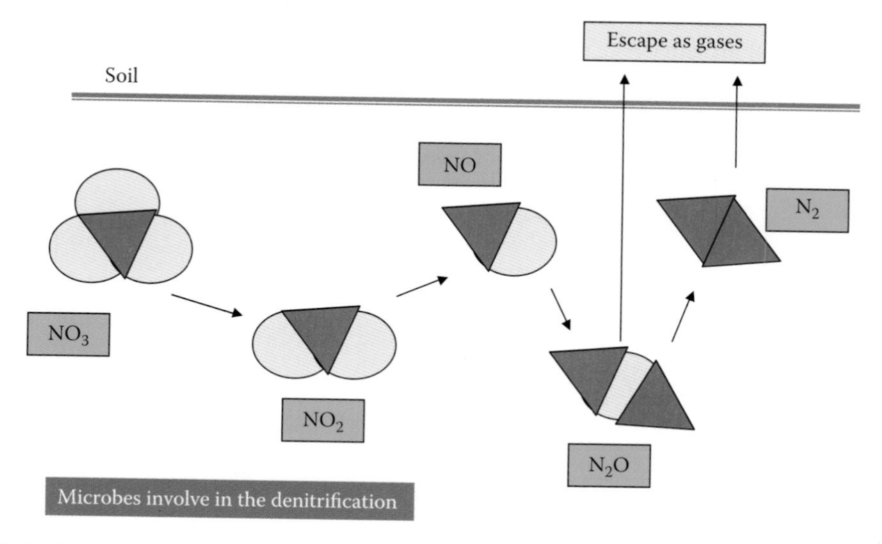

FIGURE 5.7 Denitrification cycle.

all nitrogen-fixing organisms must possess some mechanism to keep the concentration of oxygen low. Some examples of these mechanisms include

- Heterocyst formation (cyanobacteria, e.g., *Anabaena*) where one cell does not photosynthesize but instead fixes nitrogen for its neighbors, which in turn provide it with energy
- Root nodule symbioses (e.g., *Rhizobium*) with plants that supply oxygen to the bacteria bound to molecules of leghemoglobin
- Anaerobic lifestyle (e.g., *Clostridium pasteurianum*)
- Very fast metabolism (e.g., *Azotobacter vinelandii*)

The production and activity of nitrogenases is very highly regulated, both because nitrogen fixation is an extremely energetically expensive process (16–24 ATP are used per N_2 fixed) and due to the extreme sensitivity of the nitrogenase to oxygen.

5.4 MICROBIAL GROWTH

Bacterial growth is the division of one bacterium into two daughter cells in a process called *binary fission*. Providing no mutational event occurs, the resulting daughter cells are genetically identical to the original cell. Hence, "local doubling" of the bacterial population occurs. Both daughter cells from the division do not necessarily survive. However, if the number of surviving cells exceeds unity on average, the bacterial population undergoes exponential growth. The measurement of an exponential bacterial growth curve in batch culture was traditionally a part of the training of all microbiologists. The basic means requires bacterial enumeration (cell counting) by direct and individual (microscopic, flow cytometry, direct and bulk (biomass), indirect and individual (colony counting), or indirect and bulk (most probable number, turbidity, nutrient uptake) methods. Models reconcile theory with the measurements.

5.4.1 Phases of Microbial Growth

The bacterial growth in batch culture can be modeled with four different phases: lag phase, exponential or log phase, stationary phase, and death phase (Figure 5.8).

- During the *lag phase*, bacteria adapt themselves to growth conditions. It is the period where the individual bacteria are maturing and not yet able to divide. During the lag phase of the bacterial growth cycle, synthesis of RNA, enzymes, and other molecules occurs.

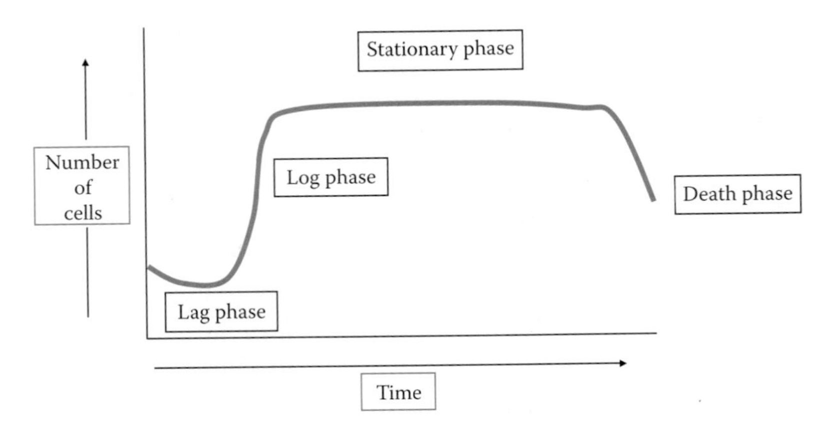

FIGURE 5.8 Phases of microbial growth.

- *Exponential phase* (sometimes called the *log phase*) is a period characterized by cell doubling. The number of new bacteria appearing per unit time is proportional to the present population. If growth is not limited, doubling will continue at a constant rate, so that both the number of cells and the rate of population increase doubles with each consecutive time period. For this type of exponential growth, plotting the natural logarithm of cell number against time produces a straight line. The slope of this line is the specific growth rate of the organism, which is a measure of the number of divisions per cell per unit time. The actual rate of this growth (i.e., the slope of the line in the figure) depends upon the growth conditions, which affect the frequency of cell division events and the probability of both daughter cells surviving. Exponential growth cannot continue indefinitely, however, because the medium is soon depleted of nutrients and enriched with wastes.
- During the *stationary phase*, the growth rate slows as a result of nutrient depletion and accumulation of toxic products. This phase is reached as the bacteria begin to exhaust the resources that are available to them. This phase is a constant value, as the rate of bacterial growth is equal to the rate of bacterial death.
- At the *death phase*, bacteria run out of nutrients and die.

In reality, even in batch cultures, the four phases are not well defined. The cells do not reproduce in synchrony without explicit and continual prompting (as in experiments with stalked bacteria) and their logarithmic phase growth is often not over a constant rate, but instead a slowly decaying rate, a constant stochastic response to pressures both to reproduce and to go dormant in the face of declining nutrient concentrations and increasing waste concentrations. Batch culture is the most common laboratory growth environment in which bacterial growth is studied, but it is only one of many. It is ideally spatially unstructured and temporally structured. The bacterial culture is incubated in a closed vessel with a single batch of medium. In some experimental regimes, some of the bacterial culture is periodically removed and added to a fresh sterile media. In extreme cases, this leads to the continual renewal of the nutrients. This is a chemostat, also known as *continuous culture*. Ideally, it is spatially unstructured and temporally unstructured, in a steady state defined by the nutrient supply rate and the reaction of the bacteria. In comparison to batch culture, bacteria are maintained in the exponential growth phase and the growth rate of the bacteria is known. Related devices include turbidostats and auxostats.

Bacterial growth can be suppressed with bacteriostats, without necessarily killing the bacteria. In a synecological, true-to-nature situation where more than one bacterial species is present, the growth of microbes is more dynamic and continual. Liquid is not the only laboratory environment for bacterial growth. Spatially structured environments such as biofilms or agar surfaces present additional complex growth models.

5.4.2 Factors That Influence Microbial Growth

As a group, microorganisms will grow under many different conditions. There are six main factors that can affect the growth of microorganisms:

- *Food*: Microorganisms grow best in foods that are high in protein or carbohydrates, such as meat, poultry, seafood, milk, rice, and eggs.
- *pH (Acid)*: The measure of acidity or alkalinity of a food also affects the growth of microorganisms. Most disease-causing bacteria multiply best at a pH of 5–8, which is near the neutral pH of 7. Fresh foods such as meat, seafood, and milk tend to have a pH of near 7 (neutral).
- *Temperature*: Food-poisoning microorganisms can multiply rapidly at temperatures between 4°C (40°F) and 60°C (140°F). This is known as the food temperature danger zone.

Hazardous foods should spend as little time as possible in the danger zone. Hot foods should be kept hot (above 60°C or 140°F) and cold foods cold (below 4°C or 40°F).

- *Time*: Microorganisms often need time to grow in the food and they can double in number every 20 min under ideal conditions.
- *Oxygen*: Some microorganisms will only grow when there is oxygen present in the food or environment (aerobic organisms). On the other hand, some microorganisms will only grow when there is NO oxygen present in the food or environment (anaerobic organisms).
- *Moisture*: Microorganisms need water to grow and multiply. However, some microorganisms can survive when there is little water, although they will not be able to grow very well.

5.5 MICROBIAL GENETICS

Most bacteria have a single circular chromosome that can range in size from only 159,662 bp in the endosymbiotic bacteria *Candidatus Carsonella ruddii*, to 13033,779 bp in the soil-dwelling bacteria *Sorangium cellulosum* (Figure 5.9). Spirochaetes of the genus *Borrelia* are a notable exception to this arrangement, with bacteria such as *Borrelia burgdorferi*, the cause of Lyme disease, containing a single linear chromosome. The genes in bacterial genomes are usually a single continuous stretch of DNA, and although several different types of introns do exist in bacteria, these are much rarer than in eukaryotes. Its length is found to be 1100 μ and its molecular weight is 2.6 × 109 Da. In addition to this major chromosome, an *E. coli* cell often possesses one or more minor chromosomes, each called a plasmid, which may contain 0.5%–2% of the DNA of the cell. Plasmids usually maintain a distinct existence from the main chromosome and replicate independently of it. Thus, in *E. coli*, the transmission of more than one genetic element from parent to offspring must frequently be followed. Finally, though bacteria reproduce chiefly by asexual reproductive means, there are a number of different avenues by which DNA from one bacterial cell can undergo genetic exchange with the DNA from another bacterial cell.

5.5.1 MUTATIONS

Mutation is a natural phenomenon resulting in variations within any population of cells. This is a change in the DNA sequence of a gene and is said to lead to a change in the genotype of the organism. There are various types of mutants found in microorganisms (Figure 5.10).

5.5.1.1 Auxotropic Mutant

Microbes such as *E. coli* can grow on a medium containing a single carbon and energy source. They are called *prototrophs*. If a mutation occurs which results in the loss of the ability to synthesize an

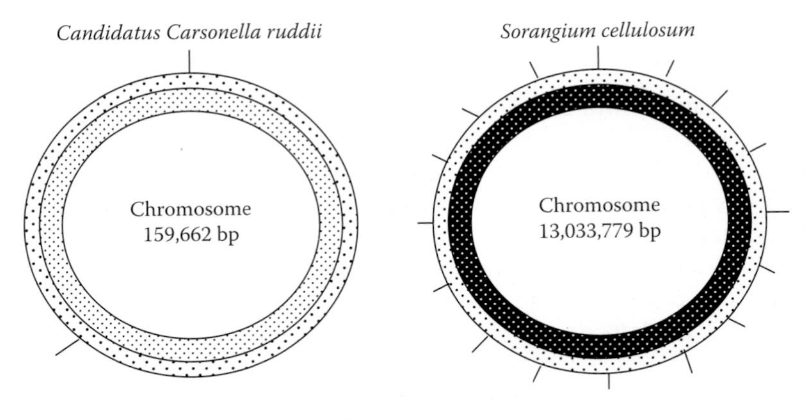

FIGURE 5.9 Chromosomes in different microbes.

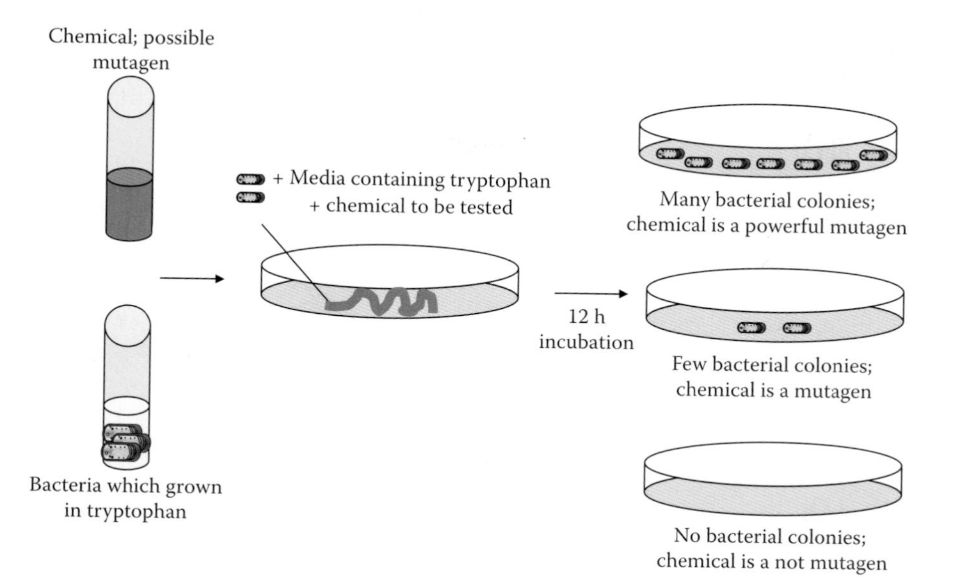

FIGURE 5.10 Mutagenic test using microbes.

essential metabolite such as an amino acid or a growth factor, it will express itself in a nutritional requirement for that substance. Such mutants are called *auxotrophs*. These proved very useful in establishing the relationship between genes and enzymes as markers for genetic experiments in construction of genetic maps.

5.5.1.2 Resistant Mutant

Bacteria may develop resistance to antibiotics and phages spontaneously, through a range of mechanisms such as loss of cell surface components that act as phage receptors and acquisition of enzymes that are able to metabolize the antibiotic.

5.5.1.3 Metabolic Mutant

These mutants have lost the ability to use a particular carbon source and are usually affected in either transport or metabolism. Mutants that have lost the capacity to make specific cell components, such as the capsular polysaccharides, and some of those that exhibit altered colonial shape (arising from mutations affecting cell wall synthesis) also belong to this category.

5.5.1.4 Regulatory Mutant

In these mutants, mutation affects either the regulatory region of the promoter of the gene or the activity of a regulatory protein.

5.5.2 Spontaneous Mutations

In 1943, experiments conducted by Salvador Luria and Max Delbruck showed that mutations occur spontaneously in bacteria. These workers used an *E. coli* strain susceptible to a bacteriophage, but which would also yield phage-resistant variants. Luria and Delbruck then wanted to know whether resistance in this bacterium to the bacteriophage arose as a result of exposure to the phage, or whether such resistant clones are always present in the bacterial population but are selected only in the presence of the phage.

To verify this, one half of a dilute suspension of *E. coli* was dispensed in 1 mL aliquots in individual tubes, while the other half was left in the flask. The cultures were allowed to grow until the cell number had increased to approximately 108 mL. Each sample in the tube and the culture in

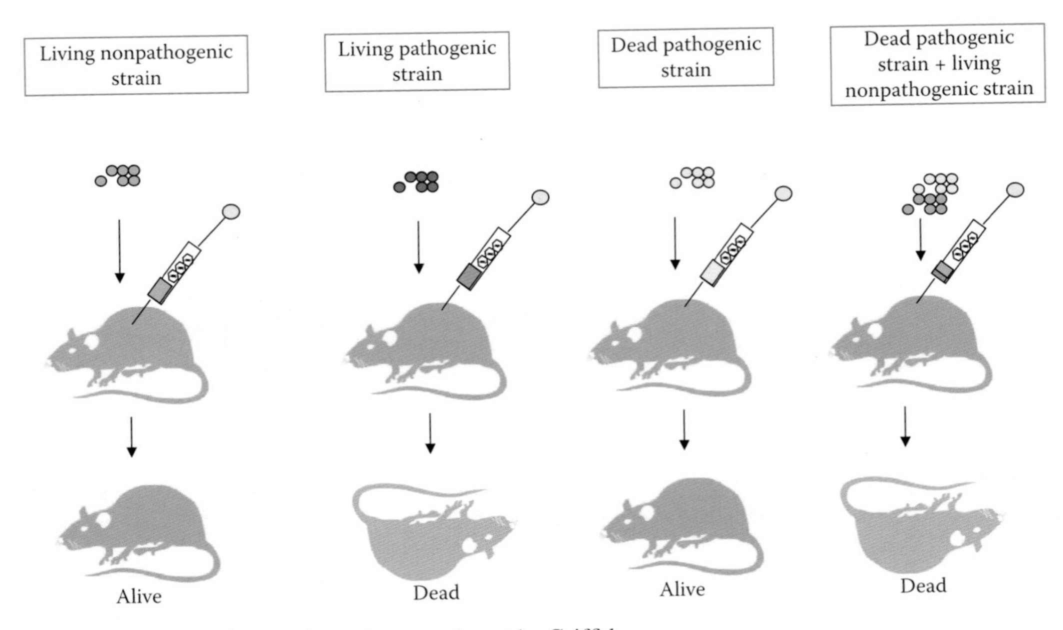

FIGURE 5.11 Genetic transformation experiment by Griffith.

the flask were then exposed to a phage and, after a while, 1 mL aliquot was plated to determine the number of phage-resistant mutants (survivors). Luria and Delbruck predicted that if resistance development to the bacteriophage occurred after the bacteria were exposed to the phage, then all the samples should contain the same number of resistant cells. Consequently, each plate should contain the same number of resistant clones. On the other hand, if resistance to the phage was the result of a spontaneous mutation, then the number of resistant colonies on each plate should vary. Luria and Delbruck found that the number of resistant mutants fluctuated from sample to sample, indicating that mutants existed in the population prior to exposure to the phage, and they concluded that mutations in bacteria occur spontaneously.

These conclusions were further corroborated by Josua Lederberg and Esther Lederberg by the use of the replica plate technique. The technique consists of first plating a small number of the test bacteria on a master plate and incubating them until growth occurs. A circular piece of wood of the size equal to the inner diameter of the Petri plate is then covered with a piece of sterile velvet cloth, and the master plate is then applied to the velvet so that the clones on the plate are transferred to the cloth. The velvet pad is then gently imprinted on several plates containing an inhibitor or a selective agent to detect resistant mutants. It was reasoned that if resistant mutants had developed on the master plate before exposure to the inhibitor, then such resistant colonies should be located at exactly the same position on each of the replica plates, while if mutations occurred as a consequence of exposure to the inhibitor, then the resistant colonies should be at different locations on different replica plates. The Lederbergs found precisely the former, proving that spontaneous mutations occur in bacteria in the absence of a selective agent. Since then, this replica plate technique has been extensively used as a basic technique for mutant detection (Figure 5.11).

5.5.3 Induction of Selective Mutations

As physical and chemical mutagens cause random mutations, obtaining a desired mutation is difficult and time consuming. The discovery of transposable genetic elements has enabled the isolation of selective mutations. This is possible because some of the transposable elements carry along with them markers that allow selection, such as resistance to antibiotics, which allows easy detection. Thus, by use of transposable elements, mutations can not only be induced, but can also be selected in any microbial function. The Mu bacteriophage has also been used as a mutagenic agent because of its ability to integrate itself with any part of the bacterial genome and thereby cause mutations. Recently, a technique by which synthetic nucleotide blocks can be inserted into different regions of gene sequences (block mutations), such as initiators, has been developed. This technique is useful in understanding the functions of DNA sequences in genes that cannot be understood otherwise.

5.5.4 Induced Mutations

Ultraviolet and nitrous acid are the most common agents used for inducing mutations. One base analog is 5-bromouracil. Other examples of mutagens are benzopyrene (in industrial soot and smoke) and aflatoxin (a fungal toxin found in animal products and foods).

5.6 GENETIC RECOMBINATION IN BACTERIA

Genetic changes due to mutations can result in the acquisition of new biological characteristics and thereby allow evolutionary change. However, evolution of the fittest organism in a particular environment can be enhanced if transfer of genes between organisms is made possible by genetic recombination. As compared with eukaryotes, where sexual recombination is of ordered nature, the process is less well developed in prokaryotes. It does not involve a true fusion of male and female gametes to produce a diploid zygote; instead, there is transfer of only some genes from the donor cell to produce a partial diploid. This is followed by recombination to restore the haploid state. There are three mechanisms by which these DNA fragments can pass from a donor to a recipient cell: transformation, transduction, and conjugation.

5.6.1 Bacterial Transformation

In the 1940s, it was recognized that inheritance in bacteria was basically governed by the same mechanisms as those in higher eukaryotic organisms. It was also realized that bacteria represent a useful tool to understand the mechanism of heredity and genetic transfer and were therefore being increasingly used in genetic studies. The first observation that bacterial properties can be changed by the use of heat-inactivated cell material was, however, discovered in 1928 by Frederick Griffith. Griffith found that in *Streptococcus pneumoniae* (earlier called *Pneumococcus*), virulence to mice was related to the presence of a capsular material, and loss of the ability to produce the capsule made the bacteria virulent.

Mutants lacking the capsular material were designated as rough (R), because colonies formed by these on solid media appeared rough, as opposed to the colonies formed by the virulent capsule-forming strains which were smooth and shining (S). Griffith's experiments involved the infection of mice with heat-killed and living preparations from two different strains of *Streptococcus pneumoniae*. When he injected the mice with either the dead S cells or a small number of living R cells, no death occurred. However, when the mice were injected with a mixture of dead S cells and a small number of live R cells, the mice died (Figure 5.11).

From these experiments, he concluded that the dead S cells which contained the capsule contributed to the killing effect by the R cells, since neither of the preparations was effective by itself. Although these observations were not well understood at that time, the term "transformation" was used to describe this phenomenon whereby one type of cells were converted by contact with the dead cells of a second strain. The material responsible for causing transformation was thought to be the capsular polysaccharide.

However, in 1943, the material responsible for bringing about this change was identified. It was left to Avery, McLeod, and McCarty in 1944 to identify the transforming principle in capsulated cells as the DNA. Their studies with purified DNA from the smooth cells of pneumonia and its ability to transform rough cells in a test tube explained the observations made by Griffith in 1928.

It was then possible to conclude that the heat-killed encapsulated cells carried the information for the synthesis of the capsule that was transferred to the live non-capsulated cells. Consequently, cells that received the genetic material for capsule formation became encapsulated and virulent.

At that time, this remarkable finding did not receive as much attention as it should have, since most believed that proteins rather than nucleic acids were the genetic elements and those proteins in the DNA preparations were responsible for bringing about transformation. Since then, however, using highly purified DNA preparations and other genetic markers, it has been shown beyond doubt that the transforming principle is DNA and not protein. The process of transformation has now also been demonstrated in several other bacteria, such as *Bacillus subtilis*, *Haemophilus influenzae*, *Rhizobium*, *E. coli*, *Streptococcus*, and *Streptomyces*. The process of transformation in all these organisms has two common features: (i) the purified donor DNA is first transported across the cell membrane into the recipient "competent cells" (cells that can take up DNA), and (ii) the DNA then undergoes recombination with the recipient DNA and is then expressed.

The uptake process is apparently not very specific, since it has been found that even calf thymus DNA can be taken in by bacterial cells, but the subsequent process of integration is highly specific. Although the double-stranded DNA is necessary for transformation, single-stranded DNA can also penetrate bacterial cells. Following uptake by the recipient cells, the transforming DNA undergoes modifications immediately and an "eclipse" period, lasting for a few minutes, is seen.

During the eclipse period, the donor DNA cannot be recovered from the recipient cells. Using isotopically labeled transforming DNA, it has been shown that during this period, the DNA exists in a single-stranded form. This is followed by the integration of DNA into the recipient DNA in an area of homology. The process of integration apparently involves recombination and the loss of a region of recipient DNA. Although many details are known about this process, our understanding of the transformation process in bacteria is yet difficult to generalize. The frequency of transformation for any single character is rather small, since the amount of DNA that is taken up by the cells and integrated is small. Nevertheless, transformations using purified DNA preparations have been useful in locating genetic loci (in genetic mapping) as well as in understanding the effect of a variety of physical and chemical treatments on the biological functioning of DNA (Figure 5.12).

5.6.2 Bacterial Transduction

Bacterial transduction is a process by which the genetic material in bacteria is transferred from one cell to another through the mediation of bacterial viruses (Figure 5.13). This process was first discovered by Norton Zinder and Joshua Lederberg in 1952 during their experiments to see whether the process of conjugation existed in *Salmonella*. In performing the "D" tube' experiments, they found that the recombinants appeared only in one arm of the tube without cell contact. Also, cell free filtrates from one culture could yield recombinants when mixed with the other. The active factor in the filtrate was, however, resistant to Dnase, and this ruled out transformation involving DNA.

Bacteriophages are viruses that parasitize bacteria and use their machinery for their own replication. During the process of replication inside the host bacteria, the bacterial chromosome or plasmid is erroneously packaged into the bacteriophage capsid. Thus, newer progeny of phages may contain fragments of the host chromosome along with their own DNA or entirely contain the host chromosome. When such a phage infects another bacterium, the bacterial chromosome in the phage also gets transferred to the new bacterium. This fragment may undergo recombination with the host chromosome and confer a new property to the bacterium. The life cycle of bacteriophages may be either lytic or lysogenic. In the former, the parasitized bacterial cell is killed with the release of mature phages, while in the latter, DNA gets incorporated into the bacterial chromosome as prophage.

It was subsequently proved that the active component was a bacteriophage that was carried by one of the strains in the prophage condition. Some bacteria have the ability to carry phage DNA within their own DNA, and such bacteria are known as *lysogenic bacteria*. In lysogenic bacteria, the

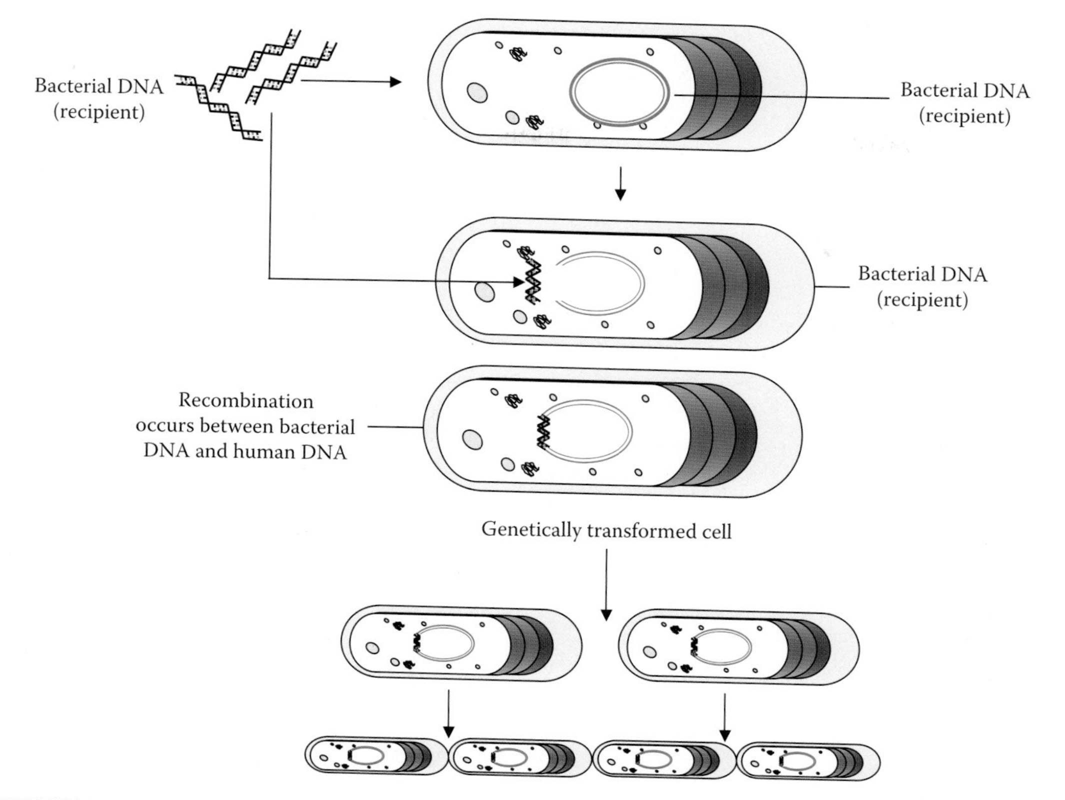

FIGURE 5.12 Mechanism of genetic transformation in bacteria.

prophage becomes active under certain conditions. It multiplies and destroys the host cell with the release of a number of phage particles. The phage particles released from a small number of bacterial cells attack sensitive cells, multiply, and release more phage particles. The lysogenic strains, however, are resistant to the same phage that they carry. Sometimes, when the prophage is released as the vegetative phage, it also carries a small fragment of the host DNA, in addition to its own DNA. These phages can infect other bacteria and carry the bacterial DNA to the recipient cells. Such phages are called *transducing phages* and act as carriers of bacterial DNA from one cell to another. The size of the DNA transferred by transduction is small when compared to either transformation or conjugation, and the amount of DNA is generally less than 1% of the bacterial genome. This technique is therefore useful only in determining the relative positions of very closely located markers and mapping regions within a gene.

5.6.2.1 Types of Transduction

In experiments performed by Zinder and Lederberg with phage P22 and *S. typhimurium* L22, it was found that the phage can carry any part of the bacterial DNA and transfer this to another *Salmonella* strain. The new bacterial DNA thus introduced into the recipient strains was subsequently integrated into the recipient cell DNA. This ability of the bacteriophage to carry with its genetic material any region of the bacterial DNA is now known as *generalized transduction* (or *unrestricted transduction*). As opposed to this, there are bacterial phages, such as the lambda phage (λ) of *E. coli,* which can carry only a specific region of the bacterial DNA to a recipient, and this is called *specialized transduction* (or *restricted transduction*). In recent years, the technique of specialized transduction has been extensively used to determine the fine structure of bacterial genes. Phage preparations obtained by the induction of lysogenic cultures possess transducing properties. However, phages that emerge from the lytic infection are inert. The generation of transducing phages involves a kind

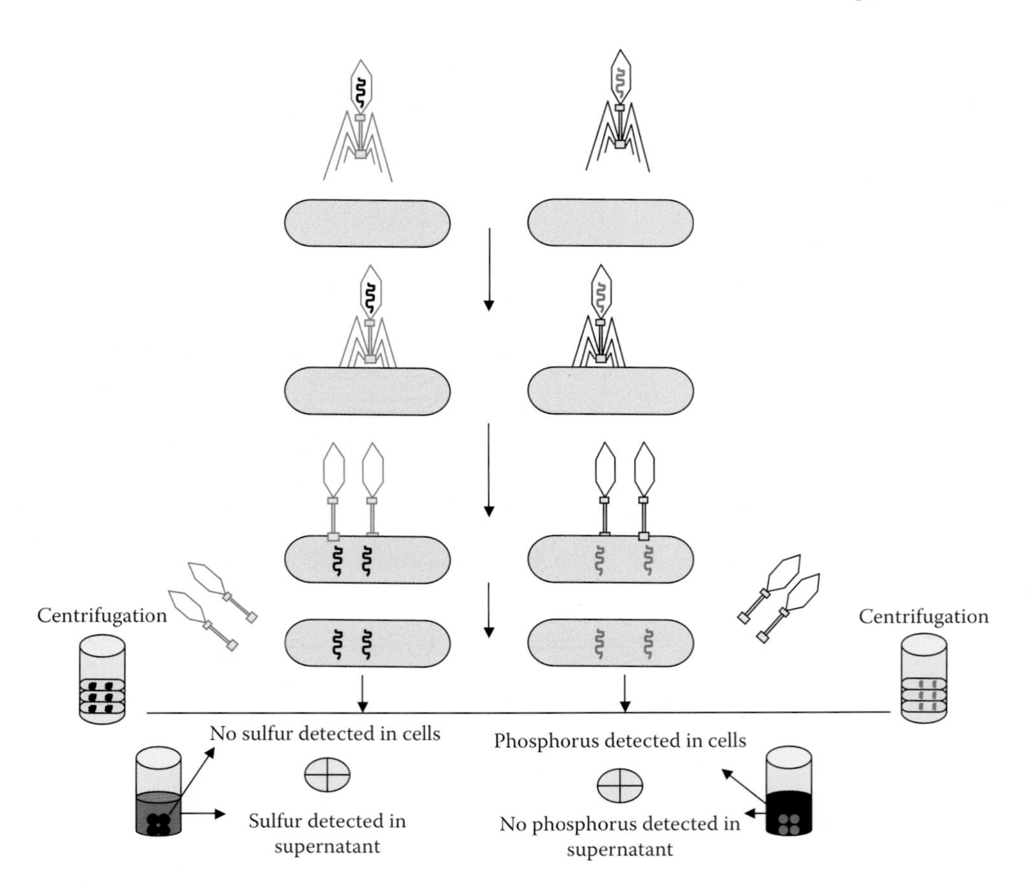

FIGURE 5.13 Genetic transformation by virus transduction.

of recombination between the bacterial and prophage chromosomes and, during later release, a part of the bacterial chromosome is exchanged for a part of the phage genome.

- *Generalized transduction*: Generalized transduction is a more common event. It is mediated by the prophages that have remained in the cytoplasm as plasmids that are not attached to the chromosome. This occurs in the PI phage and in many others. The viral DNA lies in the cytoplasm and produces copies of itself for new phage particles. In doing so, it may accidentally incorporate small chromosomal segments of bacterial DNA and incorporates these to its own DNA. Some phages may accidentally package only bacterial DNA. In most cases, normal viruses will be liberated from the cell. Occasionally, a virus contains several bacterial genes acquired in the chromosomal segments.
- If such a virus infects a new cell, it will attach to the chromosome and transduce the cell as lysogeny is established. In generalized transduction, the viral DNA enters the lytic cycle and forms new virus particles. However, tiny fragments of bacterial chromosome are sometimes incorporated into the DNA of the new viruses or may occasionally replace the viral DNA. This is a random occurrence that may involve any of the bacterial genes, hence the name "generalized transduction." Perhaps one phage in a thousand contains bacterial DNA. All bacterial genes are equally available to be picked up by the phage DNA. When the viral particles are released during lysis, the genes are carried along and, on subsequent infection, the genes enter the cytoplasm of the new host cell where they will now function. The phenomenon of lysogeny is well established in modern microbiology. Diphtheria organisms are known to contain bacteriophages that code for the toxin produced during disease. Herpes simplex viruses remain as prophages in the cytoplasm of

the body cells for many years, expressing themselves at long intervals. Certain viruses are known to attach to human chromosomes, transforming the cells to tumor cells.

- *Specialized transduction*: The bacterium may remain lysogenic for many generations, during which time the viral DNA replicates together with the bacterium. However, at some point in the future, the phage will stop coding the repressor protein, and the lytic cycle will begin. The viral DNA that was attached to the chromosome will now break free and direct the synthesis of those proteins that will yield new viruses. In detaching, however, the viral DNA may carry with it a few bacterial genes from the chromosome. The genes are then replicated along with the viral DNA and they become part of the new phage particles. When the latter are released, copies of the genes are carried along. As the cycle repeats during the next infection, phage DNA enters the new bacterial cells and inserts onto a new chromosome. However, copies of the original bacterial genes are included, and the bacterium becomes transduced. The bacterial cell now contains its own genes plus several genes from the original cell. This type of transduction is called *specialized transduction* because specific genes are removed from the bacterial chromosome, depending upon where the viral DNA was attached. This occurs in lambda phage. The removal of genes, however, is thought to be an extremely rare event.
- *Abortive transduction*: Fusion between *Enterobacteria* is common, but the formation of stable heterokaryons or diploids is not known. This is perhaps due to the difficulty in transferring the entire genome from one cell to another. After conjugation, partial zygotes (merozygotes) are formed, but these are transient and therefore are not very suitable for complementation tests. One of the systems available in bacteria to test complementation is *abortive transduction*. In this parasexual mechanism, modified temperate bacteriophages act as vectors of small fragments of bacterial DNA, transporting the DNA from a donor cell in which the phage is grown to a recipient strain that the phage can infect. However, the transduced bacterial segment fails to undergo recombination and replication, but remains functional and is transferred during cell division to only one daughter cell. In *Salmonella*, motility is one of the many characters that can be transduced by the phage P22. When P22 is grown on a motile donor strain, transducing phages which can infect nonmotile recipient bacteria can be obtained. Motile transductants can be isolated by plating the infected nonmotile recipients on soft gelatin agar in which the growth of nonmotile organism is confined and compact, while motile cells migrate outwards as they multiply to give an expanding "flare" of growth.
- Stocker, Zinder, and Lederberg noticed that in addition to flares, there are sometimes a number of linear trails of isolated colonies leading out from the outer region of confined growth. These trails were explained on the basis that the motility gene transferred through transduction does not participate in recombination, but is functional and capable of conferring motility. When such cells divide, only one daughter cell in which the gene is present is able to move away from the parent, forming tiny, compact colonies. This process of unilinear inheritance of the transferred gene continues, leaving a trail of nonmotile celled colonies until the gene is lost for reasons that are not well known. Using single cell studies, this mechanism was confirmed as *abortive transduction*. Later, in 1956, Ozeki also observed the same type of phenomenon in transduction of nutritional characters in *Salmonella*.

5.6.2.2 Stages of Transduction

The following are the stages of transduction:

- A lytic bacteriophage adsorbs to a susceptible bacterium.
- The bacteriophage genome enters the bacterium. The phage DNA directs the bacterium's metabolic machinery to manufacture bacteriophage components and enzymes.

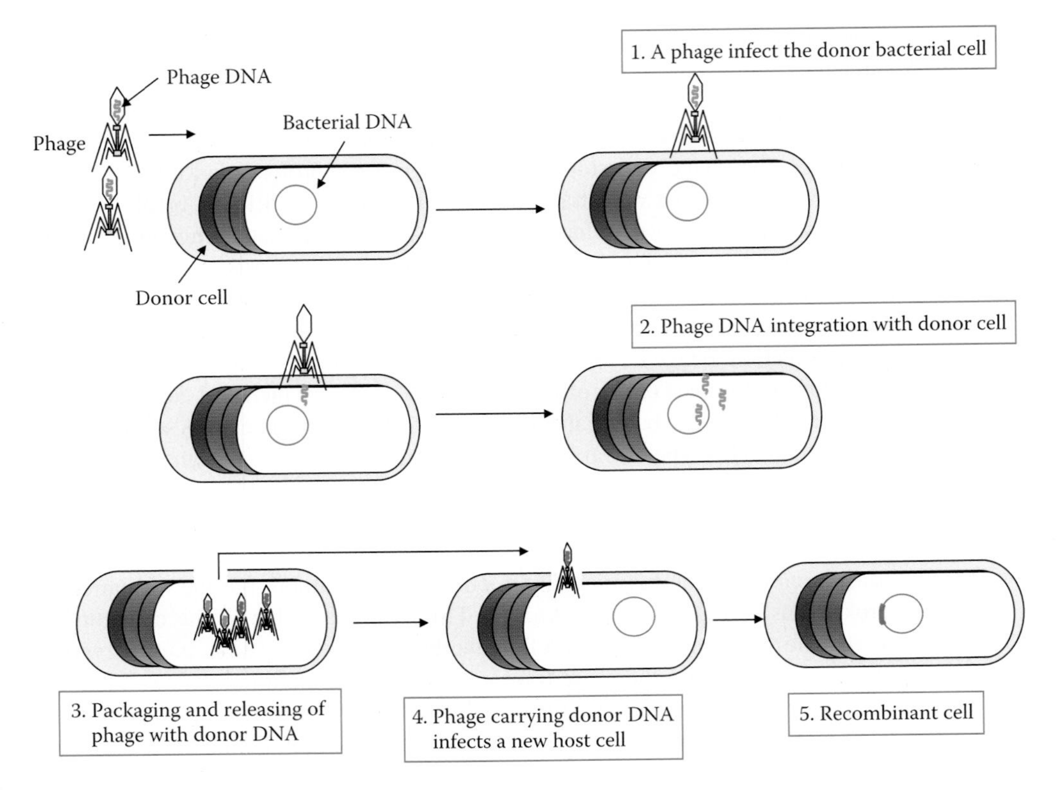

FIGURE 5.14 Genetic transformation transduction process.

- Occasionally during maturation, a bacteriophage capsid erroneously incorporates a fragment of the donor bacterium's chromosome or a plasmid instead of a phage genome.
- The bacteriophages are released with the lysis of bacterium.
- The bacteriophage carrying the donor bacterium's DNA adsorbs to another recipient bacterium.
- The bacteriophage inserts the donor bacterium's DNA that it is carrying into the recipient bacterium.
- The donor bacterium's DNA is exchanged by recombination for some of the recipient's DNA (Figure 5.14).

5.6.3 Conjugation Mechanism in Gene Recombination

Literature in bacterial morphology contains many descriptions of microscopic observations of cell pairs which were identified as indicators of mating and sexuality in bacteria. However, no confirmatory genetic evidence was available till the discovery of conjugation in *E. coli* by Lederberg and Tatum in 1946, in a study where they mixed auxotrophic mutants and selected rare recombinants. In their initial experiments, Lederberg and Tatum plated *E. coli* mutants having triple and complementary nutritional requirements (abcDEF X ABCdef) on minimal agar and obtained prototrophic bacteria (ABCDEF).

These recombinants were stable, could be propagated, and arose at a frequency of 10^{-6}, 10^{-7}. Further evidence to show that the development of prototrophic colonies required the cooperation of intact bacteria of both types was obtained by the "U" tube experiments. Neither the culture filtrates nor the cell-free culture extracts were productive, suggesting that actual cell contact was necessary. Lederberg also examined a large number of the prototrophic colonies to learn whether the process was reciprocal. He found that most colonies contained only one class of recombinants, suggesting

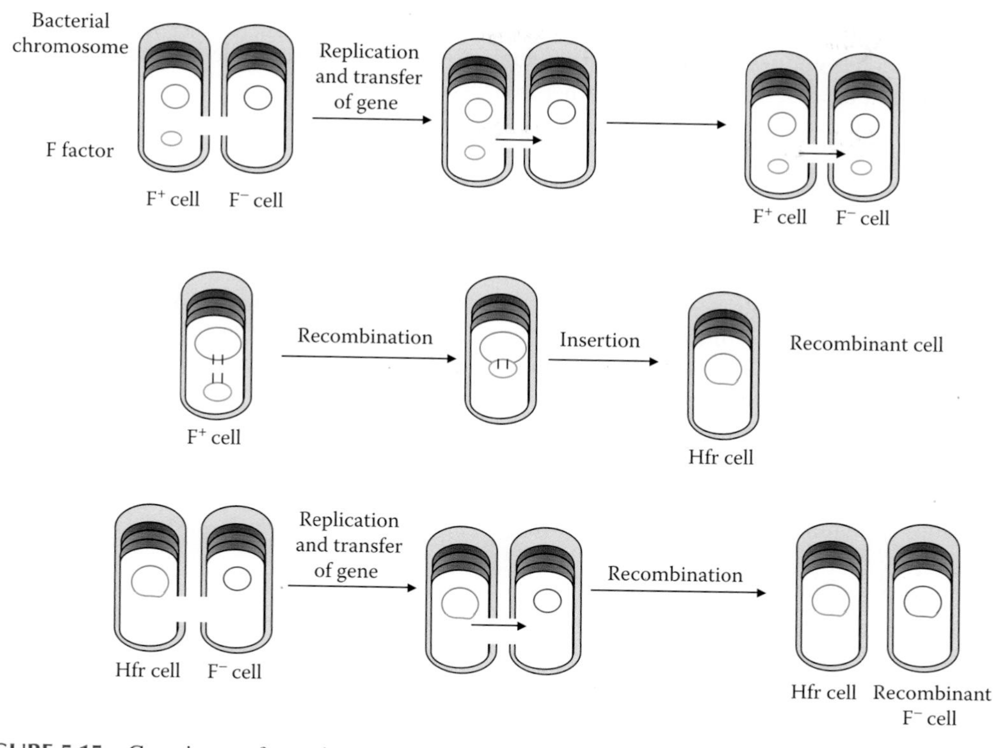

FIGURE 5.15 Genetic transformation conjugation process.

that recombination in bacteria may be of an unorthodox kind. In addition, detailed analysis of prototrophs showed an initial heterozygous nature, but later was converted to haploids.

These studies by Lederberg and his colleagues proved that bacteria possessed sex, which made them amenable to formal genetic analysis and also revealed the existence of genetic material in a chromosomal organization. Subsequent studies carried out to determine the size of the DNA fragment involved, by detecting the number of genetic markers transferred, suggested that more than one marker could be transferred at a time and, interestingly, linkage between certain markers was always seen. In this process of conjugation, it was concluded that (i) large fragments of DNA were transferred from one bacterium to another in a nonreciprocal manner, and (ii) transfer always occurred from a given point. It was also found that the size of the DNA transferred from one cell to another was much larger than in transformation, and this technique appeared to be a more useful technique for gene mapping in bacteria. The bacteria that transfer DNA are called *donor bacteria*, while those that receive the DNA are called *recipient bacteria* (Figure 5.15).

5.7 TRANSPOSABLE GENETIC ELEMENTS

Transposable genetic elements are segments of DNA that have the capacity to move from one location to another (i.e., jumping genes). Transposable genetic elements can move from any DNA molecule to any DNA of another molecule, or even to another location on the same molecule. The movement is not totally random. There are preferred sites in a DNA molecule at which the transposable genetic element will insert. The transposable genetic elements do not exist autonomously; thus, to be replicated, they must be a part of some other replicon. Transposition requires little or no homology between the current location and the new site. The transposition event is mediated by an enzyme, transposase, which is coded by the transposable genetic element. Recombination that does not require homology between the recombining molecules is called *illegitimate* or *nonhomologous recombination*. In many instances, transposition of the transposable genetic element results in the removal of the element from the original site and insertion at a new site. However, in some cases, the transposition event is accompanied by the duplication of the transposable genetic element. One copy remains at the original site and the other is transposed to the new site.

5.7.1 Types of Transposable Genetic Elements

5.7.1.1 Insertion Sequences

Insertion sequences are transposable genetic elements that carry no known genes except those that are required for transposition. Insertion sequences are small stretches of DNA that have at their ends repeated sequences, which are involved in transposition. In between the terminal repeated sequences, there are genes involved in transposition and sequences that can control the expression of the genes, but no other nonessential genes are present. The introduction of an insertion sequence into a bacterial gene will result in the inactivation of the gene. The sites at which plasmids insert into the bacterial chromosome are at or near the insertion sequence in the chromosome. In *Salmonella*, there are two genes that code for two antigenically different flagellar antigens. The expression of these genes is regulated by an insertion sequence.

5.7.1.2 Transposons

Transposons are transposable genetic elements that carry one or more other genes in addition to those which are essential for transposition. The structure of a transposon is similar to that of an insertion sequence. The extra genes are located between the terminal repeated sequences. Many antibiotic resistance genes are located on transposons. Since transposons can jump from one DNA molecule to another, these antibiotic-resistant transposons are a major factor in the development of plasmids, which can confer multiple drug resistance on a bacterium harboring such a plasmid. These multiple drug resistance plasmids have become a major medical problem.

5.8 USE OF *E. COLI* IN MICROBIAL CLONING

The microorganism *Escherichia coli* has a long history of use in the biotechnology industry and is still the microorganism of choice for most gene cloning experiments. Although *E. coli* is known to the general population for the infectious nature of one particular strain (0157:H7), few people are aware of how versatile and useful *E. coli* is to genetic research. There are several reasons *E. coli* became so widely used and is still a common host for recombinant DNA.

5.8.1 Genetic Simplicity

Bacteria make useful tools for genetic research because of their relatively small genome size compared to eukaryotes. *E. coli* cells only have about 4,400 genes, whereas the human genome project has determined that humans contain approximately 30,000 genes. Also, bacteria, including *E. coli*, live their entire lifetime in a haploid state, with no second allele to mask the effects of mutations during protein engineering experiments.

5.8.2 Growth Rate

Bacteria typically grow much faster than more complex organisms. *E. coli* grows rapidly, at a rate of one generation per 20 min under typical growth conditions. This allows for preparation of log-phase (mid-way to maximum density) cultures overnight and genetic experimental results in mere hours instead of several days, months, or years. Faster growth also means better production rates when cultures are used in scaled-up fermentation processes.

5.8.3 Safety

E. coli is naturally found in the intestinal tracts of humans and animals, where it helps provide nutrients (vitamins K and B12) to its host. There are many different strains of *E. coli* that may produce toxins or cause varying levels of infection if ingested or allowed to invade other parts of the body. Despite the bad reputation of one particularly toxic strain (O157:H7), *E. coli* is generally relatively innocuous if handled with reasonable hygiene.

5.8.4 Conjugation and the Genome Sequence

The *E. coli* genome was the first to be completely sequenced. Genetic mapping in *E. coli* was made possible by the discovery of conjugation. *E. coli* is the most highly studied microorganism, and an advanced knowledge of its protein expression mechanisms makes it simpler to use for experiments where expression of foreign proteins and selection of recombinants is essential.

5.8.5 Ability to Host Foreign DNA

Most gene cloning techniques were developed using *E. coli,* and these techniques are still more successful and effective when using *E. coli* rather than other microorganisms. *E. coli* is readily transformed with plasmids and other vectors, easily undergoes transduction, and preparation of competent cells (cells that will take up foreign DNA) is not complicated. Transformations with other microorganisms are often less successful.

5.9 PATHOGENIC BACTERIA

If bacteria form a parasitic association with other organisms, they are classed as *pathogens*. Pathogenic bacteria are a major cause of human death and disease, and cause infections such as tetanus, typhoid fever, diphtheria, syphilis, cholera, food-borne illness, leprosy, and tuberculosis. A pathogenic cause for a known medical disease may only be discovered after many years, as was the case with *Helicobacter pylori* and peptic ulcer disease. Bacterial diseases are also important in agriculture, with bacteria causing leaf spot, fire blight, and wilts in plants, as well as John's disease, mastitis, salmonellosis, and anthrax in farm animals.

Each species of pathogen has a characteristic spectrum of interactions with its human hosts. Some organisms, such as *Staphylococcus* or *Streptococcus*, can cause skin infections, pneumonia, meningitis, and even overwhelming sepsis, a systemic inflammatory response producing shock, massive vasodilation, and death. Yet these organisms are also part of the normal human flora and usually exist on the skin or in the nose without causing any disease at all. Other organisms invariably cause disease in humans, such as the *Rickettsia*, which are obligate intracellular parasites able to grow and reproduce only within the cells of other organisms. One species of *Rickettsia* causes typhus, while another causes Rocky Mountain spotted fever. *Chlamydia*, another phylum of obligate intracellular parasites, contains species that can cause pneumonia or urinary tract infection and may be involved in coronary heart disease. Finally, some species such as *Pseudomonas aeruginosa*, *Burkholderia cenocepacia*, and *Mycobacterium avium* are opportunistic pathogens and cause disease mainly in people suffering from immunosuppression or cystic fibrosis.

Bacterial infections may be treated with antibiotics, which are classified as bactericidal if they kill bacteria, or bacteriostatic if they just prevent bacterial growth. There are many types of antibiotics, and each class inhibits a process that is different in the pathogen from that found in the host. Examples of how antibiotics produce selective toxicity are chloramphenicol and puromycin, which inhibit the bacterial ribosome, but not the structurally different eukaryotic ribosome. Antibiotics are used both in treating human disease and in intensive farming to promote animal growth, where they may be contributing to the rapid development of antibiotic resistance in bacterial populations.

Infections can be prevented by antiseptic measures, such as sterilization of the skin prior to piercing it with the needle of a syringe, and by proper care of indwelling catheters. Surgical and dental instruments are also sterilized to prevent contamination by bacteria. Disinfectants such as bleach are used to kill bacteria or other pathogens on surfaces to prevent contamination and further reduce the risk of infection.

5.10 APPLICATION OF MICROBES

The economic importance of bacteria derives from the fact that bacteria are exploited by humans in a number of beneficial ways. Despite the fact that some bacteria play harmful roles, such as causing disease and spoiling food, the economic importance of bacteria includes both their useful and harmful aspects. Applications of microbes in industry are well known. Various microorganisms are used for commercial production of alcohols, acids, fermented foods, vitamins, medicines, enzymes, etc. One recent development in industrial microbiology has been the production of immobilized enzymes and cells for production of these chemicals at enhanced rates, with simultaneous recovery of the enzyme(s) involved in such processes. New strains of microbes have also been developed through recombinant DNA technology for overproduction of metabolites. Immobilized enzymes and cells could have their maximum application in industrial microbiology. Immobilized enzymes have also been utilized in medicine. In view of their various applications, microbes have become the most sought after entities for various applications.

5.10.1 Microbes and Agriculture

Besides being important in biogeochemical cycling of nutrients, microbes play vital role in maintenance of soil fertility and in crop protection. Microbes are being exploited in two important ways: as biofertilizers and for creating new nitrogen-fixing organisms. The potential of Rhizobium, Azotobacter, Beijerinckia, Azospirillum, Cyanobacteria, such as species of Aulosira, Anabaena, Nostoc, Plectonema, Scytonema, Tolypothrix, and Azalia, as biofertilizers has been exploited so as these could serve as alternatives to chemical fertilizers. Many brands of rhizobial inoculants are already in the market in the country today, and several organizations and manufacturers are producing huge quantities of Rhizobium culture. These include Micro Bac., India, Shyam Nagar, Parganas; Bacifil Inoculants, Lucknow; Govt. of Tamil Nadu; Nitro Fix Industries, Calcutta (W. Bengal); and Indian Organic Chemical Ltd., Bombay. In some other states, units are being prepared for an increase in the production of Rhizobium. Much progress has also been made with cyanobacteria in this direction. Mycorrhizae, both ecto and endomycorrhiza, help in uptake of N, P, K, and Ca. They particularly help in phosphorous nutrition.

5.10.2 Nitrogen Fixers

Through recombinant DNA technology, efforts have been made to introduce nitrogen-fixing (nif) genes into wheat, corn, rice, etc. Plasmids of the bacterium *E. coli* and yeast are being worked out for such a possibility. Hybrid *E. coli* plasmid cloned with nif genes of a nitrogen-fixing bacterium *Klebsiella pneumoniae* and hybrid yeast plasmids are then integrated.

5.10.3 Biopesticides and Bioweedicides

Several microbes are being developed as suitable biopesticides for management of insect and nematode pests. Some fungi have good potential for their use as bionematicides to control nematode pests of vegetables, fruit, and cereal crops. Some bacterial and fungal products are also in use to control diseases of roots and shoots of plants. Several fungi have been found very useful in the control of troublesome weeds in crop fields. Registered products are available for use in the market in several countries.

The bacteria are used as a Lepidopteran-specific insecticide under trade names such as Dipel and Thuricide. Because of their specificity, these pesticides are regarded as environmentally friendly, with little or no effect on humans, wildlife, pollinators, and most other beneficial insects. The bacterium *Bacillus thuringiensis* has been the most successful bioinsecticide so far. Several registered products of different strains of this microbe, such as Thuricide, are available for the control of insects, including mosquitoes—the carriers of malaria.

5.10.4 Acetone Butanol Fermentations

The *acetone butanol fermentation* is one of the oldest types of fermentation known. The fermentation is based on culturing various strains of clostridia in carbohydrate-rich media under anaerobic conditions to yield butanol and acetone. *Clostridium acetobutylicum* is the organism of choice in the production of these organic solvents. These fermentations were out of favor until very recently because of the availability of acetone and butanol from the petroleum industry. Today, there is considerable amount of interest in these fermentations. However, the concentration of end products in these fermentations is quite small, and the fermentations are a type of mixed fermentation yielding a mixture of compounds such as butyric acid, butanol, acetone, etc. Attempts to increase yields by using genetically altered strains or change in fermentation conditions have been partially successful.

5.10.5 Microbes in Recovery of Metals and Petroleum

Recently, microbes have been found very useful in enhanced recovery of metals, including uranium from low-grade ores. Through bioleaching, these microbes are able to solubilize the metals from their ores. Microbes thus play an important role in mining and recovery of metals. For instance, *Thiobacillus thiooxidans* and *T. ferrooxidans* can be used in recovery of copper. Microbes are used in tertiary recovery of petroleum. For instance, the bacterium *Xanthomonas campestris* is being exploited for this purpose. Some thiobacilli have also been found to have this potential.

5.10.6 Microbes in the Paper Industry

Mechanical pulping in the process of manufacturing paper from wood needs much energy and also does not preserve the quality of the product. Therefore, the potential of some lignin-decomposing fungi (lignolytic fungi) has been exploited for this process. Biological pulping by use of these higher fungi (*Basidiomycotina*) could find application in the paper industry. *Phanerochaete chrysosporium* has been mostly studied for its use in the biopulping process. Other lignolytic fungi found suitable for this process are *Pholiota mutabilis*, *Tremetes versicolor*, and *Phlebia spp.*

5.10.7 Microbes in Medicine

The production of antibiotics and other chemotherapeutic agents by a range of microbes is well known. Recent development in this area has been the use of microbial biotechnology in steroid transformations and biotransformation of natural penicillin G to several semi-synthetic penicillins. Penicillin acylase, produced by *Saccharomyces cerevisiae* and *Kluyvera citrophila,* is used in biotransformation of penicillin G to semi-synthetic penicillins. Microbial transformation of steroids is very important in the pharmaceutical industry. *Rhizopus nigricans* hydroxylates progesterone, forming another steroid. *Cunninghamella blakesleeana* hydroxylates cortexolone to hydrocortisone.

Microorganisms are used to produce insulin, growth hormone, and antibodies. Diagnostic assays that use monoclonal antibody, DNA probe technology, or real-time PCR are used as rapid tests for pathogenic organisms in the clinical laboratory. Microorganisms may also help in the treatment of diseases such as cancer. Research shows that clostridia can selectively target cancer cells. Various strains of nonpathogenic clostridia have been shown to infiltrate and replicate within solid tumors.

Clostridia therefore have the potential to deliver therapeutic proteins to tumors. *Lactobacillus spp.* and other lactic acid bacteria possess numerous potential therapeutic properties, including anti-inflammatory and anticancer activities.

5.10.8 Microbes in Synthetic Energy Fuels

Several microbes have been found helpful in solving the energy crisis. Some synthetic fuels produced by the activity of microbes include ethanol, methane, hydrogen, and hydrocarbons. Gasohol, a 9:1 blend of gasoline and ethanol, is a popular fuel in the United States. The most efficient microbes used for synthetic energy fuels are *Zymomonas mobilis* and *Thermoanaerobacter ethanolicus*. Methane is produced by methanogenic bacteria and biogas. A mixture of CH_4, CO_2, H_2, N_2, and O_2 is produced during fermentation of cattle dung by several bacteria, including methanogens.

5.10.9 Microbes and Environment Cleaning

Microorganisms play several key roles in the environment. Besides their well-known activities in biogeochemical cycling, soil fertility maintenance, etc., several microbes may prove to be very helpful in the maintenance of environmental quality through biodegradation of wastes (urban, municipal, and industrial) into useful products and also in biodegradation of harmful pesticides used in crop protection and public health.

DOT, lindane heptachlor, chlordane, and malathion are biodegraded by several bacteria and fungi. These microbes are efficient purifiers of the environment. Aside from this, microbes can also remove toxic heavy metals from industrial waste. Some bacteria can also metabolize the hydrocarbons in petroleum and are thus very useful in removal of oil spills and grease from water bodies. In the United States, a strain of *Pseudomonas aeruginosa* has been developed which can produce a glycolipid emulsifier that reduces the surface tension of an oil-water interface, thus removing oil from water. For removal of grease deposits, a mixture of several bacteria is used. Some microorganisms can also be used in environmental monitoring and biomonitoring. Environmental pollutants can be detected by use of appropriate strains of microbes as biosensors. *Biosensor* is a biophysical device used to detect the presence and quantify the specific substances (sugars, proteins, hormones, pollutants) in a specific environment.

5.11 FOOD MICROBIOLOGY

Most foods are excellent media for rapid growth of microorganisms. There is abundant organic matter in foods, their water content is usually sufficient, and the pH is either neutral or slightly acidic. Foods consumed by man and animals are ideal ecosystems in which bacteria and fungi can multiply. The mere presence of microorganisms in foods in small numbers need not be harmful, but their unrestricted growth may render the food unfit for consumption and can result in spoilage or deterioration. Some organisms grow and elaborate secondary metabolites that may affect food quality, which may be either desirable or undesirable. For example, the lactic fermentation of milk is a desired change and is not considered spoilage, while acidification of wine is an undesirable microbial spoilage. Some organisms may not only cause food spoilage, but also produce metabolites that may be extremely toxic to man and animals. Such examples are the production of toxins by clostridia in proteinaceous foods and the elaboration of aflatoxin by aspergilla in feeds. Generally, foods carry a variety of organisms, most of which are saprophytic. Their presence cannot be avoided since these are mostly from the environment in which the food is prepared or processed. In addition, their complete elimination is difficult. However, it is possible to reduce their number or decrease their activities by altering environmental conditions. Knowledge of the factors that either favor or inhibit their growth is therefore essential in understanding the principles of food spoilage and preservation.

5.11.1 Microbes Associated with Food Spoilage

Fruits, vegetables, meat, poultry, seafood, dairy products, and various food products differ in their biochemical composition and are therefore subject to spoilage by different microbial populations. Such changes depend upon the nature of the microbes involved in the spoilage. Thus, degradation of apple juice by yeast gives an alcoholic taste to the juice. Yeasts convert the carbohydrate into ethanol. Bacteria that attack food proteins convert these into amino acids, which are broken down again into foul-smelling end products. Digestion of cystein, for example, yields hydrogen sulfide, giving a rotten egg smell to food. Digestion of tryptophan yields indole and skatole, which give food a fecal odor. Two other products of the microbial metabolism of carbohydrates are (a) acid that causes foods to become sour and (b) gases that cause sealed cans to swell. Digestion of fats, as in spoiled butter, yields fatty acids, giving a rancid odor or taste to food. Food may become slimy due to production of capsules in bacteria. There may be pigment development, giving an odd color to foods.

5.11.2 Meat and Fish

Microorganisms that cause meat and fish spoilage are usually introduced during handling, processing, packaging, and storage. For example, if a piece of meat is ground, the surface organisms accumulate in the teeth of the grinder along with other dust-borne organisms. Bacteria from the hands of the preparer or from an errant sneeze may add more microbes. Processed meats may become contaminated during handling. For example, sausages, which are made from animal intestines, may contain residual bacteria, especially botulism spores. Organ meats such as liver and kidney spoil quickly, and may contain many bacteria trapped in their filtering tissues. Greening on meat surfaces is usually due to the Gram-positive rod, *Lactobacillus*, or the Gram-positive coccus, *Leuconostoc*.

5.11.3 Poultry and Eggs

There are reports which suggest that *Salmonella* can cause diseases in human by consuming infected chickens and turkeys or poultry or eggs. Processed foods such as potpies, egg salad, and omelet may be sources of salmonellosis. Chicken products may become contaminated due to improper handling or from the water used for cleaning the product. Eggs may become contaminated by *Proteus* causing black rot (here, hydrogen sulfide accumulates due to digestion of egg cysteine), by *Pseudomonas* causing green rot, and by *Serratia marcescens* causing red rot. The yolk is the main part of the egg prone to contamination (the white of an egg is inhibitory to Gram-positive bacteria due to the presence of the inhibitor enzyme lysozyme).

5.11.4 Breads and Bakery Products

The ingredients of bread products such as flour, egg, sugar, and salt are usually the sources of spoilage organisms. Some bacteria and molds are able to survive the baking temperatures. Some species of *Bacillus* give the bread a soft and cheesy texture, with long, stringy threads. This kind of bread is called *ropy*. Cream rolls, custards from whole eggs, and whipped cream are good media for growth of *Salmonella*, *Lactobacillus*, and *Streptococcus* species, which produce acids.

5.11.5 Other Foods

Many cereals, fruits, and vegetables are spoiled by microorganisms. The chief agents of spoilage are molds and bacteria, giving foods an unpleasant odor. Grains are spoiled by a mold, *Aspergillus flavors*, which is also present in peanut products and other foods. The mold forms a toxin called *aflatoxin*. *Claviceps purpurea* is also an agent of grain spoilage, causing ergot disease in rye, wheat, and barley grains. The mold toxin may induce convulsions and hallucinations. The drug LSD is derived from this toxin.

5.11.6 Importance of Microbes in Foods

Molds, yeasts, and bacteria play a significant part in food spoilage. It is true that molds are involved in the spoilage of many foods, but some molds are important in food manufacture, especially in mold-ripened cheeses and the preparation of oriental foods. Various fungi, such as species of *Aspergillus*, *Fusarium*, and *Penicillium,* have been found in foods, and some have been implicated in toxin production. Yeasts are both useful as well as problematic organisms, and this depends upon the food. For example, the production of wine is dependent on the growth and activity of the yeast *Saccharomyces cerevisiae*, while wine can be oxidized to CO_2 and water by wild yeast. Important yeasts in foods include species of *Saccharomyces*, which are useful in fermentations. On the other hand, species of *Zygosaccharomyces* that are osmophilic are involved in the spoilage of materials such as honey, while species of *Pichia* form pellicles in liquids such as in beer and wines. A variety of bacteria are found in foods, and the important ones are grouped based on their biochemical properties.

5.11.7 Food Fermentation

Fermentation of food results in the production of organic acids, alcohols, and esters, which not only help in preserving the food but may also generate distinctive new food products. The fermentation may be by yeast, bacteria molds, or by a combination of these organisms. Food products such as bread, beer, and wine are produced using yeast, while both yeast and bacteria are involved in the production of vinegar. Bacteria are also involved in the production of fermented milks, while molds are important in the production of oriental foods such as soy sauce and other soybean products. By virtue of their growth, these organisms bring about desirable changes in the food composition and, at the same time, bring about a certain degree of preservation. For example, in fermented pickles, the end product of fermentation, namely lactic acid, serves as a preservative. Fermented pickles, therefore, do not need the addition of any preservative.

In bread making, microorganisms are involved in gas production, which helps in the production of bread with a porous structure, and are also involved in the production of flavoring substances. During the preparation of dough, the yeast ferments the sugars to produce CO_2 and alcohol. Instead of bread yeast (baker's yeast), other gas-forming yeasts such as wild yeasts and heterolactic acid bacteria have also been used. Little growth occurs during the leavening process, but fermentation begins as soon as the dough is mixed and continues until the temperature of the oven inactivates the enzymes.

Addition of a large number of yeast cells can hasten the fermentation process and discourage growth. During fermentation, conditioning of the dough takes place, which results from the action on gluten by proteolytic enzymes in the flour by the added yeast or from the malt and a reduction in pH. Sometimes dough conditioners are also added to stimulate yeast growth. The main objective of the baker during leavening is to have enough gas produced and to have the dough in such a condition that it will hold the gas. Yeasts are also reported to contribute to the flavor of bread through products such as alcohols, acids, aldehydes, esters, etc., that are released during the fermentation. If enough time is given for growth of bacteria before baking, they may also add to the flavor. During baking, the temperature inside the loaf does not reach 100°C, but the heat serves to kill the yeast, inactivate their enzymes, and allows expansion of the gas present to give the right structure to the bread.

Beer is the principal fermented malt beverage produced worldwide from malt, hops, and yeast and malt adjuncts. The production of malt wort and fermentation has been dealt earlier. In Japan, soya sauce is prepared using *Aspergillus oryzae.* In the initial stages, this organism produces a variety of enzymes that break down soy protein and starch. In the subsequent lactic fermentation, lactic bacteria produce lactic acid. More acid production by *Pediococcus halophilus* and alcoholic fermentation by *Saccharomyces rouxi* and *Zygosaccharomyces soyae* occurs. An Indonesian food called Temphe is also prepared from soybean. Soybeans are soaked at 25°C, seed coats are then removed, and the split beans are cooked in water for 20 min. These are then cooled and inoculated with spores of *Rhizopus sp.* (*R. arrhizus* or *R. oryzae*). The mash is packed into plastic containers or rolled in banana leaves and incubated at 32°C for 20 h to allow mycelial growth. The product is then sliced, dipped in saltwater, and fried in fat before it is consumed. A variety of Japanese and Chinese foods such as miso, angkbak, and soybean cheese are also prepared by the use of fungi.

5.12 MICROBIAL BIOTECHNOLOGY

Over the last few decades, microbial biotechnology is being employed for large-scale production of a variety of biochemicals, ranging from alcohols to antibiotics, and in the processing of foods and feeds. The use of microbes to obtain a product or service of economic importance constitutes industrial microbiology or microbial biotechnology. With respect to its scope, objectives, and activities, industrial microbiology is synonymous with "fermentation," as fermentation includes any process mediated by or involving microorganisms in which a product of economic value is obtained (Casida, Jr., 1968). The goods provided by microorganisms may be entire, live, or dead microbial cells; processed microbial biomass; components of microbial cells; intracellular or extracellular enzymes; or chemicals produced by the microbes utilizing the constituents of the medium or the substrate provided. Degradation of organic wastes, detoxification of industrial wastes and toxic compounds, and degradation of oil spills are some of the services rendered by microbes. Production of biocontrol agents and biofertilizers also comes under industrial microbiology. Any activity of industrial microbiology involves isolation of microorganisms from nature, their screening to establish product formation, improvement of product yields, and maintenance of cultures, mass culture using bioreactors, and, finally, recovery of products and their purification.

5.12.1 Production of Enzymes by Microorganisms

Enzymes can be extracted from the tissues of higher organisms. It is easier and cheaper to get the enzymes from microorganisms. The first industrial production of enzymes from microorganisms dates back to 1894, when fungal taka diastase was produced in the United States to be employed as a pharmaceutical agent for digestive disorders. Enzymes have found a variety of applications in medicine, in food industries, in the textiles and leather industries, and also in analytical processes. The most commercialized microbial enzymes come from a small number of fungi and bacteria such as *Aspergillus*, *Fusarium*, *Trichoderma* and *Humicola* (all belonging to ascomycetes). *Mucor* and *Rhyzomucor* (belonging to zygomyctes) are the fungi from which biocatalysts are retrieved. *Bacillus* and *Pseudomonas* are the bacterial strains employed for the production of enzymes. The most important industrially produced microbial enzymes are alpha amylase, proteases, and lipases. Three different molecular techniques are adopted for the large-scale production of microbial enzymes: expression cloning, molecular screening, and protein engineering.

PROBLEMS

Section A: Descriptive Type

Q1. Describe the intracellular organization of microbes.
Q2. Write a brief note on pathogenic bacteria.
Q3. How can microbes be used in the field of agriculture?
Q4. Describe the significance of bacteria as material for genetic studies.
Q5. Describe mutations in bacteria.

Q6. What steps are involved in microbial transformation?
Q7. Explain the significance of *E. coli* in microbial cloning.

Section B: Multiple Choice

Q1. Microbes fossils are found to be present ________ years ago.
- a. 3 billion
- b. 3.5 billion
- c. 5 million

Q2. Bacteria were first observed by Antonio van Leeuwenhoek in the year …
- a. 1600
- b. 1676
- c. 1650
- d. 1680

Q3. Bacteria do not have a membrane-bound nucleus. True/False

Q4. Cyobacteria produce internal gas vesicles which they use to regulate …
- a. Metabolism
- b. Buoyancy
- c. Temperature
- d. Reproduction

Q5. Denitrification is the utilization of nitrate as a terminal electron acceptor. True/False

Q6. The process of bacterial growth is called …
- a. Primary fission
- b. Binary fission
- c. Tertiary fission

Q7. How many phases are there in industrial-level bacterial production?
- a. 3
- b. 4
- c. 2
- d. 5

Q8. Transduction involves the carrying over of DNA from one organism to another by an intermediate agent. True/False

Q9. The ________ genome was the first to completely sequenced.
- a. Virus
- b. *E. coli*
- c. Salmonella
- d. Nocardia

Q10. When was the first industrial production of enzymes from microorganism achieved?
- a. 1800
- b. 1850
- c. 1894

Section C: Critical Thinking

Q1. How can a bacterium be differentiated by Gram-positive and Gram-negative staining?
Q2. In order to get good bacterial growth in the bioreactor, what are the different parameters that can be regulated or modified?
Q3. Why is the lactic fermentation of milk not considered spoilage?
Q4. What would be the scenario of microbial growth in the absence of the log or exponential phase?

REFERENCES AND FURTHER READING

Demain, A.L. and Davies, J.E. *Manual of Industrial Microbiology and Biotechnology*. ASM Press, Washington, DC, 1999.

Gillor, O., Kirkup, B.C., and Riley, M.A. Colicins and microcins. The next generation antimicrobials. *Adv. Appl. Microbiol.* 54: 129–146, 2004.

Gillor, O., Nigro, L.M., and Riley, M.A. Genetically engineered bacteriocins and their potential as the next generation of antimicrobials. *Curr. Pharm. Des.* 11: 1067–1075, 2005.

Glazer, A.N. and Nikaido, H. *Microbial Biotechnology: Fundamentals of Applied Microbiology*. Cambridge University Press, Cambridge, New York, 2007.

Kirkup, B.C. Bacteriocins as oral and gastrointestinal antibiotics: Theoretical considerations, applied research, and practical applications. *Curr. Med. Chem.* 13: 3335–3350, 2006.

Lee, L.K. *Microbial Biotechnology: Principles and Applications*. World Scientific Publishing Company, Singapore, 2007.

Marquez, B. Bacterial efflux systems and efflux pumps inhibitors. *Biochimie* 87: 1137–1147, 2005.

Metlay, J.P., Camargo, C.A., MacKenzie, T. et al. Cluster-randomized trial to improve antibiotic use for adults with acute respiratory infections treated in emergency departments. *Ann. Emerg. Med.* 50: 221–230, 2007.

Spurling, G., Del Mar, C., Dooley, L., and Foxlee, R. Delayed antibiotics for respiratory infections. *Cochrane Database of Systematic Reviews (Online)* (3): CD004417. doi:10.1002/14651858.CD004417.pub3. PMID 17636757, 2007.

Toratoa, G., Funke, B., and Case, C.L. *Microbiology: An Introduction*. Pearson Benjamin Cummings, San Francisco, CA, 2007.

6 Agricultural Biotechnology

LEARNING OBJECTIVES

- Explain the significance of agricultural biotechnology
- Discuss plant-breeding techniques by the classical approach
- Discuss plant-breeding techniques by the genetic approach
- Discuss plant diseases and explain their causes
- Explain how genetically modified plants are created
- Discuss the benefits of genetically modified plants
- Cite examples of genetically modified plants

6.1 INTRODUCTION

Agricultural biotechnology is an advanced technology that allows plant breeders or farmers to make precise genetic changes in plants to impart beneficial traits which include size, yield, color, taste, and appearance. For centuries, farmers and plant breeders have labored to improve crop plants. Traditional breeding methods include selecting and sowing seeds of the strongest and most desirable plants to produce the next generation of crops. By selecting and breeding plants with characteristics such as higher yield and resistance to pests and hardiness, early farmers dramatically changed the genetic make-up of crop plants long before the science of genetics was understood. As a result, most of today's crop plants bear little resemblance to their wild ancestors. The tools of modern biotechnology allow plant breeders to select genes that produce beneficial traits and move them from one organism to another. This process is far more precise and selective than cross-breeding, which involves the transfer of tens of thousands of genes, and provides plant developers with a more detailed knowledge of the changes being made. The ability to introduce genetic material from other plants and organisms opens up a world of possibilities to benefit food production. As an example, "Bt" crops that are protected against insect damage contain selected genes found in the common soil bacteria, *Bacillus thuringiensis*. Bt genes produce proteins that are toxic to the larvae of certain plant pests but are found to be safe for humans and animals. Plants in which Bt genes are incorporated are protected from insects which eat and destroy the plants, thus improving yield, reducing the need for pesticide applications, and saving the farmer time and money. The extensive manipulation of genes in plants and microbes has revolutionized modern agricultural products. This has not only protected plants against various diseases and pests but has also improved the annual agricultural production.

6.2 PLANT BREEDING

Plant breeding has been in practice for thousands of years. It is now practiced worldwide by individuals such as gardeners and farmers, or by professional plant breeders employed by organizations such as government institutions, universities, crop-specific industry associations, or research centers. International development agencies believe that breeding new crops is important for ensuring food security by developing new varieties that are higher-yielding, resistant to pests and diseases, drought-resistant or regionally adapted to different environments and growing conditions (Figure 6.1). The extensive use of plant breeding in certain situations may lead to

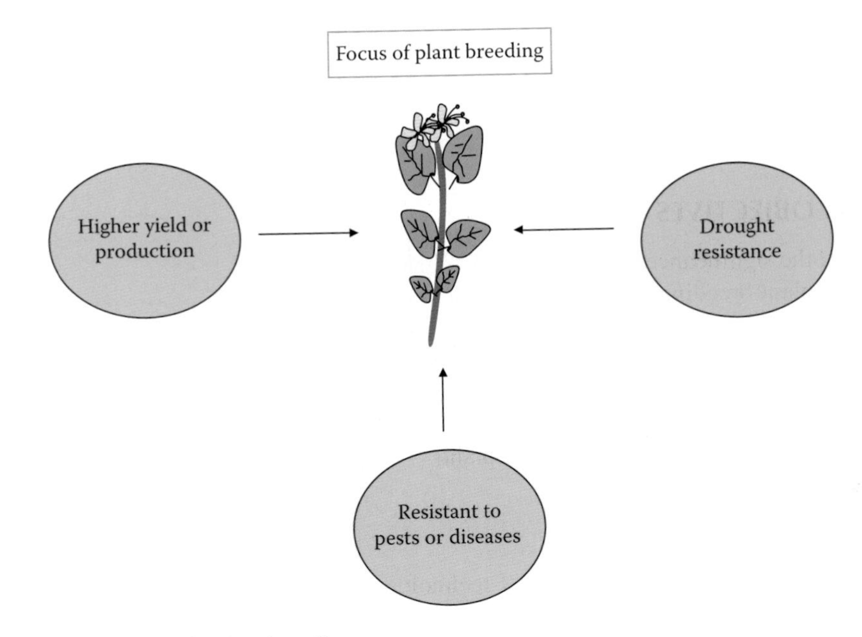

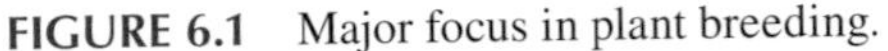
FIGURE 6.1 Major focus in plant breeding.

domestication of wild plants. Domestication of plants is an artificial selection process conducted by humans to produce plants that have more desirable traits than wild plants. Many of the crops nowadays are the result of domestication in ancient times, about 5000 years ago in the Old World and 3000 years ago in the New World. In the Neolithic period, domestication took a minimum of 1000 years and a maximum of 7000 years. Almost all the domesticated plants used today for food and agriculture were domesticated in ancient times. A plant whose origin or selection is due to human activity is called a *cultigen*, and a cultivated crop species that has evolved from wild populations due to selective pressures from traditional farmers is called a *landrace*. Landraces, which can be the result of natural forces or domestication, are plants that are ideally suited to a particular region or environment. Some examples of landraces are rice, *Oryza sativa* subspecies *indica*, which was developed in South Asia, and *Oryza sativa* subspecies *japonica*, which was developed in China.

Plant breeding is defined as identifying and selecting desirable traits in plants and combining these into one individual plant. Since 1900, Mendel's laws of genetics have provided the scientific basis for plant breeding. As all traits of a plant are controlled by genes located in chromosomes, conventional plant breeding can be considered as the manipulation of the combination of chromosomes. In general, there are three main procedures to manipulate plant chromosome combination. First, plants of a given population which show desired traits can be selected and used for further breeding and cultivation, a process called *(pure line-) selection*. Second, desired traits found in different plant lines can be combined altogether to obtain plants that exhibit all the traits simultaneously, a method termed *hybridization*. *Heterosis*, a phenomenon of increased vigor, is obtained by hybridization of inbred lines. Third, *polyploidy* (increased number of chromosome sets) can contribute to crop improvement (Figure 6.2). Plant breeding may be classified as classical breeding or modern breeding.

6.2.1 Classical Breeding

In classical plant breeding, the closely or distantly related plants are crossed to produce new crop varieties or lines with desirable properties. Plants are crossbred to introduce traits/genes from one variety or line into a new genetic background. For example, a mildew-resistant pea may be

FIGURE 6.2 Three main procedures to manipulate plant chromosome combination.

crossed with a high-yielding but susceptible pea. The goal of the crossbreeding is to introduce mildew resistance without losing the high-yield characteristics. Progeny from the cross would then be crossed with the high-yielding parent to ensure that the progeny were most like the high-yielding parent (backcrossing). The progeny from that cross would then be tested for yield and mildew resistance and high-yielding mildew-resistant plants would be further developed. Plants may also be crossed with themselves to produce inbred varieties for breeding. Classical breeding relies largely on homologous recombination between chromosomes to generate genetic diversity. The classical plant breeder may also make use of a number of *in vitro* techniques such as protoplast fusion embryo rescue or mutagenesis to generate diversity and produce hybrid plants that would not exist in nature. For over 100 years, plant breeders have tried to incorporate various traits into crop plants to increase quality and yield; to increase tolerance to environmental pressures (such as salinity, extreme temperature, drought); to increase resistance against viruses, fungi and bacteria; and to increase tolerance to pests and herbicides.

6.2.2 Plant Breeding by Traditional Techniques

6.2.2.1 Selection

In plant breeding, selection is the most ancient and basic procedure and it generally involves three distinct steps. First, a large number of selections are made from the genetically variable original population. Second, progeny rows are grown from the individual plant selections for observational purposes. After obvious elimination, the selections are grown over several years to permit observations of performance under different environmental conditions for making further eliminations. Finally, the selected and inbred lines are compared with existing commercial varieties for their yielding performance and other aspects of agronomic importance.

6.2.2.2 Hybridization

Hybridization is the most frequently employed plant-breeding technique. Its aim is to bring together desired traits found in different plant lines into one plant line via cross-pollination.

The first step in this technique is to generate homozygous inbred lines. This is normally done by using self-pollinating plants where pollen from male flowers pollinates female flowers from the same plants. Once a pure line is generated, it is out crossed or combined with another inbred line. Then the resulting progeny is selected for combination of the desired traits. If a trait (such as resistance against diseases) from a wild relative of a crop species is to be brought into the genome of the crop, a large quantity of undesired traits (like low yield, bad taste, or low nutritional value) are transferred to the crop as well. These unfavorable traits must be removed by time-consuming backcrossing, that is, by repeated crossing with the crop parent. There are two types of hybrid plants: *interspecific* and *intergeneric hybrids*. Beyond this biological boundary, hybridization cannot be accomplished due to sexual incompatibility, which limits the possibilities of introducing desired traits into crop plants. *Heterosis* is an effect which is achieved by crossing highly inbred lines of crop plants. Inbreeding of most crops leads to a strong reduction of vigor and size in the first generations. After six or seven generations, no further reduction in vigor or size is found. When such highly inbred plants are crossed with other inbred varieties, very vigorous, large-sized, and large-fruited plants may result. The most notable and successful hybrid plant ever produced is the hybrid maize. By 1919, the first commercial hybrid maize was available in the United States. Two decades later, nearly all maize was hybrid, as it is today, although farmers must buy new hybrid seed every year, because the heterosis effect is lost in the first generation after hybridization of the inbred parental lines.

6.2.2.3 Polyploidy

Plants that have three or more complete sets of chromosomes are commonly known as *polyploidy plants*. The chromosome numbers can be increased artificially by treating the plant cells with colchicine which leads to a doubling of the chromosome number. Generally, the main effect of polyploidy is increase in size and genetic variability. On the other hand, polyploid plants often have lower fertility and slower growth. Instead of relying only on the introduction of genetic variability from the wild species gene pool or from other cultivars, an alternative is the introduction of mutations induced by chemicals or radiation. The mutants obtained are tested and further selected for desired traits. The site of the mutation cannot be controlled when chemicals or radiation is used as agents of mutagenesis. Because the great majority of mutants carry undesirable traits, this method has not been widely used in breeding programs.

6.2.3 Modern Plant Breeding

Modern plant-breeding techniques basically involve molecular or genetic engineering techniques to select or to insert desirable traits into plants. In recent years, biotechnology has developed rapidly as a practical means for accelerating success in plant breeding and improving economically important crops. Some of the modern plant breeding methods used today are described below.

6.2.3.1 *In Vitro* Cultivation

In this method, plants are cultivated using *in vitro* culture from cultured isolated somatic plant cells. The parts of plants can be cultured *in vitro* and are capable of proliferation and organization into tissues and eventually into complete plants. The process of regenerating whole plants out of plant cells is called *in vitro regeneration*. Some factors affecting plant regeneration are genotype, explants source, culture conditions, culture medium, and environment. Different mixtures of plant hormones and other compounds in varying concentrations are used to achieve regeneration of plants from cultured cells and tissues. As the plant hormonal mechanisms are not yet understood completely, the development of *in vitro* cultivation and regeneration systems is still largely based on empirically testing variations of the above-mentioned factors (Figure 6.3).

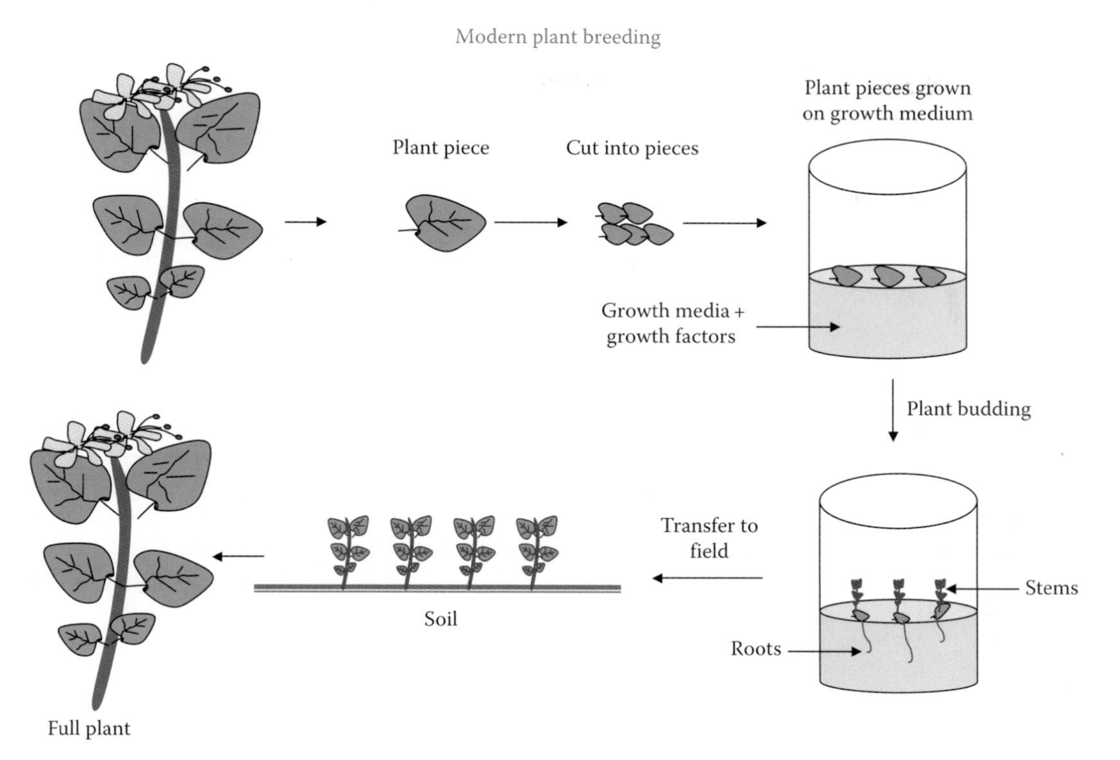

FIGURE 6.3 Plant cultivation by *in vitro* cell culture.

6.2.3.2 *In Vitro* Selection and Somaclonal Variation

Plants regenerated from *in vitro* cell cultures may exhibit phenotypes differing from their parent plants, sometimes at quite high frequencies. If these are heritable and affect desirable agronomic traits, such "somaclonal variation" can be incorporated into breeding programs. However, finding specific valuable traits by this method is largely left to chance, and hence inefficient. Rather than relying on this undirected process, *in vitro* selection targets specific traits by subjecting large populations of cultured cells to the action of a selective agent in a Petri dish. For the purpose of disease resistance, this selection can be provided by pathogens, or isolated pathotoxins that are known to have a role in pathogenesis. The selection will only allow those cells to survive and proliferate that are resistant to the challenge. Selection of cells also plays an important role in genetic engineering, where special marker genes are used to select for transgenic cells (Figure 6.4).

6.2.3.3 Somatic Hybrid Plants

Somatic hybrid plants are generated by fusion of somatic cells. Cell fusion was developed after the successful culture of a large number of plant cells that were stripped of their cell walls. The resulting cells without walls are referred to as *protoplasts*. Since protoplasts from phylogenetically unrelated species can also be fused, attempts have been made to overcome sexual incompatibility using protoplast fusion. In most cases, these attempts failed because growth and division of the fused cells did not take place when only distantly related cells were fused. Although successful fusions between sexually incompatible petunia species and between potatoes and tomatoes did not lead to economically interesting products, important contributions to the understanding of cell wall regeneration and other mechanisms were achieved (Figure 6.5).

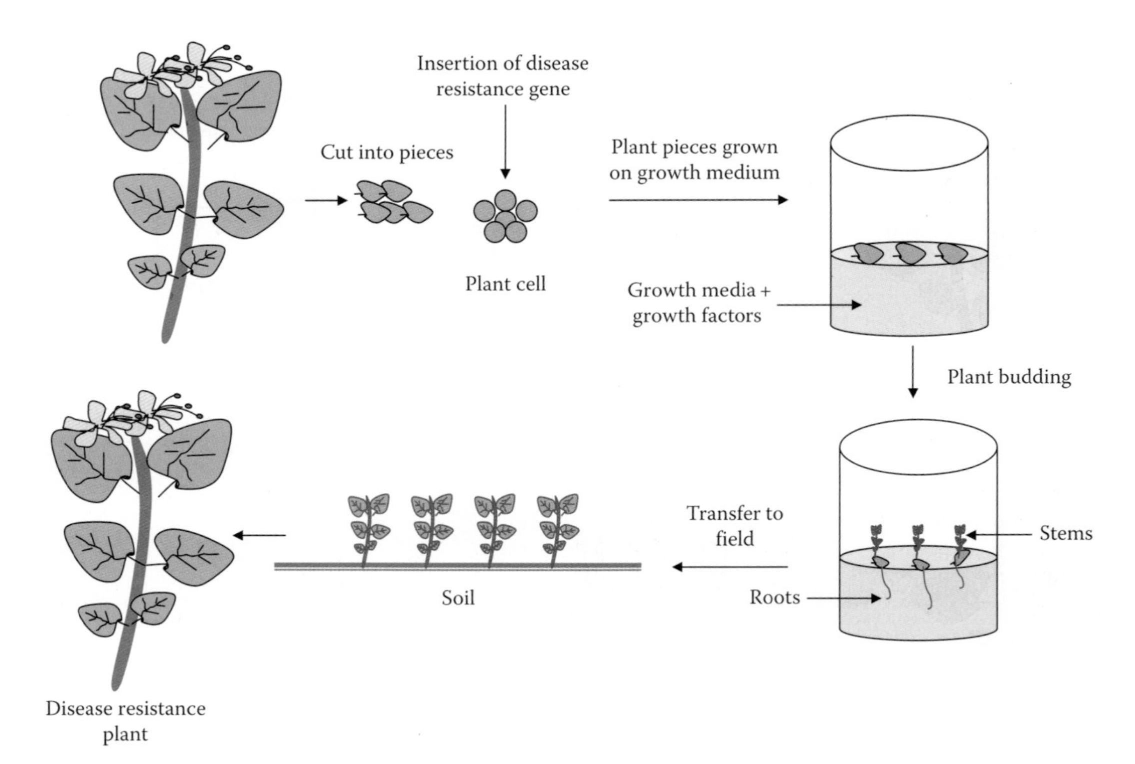

FIGURE 6.4 Plant cultivation by *in vitro* selection.

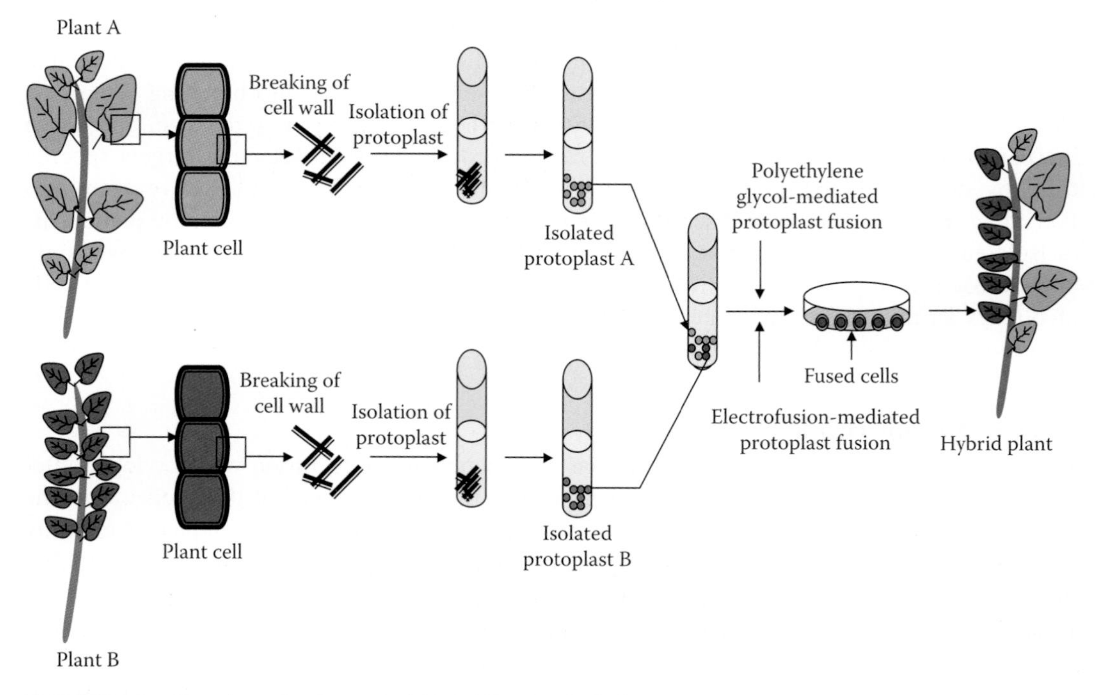

FIGURE 6.5 Plant breeding by somatic hybridization technique.

6.2.3.4 Breeding by Restriction Fragment Length Polymorphism

Traditional plant-breeding techniques are very time-consuming and sometimes a lot of undesired genes are introduced into the genome of a plant. The undesired genes have to be "sorted out" by backcrossing. The use of restriction fragment length polymorphism (RFLP) greatly facilitates conventional plant breeding, because one can progress through a breeding program much faster, with smaller populations, and without relying entirely on testing for the desired phenotype. RFLP makes use of restriction endonucleases and these enzymes recognize and cut specific nucleotide sequences in DNA. For example, the sequence GAATTC is cut by the endonuclease *EcoRl*. After treatment of a plant genome with endonucleases, the plant DNA is cut into pieces of different lengths, depending on the number of recognition sites on the DNA. These fragments can be separated according to their size by using gel electrophoresis and are made visible as bands on the gel by hybridizing the plant DNA fragments with radiolabeled or fluorescent DNA probes. As two genomes are not identical even within a given species, due to mutations, the number of restriction sites and therefore the length and numbers of DNA fragments differ, resulting in a different banding pattern on the electrophoresis gel. This variability has been termed RFLP. The closer two organisms are related, the more the pattern of bands overlap. If a restriction site lies close to or even within an important gene, the existence of a particular band correlates with the particular trait of a plant, such as disease resistance. By looking at the banding pattern, breeders can identify individuals that have inherited resistance genes, and resistant plants can be selected for further breeding. The use of this technique not only accelerates progress in plant breeding considerably but also facilitates the identification of resistance genes, thereby opening new possibilities in plant breeding.

6.2.3.5 Plant Breeding by Gene Transfer

In conventional breeding, the pool of available genes and the traits they code for is limited due to sexual incompatibility to other lines of the crops. This restriction can be overcome by using the methods of genetic engineering, which in principle allow introducing valuable traits coded for by specific genes of any organism (other plants, bacteria, fungi, animals, viruses) into the genome of any plant. The first gene transfer experiments with plants took place in the early 1980s. Normally, transgenes are inserted into the nuclear genome of a plant cell. Recently, it has become possible to introduce genes into the genome of chloroplasts and other plastids (small organelles of plant cells which possess a separate genome). Transgenic plants have been obtained using Agrobacterium-mediated DNA transfer and direct DNA transfer, the latter including methods such as particle bombardment, electroporation, and polyethylene glycol permeabilization. The majority of plants have been transformed using Agrobacterium-mediated transformation.

6.2.3.6 Agrobacterium-Mediated Gene Transfer

The Agrobacterium-mediated technique involves the natural gene transfer system in the bacterial plant pathogens of the genus Agrobacterium. In nature, *Agrobacterium tumefaciens* and *Agrobacterium rhizogenes* are the causative agents of the crown gall and the hairy root diseases, respectively. The utility of Agrobacterium as a gene transfer system was first recognized when it was demonstrated that these plant diseases were actually produced as a result of the transfer and integration of genes from the bacteria into the genome of the plant. Both Agrobacterium species carry a large plasmid (small circular DNA molecule) called Ti in *A. tumefaciens* and Ri in *A. rhizogenes*. A segment of this plasmid designated T-(for transfer) DNA is transmitted by this organism into individual plant cells, usually within wounded tissue. The T-DNA segment penetrates the plant cell nucleus and integrates randomly into the genome where it is stably incorporated and inherited like any other plant gene in a predictable, dominant Mendelian fashion. Expression of the natural genes on the T-DNA results in the synthesis of gene products that direct the observed morphological changes such as tumor or hairy root formation. In genetic engineering, the tumor-inducing genes within the T-DNA, which cause the plant disease, are removed and replaced by foreign genes. These genes

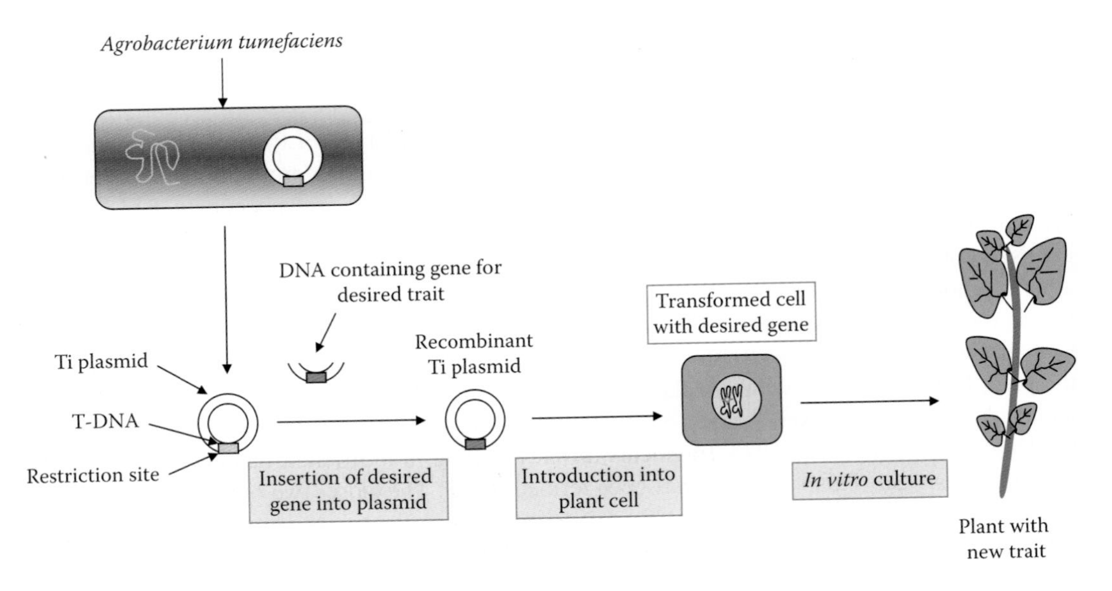

FIGURE 6.6 Agrobacterium-mediated gene transfer.

are then stably integrated into the genome of the plant after infection with the altered strain of Agrobacterium, just like the natural T-DNA. Because all tumor-inducing genes are removed, the gene transfer does not induce any disease symptoms. This reliable method of gene transfer is well suited for plants that are susceptible to infection by Agrobacterium. Unfortunately, many species, especially economically important legumes and monocotyledons such as cereals, do not respond positively to Agrobacterium-mediated transformation (Figure 6.6).

6.2.3.7 Particle Bombardment

Particle bombardment, also referred to as biolistic transformation (from biological ballistics), involves coating biologically active DNA onto small tungsten or gold particles and accelerating them into plant tissues at high velocity. The particles penetrate the plant cell wall and lodge themselves within the cell where the DNA is liberated, resulting in transformation of the individual plant cell in an explant. This technique is generally less efficient than Agrobacterium-mediated transformation, but has nevertheless been particularly useful in several plant species, most notably in cereal crops. The introduction of DNA into organized, morphogenic tissues such as seeds, embryos, or meristems has enabled the successful transformation and regeneration of rice, wheat, soybean, and maize, thus demonstrating the enormous potential of this method (Figure 6.7).

6.2.3.8 Electroporation and Direct DNA Entry into Protoplasts

Electroporation is a process whereby very short pulses of electricity are used to reversibly permeabilize lipid bilayers of plant cell membranes. The electrical discharge enables the diffusion of DNA through an otherwise impermeable plasma membrane. Because the plant cell wall will not allow the efficient diffusion of many transgene constructs, protoplasts (cells without cell walls) must be prepared. DNA uptake by plant protoplasts can also be stimulated by phosphate or calcium/ polyethylene glycol co-precipitation. However, all these methods suffer from the drawback that they use protoplasts as the recipient host which often cannot be regenerated into whole plants.

6.2.3.9 Transgene Expression

The success of transgene expression is generally based on transcription of mRNA and then into translation leading to protein synthesis. *Promoter* is a sequence of nucleic acids where RNA

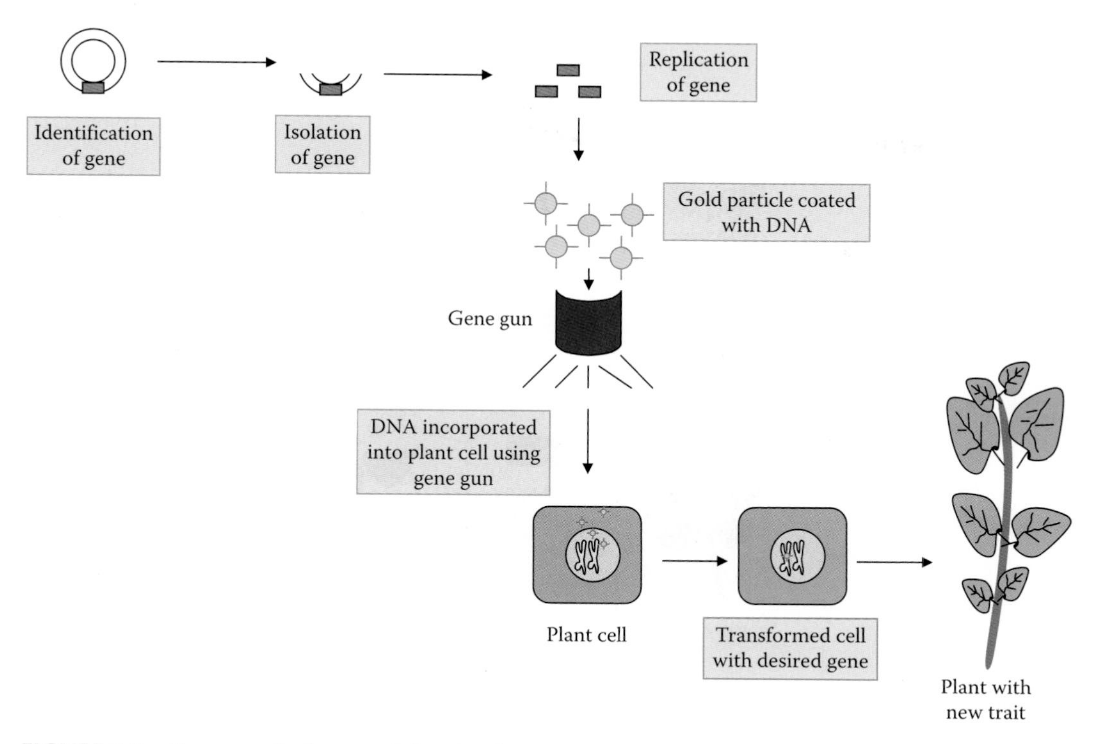

FIGURE 6.7 Gene transfer by particle bombardment.

polymerase (a complex enzyme synthesizing the mRNA transcript) attaches to the DNA template. The nature of the promoter defines (together with other expression-regulating elements) under which conditions and intensity a gene will be transcribed. The promoter of the 35S gene of cauliflower mosaic virus (CaMV) is used very frequently in plant genetic engineering. This promoter confers high-level expression of exogenous genes in most cell types from virtually all species tested. As it is often advantageous to express a transgene only in certain tissues or quantities or at certain times, a number of other promoters can also be used such as promoters inducing gene expression after wounding or during fruit ripening only. Methods of gene transfer currently employed result in the random integration of foreign DNA throughout the genome of recipient cells. The site of insertion may have a strong influence on the expression levels of the exogenous gene, resulting in different expression levels of an introduced gene, even if the same promoter/gene construct was used. The exact mechanism of this phenomenon is not yet fully understood (Figure 6.8).

6.2.3.10 Selection and Plant Regeneration

In order to select only cells that have actually incorporated the new genes, the genes coding for the desired trait are fused to a gene that allows selection of transformed cells, so-called *marker genes*. The expression of the marker gene enables the transgenic cells to grow in the presence of a selective agent, usually an antibiotic or a herbicide, while cells without the marker gene die. One of the most commonly used marker is the bacterial aminoglycoside-3′ phosphotransferase gene (APH(3′)II), also referred to as neomycin phosphotransferase II (NPTII). This gene codes for an enzyme which inactivates the antibiotics kanamycin, neomycin, and G418 through phosphorylation. In addition to NPTII, a number of other antibiotic resistance genes have been used as selective markers, such as hygromycin phosphotransferase gene conferring resistance to hygromycin. Another group of selective markers are herbicide tolerance genes. Herbicide tolerance has been obtained through the incorporation and expression of a gene that either detoxifies the herbicide in a similar manner

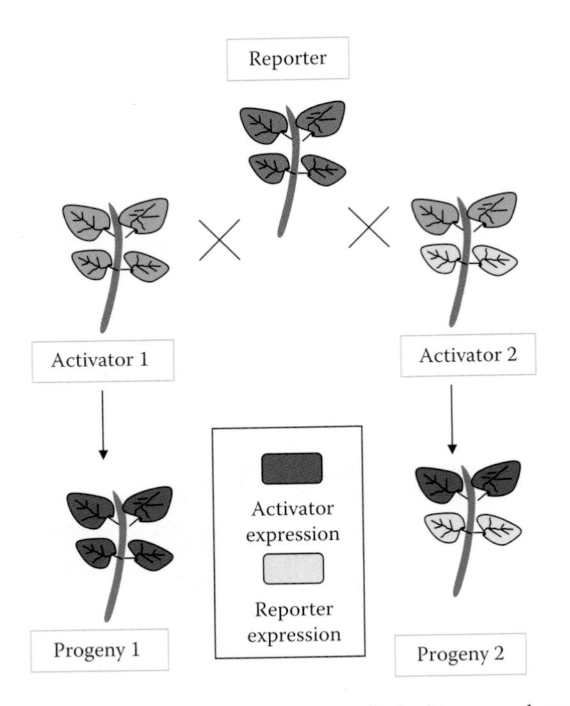

FIGURE 6.8 Transgene expression in plant: A reporter gene linked to a novel promoter is silent when first introduced into reporter plants. Transgene expression is induced by crossing to activator lines that express a heterologous transcription factor that specifically recognizes the transgene promoter. The pattern of reporter gene expression will reflect the pattern of activator expression, allowing a gene of interest to be expressed under a variety of regimes simply by crossing to an appropriate activator line.

as the antibiotic resistance gene products or a gene that expresses a product that acts like the herbicide target but is not affected by the herbicide. Herbicide tolerance may not only serve as a trait useful for selection in the development of transgenic plants but also has some commercial interest. Transformation of plant protoplasts, cells, and tissues is usually only useful if they can be regenerated into whole plants. The rate of regeneration varies greatly not only among different species but also between cultivars of the same species. Besides the ability to introduce a gene into the genome of a plant species, regeneration of intact, fertile plants out of transformed cells or tissues is the most limiting step in developing transgenic plants (Figure 6.9).

6.2.3.11 Reverse Breeding and Doubled Haploids

Reverse breeding and doubled haploids (DH) is the most efficient method to produce homozygous plants from a heterozygous starting plant, which has all desirable traits. This starting plant is induced to produce DH from haploid cells, and later on to create homozygous/DH plants from those cells. While in natural offspring recombination occurs and traits can be unlinked from each other, in DH cells and in the resulting DH plants, recombination is no longer an issue. Here, a recombination between two corresponding chromosomes does not lead to un-linkage of alleles or traits, since it just leads to recombination with its identical copy. Thus, traits on one chromosome stay linked. Selecting those offspring that have the desired set of chromosomes and crossing them will result in a final F1 hybrid plant having exactly the same set of chromosomes, genes, and traits as the starting hybrid plant. The homozygous parental lines can reconstitute the original heterozygous plant by crossing, if desired even in a large quantity. An individual heterozygous plant can be converted into a heterozygous variety (F1 hybrid) without the necessity of vegetative propagation but as the result of the cross of two homozygous/DH lines derived from the originally selected plant.

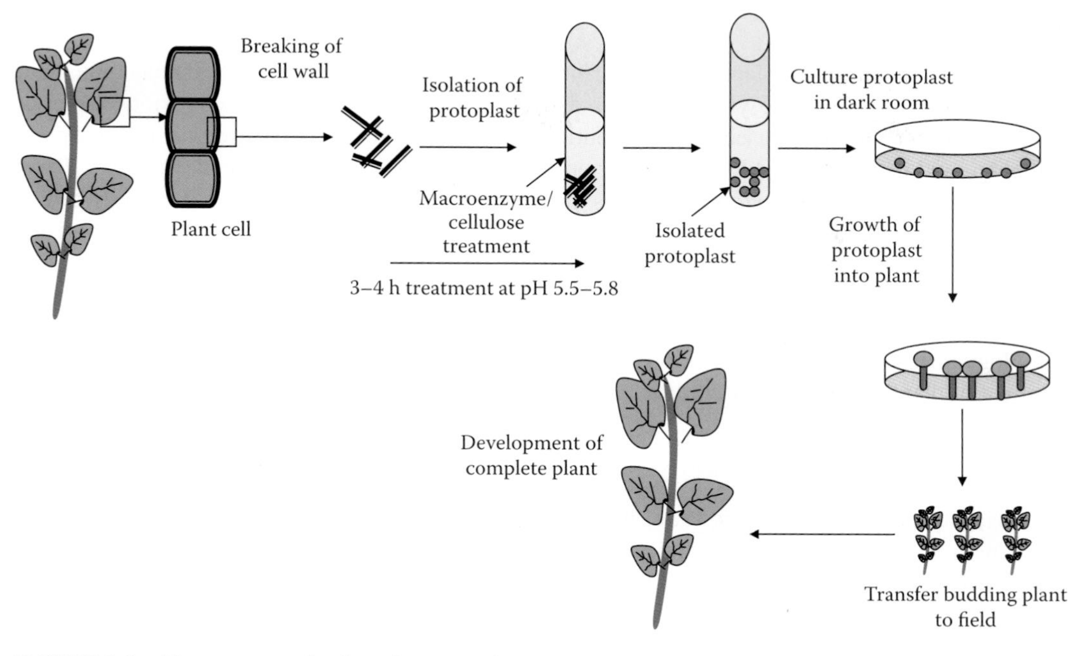

FIGURE 6.9 Plant regeneration by using protoplast.

6.2.3.12 Genetic Modification

Genetic modification of plants is achieved by adding a specific gene or genes to a plant, or by knocking out a gene with RNA interference (RNAi) technique, to produce a desirable phenotype. The plants resulting from adding a gene are often referred to as *transgenic plants*. If for genetic modification, genes of the species or of a crossable plant are used under control of their native promoter, then they are called *cisgenic plants*. Genetic modification can produce a plant with the desired trait or traits faster than classical breeding because the majority of the plant's genome is not altered. To genetically modify a plant, a genetic construct must be designed so that the gene to be added or removed will be expressed by the plant. To do this, a promoter to drive transcription and a termination sequence to stop transcription of the new gene, and the gene or genes of interest must be introduced to the plant. A marker for the selection of transformed plants is also included. In the laboratory, antibiotic resistance is a commonly used marker: plants that have been successfully transformed will grow on media containing antibiotics; plants that have not been transformed will die. In some instances markers for selection are removed by backcrossing with the parent plant prior to commercial release. The construct can be inserted in the plant genome by genetic recombination using the bacteria *A. tumefaciens* or *A. rhizogenes*, or by direct methods like the gene gun or microinjection. Using plant viruses to insert genetic constructs into plants is also a possibility, but the technique is limited by the host range of the virus. For example, CaMV only infects cauliflower and related species. Another limitation of viral vectors is that the virus is not usually passed on to the progeny, so every plant has to be inoculated. The majority of commercially released transgenic plants are currently limited to plants that have introduced resistance to insect pests and herbicides. Insect resistance is achieved through incorporation of a gene from *B. thuringiensis* (Bt) that encodes a protein that is toxic to some insects. For example, when the cotton bollworm, a common cotton pest, feeds on Bt cotton it will ingest the toxin and die. Herbicides usually work by binding to certain plant enzymes and inhibiting their action. The enzymes that the herbicide inhibits are known as the herbicides *target site*. Herbicide resistance can be engineered into crops by expressing a version of *target site* protein that is not inhibited by the herbicide. This is the method used to produce glyphosate-resistant crop plants.

6.3 PLANT DISEASES

Like humans and animals, plants are also afflicted with diseases which not only damage the plant structure but also affect the yield. Farmers and plant breeders have identified various types of diseases present in plants based on their routine checkups. In this section, we will discuss plant pathology and its causes, and learn how various diseases produce harmful effects on plants. *Plant pathology,* also known as *phytopathology,* is the scientific study of plant diseases caused by pathogens (infectious diseases) and environmental conditions (physiological factors). Some organisms that cause infectious diseases in plants are fungi, bacteria, viruses, viroids, virus-like organisms, phytoplasmas, protozoa, nematodes, and parasitic plants. Aside from these, there are also insects, mites, vertebrates, and other pests that affect plant health by consumption of plant tissues. Plant pathology also studies the identification, etiology, cycle, economic impact, epidemiology, and management of plant diseases (Figure 6.10).

6.3.1 Diseases Caused by Fungi

The majority of phytopathogenic fungi belong to the families Ascomycetes and Basidiomycetes. The fungi reproduce both sexually and asexually via the production of spores. These spores may be spread over long distances by air or water, or they may be soil borne. Many soil-borne spores, normally zoospores, are capable of living saprotrophically, carrying out the first part of their lifecycle in soil. Some examples of fungal plant pathogens are *Thielaviopsis* spp. which causes canker rot and black root rot; *Magnaporthe grisea*, which causes blast of rice and gray leaf spot in turfgrasses; *Phakospora pachyrhizi* Sydow, which causes soybean rust; and *Puccinia* species, which causes severe rusts of virtually all cereal grains and cultivated grasses.

6.3.2 Diseases Caused by Oomycetes

Oomycetes are not true fungi but are fungal-like organisms. They include some of the most destructive plant pathogens including the genus *Phytophthora*, which is the causal agent of potato late blight and sudden oak death. Despite not being closely related to fungi, *oomycetes*

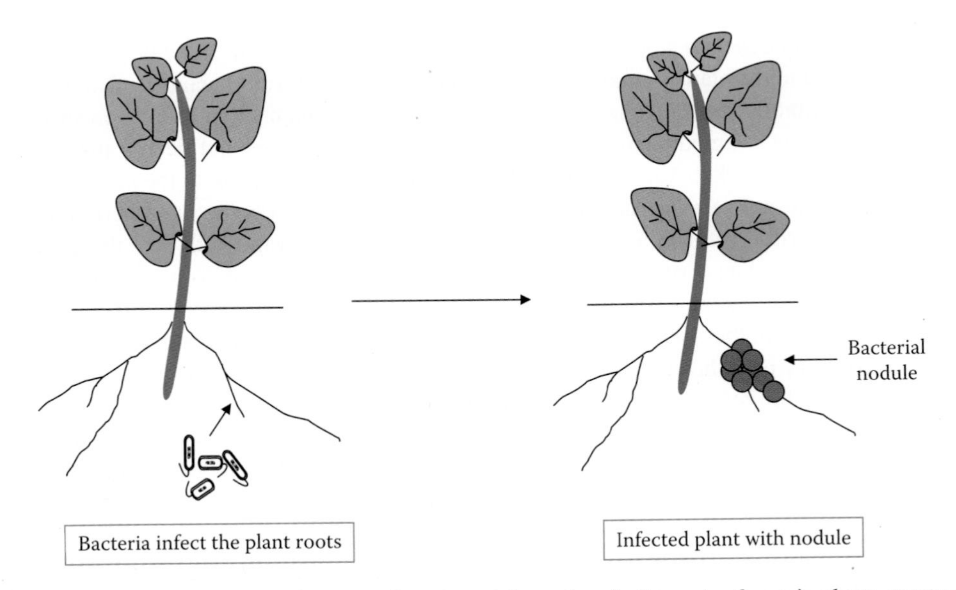

FIGURE 6.10 Plant disease: Development of a root nodule, a place in the roots of certain plants, most notably legumes (the pea family), where bacteria live symbiotically with the plant.

have developed very similar infection strategies, and so many plant pathologists group them with fungal pathogens. These pathogens are known to have caused the Great Irish Famine in 1845–1849.

6.3.3 Diseases Caused by Bacteria

Most bacteria that are associated with plants are actually saprotrophic, and do no harm to the plant itself. However, around 500 species of bacteria cause diseases. Bacterial diseases are much more prevalent in subtropical and tropical regions of the world. Most plant pathogenic bacteria are rod shaped (bacilli). Bacteria may affect plants by increasing plant cell wall, degrading enzymes, inducing toxins, or increasing the level of auxin which causes tumor formation. Some species of bacteria are also known to produce *exopolysaccharides* which block xylem vessels, often leading to the death of plants. *Phytoplasma* and *Spiroplasma* are genres of bacteria that lack cell walls and are related to the mycoplasmas, which are human pathogens. Together they are referred to as *mollicutes*. They also tend to have smaller genomes than true bacteria. They are normally transmitted by sap-sucking insects, being transferred into plants' phloem where it reproduces (Figure 6.11).

6.3.4 Diseases Caused by Plant Virus

There are many types of plant viruses, and some are even asymptomatic. Normally, plant viruses only cause yield loss. Therefore, it is not economically viable to try to control them, the exception being when they infect perennial species such as fruit trees. Most plant viruses have small, single-stranded RNA genomes. These genomes may only encode three or four proteins: a replicase, a coat protein (CP), a movement protein that allows cell to cell movement, and sometimes a protein that allows transmission by a vector.

Plant viruses must be transmitted from plant to plant by a vector. This is normally an insect, but some fungi, nematodes, and protozoa have been shown to be viral vectors.

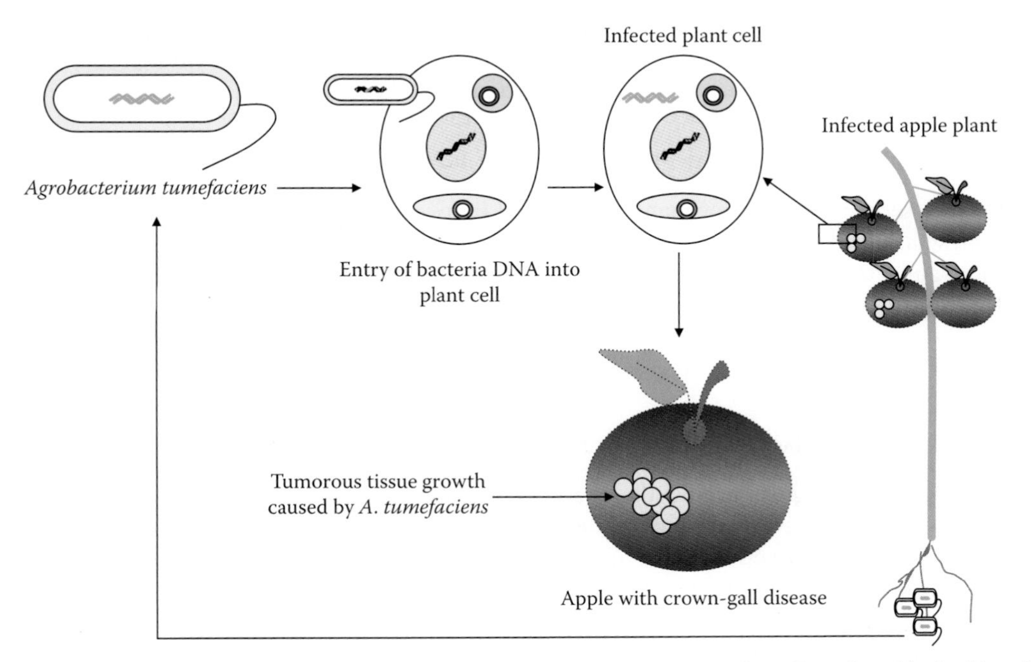

FIGURE 6.11 Plant disease caused by *Agrobacterium tumefaciens*.

6.3.5 Diseases Caused by Nematodes

Nematodes are small and multi-cellular wormlike creatures. Many live freely in the soil, but there are some species which parasitize plant roots. These nematodes cause problems in tropical and subtropical regions of the world where they infect crops. The nematodes species such as *Globodera pallida* and *Globodera rostochiensis* which caused potato cyst diseases can damage $300 million worth of potato in Europe every year. Root knot nematodes have quite a large host range, whereas cyst nematodes can infect only a few species. Nematodes can cause radical changes in root cells in order to facilitate their lifestyle.

6.3.6 Diseases Caused by Protozoa

There are a few examples of plant diseases caused by protozoa. They are transmitted as zoospores which are very durable, and may be able to survive in a resting state in soil for many years. They have also been shown to transmit plant viruses. When the motile zoospores come into contact with a root hair, they produce a plasmodium and invade the roots.

6.3.7 Diseases Caused by Parasitic Plants

Parasitic plants such as mistletoe and dodder are known to cause phytopathology in various plants. Dodder, for example, is used as a conduit for the transmission of viruses or virus-like agents from a host plant to a plant that is not typically a host or for an agent that is not graft transmissible.

6.4 APPLICATIONS OF MOLECULAR AND GENETIC TOOLS IN AGRICULTURE

Recombinant DNA technology has not only enhanced the health of humans but also contributed to exciting developments in agricultural biotechnology. Using rDNA methods, transgenic plants and animals with desirable properties such as resistance to diseases/herbicides have been developed. Flowers with exotic shapes and colors have been genetically engineered by transgenic expression of pigment genes. Recombinant growth hormones are now available for farm animals, resulting in leaner meat, improved milk yield, and more efficient feed utilization. In the future, transgenic plants and animals may serve as bioreactors for the production of medicinal or protein pharmaceuticals.

6.4.1 Expression of Viral Coat Protein to Resist Infection in Agriculture

Viruses are a serious problem for many agricultural crops and animals. Infections can result in reduced growth, less yield, and low quality. Through a standard genetic trick termed cross-protection, infection of a plant/animal with a strain of virus that produces only mild effect protects the plants/animals against infection by more damaging strains. The same principle/mechanism when applied to animals is called *vaccination*. Although the mechanism of cross-protection is not entirely known, it is thought that a particular viral-encoded protein is responsible for the protective effect.

6.4.2 Expression of Bacterial Toxin in Agriculture Using Molecular Techniques

Currently, the major weapons against the attackers of plants are chemical insecticides. However, chemicals have some impact on the environment. Natural microbial pesticides, such as the species *B. thuringiensis* (Bt) have been used in a limited manner for over 38 years. Upon sporulation, these bacteria produce a crystallized protein that is toxic to the larvae of a number of insects. The toxic protein does not harm nonsusceptible insects and has no effect on vertebrates. The crystal protein is normally expressed as a large, inactive protoxin about 1200 amino acids in length and with a molecular weight of 1,20,000 Da. The toxin acts by binding to receptors on the surface of midgut cells and blocking the functioning of these cells.

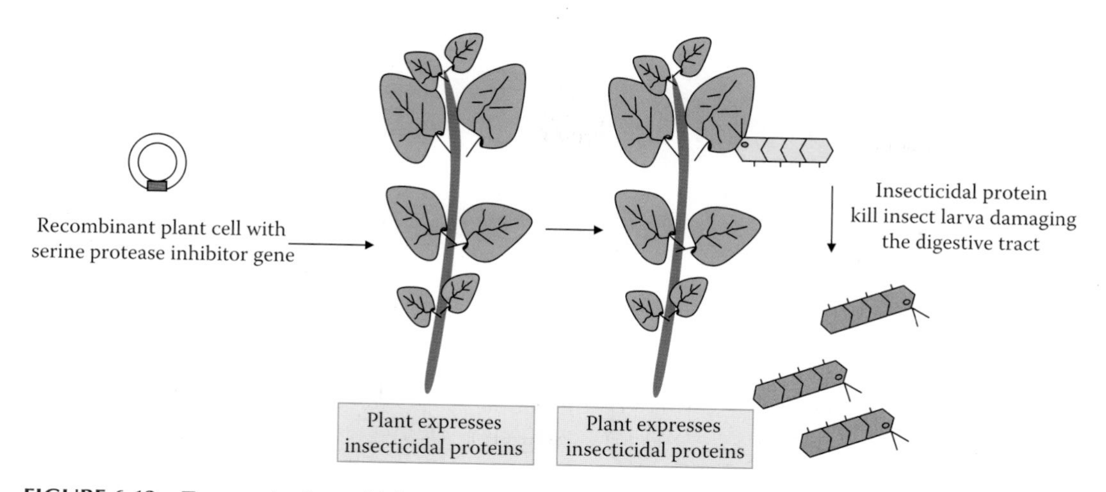

FIGURE 6.12 Transgenic plant with insecticide gene expression.

A second approach to the development of insect-resistant plants has been through the transgenic expression of serine protease inhibitors. These proteins are present in a number of plants and function to deter insects by inhibiting serine proteases in the insect digestive system (Figure 6.12).

6.5 HERBICIDE-TOLERANT PLANTS

The presence of weeds in a crop field reduces the yield. Weed killers or herbicides are not very selective and their current use relies on differential uptake between the weed and the crop plant or on application of the herbicide before planting a field, altering the food content of plants. With the ability to introduce DNA into plants, researchers are trying to create herbicide-tolerant crops using three strategies: (1) by increasing the level of the target enzyme for a particular herbicide, (2) by expressing a mutant enzyme that is not affected by the compound, and (3) by expressing an enzyme that detoxifies the herbicide. Of the large number of herbicides in use today, only a few of the cellular targets have been characterized. One strategy is to clone the cellular target genes into the plant so that they are produced in large amounts. Another strategy for creating herbicide-tolerant plants is by using mutant forms of bacterial EPSPS enzymes. Genes encoding these mutant enzymes have been derived from glycophosphate-resistant bacteria and expressed in plants. These enzymes have lesser inhibition effect by herbicides.

A third strategy is by transgenic expression of enzymes that convert the herbicide to a form that is not toxic to the plant. Some plants have developed their own detoxifying system for certain herbicides. However, these activities in plants are encoded by a complex set of genes that has not yet been fully characterized (Figure 6.13).

6.6 PIGMENTATION IN TRANSGENIC PLANTS

Plants are widely used for ornamental purposes, so it is not surprising that considerable attempts have been made to develop varieties that have flowers of new colors, shapes, and growth properties. Pigmentation in flowers is mainly due to three classes of compounds: the flavonoids, the carotenoids, and the betalains. Of these, the flavonoids are the best characterized with much information now available concerning their chemistry, biochemistry, and molecular genetics. Experiments are underway to expand the spectrum of coloring of certain floral species by introducing genes for the entire pigment biosynthesis pathway. A blue rose was never obtained because rose plants lack the enzymes that synthesize the pigment for blue flower coloration. However, introducing the genes for blue color gave very few successful results. This is due to a phenomenon called *co-suppression*, in

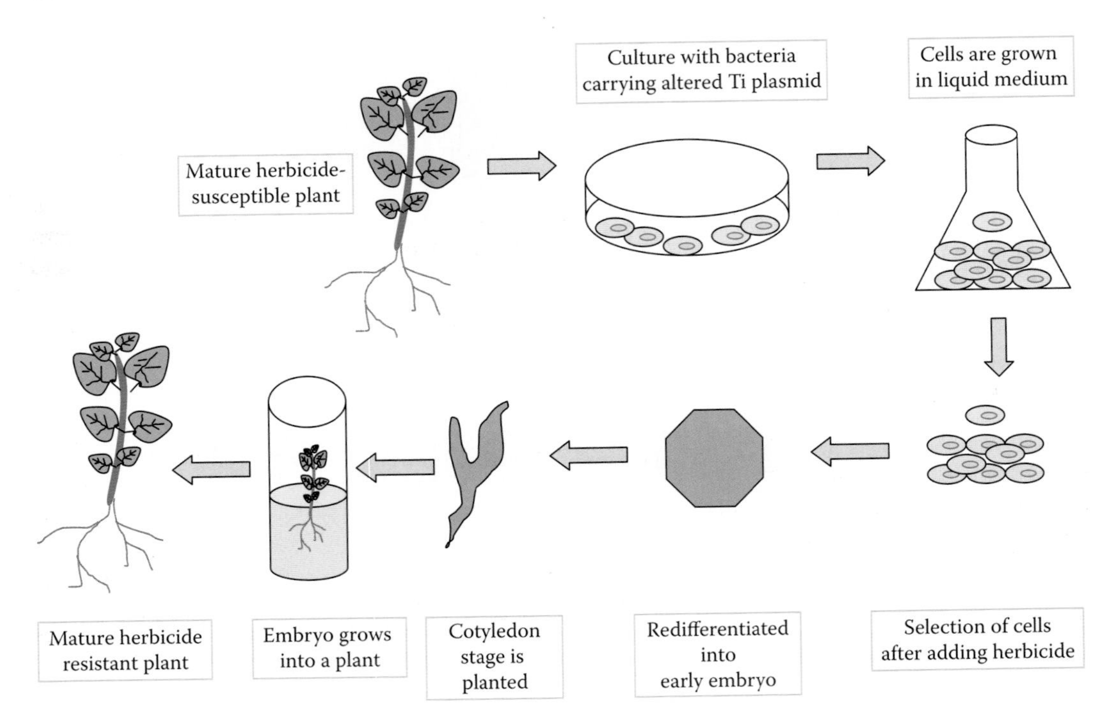

FIGURE 6.13 Herbicide-resistant plant.

which an extra copy of the gene suppresses the expression of the endogenous genes. An experiment was performed in which a second copy of a petunia pigment gene was introduced into a petunia plant with colored flowers. It is expected that increased production of the encoded enzyme might produce flowers with a deeper purple color. However, white colored flowers were produced due to co-suppression. Co-suppression has now been demonstrated in numerous other systems. It does not appear to be a dosage effect resulting from competition for transcription factors, nor is it a result of a system that detects specific duplicate plant genes. Rather, it appears to be the result of a homology-dependent interaction between homologous sequences.

6.7 ALTERING THE FOOD CONTENT OF PLANTS

Starch is the major storage carbohydrate in higher plants. A wide range of different starches are used by the food and other industries. These are obtained by sourcing starch from different plant varieties coupled with new enzymatic or chemical approaches to create novel starches with new functional properties. Higher plants produce over 200 kinds of fatty acids, some of which have value as food. However, many are likely to have industrial (not good) uses of higher value than edible fatty acids. These are widely used in detergent synthesis. *Phytate* is the main storage form of phosphorus in many plant seeds but bound in this form, it is a poor nutrient for monogastric animals. Plants with phytate gene will produce seeds with lower phytate content and higher phosphorus content. Supplementation of broiler diets with transgenic seeds resulted in improved growth rate comparable to diets supplemented with phosphate or fungal phytase.

6.8 GENE TRANSFER METHODS IN PLANTS

Recall that for production of transgenic animals, DNA is usually microinjected into pronuclei of embryonic cells at a very early stage after fertilization, or alternatively, gene targeting of embryo stem (ES) cells is employed. This is possible in animals due to the availability of specialized *in vitro* fertilization technology, which allows manipulation of ovule, zygote, or early embryo. Such techniques are not available in plants. In contrast to this in higher plants, cells or protoplasts can be cultured and used for regeneration of whole plants. Therefore, these protoplasts can be used for gene transfer followed by regeneration, leading to the production of transgenic plants. Besides cultured cells and protoplasts, other meristem cells (immature embryos or organs), pollens, or zygotes can also be used for gene transfer in plants. The enormous diversity of plant species and the availability of diverse genotypes in a species made it necessary to develop a variety of techniques, suiting different situations. These different methods of gene transfer in plants will be discussed in this chapter.

6.9 TARGET CELLS FOR GENE TRANSFORMATION

The first step in gene transfer technology is to select cells that are capable of giving rise to whole transformed plants. Transformation without regeneration and regeneration without transformation are of limited value. In many species, identification of these cell types is difficult. This is unlike the situation in animals, because plant cells are totipotent and can be stimulated to regenerate into whole plants *in vitro* via *organogenesis* or *embryogenesis*. However, *in vitro* plant regeneration imposes a degree of "genome stress," especially if plants are regenerated via a callus phase. This may lead to chromosomal or genetic abnormalities in regenerated plants, a phenomenon referred to as *somaclonal variation*. In contrast to this, gene transfer into pollen (or possibly egg cells) may give rise to genetically transformed gametes, which if used for fertilization (*in vivo*) may give rise to transformed whole plants. Similarly, insertion of DNA into zygote (*in vivo* or *in vitro*) followed by embryo rescue, may also be used to produce transgenic plants. An alternative approach is the use of individual cells in embryos or meristems, which may be grown *in vitro* or may be allowed to develop normally for the production of transgenic plants.

6.10 VECTORS FOR GENE TRANSFER

One common feature of vectors used for transformation is that they carry marker genes, which allow recognition of transformed cells (other cells die due to the action of an antibiotic or herbicide) and are described as selectable markers. Among these marker genes, the most common selectable marker is npt II which provides kanamycin resistance. Other common features of suitable transformation vector are as follows: (i) multiple unique restriction sites (a synthetic polylinker) and (ii) bacterial origins of replication (e.g., ColE1). Vectors having these properties may not necessarily have features that facilitate their transfer to plant cells or integration into the plant nuclear genome. Therefore, Agrobacterium Ti plasmid is preferred over all other vectors because of the wide host range of this bacterial system and the capacity to transfer genes due to the presence of T-DNA border sequences.

6.10.1 Structure and Functions of Ti and Ri Plasmids

The most commonly used vectors for gene transfer in higher plants are based on tumor-inducing mechanism of the soil bacterium *A. tumefaciens*, which is the causal organism for crown gall disease. On the other hand, a closely related species *A. rhizogenes* causes hairy root disease. An understanding of the molecular basis of these diseases led to the utilization of these bacteria for developing gene transfer systems. It has been shown that the disease is caused due to the transfer of a DNA segment from the bacterium to the plant nuclear genome. The DNA segment, which is transferred is called *T-DNA* and is part of a large *Ti* (tumor-inducing) plasmid found in virulent strains of *A. tumefaciens*. Similarly, *Ri* (root-inducing) megaplasmids are found in the virulent strains of *A. rhizogenes*. The Ti and Ri plasmids, inducing crown gall disease and hairy root disease, respectively, have been studied in great detail during the last decade. However, we will discuss only those aspects of these plasmids which are relevant to the design of vectors for gene transfer in higher plants. Most Ti plasmids have four regions in common:

1. Region A, comprising T-DNA, is responsible for tumor induction so that mutations in this region lead to the production of tumors with altered morphology. Sequences homologous to this region are always transferred to plant nuclear genome so that the region is described as T-DNA (transferred DNA);
2. Region B is responsible for replication;
3. Region C is responsible for conjugation;
4. Region D is responsible for virulence so that mutation in this region abolishes virulence and plays a crucial role in the transfer of T-DNA into the plant nuclear genome. The components of this Ti plasmid have been used for developing efficient plant transformation vectors (Figure 6.14).

The T-DNA consists of an *oncogenic region* and an *opine synthesis* (*os*) *region*. An oncogenic (onc) region consists of three genes (two genes, tms1 and tms2, representing "shooty locus" and one gene, tmr, representing "rooty locus") responsible for the biosynthesis of two phytohormones, namely indole acetic acid or IAA (an auxin) and isopentyladenosine 5′-monophosphate (a cytokinin). These genes encode the enzymes responsible for the synthesis of these phytohormones so that the incorporation of these genes in plant nuclear genome leads to the synthesis of these phytohormones in the host plant. The phytohormones in their turn alter the developmental program, leading to the formation of crown gall. An os region is responsible for the synthesis of unusual amino acid or sugar derivatives, which are collectively given the name *opines*. Opines are derived from a variety of compounds (such as arginine + pyruvate) that are found in plant cells. Two most common opines are octopine and nopaline. For the synthesis of octopine and nopaline, the corresponding enzymes

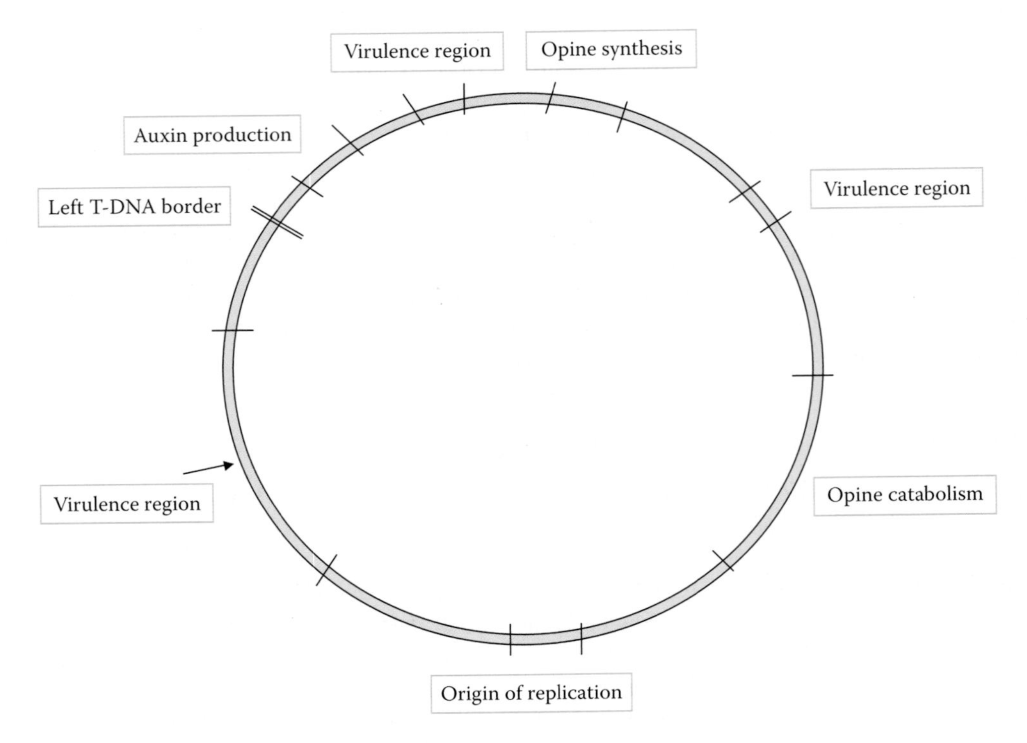

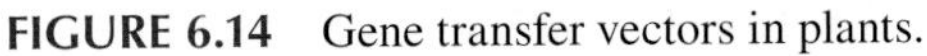
FIGURE 6.14 Gene transfer vectors in plants.

octopine synthase and nopaline synthase are coded by T-DNA. Depending upon whether the Ti plasmid encodes octopine or nopaline, it is described as *octopine-type Ti plasmid* or *nopaline-type Ti plasmid*. Many organisms including higher plants are incapable of utilizing opines, which can be effectively utilized by Agrobacterium. Outside the T-DNA region, Ti plasmid carries genes that catabolize the opines which are utilized as a source of carbon and nitrogen.

The T-DNA regions on all Ti and Ri plasmids are flanked by almost perfect 25 bp direct repeat sequences, which are essential for T-DNA transfer, acting only in *cis* orientation. It has also been shown that any DNA sequence flanked by these 25 bp repeat sequences in the correct orientation can be transferred to plant cells, an attribute that has been successfully utilized for Agrobacterium-mediated gene transfer in higher plants leading to the production of transgenic plants. Besides 25 bp flanking border sequences (with T-DNA), vir region is also essential for T-DNA transfer. While border sequences function in *cis* orientation with respect to T-DNA, vir region is capable of functioning even in transorientation. Consequently, physical separation of T-DNA and vir region into two different plasmids does not affect T-DNA transfer provided both the plasmids are present in the same Agrobacterium cell. This property played an important role in designing vectors for gene transfer in higher plants, as is discussed later.

The vir region (approximately 35 kbp) is organized into six operons, namely vir A, vir B, vir C, vir D, vir E, and vir G, of which four operons (except vir A and vir G) are polycistronic. Genes vir A, B, D, and G are absolutely required for virulence, while the remaining two genes vir C and E are required for tumor formation. The vir A locus is expressed constitutively under all conditions. The vir G locus is expressed at low levels in vegetative cells, but is rapidly induced to higher expression levels by exudates from wounded plant tissue. The vir A and vir G gene products regulate the expression of other vir loci. The vir A product (Vir A) is located on the inner membrane of Agrobacterium cells and is probably a *chemoreceptor*, which senses the presence of phenolic compounds (found in exudates of wounded plant tissue), such as acetosyringone and β-hydroxyaceto syringone. Signal transduction proceeds via activation (possibly phosphorylation) of Vir G (product of gene vir G), which in its turn induces expression of other vir genes.

6.11 TRANSFORMATION TECHNIQUES USING AGROBACTERIUM

Agrobacterium infection (utilizing its plasmids as vectors) has been extensively utilized for transfer of foreign DNA into a number of dicotyledonous species. The only important species that have not responded well are major seed legumes, even though transgenic soybean (Glycine mar) plants have been obtained. The success in this approach for gene transfer has resulted from improvement in tissue culture technology. However, monocotyledons could not be successfully utilized for Agrobacterium-mediated gene transfer except a solitary example of Asparagus. The reasons for this are not fully understood because T-DNA transfer does occur at the cellular level. It is possible that the failure in monocots lies in the lack of wound response of monocotyledonous cells.

6.11.1 Requirements of Transgenic Plants

The important requirements for Agrobacterium-mediated gene transfer in higher plants include the following:

1. The plant explants must produce acetosyringone or other active compounds in order to induce vir genes for virulence. Alternatively, Agrobacterium may be preinduced with synthetic acetosyringone.
2. The induced agrobacteria should have access to cells that are competent for transformation. For gene transfer to occur, cells must be replicating DNA or undergoing mitosis (wounded and dedifferentiated cells, fresh explants, or protoplasts have these properties).

3. Often, transformed tissues or explants do not regenerate and it is difficult to combine transformation competence with totipotency (regeneration ability). Therefore, the transformation competent cells should be able to regenerate in whole plants, a combination that can be easily achieved only in some species such as tobacco. In some cases, undifferentiated cells of embryos may undergo transformation, so that the embryos may develop into chimeric plants.

6.11.2 Explants Used for Transformation

The explants used for inoculation or cocultivation with Agrobacterium carrying the vector include protoplasts, suspension cultured cells, callus cell clumps (undifferentiated and proembryogenic), thin cell layers (epidermis), tissue slices, whole organ sections (such as leaf discs, sections of roots, stems or floral. tissues), etc. Wounding and inoculation of whole plants may also be used.

6.11.3 Marker Genes for Selection and Scoring of Cells

After explants are inoculated with Agrobacterium carrying the requisite vector having the gene of interest, transformed cells/tissues are then needed to be selected. This is facilitated by the presence of selectable marker genes available in the vector. The selectable marker genes enable the transformed cells to survive in media containing toxic levels of the selection agent, which is usually an antibiotic or an herbicide. Tobacco is used as a model transformation system, where explants (such as leaf discs) are placed on regeneration medium containing an antibiotic like kanamycin and transformed shoots can be obtained directly. Any cells, which are not transformed, die due to the presence of kanamycin. Other antibiotics and herbicides may require more judicious use, since even low concentrations can cause rapid cell death. In some cases, selection is exercised only after the regeneration is achieved because adventitious root formation is sensitive to antibiotics.

6.11.4 Neomycin Phosphotransferase Gene

This gene is used as both a selectable and a score-able marker in experiments involving transfer of genes leading to the production of transgenic plants. It imparts kanamycin resistance, so that the transformed tissue can be selected on kanamycin. An assay for NPT II enzyme is also used to detect its presence in transformed tissue or transgenic plants. The gene for NPT II enzyme is often used with nos promoter, which drives its synthesis. In some cases, npt II gene had an adverse effect on the expression of the desirable gene introduced (such as bt2 gene for insect resistance), so that alternative approaches for improving its expression had to be used. To assay an NPT II enzyme, the enzyme is first fractionated using non-denaturing poly acrylamide gel electrophoresis (PAGE). Since the enzyme detoxifies kanamycin by phosphorylation, radioactively labeled ATP (p32) is used with kanamycin in an agar layer, which is used to cover the gel containing the enzyme. The whole set is incubated at 35°C and the phosphorylation leading to incorporation of 32p in kanamycin can be detected by autoradiography. The filter with dot blots is incubated with the substrates and is then subjected to autoradiography to detect the presence NPT II enzyme.

6.12 β-GLUCURONIDASE GUS GENE

The enzyme β-glucuronidase, popularly described as GUS, breaks down glucuronidase giving a colored reaction, so that its presence can be detected *in situ* (inside the plant tissue), used either as thin section or in any other form. Several glucuronidase, which can be used as substrates, include p-nitro phenyl glucuronide (PNPG), 5-bromo, 4-chloro, 3-indolyl/glucuronide (BCIG), naphthol AS-B1 glucuronide (NAG), and resorufin glucuronide (REG). The enzyme GUS is coded by a gene gus, first isolated from *Escherichia coli*. The major advantage of using this reporter gene (score-able marker) lies in its assay, which requires no DNA extraction, electrophoresis, or autoradiography.

6.13 AGROINFECTION AND GENE TRANSFER

Agroinfection is a phenomenon in which a virus infects a host as a part of T-DNA of Ti plasmid carried by Agrobacterium. Viral DNA can be integrated into the T-DNA and can be delivered into plant cells with the normal Agrobacterium T-DNA transfer process. After infection, viral DNA is released to form a functional virus that replicates and spreads systemically. Agroinfection may also lead to the integration of viral DNA so that transgenic plants containing integrated viral DNA can be produced. In maize, agroinfection with maize streak virus has been demonstrated. This suggested that Agrobacterium-based vector system can be used for genetic engineering in cereals, although ordinarily Agrobacterium does not infect monocotyledons. Thus, agroinfection can lead to the production of transgenic plants, even though it has no better chances of yielding transgenic cereals than does Agrobacterium infection alone. However, agroinfection has great potential for studies in virus biology, because it can transfer deletion mutations or even single viral genes.

6.14 DNA-MEDIATED GENE TRANSFER

Agrobacterium-mediated gene transfer has been the most commonly used method of gene transfer in plants but cereals, comprising the most important food crops, are not amenable to this method of gene transfer. Furthermore, in many crops including cereals and legumes, tissue culture techniques for regeneration are not very successful. These two limitations forced intensive search for alternative methods for gene transfer. Physical delivery of DNA or DNA-mediated gene transfer (DMGT), as it is often described, employs methods which can be grouped according to the type of target cell.

For instance, chemically stimulated endocytosis of plasmids or DNA loaded liposomes and electroporation are employed for delivery of DNA to protoplasts only. On the other hand, techniques like microinjection, macroinjection, and shooting with microprojectiles can be used with a variety of explants (such as immature embryos, organ meristems, gametes, zygotes). The latter techniques achieve significance because regeneration of transformed plants from protoplast-derived tissues is still difficult in many species particularly in cereals (although encouraging results have been recently obtained in case of maize rice, wheat, oats, etc.).

6.14.1 Microinjection and Macroinjection

Plant regeneration from transformed protoplasts still remains a problem. Therefore, cultured tissues that encourage the continued development of immature structures provide alternate cellular targets for transformation. These immature structures may include immature embryos, meristems, immature pollens, germinating pollens, isolated ovules, and embryogenic suspension cultured cells. The main disadvantage of this technique is the production of chimeric plants with only a part of the plant transformed.

However, from this chimeric plant, transformed plants of single cell origin can be subsequently obtained. Utilizing this approach, transgenic chimeras have actually been obtained in oilseed rape (*Brassica napus*).When cells or protoplasts are used as targets in the technique of microinjection, glass micropipettes with 0.5–10 μm diameter tip are used for transfer of macromolecules into the cytoplasm or the nucleus of a recipient cell or protoplast. The recipient cells are immobilized on a solid support (such as cover slip or slide), artificially bound to a substrate, or held by a pipette under suction (as done in animal systems). Often, a specially designed micromanipulator is employed for microinjecting the DNA. Although this technique gives a high rate of success, the process is slow, expensive, and requires highly skilled and experienced personnel.

DNA macroinjection employing needles with diameters greater than cell diameter has also been tried. In rye (*Secale cereale*), a marker gene was macroinjected into the stem below the immature floral meristem so as to reach the sporogenous tissue, leading to successful production of transgenic plants. Unfortunately, this technique could not be successfully repeated with any other cereal when tried in several laboratories. Therefore, doubt is expressed about the validity of earlier experiments conducted.

6.15 ELECTROPORATION FOR GENE TRANSFER

This method is based on the use of short electrical impulses of high field strength. These impulses increase the permeability of protoplast membrane and facilitate entry of DNA molecules into the cells, if the DNA is in direct contact with the membrane.

In view of this, for delivery of DNA to protoplasts, electroporation is one of the several routine techniques for efficient transformation. However, since regeneration from protoplasts is not always possible, cultured cells or tissue explants are often used. Consequently, it is important to test whether electroporation could transfer genes into walled cells. In most of these cases, no proof of transformation was available.

The electroporation pulse is generated by discharging a capacitor across the electrodes in a specially designed electroporation chamber. Either a high voltage (1.5 kV) rectangular wave pulse of short duration or a low voltage (350 V) pulse of long duration is used. The latter can be generated by a home-made machine. Protoplasts in an ionic solution containing the vector DNA are suspended between the electrodes, electroporated, and then plated as usual. Transformed colonies are selected as described earlier. Using electroporation method, successful transfer of genes was achieved with the protoplasts of tobacco, petunia, maize, rice, wheat, and sorghum. In most of these cases, cat gene associated with a suitable promoter sequence was transferred. Transformation frequencies can be further improved by using field strength of 1.25 kV/cm, adding PEG after adding DNA, heat shocking protoplasts at 45°C for 5 min before adding DNA, and by using linear instead of circular DNA.

6.16 LIPOSOME-MEDIATED GENE TRANSFER

Liposomes are small lipid bags in which a large number of plasmids are enclosed. They can be induced to fuse with protoplasts using devices like PEG, and therefore have been used for gene transfer. This technique offers various advantages which include protection of DNA and RNA from nuclease digestion, low cell toxicity, stability, and storage of nucleic acids due to encapsulation in liposomes, high degree of reproducibility, and applicability to a wide range of cell types. In this technique, DNA enters the protoplasts due to endocytosis of liposomes, involving adhesion of the liposomes to the protoplast surface, and fusion of liposomes at the site of adhesion and then release of plasmids inside the cell. The technique has been successfully used to deliver DNA into the protoplasts of a number of plant species (such as tobacco, petunia, and carrot).

6.17 GENE TRANSFORMATION USING POLLEN

There has been a hope that DNA can be taken up by the germinating pollen and can integrate either into sperm nuclei or reach the zygote through the pollen tube pathway. Both these approaches have been tried and interesting phenotypic alterations suggesting gene transfer have been obtained. In no case, however, unequivocal proof of gene transfer has been available. In a number of experiments, when marker genes were used for transfer, only negative results were obtained. Several problems exist in this method and these include the presence of cell wall, nucleases, heterochromatic state of acceptor DNA, callose plugs in pollen tube, etc. Transgenic plants have never been recovered using this approach and this method, though very attractive, seems to have little potential for gene transfer.

6.18 APPLICATION OF TRANSGENIC PLANTS

Transgenic plants have proved to be extremely valuable tools in studies on plant molecular biology, regulation of gene action, and identification of regulatory/promoter sequences. Specific genes have been transferred into plants to improve their agronomic and other features. Genes for resistance to various biotic stresses have been engineered to generate transgenic plants resistant to insects, viruses, etc. Several gene transfers have been aimed at improving the produce quality. Transgenic plants are being used to produce novel biochemicals such as hirudin which are not produced by normal plants. Also, transgenic plants are now used in making vaccines for immunization against pathogens. In the following sections, we will learn some of the major applications of transgenic plants with suitable examples.

6.18.1 Detoxification or Degradation of Herbicides

Several detoxifying enzymes have been identified in plants as well as in microbes such as glutathione-S-transferase enzyme (found in maize and other plants) which detoxifies the herbicide atrazine, and nitrilase (encoded by gene bxn from *Klebsiella pneumoniae*) which detoxifies the herbicide bromoxynil. Similarly, phosphinothricin acetyltransferase encoded by bar gene from Streptomyces spp. detoxifies the herbicide L-phosphinothricin. Transgenic tomato, potato, oilseed rape (*Brassica napus*) and sugar beet plants expressing the bar gene from Streptomyces have been obtained. These were found to be resistant to the herbicide phosphinothricin. Similarly, transgenic tomato plants expressing the bxn gene from Klebsiella were resistant to the herbicide bromoxynil. Field trials with transgenic herbicide–resistant crops have been very successful, and several such varieties are in commercial cultivation especially in the United States.

6.18.2 Crystal (Cry) Proteins

The cry gene of *B. thuringiensis* produces a protein which forms crystalline inclusions in the bacterial spores. These crystal proteins are responsible for the insecticidal activities of this bacterium. The cry genes (or Cry proteins) have so far been grouped into 16 distinct groups, which either code for a 130kDa or a 70kDa protein. These proteins are solubilized in the alkaline environment of insect midgut and are then prototypically processed to yield a 60kDa toxic, core fragment (except in the case of cry IVD). The toxin function is localized in the N-terminal half of tile 130kDa proteins; the C-terminal half of these proteins is highly conserved and is most likely involved in crystal formation. The Cry I proteins are insecticidal to Lepidopteron insects. All the proteins, even the Cry IA subfamily, have a distinctive insecticidal spectrum. The Cry IIA proteins are active against both Lepidoptera and Diptera, while Cry IIB is specific to Diptera. The Cry III proteins are active against Coleoptera species, while Cry IV proteins are specific to Diptera. But the CytA protein does not show any insecticidal activity. It is cytolytic for a variety of vertebrate and invertebrate cells, and exhibits no homology with other Cry proteins.

6.18.3 Toxic Action of Cry Proteins

When Cry proteins are ingested by insects, they are dissolved in the alkaline juices present in the midgut lumen. The gut proteases process them hydrolytically to release the core toxic fragments. The toxic fragments are believed to bind to specific high-affinity receptors present in the brush border of midgut epithelial cells. As a result, the brush border membranes develop pores, most likely nonspecific in nature, permitting influx into the epithelial cells of ions and water which causes their swelling and eventual lysis. The presence of specific receptors in the midgut epithelium is most likely the chief reason for Cry toxin specificity. The specificity seems to be lost upon reduction of the cysteine residues of the protoxin, but can be restored by reoxidation of these residues.

6.18.4 Expression of Cry Genes in Plants

The Cry genes have been successfully transferred into tobacco, potato, and tomato. Typically, truncated cry genes are used for the production of transgenic plants since the level of expression of the complete genes in the transgenic plants is extremely low. In 1987, the first report on the response of transgenic tobacco plants to the insects *Manduca sexta* and *Heliothis virescens* was published. The transgenic tobacco plants expressing Cry protein at about 0.004% of their total leaf protein killed all *M. sexta* larvae within 6 days. Similarly, fruits of transgenic tomatoes grown in the field showed less fruit damage even under heavy infestation by fruitworm and pinworm larvae. Field tests have been quite successful with transgenic cotton as well. This crop requires about 60% of the total insecticides used in agriculture for a successful cultivation.

6.18.5 Insect Resistance to Cry Proteins

There were some reports on the development of resistance in some insects to Cry proteins. For example, *Plodia interpunctella* and *Cadra cautella* were selected by continuous exposure to high levels of *B. thuringiensis*. There was a >250-fold increase in resistance in *P. interpunctella* after 36 generations, and a sevenfold increase in *C. cautella* after 21 generations. The problem of development of insect resistance to Cry proteins may be managed by combining or alternating two or more kinds of these proteins and by reducing the selection pressure on insects by limiting cry gene expression to only the economically important plant parts. Several important insect pests are not susceptible to the currently available Cry proteins. For such insects, alternative insecticidal proteins will be needed such as inhibitors of digestive enzymes (cowpea trypsin inhibitor or CpTI), serine protease inhibitor (aprotinin), cystein protease inhibitor, proteinase inhibitor II, and lectins. These genes have been transferred into certain crop plants where they produce resistance to different insects such as members of Lepidoptera and Coleoptera species.

6.18.6 Virus Resistance

Several approaches have been used to engineer plants for virus resistance, which include CP gene, cDNA of satellite RNA, defective viral genome, antisense RNA approach, and ribozyme-mediated protection. Of these strategies, the use of CP gene has been the most successful. Transgenic plants having virus CP gene linked to a strong promoter have been produced by many crop plants such as tobacco, tomato, alfalfa, sugar beet, and potato. The first transgenic plant of this type was tobacco produced in 1986. It contained the CP gene of tobacco mosaic virus (TMV) strain U I. When these plants were inoculated with TMV U I, symptoms either failed to develop or were considerably delayed. Furthermore, there was a much less accumulation of virus than in the control plants in both inoculated and systemically infected leaves. In addition, these plants showed delayed expression of disease symptoms when inoculated with the related tomato mosaic virus (ToMV) and with tobacco mild green mosaic virus (TMGMV).

It appears to be a common feature that expression of a virus CP gene not only confers resistance to the concerned virus but also gives a measure of resistance to related viruses. The effectiveness of CP gene in conferring virus resistance can be affected by both the amount of CP produced in transgenic plants and by the concentration of virus inoculum. Most likely the resistance generated by CP is due to the blocking of the process of uncoating of virus particles, which is necessary for viral genome replication as well as expression. However, other effects seem to be involved in producing CP–mediated virus resistance. One such mechanism appears to be the prevention or delay of systemic spread of the viruses. At least in some cases, the resistance mechanism does not involve the CP itself since CP genes even in antisense orientation produce resistance to the virus.

6.18.7 Drought Resistance

A number of drought-resistant genes have been identified, isolated, cloned, and expressed in plants. These potential sources of resistance to abiotic stresses include Rab (responsive to abscisic acid) and SalT (induced in response to salt stress) genes of rice, genes for enzymes involved in proline biosynthesis in bacteria (proBA and proC in *E. coli*) and plants, and spinach genes which are involved in betaine synthesis. In plants, proline is preferentially produced from ornithine under normal conditions. However, under stress, it is made directly from glutamate, the first two reactions of the pathway being catalyzed by a single enzyme Δ^1-pyrroline 5-carboxylate synthetase (P5CS). The gene encoding P5CS has been isolated from soybean and moth bean, and cloned. The moth bean P5CS gene has been transferred and overexpressed in tobacco. The transgenic plants produced 10–18-fold more proline than the control plants. The leaves of transgenic plants retained a higher osmotic potential and showed a greater root biomass under water stress than did the control plants.

These findings indicate that overexpression of P5CS in plants enhances their tolerance to osmotic stress. The primary function of accumulation of proline and other solutes such as glycine betaine appears to be the regulation of intracellular water activity. Under water stress, they may induce the formation of strong H-bonded water around proteins, thereby preserving the native state of cell biopolymers. However, it should be kept in mind that accumulation of proline is only one of the factors that enable plants to sustain growth under water stress. Other factors also allow plants to overcome osmotic stress. For example, expression of *E. coli* gene mtl1D in plants leads to mannitol accumulation and some degree of enhanced growth under stress. An important aspect of such manipulations, however, remains that the basal metabolism of the plant should be able to sustain a high rate of accumulation of the concerned osmolytes without too much of a "cost" to the plants.

6.18.8 Modification of Seed Protein Quality

Cereal seed proteins are deficient in lysine, while those of pulses are deficient in sulfur containing amino acids such as methionine and tryptophan. This limits their nutritional value since these amino acids are essential for man. Therefore, improvement of seed storage protein quality is an important and seemingly feasible objective. The approaches to achieve this objective may be grouped into the following two broad categories which includes introduction of an appropriate transgene, and modification of the endogenous protein-encoding gene.

6.18.8.1 Introduction of an Appropriate Transgene

In this approach, a new gene encoding a storage protein, which is rich in the deficient amino acids, is introduced into the crop to correct its amino acid deficiency. The transgene is linked to a seed-specific promoter to ensure its expression only in seeds.

Vicilin is the major seed storage protein of pea. It contains 7% lysine but no sulfur containing amino acid (methionine and cysteine). Therefore, pea seed protein is generally low in sulfur containing amino acids, which needs to be ameliorated. In contrast, a sunflower seed storage protein, sunflower albumin 8 (SF A8), contains 23% methionine plus cysteine. The gene coding for SF A8 has been isolated. The SF A8 gene has been fused with the vicilin gene promoter and expressed in tobacco, which showed the accumulation of the protein in the seeds of transgenic tobacco.

6.18.8.2 Modification of Endogenous Genes

This approach to improve seed storage protein quality is based on the isolation and modification of the concerned protein-encoding gene sequence either by replacing one or few codons with the selected codons or by inserting one or few selected additional codons at appropriate sites. For example, prolamine storage proteins, such as zein of cereals, are deficient in the essential amino acids lysine and tryptophan. Single lysine replacements in the N-terminal coding sequence as well as within and between the peptide repeats, and double lysine replacement constructs of prolamine genes have been prepared. In addition, short oligonucleotides encoding lysine- and tryptophan-rich peptides were inserted separately at several different points in the coding sequence. These constructs were shown to express well in *Xenopus oocytes*, and their polypeptides were able to form normal aggregates of protein bodies.

6.18.8.3 Successful Examples

In a recent study, the 7 S legume seed storage protein, β-phaseolin, gene (driven by rice storage protein gene gtl or glutein 1) promoter was transferred in rice. Transgenic rice plants expressed the gene in their endosperm, and some plants showed up to 4% of their total proteins to be β-phaseolin. The 11 S legumin protein gene driven by gtl promoter has also been transferred, and expressed in rice endosperm. In another study, Du Pont (USA) scientists have synthesized and patented a gene encoding a protein called CP 3–5 which contains 35% lysine and 22% methionine. The CP 3–5 gene was coupled with seed-specific promoters and transferred into maize. Rice gtl gene encodes the major rice seed storage protein. It has been modified to encode higher levels of lysine, tryptophan, and methionine. The modified gtl gene, driven by its own promoter, was transferred into rice protoplasts. The resulting transgenic rice plants express the modified gene in their developing endosperm. Similarly, a modified zein protein gene encodes a protein having improved methionine. When this gene was introduced in maize, rice, and wheat, the transgenic plants showed up to 3.8% methionine in their seed proteins.

6.18.9 Co-Suppression of Genes

In case of many endogenous plant genes, an overexpression of the sense RNA or mRNA surprisingly leads to a drastic reduction in the level of expression of the genes concerned; this is called *co-suppression*. One way of achieving an overexpression of the mRNA is to introduce a homologous sense construct of the gene concerned so that it also produces sense RNA or mRNA (in addition to the endogenously present gene). The efficiency of co-suppression seems to vary among plant genes. Co-suppression has never been observed for the petunia chalcone isomerase gene, while tobacco glutamine synthetase nuclear gene is always co-repressed. CHS gene (petunia) represents the intermediate situation. The mechanism of co-suppression is not understood. According to a threshold model, when RNA transcripts of a *getle* accumulate beyond a critical, threshold level, they are selectively degraded by RNases. An accumulation of high levels of RNA transcripts of a gene may lead to the production of aberrant sense RNA transcripts of the transgene.

An accumulation of aberrant RNA transcripts is proposed to activate RNA-dependent RNA polymerase of plant origin, which transcribes the RNA transcripts to produce antisense RNA. The antisense RNA transcripts would associate with the accumulated normal and aberrant RNA transcripts of the transgene as well as the endogenous gene. This will produce RNA duplexes, which present targets for double-stranded RNA-specific RNases like RNase H. Degradation of the RNA transcripts of a gene is postulated to somehow lead to a hypermethylation of the DNA sequences homologous to the degraded RNA sequences. This often leads to a drastic reduction in the level of expression of the transgene in question and also of homologous endogenous gene(s), if any; this is called gene silencing.

Ethylene is an important phytohormone and is involved, among other things, in fruit ripening, leaf abscission, and flower senescence. It is produced from amino acid methionine, the terminal two reactions of ethylene biosynthesis being as follows. A reduced ethylene production results in delayed petal senescence in carnation and slow ripening of tomato fruits. Drastically reduced ethylene production has been achieved by (i) expression of antisense constructs of ACC synthase or ACC oxidase, (ii) co-suppression of either of these enzymes, and (iii) expression of enzymes that metabolize S-adenosyl methionine (SAM) such as SAM hydrolase from bacteriophase T3 (in tomato), or ACC such as ACC deaminase (overexpression in tomato). A carnation variety with longer vase life has its ACC synthase gene co-suppressed. A similar co-suppression approach has been used to block the onset of fruit ripening in tomatoes.

6.18.10 RNA-Mediated Interference

Silencing of homologous gene expression triggered by double-stranded RNA (dsRNA) is called *RNA-mediated interference* or RNAi. Introduction of long dsRNA into the cells of plants, invertebrates, as well as mammals leads to a sequence-specific degradation of the homologous gene transcripts. The long dsRNA molecules are cleaved by an RNase III enzyme called *Dicer*. This generates small 21–23 nucleotide long dsRNA molecules called *small interfering RNAs* (siRNAs). The siRNA molecules bind to a protein complex called *RNA-induced silencing complex*. This complex contains a helicase activity that unwinds the two strands of RNA molecules.

The antisense RNA strands so generated pair with the target RNA molecules, and an endonuclease activity then hydrolyzes the target RNA at the site where the antisense strand is bound. The RNAi is a recent but potent technology and is rapidly gaining wide acceptance. The main applications of RNAi are that (i) it serves as an antiviral defense mechanism; (ii) it is becoming a powerful and widely used tool for the analysis of gene function in invertebrates, plants, and mammals; and (iii) DNA vector–based strategy allows the suppression of endogenous genes and to produce transgenic lines with suitably modified traits. RNAi has been used to produce low caffeine coffee.

6.18.11 Biochemical Production in Plants

Production of biomass (or the culture filtrate for externally secreted products) from which the desired biochemical is to be purified is called *upstream production*.

Upstream production costs are much lower in case of plants than in microorganisms. The cellular environment of plants, on the other hand, is as good as that of animals for posttranslational modification and folding of the proteins so that animal proteins recovered from plants are usually at least as good as the native animal proteins. Plant seeds could be stored under ambient conditions from which the proteins may be isolated as per need. This would greatly reduce storage costs of the products. Biochemical production from transgenic animals raises several public and ethical concerns, while no such problem is associated with plants. However, in most cases, low levels of production are the chief problem. Purification of the desired biochemical from the biomass or the culture medium is called *downstream processing*. Generally, downstream processing from plants is more difficult and costlier than that from microorganisms mainly due to the low concentration of recombinant protein in the total biomass. Sometimes, an accumulation of a transgene product may adversely affect the performance of plants.

6.18.12 Plant-Derived Vaccines

Conventional vaccines consist of attenuated or inactivated pathogens. In case of many pathogens, the gene encoding a critical antigen has been isolated and expressed in bacteria/animals, and the recombinant protein so produced is used as a vaccine. Such vaccines are called *recombinant vaccines*. Recombinant vaccines are produced through bacterial fermentation or in animal cell cultures, which often makes their cost prohibitively high. In addition, storage and transport of vaccines especially in developing countries presents many problems such as cold storage. Therefore, plants are being developed as an alternative vaccine production and delivery systems. There are two basic methods by which these vaccines are being developed which include development of edible vaccines and production of recombinant antigenic proteins to be used as vaccines.

6.18.12.1 Edible Vaccines

Antigens of several pathogens, such as enteric pathogens, produce immunogenic response when delivered orally. Such antigens are good candidates for edible vaccines. The strategy, in simple terms, is as follows. The gene encoding the orally active antigenic protein is isolated from the pathogen, and a suitable construct for constitutive or tissue-specific expression of the gene is prepared. The gene contract is introduced and stably integrated into the genome of selected plant species and is expressed to produce the antigen. The appropriate plant parts containing the antigen may be fed raw to animals or humans to bring about immunization. For animals, crops used as feed, such as alfalfa, are suitable for the expression of such antigens, while for humans fruits like banana, which are consumed raw, have to be used. Thus, edible vaccines are produced from transgenic plants in which an orally active antigen of the target pathogen is expressed and accumulated, and which is fed to animals/humans for immunization against the pathogen.

The edible vaccines are expected to alleviate storage problems, offer an easy delivery system by feeding, and incur much lower costs than the recombinant vaccines produced by bacterial fermentation. The development of edible vaccines for animals is the first objective, followed by those for humans. It may be pointed out that animals develop a tolerance to the components of their routine food, so that these become nonimmunogenic in them. Therefore, edible vaccines cannot be used as a regular component of animal/human food. Several genes encoding antigenic proteins have been expressed in plants where they are produced in their native immunogenic forms.

An example of an edible vaccine is provided by the *E. coli* heat labile enterotoxin (LT) B. subunit (LT-B) expressed in potato. Potato tubers containing LT-B were fed raw to mice. Only four feedings of 5 g tuber each induced immune response in mice, while mice fed on normal nontransgenic tubers did not show any immune response.

6.18.12.2 Recombinant and Subunit Vaccines

Plants can serve as efficient production systems for antigenic recombinant subunit proteins, which can be purified and used as vaccines. There are two distinct strategies for the production of recombinant antigens in plants which include integration of the transgene into plant genome and expression as a CP fusion of a plant virus. The transgene encoding the antigenic protein may be integrated into the plant genome, and the recombinant protein produced by the plant is purified for use as a vaccine. The antigens need to be neither active nor used orally; they can be administered parenterally. Alternatively, the DNA sequence representing the epitope of the concerned antigen can be fused with the CP gene of a plant virus such as TMV or cowpea mosaic virus (CPMV). The recombinant virus is used to infect plants where the virus multiplies and spreads systemically in the plants. The virus particles are purified from the infected plants and may be used for oral or parenteral immunization. The antigen epitope is displayed on the CP and is present in a particulate form (virions), which is highly immunogenic.

6.18.13 Hirudin A Polypeptide

Hirudin is encoded by a synthetic gene and is expressed in fusion with the oil body protein olesin which greatly facilitates the purification of hirudin. The seed tissue expressing hirudin fused with olesin is extracted with water and the extract is centrifuged. Oil bodies containing the olesin fusion protein float on the surface and are easily separated from the rest of the seed proteins. The hirudin is cleaved from olesin at a protease recognition site located at their junction. Thus, hirudin provides a successful example of transgene expression in plants for the isolation of a polypeptide of interest at a commercial scale.

6.18.14 Phytase as an Enzyme

An example of a promising enzyme produced in plants is provided by phytase. Phytase is encoded by ao gene from the fungus *Aspergillus niger*. It enhances phosphorus utilization by chickens from their feed to the extent that phosphate supplement in feed becomes unnecessary. Transgenic tobacco seeds expressing the gene encoding phytase were fed to broiler chickens for a period of 4 weeks. This produced a gain in body weight that was comparable to those obtained with feed supplemented with phosphate or with *A. niger* phytase. This raises the feasibility of a direct use of proteins/enzymes (without purification) expressed in plants either as food/feed or for industrial applications.

6.18.15 Polyhydroxybutyrate Biodegradable Plastic Substrate

Polyhydroxy alkanoates (PHAs), such as polyhydroxybutyrate (PHB), are synthesized from acetyl-CoA used as precursor, and are used for the synthesis of biodegradable plastics with thermoplastic properties. At present, PHAs are produced by bacterial fermentation, and the cost of biodegradable plastic is substantially higher than that of synthetic plastics. Attempts are being made to produce PHAs in transgenic plants to reduce the cost. Genes encoding the two enzymes, aceto-acetyl-CoA reductase (PhbB) and PHB synthase (phbC), involved in the PHB synthesis from the precursor acetyl-CoA have been transferred from the bacterium *Alcaligenes eutrophus* and expressed in *Arabidopsis thaliana*. When the two enzymes were targeted into the plastids, PHB accumulated in leaves. PHB production by transgenic plants provides an example of a novel compound synthesized in plants. Transgenic trees like popular expressing phbB and phbC accumulate PHB in their leaves. The leaves are collected and used for PHB extraction.

6.19 HOW SAFE ARE TRANSGENIC PLANTS?

The main concerns while taking transgenic plants to the field relate to the possibilities of their becoming persistent weeds, gene transfers from them to other plants making the latter more persistent or invasive, and their being detrimental to the environment. In general, testing of transgenic plants should progress in a stepwise manner from laboratory to growth chamber, to greenhouse, to limited field testing, to large-scale field testing. Many countries have developed their own procedures and policies regulating field tests of such plants. In India, DBT, New Delhi, is concerned with the regulation of field testing of transgenic plants.

The antibiotic resistance genes used for the selection of transformed cells is expressed in every cell of the resulting transgenic plants. It has been argued that such genes and their protein products could cause problems to human health and the environment. The protein products of such genes could be toxic to humans/animals. The food from transgenic crops will contain the antibiotic resistance gene. When such food is consumed, the bacteria present in human intestine could acquire the antibiotic resistance gene present in the food. This would make the bacteria resistant to the antibiotics concerned, and they may become difficult to manage. The antibiotic resistance gene could be passed on from the transgenic crops to some other organisms in the environment, and this could damage the ecosystem.

There are two approaches to resolve these issues which include excision of the antibiotic resistance gene following transformation and selection, and use of nonantibiotic resistance selectable reporter genes. Herbicide resistance markers are similar in action to antibiotic resistance markers in that they save the transformed cells from the killing action of the selection agent.

The strategies of selection based on growth regulator autotrophy or substrate utilization, in contrast, provide the transformed cells with a metabolic advantage over the nontransformed ones, which are not killed by the selection agent. These strategies are based on the consideration that explants of most plant species are in an auxotrophic state, and they are unable to regenerate and grow in the absence of an external supply of several substances.

6.20 BIOENGINEERED PLANTS

After learning about plant-breeding techniques and plant pathology, it becomes important to eradicate various types of plant diseases. The advances in modern molecular tools make it possible to create disease-free plants by manipulating their genes. In this section, we will closely look into famous examples of bioengineered agricultural products that have revolutionized the modern crop cultivation.

6.20.1 Golden Rice

Dietary micronutrient deficiencies, such as the lack of vitamin A, iodine, iron, or zinc, are a major source of morbidity (increased susceptibility to disease) and mortality worldwide. These deficiencies affect particularly children, impairing their immune systems and normal development, causing disease and ultimately death. The best way to avoid micronutrient deficiencies is by creating genetically modified (GM) crops, and golden rice is the best example of such crop. Golden rice was created by Ingo Potrykus of the Institute of Plant Sciences at the Swiss Federal Institute of Technology together with Peter Beyer of the University of Freiburg. The project started in 1992 and at the time of publication in 2000, golden rice was considered a significant breakthrough in biotechnology as the researchers had engineered an entire biosynthetic pathway. Golden rice was designed to produce beta-carotene, a precursor of vitamin A, in the part of rice that people eat, the endosperm. The rice plant can naturally produce beta-carotene, which is a carotenoid pigment that occurs in the leaves and is involved in photosynthesis. However, the plant does not normally produce this pigment in the endosperm since photosynthesis does not occur in the endosperm (Figure 6.15).

Three transgenes providing phytoene synthase, phytoene desaturase, zeta-carotene desaturase, and lycopene cyclase activities were transferred into rice by Agrobacterium-mediated transformation. All the transgenes were introduced together in a single co-transformation experiment. The resulting transgenic rice, popularly called "golden rice" contains good quantities of beta-carotene, which gives the grains a golden color. In one transgenic line, the beta-carotene content was as high as 85% of the total carotenoids present in the grain. The original golden rice was called SGR1, and under greenhouse conditions, it produced 1.6 μg/g of carotenoids. Golden rice has been bred with local rice cultivars in the Philippines, Taiwan, and with the American rice cultivar "Cocodrie." The first field trials of these golden rice cultivars were conducted by the Louisiana State University Agricultural Center in 2004. Field testing allowed a more accurate measurement of the nutritional value of golden rice and enabled feeding tests. Preliminary results from the field tests had shown that field-grown golden rice produces 4–5 times more beta-carotene than that grown under greenhouse conditions. In 2005, a team of researchers at the biotechnology company Syngenta produced

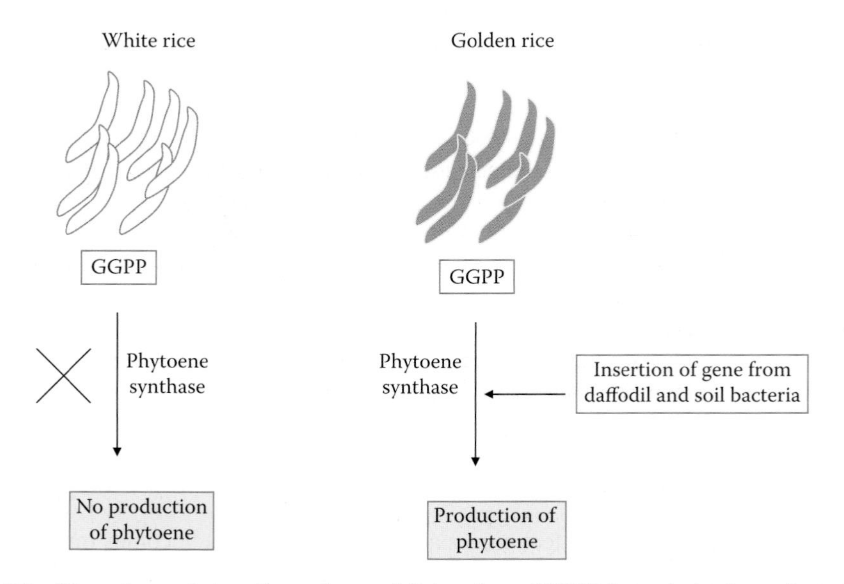

FIGURE 6.15 Bioengineered rice: Geranylgeranyldiphosphate (GGPP) is made in the grain endosperm of white rice but is not converted to phytoene because the rice phytoene synthase is not expressed in this part of the plant. The original golden rice used a phytoene synthase from daffodil (*Narcissus psuedonarcissus*) and a carotene desaturase (CrtI) from the soil bacterium *Erwiniauredovora* to convert GGPP to colored carotenoids and yielded 1.6 g of total carotenoids per gram of dry weight.

a variety of golden rice called "Golden Rice 2." They combined the phytoene synthase gene from maize with *crtI* from the original golden rice. Golden Rice 2 produces 23 times more carotenoids than golden rice (up to 37 μg/g), and preferentially accumulates beta-carotene (up to 31 μg/g of the 37 μg/g of carotenoids). In June 2005, researcher Peter Beyer received funding from the Bill and Melinda Gates Foundation to further improve golden rice by increasing the levels of or the bioavailability of provitamin A, vitamin E, iron, and zinc, and to improve protein quality through genetic modification.

Critics of genetically engineered crops have raised various concerns. One of these is that golden rice originally did not have sufficient vitamin A, but new strains were developed that solve this problem. However, there are still doubts about the speed at which vitamin A degrades once the plant is harvested, and how much would remain after cooking. In particular, since carotenes are hydrophobic, there needs to be sufficient amount of fat present in the diet for golden rice (or most other vitamin A supplements) to be able to alleviate vitamin A deficiency. In that respect, it is significant that vitamin A deficiency is rarely an isolated phenomenon, but usually coupled to a general lack of a balanced diet. Hence, assuming a bioavailability on par with other natural sources of provitamin A, Greenpeace estimated that adult humans would need to eat about 9 kg of cooked golden rice of the first breed to receive their RDA of beta-carotene, while a breast-feeding woman would need twice the amount; the effects of an unbalanced (fat-deficient) diet were not fully accounted for. In other words, it would probably have been both physically impossible to grow enough as well as to eat enough of the original golden rice to alleviate debilitating vitamin A deficiency. This claim, however, referred to a prototype cultivar of golden rice. More recent versions have considerably higher quantities of vitamin A in them.

6.20.2 Tomato "Flavr Savr"

The Flavr Savr tomato was the first commercially grown genetically engineered food to be granted a license for human consumption. It was produced by the California-based company Calgene, and submitted to the U.S. Food and Drug Administration (FDA) in 1992. It was first sold in 1994, and was only available for a few years before production ceased. Calgene made history but mounting costs prevented it from becoming profitable, and it was eventually acquired by Monsanto. The Flavr Savr tomato was created by using antisense technology. In any gene, the DNA strand oriented as $3' \rightarrow 5'$ in relation to its promoter is transcribed; this strand is called the *antisense strand*. The mRNA base sequence, therefore, is complementary to that of the antisense strand. The remaining DNA strand of the gene, called *sense strand*, is naturally complementary to the antisense strand of the gene. Therefore, the base sequence of sense strand of a gene is the same as that of the mRNA produced by it. Hence, the mRNA produced by a gene in normal orientation is also known as *sense RNA*. However, an antisense gene is produced by inverting or reversing the orientation of the protein-encoding region of a gene in relation to its promoter and as a result, the natural sense strand of the gene becomes oriented in the $3' \rightarrow 5'$ direction with reference to its promoter, and is transcribed. The normal antisense strand is not transcribed since now its orientation is $5' \rightarrow 3'$. The RNA produced by this gene has the same sequence as the antisense strand of the normal gene (except for T in DNA in the place of U in RNA), and is therefore known as *antisense RNA*. When an antisense gene is present in the same nucleus as the normal endogenous gene, transcription of the two genes yields antisense and sense RNA transcripts, respectively (Figure 6.16).

Since the sense and the antisense RNAs are complementary to each other, they would pair to produce dsRNA molecules. This event makes the mRNA unavailable for translation. At the same time, the RNA double strand is attacked and degraded by dsRNA-specific RNases. Finally, these events may somehow lead to the methylation of the promoter and coding regions of the normal gene, resulting in the silencing of the endogenous gene. The application of antisense RNA technology is explained using the slow ripening of tomato as an example. In tomato, enzyme polygalacturonase (PG) degrades pectin which is the major component of fruit cell wall. This leads to the

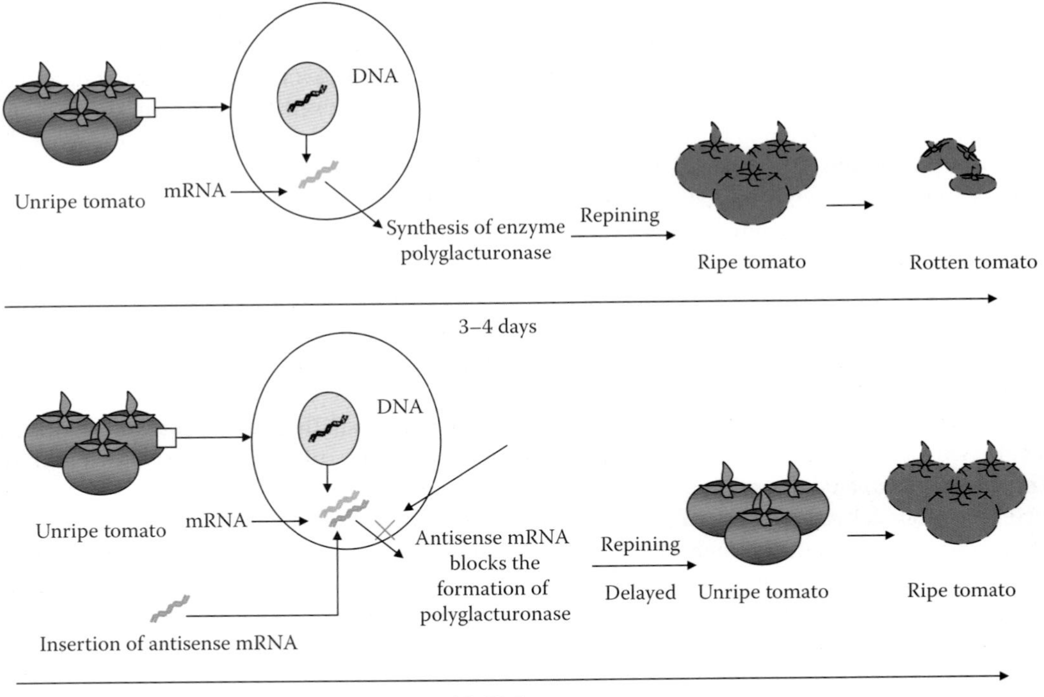

FIGURE 6.16 Making of transgenic tomato.

softening of fruits and deterioration in fruit quality. Transgenic tomatoes have been produced which contain antisense construct of the gene encoding PG. These transgenics show a drastically reduced expression of PG and markedly slower ripening and fruit softening. This has greatly improved the shelf life and the general quality of tomato fruits. Such tomatoes are being marketed in the United States under the name "Flavr Savr."

6.21 GENETICALLY MODIFIED MAIZE

Genetically engineered maize or transgenic maize (Maize) is created by incorporating desirable traits such as herbicide and pest resistance. Transgenic maize is currently grown commercially in the United States. Maize varieties resistant to glyphosate isopropylamine (salt) (Liberty) herbicides and Roundup have been produced. There are also maize hybrids with tolerance to imidazoline herbicides marketed by Pioneer Hi-Bred under the trademark Clearfield, but in these the herbicide tolerance trait was bred without the use of genetic engineering. Consequently, the regulatory framework governing the approval, use, trade, and consumption of transgenic crops does not apply for imidazoline-tolerant maize. Herbicide-resistant GM maize is grown in the United States. A variation of herbicide-resistant GM maize was approved for import into the European Union in 2004. Such imports remain highly controversial (The Independent, 2005).

The European maize borer, *Ostrinia nubilalis*, destroys maize crops by burrowing into the stem, causing the plant to fall over. Bt maize is a variant of maize, genetically altered to express the bacterial Bt toxin, which is poisonous to insect pests. In the case of maize, the pest is the European maize borer. Expressing the toxin was achieved by inserting a gene from the Lepidoptera pathogen microorganism *B. thuringiensis* into the maize genome. This gene codes for a toxin that causes the formation of pores in the larval digestive tract. These pores allow naturally occurring enteric bacteria such as *E. coli* and Enterobacter to enter the hemocoel where they multiply and cause sepsis.

This is contrary to the common misconception that Bt toxin kills the larvae by starvation. In 2001, Bt176 varieties were voluntarily withdrawn from the list of approved varieties by the United States Environmental Protection Agency when it was found to have little or no Bt expression in the ears and was not found to be effective against second-generation maize borers.

One of the interesting applications of maize is in making ethanol which is being used as biofuel especially in the United States and considered to be as environment friendly with economical cost. In view of its great potential as biofuel, farmers has extensively grown maize and earned huge profits. A biomass gasification power plant in Strem near Güssing, Burgenland, Austria was begun in 2005. Research is being done to make diesel out of biogas by the Fischer Tropsch method. Increasingly, ethanol is being used at low concentrations (10% or less) as an additive in gasoline (gasohol) for motor fuels to increase the octane rating, lower pollutants, and reduce petroleum use. The U.S. federal government announced that production of biofuel may reach a target of 35 billion gallons by 2017.

In the year 2001, the scientific journal Proceedings of the National Academy of Sciences (PNAS USA) published six comprehensive studies that showed that Bt maize pollen does not pose a risk to monarch populations. Monarch populations in the United States during 1999 had increased by 30%, despite Bt maize accounting for 30% of all maize grown in the United States that year. The beneficial effects of Bt maize on Monarch populations can be attributed to reduced pesticide use. Numerous scientific studies continue to investigate the potential effects of Bt maize on a variety of nontarget invertebrates. A synthesis of data from many such field studies found that the measured effect depends on the standard of comparison. The overall abundance of nontarget invertebrates in Cry 1Ab variety Bt maize fields is significantly higher compared to non-GM maize fields treated with insecticides, but significantly lower compared to insecticide-free non-GM maize fields. Abundance in fields of another variety, Cry 3Bb maize, is not significantly different compared to non-GM maize fields either with or without insecticides. By law, farmers in the United States who plant Bt maize must plant non-Bt maize nearby. These nonmodified fields are to provide a location to harbor pests. The theory behind these refuges is to slow the evolution of pests to the Bt pesticide. Doing so enables an area of the landscape where wild-type pests will not be immediately killed.

6.22 TERMINATOR TECHNOLOGY

Genetic use restriction technology (GURT), also known as *terminator technology*, is the name given to proposed methods for restricting the use of GM plants by causing second-generation seeds to be sterile (Figure 6.17). The technology was developed under a cooperative research and development agreement between the Agricultural Research Service of the USDA and the Delta and Pine Land Company in the 1990s, but it is not yet commercially available. Because some stakeholders expressed concerns that this technology might lead to dependence for poor smallholder farmers, Monsanto Company, an agricultural products company and the world's biggest seed supplier, pledged not to commercialize the technology in 1999. Late in 2006, it acquired Delta and Pine Land Company. The technology was discussed during the *8th Conference of the Parties to the UN's Convention on Biological Diversity* in Curitiba, Brazil, March 20–31, 2006.

6.22.1 Types of Terminator Technology

6.22.1.1 V-GURT

This type of GURT produces sterile seeds, which means that a farmer who had purchased seeds containing V-GURT technology could not save the seed from this crop for future planting. This would not have an immediate impact on the large number of primarily western farmers who use hybrid seeds, as they do not produce their own planting seeds and instead buy specialized hybrid

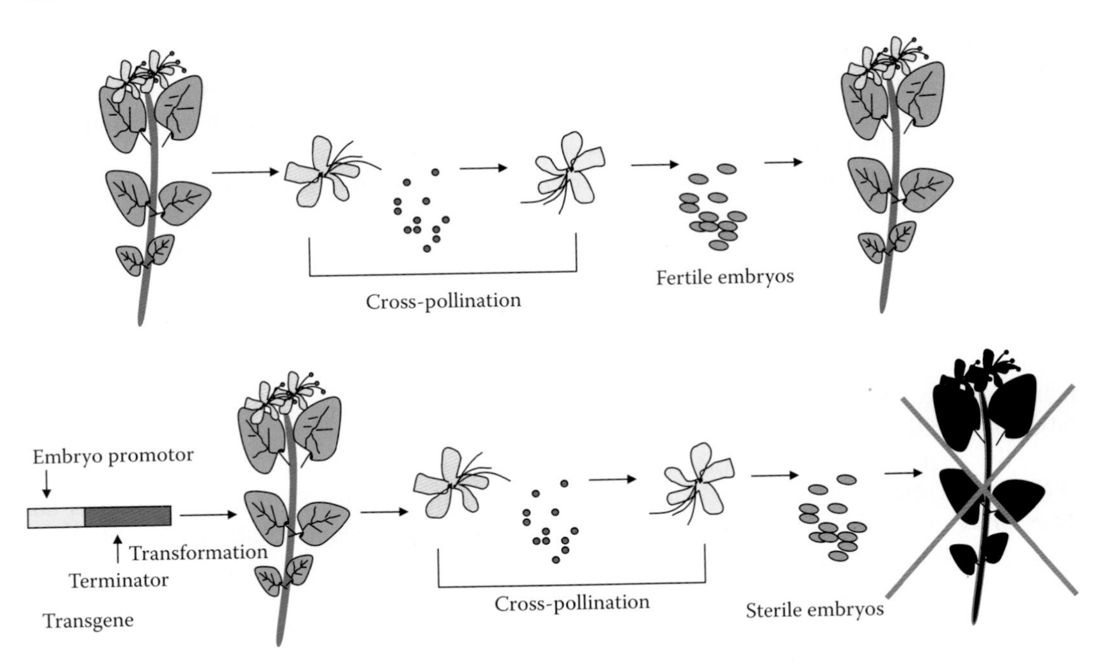

FIGURE 6.17 Terminator technology.

seeds from seed production companies. However, currently around 80% of farmers in both Brazil and Pakistan grow crops based on saved seeds from previous harvests. Consequentially, resistance to the introduction of GURT technology into developing countries is strong. The technology is restricted at the plant variety level, hence the term V-GURT. Manufacturers of genetically enhanced crops would use this technology to protect their products from unauthorized use.

6.22.1.2 T-GURT

This type of GURT modifies a crop in such a way that the genetic enhancement engineered into the crop does not function until the crop plant is treated with a chemical that is sold by the biotechnology company. Farmers can save seeds for use each year. However, they do not get to use the enhanced trait in the crop unless they purchase the activator compound. The technology is restricted at the trait level, hence the term T-GURT.

6.22.2 Benefits of Terminal Technology

Where effective intellectual property protection systems do not exist or are not enforced, GURTs could be an alternative to stimulate plant-developing activities by biotech firms.

Nonviable seeds produced on V-GURT plants may reduce the propagation of volunteer plant. Volunteer plants can become an economic problem for larger-scale mechanized farming systems that incorporate crop rotation. Under warm, wet harvest conditions non-V-GURT grain can sprout, which lowers the quality of grain produced. It is likely that this problem would not occur with the use of V-GURT grain varieties. Use of V-GURT technology could prevent escape of transgenes into wild relatives and prevent any impact on biodiversity. Crops modified to produce nonfood products could be armed with GURT technology to prevent accidental transmission of these traits into crops destined for foods.

6.22.3 Concerns of Terminal Technology

There is a concern that V-GURT plants could cross-pollinate with nongenetically modified plants, either in the wild or in the fields of farmers who do not adopt the technology. Though the V-GURT plants are supposed to produce sterile seeds, there is concern that this trait will not be expressed in the first generation of a small percentage of these plants, but be expressed in later generations. This does not seem to be much of a problem in the wild, as a sterile plant would naturally be selected out of a population within one generation of trait expression. The food safety of GURT technology would need to be assessed if a commercial release of a GURT-containing crop were proposed.

Initially developed by the U.S. Department of Agriculture and multinational seed companies, "suicide seeds" have not been commercialized anywhere in the world due to an avalanche of opposition from farmers, indigenous peoples, and civil society. In 2000, the United Nations Convention on Biological Diversity recommended a *de facto* moratorium on field testing and commercial sale of terminator seeds. The moratorium was reaffirmed in 2006. India and Brazil have already passed national laws to prohibit the technology.

PROBLEMS

Section A: Descriptive Type

Q1. Briefly describe classical breeding.
Q2. How are plants cultivated using somatic hybrid techniques?
Q3. How is plant breeding improved by using RFLP method?
Q4. Describe various diseases caused by bacteria.
Q5. How is the food content of plants altered?
Q6. How are drought-resistant plants generated?
Q7. Describe biochemical production in plants.
Q8. How are vaccines synthesized from plants?
Q9. How safe are transgenic plants?
Q10. Write an assay on golden rice.

Section B: Multiple Choice

Q1. Domestication of plants is a ________ selection process to get desirable traits of plants.
  a. Natural
  b. Artificial
  c. Classical

Q2. A plant whose origin or selection is due to human activity is known as …
  a. Mutagen
  b. Cultigen
  c. Agrogen

Q3. Landrace is a crop that evolved from wild population due to selective breeding. True/False

Q4. Classical breeding relies largely on ________ recombination between chromosomes to generate genetic diversity.
  a. Heterologous
  b. Homologous
  c. Phytologous
  d. None of the above

Q5. The chromosome numbers can be increased artificially by treating the plant with …
  a. Antibiotic
  b. Growth hormone
  c. Colchicine
  d. Oxytocin

Q6. Potato disease in Europe and South America is caused due to …
  a. Bacteria
  b. Virus

c. Nematodes
d. Oomycetes

Q7. Starch is the major storage of carbohydrates in higher plants. True/False

Q8. The most commonly used vector for gene transfer in higher plant is …
a. Nocardia
b. Salmonella
c. *Agrobacterium tumefaciens*
d. None of the above

Q9. The T-DNA regions on all Ti and Ri plasmids are flanked by almost 25 bp direct repeat sequences, which are essential for T-DNA transfer. True/False

Q10. The major advantage of using ________ gene lies in its assay, which requires no DNA extraction, electrophoresis, or autoradiography.
a. GUS gene
b. GFAP gene
c. Bt-2 gene
d. none of the above

Q11. Golden rice was created by …
a. Ingo Potrykus
b. Watson
c. Ian Wilmut
d. Fleming

Q12. "Flavr Savr" is a genetically modified …
a. Orange
b. Apple
c. Potato
d. Tomato

Section C: Critical Thinking

Q1. Explain how transformation without regeneration and regeneration without transformation are of limited value.

Q2. How would you distinguish a natural plant from a genetically modified plant?

Q3. Do you agree that genetically modified plants which are resistant to pest are harmful to human health? Why or why not?

ASSIGNMENT

Can agriculture biotechnology assist in meeting the food demands of a growing global population? Prepare a report indicating the global status of agricultural products developed from genetically modified plants.

REFERENCES AND FURTHER READING

Briggs, F.N. and Knowles, P.F. *Introduction to Plant Breeding*. Reinhold Publishing Corporation, New York, 1967.

Chavarro, J.E., Toth, T.L., Sadio, S.M., and Hauser, R. Soy food and isoflavone intake in relation to semen quality parameters among men from an infertility clinic. *Hum. Reprod.* 2: 2584–2590, 2008.

Chilcutt, C.F. and Tabashnik, B.E. Contamination of refuges by Bacillus thuringiensis toxin genes from transgenic maize. *Proc. Natl. Acad. Sci. USA* 101: 7526–7529, 2004.

Crawford, G.W. East Asian plant domestication, in: *Archaeology of East Asia*, Stark M. (ed.). Blackwell, Oxford, U.K., p. 81, 2006.

Dawe, D., Robertson, R., and Unnevehr, L. Golden rice: What role could it play in alleviation of vitamin A deficiency? *Food Policy*. 27: 541–560, 2002.

de Lemos, M.L. Effects of soy phytoestrogens genistein and daidzein on breast cancer growth. *Ann. Pharmacother*. 5: 1118–1121, 2001.

Derbyshire, E. et al. Review: Legumin and vicilin, storage proteins of legume seeds. *Phytochemistry* 15: 3–24, 1976.

Dillingham, B.L., McVeigh, B.L., Lampe, J.W., and Duncan, A.M. Soy protein isolates of varying isoflavone content exert minor effects on serum reproductive hormones in healthy young men. *J. Nutr*. 135: 584–591, 2005.

Fradin, M.S. and Day, J.F. Comparative efficacy of insect repellents against mosquito bites. *N. Engl. J. Med*. 34: 13–18, 2002.

Gepts, P. A comparison between crop domestication, classical plant breeding, and genetic engineering. *Crop Sci*. 42: 1780–1790, 2002.

Giampietro, P.G., Bruno, G., Furcolo, G. et al. Soy protein formulas in children: No hormonal effects in long-term feeding. *J. Pediatr. Endocrinol. Metab*. 17: 191–196, 2004.

Gottstein, N., Ewins, B.A., Eccleston, C. et al. Effect of genistein and daidzein on platelet aggregation and monocyte and endothelial function. *Br. J. Nutr*. 89: 607–616, 2003.

Heald, C.L., Ritchie, M.R., Bolton-Smith, C., Morton, M.S., and Alexander, F.E. Phyto-oestrogens and risk of prostate cancer in Scottish men. *Br. J. Nutr*. 98: 388–396, 2007.

Hirschberg, J. Carotenoid biosynthesis in flowering plants. *Curr. Opin. Plant Biol*. 4: 210–218, 2001.

Hogervorst, E., Sadjimim, T., Yesufu, A., Kreager, P., and Rahardjo, T.B. High tofu intake is associated with worse memory in elderly Indonesian men and women. *Dement. Geriatr. Cogn. Disord*. 26: 50–57, 2008.

Humphrey, J.H., West, K.P. Jr., and Sommer, A. Vitamin A deficiency and attributable mortality in under-5-year-olds. *WHO Bulletin* 70: 225–232, 1992.

International Rice Research Institute. 2005. Program 3, Annual Report of the Director General, 2004–2005. www.irri.org

King, D. and Gordon, A. Contaminant found in Taco Bell taco shells. Food safety coalition demands recall (press release), 2001. Friends of the Earth, Washington, DC 2000. Available: http://www.foe.org/act/getacobellpr.html

Martineau, B. First fruit, in: *The Creation of the Flavr Savr Tomato and the Birth of Biotech Food*. McGraw-Hill, New York, 2001.

Maskarinec, G., Morimoto, Y., Hebshi, S., Sharma, S., Franke, A.A., and Stanczyk, F.Z. Serum prostate-specific antigen but not testosterone levels decrease in a randomized soy intervention among men. *Eur. J. Clin. Nutr*. 60: 1423–1429, 2006.

Meagher, R.B. Phytoremediation of toxic elemental and organic pollutants. *Curr. Opin. Plant Biol*. 3: 153–162, 2000.

Merritt, R.J. and Jenks, B.H. Safety of soy-based infant formulas containing isoflavones: The clinical evidence. *J. Nutr*. 134: 1220S–1224S, 2004.

Messina, M., McCaskill-Stevens, W., and Lampe, J.W. Addressing the soy and breast cancer relationship: Review, commentary, and workshop proceedings. *J. Natl. Cancer Inst*. 98: 1275–1284, 2006.

Paine, et al. Improving the nutritional value of golden rice through increased pro-vitamin A content. *Nat. Biotechnol*. 23, 4: 482, 2005.

Potrykus, I. Golden rice and beyond. *Plant Physiol*. 125: 1157–1161, 2001.

Schaub, P. et al. Why is golden rice golden (yellow) instead of red? *Plant Physiol*. 138: 441–450, 2005.

Sears, M.K. et al. Impact of Bt corn pollen on monarch butterfly populations: A risk assessment, in: *Proceedings of the National Academy of Sciences*. 98: 11937–11942, October 9, 2001.

Winston, L.R. and Koffler, H. Corn steep liquor in microbiology. *Bacteriol Rev*. 12: 297–311, 1948.

7 Animal Biotechnology

LEARNING OBJECTIVES

- Discuss the significance of use of animals in research
- Explain why animal research is important on historical perspectives
- Discuss about animal biotechnology and its significance
- Discuss about animal testing in pharmaceutical biotechnology companies
- Discuss about animal cloning and transgenic animals
- Discuss about commercial aspects of animal biotechnology

7.1 INTRODUCTION

We know that animals play an important role in human life and in the ecosystem, but not all of us know that many drugs for humans were first tested in animals. Throughout history, scientists have been solving medical problems and developing treatments and cures for diseases, all by using animals in biomedical research. One of the most frequently asked questions about animals is generally related to the use of animals in research, and the general impression is that scientists kill animals for no reason. That is why it is necessary to put a balance in perspectives about the use of animals to humans to understand the ethical as well as the scientific points of view. The following are some of the reasons why scientists use animals in their researches:

- The principles of anatomy and physiology are the same for humans and animals, especially mammals. When the scientists learned that animals are similar to humans in physiology and anatomy, it became preferable to use animals rather than humans for their preliminary researches.
- Certain strains or breeds of animals get the same diseases or conditions as humans. "Animal models" are frequently critical in understanding a disease and in developing appropriate treatments.
- Research meant introducing one variable and observing the results of that one item. With animals, we can control their environment (temperature, humidity, etc.), and shield them from diseases or conditions not related to the research (control their health). Although humans and animals get the disease that may be the subject of a research investigation, the different lifestyles or living conditions make them poor subjects until preliminary research under controlled conditions has been done.
- We can use scientifically valid numbers of animals. Data from one animal or human are not considered a reliable data in research. In order to make reliable data to scientifically test a hypothesis, an adequate number of animals must be used to statistically test the results of the research (Figure 7.1).

All the above points clearly suggest the significance of animals in research and experimentation.

Animals are used as models to test drugs and medical procedures to treat various human diseases. For a new therapy to be developed, it is first being tested in animals to check its efficacy. If animal studies are positive, then the drug is finally tested in human volunteers. Some individuals claim that we should use humans or animals that already have a disease to study that disease.

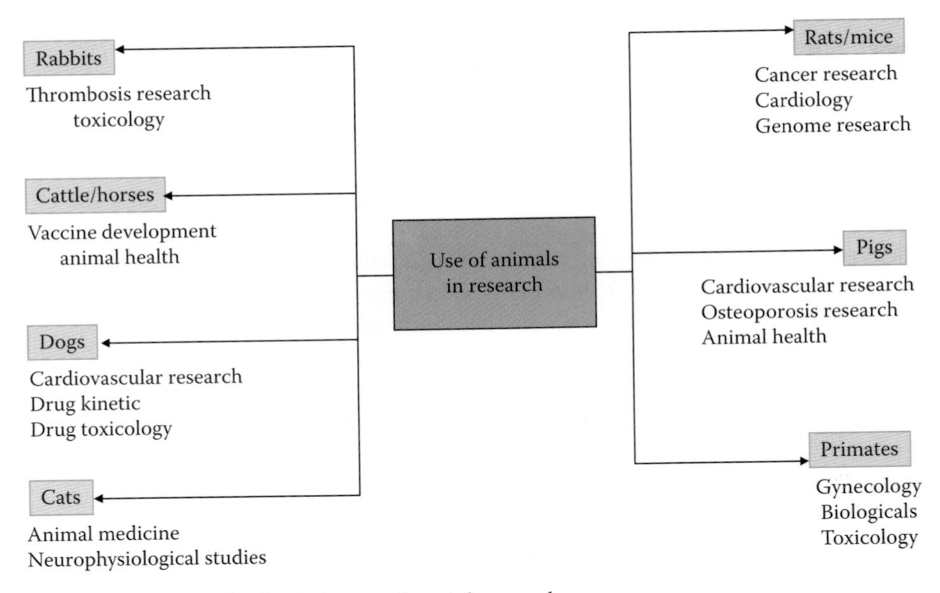

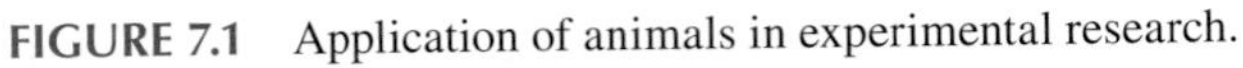
FIGURE 7.1 Application of animals in experimental research.

Certainly, epidemiological studies tracking the occurrence of a disease or condition have provided many important insights into the cause of a disease or a condition, especially when an environmental aspect is responsible. However, epidemiological studies are successful in only a limited number of situations. As noted earlier, the study of a disease is severely hindered or not possible when the research subjects have been/are exposed to a variety of environmental factors. It is important to note that, according to the American Medical Association, humans are the most frequently used animal in research. However, research studies conducted on humans follow preliminary studies conducted in animals. These animal studies make human studies a reasonable risk. The animal studies are not a guarantee of success, but they do tell us that the human research has a reasonable probability of success. Mice, dogs, rats, rabbits, cats, sheep, pigs, monkeys, and horses are some of the most widely used animals in drug testing. On different note, we should not forget the sacrifice of thousands of laboratory animals who have been equally contributing in the development of new therapies or new drug molecules for so many years. Table 7.1 illustrates the pros and cons of animal testing.

TABLE 7.1
Pros and Cons of Animal Testing

Pros	Cons
Animal testing helps in making safe products. These products were tested first in animals before they are launched in the market for human use.	The side against the animal tests proclaims that the main reason for conducting animal tests is to earn profit.
The decrease in human deaths due to cardiovascular diseases is because of the knowledge gained out of animal testing and research.	Most of the time, animals are used in research of diseases that are not found in them naturally, such as different type of cancers to which humans falls prey, or HIV.
The animals are used in research, only when they are indispensable.	Animal rights activists believe that animals have an equal right to live like that of humans.
Animal testing helps preserve the environment.	They also propose to use available alternatives to animal experimentation such as chemical assay tests, tissue, cell and organ culture systems, cloned human skin cells, and computer and mathematical models.
During research, the animals are treated humanely.	

In the next section, we will first examine historical perspective of animal use in research and how animal testing helped in the development of the drug molecules and other milestones in chronological sequence.

7.2 HISTORY OF THE USE OF ANIMALS IN RESEARCH

It may appear that animal use in research is only happening in modern times but you may be surprised to know that the information pertaining to animal testing is found in ancient scripts. Aristotle (384–322 BCE) and Erasistratus (304–258 BCE) were among the first to perform experiments on living animals. In the 1880s, Louis Pasteur convincingly demonstrated the germ theory of medicine by inducing anthrax in sheep. In the 1890s, Ivan Pavlov famously used dogs to describe classical conditioning. Insulin was first isolated from dogs in 1922, and revolutionized the treatment of diabetes. In 1957, a Russian dog, Laika, became the first of many animals to orbit the earth. In the 1970s, antibiotic treatments and vaccines for leprosy were developed using armadillos, then given to humans. The ability of humans to change the genetics of animals took a large step forward in 1974 when Rudolf Jaenisch was able to produce the first transgenic mammal by integrating DNA from the SV40 virus into the genome of mice. This genetic research progressed rapidly and, in 1996, Dolly the sheep was born, the first mammal to be cloned from an adult cell.

The controversy surrounding animal testing dates back to the seventeenth century. There were also objections on an ethical basis, contending that the benefit to humans did not justify the harm to animals. Early objections to animal testing also came from another angle. Many people believed that animals were inferior to humans and so different that results from animals could not be applied to humans. On the other side of the debate, those in favor of animal testing held that experiments on animals were necessary to advance medical and biological knowledge. Claude Bernard, and his wife, Marie Françoise Martin, founded the first antivivisection society in France in 1883 famously wrote in 1865 that "the science of life is a superb and dazzlingly lighted hall which may be reached only by passing through a long and ghastly kitchen." Arguing that "experiments on animals are entirely conclusive for the toxicology and hygiene of man...the effects of these substances are the same on man as on animals, save for differences in degree." In 1883, Claude Bernard, the Father of Physiology, established animal experimentation as part of the standard scientific method.

7.3 DRUG TESTING IN ANIMALS IS MANDATORY

In the nineteenth century, laws regulating drugs were more relaxed in the United States and the federal government could only ban a drug after a company had been prosecuted for selling products that harmed customers. However, in response to a tragedy in 1937 where a drug labeled "Elixir of Sulfanilamide" killed more than 100 people, the U.S. congress passed laws that required safety testing of drugs on animals before they could be marketed and prescribed for human use. Other countries enacted similar legislation. In the 1960s, in reaction to the thalidomide tragedy, further laws were passed requiring safety testing on pregnant animals before a drug can be sold. Today, the majority of people in our society do agree with the idea of the humane and responsible use of animals in research. In fact, today's regulatory guidelines pertaining to use of animal as experiment models have been drafted and followed in most of the countries in the world.

7.4 MOST COMMONLY USED ANIMALS IN RESEARCH

One of the major problems in animal testing is the availability of a large number of laboratory animals especially those that have long lifespan. Because of this problem, scientists prefer to use laboratory animals with a short lifespan and in a short duration of time. Table 7.2 lists some of the most commonly used animals in laboratory research experimentation.

TABLE 7.2
Use of Animals in Research and Product Testing

U.S. Statistics	Europe Statistics
Approximately 17–22 million of animals are used in research each year. Rats, mice, and other rodents make up 85%–90% of all research animals. Only 1%–1.5% of research animals are dogs and cats. Only 0.5% of research animals are NHPs. There has been a 40% decrease in the number of animals used in biomedical research and testing in the United States since 1968. The animal test statistics points to the fact that 50% of all animals used in cosmetic testing die 2–3 weeks after the experiments.	Approximately 12 million of animals are used in research each year. Most animals used in the United Kingdom are mice. In 2005, Finland and Ireland both used a lesser number of animals. In contrast, Sweden, Spain, and Greece all increased their use of animals, either doubling or near-doubling their use. In the United Kingdom, reptiles are generally the least used animals because their anatomy tends to be incompatible with most of the research performed, which is primarily biomedical and toxicology based. France used 2.3 million animals in 2005 while Germany used 1.8 million animals in that same year for testing purposes. National U.K. statistics contrasted somewhat because they showed that for 2004, France used 2.3 million animals while in 2005, Germany used 2.4 million. Huntingdon Life Sciences is one of Europe's biggest testers of animals. They kill approximately 75,000 animals each year, with 87% of these animals being rodents. Across all of Europe, there are approximately 12.1 million animal testing experiments performed each year. Britain is the top user of animals with its use of nearly 3 million animal experiments each year. France is a very close second and generates a large amount of debate given that L'Oreal—a major global cosmetics company—is based in France and still tests cosmetics on animals. France is also the biggest critic of the near-total ban on cosmetics testing that took effect last 2009. Europe's overall laboratory use of animals has actually increased very recently by 3.2%. This contrasts with the fall in animal testing over the last few decades. The United Kingdom has banned cosmetic testing since 1990. France, which is one of the most influential nations in Europe, is against the ban on animal testing.

7.4.1 Invertebrates

Invertebrates are extensively used in basic research than applied research. Most of the experiments based on invertebrates are largely conducted without getting animal ethical approvals and are largely unregulated by law. The most used invertebrate species are *Drosophila melanogaster*, a fruit fly, and *Caenorhabditis elegans*, a nematode worm. In the case of *Caenorhabditi elegans*, the worm's body is completely transparent and the precise lineage of all the organism's cells is known, while studies in the fly *Drosophila melanogaster* can use an amazing array of genetic tools. These animals offer great advantages over vertebrates, including their short life cycle and the ease with which large numbers may be studied, with thousands of flies or nematodes fitting into a single room. However, the lack of an adaptive immune system and their simple organs prevent worms from being used in medical research such as vaccine development. Similarly, flies are not widely used in applied medical research, as their immune system differs greatly from that of humans, and diseases in insects can be very different from diseases in more complex animals.

7.4.2 Fish and Amphibians

Besides rodents, fish and amphibians are extensively used in biological education, training, and research. In the United Kingdom alone, nearly 200,000 fish and 20,000 amphibians were used in research experimentations each year. The main species used in research is the zebrafish, *Danio rerio*, because it is very easy to study their developmental stages in great details. Secondly, the body of zebrafish remains translucent during their embryonic stage. On the other hand, the African clawed frog, *Xenopus laevis*, is used to study regeneration. Another striking difference with vertebrate models is that these animals do not come under strict ethical and animal guidelines and provide an easy access for research experimentation.

7.4.3 Vertebrates

7.4.3.1 Rodents

Among vertebrates, rats, mice, and rabbits are the most commonly used animals in biomedical researches. In the United States alone, the number of rats and mice used is estimated at 20 million/year. Other rodents commonly used are guinea pigs, hamsters, and gerbils. Of these rodents, mice are the most commonly used because of their size, low cost, ease of handling, and fast reproduction rate. Mice are widely considered to be the best model of inherited human disease and share 99% of their genes with humans. With the advent of genetic engineering technology, genetically modified mice are generated to provide models for a range of human diseases. Rats are also widely used for physiology, toxicology, and cancer research, but genetic manipulation is much harder in rats than in mice, which limits the use of these rodents in basic science. Albino rabbits are used in eye irritancy tests because rabbits have less tear flow than other animals, and the lack of eye pigment in albinos make the effects easier to visualize. On the other hand, mice are used in monoclonal antibody production, while rabbits are frequently used for the production of polyclonal antibodies.

7.4.3.2 Cats and Dogs

Although rats, mice, and other small animals provide basic information on human diseases, they do not have the same neurological characteristics as that of humans. In order to study the pathological conditions associated with neurological diseases, researchers prefer to work on higher vertebrates such as cats and dogs. For neurological pathology associated with humans, the favored animal models are cats. According to the American antivivisection society, about 25,500 cats were used in neurological researches in the year 2000 in the United States alone. Most of them were used to study the potential cause of pain and/or distress. On the other hand, dogs are widely used in biomedical research, testing, and education particularly beagles because they are gentle and easy to handle. Dogs are commonly used animal model for human diseases in cardiology, endocrinology, and bone and joint studies, researches that tend to be highly invasive according to the Humane Society of the United States. The U.S. Department of Agriculture's Animal Welfare Report shows that about 66,000 dogs were used in USDA-registered facilities in the year 2005. In the United States, some of the dogs are purpose-bred, while most are supplied by so-called Class B dealers licensed by the USDA.

7.4.3.3 Primates and Nonprimates

Researchers find it difficult to do experiments on human cognition and behaviors using small animals. In fact, it is very difficult to create human-like behaviors in animals, but there are higher-order animals close to humans that are being considered for such studies. It has been estimated that around 65,000 primates are being used in research projects in each year in the United States and Europe. Nonhuman primates (NHPs) are used in toxicology tests, studies of AIDS and hepatitis, studies of neurology, behavior and cognition, reproduction, genetics, and xenotransplantation. They are caught in the wild or purpose-bred. In the United States and China, most primates are domestically purpose-bred, whereas in Europe, the majority is imported purpose-bred such as rhesus monkeys, cynomolgus monkeys, squirrel monkeys, and owl monkeys. Around 12,000–15,000 monkeys are imported into the United States, annually. In total, around 70,000 NHPs are used each year in the United States and European Union (EU) countries. Most of the NHPs used are macaques, although marmosets, spider monkeys, squirrel monkeys, baboons, and chimpanzees are also used in the United States. In 2006, there were 1133 chimpanzees in U.S. primate centers. The first transgenic primate was produced in 2001, with the development of a method that could introduce new genes into a rhesus macaque. This transgenic technology is now being applied in the search for a treatment for the genetic disorder Huntington's disease. In addition, NHPs have been part of the polio vaccine development.

7.5 APPLICATION OF ANIMAL MODELS

In the previous section, we have learned the uses of different types of animals in research and testing purpose. With the discovery of molecular and genetic tools, it became possible to create specific kinds of animals by manipulating their genes using genetic engineering technology or cloning. These genetically designed animals are regularly being used in laboratory research and drug testing purpose. In this section, we will learn various applications of animals in laboratory research. We have broadly classified the application of animals into two subcategories, the use of animals in basic research and the use of animals in applied or industry research.

7.5.1 Use of Animals in Basic Research

Animals are extensively used in researches that have been mostly carried out by university laboratories and research centers funded by the federal government or agencies. The scientists in the universities usually work on animals to know the cause or progression of diseases. When they come up with any new information, they publish their research in scientific journals. In basic research, scientists use large numbers and a greater variety of animals than applied research. Fruit flies, nematode worms, mice, and rats together account for the vast majority, though small numbers of other species are used, ranging from sea slugs through to armadillos. We have listed a few examples of types of animals and experiments used in basic research: (1) Animals are used to study *embryogenesis* and developmental aspects of human body or tissues. Researchers have created genetically designed fruit fly by adding or deleting genes. By studying the changes in development that these changes produce, scientists aim to understand both how organisms normally develop, and what can go wrong in this process. These studies are particularly powerful since the basic controls of development, such as the homeobox genes, have similar functions in organisms as diverse as fruit flies and man. (2) Animals are used to understand how organisms detect and interact with each other and their environment in which fruit flies, worms, mice, and rats are all widely used. Studies of brain function, such as memory and social behavior, often use rats and birds. For some species, behavioral research is combined with enrichment strategies for animals in captivity because it allows them to engage in a wider range of activities. (3) Breeding experiments on animals are used to study *evolution* and *genetics*. Laboratory mice, flies, fish, and worms are inbred through many generations to create strains with defined characteristics. These provide animals of a known genetic background, an important tool for genetic analyses. Larger mammals are rarely bred specifically for such studies due to their slow rate of reproduction, though some scientists take advantage of inbred domesticated animals, such as dog or cattle breeds, for comparative purposes. Scientists also used animals to check genetic mutations in the animal population. One example are the sticklebacks, which are now being used to study how many and which types of mutations are selected to produce adaptations in animals' morphology during the evolution of new species.

7.5.2 Use of Animals in Applied Research

Interestingly, most of the industry researches for new treatments or therapies are in fact based on the outcomes of basic researches. In this section, we discuss some of the areas of industrial researches where animal models have been extensively used.

7.5.2.1 Genetic Diseases

Genetic diseases are caused by mutation in a specific type of gene(s). To make a treatment or therapy for a genetic disease, it is very important to know the cause of the disease. The cause of a genetic mutation can be studied by creating similar mutations in animals. With the advancement of modern genetic engineering tools, it is now possible to add or delete a specific gene and induce mutations in animals. These animals are known as *transgenic animals*. Transgenic animals have specific genes inserted, modified, or removed to mimic specific conditions such as single gene disorders like the Huntington's disease. Other models mimic complex, multifactor diseases with genetic components such as diabetes, or even transgenic mice that carry the same mutations that occur during the development of cancer. These models allow investigations on how and why the disease develops. They also provide ways to develop and test new treatments. The vast majority of these transgenic models of human diseases are lines of mice and mammalian species in which genetic modification is most efficient. Smaller numbers of other animals are also used, including rats, pigs, sheep, fish, birds, and amphibians.

7.5.2.2 Virology

Certain domestic and wild animals have a natural propensity or predisposition for certain conditions that are also found in humans. Cats are used as a model to develop immunodeficiency virus vaccines and to study leukemia because their natural predisposition to feline leukemia virus. Certain breeds of dog suffer from narcolepsy making them the major model used to study the human condition. Armadillos and humans are among only a few animal species that naturally suffer from leprosy, as the bacteria responsible for this disease are yet to be grown in culture. Armadillos are the primary source of bacilli used in leprosy vaccines.

7.5.2.3 Neurological Disorders

Researchers are trying to find a dream drug or therapy for all neurological disorders, which are not curable now. In order to make the dream drug, researchers have to first create neurological disorder-like conditions in the animals and test the drug or cells in these animal models. In the last few years, researchers are now able to model human diseases in animals. The stroke model in animals is created by restricting blood flow to the brain, and Parkinson-like syndrome can be created by injecting neurotoxins into the substantia nigra region of the midbrain, which are known to involve in Parkinson's disease. Such studies can be difficult to interpret, and it is argued that they are not always comparable to human diseases. For example, although such models are now widely used to study Parkinson's disease, the British antivivisection (BUAV) interest group argues that these models only superficially resemble the disease symptoms, without the same time course or cellular pathology. In contrast, scientists assess the usefulness of animal models of Parkinson's disease, as well as the medical research charity. The Parkinson's appeal states that these models were invaluable and that they led to improved surgical treatments such as pallidotomy, new drug treatments such as levodopa, and later, deep brain stimulation.

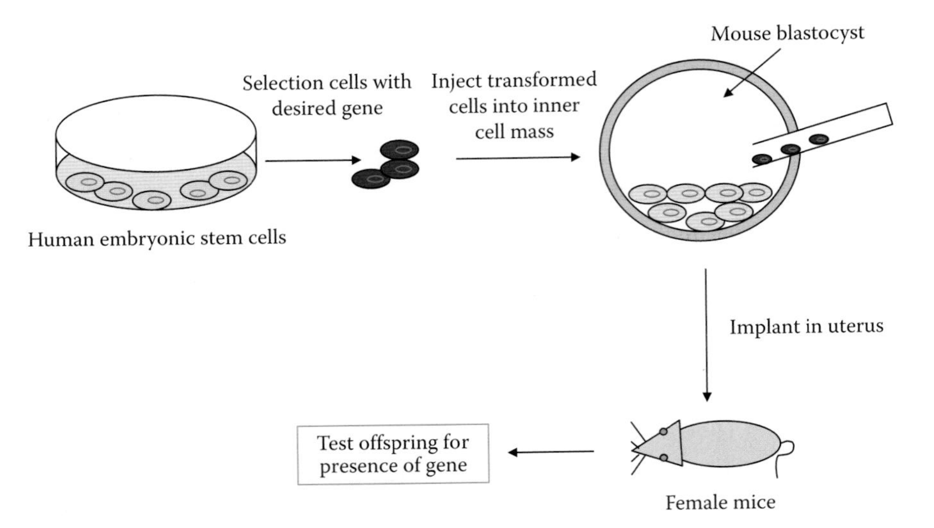

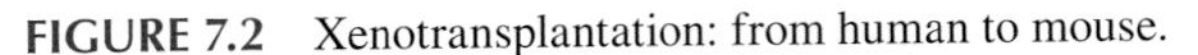
FIGURE 7.2 Xenotransplantation: from human to mouse.

7.5.2.4 Organ Transplantation

With the rampant increase in organ failures in human, there is a tremendous demand of human body organ for transplantation such as kidney and heart. Getting these organs are highly difficult and hundreds are patients are dying because of nonavailability of these organs. In this background, researchers are working to find suitability of using animal organs in place of human organs. The science of transplanting an organ of one species to another, in this case from animal to human, is called *xenotransplantation*. Xenotransplantation research involves transplanting tissues or organs from one species to another as a way to overcome the shortage of human organs for use in organ transplants. Current research involves the use of primates as the recipients of organs from pigs that have been genetically modified to reduce the primates' immune response against the pig tissue. Although transplant rejection remains a problem, recent clinical trials that involved implanting pig insulin-secreting cells into diabetic patients did reduce these people's need for insulin. As success rate is a bit low at the moment, xenotransplantation is not yet safe in human and extensive clinical trials must be conducted first to achieve complete success. In the year 1999, the British Home Office released figures which showed that 270 monkeys had been used in xenotransplantation research during 1995–1999. Scientists used wild baboons imported from Africa for xenotransplantation by grafting pigs' hearts and kidneys. Unfortunately, some baboons died after suffering strokes, vomiting, diarrhea, and paralysis (Figure 7.2).

7.5.2.5 Drug Efficacy Testing

All pharmaceutical companies ensure that drugs must be properly tested in animals before used in humans. *Drug efficacy* is the effect of drug on body or body organ(s) at various time intervals. It also checks whether or not the drug has reached the target site and produced desirable effects. A drug efficacy test is commonly a technical examination of urine, hair, blood, sweat, or oral fluid samples to determine the presence or absence of specified drugs or their metabolized traces. In the early twentieth century, laws regulating drugs were not so stringent, but nowadays all new pharmaceuticals undergo rigorous animal testing before being used in humans. The following are some tests done on pharmaceutical products:

- Metabolic tests or investigating pharmacokinetics—how drugs are absorbed, metabolized, and excreted by the body when introduced orally, intravenously, intraperitoneally, intramuscularly, or transdermally.

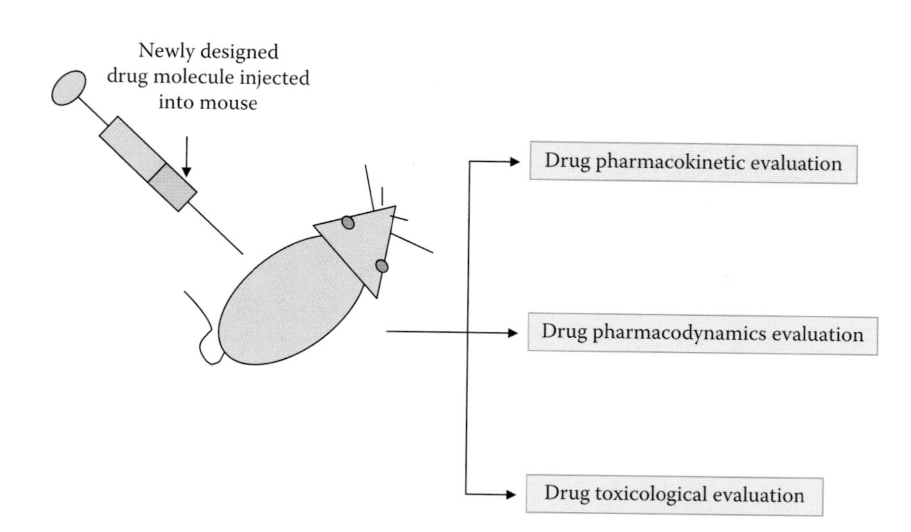

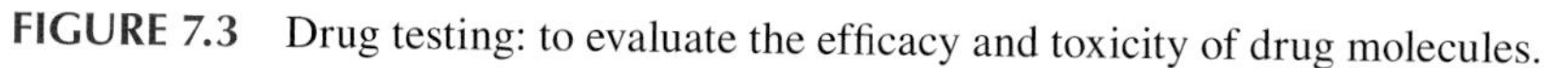
FIGURE 7.3 Drug testing: to evaluate the efficacy and toxicity of drug molecules.

- Efficacy studies that test whether experimental drugs work by inducing the appropriate illness in animals. The drug is then administered in a double-blind controlled trial, which allows researchers to determine the effect of the drug and the dose–response curve.
- Specific tests on reproductive function, embryonic toxicity, or carcinogenic potential can all be required by law, depending on the result of other studies and the type of drug being tested.

All the above tests are very critical to make a successful drug (Figure 7.3).

7.5.2.6 Toxicological Analysis

As per international drug testing guidelines, it is mandatory to know any toxic effect(s) of a new drug molecule before testing in humans, and newly synthesized drug must be tested in animals for checking any undesirable effects. *Toxicology* is the study of the adverse effects of drugs or chemicals on living organisms. It is the study of symptoms, mechanisms, treatments, and detection of toxic effects associated with drug or chemical consumption. Pharmaceutical and biotechnology companies normally conduct almost all the toxicological testing in animals. According to 2005 EU figures, around 1 million animals are used every year in Europe in toxicology tests, which are about 10% of all procedures. The toxicological tests are conducted without anesthesia because interactions between drugs can affect low animals detoxity chemicals and may interfere with the results. Toxicology tests are required for the products such as pesticides, medications, food additives, packing materials, and air freshener, or their chemical ingredients. Most tests involve testing ingredients rather than finished products. The substances are applied to the skin or dripped into the eyes; injected intravenously, intramuscularly, or subcutaneously; inhaled either by placing a mask over the animals and restraining them, or by placing them in an inhalation chamber; or administered orally, through a tube into the stomach, or simply in the animal's food.

There are several different types of acute toxicity tests. The lethal dose 50 (LD50) test is used to evaluate the toxicity of a substance by determining the dose required to kill 50% of the test animal population. This test was removed from OECD international guidelines in 2002, replaced by methods such as the fixed-dose procedure, which uses fewer animals and causes less suffering. The Humane Society of the United States writes that the procedure can cause redness, ulceration, hemorrhaging, cloudiness, or even blindness in animals. The most stringent tests are reserved for drugs and foodstuffs. For these, a number of tests are performed, lasting less than a month (acute), 1–3 months (subchronic), and more than 3 months (chronic) to test general toxicity (damage to organs),

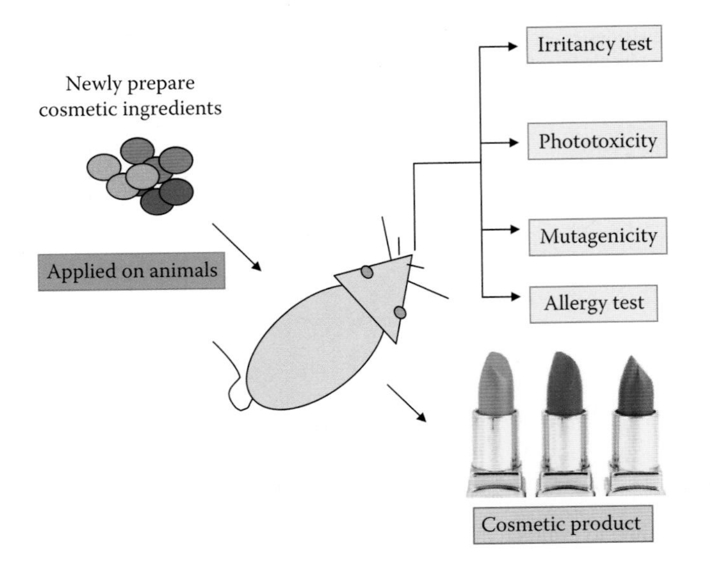

FIGURE 7.4 Cosmetic testing by using animals.

eye and skin irritancy, mutagenicity, carcinogenicity, teratogenicity, and reproductive problems. The cost of the full complement of tests is several million dollars per substance and it may take 3 or 4 years to complete.

7.5.2.7 Cosmetics Testing

There was a surge in cosmetic products and cosmetic-based industries around the world and, like pharmaceutical products, cosmetic products are also undergoing toxicological test before use in humans. These cosmetic products are usually made of chemicals or composition of chemicals derived either from natural sources or by synthetic sources. The cosmetic products are first tested in animals to check their efficacy and safety. Using animal testing in the development of cosmetics may involve testing either a finished product or the individual ingredients of a finished product on animals, often rabbits, but also mice, rats, and other animals. In some cases, the products or ingredients are applied to the mucous membranes of the animal, including eyes, nose, and mouth, to determine whether they cause allergic or other reactions (Figure 7.4). The cosmetics testing on animals is a controversial test but such tests are still conducted in the United States, involving general toxicity, eye and skin irritancy, phototoxicity (toxicity triggered by ultraviolet light), and mutagenicity. Cosmetics testing are banned in the Netherlands, Belgium, and the United Kingdom. In 2002, after 13 years of discussion, the EU agreed to phase in a near-total ban on the sale of animal-tested cosmetics throughout the EU from 2009, and to ban all cosmetics-related animals testing. France, which is home to the world's largest cosmetics company, L'Oreal, has protested the proposed ban by lodging a case at the European Court of Justice in Luxembourg, asking that the ban be quashed. The ban is also opposed by the European Federation for Cosmetics Ingredients, which represents 70 companies in Switzerland, Belgium, France, Germany, and Italy.

7.6 ANIMAL MODELS

Researchers always wanted an ideal situation where they can induce human-like disease conditions and try to reverse it by drug therapy. In particular, model organisms are widely used to explore potential causes and treatments for human disease when human experimentation would be unfeasible or unethical. This strategy is made possible by the common descent of all living organisms, and the conservation of metabolic and developmental pathways and genetic material over the course of evolution. Studying animal models can be informative, but care must be taken when generalizing from one organism to another. There are many animal models. One of the first model systems for molecular biology was the bacterium *Escherichia coli*, a common constituent of the human digestive system. Several of the bacterial viruses (bacteriophage) that infect *E. coli* also have been very useful for the study of gene structure and gene regulation (e.g., phages lambda and T4). However, bacteriophages are not organisms because they lack metabolism and depend on functions of the host cells for propagation. In eukaryotes, several yeasts, particularly *Saccharomyces cerevisiae* have been widely used in genetics and cell biology, largely because they are quick and easy to grow. The cell cycle in yeast is simple, and is very similar to the cell cycle in humans and is regulated by homologous proteins. In the next section, we will study some of the most widely used animal models in details.

7.6.1 *Caenorhabditis elegans*

Caenorhabditis elegans (*C. elegans*) is a free-living, transparent nematode (roundworm), about 1 mm in length, which lives in temperate soil environments. Research into the molecular and developmental biology of *C. elegans* was begun in 1974 by Sydney Brenner and it has since been used extensively as a model organism. *C. elegans* is studied as a model organism for a variety of reasons. One of the main advantages of *C. elegans* is that these organisms are cheap to breed and can be frozen for further study. *C. elegans* has proven especially useful for studying cellular differentiation, because the complete cell lineage of the species has been determined. From a research perspective, *C. elegans* has the advantage of being a multicellular eukaryotic organism that is simple enough to be studied in detail. In addition, it is transparent, facilitating the study of developmental processes in the intact organism (Figure 7.5). In addition, *C. elegans* is one of the simplest organisms with a nervous system. In the hermaphrodite, this comprises 302 neurons whose pattern of connectivity has been completely mapped out, and shown to be a small-world network. Research has explored the neural mechanisms responsible for several of the more interesting behaviors shown by *C. elegans*, including chemotaxis, thermotaxis, mechanotransduction, and male mating behavior. A useful feature of *C. elegans* is that it is relatively straightforward to disrupt the function of specific genes by RNA interference (RNAi). Silencing the function of a gene in this way can sometimes allow a researcher to infer what the function of that gene may be. The nematode can either be soaked in (or injected with) a solution of double stranded RNA, the sequence of which is complementary to the sequence of the gene that the researcher wishes to disable. Alternatively, worms can be fed on genetically transformed bacteria, which express the double-stranded RNA of interest.

C. elegans has also been useful in the study of meiosis. As sperm and egg nuclei move down the length of the gonad, they undergo a temporal progression through meiotic events. This progression means that every nucleus at a given position in the gonad will be at roughly the same step in meiosis, eliminating the difficulties of heterogeneous populations of cells. The organism has also been identified as a model for nicotine dependence as it has been found to experience the same symptoms humans experience when they quit smoking. As for most model organisms, there is a dedicated online database for the species that is actively curated by scientists working in this field. The WormBase database attempts to collate all published information on *C. elegans* and other related nematodes.

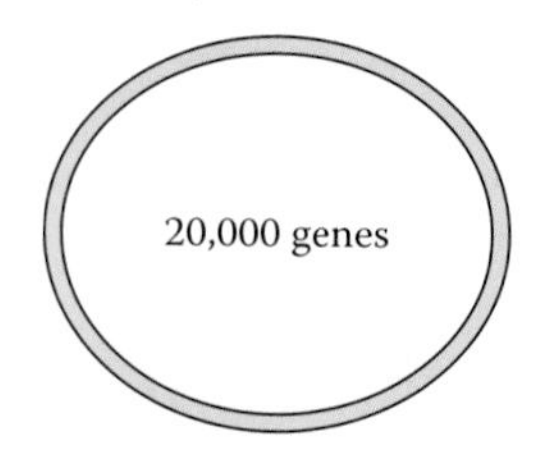

FIGURE 7.5 *Caenorhabditis elegans* genome.

C. elegans was the first multicellular organism to have its genome completely sequenced. The finished genome sequence was published in 1998, although a number of small gaps were present. The *C. elegans* genome sequence contains approximately 100 million bp and approximately 20,000 genes. The vast majority of these genes encode for proteins but there are likely to be as many as 1000 RNA genes. Scientific curators continue to appraise the set of known genes such that new gene predictions continue to be added and incorrect ones modified or removed. In 2003, the genome sequence of the related nematode *C. briggsae* was also determined, allowing researchers to study the comparative genomics of these two organisms. Work is now ongoing to determine the genome sequences of more nematodes from the same genus such as *C. remanei*, *C. japonica*, and *C. brenneri*. These newer genome sequences are being determined by using the whole genome shotgun technique, which means that the resulting genome sequences are likely to not be as complete or accurate as *C. elegans* (which was sequenced using the "hierarchical" or clone-by-clone approach). The official version of the *C. elegans* genome sequence continues to change as and when new evidence reveals errors in the original sequencing, and we have to remember that DNA sequencing is not an error-free process.

7.6.2 *Drosophila melanogaster*

Drosophila melanogaster, also known as fruit fly or vinegar fly, is one of the most studied organisms in biological research, particularly in genetics and developmental biology (Figure 7.6). This is so because it is small and easy to grow in the laboratory, and their morphology is easy to identify once they are anesthetized usually with ether, carbon dioxide gas, or by cooling them. It has a short lifespan of about 10 days at room temperature, so several generations can be studied within a few weeks. It has a high fecundity, females can lay more than 800 eggs in a lifetime. Males and females are readily distinguished and virgin females are easily isolated, facilitating genetic crossing. The mature larvae show giant chromosomes in the salivary glands called *polytene chromosomes*. It has only four pairs of chromosomes: three autosomes and one sex chromosome. Males do not show meiotic recombination, facilitating genetic studies. Recessive lethal "balancer chromosomes" carrying visible genetic markers can be used to keep stocks of lethal alleles in a heterozygous state without recombination due to multiple inversions in the balancer. Genetic transformation techniques have been available since 1987. *Drosophila* genes are traditionally named after the phenotype they cause when mutated. For example, the absence of a particular gene in *Drosophila* will result in a mutant embryo that does not develop a heart. Scientists have thus called this gene *tinman*, named after the Oz character of the same name.

About 75% of known human disease genes have a recognizable match in the genetic code of fruit flies and 50% of fly protein sequences have mammalian analogues. An online database called Homophila is available to search for human disease gene homologues in flies and vice versa. *Drosophila* is being used as a genetic model for several human diseases including the neurodegenerative disorders Parkinson's, Huntington's, spinocerebellar ataxia, and Alzheimer's disease. The fly is also being used to study mechanisms underlying aging and oxidative stress, immunity, diabetes, and cancer, as well as drug abuse. In 1971, Ron Konopka and Seymour Benzer published

FIGURE 7.6 *Drosophila melanogaster* genome.

"Clock mutants of *Drosophila melanogaster*," a paper describing the first mutations that affected an animal's behavior. Wild-type flies show an activity rhythm with a frequency of about a day. They found mutants with faster and slower rhythms as well as broken rhythms—flies that move and rest in random spurts. Work over the following 30 years has shown that these mutations (and others like them) affect a group of genes and their products that comprise a biochemical or biological clock. This clock is found in a wide range of fly cells, but the clock-bearing cells that control activity are several dozen neurons in the fly's central brain. Since then, Benzer and others have used behavioral screens to isolate genes involved in vision, olfaction, audition, learning/memory, courtship, pain, and other processes such as longevity. The first learning and memory mutants (*dunce*, *rutabaga*, etc.) were isolated by William "Chip" Quinn while in Benzer's lab, and were eventually shown to encode components of an intracellular signaling pathway involving cyclic AMP, protein kinase A, and a transcription factor known as CREB. These molecules were shown to be also involved in synaptic plasticity in *Aplysia* and mammals. Furthermore, *Drosophila* has been used in neuropharmacological research, including studies of cocaine and alcohol consumption.

7.6.3 Laboratory Mouse

Mice are the most commonly used animal model with hundreds of established inbred, outbred, and transgenic strains. Mice are common experimental animals in biology and psychology primarily because they are mammals, and thus share a high degree of homology with humans. The mouse genome has been sequenced, and virtually all mouse genes have human homologs. They can also be manipulated in ways that would be considered unethical to do with humans. Mice are a primary mammalian model organism, as are rats. There are many additional benefits of mice in laboratory research. Mice are small, inexpensive, easily maintained, and can reproduce quickly. Several generations of mice can be observed in a relatively short period of time. Some mice can become docile if raised from birth and given sufficient human contact. However, certain strains have been known to be quite temperamental.

Most laboratory mice are hybrids of different subspecies, most commonly of *Mus musculus domesticus* and *Mus musculus musculus*. Laboratory mice come in a variety of coat colors including agouti, black, and albino. Many (but not all) laboratory strains are inbred, so as to make them genetically almost identical. The different strains are identified with specific letter-digit combinations, such as C57BL/6 and BALB/c. The first such inbred strains were produced by Clarence Cook Little in 1909. The sequencing of the mouse genome was completed in late 2002. The haploid genome is about 3 billion bases long (3000 MB distributed over 20 chromosomes) and therefore equal to the size of the human genome. Estimating the number of genes contained in the mouse genome is difficult, in part because the definition of a gene is still being debated and extended. The current estimated gene count is 23,786. This estimate takes into account knowledge of molecular biology as well as comparative genomic data. For comparison, humans are estimated to have 23,686 genes (Figure 7.7).

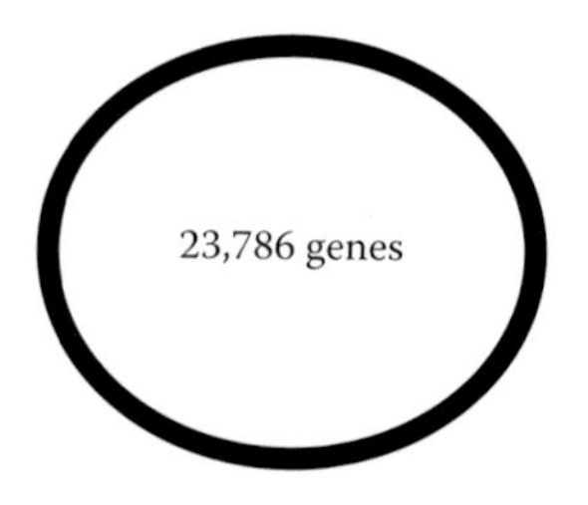

FIGURE 7.7 Mouse genome.

7.6.4 Rhesus Monkey

The rhesus macaque (*Macaca mulatta*), often called the rhesus monkey, is one of the best known species of Old World monkeys. It typically has a lifespan of about 25 years. The species is native to northern India, Bangladesh, Pakistan, Burma, Thailand, Afghanistan, southern China, and some neighboring areas. The rhesus macaque is well known to science owing to its relatively easy upkeep in captivity, and has been used extensively in medical and biological research. It has given its name to the Rhesus factor, one of the elements of a person's blood group, by the discoverers of the factor, Karl Landsteiner and Alexander Wiener. The rhesus macaque was also used in the well-known experiments on maternal deprivation carried out in the 1950s by comparative psychologist Harry Harlow. In January 2000, the rhesus macaque became the first cloned primate with the birth of Tetra. January 2001 saw the birth of ANDi, the first transgenic primate. ANDi carries foreign genes originally from a jellyfish. The rhesus macaque, Tetra, was recently cloned with the technique of embryo spitting. Work on the genome of the rhesus macaque was completed in 2007, making rhesus macaque the second NHP to have its genome sequenced. The study shows that humans and macaques share about 93% of their DNA sequence and shared a common ancestor roughly 25 million years ago (Figure 7.8).

7.6.5 *Xenopus laevis*

The *Xenopus laevis* also known as African clawed frog is a species of South African aquatic frog of the genus *Xenopus*. It can grow up to 12 cm long with a flattened head and body, but no external ear or tongue. The species is found throughout much of Africa, and in isolated, introduced populations in North America, South America, and Europe. Although *X. laevis* does not have the short generation time and genetic simplicity generally desired in genetic model organisms, it is an important model organism in developmental biology. *X. laevis* takes 1–2 years to reach sexual maturity and, like most of its genus, it is tetraploid. However, it does have a large and easily manipulable embryo. The ease of manipulation in amphibian embryos has given them an important place in historical and modern developmental biology. A related species, *X. tropicalis*, is now being promoted as a more viable model for genetics. Roger Wolcott Sperry used *X. laevis* for his famous experiments describing the development of the visual system. These experiments led to the formulation of the Chemoaffinity hypothesis (Figure 7.9).

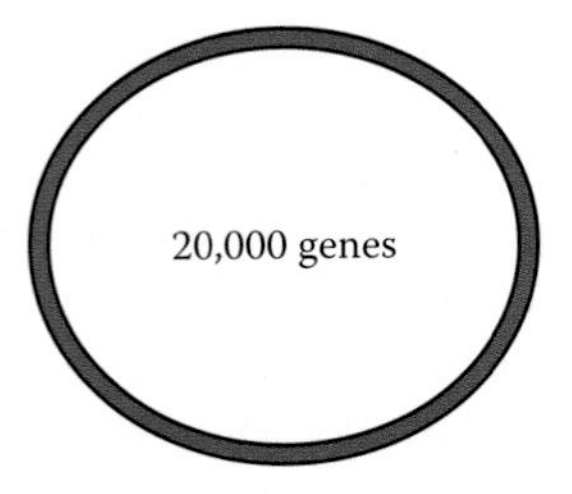

FIGURE 7.8 Rhesus monkey genome.

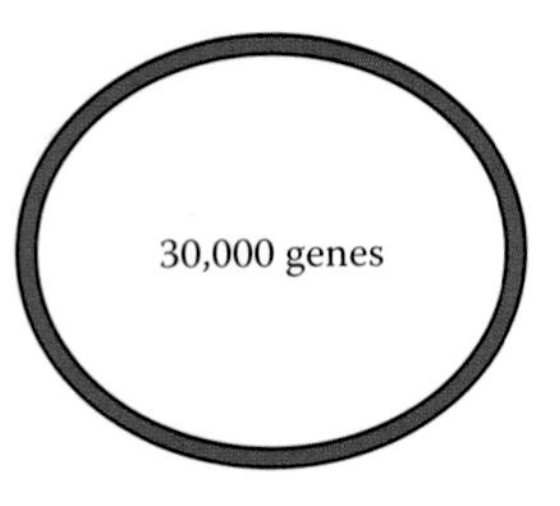

FIGURE 7.9 *Xenopus laevis* genome.

Xenopus oocytes provide an important expression system for molecular biology. By injecting DNA or mRNA into the oocyte or developing embryo, scientists can study the protein products in a controlled system. This allows rapid functional expression of manipulated DNAs (or mRNA). This is particularly useful in electrophysiology, where the ease of recording from the oocyte makes expression of membrane channels attractive. One challenge of oocyte work is in eliminating native proteins that might confound results, such as membrane channels native to the oocyte. Translation of proteins can be blocked or splicing of pre-mRNA can be modified by injection of Morpholino antisense oligos into the oocyte (for distribution throughout the embryo) or early embryo (for distribution only into daughter cells of the injected cell). *X. laevis* is also notable for its use as the first well-documented method of pregnancy testing when it was discovered that the urine from pregnant women induced *X. laevis* oocyte production. Human chorionic gonadotropin (HCG) is a hormone found in substantial quantities in the urine of pregnant women. Today, commercially available HCG is injected into *Xenopus* males and females to induce mating behavior and breed these frogs in captivity at any time of the year.

7.6.6 Zebrafish

The zebrafish, *Danio rerio*, is a tropical freshwater fish belonging to the minnow family (Cyprinidae). It is a popular aquarium fish, frequently sold under the trade name zebra danio, and is an important vertebrate model organism in scientific research. *D. rerio* are a common and useful model organism for studies of vertebrate development and gene function. They may supplement higher vertebrate models, such as rats and mice. Pioneering work of George Streisinger at the University of Oregon established the zebrafish as a model organism; large-scale forward genetic screens consolidated its importance. Zebrafish embryonic development provides advantages over other vertebrate model organisms. Although the overall generation time of zebrafish is comparable to that of mice, zebrafish embryos develop rapidly, progressing from eggs to larvae in less than 3 days. The embryos are large, robust, and transparent and develop externally to the mother, all characteristics that facilitate experimental manipulation and observation. Their nearly constant size during early development facilitates simple staining techniques, and drugs may be administered by adding directly to the tank.

Despite the complications of the zebrafish genome (Figure 7.10), a number of commercially available global platforms for analysis of both gene expression by microarrays and promoter regulation using ChIP on-chip exist. Zebrafish have the ability to regenerate fins, skin, heart, and the brain. Zebrafish have also been found to regenerate photoreceptors and retinal neurons following injury. The mechanisms of this regeneration are unknown, but are currently being studied. Researchers frequently cut the dorsal and ventral tail fins and analyze their regrowth to test for mutations. This research is leading the scientific community in the understanding of healing/repair mechanisms in vertebrates. In December 2005, a study of the *golden* strain identified the gene responsible for the unusual pigmentation of this strain as SLC24A5, a solute carrier that appeared to be required for melanin production, and confirmed its function with a Morpholino knockdown.

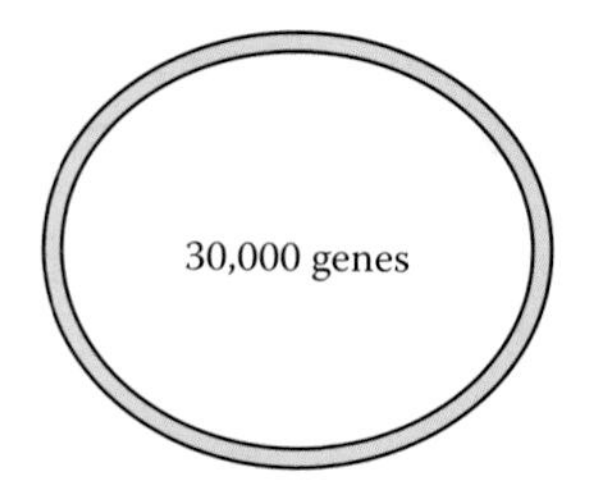

FIGURE 7.10 Zebrafish genome.

The orthologous gene was then characterized in humans and a 1 bp difference was found to segregate strongly between fair-skinned Europeans and dark-skinned Africans. This study featured on the cover of the academic journal *Science* and demonstrates the power of zebrafish as a model organism in the relatively new field of comparative genomics. In January 2007, Chinese researchers at Fudan University raised genetically modified fish that can detect estrogen pollution in lakes and rivers, showing environmental officials what waterways need to be treated for the substance, which is linked to male infertility. Song Houyan and Zhong Tao, professors at Fudan's molecular medicine lab, spent 3 years cloning estrogen-sensitive genes and injecting them into the fertile eggs of zebrafish. The modified fish turn green if they are placed into water that is polluted by estrogen.

Researchers at the University College London grew a type of zebrafish adult stem cell found in the eyes of fish and mammals that develops into neurons in the retina, the part of the eye that sends messages to the brain. These cells could be injected in the eye to treat all diseases where the retinal neurons are damaged, nearly all diseases of the eye including macular degeneration, glaucoma, and diabetes-related blindness. Damage to the retina is responsible for most cases of sight loss. The researchers studied Müller glial cells in the eyes of humans aged from 18 months to 91 years and were able to develop them into all types of neurons found in the retina. They were also able to grow them easily in the lab. The results of their experiments were reported in the journal Stem Cells. The cells were tested in rats with diseased retinas, where they successfully migrated into the retina and took on the characteristics of the surrounding neurons. Now, the team is working on the same approach in humans. In February 2008, researchers at Children's Hospital, Boston, reported in the journal *Cell Stem Cell* the development of a new strain of zebrafish, named Casper, with see-through body. This allows for detailed visualization of individual blood stem cells and metastasizing (spreading) cancer cells within a living adult organism. Because the functions of many genes are shared between fish and humans, this tool is expected to yield insight into human diseases such as leukemia and other cancers.

7.6.7 Bioengineered Mosquito

Malaria infects about 400 million and manages to kill more than a million. More disturbingly, the combination of pesticide and drug immunities along with the rise of global transportation and current climate changes has resulted in mosquito-vectored ailments appearing in places where they have not been seen before. In the United States, dengue appeared for the second time in Hawaii and the first time in the Gulf States, and there were about 4000 reported cases of West Nile and about 300 deaths. Malaria has so far failed to make serious new inroads into the United States, but in the slightly purple words of the Malaria Foundation International, "a plague is coming back and we have only ourselves to blame." To address these concerns, during the past 15 years scientists have been trying to move beyond the chemical paradigm and into a genetic one. The dream has been to build a genetically modified insect, a transgenic mosquito that is unable to transmit such diseases. This new insect would then be introduced in the wild, thus supplanting malaria carriers with a harmless imposter. Seven teams, both in America and in Europe, demonstrating a collaborative spirit not often found in modern science, have been at work on the project.

7.7 ANIMAL BIOTECHNOLOGY

7.7.1 Use of Animals in Antibody Production

In general, an antigen molecule has antigenic determinants of more than one specificity, this means different determinants will interact with different antibodies. Each distinct antigenic determinant of the antigen will bind to a distinct mature B cell whose sIg matches the specificity presented by the concerned determinant. As a result, such a single antigen will activate B cells of more than one sIg specificity. Activated B cells of each sIg specificity will divide and differentiate to give rise to clones of plasma cells producing antibodies of the same specificity. Thus, a single antigen would induce more than one distinct clone of plasma cells, which will produce antibodies of different specificities. Therefore, the serum of an animal immunized by a single antigen will contain antibodies of different specificities but reacting to the same antigen. These are called *polyclonal antibodies*.

In contrast, a hybridoma clone produces antibodies of a single specificity since the clone is derived from the fusion of a single differentiated (antibody producing) B cell with a myeloma cell, that is, a clone of a single B cell. Therefore, such antibodies are called *monoclonal antibodies* (MABs). Obviously, all the molecules of a MAB will have the same specificity. The chief advantage of MABs is that all the antibody molecules in a single preparation react with a single epitope or antigenic determinant. Therefore, the results obtained by using MABs are clearcut as there is no background confusion that arises due to the presence of antibodies of other specificities in the case of conventionally used antisera. The multitude of various applications of MABs may be grouped into the three categories such as diagnostic, therapeutic, and purification.

7.7.1.1 Monoclonal Antibodies in Diagnostic Applications

MABs are used to detect the presence of a specific antigen or of antibodies specific to an antigen in a sample or samples—this constitutes a diagnostic application. The presence of antigen is detected by assaying the formation of antigen antibody complex (Ag–Ab complex) for which a number of assay techniques have been devised. These assays are highly precise, extremely efficient, rapid, and surprisingly versatile for a large variety of applications. Some examples of diagnostic applications are (1) MABs are available for unequivocal classification of blood groups such as ABO and Rh and (2) MABs are applied for a clear and decisive detection of pathogens involved in diseases (disease diagnosis).

7.7.1.2 Monoclonal Antibodies in Therapeutic Applications

The MABs are used for either the treatment of or protection from a disease. Antibodies specific to a cell type, for example, tumor cells, can be linked with a toxin polypeptide to yield a conjugate molecule called *immunotoxin*. The antibody component of immunotoxin will ensure that it is bound specifically and only to the target cells, and the attached toxin will kill such cells. Immunotoxins, having ricin, have been prepared and evaluated for killing of tumor cells with considerable success. *Ricin* is a natural toxin found in the endosperm of castor (*Ricinus communis*). It has two polypeptides called A (toxin peptide) and B (a cell binding polypeptide, lectin). Ricin A polypeptide enzymatically and irreversibly modifies the larger subunit of ribosomes (in fact, their EF2 binding site) making them incapable of protein synthesis. This toxin is effective against both dividing and nondividing cells as it inhibits protein synthesis.

Antibody-Ricin A conjugate has been shown to reduce protein synthesis in mouse B cell tumors. The antibody used in the conjugate was specific to the antigen molecules present on the surface of target tumor cells. It is noteworthy that this immunotoxin did not bind to either other tumor cells or the normal cells. The same principle has been used to deliver radioactivity, specifically to target tumor cells. In such cases, radioactivity due to 1311 (iodine), 90y (yitrium), 67Cu, 212Pb, etc., is incorporated into the tumor-specific antibody in the place of toxin. Radiolabeled antibodies have been used in patients having hepatoma, human T-cell leukemia/lymphoma virus-I (HTL V-1), adult-T cell leukemia (ATL), etc., B-lymphocyte proliferation, maturation, and antibody secretion are dependent on interleukin-2 produced by activated T-lymphocytes. Furthermore, Tc cells mediate graft rejection. Thus, an effective strategy to minimize the rejection of grafts from other individuals would be to eliminate the T cells from their bone marrow/circulatory system (blood stream) by using T-cell specific MABs. T cells exhibit several antigens of which CD3, CD4, CD8, etc., have been the preferred targets for MAB development.

In bone marrow transplantation, the cells of the recipient are inactivated by appropriate irradiation. The donor cells are treated with T cell specific antibodies to destroy the T cells present in them.

The remaining cells are then transplanted into the recipient. Experiments with mice have shown remarkable success. In order to minimize tissue-graft rejection, the T cells present in circulatory system of the recipient are eliminated prior to the transplant by an administration of T-cell specific MABs. This treatment abolishes, though temporarily, the ability of recipient to mount immune response against any foreign antigen, including those present in the graft tissue. MAB OKT3 is the most widely used for treatment of acute cases of rejection of kidney transplants. MABs can be administered to provide passive immunity against diseases. In case of active immunity, the immunized individual will itself produce the antibodies against the concerned pathogen. However, in case of passive immunity, antibodies produced elsewhere are introduced into the body of an individual to provide immunity against a pathogen. MABs are very useful in the purification of antigens specific to pathogens. These purified antigens are used as vaccines.

7.7.1.3 Recombinant Antibodies

When antibody molecules are modified or designed using recombinant DNA technology to suit specific applications, such antibodies are called *recombinant* or *hybrid antibodies*, and the approach itself is termed as *antibody engineering*. A recombinant antibody could be constructed so that the constant end of its heavy chain is fused to a polypeptide chain having an enzymatic function. Such an antibody is extremely useful for enzyme-linked immunosorbent assay (ELISA) as there will be no need for a second antibody. A gene producing such an antibody can be produced by fusing the heavy chain gene having the appropriate L-V-D-J and yl sequences with the sequence coding for the selected enzyme function. The heavy chain gene of an antibody specific to a tumor-specific antigen may be fused with a gene encoding a toxin polypeptide. Such hybrid antibodies will carry the toxin specifically to the tumor cells and, thereby, kill them. Gene segments encoding the variable region or V-region (involved in antigen–antibody interaction) of an antibody have been fused with the constant region or C-region of another antibody to yield a hybrid antibody. A hybrid antibody has the antigen specificity of the first antibody (which contributed the V-region segments), but its other properties are due to the second antibody (contributing the C-region segments).

7.7.2 *In Vitro* Fertilization and Embryo Transfer

The union of egg cell with sperm occurs outside the body in a culture vessel—this is known as *in vitro* fertilization. This involves collection of healthy ova and sperms from healthy females and males, respectively, and their fusion under appropriate conditions *in vitro*. The resulting zygotes may be cultured *in vitro* for a period of time to obtain young embryos, which ultimately are implanted in the uterus of healthy females to complete their development. The implementation of young embryos developed *in vitro* or obtained from the uterus of different donor females into the womb of selected females is termed as *embryo transplantation*. The techniques of *in vitro* fertilization and embryo transfer are being applied to animals for a rapid multiplication of desirable genotypes of animals, and in cases of infertility of certain types in humans (Figure 7.11).

7.7.2.1 Embryo Transfer in Cattle

Young embryos of cattle of superior genotype are collected prior to their implantation in uterus, and are implanted in the uterus of other females of inferior genotype, where they complete development; this is called *embryo transfer*. The chief objective of embryo transfer is to obtain several progeny per year from a single female of superior genotype. A country like India, most cattle are of inferior genotype with rather low productivity, while superior genotype females are limited in number and of high price. Therefore, a program of artificial insemination (AI) was widely used in an effort to improve cattle breeds. A limitation of AI is that the superior genes (50%) are contributed by only the male side, while the female contributes (50%) the inferior genes.

In contrast, in embryo transfer technique, the inferior females used as surrogate or substitute mothers do not contribute any genes to the progeny. They only serve as extremely sophisticated

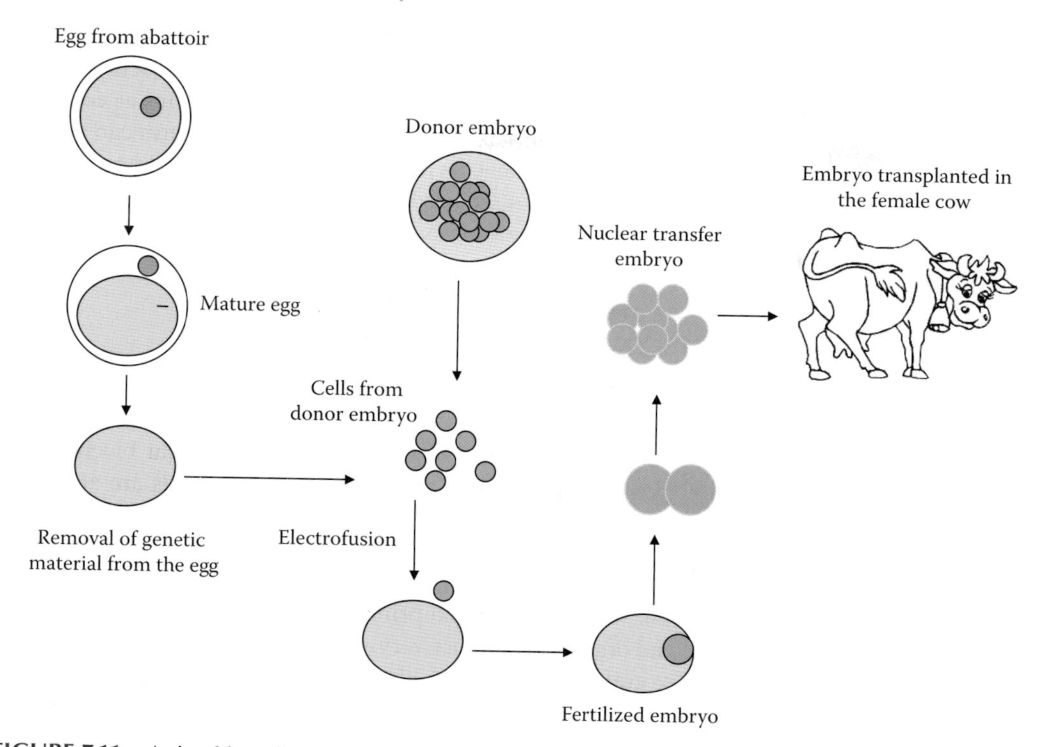

FIGURE 7.11 Animal breeding by nuclear transfer technology.

natural incubators for the normal development of young embryos. As a result, the progeny obtained by embryo transfer are of superior genotype.

In embryo transfer, a genetically superior and highly productive female serves as the donor of embryos to be transferred and healthy, young females of inferior genotype are selected to be the recipients of embryos to be transferred. These females are called *surrogate* or *substitute mothers.* The donor females are treated with appropriate doses of the selected *gonadotrophin*, a follicle stimulating hormone (FSH) or luteinizing hormone (LH), to increase the number of ova released at the time of ovulation, This process is called *superovulation*. Under optimum treatment conditions, a single female can provide up to 15 embryos in a single cycle. The chief objective of superovulation is to greatly increase the number of embryos recovered per female in a single cycle. When the donor female is in heat, it is artificially inseminated using semen from a genetically superior bull of top pedigree. The fertilized eggs/young embryos are collected by flushing the uterus of donor females with a special nutrient solution, 7 days after insemination. The embryos are examined under a stereoscopic microscope and normal looking healthy embryos are selected. The selected embryos are incubated in a special nutrient medium at 37°C until they are transferred into the surrogate mothers. Alternatively, they may be frozen and stored in liquid nitrogen for future use. A single embryo is transferred into the uterus of each surrogate mother. It is important that the estrus cycles of donor and surrogate mothers are synchronized by administering prostaglandins to provide the optimum uterine environment for survival, establishment, and normal development of the young embryos.

This technology achieves a surprisingly rapid rate of multiplication of animals of the selected superior genotype. In natural course, a single female will produce a single progeny in about a year. However, using superovulation and embryo transfer technology, it is feasible to collect around 36 embryos from one female in 1 year. Assuming an average success rate of 50% in the embryo transfer, an average of 18 progeny can be derived from one superior female in 1 year. Each young embryo can be split into 2–4 parts, each of which would develop into a separate progeny. This process is called *embryo splitting*. By combining embryo splitting with superovulation, the rate of multiplication can be further increased. The young embryos can be frozen and stored in liquid nitrogen (at −196°C; cryopreservation) for up to 10 years or more and used at a subsequent date. The frozen embryos are far easier to transport, and present negligible quarantine problems as compared to the animals themselves. Superior cows that are unfit to carry the fetus for full term can serve as donors of the young embryos. In spite of various benefits of embryo transfer technology, there are also some limitations. A high degree of expertise is required for an efficient and successful embryo transfer operation. The cost of producing each progeny is several-fold higher than that from the natural process. The donor females are removed from production for the period they are used as donors of young embryos.

7.7.3 Animal Cell Culture Products

Animal cell cultures are used to produce virus vaccines, as well as a variety of useful biochemicals, which are mainly high molecular weight proteins like enzymes, hormones, cellular biochemicals like interferon's, and immunobiological compounds including MABs. Animal cells are also good hosts for the expression of recombinant DNA molecules and a number of commercial products have been/are being developed. Initially, virus vaccines were the dominant commercial products from cell cultures, but at present, monoclonal antibody production is the chief commercial activity. It is expected that recombinant proteins would become the prime product from cell cultures in the near future. Transplantable tissues and organs are another very valuable product from cell cultures. Artificial skins are already in use for grafting in burn and other patients, and efforts are focused on developing transplantable cartilage and other tissues.

7.7.4 Animal Cloning

Animal *cloning* is the process by which an entire organism is reproduced from a single cell taken from the parent organism and in a genetically identical manner. Cloning in biotechnology refers to processes used to create copies of DNA fragments (molecular cloning), cells (cell cloning), or organisms. Scientists have been attempting to clone animals for a very long time. Many of the early attempts came to nothing. The first successful result in animal cloning was seen when tadpoles were cloned from frog embryonic cells in 1952. This was the first vertebrate to be cloned. This cloning was done by the process of nuclear transfer. The tadpoles, which were created by this method, did not survive to grow into mature frogs, but it was a major breakthrough nevertheless. After this, using the process of nuclear transfer on embryonic cells, scientists managed to produce clones of mammals. Again, the cloned animals did not live very long (Figure 7.12).

The first successful instance of mammal cloning was that of Dolly the Sheep. Dolly (05-07-1996–14-02-2003), a Finn Dorsett ewe, was the first mammal to have been successfully cloned from an adult cell. Dolly was cloned at the Roslin Institute in Scotland and lived there until her death when she was six. On April 9, 2003, her stuffed remains were placed at Edinburgh's Royal Museum, part of the National Museums of Scotland. Dolly was publicly significant because the effort showed that the genetic material from a specific adult cell, programmed to express only a distinct subset of its genes, can be reprogrammed to grow an entire new organism. Cloning Dolly had a low success rate per fertilized egg. She was born after 237 eggs were used to create 29 embryos, which only produced three lambs at birth, only one of which lived. Seventy calves have been created from 9000 attempts and one-third of them died young. Prometea took 328 attempts. Notably, although the first clones were frogs, no adult cloned frog has yet been produced from a somatic adult nucleus donor cell. There were early claims that Dolly the Sheep had pathologies resembling accelerated aging. Scientists speculated that Dolly's death in 2003 was related to the shortening of telomeres, DNA-protein complexes that protect the end of linear chromosomes. However, other researchers, including Ian Wilmut who led the team that successfully cloned Dolly, argued that Dolly died due to respiratory infection. We have listed the animals which have been cloned in Table 7.3.

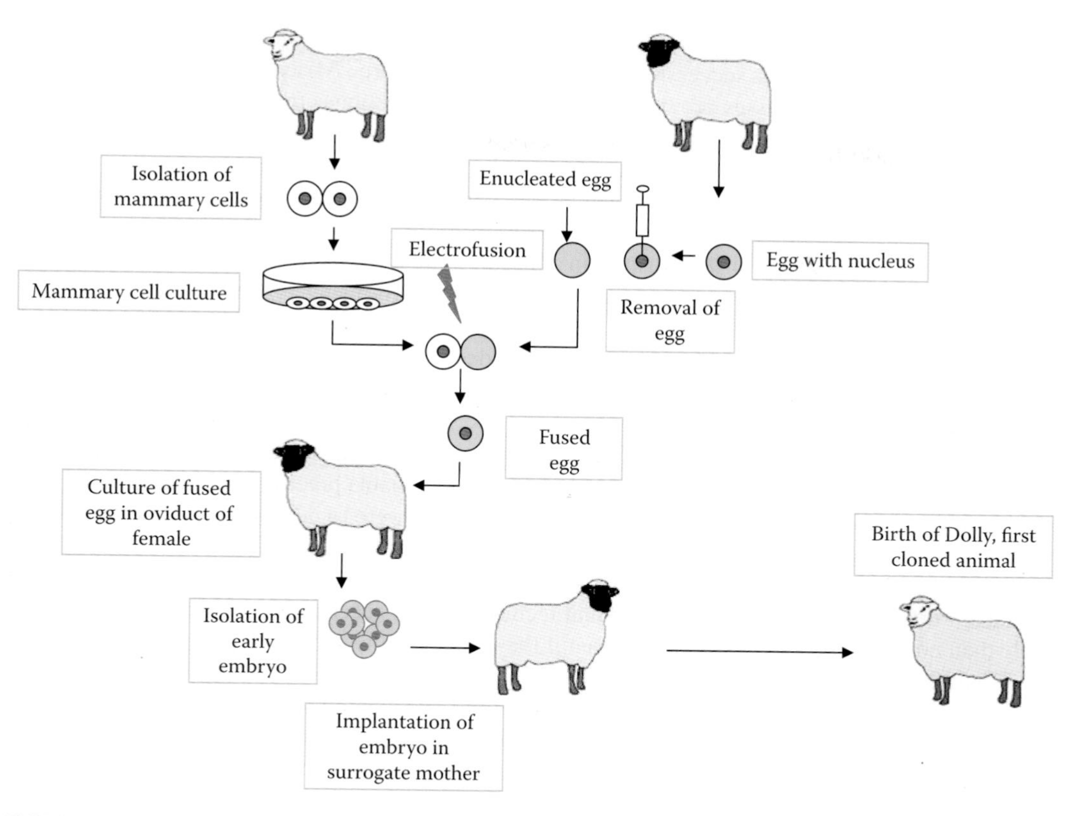

FIGURE 7.12 Animal cloning.

TABLE 7.3
List of Cloned Animals

Name of Animals Cloned	Year of Cloning Done	Country
Carp	1963	China
Mice	1986	Russia
Sheep: Dolly	1997	Scotland
Gaur	2001	United States
Cattle: Alpha and Beta	2001 and 2005	Brazil
Dog: Snuppy	2005	United States
Rat: Ralph	2003	China and France
Horse: Prometa	2003	United States
Mule: Idahu Gem	2003	United States
CopyCat	2001	United States
Water Buffalo: Samrupa	2009	India
Camel: Injaz	2009	U.A.E.

7.7.4.1 Cloning of Extinct and Endangered Species

Cloning techniques have been in use for many years to save extinct species. The possible implications of animal cloning were dramatized in the best-selling novel by Michael Crichton and high budget Hollywood thriller *Jurassic Park*. In real life, one of the most anticipated targets for cloning was once the Woolly Mammoth, but attempts to extract DNA from frozen mammoths have been unsuccessful, though a joint Russo-Japanese team is currently working toward this goal. In 2001, a cow named Bessie gave birth to a cloned Asian gaur, an endangered species, but the calf died after 2 days. In 2003, a banteng was successfully cloned, followed by three African wildcats from a thawed frozen embryo. These successes provided hope that similar techniques (using surrogate mothers of another species) might be used to clone extinct species. Anticipating this possibility, tissue samples from the last *bucardo* (Pyrenean Ibex) were frozen immediately after it died. Researchers are also considering cloning endangered species such as the giant panda, ocelot, and cheetah. The "Frozen Zoo" at the San Diego Zoo now stores frozen tissue from the world's rarest and most endangered species. In 2002, geneticists at the Australian Museum announced that they had replicated DNA of the Thylacine (Tasmanian tiger), extinct about 65 years ago, using polymerase chain reaction (PCR). However, in the year 2005, the museum announced that it was stopping the project after tests showed that the specimens' DNA had been too badly degraded by the (ethanol) preservative. In the year 2005, it was announced that the Thylacine project would be revived, with new participation from researchers in New South Wales and Victoria.

One of the continuing obstacles in the attempt to clone extinct species is the need for nearly perfect DNA. Cloning from a single specimen could not create a viable breeding population in sexually reproducing animals. Furthermore, even if males and females were to be cloned, the question would remain open whether they would be viable at all in the absence of parents that could teach or show them their natural behavior. Essentially, if cloning an extinct species were successful, it must be considered that cloning is still an experimental technology that succeeds only by chance. It is more likely than not that any resulting animals, even if they were healthy, would be little more than curios or museum pieces. Cloning endangered species is a highly ideological issue. Many conservation biologists and environmentalists vehemently oppose cloning endangered species mainly because they think it may deter donations to help preserve natural habitat and wild animal populations. The "rule-of-thumb" in animal conservation is that, if it is still feasible to conserve habitat and viable wild populations, breeding in captivity should not be undertaken in isolation. In a 2006 review, David Ehrenfeld concluded that cloning in animal conservation is an experimental technology that, at its state in 2006, could not be expected to work except by pure chance and utterly failed a cost-benefit analysis.

7.7.5 Transgenic Animals

Transgenic animals are those animals whose genes have been deliberately modified in order to change their morphological appearance or their physiological functions. A transgenic animal is one whose genome has been changed to carry genes from other species. The nucleus of all cells in every living organism contains genes made up of DNA. These genes store information that regulates how our bodies form and function. Genes can be altered artificially, so that some characteristics of an animal are changed. For example, an embryo can have an extra, functioning gene from another source artificially introduced into it, or a gene introduced that can knock out the functioning of another particular gene in the embryo. The majority of transgenic animals produced so far are mice, the animal that pioneered the technology. The first successful transgenic animal was a mouse. A few years later, it was followed by rabbits, pigs, sheep, and cattle (Figure 7.13).

Transgenic animals are useful as disease models and producers of substances for human welfare. Some transgenic animals are produced for specific economic traits. For example, transgenic cattle were created to produce milk containing particular human proteins, which may help in the treatment of human emphysema. Other transgenic animals are produced as disease models (animals genetically manipulated to exhibit disease symptoms so that effective treatment can be studied). For example, Harvard scientists made a major scientific breakthrough when they received a U.S. patent, when the company DuPont holds exclusive rights to its use for a genetically engineered mouse, called OncoMouse®, or the Harvard mouse, which carries a gene that promotes the development of various human cancers.

Transgenic mice have also been used in scientific researches. Normal mice cannot be infected with polio virus. They lack the cell-surface molecule that, in humans, serves as the receptor for the virus.

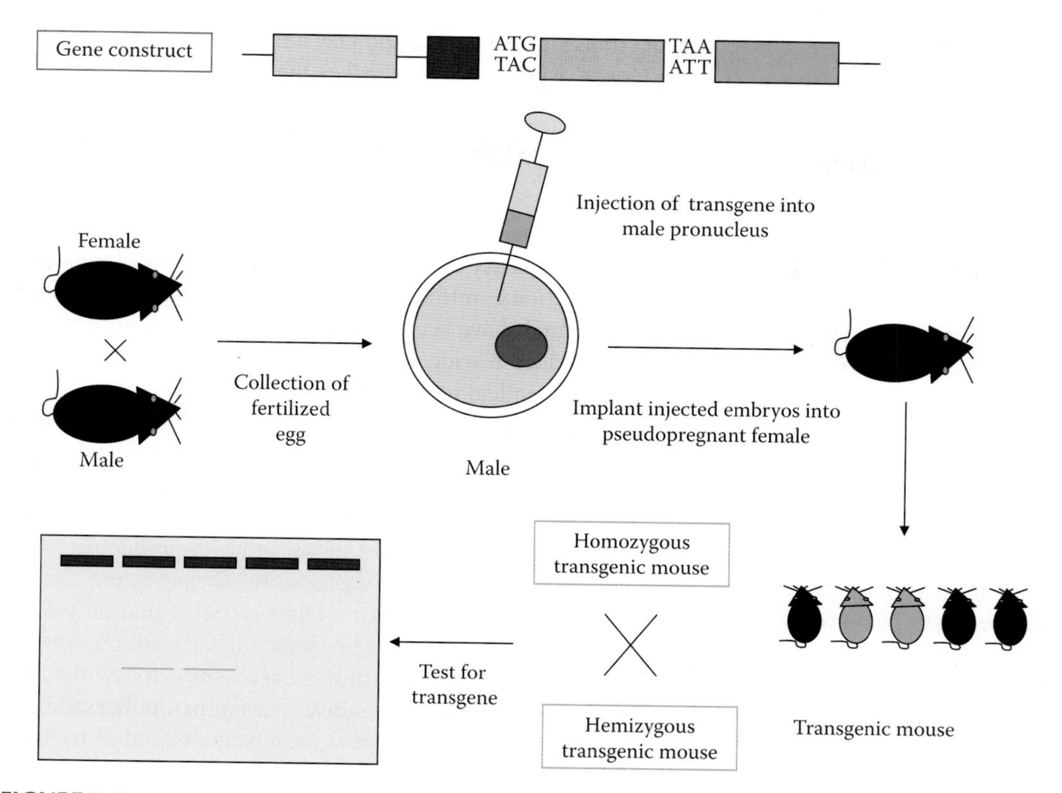

FIGURE 7.13 Generation of transgenic mouse.

Henceforth, normal mice cannot serve as an inexpensive, easily manipulated model for studying the disease. However, transgenic mice expressing the human gene for the polio-virus receptor can be infected by polio virus and even develop paralysis and other pathological changes characteristic of the disease in humans.

Farmers have always used selective breeding to produce animals that exhibit desired traits (such as increased milk production or high growth rate). Traditional breeding is a time-consuming, difficult task. When technology using molecular biology was developed, it became possible to develop traits in animals in a shorter time and with more precision. In addition, it offers the farmer an easy way to increase yields. Transgenic cows exist that produce more milk or milk with less lactose or cholesterol.

In the past, farmers used growth hormones (GH) to spur the development of animals but this technique was problematic, especially since residue of the hormones remained in the animal product. Products such as insulin, GH, and blood anticlotting factors may soon be or have already been obtained from the milk of transgenic cows, sheep, or goats. Research is also underway to manufacture milk through transgenesis for treatment of debilitating diseases such as phenylketonuria (PKU), hereditary emphysema, and cystic fibrosis. In 1997, the first transgenic cow, Rosie, produced human protein-enriched milk at 2.4 g/L. This transgenic milk is a more nutritionally balanced product than natural bovine milk and could be given to babies or the elderly with special nutritional or digestive needs. Rosie's milk contains the human gene alpha-lactalbumin.

Patients die every year for lack of a replacement heart, liver, or kidney. For example, about 5000 organs are needed each year in the United Kingdom alone. Transgenic pigs may provide the transplant organs needed to alleviate the shortfall. Currently, a pig protein that can cause donor rejection hampers xenotransplantation, but research is underway to remove the pig protein and replace it with a human protein. Human gene therapy involves adding a normal copy of a gene (transgene) to the genome of a person carrying defective copies of the gene. The potential for treatments for the

5000 named genetic diseases is huge and transgenic animals could play a role. For example, the A. I. Virtanen Institute in Finland produced a calf with a gene that makes the substance that promotes the growth of red cells in humans.

Besides agricultural, dairy, and medical fields, the transgenic animals have also been used in industrial applications. In 2001, two scientists at Nexia Biotechnologies in Canada spliced spider genes into the cells of lactating goats. The goats began to manufacture silk along with their milk and secrete tiny silk strands from their body by the bucketful. By extracting polymer strands from the milk and weaving them into thread, the scientists can create a light, tough, flexible material that could be used in such applications as military uniforms, medical microsutures, and tennis racket strings. Toxicity-sensitive transgenic animals have been produced for chemical safety testing. Microorganisms have been engineered to produce a wide variety of proteins, which in turn can produce enzymes that can speed up industrial chemical reactions.

To date, there are three basic methods of producing transgenic animals: DNA microinjection, retrovirus-mediated gene transfer, and embryonic stem (ES) cell-mediated gene transfer. Gene transfer by microinjection is the predominant method used to produce transgenic farm animals. Since the insertion of DNA results in a random process, transgenic animals are mated to ensure that their offspring acquire the desired transgene. However, the success rate of producing transgenic animals individually by these methods is very low and it may be more efficient to use cloning techniques to increase their numbers. For example, gene transfer studies revealed that only 0.6% of transgenic pigs were born with a desired gene after 7000 eggs were injected with a specific transgene. Although we benefit a lot from transgenic animals, still there are issues on how they are created, and these issues need to be resolved. The biotechnology industry, scientists, policy-makers, and the public cannot ignore thoughtful ethics, and ethical concerns must be addressed as technology grows, including the issue of the welfare of laboratory animal. Interestingly, the creation of transgenic animals has resulted in a shift in the use of laboratory animals from the use of higher-order species such as dogs to lower-order species such as mice and has decreased the number of animals used in such experimentation, especially in the development of disease models. This is certainly a good turn of events since transgenic technology holds great potential in many fields such as agriculture, medicine, and industry. In the next sections, we have listed different types of transgenic animals that are generated and used in various applications.

7.7.5.1 Transgenic Cow

In earlier attempts (in Canada) for the production of transgenic cows, embryos or fertilized oocytes produced *in vivo* were utilized. Fertilized oocytes or proembryos were surgically retrieved from superovulated and artificially inseminated cows. Microinjected zygotes were then transferred by surgery either directly into the oviduct of recipient cows or into temporary hosts like sheep or rabbits. In view of two surgical operations, this method is labor intensive and more expensive. In the Netherlands, recently (1991), a technique has been developed for *in vitro* embryo production. In this new procedure, oocytes obtained from the ovaries of slaughter house cows, were matured and fertilized *in vitro*. Their pronuclei were microinjected with a construct containing a bovine alpha-SI-casein promoter (bovine = ox) driving a cDNA encoding the antibacterial human iron binding protein, "lactoferrin." The embryos were cultured to morula/blastula stage and, then nonsurgically transferred to recipient females. Two of the 19 calves born from 103 transferred zygotes were transgenic (one male and other female). This procedure may facilitate the use of cows as bioreactors at the commercial level.

The transgenes introduced into sheep were inserted randomly in the genome and often worked poorly. However, in July 2000, success at inserting a transgene into a specific gene locus was reported. The gene was the human gene for alpha1-antitrypsin, and two of the animals expressed large quantities of the human protein in their milk. This is how it was done: Sheep fibroblasts (connective tissue cells) growing in tissue culture was treated with a vector that contained these segments of DNA: two regions homologous to the sheep *COL1A1* gene. This gene encodes Type 1 collagen.

(Its absence in humans causes the inherited disease osteogenesis imperfecta.) This locus was chosen because fibroblasts secrete large amounts of collagen and thus one would expect the gene to be easily accessible in the chromatin. A neomycin-resistance gene was used to aid in isolating those cells that successfully incorporated the vector.

7.7.5.2 Transgenic Pigs

The efficiency of the production of transgenic pigs is still very low compared to that of the production of transgenic mice. In mice, 2.5%–6% of the microinjected eggs developed into transgenic mice, but in pigs this frequency was as low as 0.6% even when as many as 7000 eggs were injected. Despite this low frequency, transgenic pigs carrying GH gene from bovine (of "ox" origin) or human, and sheep globin gene have been produced (by V.G. Purse) at Agriculture Research Service, Beltsville, United States. The pigs carrying hGH gene showed different levels of expression and only 66% of these animals showed detectable levels of hGH and bGH in their plasma. The animals grew a little faster but did not become large; similarly pigs with sheep globin gene did not show any expression of the transgene for unknown reasons.

In these transgenic pigs, however, a modest increase of 10%–15% in daily weight and 16%–18% in feed efficiency was observed, which, though lower to those in mice, is comparable to those obtained due to daily injection with pig GH. It was also observed that there was a marked reduction in the subcutaneous fat in some of these transgenic pigs suggesting the possibility of producing leaner meat with lower fat content. These results may have a significant impact on the 9.5 billion dollar annual pig industry in United States. It is also reported that a long-term elevation of GH was generally detrimental to health. The pigs had incidence of gastric ulcers, arthritis, and several other diseases. Therefore, techniques will have to be developed to manipulate better the transgene expression by a variety of methods (e.g., changing genetic background or modifying the husbandry regiment).

7.7.5.3 Transgenic Goat

Transgenic goats were successfully produced recently by groups headed by John McPherson and Karl Ebert both in the Untied States. These transgenic goats expressed a heterologous protein (a variant of human tissue type plasminogen activator = LAtPA) in their milk. This protein is used for dissolving blood clots, that is, for treatment of coronary thrombosis. A cDNA representing LAtPA was linked with either the murine whey acid promoter (WAP) or a, f3 casein promoter in an expression vector and used for injecting early embryos obtained surgically from the oviducts of superovulated dairy goats.

These injected embryos were either immediately transferred to the oviducts of recipient females (surrogate mothers) or cultured for 72 h (blocked at 8–16 cell stage), before transfer to the uterus of recipient females. Of the 29 offspring from 36 recipients, one male and one female contained the transgene. The transgenic female delivered five offspring, one of which was transgenic showing expression of LAtPA at a low level of few milligrams per liter of milk. In another case of a transgenic goat, few grams of LAtPA per liter of milk could be obtained. At this concentration, the dairy goat may become an economically viable bioreactor for human pharmaceuticals.

7.7.5.4 Transgenic Sheep

The rate of transgenesis in sheep is very low (0.1%–0.2%). This can be improved, if only transgenic viable embryos (after necessary checking) are transferred to surrogate ewes (female sheep). Embryos at 8–16 cell stage can be split into two parts, one for continued culture and the other for detection of integrated genes using PCR. Although microinjection is the most common method for DNA delivery, gene targeting may be increasingly used in future. In this approach, ES cells in culture are transfected with a vector, which targets the gene to a particular site by homologous recombination. This technique, though successfully used in mice, has yet to be applied to sheep, where ES cells will have to be isolated first. The first reports of transgenic sheep were published by J.P. Simons (1988) of Edinburgh.

Two transgenic ewes were produced, each. carrying about 10 copies of human antihemophilic factor IX gene (cDNA) fused with the 10.5 kb BLG gene (BLG = p-lactoglobulin). BLG gene was used, because it is necessary for specific expression of gene in mammary glands. Consequently, the gene had a tissue specific expression and ewes secreted human factor IX (or alpha-1 antitrypsin) into their milk; this human factor IX is active, even though the expression of transgene is low. The transgenic ewes were born in early summer of 1986 and were successfully mated same year in December.

In 1987, each ewe gave birth to a single lamb. Each lamb inherited BLG-F IX transgene and secreted factor IX in the milk. This program of the production of transgenic animals by J.P. Simons at Edinburgh' was funded by Pharmaceutical Proteins Ltd. (Cambridge, United Kingdom), due to its commercial appeal. In another report (published in 1991), also from Edinburgh, five transgenic sheep were produced (Alan Colman and colleagues). In all these cases, transgene involved fusion of the ovine β lactoglobulin gene promoter fused to the human at antitrypsin (hα1AT) gene.

Four of these animals were female and one male. In one female, the protein hα1AT reached a level of 35 g/L of milk. The protein purified from milk had a biological activity indistinguishable from human plasma derived antitrypsin. The deficiency for hα1AT leads to a lethal disease emphysema, which is a common hereditary disorder among Caucasian males of European descent. Therefore, any strategy giving high yield of this protein economically will be most welcome. In view of this, transgenic sheep with hα1AT gene will prove very useful as a bioreactor.

Recombinant DNA technique can also be used to increase the ability of sheep for wool growth. For this purpose, genes essential for synthesis of some important amino acids found in keratin proteins of wool, have been cloned and introduced in embryos to produce transgenic sheep. For instance, genes (cysE and cysM) for two enzymes (serine acetyltransferase = SAT and O-acetylserine sulphydrylase = OAS), involved in cysteine biosynthesis, were isolated from bacteria and cloned in a vector. These genes were introduced in sheep cells, ultimately leading to the production of transgenic sheep, where these genes are expressed. GH genes have also been introduced and can be used to promote body weight. Other genes involved in wool production have also been cloned and, will be used for transgenesis, thus increasing the potential of wool production through genetic engineering.

7.7.5.5 Transgenic Chickens

Chickens grow faster than sheep and goats, and large numbers can be grown in close quarters. They can also synthesize several grams of protein in the "white" of their eggs. Preliminary results from both methods indicate that it may be possible for chickens to produce as much as 0.1 g of human protein in each egg that they lay. Not only should this cost less than producing therapeutic proteins in culture vessels, but chickens will probably add the correct sugars to glycosylated proteins, something that *E. coli* cannot do.

7.7.5.6 Transgenic Primates

In the May 28, 2009, issue of Nature, Japanese scientists report success in creating transgenic marmosets. Marmosets are primates, and thus our closest relatives (so far) to be genetically engineered. In some cases, the transgene (for green fluorescent protein [GFP]) was incorporated into the germline and passed on to the animal's offspring. The hope is that these transgenic animals will provide the best model yet for studying human disease and possible therapies.

7.8 BIOTECHNOLOGY AND FISH FARMING

Aquaculture is the farming of freshwater and saltwater organisms such as finfish, mollusks, crustaceans, and aquatic plants. Also known as aquafarming, aquaculture involves cultivating aquatic organisms under controlled conditions. One-half of the world's commercial production of fish and shellfish that is directly consumed by humans comes from aquaculture. Aquaculture is a very old fish farming technique and it has been in practice since 2500 BC.

7.8.1 Mariculture

Mariculture is a specialized branch of aquaculture involving the cultivation of marine organisms in the open ocean, an enclosed section of the ocean, or in tanks, ponds, or raceways, which are filled with seawater. An example of the latter is the farming of marine fish, prawns, or oysters in saltwater ponds. Nonfood products produced by mariculture include fish meal, nutrient agar, jewelry (such as cultured pearls), and cosmetics. With fishery catches on the decline, and aquaculture products on the incline, mariculture holds a great promise for the future, both economically and environmentally. The broad perception by a large majority of fish consumers is that fish is healthy and nutritious, and an advantage to mariculture. Another advantage is that the naturalness that farmed fish possess is comparable to those that are harvested from the ocean. Mariculture farming helps the species that may be depleting in the wild such as trout, sea bass, and salmon. The consistency of supply all year round and more routine quality control has enabled mariculture supply to be integrated in other food market channels. These benefits have also been able to reach different socioeconomic classes who may not have been able to purchase fish because of high prices.

7.8.2 Polyculture

Polyculture is agriculture using multiple crops in the same space, in imitation of the diversity of natural ecosystems, and avoiding large stands of single crops, or monoculture. It includes crop rotation, multicropping, intercropping, companion planting, beneficial weeds, and alley cropping. Polyculture, though it often requires more labor, has several advantages over monoculture. The diversity of crops avoids the susceptibility of monocultures to disease. For example, a study in China, reported in *Nature*, showed that planting several varieties of rice in the same field increased yields by 89%, largely because of a dramatic (94%) decrease in the incidence of disease, which made pesticides redundant. Also, the greater variety of crops provides habitat for more species, increasing local biodiversity. This is one example of reconciliation ecology, or accommodating biodiversity within human landscapes.

7.8.3 Aquatic Biotechnology

With the advancement of molecular and genetic tools, there is also development of transgenic aquatic animals to understand their genetic information. The genetic information received from aquatic animals is used to study how the animals live and survive in both extreme cold water and hot water. In this section, we will discuss how different useful genes are isolated from aquatic animals, which are being later used in various applications.

7.8.3.1 Transgenic Fish

Attempts to produce transgenic fish started in 1985, and some encouraging results have been obtained. The genes that have been introduced by microinjection in fish are (1) human or rat gene for GH, (2) chicken gene for delta crystalline protein, (3) *E. coli* gene for β-galactosidase, (4) *E. coli* gene for neomycin resistance, (5) winter flounder gene for antifreeze protein (AFP), and (6) rainbow trout gene for GH. The technique of microinjection has been successfully used to generate transgenic fish in many species such as common carp, catfish, goldfish, loach, medaka, salmon, tilapia, rainbow trout, and zebrafish.

In other animals (such as mice, cows, pigs, sheep, and rabbits), usually direct microinjection of cloned DNA into male pronuclei of fertilized eggs has proved very successful, but in most fish species studied so far, pronuclei cannot be easily visualized (except in medaka); therefore, the DNA needs to be injected into the cytoplasm. Eggs and sperms from mature individuals are collected and placed into a separate dry container. Fertilization is initiated by adding water and sperm to the eggs, with gentle stirring to facilitate the fertilization process. Egg shells are hardened in water.

About 106–108 molecules of linearized DNA in a volume of 20 mL or less is microinjected into each egg (1–4 cells stage) within the first few hours after fertilization. Following microinjection, eggs are incubated in appropriate hatching trays and dead embryos are removed daily. Since in fish, fertilization is external, *in vitro* culturing of embryos and their subsequent transfer into foster mothers (required in mammalian systems) is not required. Further, the injection into the cytoplasm is not as harmful as that into the nucleus; therefore, the survival rate in fish is much higher (35%–80%). Human GH genes transferred to transgenic fish allowed growth that was twice the size of their corresponding nontransgenic fish (goldfish, rainbow trout, and salmon).

7.8.3.2 Green Fluorescent Protein

GFP is a protein isolated from *coelenterates* such as the Pacific jellyfish, *Aequoria victoria*, or from the sea pansy, *Renilla reniformis*. Although many other marine organisms have similar GFPs, GFP traditionally refers to the protein first isolated from the jellyfish *Aequorea victoria*. The GFP from *A. victoria* has a major excitation peak at a wavelength of 395 nm and a minor one at 475 nm. Its emission peak is at 509 nm, which is in the lower green portion of the visible spectrum. The GFP from the sea pansy (*Renilla reniformis*) has a single major excitation peak at 498 nm. In cell and molecular biology, the GFP gene is frequently used as a reporter of expression. In modified forms, it has been used to make biosensors, and many animals have been created that express GFP as a proof-of-concept that a gene can be expressed throughout a given organism. The GFP gene can be introduced into organisms and maintained in their genome through breeding, injection with a viral vector, or cell transformation. To date, the GFP gene has been introduced and expressed in many bacteria, yeast and other fungi, fish (such as zebrafish), plant, fly, and mammalian cells, including human. Martin Chalfie, Osamu Shimomura, and Roger Y. Tsien were awarded the 2008 Nobel Prize in chemistry on October 10, 2008, for their discovery and development of the GFP.

7.8.3.3 Antifreeze Proteins

Scientists found out that aquatic animals living in extreme cold conditions (such as whales and sharks), without having any problems, produce some kind of proteins that antagonize the effect of cold. They isolated the protein and gene responsible for the antifreezing properties and called this specialized protein as AFPs or ice-structuring proteins (ISPs). These AFPs refer to a class of polypeptides produced by certain vertebrates, plants, fungi, and bacteria that permit their survival in subzero environments. AFPs bind to small ice crystals to inhibit growth and recrystallization of ice that would otherwise be fatal. There is also increasing evidence that AFPs interact with mammalian cell membranes to protect from cold damage. In the 1950s, Canadian scientist Scholander set out to explain how Arctic fish can survive in water colder than the freezing point of their blood. His experiments led him to believe that there was "antifreeze" in the blood of Arctic fish. Then, in the late 1960s, animal biologist Arthur DeVries was able to isolate the AFP through his investigation of Antarctic fish. Antifreeze glycoproteins (AFGPs) were the first AFPs to be discovered. At that time, they were called "glycoproteins as biological antifreeze agents." These proteins were later called AFGPs or antifreeze glycopeptides to distinguish them from newly discovered nonglycoprotein biological AFPs.

The following are some of the vast commercial applications of AFPs: (1) increases freeze tolerance of crop plants and extend the harvest season in cooler climates, (2) improves farm fish production in cooler climates, (3) lengthens shelf life of frozen foods, (4) improves cryosurgery, and (5) enhances preservation of tissues for transplant or transfusion in medicine. Currently, two companies market AFPs like A/F Protein, Inc., Waltham and Ice Biotech, Inc., Ontario, Canada. One recent successful business endeavor has been the introduction of AFPs into ice cream and yogurt products. This ingredient labeled as ISP has been approved by the U.S. Food and Drug Administration (USFDA). Currently, Unilever incorporates AFPs into some of its products, including some popsicles and a new line of Breyers Light Double Churned ice cream bars. In ice cream, AFPs allow the production of very creamy, dense, reduced fat ice cream with fewer additives.

They control ice crystal growth brought on by thawing on the loading dock or kitchen table, which drastically reduces texture quality. In November 2009, the Proceedings of the National Academy of Sciences published the discovery of a molecule in an Alaskan beetle that behaves like AFPs, but is composed of saccharides and fatty acids. Similarly, AFP gene was transferred in several cases and its expression as studied in transgenic salmon. It was shown that the level of AFP gene expression is still too low to provide protection against freeze.

7.8.3.4 Transgenic Salmon

Two potential ways in which transgenic technologies can be used to solve the problem of overwintering salmon in sea cages in Atlantic Canada are (1) Produce freeze-resistant salmon by giving them a set of AFP genes, and (2) enhance growth rates by GH gene transfer so that overwintering may not be necessary. Many commercially important fish (such as salmon and halibut) lack these proteins and their genes and, as a consequence, they will not survive if cultured in icy sea water. In 1982, transgenic studies were initiated by injecting winter flounder antifreeze genes into the fertilized eggs of Atlantic salmon. A full length gene encoding the major liver secretory AFP was used and the AFP transgene was successfully integrated into the salmon chromosomes, expressed, and found to exhibit Mendelian inheritance over five generations to date. Expressed levels of AFP in the blood of these fish are quite low (100–400 μg/mL) and is insufficient to confer any significant increase in freeze resistance to the salmon. However, the "proof-of-the-concept" that salmon and other fish species can be rendered more freeze resistant by gene transfer has been established. The challenge now is to design an antifreeze gene construct(s) that will result in enhanced expression in appropriate tissues—epithelia and liver.

7.8.3.5 Transgenic Mussel

Byssus generally refers to a filament created by certain kinds of marine and freshwater bivalve mollusks, which it uses to attach themselves to rocks, substrates, or sea beds. In edible mussels, the inedible byssus is commonly known as the "beard," and is removed before cooking. Byssus specifically refers to the long, fine, silky threads secreted by the large Mediterranean pen shell, *Pinna nobilis*. The byssus threads from this *Pinna* species can be up to 6 cm in length and have historically been made into cloth. Many species of mussels secrete byssus threads to anchor themselves on hard surfaces, with families including the Arcidae, Mytilidae, Anomiidae, Pinnidae, Pectinidae, Dreissenidae, and Unionidae. When a mussel's foot encounters a crevice, it creates a vacuum chamber by forcing out the air and arching up, similar to a plumber's plunger unclogging a drain. The byssus, which is made of keratin, quinone-tanned proteins, and other proteins, is spewed into this chamber in liquid form, and bubbles into sticky foam. By curling its foot into a tube and pumping the foam, the mussel produces sticky threads about the size of a human hair. The mussel then varnishes the threads with another protein, resulting in an adhesive. Byssus is a remarkable adhesive, one that is neither degraded nor deformed by water, as are synthetic adhesives. This property has spurred genetic engineers to insert mussel DNA into yeast cells for translating the genes into the appropriate proteins. Byssal fiber proteins have been used in several applications from tyre, bone, and tooth manufacturing industries.

7.8.3.6 Fugu Fish

Fugu fish contains lethal amounts of the poison tetrodotoxin (TTX) in the organs, especially the liver and ovaries, and also the skin. TTX is a very potent neurotoxin that shuts down electrical signaling in nerves by binding to the pores of sodium channel proteins in nerve cell membranes. Currently, there is no known antidote, and the standard medical approach is to try to support the respiratory and circulatory systems until the poison wears off. Fugu has been consumed in Japan for centuries, although its historic origins are unclear and TTX is not affected by cooking. It does not cross the blood-brain barrier, leaving the victim fully conscious while paralyzing the remainder of the body. There is no known antidote, and treatment consists of emptying the stomach, feeding the victim with activated charcoal to bind the toxin, and taking standard life-support measures to keep the victim alive until the poison has worn off.

Scientists at Nagasaki University have reportedly succeeded in creating a nontoxic variety of torafugu by restricting the fish's diet. With over 4800 fish raised and found to be nontoxic, they are fairly certain that the fish's diet and digestive process are what actually produce the toxins that make it deadly. The nontoxic version is said to taste the same, but is completely safe for consumption. Some skeptics say that the species being offered as nontoxic may be of a different species and that the toxicity has nothing to do with the diet of the pufferfish. Recent evidence has shown that TTX is produced by certain bacteria such as *Pseudoalteromonas tetraodonis*, certain species of Pseudomonas and Vibrio, as well as some others. *Fugu rubripes* is a commonly used genetic model organism, particularly useful to bioinformaticians. The *Fugu* genome is unusually small for an organism of its complexity, and contains very little junk DNA. This compactness makes its genome sequence very useful for identifying conserved functional elements. Furthermore, the sodium channel blocking properties of TTX means that it is widely used in the study of ion channels in neuropharmacology.

7.8.3.7 *Squalus acanthias*

The spiny dogfish, spurdog, mud shark, or piked dogfish, *Squalus acanthias*, is one of the best known of the dogfish, which are members of the family Squalidae in the order Squaliformes. While these common names may apply to several species, *Squalus acanthias* is distinguished by having two spines (one anterior to each dorsal fin) and lacks an anal fin. It is found mostly in shallow waters and further offshore in most parts of the world, especially in temperate waters. Squalamine is a strong naturally derived broad-spectrum antibiotic that is predominantly derived from the livers of dogfish and other shark species. Often mistaken for shark liver oil (squalene), it is instead an entirely different substance derived from the livers of these species.

While used first as an effective broad-spectrum antibiotic, it has been found recently that it can perform a powerful antiangiogenesis function, that is, it inhibits the growth of blood vessels. Because of this function, squalamine in its intravenous form, squalamine lactate, is in the process of being tested as a treatment of fibrodysplasia ossificans progressiva, a rare disease where connective tissue will ossify when damaged. Squalamine is also undergoing trials for treatment of nonsmall cell lung cancer (Stage I/IIA) as well as general Phase I pharmacokinetic studies. In 2005, the Food and Drug Administration (FDA) granted squalamine Fast Track status for approval for treatment of age-related macular degeneration. However, the Genaera Corporation, the company that has done the most work with squalamine, discontinued trials for its use in treating prostate cancer and wet age-related macular degeneration in 2007. The compound is also being marketed under the brand name of Squalamax as a dietary supplement, but it has not been approved as a drug.

7.8.3.8 *Limulus polyphemus*

The horseshoe crab or Atlantic horseshoe crab (*Limulus polyphemus*) is a marine chelicerate arthropod. Despite its name, it is more closely related to spiders, ticks, and scorpions than to crabs. Horseshoe crabs are most commonly found in the Gulf of Mexico and along the northern Atlantic coast of North America. A main area of annual migration is Delaware Bay, although stray individuals are occasionally found in Europe. The blood of horseshoe crabs as well as that of most mollusks, including cephalopods and gastropods, contains the copper-containing protein, hemocyanin, at concentrations of about 50 g/L. These creatures do not have hemoglobin (iron-containing protein), which is the basis of oxygen transport in vertebrates. Hemocyanin is colorless when deoxygenated and dark blue when oxygenated. The blood in the circulation of these creatures, which generally live in cold environments with low oxygen tensions, is gray-white to pale yellow, and it turns dark blue when exposed to the oxygen in the air, as seen when they bleed.

The blood of horseshoe crabs contains one type of blood cell, the amebocytes, and these cells play an important role in the defense against pathogens. Amebocytes contain granules with a clotting factor known as *coagulogen*. This is released outside the cell when bacterial endotoxin is encountered. The resulting coagulation is thought to contain bacterial infections in the animal's semiclosed circulatory system. This is known as the Limulus amebocyte lysate (LAL) test. Horseshoe crabs are valuable as a species to the medical research community, and in medical testing. The above-mentioned clotting reaction of the animal's blood is used in the LAL test to detect bacterial endotoxins in pharmaceuticals and to test for several bacterial diseases. In an another application, enzymes from horseshoe crab blood are used by astronauts in the International Space Station to test surfaces for unwanted bacteria and fungi. A protein from horseshoe crab blood is also under investigation as a novel antibiotic.

7.9 REGULATIONS OF ANIMAL TESTING

Animal testing is one of the most strictly regulated affairs where researchers need to get approval from regulators and ethical organizations first before they are allowed to carry out research work. Different agencies regulate broad range of products from cosmetics to drugs to food additives that routinely undergo animal testing by manufacturers before they can be used by people. Failure to follow the regulations means that test results may be faulty, the data may be rejected, and whatever the submitter was seeking will be denied. The regulations call for humane treatment of the animals, with close attention to housing, bedding, food, and water. They stress that animals must be carefully identified and that identity maintained, since losing track of test animals can void a study. Animals from one study must not be mingled with others, and no more animals that are needed are to be used. Regulatory authorities generally ensure that researchers or companies should use the animals with great care and they insist that animals should be properly housed, fed, and kept under healthy and hygienic conditions. They also make sure the number of animals used in the study is properly justified and required. The USFDA has a responsibility to assure that products are safe and effective for the public. A heated debate has developed over the use of animal testing to ensure the safety of human beings. The FDA has placed requirements on the use of animal testing and alternative methods. The U.S. Food, Drug, and Cosmetic Act (FD&C Act) does not require the use of animals in testing cosmetic products for safety, nor does it require premarket approval of cosmetics by the FDA. The agency advises cosmetics manufacturers to use whatever form of testing is appropriate and effective for proving the safety of their products.

7.9.1 Alternatives to Animal Testing

With a view to minimize the access use of animals in research, constant efforts have been made over the past 50 years to come up with alternative strategy. In 1954, Charles Hume, founder of the Universities Federation for Animal Welfare (UFAW), made an original proposal based on reduction, refinement, and replacement concepts to minimize the access use of animals. The replacement methods enable researchers to obtain comparable levels of information from fewer animals, or to obtain more information from the same number of animals. The refinement methods alleviate or minimize potential pain, suffering, or distress, and enhance animal welfare for the animals still used. The replacement method is a nonanimal method over animal method whenever it is possible to achieve the same scientific aim. We have described several approaches on similar lines, which have proposed in recent times to minimize the use of animals.

7.9.1.1 *In Vitro* Cell Culture Technique

The *in vitro* cell culture technique is currently the most successful and promising alternative to animal use. Prior to use of *in vitro* cell culture technique, animals have been used in production of MABs, which required animals to undergo a procedure likely to cause pain and distress.

Historically, animals were used for purposes such as vaccine development and creation. Throughout the 1970s, in the Netherlands, for instance, thousands of monkeys were used to formulate the polio vaccine. Today, a kidney cell culture can be used to make polio vaccine. The added benefit of using cell cultures is that any vaccines from cell cultures are in a form that is particularly pure compared to one derived directly from the animal. This means that the usual safety testing that must occur on the vaccines can essentially be bypassed.

Human skin equivalent tests are being used to replace animal-based corrosive studies. Two products, EpiDerm and EpiSkin are derived from human skin cells, which have been cultured to produce a model of human skin. These methods are currently accepted replacements in Canada and the EU. In another synthetic replacement method, protein membrane is used to simulate a skin barrier and is approved as a partial replacement by the U.S. Department of Transportation as well as by EU. Several tissue culture methods, which measure the rate of chemical absorption by the skin, have been approved by the Organization for Economic Cooperation and Development (OECD), although they have not yet been approved as a replacement in the United States. In order to check the phototoxicity caused by exposure to light following exposure to a chemical, the OECD approved 3T3 Neutral Red Uptake (NRU) phototoxicity test to detect the viability of 3T3 cells after exposure to a chemical in the presence or absence of light. Similarly, skin patch test has been designed and is used in Canada to measure development of rashes, inflammation, and swelling or abnormal tissue growth on human volunteers. Unlike corrosives, irritants cause only reversible skin damage. Pyrogens are most often pharmaceutical products or intravenous drugs that may cause inflammation or fever when they interact with immune system cells. This interaction can be quickly and accurately tested *in vitro* using donated human blood.

7.9.1.2 Synthetic Membranes

By growing cells via artificial means, many biomedical tests can avoid the use of animals. In one case, an organization in the United States began using synthetic membranes in the early 1990s, and the techniques for creating such mediums have vastly improved since that time. These synthetic membranes are substituted for animals and they are used to demonstrate the effects of chemicals or topical treatments on skin. This contrasts greatly with the traditional tests where an animal's fur would be shaved, and then a corrosive chemical would be applied to its back to observe the effects.

7.9.1.3 Statistics Instead of Animal Testing

Statistical procedures can allow researchers to use comprehensive data sets to better gauge how a disease can spread. They also make use of data previously obtained from animal testing. This allows them to avoid using animals, which ultimately reduces the number of animals used in testing.

7.9.1.4 Newer Scanning Techniques

Some of the newer technologies entail improved scans such as magnetic resonance imaging (MRI). This enables researchers to actually investigate disease through human scans rather than performing animal testing. Another important alternative is the use of an autopsy to provide information relevant to biomedical research. While this cannot replace animals completely, it does reduce the numbers used.

7.9.1.5 Computer Models

Computer models are effective tools to simulate the response to a specific research question or experiment. While they still do not replace an entire animal organism, they have proved useful as a substitute for animals in some cases. Examples of computer simulations available include models of diabetes, asthma, and drug absorption, though potential new medicines identified using these techniques are still currently required to be verified in animal and human tests before licensing. Computer-operated mannequins, also known as *crash test dummies*, complete with internal sensors and video, have replaced live animal trauma testing for automobile crash testing. The first of these was "Sierra Sam" built in 1949 by Alderson Research Labs (ARL) Sierra Engineering.

These dummies continue to be refined. Prior to this, live pigs were used as test subjects for crash testing. Other nonanimal simulators have been developed for military use to mimic battlefield induced traumas. TraumaMan and the Combat Trauma Patient Simulator can be used to simulate hemorrhaging, fractures, amputations, and burns. Previously, animals were intentionally subjected to various traumas to provide military training. TraumaMan is also now used for training medical students. Several virtual humans have been constructed by creating mathematical models of a human based on known human reactions. Computer models have been constructed to model human metabolism to study plaque buildup and cardiovascular risk and to evaluate toxicity of drugs, tasks for which animals are also used.

PROBLEMS

Section A: Descriptive Type

Q1. Why do scientists prefer to work on animal models than humans?
Q2. Do you think selling drugs without being tested in animals is unlawful? Why or why not?
Q3. What are the most commonly used animals in research? Briefly explain their uses.
Q4. How are animals used in basic research?
Q5. Why do animal models of neurological disorder fail to create same syndromes as that in humans?
Q6. Is it mandatory to know the toxic effect(s) of a new drug molecule before being tested in humans? Why or why not?
Q7. Explain briefly the use of zebrafish as a research model.
Q8. What are the benefits of animal cloning?
Q9. How are transgenic animals produced?
Q10. What are the ethical concerns related to transgenic animals?
Q11. What is LAL test?
Q12. How are antifreezing proteins isolated from fish?

Section B: Multiple Choices

Q1. How many reasons scientists will give in favor of use of animals in research?
a. 4
b. 5
c. 7
d. 10

Q2. On which year did the U.S. government made animal testing mandatory for all medicines before it is used in humans?
a. 1937
b. 1940
c. 1945
d. 1949

Q3. Which vertebrate animal is the most commonly used in research?
a. Rodents
b. Monkeys
c. Cats
d. Dogs

Q4. Which is the most commonly used animal for neurological research?
a. Dog
b. Cat
c. Fish
d. Mouse

Q5. Drug efficacy is the effect of ________ at various time intervals.
   a. Drug on body
   b. Body on drug
   c. None of these
Q6. What is LD50?
   a. Dose at which 50% of test animals were killed
   b. Dose at which 50% of test animals survived
   c. Dose at which 50 test animals were killed
   d. None of the above
Q7. Make the right pairing.

a. Chronic	i. effect lasting for less than 1 month
b. Acute	ii. effect lasting for 1–3 months
c. Subacute	iii effect lasting for more than 3 months

Q8. In 2002, cosmetic testing on animals was banned in these countries except …
   a. Belgium
   b. United Kingdom
   c. United States
   d. Netherlands
Q9. What was the first multicellular organism to have its genome completely sequenced?
   a. *Drosophila melanogaster*
   b. *C. elegans*
   c. Zebrafish
   d. *C. brenneri*
Q10. ________ was the first learning and memory mutant fruitfly.
   a. Dunce
   b. Tetra
   c. Pheta
   d. TTcc
Q11. The first successful instance of animal cloning was the …
   a. Goat
   b. Cat
   c. Sheep
   d. Mule

Section C: Critical Thinking

Q1. Is it possible to test a drug without using animals? Explain.
Q2. Do you agree with the ban imposed by some European countries on the use of animals for cosmetics testing? Why or why not?

ASSIGNMENT

Organize a class debate on the pros and cons of using transgenic animals for various applications.

REFERENCES AND FURTHER READING

Butler, M. *Animal Cell Technology: From Biopharmaceuticals to Gene Therapy*. T & F Books, Oxford, U.K., 2009.

Dahm, R. The zebrafish exposed. *Am. Sci.* 94: 446–453, 2006.

Kimmel, C.B. and Law, R.D. Cell lineage of zebrafish blastomeres. I. Cleavage pattern and cytoplasmic bridges between cells. *Dev. Biol.* 10: 78–85, 1985.

Kimmel, C.B. and Law, R.D. Cell lineage of zebrafish blastomeres. III. Clonal analyses of the blastula and gastrula stages. *Dev. Biol.* 108: 94–101, 1985.
Lamason, R.L., Mohideen, M.A., Mest, J.R. et al. SLC24A5, a putative cation exchanger, affects pigmentation in zebrafish and humans. *Science* 310: 1782–1786, 2005.
Mills, D. *Eyewitness Hnbk Aquarium Fish*. Harper Collins, Australia, 1993. ISBN 0-7322-5012-9.
Portner, R. *Animal Cell Biotechnology: Methods and Protocols*. Humana Press, Totowa, NJ, 2007.
Twine, R. *Animals as Biotechnology: Ethics, Sustainability and Critical Animal Studies*. Earthscan Publications, Ltd., London, U.K., 2010.
Watanabe, M.M., Iwashita, M., Ishii, M. et al. Spot pattern of leopard Danio is caused by mutation in the zebrafish connexin 41.8 gene. *EMBO Rep.* 7: 893–897, 2006.
White, R.M., Sessa, A., Burke, C. et al. Transparent adult zebrafish as a tool for in vivo transplantation analysis. *Cell Stem Cell* 2: 183–189, 2008.

8 Environmental Biotechnology

LEARNING OBJECTIVES

- Discuss the importance of a healthy environment
- Discuss some factors affecting the environment
- Explain the role of biotechnology in maintaining the planet
- Explain how bioremediation helps clean wastewater
- Discuss some environmentally friendly products such as biofuel, biodegradable plastic, oil-eating bacteria, biostimulation, bioleaching, vermitechnology, and biosorption

8.1 INTRODUCTION

It is very important to know the environment and its components before we discuss the factors affecting it. The *natural environment*, or simply *environment*, is a term that encompasses all living and nonliving things occurring naturally on earth or on some of its regions. The concept of the natural environment is classified into two major components. The first component pertains to natural systems without massive human intervention, such as vegetation, animals, microorganisms, soil, rocks, the atmosphere, and the natural phenomena that occur within their boundaries. The second component pertains to universal natural resources and physical phenomena that lack clear-cut boundaries, such as air, water, and the climate, as well as energy, radiation, electric charge, and magnetism that do not originate from human activity (Figure 8.1).

8.1.1 Ecosystem

An *ecosystem* is a natural unit consisting of all plants, animals, and microorganisms in an area functioning together with all the nonliving physical (abiotic) factors of the environment. The living organisms are continually engaged in a highly interrelated set of relationships, with every other element constituting the environment in which they exist. The human ecosystem concept is then grounded in the deconstruction of the human/nature dichotomy and the emergent premise that all species are ecologically integrated with each other, as well as with the abiotic constituents of their biotope. Ecosystems can be grouped and discussed with tremendous variety of scope. An ecosystem can also describe any situation where there is relationship between organisms and their environment. If humans are part of the organisms, one can speak of a "human ecosystem." As virtually no surface of the earth today is free of human contact, all ecosystems can be more accurately considered human ecosystems or—more neutrally—human-influenced ecosystems.

8.1.2 Biomes

Biomes are terminologically similar to the concept of ecosystems and are climatically and geographically defined areas of ecologically similar climatic conditions, such as communities of plants, animals, and soil organisms. Biomes are defined based on factors such as plant structure (such as trees, shrubs, and grasses), leaf types (such as broadleaf and needle leaf), plant spacing (forest, woodland, savanna), and climate. Unlike *ecozones*, biomes are not defined by genetic, taxonomic, or historical similarities. Biomes are often identified with particular patterns of ecological succession and climate vegetation.

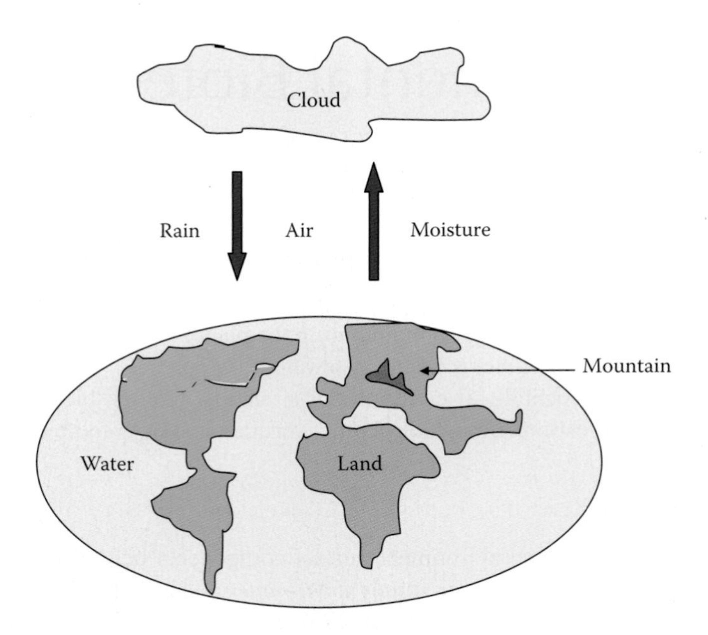

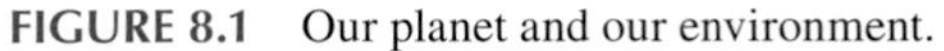

FIGURE 8.1 Our planet and our environment.

8.1.3 Wilderness

Wilderness is generally defined as a natural environment on earth that has not been significantly modified by human activity. The WILD Foundation defines wilderness as "the most intact, undisturbed wild natural areas left on our planet—those last truly wild places that humans do not control and have not developed with roads, pipelines, or other industrial infrastructure." Wilderness areas and protected parks are considered important for the survival of certain species, ecological studies, conservation, solitude, and recreation. Wilderness is deeply valued for cultural, spiritual, moral, and aesthetic reasons. Some nature writers believe wilderness areas are vital for the human spirit and creativity. The word "wilderness" was derived from the notion of wildness or, in other words, that which is not controllable by humans. The mere presence or activity of people does not disqualify an area from being a "wilderness." Many ecosystems that are, or have been, inhabited or influenced by activities of people may still be considered "wild." This way of looking at wilderness includes areas within which natural processes operate without very noticeable human interference.

8.1.4 Geological Activity

Another important feature of our planet is its geological activity such as volcanic fissures and lava eruption. The earth's crust (or continental crust) is the outermost solid land surface of the earth. It is chemically and mechanically different from the underlying mantle, and has been generated largely by igneous processes in which magma (molten rock) cools and solidifies to form solid land. Plate tectonics, mountain ranges, volcanoes, and earthquakes are geological phenomena that can be explained in terms of energy transformations in the earth's crust, and might be thought of as the process by which the earth resurfaces itself. Beneath the earth's crust lies the mantle, which is heated by the radioactive decay of heavy elements. The mantle is not quite solid, and consists of magma, which is in a state of semi-perpetual convection. This convection process causes the lithospheric plates to move, although slowly. The resulting process is known as *plate tectonics*. Volcanoes result primarily from the melting of subducted crust material. Crust material that is forced into the asthenosphere melts, and some portion of the melted material becomes light enough to rise to the surface, giving birth to volcanoes.

8.1.5 Oceanic Activity

Approximately 71% of the earth's surface (an area of some 361 million square kilometers) is covered by oceans, a continuous body of water that is customarily divided into several principal oceans and smaller seas. More than half of this area is over 3000 m (9800 ft) deep. Average oceanic salinity is around 35 parts per thousand (ppt) (3.5%), and nearly all seawater has a salinity in the range of 30–38 ppt. Though generally recognized as several "separate" oceans, these waters comprise one global interconnected body of salt water, often referred to as the World Ocean or global ocean. This concept of a global ocean as a continuous body of water with relatively free interchange among its parts is of fundamental importance to oceanography. The major oceanic divisions are defined in part by the continents, various archipelagos, and other criteria. These divisions are (in descending order of size) the Pacific Ocean, the Atlantic Ocean, the Indian Ocean, the Southern Ocean (which is sometimes subsumed as the southern portions of the Pacific, Atlantic, and Indian Oceans), and the Arctic Ocean (which is sometimes considered a sea of the Atlantic). The Pacific and the Atlantic oceans may be further subdivided by the equator into northerly and southerly portions. Smaller regions of the oceans are called seas, gulfs, bays, and by other names. There are also salt lakes, which are smaller bodies of landlocked saltwater that are not interconnected with the World Ocean. Two notable examples of salt lakes are the Aral Sea and the Great Salt Lake.

8.2 FACTORS AFFECTING THE ENVIRONMENT

Although the industrial revolution created useful products for humans, it also created problems like global warming; contamination of air, water, and land; and destruction of forests, to name a few. The initiation of large-scale projects such as dams and power plants pose special and growing challenges and risks to the natural environment. In this section, we will discuss some factors that negatively affect our environment, directly or indirectly.

8.2.1 Global Warming

Global warming is a serious threat to the earth. The potential dangers of global warming are being increasingly studied by a wide global consortium of scientists, who are concerned about the potential long-term effects of global warming on our natural environment and on the earth in general. One particular concern is how climate change and global warming caused by anthropogenic or human-made releases of greenhouse gases (GHGs), most notably carbon dioxide (CO_2), can act interactively and have adverse effects upon the planet, it's natural environment, and human existence. Efforts have been increasingly focused on the mitigation of GHGs that are causing climatic changes and on developing adaptive strategies to assist humans, animal and plant species, ecosystems, regions, and nations to adjust to the effects of global warming (Figure 8.2).

8.2.1.1 Carbon Footprint

A *carbon footprint* is "the total set of GHG emissions caused directly and indirectly by an individual, organization, event, or product" (U.K. Carbon Trust 2008). The carbon footprint of an individual, a nation, or an organization is measured by undertaking a GHG emissions assessment. Once the size of the carbon footprint is known, a strategy can be devised to reduce it. Carbon offsets, or the mitigation of carbon emissions through the development of alternative projects such as solar or wind energy or reforestation, represent one way of managing a carbon footprint. The term and concept of the carbon footprint originates from the ecological footprint discussion. The carbon footprint is a subset of the ecological footprint.

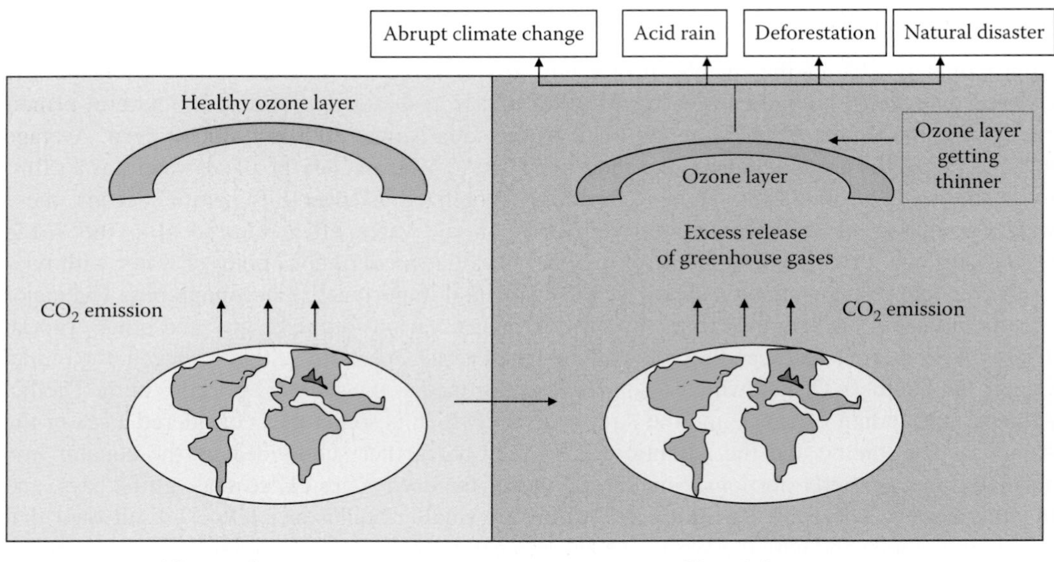

FIGURE 8.2 Global warming.

8.2.1.2 Destruction of Forests

Deforestation can occur in two ways: by natural means or by human intervention. Deforestation happens naturally from time to time via wildfires. Interestingly, trees, plants, and animals all recover from such events naturally, so there are indeed some benefits from a wildfire. In addition, birds such as the black-backed woodpecker thrive only in freshly burned areas, where they eat insects that bore into the burned trees. Some trees, such as the lodgepole pine, produce serrotonous cones that only open when heat cooks the cone, thereby spreading the seeds into a freshly burned area with little other competition. Over time, burned areas regrow into forests. This has been observed in the case of Yellowstone National Park, where wildfires burnt all the trees and 20 years later the park was once again filled with medium-height lodgepole pines. Table 8.1 shows more facts about deforestation.

In the second scenario, people cause deforestation for a number of reasons. The action is not always permanent; some countries are better at replanting forests than others. Trees can be converted into paper or wood, two products that human civilization uses daily. In some parts of the world, wood is still a major source of fuel for cooking and heating. Chances are that wood is a major component of your home's structure, the main use for timber in all countries. Wood is also used to

TABLE 8.1
Facts about Deforestation

- Tropical forests are disappearing at a rate of 4.6 million hectares (17,760 square miles) a year (Asia, Latin America, the Caribbean, and Africa).
- Moist deciduous forests lose approximately 6.1 million hectares (23,552 square miles) every year (Latin America and the Caribbean).
- Dry deciduous forests lose more than 1.8 million hectares (6,949 square miles) every year (Sudan, Paraguay, Brazil, and India).
- Very dry forests lose approximately 341,000 hectares (1,316 square miles) annually (Sudan and Botswana).
- Desert forests lose an estimated 82,000 hectares (316 square miles) annually (Mexico and Pakistan).
- Hills and mountains lose about 2.5 million hectares (9,652 square miles) of forest annually (Brazil, Mexico, and Indonesia).

build furniture, cabinetry, and other products. Paper is used in many ways. Trees may also be cut down for forest management reasons. One reason is to limit a wildfire's ability to spread. This may be done in an emergency to combat an active fire, or in a methodically planned long-term harvest. Typically, the forest is allowed to recover and it does so in a period of decades. Certain types of wildlife benefit from recovering forests after a harvest or wildfire. Many animals benefit from an edge habitat created by responsible logging or smaller-scale fires. Plants and animals in the natural world typically benefit from one type of habitat or another, or else benefit from living along the boundaries between two habitat types (known as an "edge" habitat). Animals that are specifically adapted to living in the forest cannot usually survive if their habitat is taken away. However, deforestation may benefit certain other animals, particularly grazing animals. It is for this reason that humans clear forests such as the Amazon for cattle grazing. Many birds benefit from having two habitats next to each other; the forest provides security but little food, while the open field provides food but relatively little security. Living along the boundary allows these types of animals to benefit from the strengths of both habitats. Deer are another example of an animal that may benefit from an edge habitat.

8.2.1.3 Air Pollution

Air pollution is the introduction of chemicals, particulate matter, or biological materials that cause harm or discomfort to humans or other living organisms, or that damage the natural environment, into the atmosphere. The atmosphere is a complex, dynamic natural gaseous system that is essential to support life on earth. Stratospheric ozone depletion due to air pollution has long been recognized as a threat to human health as well as to the earth's ecosystems. According to the Environmental Science Engineering Program at the Harvard School of Public Health, approximately 4% of deaths in the United States can be attributed to air pollution. The pollutants can be classified into two major categories: primary and secondary.

Primary pollutants produced by human activity include toxic metals (such as lead, cadmium, and copper), odors (from garbage, sewage, and industrial processes), and radioactive pollutants produced by nuclear explosions, war explosives, and natural processes such as the radioactive decay of radon. Secondary pollutants include particulate matter formed from gaseous primary pollutants and compounds in photochemical smog (Table 8.2). Smog is a kind of air pollution; the word "smog" is a portmanteau of smoke and fog. Classic smog results from large amounts of coal burning in an area and is caused by a mixture of smoke and sulfur dioxide. Modern smog does not usually come from coal, but from vehicular and industrial emissions that are acted on in the atmosphere by sunlight to form secondary pollutants that also combine with the primary emissions to form photochemical smog (Figure 8.3).

8.2.1.4 Major Corbon Dioxide Emission Countries

The following is a list of the world's top CO_2 emitting countries. The data itself were collected in 2007 by the Carbon Dioxide Information Analysis Center (CDIAC) for the United Nations (UN). The data considers only CO_2 emissions from the burning of fossil fuels, but not emissions from deforestation, fossil fuel exporters, etc. These statistics are rapidly dated due to the huge recent growth of emissions in Asia. The United States is the tenth largest emitter of CO_2 emissions per capita as of 2004. According to preliminary estimates, China has had a higher total emission since 2006 due to its much larger population and an increase of emissions from power generation (Table 8.3).

8.2.1.5 Greenhouse Effect

GHGs are gases in an atmosphere that absorb and emit radiation within the thermal infrared range. This process is the fundamental cause of the greenhouse effect. Common GHGs in the earth's atmosphere include water vapor, CO_2, methane, nitrous oxide, ozone (O_3), and chlorofluorocarbons (CFCs). In our solar system, the atmospheres of Venus, Mars, and Titan also contain gases that cause greenhouse effects. GHGs, mainly water vapor, are essential to help determine the temperature of the earth. Without them, this planet would likely be so cold as to be uninhabitable. Although many

TABLE 8.2
Impact of Air Pollutants on the Environment

Pollutants Released	Environmental Implications
Sulfur oxides (SO_x)	Sulfur dioxide (SO_2) is produced by volcanoes and various industrial processes. Since coal and petroleum often contain sulfur compounds, their combustion generates sulfur dioxide. Further oxidation of SO_2, usually in the presence of a catalyst such as NO_2, forms H_2SO_4 and thus acid rain.
Nitrogen oxides (NO_x)	Nitrogen dioxide (NO_2) is emitted from high-temperature combustion. It can be seen as the brown haze dome above, or plume downwind of, cities. This reddish-brown toxic gas has a characteristic sharp, biting odor. NO_2 is one of the most prominent air pollutants.
Carbon monoxide (CO)	Carbon monoxide is a colorless, odorless, nonirritating, but very poisonous gas. It is produced by incomplete combustion of fuel such as natural gas, coal, or wood. Vehicular exhaust is a major source of carbon monoxide.
Carbon dioxide (CO_2)	Carbon dioxide is GHG emitted from combustion, but is also a gas vital to living organisms. It is a natural gas in the atmosphere.
Volatile organic compounds (VOCs)	Volatile organic compounds are an important outdoor air pollutant. In this field, they are often divided into separate categories: methane (CH_4) and non-methane (NMVOCs). Methane is an extremely efficient GHG which contributes to enhance global warming. Other hydrocarbon VOCs are also significant GHGs via their role in creating ozone and in prolonging the life of methane in the atmosphere, although the effect varies depending on local air quality. Within the NMVOCs, the aromatic compounds benzene, toluene, and xylene are suspected carcinogens and may lead to leukemia through prolonged exposure. 1,3-butadiene is another dangerous compound that is often associated with industrial uses.
Particulate matter (PM)	Particulate matter, alternatively referred to as fine particles, consists of tiny particles of solid or liquid suspended in a gas. In contrast, aerosol refers to particles and the gas together. Sources of particulate matter can be man-made or natural. Some particulates occur naturally, originating from volcanoes, dust storms, forest and grassland fires, living vegetation, and sea spray. Human activities, such as the burning of fossil fuels in vehicles, power plants, and various industrial processes also generate significant amounts of aerosols. Increased levels of fine particles in the air are linked to health hazards such as heart disease, altered lung function, and lung cancer.
Chlorofluorocarbons (CFCs)	Chlorofluorocarbons are harmful to the ozone layer. They are emitted from products currently banned from use.
Ammonia (NH_3)	Ammonia is a compound emitted from agricultural processes. It is normally encountered as a gas with a characteristic pungent odor. Ammonia contributes significantly to the nutritional needs of terrestrial organisms by serving as a precursor to foodstuffs and fertilizers. Ammonia, either directly or indirectly, is also a building block for the synthesis of many pharmaceuticals. Although in wide use, ammonia is both caustic and hazardous.

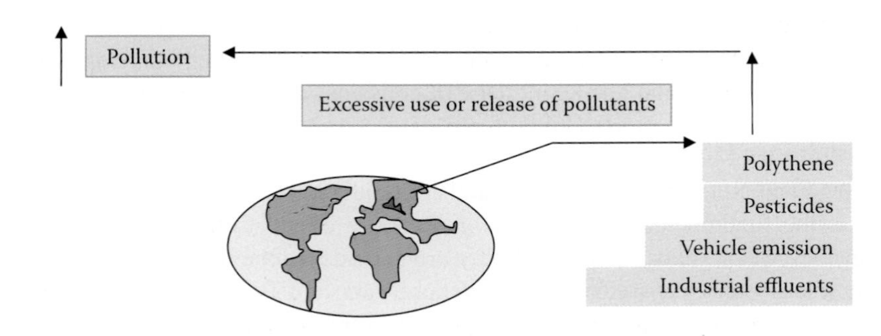

FIGURE 8.3 Global pollution issue.

TABLE 8.3
World Top 10 CO_2 Emission Countries

Rank	Country	Annual CO_2 Emissions (in Thousands of Metric Tons)	Percentage of Global Total (%)
1	China	6,538,367.00	22.30
2	United States	5,838,381.00	19.91
3	India	1,612,362.00	5.50
4	Russia	1,537,357.00	5.24
5	Japan	1,254,543.00	4.28
6	Germany	787,936.00	2.69
7	Canada	557,340.00	1.90
8	United Kingdom	539,617.00	1.84
9	South Korea	503,321.00	1.72
10	Iran	495,987.00	1.69

factors such as the sun and the water cycle are responsible for the earth's weather and energy balance, if all else were held equal and stable, the planet's average temperature should be considerably lower without GHGs. Human activities have an impact upon the levels of GHGs in the atmosphere, which has other effects upon the system, with their own possible repercussions. The most recent assessment report compiled by the Intergovernmental Panel on Climate Change (IPCC) observed that "changes in atmospheric concentrations of GHGs and aerosols, land cover, and solar radiation alter the energy balance of the climate system," and concluded that "increases in anthropogenic GHG concentrations is very likely to have caused most of the increases in global average temperatures since the mid-twentieth century."

8.2.1.6 Acid Rain

The term *acid rain* refers to what scientists call acid deposition. It is caused by airborne acidic pollutants and has highly destructive results. Scientists first discovered acid rain in 1852, when the English chemist Robert Agnus invented the term. From then until now, acid rain has been an issue of intense debate among scientists and policy makers. Acid rain, one of the most important environmental problems of all, cannot be seen. The invisible gases that cause acid rain usually come from automobiles or coal-burning power plants. Acid rain moves easily, affecting locations far beyond those that emit pollution. As a result, this global pollution issue causes great debate between countries that fight over polluting each other's environments.

For years, science studied the true causes of acid rain. Some scientists concluded that human production was primarily responsible, while others cited natural causes as well. Recently, more intensive research has been conducted so that countries have the information they need to prevent acid rain and its dangerous effects. The levels of acid rain vary from region to region. In Third World nations without pollution restrictions, acid rain tends to be very high. In Eastern Europe, China, and the Soviet Union, acid rain levels have also risen greatly. However, because acid rain can move about so easily, the problem is definitely a global one.

8.2.1.7 Ocean Acidification

Ocean acidification is the name given to the ongoing decrease in the pH of the earth's oceans, caused by their uptake of anthropogenic CO_2 from the atmosphere. One of the first detailed data sets examining temporal variations in pH at a temperate coastal location found that acidification was occurring at a rate much higher than that previously predicted, with consequences for near-shore benthic ecosystems. A December 2009 National Geographic report quoted Thomas Lovejoy, former chief biodiversity advisor to the World Bank, on recent research suggesting "the acidity of the oceans will more than double in the next 40 years. This rate is 100 times faster than any changes in ocean acidity in the last 20 million years, making it unlikely that marine life can somehow adapt to the changes." According to research from the University of Bristol, published in the journal *Nature Geoscience* in February 2010 and comparing current rates of ocean acidification with the greenhouse event at the Paleocene-Eocene boundary, surface ocean temperatures rose by 5°C–6°C approximately 55 million years ago. During this time, no catastrophe is seen in surface ecosystems, yet bottom-dwelling organisms in the deep ocean experienced a major extinction. The study concluded that the current acidification is headed to reach levels higher than any seen in the last 65 million years. The study also found that the current rate of acidification is "10 times the rate that preceded the mass extinction 55 million years ago."

The main cause of acidification is excess CO_2 levels in the atmosphere. The carbon cycle describes the fluxes of CO_2 between the oceans, terrestrial biosphere, lithosphere, and the atmosphere. Human activities such as land-use changes, the combustion of fossil fuels, and the production of cement have led to a new flux of CO_2 into the atmosphere. Some of this has remained there, some has been taken up by terrestrial plants, and some has been absorbed by the oceans. The carbon cycle comes in two forms: the organic carbon cycle and the inorganic carbon cycle. The inorganic carbon cycle is particularly relevant when discussing ocean acidification, for it includes the many forms of dissolved CO_2 present in the earth's oceans. When CO_2 dissolves, it reacts with water to form a balance of ionic and nonionic chemical species: dissolved free CO_2, carbonic acid, bicarbonate, and carbonate. The ratio of these species depends on factors such as seawater temperature and alkalinity.

8.2.1.8 Health Hazards due to Pollution

The World Health Organization states that 2.4 million people die each year from causes directly attributable to air pollution, with 1.5 million of these deaths attributable to indoor air pollution. "Epidemiological studies suggest that more than 500,000 Americans die each year from cardiopulmonary disease linked to breathing fine particle air pollution." A study by the University of Birmingham has shown a strong correlation between pneumonia-related deaths and air pollution from motor vehicles. The worst short-term civilian pollution crisis in India was the 1984 Bhopal disaster. Leaked industrial vapors from the Union Carbide Factory, belonging to Union Carbide, Inc., United States, killed more than 2000 people outright and injured anywhere from 150,000–600,000 others, some 6000 of whom would later die from their injuries. The United Kingdom suffered its worst air pollution event when the Great Smog of 1952 formed over London. In 6 days, more than 4000 died, and 8000 more died within the following months. The worst single incident of air pollution to occur in the United States occurred in Donora, Pennsylvania, in late 1948, when 20 people died and over 7000 were injured. Diesel exhaust (DE) is a major contributor to combustion-derived particulate matter air pollution. In several human experimental studies using a well-validated exposure chamber setup, DE has been linked to acute vascular dysfunction and increased thrombus formation. This serves as a plausible mechanistic link between the previously described association between particulate matter air pollution and increased cardiovascular morbidity and mortality. Table 8.2 shows the impact of pollutants on our environment.

8.3 ENVIRONMENT PROTECTION BY BIOTECHNOLOGY

In the previous section, we learned about the environment and its various attributes. We also learned how different kinds of pollution are responsible for deteriorating the quality of water, air, and land, thus causing health hazards. Air and water pollution released from industrial, agricultural, and household wastes has contaminated our environment to such an extent that it definitely requires a rescue plan to save our planet. Toxic affluents have not only affected human life, but also affected other living organisms, and quite a few species are now on the verge of extinction. It is true that

TABLE 8.4
Ways to Protect Our Environment

- Reduction and cleanup of pollution, with future goals of zero pollution
- Cleanly converting nonrecyclable materials into energy through direct combustion or after conversion into secondary fuels
- Reducing societal consumption of nonrenewable fuels
- Development of alternative, green, low-carbon, or renewable energy sources
- Conservation and sustainable use of scarce resources such as water, land, and air
- Protection of representative or unique or pristine ecosystems
- Preservation of threatened and endangered species
- Establishment of nature and biosphere reserves under various types of protection
- Protection of biodiversity and ecosystems upon which all human and other life on earth depends

humans are responsible for deteriorating the quality of water, air, and land, but they themselves find ways to solve the problems they have created. In the process of solving these problems, scientists such as biotechnologists play a vital role in providing arrays of solutions to reduce pollution. In this section we will discuss some tools of biotechnology which are used to create environmentally friendly technology and products (Table 8.4).

8.3.1 Bioremediation

Bioremediation can be defined as any process that uses microorganisms such as fungi, green plants, or their enzymes to return the natural environment altered by contaminants to its original condition. Bioremediation may be employed to attack specific soil contaminants, such as degradation of chlorinated hydrocarbons by bacteria. An example of a more general approach is the cleanup of oil spills by the addition of nitrate and/or sulfate fertilizers to facilitate the decomposition of crude oil by indigenous or exogenous bacteria. Naturally occurring bioremediation and phytoremediation have been used for centuries. For example, desalination of agricultural land by phyto-extraction has a long tradition.

Bioremediation technology using microorganisms was reportedly invented by George M. Robinson. Bioremediation technologies can be generally classified as in situ or ex situ. In situ bioremediation involves treating the contaminated material at the site, while ex situ bioremediation involves the removal of the contaminated material to be treated elsewhere. Some examples of bioremediation technologies are bioventing, land farming, bioreactor, composting, bio-augmentation, rhizofiltration, and biostimulation. However, not all contaminants are easily treated by bioremediation using microorganisms. For example, heavy metals such as cadmium and lead are not readily absorbed or captured by organisms. The assimilation of metals such as mercury into the food chain may worsen matters. Phytoremediation is useful in these circumstances, because natural plants or transgenic plants are able to bioaccumulate these toxins in their above-ground parts, which are then harvested for removal. The heavy metals in the harvested biomass may be further concentrated by incineration or even recycled for industrial use. The elimination of a wide range of pollutants and wastes from the environment requires increasing our understanding of the relative importance of different pathways and regulatory networks to carbon flux in particular environments and for particular compounds, and they will certainly accelerate the development of bioremediation technologies and biotransformation processes (Figure 8.4).

The use of genetic engineering to create organisms specifically designed for bioremediation has great potential. The bacterium *Deinococcus radiodurans* has been modified to consume and digest toluene and ionic mercury from highly radioactive nuclear waste. There are a number of cost-efficiency advantages to bioremediation that can be employed in areas that are inaccessible without excavation. For example, hydrocarbon spills (specifically, petrol spills) or certain chlorinated

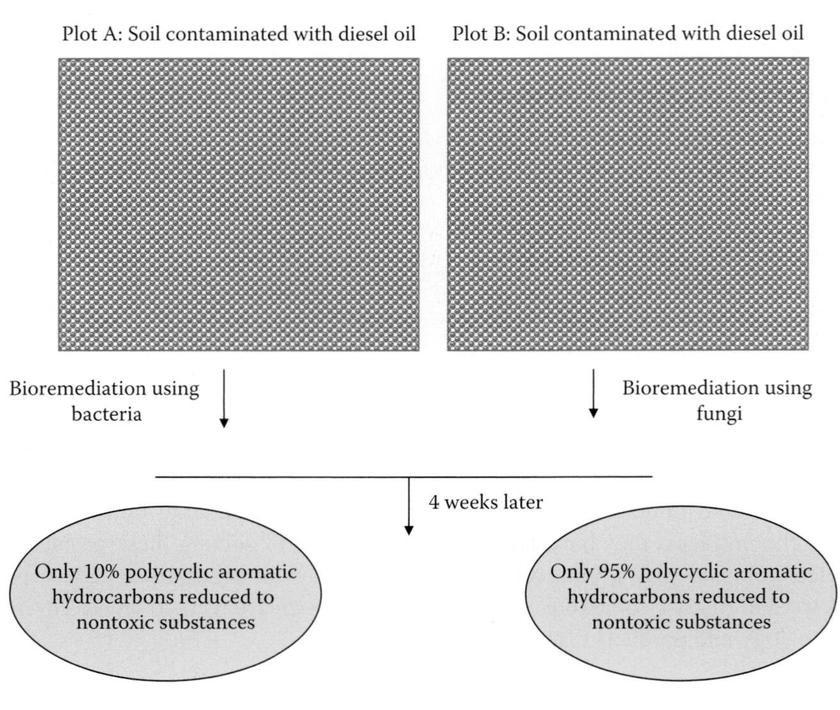

FIGURE 8.4 Bioremediation using fungi.

solvents may contaminate groundwater, and introducing the right electron acceptor or electron donor amendment, as appropriate, may significantly reduce contaminant concentrations after a lag time allowing for acclimation. This is typically much less expensive than excavation followed by disposal elsewhere, incineration, or other ex situ treatment strategies. It also reduces or eliminates the need for "pump and treat," a common practice at sites where hydrocarbons have contaminated clean groundwater.

8.3.1.1 Mycoremediation

Mycoremediation is a form of bioremediation, the process of using fungi to return an environment contaminated by pollutants to a less contaminated state. The term *mycoremediation* refers specifically to the use of fungal mycelia in bioremediation. One of the primary roles of fungi in the ecosystem is decomposition, which is performed by the mycelium. The mycelium secretes extracellular enzymes and acids that break down lignin and cellulose, the two main building blocks of plant fiber. These are organic compounds composed of long chains of carbon and hydrogen, structurally similar to many organic pollutants. The key to mycoremediation is determining the right fungal species to target a specific pollutant. Certain strains have been reported to successfully degrade certain nerve gases. With a view to establishing the bioremediation capability of fungi, scientists conducted an experiment where a plot of soil contaminated with diesel oil was inoculated with mycelia of oyster mushrooms. Traditional bioremediation techniques (bacteria) were used on control plots. After 4 weeks, more than 95% of many of the polycyclic aromatic hydrocarbons (PAHs) had been reduced to nontoxic components in the mycelial-inoculated plots. It appears that the natural microbial community participates with the fungi to break down contaminants, eventually into CO_2 and water. Moreover, wood-degrading fungi are particularly effective in breaking down aromatic pollutants such as toxic components of petroleum, as well as chlorinated compounds and certain persistent pesticides. The process of bioremediation can be monitored indirectly by measuring the oxidation reduction potential, or redox, in soil and groundwater, together with pH, temperature, oxygen content, electron acceptor/donor concentrations, and concentration of breakdown products such as CO_2.

8.3.2 Waste Water Treatment

Sewage is created by residences, institutions, hospitals, and commercial and industrial establishments. Raw influent includes household waste liquid from toilets, baths, showers, kitchens, sinks, and so forth, that is disposed of via sewers. In many areas, sewage also includes liquid waste from industry and commerce. The separation and draining of household waste into gray water and black water is becoming more common in the developed world, with gray water being permitted to be used for watering plants or recycled for flushing toilets (Figure 8.5). Sewage water can be treated close to where it is created, in septic tanks, biofilters, or aerobic treatment systems, or can be collected and transported via a network of pipes and pump stations to a municipal treatment plant. Sewage collection and treatment is typically subject to local, state, and federal regulations and standards. Industrial sources of wastewater often require specialized treatment processes. Conventional sewage treatment involves three stages: primary, secondary, and tertiary treatments. First, the solids are separated from the wastewater stream. Then, dissolved biological matter is progressively converted into a solid mass by using indigenous, waterborne microorganisms. Finally, the biological solids are neutralized, then disposed of or reused, and the treated water may be disinfected chemically or physically—for example, by lagoons and microfiltration. The final effluent can be discharged into a stream, river, bay, lagoon, or wetland, or it can be used for the irrigation of golf courses, greenways, or parks. If it is sufficiently clean, it can also be used for groundwater recharge or for agricultural purposes. Wastewater treatment is a critical component of the health and hygienic status of any city or town; therefore, the different phases of sewage treatment have been described in detail.

8.3.2.1 Pretreatment Phase

Pretreatment removes the materials that can be easily collected from the raw wastewater and disposed of. The typical materials that are removed during primary treatment include fats, oils, and greases (also referred to as FOG); sand, gravel, and rocks; larger settleable solids; and floating materials such as rags and flushed feminine hygiene products. In developed countries, sophisticated equipment with remote operation and control are employed, while developing countries still rely on low-cost equipment for cleaning.

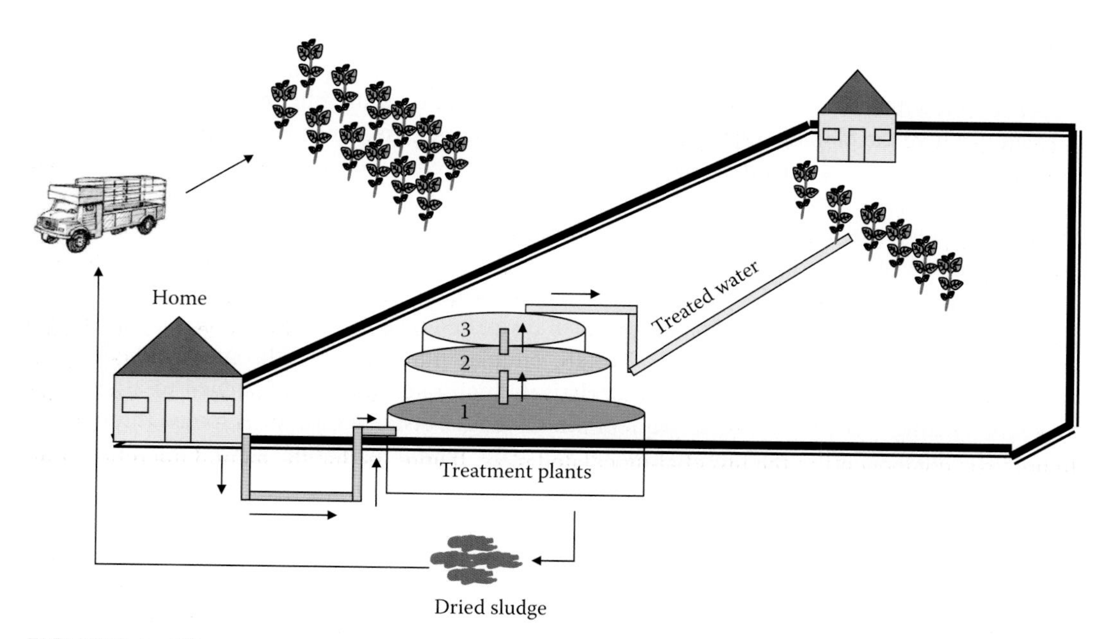

FIGURE 8.5 Wastewater treatment process.

8.3.2.2 Screening Phase

The influent sewage water is strained to remove all large objects carried in the sewage stream, such as rags, sticks, tampons, cans, fruit, etc. This is most commonly done with a manual or automated mechanically raked bar screen. The raking action of a mechanical bar screen is typically paced according to the accumulation on the bar screen. The bar screen is used because large solids can damage or clog the equipment used later in the sewage treatment plant. The large solids can also hinder the biological process. The solids are collected and later disposed of in a landfill or incinerator. During screening phase, incoming wastewater is carefully controlled to allow sand, grit, and stones to settle, while keeping the majority of the suspended organic material in the water column. This equipment is called a *degritter* or *sand catcher.* Sand, grit, and stones need to be removed early in the process to avoid damage to pumps and other equipment in the remaining treatment stages. Sometimes, there is a sand washer followed by a conveyor that transports the sand to a container for disposal. The contents from the sand catcher may be fed into the incinerator in a sludge processing plant, but in many cases, the sand and grit are sent to a landfill.

8.3.2.3 Sedimentation Phase

In this phase, sewage flows through large tanks, commonly called *primary clarifiers* or *primary sedimentation tanks.* The tanks are large enough for sludge to settle and for floating materials (grease and oils) to be skimmed off. The main purpose of the primary sedimentation stage is to produce both a generally homogeneous liquid capable of being treated biologically and a sludge that can be separately treated or processed. Primary settling tanks are usually equipped with mechanically driven scrapers that continually drive the collected sludge toward a hopper in the base of the tank, from where it can be pumped to where subsequent sludge treatment stages are carried out.

8.3.2.4 Secondary Treatment Phase

This phase of treatment is designed to substantially degrade the biological content of the sewage that is derived from human waste, food waste, soaps, and detergents. The majority of municipal plants treat the settled sewage liquor using aerobic biological processes. For this to be effective, the biota requires both oxygen and a substrate on which to live. There are a number of ways in which this is done. In all these methods, the bacteria and protozoa consume biodegradable soluble organic contaminants (such as sugars, fats, and organic short-chain carbon molecules) and bind much of the less soluble fractions into flock. Secondary treatment systems are classified as fixed film or suspended growth. The fixed-film treatment process includes trickling filter and rotating biological contactors where the biomass grows on media and the sewage passes over its surface. In suspended growth systems, such as activated sludge, the biomass is well mixed with the sewage and can be operated in a smaller space than fixed-film systems that treat the same amount of water. However, fixed-film systems are more able to cope with drastic changes in the amount of biological material, and can provide higher removal rates for organic material and suspended solids than suspended growth systems. Roughing filters are intended to treat particularly strong or variable organic loads (typically, industrial loads), to allow them to then be treated by conventional secondary treatment processes. Characteristics include typically tall, circular filters filled with open synthetic filter media to which wastewater is applied at a relatively high rate. They are designed to allow high hydraulic loading and a high flow-through of air. On larger installations, air is forced through the media using blowers. The resultant wastewater is usually within the normal range for conventional treatment processes (Figure 8.6).

8.3.2.5 Activated Sludge

During treatment, activated sludge plants involve processes that use dissolved oxygen to promote the growth of biological flock that substantially removes organic material. Most biological oxidation processes for treating industrial wastewaters have in common the use of oxygen (or air)

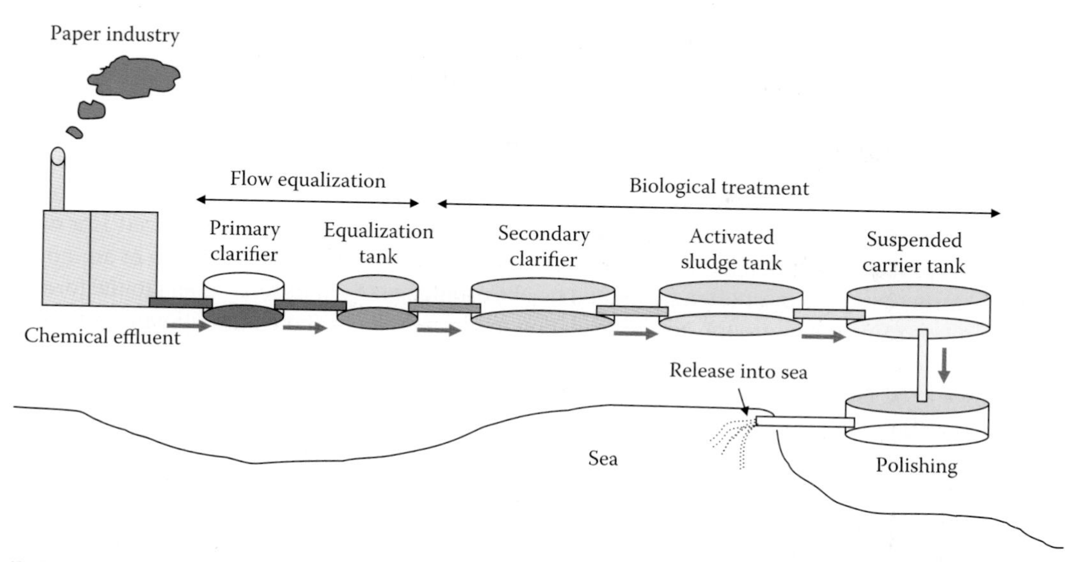

FIGURE 8.6 Industrial wastewater treatment plant.

and microbial action. Surface-aerated basins achieve 80%–90% removal of biochemical oxygen demand with retention times of 1–10 days.

8.3.2.6 Filter Beds

In older plants and in plants receiving more variable loads, trickling filter beds are used, where the settled sewage liquor is spread onto the surface of a deep bed made up of coke or carbonized coal, limestone chips, or specially fabricated plastic media. Such media must have high surface areas to support the biofilms that form. The liquor is distributed through perforated rotating arms radiating from a central pivot. The distributed liquor trickles through this bed and is collected in drains at the base. These drains also provide a source of air, which percolates up through the bed, keeping it aerobic. Biological films of bacteria, protozoa, and fungi form on the media's surfaces and eat or otherwise reduce the organic content. This biofilm is grazed by insect larvae and worms, which helps maintain an optimal thickness. Overloading of beds increases the thickness of the film, leading to clogging of the filter media.

8.3.2.7 Biological Aerated Filters

A biological aerated filter (BAF) is generally used to provide access to oxygen and induce carbon reduction and nitrification. A BAF usually includes a reactor filled with a filter media; the media is either in suspension or supported by a gravel layer at the foot of the filter. The dual purpose of this media is to support the highly active biomass that is attached to it and to filter suspended solids. Carbon reduction and ammonia conversion occurs in aerobic mode and is sometimes achieved in a single reactor, while nitrate conversion occurs in anoxic mode. A BAF is operated either in upflow or downflow configuration, depending on the design specified by the manufacturer.

8.3.2.8 Nutrient Removal

One of the major problems with wastewater treatment is handling high levels of the nutrients nitrogen and phosphorus. Their excessive release into the environment can lead to a buildup of nutrients, called *eutrophication*, which can in turn encourage the overgrowth of weeds, algae, and cyanobacteria. This may cause an algal bloom, a rapid growth in the population of algae. The algae numbers are unsustainable, and eventually most of them die. The decomposition of the algae by bacteria uses up so much of oxygen in the water that most or all of the animals die, which creates more organic matter for the bacteria to decompose. In addition to causing deoxygenating, some algal species produce toxins that contaminate drinking water supplies. Different treatment processes are required to remove nitrogen and phosphorus.

8.3.2.9 Nitrogen Removal

The removal of nitrogen is effected through the biological oxidation of nitrogen from ammonia to nitrate (nitrification), followed by *denitrification*, the reduction of nitrate to nitrogen gas. Nitrogen gas is released into the atmosphere and thus removed from the water. Nitrification itself is a two-step aerobic process, each step facilitated by a different type of bacteria. The oxidation of ammonia (NH_3) to nitrite (NO_2^-) is most often facilitated by Nitrosomonas spp. ("nitroso" refers to the formation of a nitroso functional group). Nitrite oxidation to nitrate (NO_3^-), though traditionally believed to be facilitated by Nitrobacter spp. ("nitro" refers to the formation of a nitro functional group), is now known to be facilitated in the environment almost exclusively by Nitrospira spp.

Denitrification requires anoxic conditions to encourage the appropriate biological communities to form. It is facilitated by a wide diversity of bacteria. Sand filters, lagooning, and reed beds can all be used to reduce nitrogen, but the activated sludge process can do the job most easily. Since denitrification is the reduction of nitrate to dinitrogen gas, an electron donor is needed. Depending on the wastewater, this can be organic matter from feces, sulfide, or an added donor-like methanol. Sometimes, the conversion of toxic ammonia to nitrate alone is referred to as tertiary treatment.

8.3.2.10 Phosphorus Removal

The removal of phosphorus is an important step, as it limits nutrients required for algae growth in many freshwater systems. It is also particularly important for water reuse systems, such as reverse osmosis, where high phosphorus concentrations may lead to fouling of downstream equipment. Phosphorus can be removed biologically in a process called *enhanced biological phosphorus removal*. In this process, specific bacteria called polyphosphate-accumulating organisms, are selectively enriched and accumulate large quantities of phosphorus within their cells. When the biomass enriched in these bacteria is separated from the treated water, these biosolids have a high fertilizer value. Phosphorus removal can also be achieved by chemical precipitation, usually with salts of iron, aluminum, or lime. This may lead to excessive sludge production, as hydroxide precipitates, and the added chemicals can be expensive. Despite this, chemical phosphorus removal requires significantly smaller equipment footprint than does biological removal. It is easier to operate and is often more reliable than biological phosphorus removal. Once removed, phosphorus, in the form of a phosphate-rich sludge, may be used in landfills or as fertilizer in agricultural fields.

8.3.2.11 Disinfection of Wastewater

Before releasing the treated wastewater back into the environment, the wastewater needs to be disinfected to substantially reduce the number of microorganisms in the water. Common methods of disinfection include Ozone (O_3), chlorine, or ultraviolet (UV) light. Chloramine, which is used for drinking water, is not used in wastewater treatment because of its persistence. Chlorination remains the most common form of wastewater disinfection in North America due to its low cost and long-term history of effectiveness. One disadvantage is that chlorination of residual organic material can generate chlorinated organic compounds that may be carcinogenic or may be harmful to the environment. Residual chlorine or chloramines may also be capable of chlorinating organic material in the natural aquatic environment. Further, because residual chlorine is toxic to aquatic species, the treated effluent must also be chemically dechlorinated, adding to the complexity and cost of treatment. UV light can be used instead of chlorine, iodine, or other chemicals. Because no chemicals are used, the treated water has no adverse effect on organisms that later consume it, as may be the case with other methods. The cities of Edmonton and Calgary in Alberta, Canada, also use UV light for their water treatment. O_3 is generated by passing oxygen through a high voltage potential, resulting in a third oxygen atom becoming attached and forming O_3. O_3 is very unstable and reactive and oxidizes most organic material it comes in contact with, thereby destroying many pathogenic microorganisms. O_3 is considered to be safer than chlorine because, unlike chlorine, which has to be stored on site, O_3 is generated onsite as needed. O_3 also produces fewer disinfection by-products than chlorination. A disadvantage of O_3 disinfection is the high cost of the O_3 generation equipment and the requirement for special operators.

8.3.2.12 Sludge Disposal

The liquid sludge is produced after various treatments are further processed to make it suitable for final disposal. Typically, sludge is thickened to reduce the volumes transported offsite for disposal. There is no process that completely eliminates the need to dispose of biosolids. There is, however, an additional step some cities are taking to superheat the wastewater sludge and convert it into small pellets that are high in nitrogen and other organic materials. In New York City, for example, several sewage treatment plants have dewatering facilities that use large centrifuges along with the addition of chemicals such as polymer to further remove liquid from the sludge. The removed fluid is typically reintroduced into the wastewater process. The product that is left is called "cake," which is picked up by companies that turn it into fertilizer pellets. This product is then sold to local farmers and turf farms as a soil fertilizer.

8.3.3 Biofuels

The world's fossil fuel reservoirs are fast diminishing and may be exhausted within the next 100 years. In recent years, it has become apparent that reliance on the limited sources of fossil fuel could easily be overcome by the use of biofuel. A biofuel is difficult to define precisely; generally, however, biofuel is liquid or gaseous fuel obtained from relatively recently lifeless biological material and is different from fossil fuels, which are derived from long-dead biological material. In addition, various plants and plant-derived materials are used for biofuel manufacturing. Globally, biofuels are most commonly used to power vehicles, to heat homes, and for cooking.

Most of the fossil fuels we use are biological in nature. Perhaps we have to say that a biofuel is one that does not add to the stock of total CO_2 in the atmosphere. These are plant forms that typically remove CO_2 from the atmosphere and give up the same amount when burnt. Naturally, all the other fossil fuels have done the same, but we are talking about a time scale of 1 or 2 years for the biofuels, whereas the fossil fuels can only be considered on a time scale measured in millions of years. Biofuel industries are expanding in Europe, Asia, and the United States. Recent technology developed at Los Alamos National Lab even allows for the conversion of pollution into renewable biofuel. *Agrofuels* are biofuels which are produced from specific crops, rather than from waste processes such as landfill off-gassing or recycled vegetable oil.

There are two common strategies for producing liquid and gaseous agrofuels. The first strategy is to grow crops high in sugar or starch and then use yeast fermentation to produce ethyl alcohol. The second strategy is to grow plants that contain high amounts of vegetable oil, such as oil palm, soybean, algae, jatropha, or pongamia pinnata. When these oils are heated, their viscosity is reduced, and they can either be burned directly in a diesel engine or they can be chemically processed to produce fuels such as biodiesel. Wood and its byproducts can also be converted into biofuels such as wood gas, methanol, or ethanol fuel. It is also possible to make cellulosic ethanol from nonedible plant parts, but this may not be economically viable.

Biomass or biofuel is material derived from recently living organisms. This includes plants, animals, and their by-products; for example, manure, garden waste, and crop residues are all sources of biomass. It is a renewable energy source based on the carbon cycle, unlike other natural resources such as petroleum, coal, and nuclear fuels. Biomass is made from many types of waste organic matter such as crop stalks, tree thinning, wooden pallets, construction waste, chicken and pig waste, and agricultural waste and lawn trimmings. It is used to produce power, heat and steam, and fuel, through a number of different processes. Although renewable, biomass often involves a burning process that produces emissions such as sulfur dioxide, nitrogen oxide, and CO_2, but fortunately in

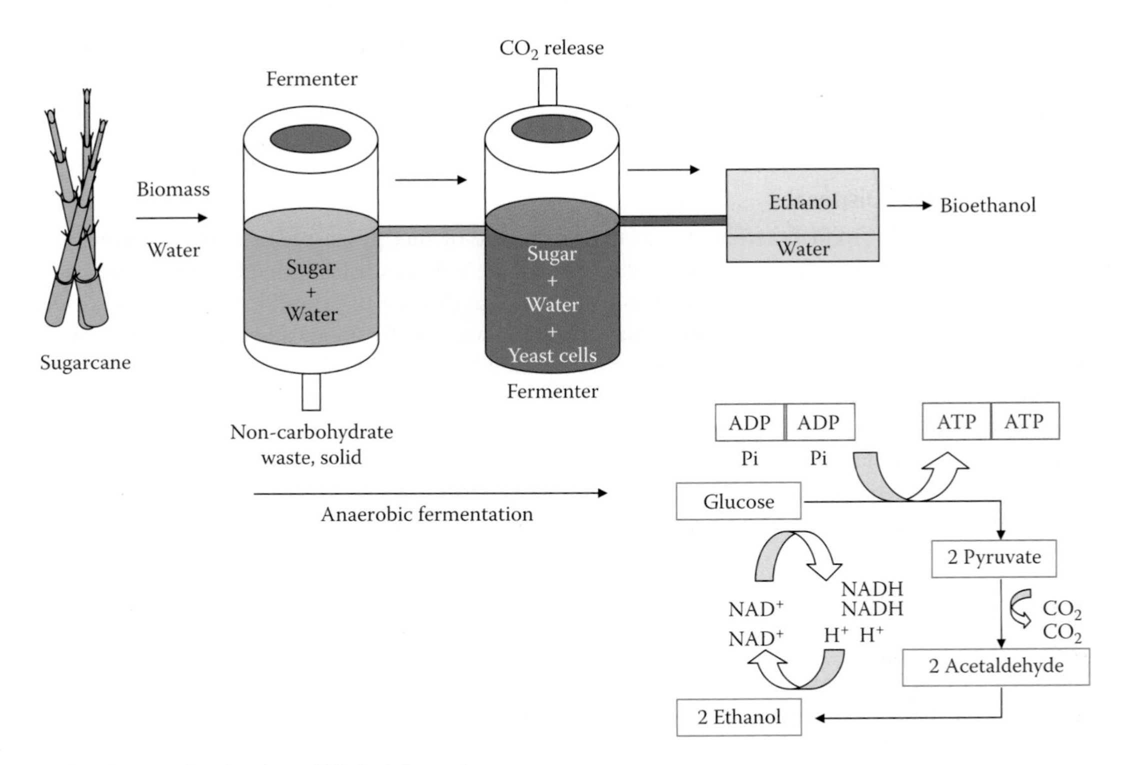

FIGURE 8.7 Production of biofuel from plants.

quantities far less than those emitted by coal plants. However, proponents of coal plants feel that their way of doing it is a lot cheaper and there is a lot of dispute over this.

Animal waste is a persistent and unavoidable pollutant produced primarily by the animals housed in industrial-sized farms. Researchers from Washington University have figured out a way to turn manure into biomass. In April 2008, with the help of imaging technology, they noticed that vigorous mixing helps microorganisms turn farm waste into alternative energy, providing farmers with a simple way to treat their waste and convert it into energy. There are also agricultural products specifically grown for biofuel production, including corn, switch grass, and soybeans, primarily in the United States; rapeseed, wheat, and sugar beet, primarily in Europe; sugar cane in Brazil; palm oil and miscanthus in South East Asia; sorghum and cassava in China; jatropha and pongamia pinnata in India; and pongamia pinnata in Australia and the tropics. Hemp has also been proven to work as a biofuel. Biodegradable outputs from industry, agriculture, forestry, and households can be used for biofuel production. Biomass can come from waste plant material. The use of biomass fuels can therefore contribute to waste management as well as fuel security and help to prevent global warming, though alone they are not a comprehensive solution to these problems (Figure 8.7).

8.3.3.1 Biodiesel

Biodiesel was probably the first of the alternative fuels to really become known to the public. The great advantage of biodiesel is that it can be used in existing vehicles with little or no adaptation necessary. Biodiesel is, naturally, a compromise for this reason, but still balances positively on the energy scales. There are energy plants available that will produce a higher yield in kW h per area, but the simplicity of having a fuel that is fully compatible with present fuel and engine technology makes it very attractive. Cars running on bioethanol, which is produced from agricultural crops, sugar cane, or biomass, are governed by the same law of physics as those using gasoline. This means that although both emit CO_2 as an inevitable consequence of the combustion process, there is a crucial difference: burning ethanol, in effect, recycles the CO_2 because it has already been removed from the atmosphere by photosynthesis during the natural growth process. In contrast, the use of gasoline or diesel injects into the atmosphere additional new quantities of CO_2 that have lain fixed underground in oil deposits for millions of years.

Biodiesel is the most common biofuel in Europe. It is produced from oils or fats using transesterification and is a liquid similar in composition to fossil/mineral diesel. Its chemical name is fatty acid methyl (or ethyl) ester (FAME). Oils are mixed with sodium hydroxide and methanol and the chemical reaction produces biodiesel (FAME) and glycerol. One part glycerol is produced for every 10 parts biodiesel. Feedstock for biodiesel includes animal fats, vegetable oils, soy, rapeseed, jatropha, mahua, mustard, flax, sunflower, palm oil, hemp, field pennycress, pongamia pinnata, and algae. Since biodiesel is an effective solvent and cleans residues deposited by mineral diesel, engine filters may need to be replaced more often, as the biofuel dissolves old deposits in the fuel tank and pipes. It also effectively cleans the engine combustion chamber of carbon deposits, helping to maintain efficiency. In many European countries, a 5% biodiesel blend is widely used and is available at thousands of gas stations. Biodiesel is also an *oxygenated fuel*, meaning that it contains a reduced amount of carbon and higher hydrogen and oxygen content than fossil diesel. This improves the combustion of fossil diesel and reduces the particulate emissions from unburnt carbon.

8.3.3.2 Biogas

Biogas is becoming increasingly interesting as an alternative to natural gas. It is especially useful that the composition of biogas and natural gas is practically identical, so the same burners can be used for both fuels. Biogas can be produced from plant or animal waste or a combination of both. A mixture of both has proven to be the best method. There are many different methods used, depending on the starting material and quantity involved. Animal waste produces the nitrogen needed for growth of the bacteria, and vegetable waste supplies most of the carbon and hydrogen necessary.

8.3.3.3 Bioalcohols

Bioalcohols are produced by the action of microorganisms and enzymes through the fermentation of sugars or starches (which is easiest), or cellulose (which is more difficult). Biobutanol (also called *biogasoline*) is often claimed to provide a direct replacement for gasoline, because it can be used directly in a gasoline engine (in a similar way to biodiesel in diesel engines). Ethanol fuel is the most common biofuel worldwide, particularly in Brazil. Alcohol fuels are produced by fermentation of sugars derived from wheat, corn, sugar beets, sugar cane, molasses, and any sugar or starch that alcoholic beverages can be made from, like potato and fruit waste. The ethanol production methods used are enzyme digestion (to release sugars from stored starches), fermentation of the sugars, distillation, and drying. The distillation process requires significant energy input for heat (often unsustainable natural gas fossil fuel, but cellulosic biomass such as bagasse, the waste left after sugar cane is pressed to extract its juice, can also be used and is more sustainable). Ethanol can be used in petrol engines as a replacement for gasoline. It can be mixed with gasoline to any percentage. Most existing automobile petrol engines can run on blends of up to 15% bioethanol with petroleum/gasoline. Gasoline with ethanol added has higher octane, which means that an engine can typically burn hotter and more efficiently. In high-altitude (thin air) locations, some states mandate a mix of gasoline and ethanol as a winter oxidizer to reduce atmospheric pollution emissions.

Many car manufacturers are now producing flexible-fuel vehicles (FFVs), which can safely run on any combination of bioethanol and petrol, up to 100% bioethanol. They dynamically sense exhaust oxygen content and adjust the engine's computer systems, spark, and fuel injection accordingly. Alcohol mixes with both petroleum and with water, so ethanol fuels are often diluted after the drying process by absorbing environmental moisture from the atmosphere. Water in alcohol-mix fuels reduces efficiency, makes engines harder to start, causes intermittent operation (sputtering), oxidizes aluminum (in carburetors), and rusts steel components. Methanol is currently produced from natural gas, a nonrenewable fossil fuel. It can also be produced from biomass, as biomethanol.

The methanol economy is an interesting alternative to the hydrogen economy compared to today's hydrogen produced from natural gas, but not hydrogen production directly from water and state-of-the-art clean solar thermal energy processes.

8.3.3.4 Bioethers

Bioethers, also referred to as fuel ethers, are cost-effective compounds that act as octane enhancers. They also enhance engine performance, whilst significantly reducing engine wear and toxic exhaust emissions. Greatly reducing the amount of ground-level O_3, they contribute to the quality of the air we breathe.

8.3.3.5 Syngas

Syngas, a mixture of carbon monoxide and hydrogen, is produced by partial combustion of biomass—that is, combustion with an amount of oxygen that is not sufficient to convert the biomass completely to CO_2 and water. Before partial combustion, the biomass is dried, and sometimes pyrolyzed. The resulting gas mixture, syngas, is itself a fuel. Using syngas is more efficient than direct combustion of the original biofuel; more of the energy contained in the fuel is extracted. Syngas may be burned directly in internal combustion engines or turbines. The wood gas generator is a wood-fueled gasification reactor mounted on an internal combustion engine. Syngas can be used to produce methanol and hydrogen, or converted via the Fischer–Tropsch process to produce a synthetic diesel substitute or a mixture of alcohols that can be blended into gasoline. Gasification normally relies on temperatures >700°C. Lower temperature gasification is desirable when co-producing biochar, but results in a syngas polluted with tar.

8.3.3.6 Solid Biofuels

Some examples of solid biofuels are wood, sawdust, grass cuttings, domestic refuse, charcoal, agricultural waste, nonfood energy crops, and dried manure. When raw biomass is already in a suitable form (such as firewood), it can burn directly in a stove or furnace to provide heat or raise steam. When raw biomass is in an inconvenient form (such as sawdust, wood chips, grass, agricultural wastes), another option is to pelletize the biomass with a pellet mill. The resulting fuel pellets are easier to burn in a pellet stove. A problem with the combustion of raw biomass is that it emits considerable amounts of pollutants such as particulates and PAHs. Even modern pellet boilers generate far more pollutants than oil or natural gas boilers. Pellets made from agricultural residues are usually worse than wood pellets, producing much larger emissions of dioxins and chlorophenols. Another solid biofuel is biochar, which is produced by biomass pyrolysis. Biochar pellets made from agricultural waste can substitute for wood charcoal. In countries where charcoal stoves are popular, this can reduce deforestation.

8.3.3.7 Second-Generation Biofuels

Supporters of biofuels claim that a more viable solution is to increase political and industrial support for (and rapidity of) second-generation biofuel implementation from nonfood crops, including cellulosic biofuels. Second-generation biofuel production processes can use a variety of nonfood crops. These include waste biomass, the stalks of wheat, corn, wood, and special energy or biomass crops (e.g., Miscanthus). Second-generation biofuels use biomass-to-liquid technology, including cellulosic biofuels from nonfood crops. Many second-generation biofuels are under development, such as biohydrogen, biomethanol, 2,5-dimethylfuran (DMF), dimethylether (Bio-DME), Fischer–Tropsch diesel, biohydrogen diesel, mixed alcohols, and wood diesel.

8.3.3.8 Third-Generation Biofuels

Algae fuel, also called *oilgae* or *third-generation biofuel*, is a biofuel from algae. Algae are low-input, high-yield feedstock that produce biofuels. They produce 30 times more energy per acre than land crops such as soybeans. With the higher prices of fossil fuels (petroleum), there is much interest in algaculture (farming algae). One advantage of many biofuels over most other fuel types is that they are biodegradable, and so relatively harmless to the environment if spilled. The U.S. Department of Energy estimated that if algae fuel replaced all the petroleum fuel in the United States, it would require 15,000 square miles, which is roughly the size of Maryland. Algae, such as *Botryococcus braunii* and *Chlorella vulgaris*, are relatively easy to grow, but the algal oil is hard to extract. There are several approaches, some of which work better than others.

8.3.3.9 International Biofuel Efforts

Recognizing the importance of implementing bioenergy, there are international organizations such as International Energy Agency (IEA) Bioenergy, established in 1978 by the Organization for OECD, with the aim of improving cooperation and information exchange between countries that have national programs in bioenergy research, development, and deployment. The UN International Biofuels Forum, has been formed by Brazil, China, India, South Africa, the United States, and the European Commission. The world leaders in biofuel development and use are Brazil, the United States, France, Sweden, and Germany. There are various current issues with biofuel production and use, which are presently being discussed in the popular media and scientific journals. These include the effect of moderating oil prices, the "food versus fuel" debate, carbon emission levels, sustainable biofuel production, deforestation and soil erosion, impact on water resources, human rights issues, poverty reduction potential, biofuel prices, energy balance and efficiency, and centralized versus decentralized production models.

8.3.3.10 Future of Biofuels

A number of features of biofuels are distinctly disadvantageous. Some of these are that very large volumes of the product are required, so that the production has to be on a very large scale, and usually near to the site of use. As a consequence, the substrate requirement is equally large, for which large areas of land would be needed, and often transport of this material will also be necessary. The product is generally of low value and has rather a low profit margin, the cost of production being more than 75% of the sale price. This has prevented a more rapid industrialization of biofuel production. More efficient processes, especially continuous processes, must be developed to reduce the cost of production. To achieve this, organisms with higher product yields and greater product tolerance must be developed. Efficient processes to utilize low-cost substrates for biofuel production are urgently needed. The sources of biomass to be used as substrates should be identified and their cheap and abundant supply should be ensured.

8.3.4 Biodegradable Plastic

Plastics are durable and degrade very slowly, and the molecular bonds that make plastic so durable make it equally resistant to natural processes of degradation. Since the 1950s, one billion tons of plastic have been discarded and may persist for hundreds or even thousands of years. In some cases, burning plastic can release toxic fumes. In addition, manufacture of plastics often creates large quantities of chemical pollutants as well.

Prior to the ban on the use of CFCs in extrusion of polystyrene (and general use, except in life-critical fire suppression systems), the production of polystyrene contributed to the depletion of the O_3 layer. However, non-CFCs are currently used in the extrusion process. By 1995, plastic recycling programs were common in the United States and elsewhere (Figure 8.8).

On one hand, researchers have been working to find a method to degrade the existing plastic; on the other hand, they are also working on the development of biodegradable plastic. Recently, there have been reports of making biodegradable plastics using polyhydroxy alkanoates (PHAs), such as polyhydroxybutyrate (PHB), which are synthesized from acetyl-CoA used as a precursor, and are used for the synthesis of biodegradable plastics with thermoplastic properties. At present, PHAs are produced by bacterial fermentation, and the cost of biodegradable plastic is substantially higher

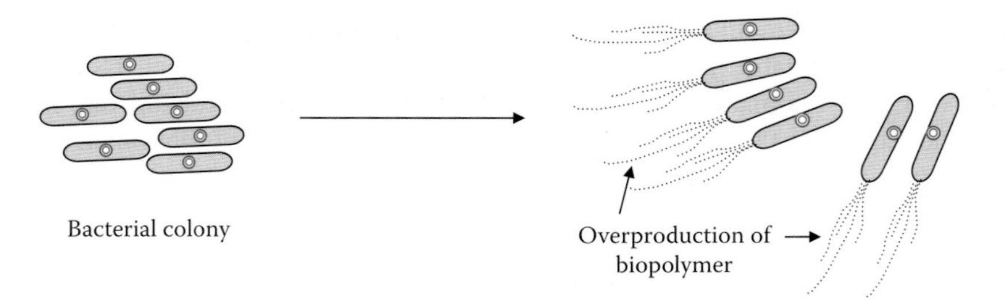

FIGURE 8.8 Bioengineered bacteria overproducing biopolymer.

than that of synthetic plastics. To reduce the cost, attempts are being made to produce PHAs in transgenic plants. Genes encoding the two enzymes—aceto-acetyl-CoA reductase (PhbB) and PHB synthase (phbC)—which are involved in the PHB synthesis from the precursor acetyl-CoA have been transferred from the bacterium *Alcaligenes eutrophus* and expressed in *Arabidopsis thaliana*. When the two enzymes were targeted into the plastids, PHB accumulated in leaves. PHB production by transgenic plants provides an example of a novel compound synthesized in plants. Transgenic trees, such as poplars, expressing phbB and phbC accumulate PHB in their leaves. The leaves are then collected and used for PHB extraction.

8.3.5 Biodegradation by Bacteria

Pseudomonas putida F1 is a Gram-negative rod-shaped saprotrophic soil bacterium. This bacterium was isolated from a polluted creek in Urbana, Illinois, by enrichment culture, with ethylbenzene as the sole source of carbon and energy. *Pseudomonas putida* F1 is one of the most well-studied aromatic hydrocarbon degrading bacterial strains. Over 200 articles have been written about various aspects of Pseudomonas F1 physiology, enzymology, and genetics by microbiologists and biochemists, in addition to more applied studies by chemists and environmental engineers utilizing Pseudomonas F1 and its enzymes for green chemistry application and bioremediation. Strain F1 grows well with benzene, toluene, ethylbenzene, and *p*-cymene. Mutants of strain F1 that are capable of growth with *n*-propylbenzene, *n*-butylbenzene, isopropylbenzene, and biphenyl are easily obtained. In addition to aromatic hydrocarbons, the broad substrate toluene dioxygenase in strain F1 can oxidize trichloroethylene (TCE), indole, nitrotoluenes, chlorobenzenes, chlorophenols, and many other aromatic substrates. Although Pseudomonas F1 cannot use TCE as a source of carbon and energy, it is capable of degrading and detoxifying TCE in the presence of an additional carbon source. The ability of Pseudomonas F1 to degrade benzene, toluene, and ethylbenzene has a direct bearing on the development of strategies for dealing with environmental pollution. For example, many underground gasoline storage tanks leak and contaminate groundwater supplies with benzene, toluene, ethylbenzene, and xylenes (BTEX). Pseudomonas F1 can oxidize all of these compounds, and thus a detailed knowledge of the physiology and biodegradation capabilities of this organism will be essential in providing the scientific foundations for the emerging bioremediation industry (Figure 8.9).

The sequence of F1 will provide an opportunity for comparative genomics of Pseudomonas KT2440 and F1. As far as we know, *P. putida* KT2440 does not have the ability to grow on any aromatic hydrocarbons, although it has pathways for the degradation of numerous aromatic acids. Based on the significantly wider range of growth substrates known for Pseudomonas F1, we expect to identify islands of catabolic diversity interspersed among standard Pseudomonas genes. The KT2440 genome should provide a useful scaffold for the Pseudomonas F1 genome sequence, and this should speed the process of genome analysis and annotation. Regions of similarity should be immediately obvious, and annotation and functional genomic analysis can focus on unique regions of the sequence.

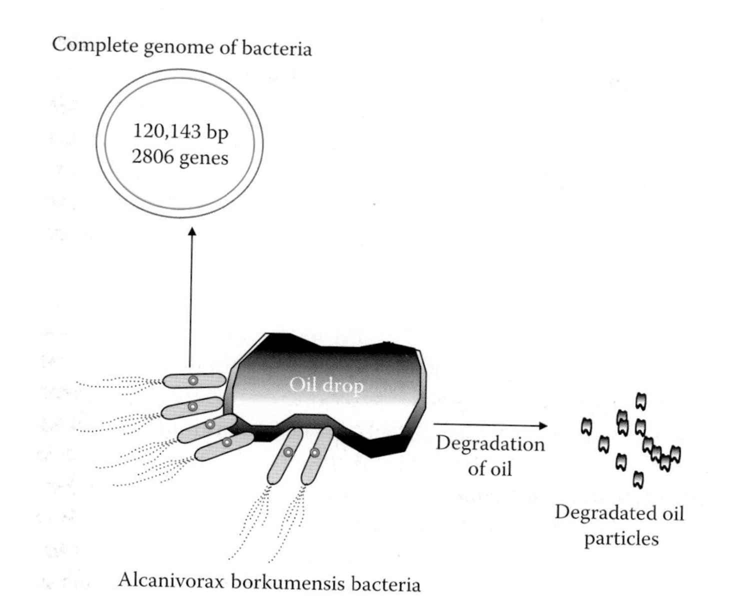

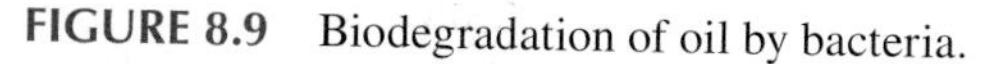
FIGURE 8.9 Biodegradation of oil by bacteria.

8.3.6 Oil-Eating Bacteria

The extensive use and transportation of oil also caused the incidence of oil spillage around the world. An *oil spill* is a release of a liquid petroleum hydrocarbon into the environment due to human activity and is a form of pollution. The term often refers to marine oil spills, where oil is released into the ocean or coastal waters. Oil spills include releases of crude oil from tankers, offshore platforms, drilling rigs and wells, as well as spills of refined petroleum products (such as gasoline, diesel) and their by-products, and heavier fuels used by large ships such as bunker fuel, or the spillage of any oily refuse or waste oil. The oil penetrates into the structure of the plumage of birds, reducing its insulating ability, thus making the birds more vulnerable to temperature fluctuations and much less buoyant in the water. It also impairs birds' flight abilities to forage and escape from predators. As they attempt to preen, birds typically ingest the oil that covers their feathers, causing kidney damage, altered liver function, and digestive tract irritation. This and the limited foraging ability quickly cause dehydration and metabolic imbalances. Hormonal balance alteration, including changes in luteinizing protein, can also result in some birds that have been exposed to petroleum. Most birds affected by an oil spill die unless there is human intervention. Marine mammals exposed to oil spills are affected in similar ways as seabirds. Oil coats the fur of sea otters and seals, reducing their insulation abilities and leading to body temperature fluctuations and hypothermia. Ingestion of the oil causes dehydration and impaired digestion. Because oil floats on top of water, less sunlight penetrates into the water, limiting the photosynthesis of marine plants and phytoplankton. This, as well as decreasing fauna populations, affects the food chain in the ecosystem.

Besides natural calamities, companies that drill oil from the seas also get into trouble because of oil spillage, and one such recent example was the oil spillage that killed 11 people and that contaminated a large area of the U.S. marine environment along the Gulf of Mexico. BP's bill for containing and cleaning up the oil spill has reached nearly $10 bn (£6.4 bn), as the U.S. government declared that the blown-out well has finally been plugged, 5 months after the explosion on the Deepwater Horizon rig. The beleaguered oil company revealed that its total cost of the spill had climbed to $9.5 bn. BP also said payouts to people affected by the spill, such as fishermen, hoteliers, and retailers, had dramatically increased since it handed over authority for dispensing funds to a White House appointee. BP has set up a $20 bn compensation fund, which has so far paid out

19,000 claims totaling more than $240 m. The fund is run by lawyer Kenneth Feinberg, the Obama administration's former executive pay tsar.

Cleanup and recovery from an oil spill is difficult and depends upon many factors, including the type of oil spilled, the temperature of the water (affecting evaporation and biodegradation), and the types of shorelines and beaches involved. Among various cleanup methods, bioremediation has attracted lot of attention and has been found to be the best method to clean up oil spills. Bioremediation can be defined as any process that uses microorganisms, fungi, green plants, or their enzymes, to return the natural environment altered by contaminants to its original condition. Bioremediation may be employed to attack specific soil contaminants, such as degradation of chlorinated hydrocarbons by bacteria. An example of a more general approach is the cleanup of oil spills by the addition of nitrate and/or sulfate fertilizers to facilitate the decomposition of crude oil by indigenous or exogenous bacteria. The use of genetic engineering to create organisms specifically designed for bioremediation has great potential. The bacterium *Deinococcus radiodurans* (the most radio-resistant organism known) has been modified to consume and digest toluene and ionic mercury from highly radioactive nuclear waste.

Dr. Anand Mohan Chakraborty, an Indian-born American scientist, and his coworkers produced an oil-eating bug or super bug by introducing plasmids from different strains into a single cell of *Pseudomonas putida*. This super bug can degrade all the four types of substrates for which four separate plasmids were required. The U.S. government allowed Dr. Chakraborthy, then with General Electric Company, to use this super bug for cleaning up an oil spill in Texas. Large-scale production of the super bugs was carried out in the laboratory; they were mixed with straw and dried. The bacteria-laden straw was stored for future use. In the event of an oil slick occurring, the bacteria-laden straw is spread over it. The straw soaks up the oil, and the bacteria present on it break the oil into nonpolluting, harmless products. Several other microbes for the treatment of oil spills are also being developed through genetic engineering (Table 8.5).

8.3.7 Biostimulation and Bioaugmentation

A xenobiotic compound, if it is to be degraded by a microorganism, should be available in an acceptable concentration below toxic levels. A constant supply of nutrients and co-metabolites should be available for selective maintenance of microbes capable of its degradation. Most important is the provision of a microbial population or inoculum. Natural evolution of the microbial population is a time-consuming process. The main way of initiating a biological treatment system is to inoculate the system with a suitable population of microbes having the specific degrading capability. Biostimulation makes use of existing microorganisms by providing favorable conditions for their action. Nutrients, pH, temperature, growth factors, etc., are adjusted so as to stimulate the growth of microbes.

Bioaugmentation is the process of introducing, from outside, cultures which have the specific degradation capacities. The introduced microorganisms must remain viable and should compete with the microorganisms already existing in the system. A number of inoculants which specifically degrade various xenobiotic compounds are commercially available. These are naturally occurring species. They are isolated, purified, and characterized for features, including capability of xenobiotic degradation. These isolated stocks are air-dried or freeze-dried and are preserved. They are usually available as dry powders and are easy to store, transport, and reconstitute on the spot by adding water.

8.3.8 Bioleaching

Microbial leaching involves the process of dissolution of metals from ore-bearing rocks using microorganisms. Recently, bacterial activity has been implicated in the weathering, leaching, and deposition of mineral ores. The new discipline created by the marriage between biotechnology and metallurgy is known as *biohydrometallurgy*, *bioleaching*, or *biomining*. Conventional

TABLE 8.5
Oil Spills of Over 30 Million U.S. Gallons

Spill/Tanker	Location	Date	Volume of Crude Oil
Kuwaiti oil fires	Kuwait	January 1991–November 1991	136,000,000–205,000,000
Kuwaiti oil lakes	Kuwait	January 1991–November 1991	3,409,000–6,818,000
Lakeview gusher	United States, Kern County, California	May 14, 1910–September 1911	1,200,000
Gulf War oil spill	Iraq, Persian Gulf, and Kuwait	January 19, 1991–January 28, 1991	818,000–1,091,000
Deepwater Horizon	United States, Gulf of Mexico	April 20, 2010–July 15, 2010	560,000–585,000
Ixtoc I	Mexico, Gulf of Mexico	June 3, 1979–March 23, 1980	454,000–480,000
Atlantic Empress/Aegean Captain	Trinidad and Tobago	July 19, 1979	287,000
Fergana valley	Uzbekistan	March 2, 1992	285,000
Nowruz field platform	Iran, Persian Gulf	February 4, 1983	260,000
ABT Summer	Angola, 700 nmi (1300 km; 810 mi) offshore	May 28, 1991	260,000
Castillo de Bellver	South Africa, Saldanha Bay	August 6, 1983	252,000
Amoco Cadiz	France, Brittany	March 16, 1978	223,000
MT Haven	Italy, Mediterranean Sea, near Genoa	April 11, 1991	144,000
Odyssey	Canada, 700 nmi (1300 km; 810 mi) off Nova Scotia	November 10, 1988	132,000
Sea Star	Iran, Gulf of Oman	December 19, 1972	115,000
Irenes Serenade	Greece, Pylos	February 23, 1980	100,000
Urquiola	Spain, A Coruña	May 12, 1976	100,000
Torrey Canyon	United Kingdom, Scilly Isles	March 18, 1967	80,000–119,000
Greenpoint oil spill	United States, Brooklyn, New York City	1940–1950s	55,000–97,000

metallurgy involves smelting of ores at high temperatures. This involves high energy costs and also leads to pollution. That copper could be leached from its ores by the activity of a bacterium, *Acidithiobacillus ferroxidans*, was discovered in 1947. This discovery opened the way to biohydrometallurgy for extraction of uranium in Canada and gold in South Africa. This technology is now commercially exploited in several countries for extraction of copper, arsenic, nickel, zinc, etc. Microbes useful for metal recovery depend upon the temperature of the recovery process. Microbial technology is helpful in recovery of ores that cannot be economically processed with chemical methods, since they contain only low-grade ores. During separation of high-grade ores, large quantities of low-grade ore are produced and are discarded into waste heaps, from where they reach the environment.

Microbial leaching plays its role here, to retrieve lead, nickel, and zinc ores, which are there in significant amounts. Large-scale chemical processing causes environmental problems when the dump is not managed properly. Leach fluid containing a large quantity of metals at very low pH seeps into nearby natural water supplies and ground water. Biomining is thus an economically sound hydrometallurgical process that creates fewer environmental problems than conventional chemical processing. The vast unexploited mineral treasure of India is the treasure arena of action for biomining technology.

8.3.9 Single-Cell Protein and Biomass from Waste

In highly populated countries like India, the high quantity of waste that is generated can be effectively utilized as the growth medium for single-cell protein (SCP). *SCP* typically refers to sources of mixed protein extracted from pure or mixed cultures of algae, yeasts, fungi, or bacteria (grown on agricultural wastes) used as a substitute for protein-rich foods in human and animal feeds. In the 1960s, researchers at British Petroleum developed what they called "proteins-from-oil process," a technology for producing SCP by yeast fed by waxy n-paraffins, a product produced by oil refineries. Initial research work was done by Alfred Champagnat at BP's Lavera Oil Refinery in France. A small pilot plant there started operations in March 1963, and the construction of the second pilot plant was authorized at Grangemouth Oil Refinery in Britain.

SCPs develop when microbes ferment waste materials (including wood, straw, cannery, and food processing wastes, residues from alcohol production, hydrocarbons, or human and animal excreta). The problem with extracting SCPs from the wastes is the dilution and cost. They are found in very low concentrations, usually less than 5%. Engineers have developed ways to increase the concentrations, including centrifugation, flotation, precipitation, coagulation and filtration, or the use of semipermeable membranes. SCPs need to be dehydrated to approximately 10% moisture content and/or acidified to aid in storage and prevent spoilage. The methods to increase the concentrations to adequate levels and the dewatering process require equipment that is expensive and not always suitable for small-scale operations. It is economically prudent to feed the product locally and shortly after it is produced. Microbes employed include yeasts (*Saccharomyces cerevisiae*, *Candida utilis Torulopsis and Geotrichum candidum*, *Oidium lactis*), other fungi (*Aspergillus oryzae*, *Sclerotium rolfsii*, *Polyporus*, *and Trichoderma*), and bacteria (*Rhodopseudomonas capsulata*) and algae (*Chlorella and Spirulina*).

8.3.10 Vermitechnology

Vermitechnology or *vermicomposting* is an easy and effective way to recycle agricultural waste, city garbage, kitchen waste, etc., by converting these organic waste materials into nutritious compost by the activity of earthworms. Worms transform the waste into high-quality fertilizer. Vermicast is a valuable soil amendment and can replace chemical fertilizers to some extent. Vermicompost is potential organic manure, rich in plant nutrients. It also contains micronutrients, certain hormones and enzymes, etc., which have stimulatory effects on plant growth. It also lodges many beneficial bacteria and actinomycetes.

Vermicompost containing large number of earthworm eggs that hatch out within a month is equivalent to a mini-fertilizer factory in the soil. The earthworms eat biomass and excrete it in a digested form, generally called *vermicompost*. Vermicompost increases water-holding capacity and reduces the irrigation water requirement of crops. The worm species most commonly used are *Eisenia foetida*, *Lubricus rubellus*, *Eudrilus eugeniae*, *Perionyx excavatus*, *P. arbricola*, and *P. sansibaricus*. Vermicompost influences the physicochemical as well as the biological properties of the soil, which in turn improves its fertility. It is rich in micronutrients like magnesium, iron, molybdenum, boron, copper, zinc, and also some growth regulators. It enhances the water-holding capacity of light-textured sandy soil. Vermicompost is rich in several microflora, several enzymes, auxins, and complex growth regulators like gibberellins, which are present in the earthworm castings. It also has a buffering action and so neutralizes soil pH and helps in making minerals and trace elements more easily available to crops. It also enhances soil fertility and reduces toxicity.

The quality, shelf life, and nutritive value of horticultural crops are enhanced, enabling value addition to the produce. Vermitechnology is an essential tool of organic farming that increases crop productivity in a sustainable manner, thus resulting in quality produce, and reduces the cost of agricultural inputs, in addition to improving the inherent capacity of the soil without deleterious effects on the environment.

8.3.11 Biosorption

Pollution interacts naturally with biological systems. It is currently uncontrolled, seeping into any biological entity within the range of exposure. The most problematic contaminants include heavy metals, pesticides, and other organic compounds that can be toxic to wildlife and humans in small concentrations. There are existing methods for remediation, but they are expensive or ineffective. However, an extensive body of research has found that a wide variety of commonly discarded waste, including eggshells, bones, peat, fungi, seaweed, yeast, and carrot peels can efficiently remove toxic heavy metal ions from contaminated water. Ions from metals like mercury can react in the environment to form harmful compounds such as methyl mercury, a compound known to be toxic to humans. In addition, adsorbing biomass, or biosorbents, can also remove other harmful metals such as arsenic, lead, cadmium, cobalt, chromium, and uranium. Biosorption is a physiochemical process that occurs naturally in certain biomass, which allows it to passively concentrate and bind contaminants onto its cellular structure. Biosorption may be used as an environmentally friendly filtering technique. There is no doubt that the world would benefit from more rigorous filtering of harmful pollutants created by industrial processes and pervasive human activity. Bacteria, yeast, algae, and fungi are found to be most effective in bioabsorption due to the charged groups, cell walls, and envelopes. (Figure 8.10).

8.3.11.1 Bacteria

Bacteria make excellent biosorbents because of their high surface-to-volume ratios and a high content of potentially active chemosorption sites, such as on teichoic acid, in their cell walls. It has been shown that the Gram-negative strains *Escherichia coli* K-12 and *Pseudomonas aeruginosa* and a Gram-positive strain *Micrococcus luteus* are excellent biosorbents for various chemicals,

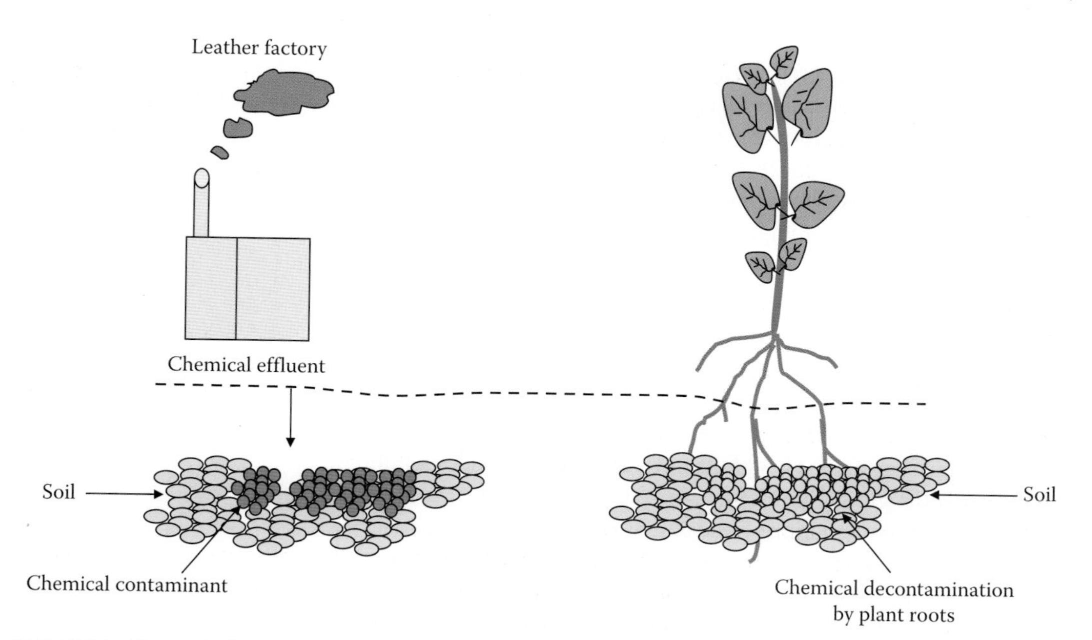

FIGURE 8.10 Pollution removal by plants.

which include copper, chromium, cobalt, and nickel. Their sorption binding constants suggested that *Escherichia coli* cells were the most efficient at binding copper, chromium, and nickel, and *Micrococcus luteus* absorbed cobalt most efficiently. The most relevant work on true bacterial biosorption has been done by the Brierleys, who took the metal biosorption concept all the way to the commercial stage. The bacterium *Pseudomonas fluorescens* was found to be effective in removing lead, zinc whereas *Pseudomonas aeruginosa* was found to be effective in removing plutonium with 75%–80% efficiency. However, this technique is not used on an industrial scale, and *Streptomyces virido* chromogens immobilized in polyacrylamide gel can also remove uranium.

8.3.11.2 Fungi

Fungal biomass can be used to adsorb metal particulates, elemental sulfur, insoluble sulfides, clay, charcoal, magnetite, etc. Particulate adsorption by fungal cell walls occurs independent of cellular metabolism. Adsorption of metallic cations takes place on the anionic sites such as amines, carboxyl, hydroxyl, and phosphate groups on the cell walls of fungi. *Rhizopus arrhisus* adsorbs 170 mg radionuclides per gram dry weight and 30–130 mg cadmium per gram dry weight. *Aspergillus niger* can adsorb silver, but it cannot desorb it easily. *Aspergillus oryzae* can recover cadmium with 99% efficiency. *Penicillium chrysogenum* is used for the removal of uranium and radium.

8.3.11.3 Algae

Cells of *Chlorella vulgaris* immobilized on polyacrylamide gel have been used to remove uranium from seawater. It is also used to remove gold, silver, and mercury at low pH (2–3). Considerable amounts of heavy metals can be bound to the algal biomass. Zinc and copper can be released by an acetate buffer at pH 5 and 3.5, respectively. Mercury can be released by the same buffer at pH 2. With the help of algae, simple and cheap recovery of metals is possible.

8.3.12 Genetically Engineered Organisms

Genetically engineered microbes (GEMs) are microorganisms into which a gene or genes have been introduced by using recombinant DNA technology. Polyaromatic hydrocarbons like naphthalene, phenanthrene, and anthracene occur in soil due to leakage of fossil fuels or petroleum products. *Pseudomonas flourescens* were isolated from PAHs-contaminated soils in the United States. These were genetically engineered with lux genes from *Vibrio fisberi*, which is a bacterium that lives in the light, producing organs in deep-sea fishes. The lux gene was recombined with a promoter of the naphthalene degradation gene. The *Pseudomonas flourescens* HK44 thus modified responds to naphthalene by luminescence. The luminescence can be detected by light-sensing probes. Strains of *Gliocladium virens*, *Pseudomonas chrysosporium*, and *Gliocladium virens* have the potential for bioremediation of contaminated soil.

PROBLEMS

Section A: Descriptive Type

Q1. What are the effects of global warming?
Q2. Describe the various health hazards due to man-made pollution.
Q3. What is the greenhouse effect?
Q4. What is bioremediation? Describe its significance.
Q5. Describe the various steps involved in the waste water treatment process.
Q6. Can biofuel replace fossil fuel in the future? Describes the advantages and disadvantages of biofuel.
Q7. How is energy converted from biowaste?
Q8. Do you think it is possible to make biodegradable plastics?

Section B: Multiple Choice

Q1. A carbon footprint is …
   a. Greenhouse gas emission
   b. Hydrogen gas emission
   c. Methane gas emission

Q2. Greenhouse gases are gases in the atmosphere that absorb and emit radiation within the thermal infrared range. True/False

Q3. Acid rain is caused due to a reaction between ________ in the atmosphere.
   a. Hydrogen and nitrogen
   b. Carbon and hydrogen
   c. Sulfur and nitrogen
   d. Hydrogen and sulfur

Q4. The bacterium ________ has been modified to consume and digest toluene and ionic mercury from highly radioactive nuclear waste.
   a. *Deinococus radiodurans*
   b. Salmonella
   c. Nocardia
   d. *E. coli*

Q5. The term mycoremediation refers to …
   a. Bacterial bioremediation
   b. Yeast bioremediation
   c. Fungal bioremediation
   d. None of the above

Q6. The secondary treatment phase of wastewater treatment deals with the …
   a. Degradation of biological waste
   b. Degradation of chemical waste
   c. Treatment of microbes
   d. None of the above

Q7. Which among the following is not a biofuel?
   a. Ethanol
   b. Methanol
   c. Butanol
   d. Butane

Q8. FFV is a …
   a. Fast-fuel vehicle
   b. Flexible-fuel vehicle
   c. Front-fuel vehicle
   d. None of the above

Q9. Syngas is a mixture of carbon monoxide and hydrogen. True/False

Q10. Third-generation biofuel refers to …
   a. Ethanol fuel
   b. Algal fuel
   c. Fungal fuel
   d. Bacterial fuel

Q11. *P. putida* F1 is one of the most well-studied aromatic hydrocarbons degrading bacterial strains. True/False

Section C: Critical Thinking

Q1. What efforts should we make to prevent the access release of greenhouse gases?

Q2. Do you think biofuel will ever replace fossil fuel? If yes, explain how.

Debate

Organize a debate on the topic "Global warming: myth or reality?" The students may speak about global warming based on scientific data and research.

Field Visit

With the help of your course instructor, arrange a field visit to your city wastewater treatment plant and learn how wastewater is being treated and cleaned in the plant and how the biowaste water or sludge is being used for irrigation and to generate energy. Also, write a brief report based on your visual experience of the sewage treatment plant and submit your report to your course instructor.

REFERENCES AND FURTHER READING

Adams, S. and Lambert, D. *Earth Science: An Illustrated Guide to Science*, Chelsea House, New York p. 20, 2006.

ADM Biodiesel. 2010. Hamburg, Leer, Mainz (http://wwwbiodisel.de).

Bent, F., Bruzelius, N., and Rothengatter, W. *Megaprojects and Risk: An Anatomy of Ambition*, Cambridge University Press, Cambridge, U.K., 2003.

Beychok, M.R. Performance of surface-aerated basins. *Chem. Eng. Progr. Symp. Ser.* 67: 322–339, 1971.

Bounds, A. OECD warns against biofuels subsidies. *Financial Times*. http://www.ft.com/cms/s/0/e780d216-5fd5-11dc-b0fe-0000779fd2ac.html (retrieved on March 7, 2008), 2007.

Caldeira, K. and Wickett, M.E. Anthropogenic carbon and ocean pH. *Nature* 425: 365–365, 2003.

Choi, E.N., Cho, M.C., Kim, Y., Kim, C.-K., and Lee, K. Expansion of growth substrate range in *Pseudomonas putida* F1 by mutations in both *cymR* and *todS*, which recruit a ring-fission hydrolase CmtE and induce the *tod* catabolic pathway, respectively. *Microbiology* 149: 795–805, 2003.

Earth's Energy Budget. Oklahoma climatological survey. 1996–2004. http://okfirst.mesonet.org/train/meteorology/EnergyBudget.html (retrieved on November 17, 2007), 2007.

Eaton, R.W. P-Cymene catabolic pathway in *Pseudomonas putida* F1: Cloning and characterization of DNA encoding conversion of p-cymene to p-cumate. *J. Bacteriol.* 179: 3171–3180, 1997.

EurekAlert. Fifteen new highly stable fungal enzyme catalysts that efficiently break down cellulose into sugars at high temperatures, 2009 (www.eurekalert.org).

Evans, J. Biofuels aim higher. Biofuels, bioproducts and biorefining (BioFPR). http://www.biofpr.com/details/feature/102347/Biofuels_aim_higher.html (retrieved on December 3, 2008), 2008.

Farrell, A.E. et al. Ethanol can contribute to energy and environmental goals. *Science* 311: 506–508, 2006.

Gibson, D.T., Koch, J.R., and Kallio, R.E. Oxidative degradation of aromatic hydrocarbons by microorganisms I. Enzymatic formation of catechol from benzene. *Biochemistry* 7: 2653–2661, 1968.

Globeco biodegradable bio-diesel. 2008. (www.globeco.co.uk).

Greenfuelonline.com. 2011. (www.greenfuelonline.com).

Hammerschlag, R. Ethanol's energy return on investment: A survey of the literature 1999-present. *Environ. Sci. Technol.* 40: 1744–1750, 2006.

Hartman, E. A Promising oil alternative: Algae energy. *Washington Post*. http://www.washingtonpost.com/wpdyn/content/article/2008/01/03/AR2008010303907.html (retrieved on January 15, 2008), 2008.

Hudlicky, T., Gonzalez, D., and Gibson, D.T. Enzymatic dihydroxylation of aromatics in enantioselective synthesis: Expanding asymmetric methodology. *Aldrichim. Acta* 32: 35–62, 1999.

IEA Bioenergy. 2010. (www.ieabioenergy.com).

John, F.K., Hamilton, T.G., Bra, M., and Röckmann, T. Methane emissions from terrestrial plants under aerobic conditions. *Nature* 439: 187–191, 2006.

Karl, T.R. and Trenberth, K.E. Modern global climate change. *Science* 302: 1719–1723, 2003.

Key, R.M., Kozyr, A., Sabine, C.L. et al. A global ocean carbon climatology: Results from GLODAP. *Global Biogeochem. Cycles* 18: GB4031, 23, 2004.

Kiehl, J.T., Kevin, E., and Trenberth. Earth's annual global mean energy budget. *Bull. Amer. Meteor. Soc.* 78(2): 197–208, 1997.

Kyoto Protocol from United Nations Framework Convention on Climate Change 2008 (retrieved August 2008) (www.unfccc.int).

Lau, P.C.K., Wang, Y., Patel, A., Labbé, D. et al. A bacterial basic region leucine zipper histidine kinase regulating toluene degradation. *Proc. Natl. Acad. Sci. USA* 94: 1453–1458, 1997.

Le Treut, H., Somerville, R., and Cubasch, U. et al. Historical overview of climate change science. in: *Climate Change 2007: The Physical Science Basis*, Contribution of Working Group I to the Fourth Assessment Report of the Intergovernmental Panel on Climate Change, S. Solomon, D. Qin, M. Manning (eds.). et al. Cambridge University Press, Cambridge, U.K., 2007.

Low cost algae production system introduced. 2011. (www.energy-arizona.org).

Non-CO_2 gases economic analysis and inventory 2007: Global warming potentials and atmospheric lifetimes, U.S. *Environmental Protection Agency*, (accessed August 31, 2007).

Ocean. *The Columbia Encyclopedia*. Columbia University Press, New York, 2002.

Odum, E.P. *Fundamentals of Ecology*, 3rd edn. Saunders, New York, 1971.

Oldroyd, D. *Earth Cycles: A Historical Perspective*. Greenwood Press, Westport, CT, 2006.

Orr, J.C., Fabry, V.J., Aumont, O. et al. Anthropogenic ocean acidification over the twenty-first century and its impact on calcifying organisms. *Nature* 437: 681–686, 2005.

Parales, R.E., Ditty, J.L., and Harwood, C.S. Toluene-degrading bacteria are chemotactic to the environmental pollutants benzene, toluene, and trichoroethylene. *Appl. Environ. Microbiol.* 66: 4098–4104, 2000.

Pelletier, J.D. Natural variability of atmospheric temperatures and geomagnetic intensity over a wide range of time scales. *Proc. Natl. Acad. Sci.* 99: 2546–2553, 2002.

Robert, W.C. *Geosystems: An Introduction to Physical Geography*, Prentice Hall, Inc., Upper Saddle River, NJ, 1996.

Simison, W.B. The mechanism behind plate tectonics. 2007. http://www.ucmp.berkeley.edu/geology/tecmech.html (retrieved on November 17, 2007).

Smith, G.A. and Pun, A. *How Does the Earth Work*? Pearson Prentice Hall, Upper Saddle River, NJ, 2006.

Somerville, C. Development of cellulosic biofuels. U.S. Department of Agriculture. http://www.usda.gov/oce/forum/2007%20Speeches/PDF%20PPT/CSomerville.pdf (retrieved on January 15, 2008), 2008.

Spilhaus, A.F. Distribution of land and water on the planet, in: *UN Atlas of the Oceans, Maps of the whole world ocean, Geogr. Rev*. 32(3): 431–435, 1942.

UN Biofuels Report. 2007. (http://esa.in.org).

Wackett, L.P. and Gibson, D.T. Degradation of trichloroethylene by toluene dioxygenase in whole cell studies with *Pseudomonas putida* F1. *Appl. Environ. Microbiol.* 54: 1703–1708, 1988.

Welcome to Biodiesel Filling Stations. 2009. (http://biodieselrefillingstations.co.uk).

Western Climate Initiative. 2008. (www.westernclimateinitiative.org).

Yunqiao, P., Dongcheng, Z., Singh, P.M., and Ragauskas, A.J. The new forestry biofuels sector. *Biofuel Bioprod. Biorefin*. 2(1): 58–73, 2007.

Zylstra, G.J. and Gibson, D.T. Toluene degradation by *Pseudomonas putida* F1: Nucleotide sequence of the todC1C2BADE genes and their expression in *E. coli*. *J. Biol. Chem*. 264: 14940–14946, 1989.

Zylstra, G.J. and Gibson, D.T. Aromatic hydrocarbon degradation: A molecular approach. *Genet. Eng*. 13: 183–203, 1991.

9 Medical Biotechnology

LEARNING OBJECTIVES

- Explain the role of biotechnology in medicine
- Discuss how vaccines are manufactured and explain the related issues
- Discuss the process of making antibodies and hybridoma technology
- Explain the significance of therapeutic protein and its production
- Discuss the significance of stem cell therapy and the related issues
- Discuss the success and failure of gene therapy and organ transplantation
- Discuss some tools and techniques used in medical biotechnology

9.1 INTRODUCTION

Biotechnology has found a wide range of applications in medicine. While dealing with diseases, applications of biotechnology include prevention, diagnosis, and cure of diseases. Through human genetics, biotechnology has found use in genetic counseling, antenatal diagnosis, and gene therapy. In forensic medicine, it has already been used for identification of individuals who could be criminals. The major applications include animal and human healthcare, genetic counseling, and forensic medicine. Biotechnology may have applications in human healthcare either directly or indirectly. Its direct and more important use involves efforts to overcome the menace due to a variety of diseases. Biotechnology is useful in providing immunity against a disease through development of a vaccine and in diagnosis of a disease at an early stage of its onset. Table 9.1 shows a list of the various applications of medical biotechnology.

9.1.1 Vaccine Development

A *vaccine* is a biological preparation that improves immunity to a particular disease. It typically contains an agent that resembles a disease-causing microorganism and is often made from weakened or killed forms of the microbe. The agent stimulates the body's immune system to recognize the agent as foreign, destroy it, and "remember" it, so that the immune system can more easily recognize and destroy any of these microorganisms that it later encounters. Sometime during the 1770s, Edward Jenner heard a milkmaid boast that she would never have the often-fatal or disfiguring disease smallpox because she had already had cowpox, which has a very mild effect in humans. In 1796, Jenner took pus from the hand of a milkmaid with cowpox, inoculated an 8-year-old boy with it, and, 6 weeks later, variolated the boy's arm with smallpox, afterward observing that the boy did not catch smallpox. Since vaccination with cowpox was much safer than smallpox inoculation, the latter, though still widely practiced in England, was banned in 1840.

Louis Pasteur generalized Jenner's idea by developing what he called a rabies vaccine (now termed an antitoxin), and in the nineteenth century vaccines were considered a matter of national prestige, and compulsory vaccination laws were passed. The twentieth century saw the introduction of several successful vaccines, including those against diphtheria, measles, mumps, and rubella. Major achievements included the development of the polio vaccine in the 1950s and the eradication of smallpox during the 1960s and 1970s. As vaccines became more common, many people began taking them for granted. However, vaccines remained elusive for many important diseases,

TABLE 9.1
Applications of Medical Biotechnology

Contribution	Technology
Monoclonal antibodies (MABs) for disease diagnosis	Hybridoma technology
DNA probes for disease diagnosis	Genetically engineered microbes
Recombinant vaccines	Genetically engineered microbes
Drugs such as human insulin, interferon, growth hormone, etc.	Genetically engineered microbes
Gene therapy for the cure of genetic diseases	Replacement of defective genes by normal genes, artificially cloned
Babies of specified sex	Artificial insemination followed by separation of sperms containing X and Y chromosomes
Identification of parents, siblings, and criminals	DNA fingerprinting
Test tube babies	*In vitro* fertilization and embryo transfer

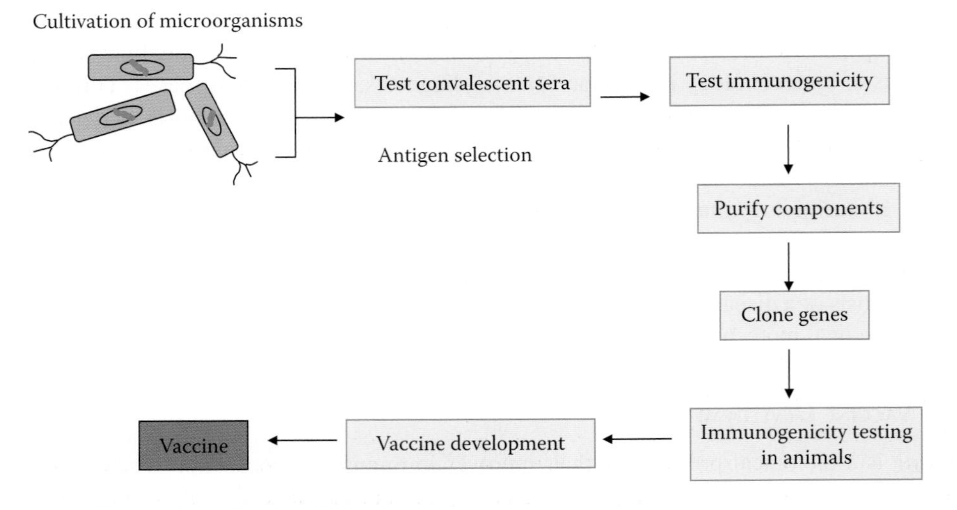

FIGURE 9.1 Conventional vaccine development.

including malaria and HIV. There are several types of vaccines currently in use. These represent different strategies used to reduce the risk of illness, while retaining the ability to induce a beneficial immune response (Figure 9.1).

9.1.1.1 Killed Vaccines

Some vaccines contain killed microorganisms. These microorganisms are previously virulent microorganisms that have been killed with chemicals or heat. Some examples are vaccines against influenza, cholera, the bubonic plague, polio, and hepatitis A.

9.1.1.2 Attenuated Vaccines

Some vaccines contain live, attenuated virus microorganisms. These are live microorganisms that have been cultivated under conditions that disable their virulent properties, or which use closely-related but less dangerous organisms to produce a broad immune response. They typically provoke more durable immunological responses and are the preferred type for healthy adults. Examples include vaccines against yellow fever, measles, rubella, and mumps. The live Mycobacterium tuberculosis vaccine, developed by Calmette and Guérin, is not made of a contagious strain, but contains a virulently modified strain called "BCG," used to elicit immunogenicity to the vaccine.

9.1.1.3 Toxoid Vaccines

Toxoids are inactivated toxic compounds in cases where these (rather than the microorganism itself) cause illness. Examples of toxoid-based vaccines include tetanus and diphtheria. Not all toxoids are for microorganisms; for example, *Crotalus atrox* toxoid is used to vaccinate dogs against rattlesnake bites.

9.1.1.4 Subunit Vaccines

Rather than introducing an inactivated or attenuated microorganism to an immune system (which would constitute a "whole-agent" vaccine), a fragment of it can create an immune response. These are called *subunit vaccines*. Characteristic examples include the subunit vaccine against the hepatitis B virus, which is composed of only the surface proteins of the virus (produced in yeast), and the virus-like particle (VLP) vaccine against human papillomavirus (HPV), which is composed of the viral major capsid protein.

9.1.1.5 Conjugate Vaccines

Certain bacteria have polysaccharide outer coats that are poorly immunogenic. By linking these outer coats to proteins (such as toxins), the immune system can be led to recognize the polysaccharide as if it were a protein antigen. This approach is used in the *Haemophilus influenzae* type B vaccine.

9.1.1.6 Experimental Vaccines

There are a number of innovative vaccines currently in the developmental stage or which are already in use, such as recombinant vector vaccines, DNA vaccines, and T-cell receptor peptide vaccines. Recall that in recombinant vector vaccines, combining the physiology of one microorganism and the DNA of another can create immunity against diseases that have complex infection processes. On the other hand, a new type of vaccine, called a *DNA vaccine*, has been created recently from an infectious agent's DNA. It works by insertion (and expression, triggering immune system recognition) of viral or bacterial DNA into human or animal cells. Some cells of the immune system that recognize the proteins expressed will mount an attack against these proteins and cells expressing them. Because these cells live for a very long time, if the pathogen that normally expresses these proteins is encountered at a later time, they will be attacked instantly by the immune system. One advantage of DNA vaccines is that they are very easy to produce and store. As of 2006, DNA vaccination is still experimental. Likewise, T-cell receptor peptide vaccines are under development for several diseases, using models of valley fever, stomatitis, and atopic dermatitis. These peptides have been shown to modulate cytokine production and improve cell-mediated immunity.

Targeting of identified bacterial proteins that are involved in complement inhibition would neutralize the key bacterial virulence mechanism. While most vaccines are created using inactivated or attenuated compounds from microorganisms, synthetic vaccines are composed mainly or wholly of synthetic peptides, carbohydrates, or antigens.

9.1.1.7 Valence Vaccines

Valence vaccines are monovalent (also called *univalent*) or multivalent (also called *polyvalent*) in nature. A monovalent vaccine is designed to immunize against a single antigen or single microorganism. A multivalent vaccine is designed to immunize against two or more strains of the same microorganism or against two or more microorganisms. In certain cases, a monovalent vaccine may be preferable for rapidly developing a strong immune response.

9.1.1.8 Vaccine Production

One challenge in vaccine development is the economic factor. Many diseases, such as HIV, malaria, and tuberculosis—which exist principally in poor countries—demand a vaccine. Pharmaceutical firms and biotechnology companies have little incentive to develop vaccines for these diseases because there is little revenue potential. Even in more affluent countries, financial returns are usually minimal and the financial and other risks are great. Most vaccine development to date has relied on "push" funding by governments, universities, and nonprofit organizations. Many vaccines have been highly cost-effective and beneficial for public health. The number of vaccines actually administered has risen dramatically in recent decades.

Vaccine production has several stages. First, the antigen itself is generated. Viruses are grown either on primary cells such as chicken eggs (e.g., for influenza), or on continuous cell lines such as cultured human cells (e.g., for hepatitis A). Bacteria such as *Haemophilus influenzae* type b are grown in bioreactors. Alternatively, a recombinant protein derived from the viruses or bacteria can be generated in yeast, bacteria, or cell cultures. After the antigen is generated, it is isolated from the cells used to generate it. A virus may need to be inactivated, possibly with no further purification required. Recombinant proteins need many operations involving ultrafiltration and column chromatography. Finally, the vaccine is formulated by adding adjuvant, stabilizers, and preservatives, as needed. The adjuvant enhances the immune response of the antigen, the stabilizers increase the storage life, and the preservatives allow the use of multidose vials. Combination vaccines are harder to develop and produce because of potential incompatibilities and interactions among the antigens and other ingredients involved.

Vaccine production techniques are evolving. Cultured mammalian cells are expected to become increasingly important, compared to conventional options such as chicken eggs, due to greater productivity and few problems with contamination. Recombination technology that produces genetically detoxified vaccines is expected to grow in popularity for the production of bacterial vaccines that use toxoids. Combination vaccines are expected to reduce the quantities of antigens they contain, and thereby decrease undesirable interactions, by using pathogen-associated molecular patterns. Many vaccines need preservatives to prevent serious adverse effects such as the *Staphylococcus* infection that, in one 1928 incident, killed 12 of 21 children inoculated with a diphtheria vaccine that lacked a preservative. Several preservatives are available, including thiomersal, phenoxyethanol, and formaldehyde. Thiomersal is more effective against bacteria, has better shelf life, and improves vaccine stability, potency, and safety. However, in the United States, the European Union, and a few other affluent countries, it is no longer used as a preservative in childhood vaccines, as a precautionary measure due to its mercury content. Although controversial claims have been made that thiomersal contributes to autism, no convincing scientific evidence supports these claims.

There are several new delivery systems in development, which will make vaccines more efficient to deliver. Possible methods include liposomes and *immune stimulating complex* (ISCOM). The latest developments in vaccine delivery technologies have resulted in oral vaccines. A polio vaccine was developed and tested by volunteer vaccinators with no formal training. The results were very positive, in that the ease of the vaccines increased dramatically. With an oral vaccine, there is no risk of blood contamination. Oral vaccines are likely to be solid, and have proven to be more stable and less likely to freeze. This stability reduces the need for a "cold chain" (the resources required to keep vaccines within a restricted temperature range from the manufacturing stage to the point of administration), which in turn will decrease the cost of vaccines. Finally, a microneedle approach, which is still in the developmental stage, seems to be the vaccine of the future. The microneedle has pointed projections fabricated into arrays that can create vaccine delivery pathways through the skin. The use of plasmids has been validated in preclinical studies as a protective vaccine strategy for cancer and infectious diseases. However, the crossover application into human studies has met with poor results, based on the inability to provide clinically relevant benefits. The overall efficacy of plasmid DNA immunization depends on increasing the plasmid's immunogenicity while also correcting for factors involved in the specific activation of immune effecter cells.

9.1.1.9 Making of Influenza Vaccines

Influenza, commonly referred to as the flu, is an infectious disease caused by RNA viruses of the family *Orthomyxoviridae* (the influenza viruses), which affects birds and mammals. The most common symptoms of the disease are chills, fever, sore throat, muscle pains, severe headache, coughing, weakness, and general discomfort. Sore throat, fever, and coughs are the most frequent symptoms. In more serious cases, influenza causes pneumonia, which can be fatal, particularly for the young and the elderly. Although it is often confused with other influenza-like illnesses, especially the common cold, influenza is a much more severe disease than the common cold and is caused by a different type of virus. Influenza may produce nausea and vomiting, particularly in children, but these symptoms are more common in the unrelated gastroenteritis, which is sometimes called stomach flu or 24 h flu.

Typically, influenza is transmitted through the air by coughs or sneezes, creating aerosols containing the virus. Influenza can also be transmitted by direct contact with bird droppings or nasal secretions or through contact with contaminated surfaces. Airborne aerosols have been thought to cause most infections, although which means of transmission is most important is not entirely clear. Influenza viruses can be inactivated by sunlight, disinfectants, and detergents. As the virus can be inactivated by soap, frequent hand washing reduces the risk of infection. Influenza spreads around the world in seasonal epidemics, resulting in the deaths of hundreds of thousands worldwide annually, and millions in pandemic years. On the average, between 1979 and 2001, approximately 41,400 people in the United States died from influenza each year. Three influenza pandemics occurred in the twentieth century and killed tens of millions of people, with each of these pandemics being caused by the appearance of a new strain of the virus in humans. Often, these new strains appear when an existing flu virus spreads to humans from other animal species or when an existing human strain picks up new genes from a virus that usually infects birds or pigs. An avian strain named H5N1 raised the concern of a new influenza pandemic after it emerged in Asia in the 1990s, but it has not evolved to a form that spreads easily between people. In April 2009, a novel flu strain evolved that combined genes from human, pig, and bird flu, initially dubbed "swine flu" and also known as influenza A/H1N1, and emerged in Mexico, the United States, and several other nations. The World Health Organization officially declared the outbreak to be a pandemic on June 11, 2009.

Vaccinations against influenza are usually given to people in developed countries and to farmed poultry. The most common human vaccine is the trivalent influenza vaccine (TIV), which contains purified and inactivated material from three viral strains. Typically, this vaccine includes material from two influenza A virus subtypes and one influenza B virus strain. The TIV carries no risk of transmitting the disease, and it has very low reactivity. A vaccine formulated for 1 year may be ineffective in the following year, since the influenza virus evolves rapidly and new strains quickly replace the older ones. Antiviral drugs can be used to treat influenza, with neuraminidase inhibitors being particularly effective. The following lists the steps in producing influenza vaccine.

Step 1: A survey must first be done to find out which kinds of influenza strains are dominantly present in particular geographical locations throughout the year.

Step 2: The surveillance data, based on the presence of dominant influenza strains, are analyzed by researchers and the selected influenza strains are submitted to the Food and Drug Administration (FDA) to recommend which three to include. The FDA distributes seed viruses to manufacturers to begin the vaccine production.

Step 3: The seed viruses received from the FDA are processed for production in specially designed labs. Each virus strain is produced separately and later combined to make one vaccine. Millions of specially prepared chicken eggs are used to produce the vaccine. For 7 months, fertilized eggs are delivered to the manufacturer. Each egg is cleaned with a disinfectant spray and injected with one strain. The eggs are incubated for several days to allow the virus to multiply. After incubation the virus-loaded fluid is harvested.

Step 4: The virus fluid undergoes multiple purification steps and a special chemical treatment to ensure the virus is inactivated or killed. The virus is split by chemically disrupting the whole virus.

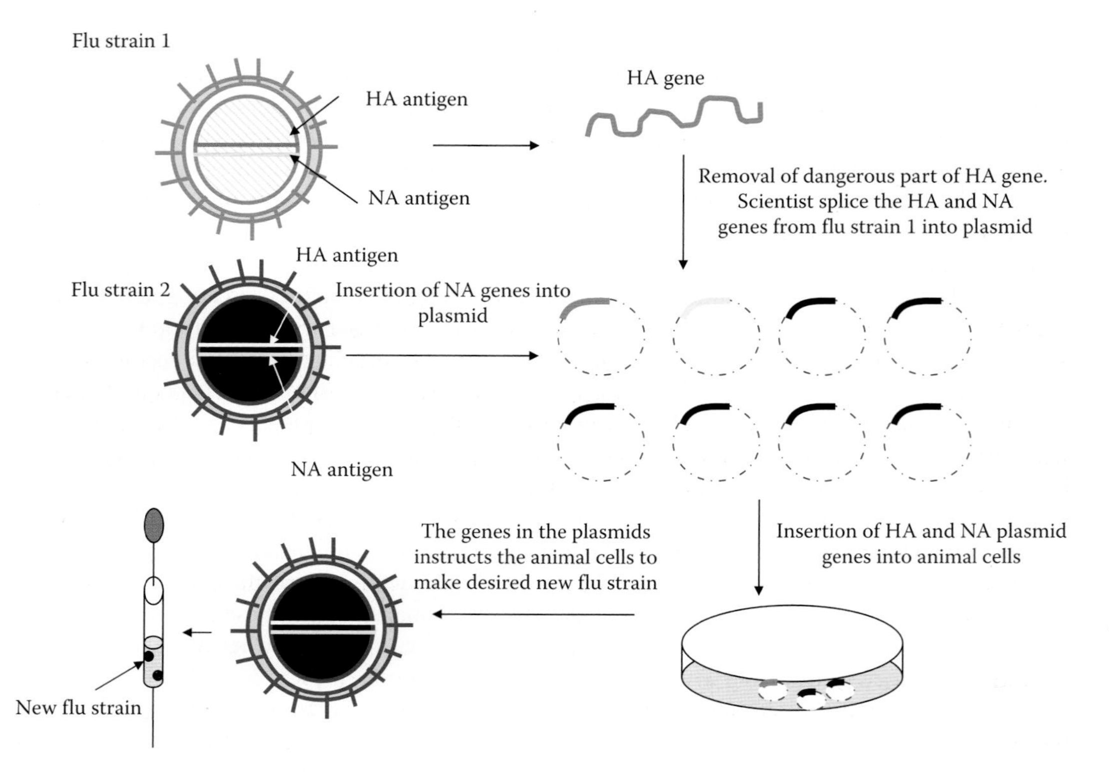

FIGURE 9.2 The process of making influenza vaccine.

Viral fragments from all the virus strains are collected from different batches and combined after completion of quality control tests. Manufacturers and the FDA test the vaccine concentrates to determine the amount and yield of the virus to ensure the concentrate is adequate for immunization.

Step 5: Upon receiving FDA approval and licensing, the vaccine is released for distribution in time for immunization. Manufacturers begin filling the doses into vials and syringes, which are then sealed and carefully inspected before labels are applied to show the vaccine batch, lot numbers, and expiration dates. Each lot must be specially released by the FDA before the manufacturer can ship it.

Step 6: Vaccine shipments typically begin in August/September and continue till November each year. Immunization generally begins in the month of October or as soon as the vaccine becomes available, continues throughout the influenza season, and typically ends in March. Immunity develops approximately 2 weeks following vaccination (Figure 9.2).

9.1.1.10 Large-Scale Production of Vaccines

The first step in making a vaccine is to store the virus of interest under frozen conditions that prevent the virus from becoming either stronger or weaker than desired. After defrosting and warming the seed virus under carefully specified conditions (i.e., at room temperature or in a water bath), a small amount of virus cells is placed into a "cell factory," a small machine that, with the addition of an appropriate medium, allows the virus cells to multiply. Each type of virus grows best in a medium specific to it, established in premanufacturing laboratory procedures, but all contain proteins from mammals in one form or another, such as purified protein from cow blood. Mixed with the appropriate medium, at the appropriate temperature, and with a predetermined amount of time, viruses will multiply.

The virus from the cell factory is then separated from the medium and placed in a second medium for additional growth. Early methods dating back 40 or 50 years ago used a bottle to hold the mixture, and the resulting growth was a single layer of viruses floating on the medium. It was soon discovered that if the bottle was turned while the viruses were growing, even more viruses could be produced, because a layer of viruses grew on all the inside surfaces of the bottle. An important discovery in the 1940s was that cell growth is greatly stimulated by the addition of enzymes to a medium, of which trypsin is the most commonly used. An enzyme is a protein that also functions as a catalyst in the feeding and growth of cells. In current practice, bottles are not used at all. The growing virus is kept in a container larger than but similar to the cell factory, and mixed with "beads" (near-microscopic particles to which the viruses can attach themselves). The use of the beads provides the virus with a much greater area to attach to, and consequently allows a much greater growth of virus. As in the cell factory, temperature and pH are strictly controlled. Time spent in growing viruses varies according to the type of virus being produced and is, in each case, a closely guarded secret of the manufacturer. When there are a sufficient number of viruses, they are separated from the beads in one or more ways. The broth might be forced through a filter with openings large enough to allow the viruses to pass through, but small enough to prevent the beads from passing. The mixture may be centrifuged several times to separate the virus from the beads in a container from which they can then be drawn off. The eventual vaccine will be made of either attenuated (weakened) virus or a killed virus. The choice of one or the other depends on a number of factors, including the efficacy of the resulting vaccine and its secondary effects. The vaccine based on rubella is developed almost every year in response to new variants of the causative virus, is always an attenuated vaccine. The virulence of a virus can dictate the choice. The rabies vaccine, for example, is always a killed vaccine.

If the vaccine is attenuated, the virus is usually attenuated before it goes through the production process. Carefully selected strains are cultured (grown) repeatedly in various media. There are strains of viruses that actually become stronger as they grow. These strains are clearly unusable for an attenuated vaccine. Other strains become too weak as they are cultured repeatedly, and these too are unacceptable for vaccine use. Like the porridge, chair, and bed that Goldilocks liked, only some viruses are "just right," reaching a level of attenuation that makes them acceptable for vaccine use and unchanging in strength. Recent molecular technology has made possible the attenuation of live viruses by molecular manipulation, but this method is still rare. The virus is then separated from the medium in which it has grown. Vaccines that are of several types (as most are) are combined before packaging. The actual amount of vaccine given to a patient will be relatively small compared to the medium in which it is given. Decisions about whether to use water, alcohol, or some other solution, for an injectable vaccine, for example, are made after repeated tests for safety, sterility, and stability.

9.1.1.11 Economics Involved in Vaccine Production

Influenza produces direct costs due to lost productivity and associated medical treatment, as well as indirect costs related to preventative measures. In the United States, influenza is responsible for a total cost of over $10 billion per year, while it has been estimated that a future pandemic could cause hundreds of billions of dollars in direct and indirect costs. However, the economic impacts of past pandemics have not been intensively studied, and some authors have suggested that the Spanish influenza actually had a positive long-term effect on per capita income growth, despite a large reduction in the working population and severe short-term depressive effects. Other studies have attempted to predict the costs to the U.S. economy of a pandemic as serious as the 1918 Spanish flu, where 30% of all workers became ill and 2.5% died. A 30% sickness rate and a 3-week length of illness would decrease the gross domestic product by 5%. The medical treatment of 18 million to 45 million people would incur additional costs, and the total economic costs would be approximately $700 billion. Preventative costs are also high. Governments worldwide have spent billions of U.S. dollars preparing and planning for a potential H5N1 avian influenza pandemic, with costs associated with purchasing drugs and vaccines as well as developing disaster drills and strategies for improved border controls. On November 1, 2005, U.S. President George W. Bush unveiled the National Strategy to Safeguard Against the Danger of Pandemic Influenza, backed by a request to Congress for $7.1 billion to begin implementing the plan. Internationally, on January 18, 2006, donor nations pledged $2 billion to combat bird flu at the 2-day International Pledging Conference on Avian and Human Influenza held in China. In an assessment of the impact of the 2009 H1N1 pandemic on selected countries in the southern hemisphere, data suggest that all countries experienced some time-limited and/or geographically-isolated socio/economic effects, and a temporary decrease in tourism, most likely due to fear of 2009 H1N1 disease. It is still too early to determine whether the H1N1 pandemic has caused any long-term economic impact.

9.1.1.12 Trends in Vaccine Research

Research on influenza includes how the virus produces disease–host immune responses, viral genomics, and how the virus spreads. These studies help in developing influenza countermeasures. For example, a better understanding of the body's immune system response helps vaccine development, and a detailed picture of how influenza invades cells aids the development of antiviral drugs. One important basic research program is the Influenza Genome Sequencing Project, which is creating a library of influenza sequences. This library should help clarify which factors make one strain more lethal than another, which genes most affect immunogenicity, and how the virus evolves over time. Research into new vaccines is particularly important, as current vaccines are very slow and expensive to produce and must be reformulated every year. The sequencing of the influenza genome and rDNA technology may accelerate the generation of new vaccine strains by allowing scientists to substitute new antigens into a previously developed vaccine strain. New technologies are also being developed to grow viruses in cell culture, which promises higher yields, less cost, better quality, and surge capacity. Research on a universal influenza A vaccine, targeted against the external domain of the transmembrane viral M2 protein (M2e), is being done at the University of Ghent by Walter Fiers, Xavier Saelens, and their team and has now successfully concluded Phase I clinical trials. A number of biologics, therapeutic vaccines, and immunobiologics are also being investigated for treatment of infections caused by viruses. Therapeutic biologics are designed to activate the immune response to viruses or antigens. Typically, biologics do not target metabolic pathways as do antiviral drugs, but stimulate immune cells such as lymphocytes, macrophages, and/or antigen-presenting cells in an effort to drive an immune response toward a cytotoxic effect against the virus. Influenza models such as murine influenza are convenient models to test the effects of prophylactic and therapeutic biologics. For example, Lymphocyte T-Cell Immune Modulator inhibits viral growth in the murine model of influenza.

When it comes to making an influenza vaccine, there seems to be a crack in the system. Although dependable since the 1970s, the current practice of injecting the flu virus into fertilized hens' eggs requires at least 6 months and hundreds of millions of eggs to produce a sufficient supply of vaccine for the U.S. population. The egg method is not sufficient to ensure rapid vaccine supply, and vaccine manufacturers need to arrange for egg supplies months in advance. ID Biomedical was recently awarded a $9.5 million contract by the National Institute of Allergy and Infectious Diseases (NIAID) to study an alternative method (called cell culture) to rapidly produce large quantities of the flu vaccine.

In recent times, the cell culture technique is a very interesting alternative, as it is an efficient and flexible strategy for manufacturing influenza vaccines. Instead of injecting the flu virus into eggs, the virus is overlaid onto special dog kidney cells, which, unlike eggs, grow and multiply rapidly in the lab. The inoculated cells are then incubated inside a growth chamber (called a bioreactor), adhering to small round beads (called microcarriers). As with the egg method, once the virus-containing fluid is removed from the bioreactor, the virus is purified, killed, split, and then blended with the two other viruses to make doses of flu vaccine. The advantages of the cell-culture technique are that if the need should arise for increased amounts of vaccine, a manufacturer could simply thaw out more cells and perhaps add more bioreactors. In addition, unlike eggs, the cells are naturally sterile, which helps to control product quality. Finally, the nutrients in which the cells grow contain no animal serum, which reduces the risk of microbial contamination. Currently, the research team is conducting studies to determine the optimal conditions under which the cell culture vaccine can be produced.

9.1.1.13 Issues Related to Vaccines

Producing a usable, safe antiviral vaccine involves a large number of steps; unfortunately, these steps cannot always be carried out for each and every virus. There is still much to be done and learned. The new methods of molecular manipulation have caused more than one scientist to believe that vaccine technology is only now entering a "golden age." Refinements of existing vaccines are possible in the future. The rabies vaccine, for example, produces side effects which make the vaccine unsatisfactory for mass immunization. In the United States, the rabies vaccine is now used only on patients who have contracted the virus from an infected animal and are likely to develop the fatal disease without immunization. The HIV virus, which biologists believe causes AIDS, is not currently amenable to traditional vaccine production methods. The AIDS virus rapidly mutates from one strain to another, and any given strain does not seem to confer immunity against other strains. Additionally, a limited immunizing effect of either attenuated or killed virus cannot be demonstrated either in the laboratory or in test animals. No HIV vaccine has yet been developed.

9.1.1.14 Synthetic Peptides as Vaccines

Vaccines can also be prepared through short synthetic peptide chains, which have therefore become a subject of considerable research activity. In order to synthesize peptides to be used as vaccines, the structure and function of proteins involved should be studied. Since it is the three-dimensional structure (TDS) and not the amino acid sequence which is responsible for the immunogenic response, it may be necessary to identify the protein region involved in the immunogenic response. For instance, in foot and mouth disease virus (FMDV), it is the amino acid 114–160 of the virus polypeptide that can produce antibodies neutralizing FMDV and thus provide protection.

Neutralization of FMDV was also possible through the region of 201–213 amino acids of the same protein. It has been shown that small synthetic peptides representing these regions of proteins can show immunogenic response and can therefore be used for development of a vaccine. An alternative approach to identify the immunogenic region of protein is through the study of the gene encoding the protein.

Recently, it has been shown that a cloned gene of an immunogenic protein of a pathogen (feline leukemia virus or FeLV) can be cut into fragments by DNAase, and these fragments can be cloned in lambda phages, where they may express. Phage colonies (plaques) having different cloned fragments are screened with a specific monoclonal antibody (MAB) that neutralizes the pathogen. The fragments, which react with the antibody, must be synthesizing the immunogenic peptide fragment. This cloned DNA fragment can then be sequenced. In this manner, it was possible to identify a 14 amino acid immunogen of the envelope protein of FeLV. The corresponding synthetic peptide was also found to compete with the virus for antibodies. When injected in guinea pigs, such synthetic peptides also elicited a partial immunogenic response. Therefore, there is great promise for the use of such synthetic peptides to be used as vaccines. In fact, a vaccine for malaria in the form of a synthetic peptide has already been prepared and is being tested for its suitability. This is the first example of a vaccine developed in the form of a synthetic peptide. Recently, it has been shown that the immunogenic region of protein of a pathogen can also be identified by eluting it from purified major histocompatibility complex (MHC) molecules. Different MHC allelic variants are available in cells for binding of different proteins, and they can be purified using specific T-cells. Peptides can be eluted from these purified MHC molecules and the sequence of such peptides can be determined and used for manufacturing synthetic peptides to be used as vaccines. Using the above three approaches, vaccines against several pathogens have either been produced in recent years or are expected to be produced in the near future.

9.2 ANTIBODY PRODUCTION

The body produces antibodies against antigens to fight back bacterial or viral diseases. These antibodies make our immune system very strong, but in some cases, the body fails to produce such antibodies. In that case, the body can be provided with antibodies produced in animals and bioreactors. In this section, we will discuss various MABs and their production and applications. MABs are

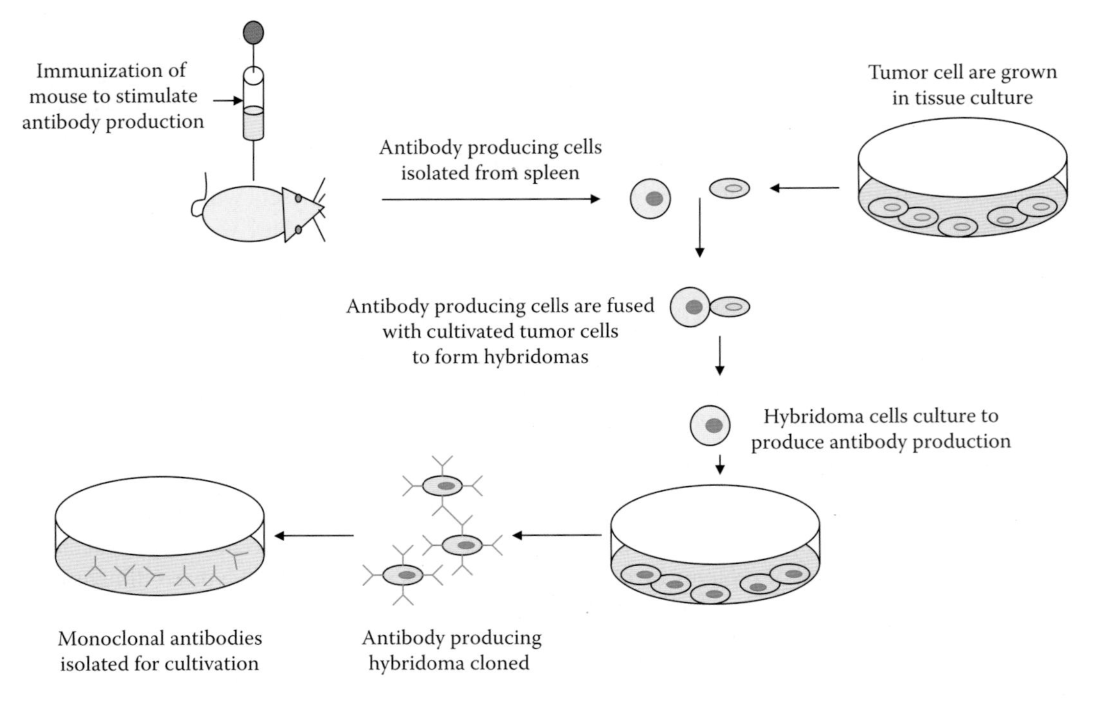

FIGURE 9.3 Generation of MABs.

mono-specific antibodies that are identical because they are produced by one type of immune cell that are all clones of a single parent cell. Given almost any substance, it is possible to create MABs that specifically bind to that substance; they can then serve to detect or purify that substance. This has become an important tool in biochemistry, molecular biology, and medicine. When used as medications, the generic drug name ends in "mab." The idea of a "magic bullet" was first proposed by Paul Ehrlich, who, at the beginning of the twentieth century, postulated that if a compound could be made that selectively targeted a disease-causing organism, then a toxin for that organism could be delivered along with the agent of selectivity. In the 1970s, the B-cell cancer, multiple myeloma, was known, and it was understood that these cancerous B-cells all produce a single type of antibody, called *paraprotein*. This knowledge was used to study the structure of antibodies, but it was not yet possible to produce identical antibodies specific to a given antigen. A process of producing MABs involving human–mouse hybrid cells was described in 1973. The invention of MABs in 1975 is generally accredited to Georges Köhler, César Milstein, and Niels Kaj Jerne. They shared the Nobel Prize in Physiology or Medicine in 1984 for the discovery of hybridoma technology. The key idea was to use a line of myeloma cells that had lost their ability to secrete antibodies, come up with a technique to fuse these cells with healthy antibody-producing B-cells, and be able to select the successfully fused cells (Figure 9.3).

In 1988, Greg Winter and his team pioneered the techniques to humanize MABs, removing the reactions that many MABs caused in some patients. The production of recombinant MABs involves technologies referred to as repertoire cloning or phage display/yeast display. Recombinant antibody engineering involves the use of viruses or yeast, instead of mice, to create antibodies. These techniques rely on rapid cloning of immunoglobulin gene segments to create libraries of antibodies with slightly different amino acid sequences, from which antibodies with desired specificities can be selected. These techniques can be used to enhance the specificity with which antibodies recognize antigens, their stability in various environmental conditions, their therapeutic efficacy, and their delectability in diagnostic applications. Fermentation chambers have been used to produce these antibodies on a large scale.

9.2.1 Applications of Monoclonal Antibodies

Although MABs are long-established as essential research tools, their therapeutic promise has taken considerably longer to realize, requiring further advances such as the humanization of mouse antibodies and recombinant production protocols. The diagnostic and therapeutic implications of MABs were immediately recognized. We will next discuss the various applications of MABs.

9.2.1.1 Diagnostic Test

MABs have been extensively used for diagnostic purposes, to identify a disease and its progression. Several antibody tests and kits have been developed for rapid detection of disease using the enzyme-linked immunosorbent assay (ELISA) technique. The western blot test or immuno-dot blot test detects the protein on a membrane. They are also very useful in immunohistochemistry, which detects antigens in fixed tissue sections, and the immunofluorescence test, which detects the substance in a frozen tissue section or in live cells.

9.2.1.2 Cancer Treatment

One possible treatment for cancer involves MABs that bind only to cancer-cell-specific antigens and induce an immunological response against the target cancer cells. Such MABs could also be modified for delivery of a toxin, radioisotope, cytokine, or other active conjugate. It is also possible to design bispecific antibodies that can bind with their fragment antigen-binding (Fab) regions, both to target antigens and to a conjugate or effector cell. In fact, every intact antibody can bind to cell receptors or other proteins with its fragment crytallazable (Fc) region. Antibody-directed enzyme pro-drug therapy (ADEPT) is a new type of cancer treatment that uses drugs called MABs. At the moment, ADEPT is being used only in clinical trials. The trials aim to find out whether ADEPT may be useful as a new type of treatment for bowel cancer. ADEPT is a type of targeted therapy. It uses a MAB to carry an enzyme directly to the cancer cells. Enzymes are proteins that control chemical reactions in the body. First, the MAB is given (with the enzyme attached). A few hours later, a second drug (the pro-drug) is given. When the pro-drug comes into contact with the enzyme, a reaction takes place. This reaction activates the pro-drug and it is then able to destroy the cancer cells. As the enzyme does not attach to normal cells, this treatment does not affect them.

9.2.2 Hybridoma Technology

We will now discuss the process of generating MABs using hybridoma technology. In hybridoma technology, MABs are typically made by fusing myeloma cells with the spleen cells from a mouse that has been immunized with the desired antigen. However, recent advances have allowed the use of rabbit B-cells. Polyethylene glycol (PEG) is used to fuse adjacent plasma membranes, but the success rate is low, so a selective medium is used in which only fused cells can grow. This is because myeloma cells have lost the ability to synthesize *hypoxanthine-guanine-phosphoribosyl transferase* (HGPRT), an enzyme necessary for the salvage synthesis of nucleic acids. The absence of HGPRT is not a problem for these cells unless the de novo purine synthesis pathway is also disrupted. By exposing cells to aminopterin, a folic acid analogue that inhibits dihydrofolate reductase (DHFR), they are unable to use the de novo pathway and become fully auxotrophic for nucleic acids, requiring supplementation to survive.

The selective culture medium is called *HAT medium*, because it contains hypoxanthine, aminopterin, and thymidine. This medium is selective for fused (hybridoma) cells. Unfused myeloma cells cannot grow because they lack HGPRT and thus cannot replicate their DNA. Unfused spleen cells cannot grow indefinitely because of their limited life span. Only fused hybrid cells, referred to as *hybridomas*, are able to grow indefinitely in the media, because the spleen cell partner supplies HGPRT and the myeloma partner has traits that make it immortal (as it is a cancer cell). This mixture of cells is then diluted, and clones are grown from single parent cells on microtiter wells.

The antibodies secreted by the different clones are then assayed for their ability to bind to the antigen (with a test such as ELISA or antigen microarray assay) or immuno-dot blot. The most productive and stable clone is then selected for future use. The hybridomas can be grown indefinitely in a suitable cell culture media. When the hybridoma cells are injected in mice (in the peritoneal cavity, the gut), they produce tumors containing an antibody-rich fluid called *ascites fluid*. The medium must be enriched during selection to further favor hybridoma growth. This can be achieved by the use of a layer of feeder fibrocyte cells or supplement medium such as briclone.

9.2.2.1 Purification of Monoclonal Antibodies

After obtaining either a media sample of cultured hybridomas or a sample of ascites fluid, the desired antibodies must be extracted. The contaminants in the cell culture sample would consist of media components such as growth factors, hormones, and transferrins. In contrast, the *in vivo* sample is likely to have host antibodies, proteases, nucleases, nucleic acids, and viruses. In both cases, other secretions by the hybridomas, such as cytokines, may be present. There may also be bacterial contamination, resulting in endotoxins being secreted by the bacteria. Depending on the complexity of the media required in cell culture, and thus the contaminants in question, one method (*in vivo* or *in vitro*) may be preferable to the other. The sample is first conditioned or prepared for purification. Cells, cell debris, lipids, and clotted material are first removed, typically by filtration with a 0.45 μm filter. These large particles can cause a phenomenon called *membrane fouling* in subsequent purification steps. Additionally, the concentration of the product in the sample may not be sufficient, especially in cases where the desired antibody is one produced by a low-secreting cell line. The sample is therefore condensed by ultrafiltration or dialysis.

Most of the charged impurities are usually anions such as nucleic acids and endotoxins. These are often separated by ion exchange chromatography. Either cation exchange chromatography is used at a low enough pH that the desired antibody binds to the column while anions flow through, or anion exchange chromatography is used at a high enough pH for the desired antibody to flow through the column while anions bind to it. Various proteins can also be separated out along with the anions, based on their isoelectric point (pI). For example, albumin has a pI of 4.8, which is significantly lower than that of most MABs, which have a pI of 6.1. In other words, at a given pH, the average charge of albumin molecules is likely to be more negative. Transferrin, on the other hand, has a pI of 5.9, so it cannot easily be separated out by this method. A difference of at least 1 pI is necessary for a good separation. Transferrin can instead be removed by size-exclusion chromatography. The advantage of this purification method is that it is one of the more reliable chromatography techniques. Since we are dealing with proteins, properties such as charge and affinity are not consistent and vary with pH as molecules are protonated and deprotonated, while size stays relatively constant. Nonetheless, this method of separation has drawbacks, such as low resolution, low capacity, and low elution times. A much quicker, single-step method of separation is Protein A/G affinity chromatography. The antibody selectively binds to Protein A/G, so a high level of purity (generally >80%) is obtained. However, this method may be problematic for antibodies that are easily damaged, as harsh conditions are generally used. A low pH can break the bonds to remove the antibody from the column. In addition to possibly affecting the product, low pH can cause Protein A/G itself to leak off the column and appear in the eluted sample. Gentle elution buffer systems that employ high salt concentrations are also available to avoid exposing sensitive antibodies to low pH. Cost is also an important consideration with this method, because immobilized Protein A/G is a more expensive resin.

To achieve maximum purity in a single step, affinity purification can be performed by using the antigen to provide exquisite specificity for the antibody. In this method, the antigen used to generate the antibody is covalently attached to an agarose support. If the antigen is a peptide, it is commonly synthesized with a terminal cysteine which allows selective attachment to a carrier protein, such as KLH, during development and to the support for purification. The antibody-containing media are then incubated with the immobilized antigen, either in batches or as the antibody is passed through a column, where it is selectively retained while impurities are removed. An elution with a low pH buffer, high salt elution buffer is then used to recover the purified antibody from the support. To further select for antibodies, the antibodies can be precipitated out using sodium sulfate or ammonium sulfate. Antibodies precipitate at low concentrations of the salt, while most other proteins precipitate at higher concentrations. The appropriate level of salt is added in order to achieve the best separation. Excess salt must then be removed by a desalting method such as dialysis. The final purity can be analyzed using a chromatogram. Any impurities will produce peaks, and the volume under the peak indicates the amount of the impurity. Alternatively, gel electrophoresis and capillary electrophoresis (CE) can be carried out. Impurities will produce bands of varying intensity, depending on how much of the impurity is present.

9.2.3 Recombinant Monoclonal Antibodies

The need to overcome the immunogenicity problem of rodent antibodies in clinical practice has resulted in a plethora of strategies to isolate human antibodies. If human antibodies are to be used, then it is necessary to understand the basis by which different isotypes interact with host effector systems and, if possible, improve on nature by engineering in desirable modifications. Recombinant antibody technology is a rapidly evolving field that enables the study and improvement of antibody properties by means of genetic engineering. Moreover, the functional expression of antibody fragments in *Escherichia coli* has formed the basis for antibody library generation and selection, a powerful method to produce human antibodies for therapy. Because *in vitro*-generated antibodies offer various advantages over traditionally produced MABs, such molecules are now increasingly used for standard immunological assays. The original humanization strategy described by Dr. Winter and his group exploited knowledge of the solved crystal structure to graft rodent complementarity determining regions (CDRs) into the defined human frameworks and judicious mutations in critical framework residues.

9.2.4 Constraints in Making Monoclonal Antibodies

There are constraints in producing MABs, which include use of animals to generate antibodies. Although murine antibodies are very similar to human antibodies, the human immune system recognizes mouse antibodies as foreign, rapidly removing them from circulation and causing systemic inflammatory effects. A solution to this problem would be to generate human antibodies directly from humans. However, this is not easy, primarily because it is generally not seen as ethical to challenge humans with antigens in order to produce antibodies. The ethics of doing the same to nonhumans is a matter of debate. Furthermore, it is not easy to generate human antibodies against human tissues. Various approaches using rDNA technology to overcome this problem have been tried since the late 1980s. In one approach, the DNA that encodes the binding portion of monoclonal mouse antibodies it is merged with human antibody-producing DNA. Mammalian cell cultures are then used to express this DNA and produce these half-mouse and half-human antibodies. Bacteria cannot be used for this purpose, since they cannot produce this kind of glycoprotein. Depending on how big a part of the mouse antibody is used, these may be referred to as chimeric antibodies or humanized antibodies. Another approach involves mice genetically engineered to produce more human-like antibodies. MABs have been generated and approved to treat cancer, cardiovascular disease, inflammatory diseases, macular degeneration, multiple sclerosis, and viral infection. In August 2006, the Pharmaceutical Research and Manufacturers of America (PhRMA) reported that U.S. companies had 160 different MABs in clinical trials or awaiting approval by the FDA.

9.3 THERAPEUTIC PROTEINS

Therapeutic proteins are proteins that are either extracted from human cells or engineered in the laboratory for pharmaceutical use. The majority of therapeutic proteins are recombinant human proteins manufactured using nonhuman mammalian cell lines that are engineered to express certain human genetic sequences to produce specific proteins. Recombinant proteins are an

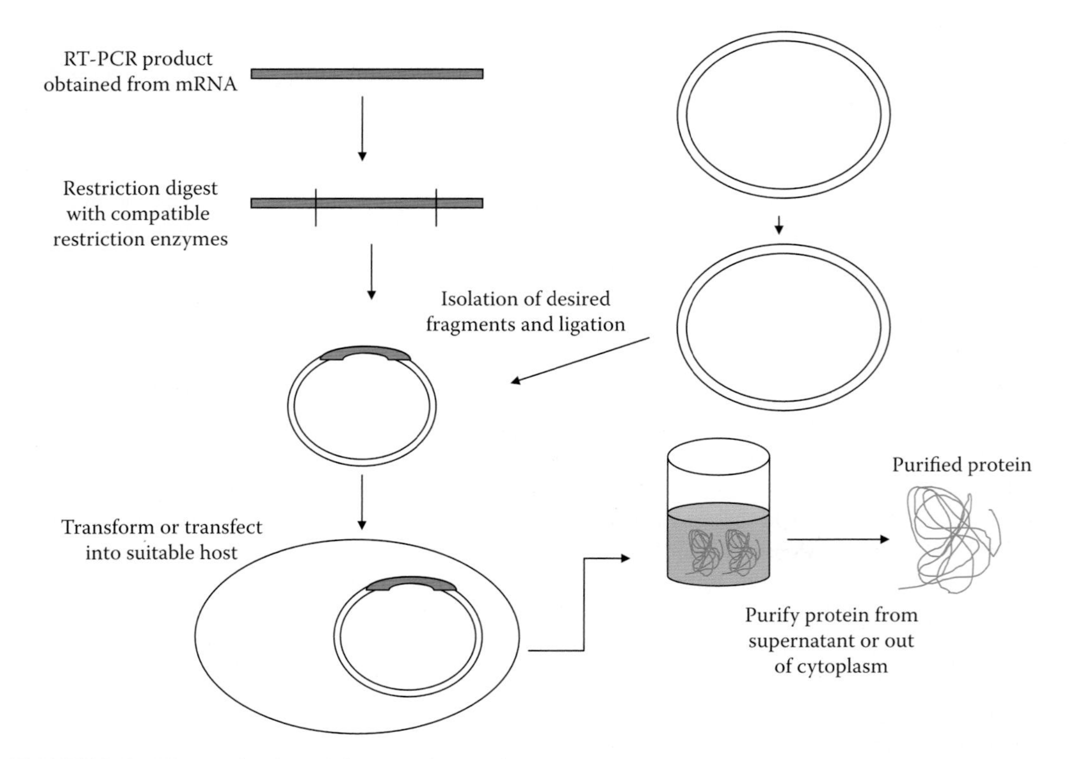

FIGURE 9.4 The production of therapeutic protein.

important class of therapeutics used to replace deficiencies in critical blood-borne growth factors and to strengthen the immune system to fight cancer and infectious disease. Therapeutic proteins are also used to relieve patients suffering from many conditions, including various cancers (treated by MABs and interferon), heart attacks, strokes, cystic fibrosis and Gaucher's disease (treated by enzymes and blood factors), diabetes (treated by insulin), anemia (treated by erythropoietins, called EPOs), and hemophilia (treated by blood clotting factors) (Figure 9.4).

The FDA has approved 75 therapeutic proteins, also known as biopharmaceuticals, and there are more than 500 additional proteins under development. Worldwide sales of therapeutic proteins were reported to be approximately $53 billion in 2005 and are expected to increase to more than $70 billion by 2008. To date, most of the growth has been in sales of EPO (used to treat anemia) and insulin (used to treat diabetes). Many of the proteins currently on the market will lose the protection of certain patent claims over the next 15 years. In addition, many marketed proteins are facing increased competition from next-generation versions or from other therapeutic proteins approved for the same disease indications. Because proteins are broken down in the gastrointestinal system, marketed therapeutic proteins must be administered by injection. Once in the bloodstream, therapeutic proteins are broken down by enzymes and cellular activity, as well as filtered out of the blood by the body's organs. Therefore, injections must be given frequently to achieve effective therapeutic levels. These therapeutic proteins are produced using rDNA technology, as already described in Chapter 4. Table 9.2 lists some available recombinant products.

9.3.1 Growth Factors

Growth factors belong to the group of proteins called cytokines. They can alter cell production, organogenesis, and disease susceptibility in animals. The effect produced by a growth factor will mainly depend on the presence of other growth factors, the target cells, and their receptors

TABLE 9.2
Some Available Recombinant Proteins

Product (Protein)	Gene Expressed in (Host)	Application
Human growth hormone (HGH)	Mouse mammary tumor cell line	Dwarfism
Tissue plasminogen activator (TPA)	Chinese hamster ovary (CHO) cell line	Thrombolysis
Erythropoietin	(CHO) cell line	Anemia
Blood clotting factor VIII	(CHO) cell line	Hemophilia
In Advanced Stages of Development		
Antithrombin VIII	(CHO) cell line	Anticoagulant
Blood factor VIII	BHK-21 cell line	Hemophilia
Blood factor IX	CHO cell line	Hemophilia-B
Interleukin-2 (11–2)	—	Cancer therapy
Human recombinant antibodies	CHO cell line and mouse myeloma lines	Infections, various cancers
Hepatitis-B surface antigen (HBSAg)	CHO cell line and C-127 cell line	As vaccine for hepatitis-B
Interferon (IFN)	—	Tumors
Plasminogen activator (including chimeric and modified)	CHO cell line and *Spodoptera frugiperda*	Thrombolysis
CD4 protein	Mammalian cell line	AIDS vaccine
gp 160 protein	Mammalian cell line	AIDS vaccine
p24 protein	Mammalian cell line	AIDS vaccine
Interleukin-l receptor antagonist	Mammalian cell line	Severe sepsis
Atrial natriuretic	Mammalian cell line	Kidney failure
Ciliary neurotropic factor	Mammalian cell line	Amyotrophic sclerosis
DNase	Mammalian cell line	Cystic fibrosis

on target cells. The nomenclature of cytokines is confusing, and terms like interleukins, growth factors, and colony-stimulating factors (CSFs) are often used for a single protein. The term interleukin is the preferred one, and new leukocyte products are designated by this name followed by a number, such as interleukin-13. The various growth factors can be grouped into the following families:

- Insulin-like growth factors (IGF, e.g., IGF-I and IGF-II),
- Nerve growth factors (NGF)
- Epidermal growth factors (EGF)
- Transforming growth factor β (TGF-β)
- Platelet-derived growth factors (PDGF)
- Fibroblast growth factors (FGFs)
- Hepatocyte growth factors (HGF)
- Hemopoietic growth factors (at least 16 cytokines; affect production and function of blood cells)

Many of the growth factors have been approved for treatment of human diseases. EPO is used on a considerable scale for the treatment of anemia. EPO is also used to stimulate red blood cell production in kidney dialysis or cancer patients. Interleukin-2, in conjunction with lymphocyte-activated killer (LAK) cells, is being used for cancer therapy. Similarly, granulocyte–macrophage CSF and granulocyte CSF (G-CSF) are used to accelerate neutrophil recovery after chemotherapy or bone marrow transplantation. Several other growth factors have been/are likely to be approved for similar and other applications.

9.4 STEM CELL TRANSPLANTATION

Stem cells are distinguished from other cell types by two important characteristics. First, they are unspecialized cells capable of renewing themselves through cell division, sometimes after long periods of inactivity. Second, under certain physiologic or experimental conditions, they can be induced to become tissue- or organ-specific cells with special functions. In some organs, such as the gut and bone marrow, stem cells regularly divide to repair and replace worn-out or damaged tissues. In other organs, however, such as the pancreas and the heart, stem cells only divide under special conditions. Stem cells can be divided into two major types: adult stem cells and embryonic stem (ES) cells.

9.4.1 Adult Stem Cells

Adult stem cells are undifferentiated cells, found throughout the body, that multiply by cell division to replenish dying cells and regenerate damaged tissues. Also known as somatic stem cells, they can be found in juvenile as well as adult animals and humans. Scientific interest in adult stem cells has centered on their ability to divide or self-renew indefinitely and generate all the cell types of the organ from which they originate, potentially regenerating the entire organ from a few cells. Unlike ES cells, the use of adult stem cells in research and therapy is not considered to be controversial as they are derived from adult tissue samples rather than destroyed human embryos. They have mainly been studied in humans and model organisms such as mice and rats. Adult stem cell research has been focused on uncovering the general molecular mechanisms that control their self-renewal and differentiation. The transcriptional repressor Bmi-1 is one of the Polycomb-group proteins that was discovered as a common oncogene activated in lymphoma and later shown to specifically regulate hematopoietic stem cells (HSCs). The role of Bmi-1 has also been illustrated in neural stem cells. The Notch pathway has been known to developmental biologists for decades. Its role in control of stem cell proliferation has now been demonstrated for several cell types, including hematopoietic, neural, and mammary stem cells. The Wnt developmental pathways are also strongly implicated as stem cell regulators. Under special conditions, tissue-specific adult stem cells can generate a whole spectrum of cell types of other tissues, even crossing germ layers. This phenomenon is referred to as *stem cell transdifferentiation* or *plasticity*. It can be induced by modifying the growth medium when stem cells are cultured *in vitro* or by transplanting them to an organ of the body different from the one they were originally isolated from. There is yet no consensus among biologists on the prevalence and physiological and therapeutic relevance of stem cell plasticity. We can classify adult stem cells into various types, based on the source from which they are isolated (Figure 9.5).

9.4.1.1 Dental Pulp–Derived Stem Cells

Stem cells that have been successfully recovered from dental pulp in the perivascular niche are known as stem cells harvested from exfoliated deciduous teeth or SHED cells. These cells have been shown to have the same cellular markers and differential abilities as *mesenchymal stem cells* (MSCs). As deciduous baby teeth are shed naturally, this is a noninvasive and painless way to harvest stem cells for either storage or future medical use.

9.4.1.2 Hematopoietic Stem Cells

HSCs are multipotent stem cells that give rise to all the blood cell types, including myeloid (monocytes and macrophages, neutrophils, basophils, eosinophils, erythrocytes, megakaryocytes/platelets, dendritic cells) and lymphoid lineages (T-cells, B-cells, NK-cells). The definition of HSCs has undergone considerable revision in the last two decades. The hematopoietic tissue contains cells with long-term and short-term regeneration capacities and committed multipotent, oligopotent, and unipotent progenitors. Recently, long-term transplantation experiments point toward a clonal diversity model of HSCs. Here, the HSC compartment consists of a fixed number of different

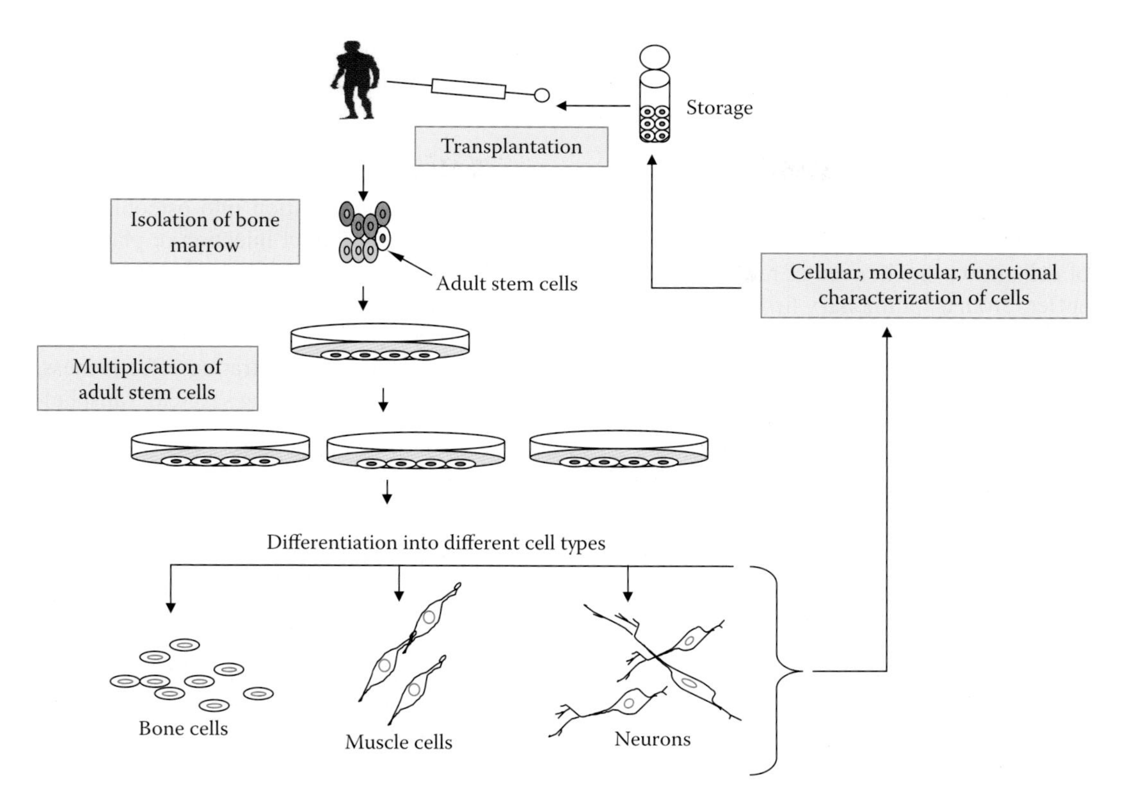

FIGURE 9.5 Adult stem cell transplantation: Bone marrow stem cells are isolated from an adult and are normally cultured to get an enriched population of stem cells. These isolated stem cells can be differentiated into neurons, muscle cells, or bone cells and can be transplanted into patients suffering from degenerative diseases.

types of HSCs, each with epigenetically preprogrammed behavior. This contradicts older models of HSC behavior, which postulated a single type of HSC that can be continuously molded into different subtypes of HSCs. HSCs constitute 1:10.000 of cells in myeloid tissue. HSCs are found in the bone marrow of adults, which includes femurs, hip, ribs, sternum, and other bones. Cells can be obtained directly by removal from the hip using a needle and syringe, or from the blood following pretreatment with cytokines such as G-CSFs, which induce cells to be released from the bone marrow compartment. Other sources for clinical and scientific use include umbilical cord blood, placenta, and mobilized peripheral blood. For experimental purposes, fetal liver, fetal spleen, and aorta-gonad-mesonephros (AGM) of animals are also useful sources of HSCs.

9.4.1.3 Mammary Stem Cells

Mammary stem cells provide the source of cells for growth of the mammary gland during puberty and gestation, and play an important role in carcinogenesis of the breast. Mammary stem cells have been isolated from human and mouse tissue, as well as from cell lines derived from the mammary gland. Single mammary stem cells can give rise to both the luminal and myoepithelial cell types of the gland, and have been shown to have the ability to regenerate the entire organ in mice.

9.4.1.4 Mesenchymal Stem Cells

MSCs are multipotent stem cells that can differentiate into a variety of cell types. Cell types that MSCs have been shown to differentiate into *in vitro* or *in vivo* include osteoblasts, chondrocytes, myocytes, adipocytes, endotheliums, and, as described lately, beta-pancreatic islets cells. In 1924, Russian-born morphologist Alexander A. Maximow used extensive histological findings to identify a singular type of precursor cell within mesenchyme which develops into different types of blood cells.

Scientists Ernest A. McCulloch and James E. Till first revealed the clonal nature of marrow cells in the 1960s. An *ex vivo* assay for examining the clonogenic potential of multipotent marrow cells was later reported in the 1970s by Friedenstein and colleagues. In this assay system, stromal cells were referred to as *colony-forming unit –fibroblasts*.

Culturing marrow stromal cells in the presence of osteogenic stimuli such as ascorbic acid, inorganic phosphate, and dexamethasone could promote their differentiation into osteoblasts. In contrast, the addition of TGF-β could induce chondrogenic markers. Direct injection or placement of cells into a site in need of repair may be the preferred method of treatment, as vascular delivery suffers from a "pulmonary first-pass effect," where intravenous injected cells are sequestered in the lungs. Clinical case reports in orthopedic applications have been published, though the number of patients treated is small and these methods still lack rigorous study demonstrating effectiveness. Wakitani has published a small case series of nine defects in five knees involving surgical transplantation of MSCs, with coverage of the treated chondral defects.

9.4.1.5 Neural Stem Cells

The existence of stem cells in the adult brain has been postulated following the discovery that the process of neurogenesis (the birth of new neurons) continues into adulthood in rats. It has since been shown that new neurons are generated in adult mice, songbirds, and primates, including humans. Normally, adult neurogenesis is restricted to two areas of the brain: the subventricular zone, which lines the lateral ventricles, and the dentate gyrus of the hippocampal formation. Although the generation of new neurons in the hippocampus is well established, the presence of true self-renewing stem cells there has been debated. Under certain circumstances, such as following tissue damage due to ischemia, neurogenesis can be induced in other brain regions, including the neocortex. Neural stem cells are commonly cultured *in vitro* as so-called *neurospheres*, floating heterogeneous aggregates of cells, containing a large proportion of stem cells. They can be propagated for extended periods of time and differentiated into both neuronal and glia cells, and therefore behave like stem cells. However, some recent studies suggest that this behavior is induced by the culture conditions in *progenitor cells*, the progeny of stem cell division that normally undergo a strictly limited number of replication cycles *in vivo*. Furthermore, neurosphere-derived cells do not behave like stem cells when transplanted back into the brain. Neural stem cells share many properties with HSCs. Remarkably, when injected into the blood, neurosphere-derived cells differentiate into various cell types of the immune system. Cells that resemble neural stem cells have been found in the bone marrow, the home of HSCs. It has been suggested that new neurons in the dentate gyrus arise from circulating HSCs. Indeed, newborn cells first appear in the dentate, in the heavily vascularized subgranular zone immediately adjacent to blood vessels.

9.4.1.6 Olfactory Adult Stem Cells

Stem cells have been successfully harvested from human olfactory mucosa cells, which are found in the lining of the nose and are involved in the sense of smell. If they are given the right chemical environment, these cells have the same ability as ES cells to develop into many different cell types. Olfactory stem cells hold the potential for therapeutic applications and, in contrast to neural stem cells, can be harvested with ease without harm to the patient. This means they can be easily obtained from all individuals, including older patients who might be most in need of stem cell therapies.

9.4.1.7 Clinical Applications of Stem Cells

Adult stem cell treatments have been used for many years to successfully treat leukemia and related bone/blood cancers utilizing bone marrow transplants. The use of adult stem cells in research and therapy is not considered as controversial as the use of ES cells, because the production of adult stem cells does not require the destruction of an embryo. Consequently, the majority of U.S. government funding for research in this field is restricted to supporting adult stem cell research. In the beginning, the regenerative applications of adult stem cells focused on intravenous delivery of HSCs. Other early commercial applications have focused on MSCs, and in both HSCs and MSCs the cells are either injected through at the site of repair or through blood capillaries. However, the injection or placement of cells into a site in need of repair is found to be the preferred method of treatment, as vascular delivery suffers from a "pulmonary first-pass effect" where intravenous injected cells are sequestered in the lungs.

In a fetus, the blood cells are created in the liver, but soon after birth they are produced exclusively in bone marrow. These stem cells can be isolated, stored, and are also cultured (rarely). If marrow is transplanted to the same individual from whom it was taken for culture, the procedure is described as *autotransplantation*; if donated to someone else, the procedure is called *allogeneic transplantation*. In 1 h, HSCs produce approximately 3–10 billion platelets, red cells, neutrophils, lymphocytes, and even macrophages of the immune system. These cells are used for transplantation in patients whose immune and blood-forming systems have been devastated by leukemia, cancer, chemotherapy, or other unknown causes. Isolation of stem cells from the mass of cells in the bone marrow has been achieved through the use of MABs that react to a cell marker or antigen called *CD34*, which is specific for primitive bone marrow stem cells.

However, all CD34-positive cells may not be the pluripotent stem cells required for bone marrow transplantation; more specialized techniques may therefore be used for further screening of CD34-positive cells. Culture systems are also available, which allow multiplication of these cells. These cultures may also be used in future for transplant therapy. Several hormones have also been identified, which stimulate replication of precursor blood cells. These include G-CSF, interleukin-6, interleukin-11, and a stem cell factor.

Bone marrow donors are classified by a method called *HLA typing*, which is used for matching the recipient. However, for autologous transplantation, a patient's own stem cells are collected, frozen, and stored outside the body, while chemotherapy or radiation therapy is administered to remove the malignant cells. The bone marrow thus stored is returned directly into a vein. However, bone marrow may be sometimes invaded by cover cells, so that healthy cells will have to be separated from malignant stem cells. MABs linked with toxins can be used for killing the malignant cells.

Chemotherapy with extracted bone marrow can also be used to destroy malignant stem cells. It has also been shown that while the malignant cells have an advantage over normal cells for growth within the body, they have a relative disadvantage outside the body. As a result, long-term culture systems have been used to remove cancerous cells from bone marrow. In general, 200×10^6 bone marrow cells are needed for 15 kg of body weight. This requires multiple extractions involving 500–1000 mL of bone marrow for one patient (given that one extraction gives 10–15 mL of bone marrow). Administering hormones like CSF and G-CSF helps stimulate blood production. Adding stem cells from peripheral blood to bone marrow may further help engraftment. In future, it may be possible for a healthy person to store his own stem cells permanently and use them whenever required.

9.4.1.8 Cancer Treatment

Research has shown that injecting neural adult stem cells into the brains of dogs has been very successful in treating cancerous tumors. With traditional techniques, brain cancer is almost impossible to treat, because it spreads so rapidly. Researchers at the Harvard Medical School induced intracranial tumors in rodents, and then injected human neural stem cells. Within days, the cells had migrated into the cancerous area and produced *cytosine deaminase*, an enzyme that converts a non-toxic pro-drug into a chemotherapeutic agent. As a result, the injected substance was able to reduce tumor mass by 81%. The stem cells neither differentiated nor turned tumorigenic. Some researchers believe that the key to finding a cure for cancer is to inhibit cancer stem cells, where the cancer tumor originates. Currently, cancer treatments are designed to kill all cancer cells, but through this method, researchers would be able to develop drugs to specifically target these stem cells.

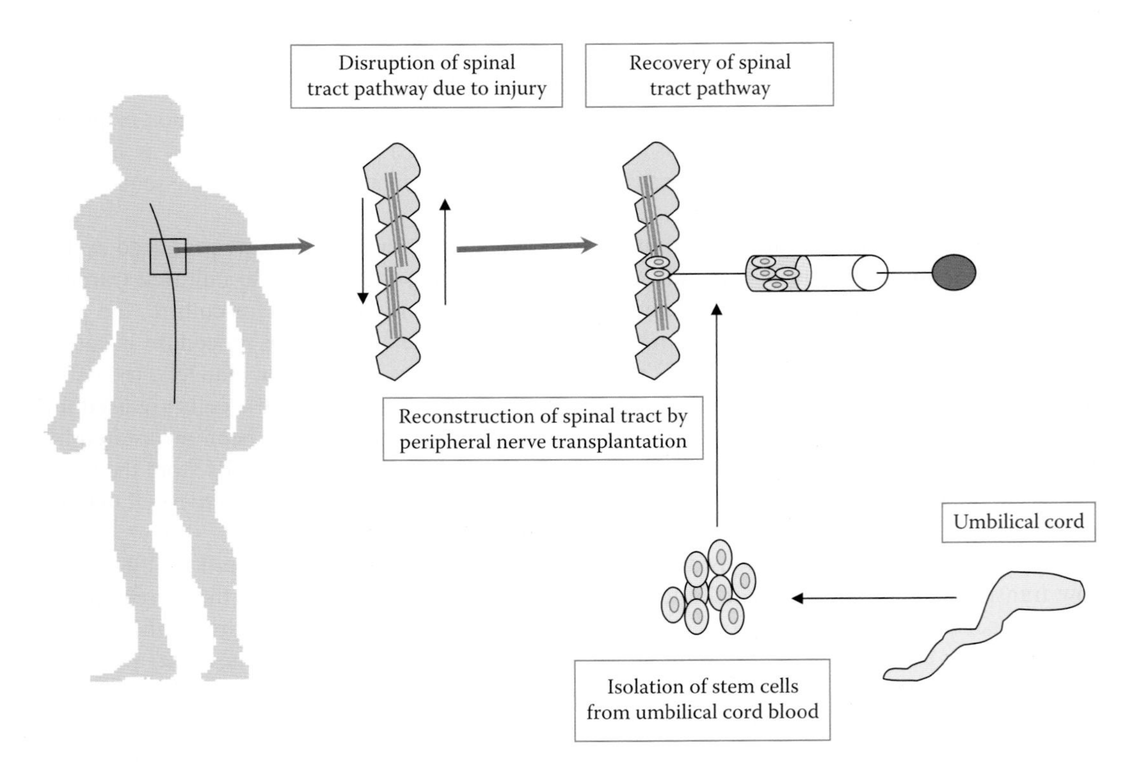

FIGURE 9.6 Repair of spinal cord injury by stem cell therapy.

9.4.1.9 Spinal Cord Injury

On November 25, 2004, a team of Korean researchers reported that they had transplanted multipotent adult stem cells from umbilical cord blood to a patient suffering from a spinal cord injury and that she could now walk on her own, without difficulty. The patient had not been able to stand up for approximately 19 years. For the unprecedented clinical test, the scientists isolated adult stem cells from umbilical cord blood and then injected them into the damaged part of the spinal cord (Figure 9.6).

9.4.1.10 Corneal Repair

In April 2005, a group of doctors in the United Kingdom transplanted corneal stem cells from an organ donor to the cornea of Deborah Catlyn, a woman who was blinded in one eye when acid was thrown in her eye at a nightclub. The cornea, which is the transparent window of the eye, is a particularly suitable site for transplants. In fact, the first successful human transplant was a cornea transplant. The cornea has a remarkable property in that it does not contain any blood vessels, making it relatively easy to transplant. The majority of corneal transplants carried out today are due to a degenerative disease called keratoconus. The University Hospital of New Jersey claims that the success rate when growing the new cells from transplanted stem cells varies from 25% to 70%. In 2009, researchers at the University of Pittsburgh Medical Center demonstrated that stem cells collected from human corneas can restore transparency without provoking a rejection response in mice with corneal damage.

9.4.2 Embryonic Stem Cells

ES cells are stem cells derived from the inner cell mass of an early-stage embryo, known as a *blastocyst*. Human embryos reach the blastocyst stage 4–5 days post fertilization, at which time they consist of 50–150 cells which are known as *inner cell mass*. One unique property of ES cells

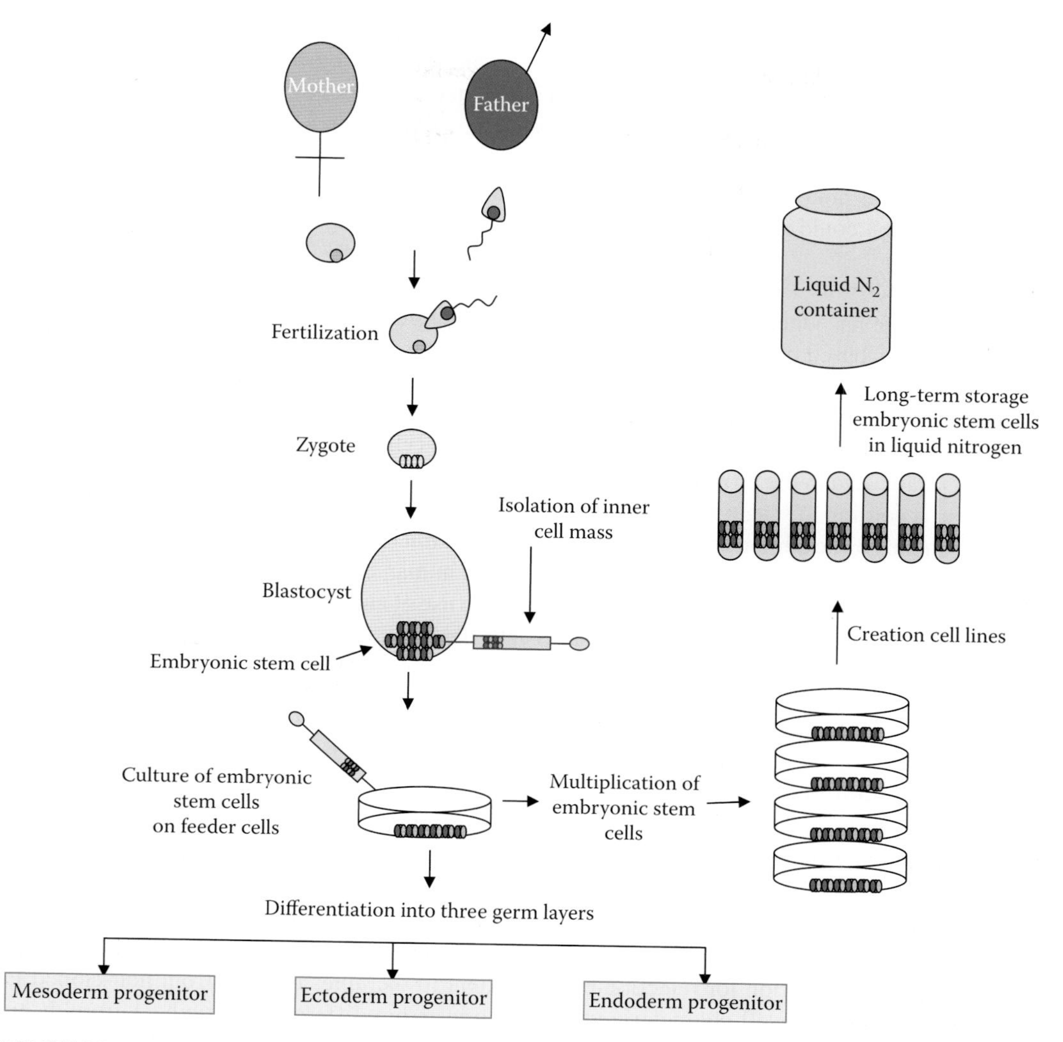

FIGURE 9.7 Generation of human ES cells.

is *pluripotentcy*. This means that the cells are able to differentiate into all derivatives of the three primary germ layers: ectoderm, endoderm, and mesoderm. These include each of the more than 220 cell types in the adult body. Pluripotency distinguishes ES cells from multipotent progenitor cells found in the adult. These only form a limited number of cell types. Under sterile and controlled *in vitro* culture conditions, ES cells can maintain pluripotency for many cell divisions. The presence of pluripotent adult stem cells remains a subject of scientific debate, although research has demonstrated that pluripotent stem cells may have therapeutic potential. Because of their plasticity and potentially unlimited capacity for self-renewal, ES cell therapies have been proposed for regenerative medicine and tissue replacement for a number of blood- and immune-system-related genetic diseases, cancers, and disorders; juvenile diabetes; Parkinson's disease; blindness; and spinal cord injuries (Figure 9.7).

9.4.2.1 How Were Embryonic Stem Cells Discovered?

The first report of stem cell identification was in 1964, when researchers had isolated a single type of cell from a *teratocarcinoma*, a form of cancer that replicated and grew in cell culture as a stem cell. Subsequently, researchers isolated a primordial embryonic germ cell (EG cell) that, after replicating and growing in cell culture as a stem cell, was capable of developing into different cell types. In 1981, ES cells were first derived from mouse embryos by Martin Evans and Matthew Kaufman, and independently by Gail R. Martin is credited with coining the term "embryonic stem cell." In 1998, a breakthrough occurred when researchers, led by James Thomson at the University of Wisconsin–Madison, first developed a technique to isolate and grow ES cells in the laboratory. After that, several mouse and human ES cell lines were isolated, cultured, and differentiated and characterized, but could not be tested in humans because of cell line contamination.

9.4.2.2 Cell Line Contamination

The research that has jolted cell transplant experts is contamination of ES cells with animal components, which was based on a research finding published in the online edition of *Nature Medicine* on January 24, 2005, which stated that the human ES cells available for federally funded research were contaminated with mouse feeder cells. It is a common technique to use mouse cells and other animal cells to maintain the pluripotency of actively dividing stem cells. The problem was discovered when nonhuman sialic acid in the growth media was found to compromise the potential uses of the ES cells in humans, according to scientists at the University of California, San Diego. On other hand, a study published in the online edition of the medical journal *The Lancet* on March 8, 2005, detailed information about a new stem cell line which was derived from human embryos under completely cell- and serum-free conditions. This means that there has been no use of animal tissue or media while growing the ES cells. This study further suggested that ES cells demonstrated the potential to form derivatives of all three EG layers, both *in vitro* and in teratomas. These properties were also successfully maintained for more than 30 passages with the established stem cell lines. These cell lines first need to be tested rigorously on animal models before use in humans.

9.4.2.3 Immunorejection

Another issue with ES cell-based therapy is immunorejection, and there is a high possibility that the host may reject the transplanted cells in humans. There is also ongoing research to reduce the potential for rejection of the differentiated cells derived from ES cells once researchers are capable of creating an approved therapy from ES cell research. One of the possibilities to prevent rejection is by creating ES cells that are genetically identical to the patient via therapeutic cloning. An alternative solution for rejection by the patient to therapies derived from non-cloned ES cells is to derive many well-characterized ES cell lines from different genetic backgrounds and use the cell line that is most similar to the patient. Treatment can then be tailored to the patient, minimizing the risk of rejection.

9.4.2.4 Alternative Approach to Creating Embryonic Stem Cells

In view of the ethical issues that have severely affected the ES cell research, attempts were made to generate ES cells by alternative methods. We will go through some examples of the recent breakthroughs in generating ES cells without killing the embryos. On August 23, 2006, the online edition of the scientific journal *Nature* published a report on generating ES cells without destroying the embryo. This technical achievement would potentially enable scientists to work with new lines of ES cells derived using public funding in the United States, where federal funding was, at the time, limited to research using ES cell lines derived prior to August 2001. In March 2009, the limitation was lifted. Professor Shinya Yamanaka had a recent breakthrough in which the skin cells of laboratory mice were genetically manipulated back to their embryonic state. This work was confirmed by two other groups, demonstrating that the addition of just four genes (Oct3/4, Sox2, Klf4, and c-Myc) could reprogram mouse skin cells into ES-like cells. These cells produced by Yamanaka, as well as by other laboratories, demonstrated all the hallmarks of ES cells, including the ability to form chimeric mice and contribute to the germ line. One issue with this work is that the mice generated from these ES lines were prone to develop cancer due to the use of Myc, which is a known oncogene.

On November 20, 2007, two research teams, one of which was headed by Professor Yamanaka and the other by James Thomson, announced a similar breakthrough with ordinary human skin cells, which were transformed into batches of cells that look and act like ES cells. This may enable the generation of patient-specific ES cell lines that could potentially be used for cell replacement therapies. In addition, this will allow the generation of ES cell lines from patients with a variety of genetic diseases and will provide invaluable models to study those diseases. There is still much work to be done before this technology can be used for the treatment of disease. First, the genes used to reprogram the skin cells into ES-like cells were added by the use of retroviruses, which can cause mutations and lead to the risk of possible cancers, although recent research by professor Yamanaka's research group has made advances in avoiding this particular problem. In addition, as shown with the mouse work, one of the genes used to reprogram, Myc, can also cause cancer. The group led by Thomson did not use Myc to reprogram and may not have this difficulty.

Future work is aimed at attempting to reprogram without permanent genetic manipulation of the cells with viruses. This could be accomplished by either small molecules or other methodologies to express these reprogramming genes. However, as a first indication that the induced pluripotent stem (iPS) cell technology can lead to new cures in rapid succession, it was used by a research team headed by Rudolf Jaenisch of the Whitehead Institute for Biomedical Research in Cambridge, Massachusetts, to cure mice of sickle cell anemia, as reported by *Science* journal's online edition on December 6, 2007. On January 16, 2008, a California-based company, Stemagen, announced that they had created the first mature cloned human embryos from single skin cells taken from adults. These embryos can be harvested for patient-matching ES cells.

9.4.2.5 Embryonic Stem Cells as Models for Human Genetic Disorders

In recent years, there have been several reports regarding the potential use of human ES cells as models for human genetic diseases. This issue is especially important due to the species-specific nature of many genetic disorders. The relative inaccessibility of human primary tissue for research is another major hindrance. Several new studies have started to address this issue. This has been done either by genetically manipulating the cells or, more recently, by deriving diseased cell lines identified by prenatal genetic diagnosis (PGD). This approach may very well prove invaluable for studying disorders such as Fragile-X syndrome, cystic fibrosis, and other genetic maladies that have no reliable model system. Yury Verlinsky, a Russian-American medical researcher who specialized in embryo and genetic cytology, developed prenatal diagnosis testing methods to determine genetic and chromosomal disorders a month and an half earlier than standard amniocentesis. The techniques are now being used by many pregnant women and prospective parents, especially those couples with a history of genetic abnormalities. In addition, by allowing parents to select an embryo without genetic disorders, they have the potential of saving the lives of siblings that already have similar disorders and diseases using cells from the disease-free offspring.

9.4.2.6 First Clinical Trial

The U.S. FDA approved the first clinical trial using ES cells in humans in 2009, and Geron Corporation is conducting the trial. The study involves the injection of neural stem cells into paraplegic mice afflicted with spinal cord injury. However, the researchers emphasize that the injections are not expected to fully cure the patients and restore all mobility. Based on the results of the mice trials, the researchers state that restoration of myelin sheaths and an increase in mobility is probable. This first trial is mainly testing the safety of these procedures, and if everything goes well, it could lead to future studies that involve people with more severe disabilities.

9.5 BIOENGINEERED SKIN

Skin is perhaps the only organ that can be artificially made from cell culture and used for grafting when skin is severely damaged due to severe burns. Of the few cell types that can be cultured are keratinocytes, which comprise 90% of the epidermis of skin. They are responsible for giving rise to dead cells making the external cornified layers of skin, and their proliferation is facilitated by the products of fibroblasts found in the dermis layer of the skin. Since fibroblasts are useful for culturing keratinocytes, fibroblast cells called 3T3 cells were used (after irradiating with lethal dose to prevent their own growth) to cover the bottom of vessel, before adding epidermal cells for culturing them. Other substances added in culture medium included epidermal growth factors, cholera toxin, and a mixture of other growth factors. Only 1%–10% epidermal cells proliferate, others having already started the process of differentiation. These cells form colonies, are separated again, and transferred to fresh culture to allow better growth. The process of separating cells from colonies and reculturing them is continued to discourage stratification of cell layers and allow the cell colonies to become confluent, forming a sheet of pure epithelium. The cells of this sheet of epithelium are linked by desmosomes. This cultured epithelium can be detached from the vessel using the enzyme dispase, washed free of extraneous protein, attached to a backing of gauze, and brought to the hospital to be used for grafting on patients having severe burns.

A meticulous preparation of wounds is required since they are often contaminated, and complete elimination of microorganisms is essential. It is important that cultured kerationcytes used to generate epidermis must come from the unburned portion of the skin of the patient; otherwise, these will soon be rejected. Starting from 3 cm^2 of the patient's skin, it can be expanded 5000-fold in 3–4 weeks to supply 1.7 m^2 of skin needed for an adult human. The entire body may need 350 grafts, each of 25 cm^2, to cover only the front or back surface. Before the cultured epithelium became available, split thickness grafts were used, which involved transfer of 0.3 mm thick skin (epidermis plus part of the dermis) from one part of the body to the other. Such grafting leads to quick recovery and normalization of skin. However, in grafting of cultured skin, if samples of regenerated skin are taken over a period of 5 years, the different elements of normal skin return at different rates, but eventually normal skin, with all its essential components, is regenerated. Although more than 500 patients throughout the world have already received cultured keratinocytes for treatment of burns, ulcers, or other conditions, this is a very small number in comparison to the number of patients who need such help.

9.6 BIOENGINEERED ORGAN TRANSPLANTATION

In 2008, the first full transplant of a human organ grown from adult stem cells was carried out by Paolo Macchiarini at the Hospital Clínic of Barcelona on Claudia Castillo, a Columbian female adult whose trachea had collapsed due to tuberculosis. Researchers from the University of Padua, the University of Bristol, and Politecnico di Milano harvested a section of trachea from a donor and stripped off the cells that could cause an immune reaction, leaving a grey trunk of cartilage. This section of trachea was then "seeded" with stem cells taken from Ms. Castillo's bone marrow, and a new section of trachea was grown in the laboratory over 4 days. The new section of trachea was then transplanted into the left main bronchus of the patient. Because the stem cells were harvested from the patient's own bone marrow, Professor Macchiarini did not think it was necessary for her to be given any immunosuppressive medication, and when the procedure was reported 4 months later in *The Lancet*, the patient's immune system was showing no signs of rejecting the transplant.

9.7 GENE THERAPY

A potential approach to the treatment of genetic disorders in humans is gene therapy. This is a technique whereby the absent or faulty gene is replaced by a working gene, so that the body can make the correct enzyme or protein and consequently eliminate the root cause of the disease. The most likely candidates for gene therapy trials are rare diseases such as Lesch–Nyhan syndrome, a distressing disease in which patients are unable to manufacture a particular enzyme. If gene therapy does become practicable, the biggest impact would be the treatment of diseases where the normal gene needs to be introduced into one organ. One such disease is phenylketonuria (PKU), which affects approximately one in 12,000 white children and, if not treated early, can result in severe mental retardation. The disease is caused by a defect in a gene producing a liver enzyme. If detected early enough, the child can be placed on a special diet for its first few years, but this is very unpleasant and can lead to many problems within the family.

The types of gene therapy described so far all have one factor in common, which is that the tissues being treated are somatic. In contrast to this is the replacement of defective genes in the germline cells that contribute to the genetic heritage of the offspring. Gene therapy in germline cells has the potential to affect not only the individual being treated, but also his or her children. Germline therapy would change the genetic pool of the entire human species, and future generations would have to live with that change. In addition to these ethical problems, a number of technical difficulties make it unlikely that germline therapy would be tried on humans in the near future.

Before treatment for a genetic disease can begin, an accurate diagnosis of the genetic defect needs to be made. It is here that biotechnology is also likely to have a great impact in the near future. Genetic engineering research has produced a powerful tool to undercover specific diseases rapidly and accurately. Short pieces of DNA called *DNA probes* can be designed to stick very specifically to certain other pieces of DNA. The technique relies upon the fact that complementary pieces of DNA stick together. DNA probes are more specific and have the potential to be more sensitive than conventional diagnostic methods, and it should be possible in the near future to distinguish between defective genes and their normal counterparts, which is an important development.

Somatic cells are nonreproductive. Somatic cell therapy is viewed as a more conservative and safer approach because it affects only the targeted cells in the patient and is not passed on to future generations. In other words, the therapeutic effect ends with the individual who receives the therapy. However, this type of therapy presents unique problems of its own. Often the effects of somatic cell therapy are short-lived. Because the cells of most tissues ultimately die and are replaced by new cells, repeated treatments over the course of the individual's life span are required to maintain the therapeutic effect. Transporting the gene to the target cells or tissue is also problematic. Regardless of these difficulties, however, somatic cell gene therapy is appropriate and acceptable for many disorders, including cystic fibrosis, muscular dystrophy, cancer, and certain infectious diseases. Clinicians can even perform this therapy in utero, potentially correcting or treating a life-threatening disorder that may significantly impair a baby's health or development if not treated before birth. In brief, the distinction is that the results of any somatic gene therapy are restricted to the actual patient and are not passed on to his or her children. To date, all gene therapy on humans has been directed at somatic cells, while germline engineering in humans remains controversial and prohibited (for instance, in the European Union). There are broad categories of somatic gene therapy, *ex vivo* gene therapy, and *in vivo* gene therapy.

9.7.1 *Ex Vivo* Gene Therapy

In *ex vivo* gene therapy, cells are modified outside the body and then transplanted back in again. In this process, cells from the patient's blood or bone marrow are removed and grown in the laboratory. The cells are then exposed to the virus that is carrying the desired gene. The virus enters the cells and inserts the desired gene into the cells' DNA. The cells grow in the laboratory and are transplanted back to the patient. *Ex vivo* gene therapy can be classified into two types: *ex vivo* gene therapy using viral vectors and *ex vivo* gene therapy using nonviral vectors (Figure 9.8).

9.7.1.1 Gene Therapy Using Viral Vectors

Efforts have also been made to repair genes by using viral vectors. Viruses are obligate intracellular parasites, designed through the course of evolution to infect cells, often with great specificity to a particular cell type. They tend to be very efficient at transfecting their own DNA into the host cell, which is expressed to produce new viral particles. By replacing genes that are needed for the replication phase of their life cycle (the nonessential genes) with foreign genes of interest, the recombinant

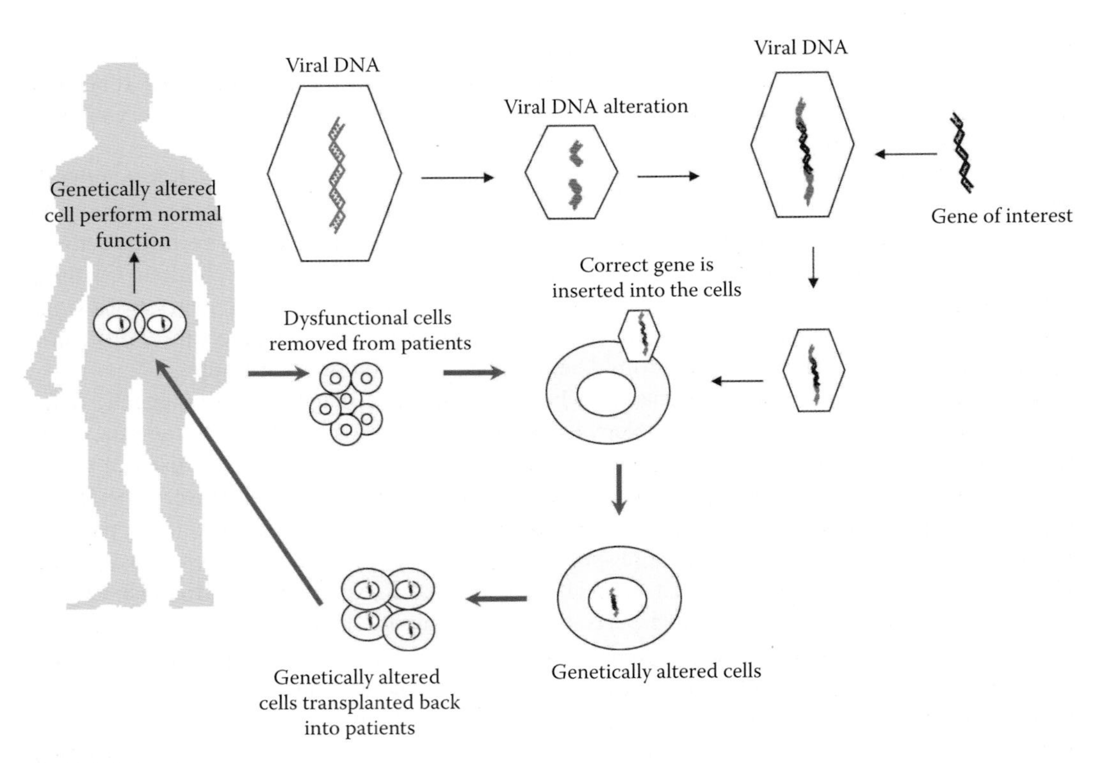

FIGURE 9.8 *Ex vivo* gene therapy.

viral vector can transduce the cell type it would normally infect. To produce such recombinant viral vectors, the nonessential genes are provided in *trans*, either integrated into the genome of the packaging cell line or on a plasmid. As viruses have evolved as parasites, they all elicit a host immune system response to some extent. Though a number of viruses have been developed, interest has centered on four types: retroviruses (including lentiviruses), adenoviruses, adeno-associated viruses (AAVs), and herpes simplex virus type 1 (HSV-1).

9.7.1.1.1 Retroviruses

Retroviruses are a class of enveloped viruses containing a single-stranded RNA molecule as the genome. Following infection, the viral genome is reverse transcribed into double-stranded DNA, integrates into the host genome, and is expressed as proteins. The viral genome is approximately 10 kb, containing at least three genes: gag (coding for core proteins), pol (coding for reverse transcriptase), and env (coding for the viral envelope protein). At each end of the genome are long terminal repeats (LTRs), which include promoter/enhancer regions and sequences involved with integration. In addition, there are sequences required for packaging the viral DNA and RNA splice sites in the env gene.

Some retroviruses contain protooncogenes which, when mutated, can cause cancers, although these are removed in the production of vectors. Retroviruses can also transform cells by integrating near to a cellular protooncogene and driving inappropriate expression from the LTR, or by disrupting a tumor suppresser gene. This event, termed *insertional mutagenesis*, though extremely rare, could still occur when retroviruses are used as vectors. Retroviral vectors are most frequently based upon the Moloney murine leukemia virus (Mo-MLV), which is an amphotrophic virus, capable of infecting both mouse cells, enabling vector development in mouse models and human cells, enabling human treatment. The viral genes (gag, pol, and env) are replaced with the transgene of interest and expressed on plasmids in the packaging cell line. Because the nonessential genes lack the packaging sequence (psi), they are not included in the virion particle. To prevent recombination resulting in replication-competent retroviruses, all regions of homology with the vector backbone should be removed and the nonessential genes should be expressed by at least two transcriptional units. Even so, replication-competent retroviruses do occur at a low frequency.

The essential regions include the 5′ and 3′ LTRs and the packaging sequence lying downstream of the 5′ LTR. Transgene expression can either be driven by the promoter/enhancer region in the 5′ LTR or by alternative viral (e.g., cytomegalovirus, Rous sarcoma virus) or cellular (e.g., beta-actin, tyrosine) promoters. Mutational analysis has shown that up to the entire gag coding sequence and the immediate upstream region can be removed without effecting viral packaging or transgene expression. To aid identification of transformed cells, selectable markers such as neomycin and beta-galactosidase can be included, and transgene expression can be improved with the addition of internal ribosome sites. The available carrying capacity for retroviral vectors is approximately 7.5 kb, which is too small for some genes, even if the complementary DNA (cDNA) is used.

A requirement for retroviral integration and expression of viral genes is that the target cells should be dividing. This limits gene therapy to proliferating cells *in vivo* or *ex vivo*, whereby cells are removed from the body, treated to stimulate replication, and then transduced with the retroviral vector before being returned to the patient. When treating cancers *in vivo*, tumor cells are preferentially targeted. However, *ex vivo* cells can be more efficiently transduced, due to exposure to higher virus titers and growth factors. Furthermore, *ex vivo*–treated tumor cells will associate with the tumor mass and can direct tumoricidal effects. Though transgene expression is usually adequate *in vitro* and initially *in vivo*, prolonged expression is difficult to attain. Retroviruses are inactivated by C1 complement protein and an anti-alpha galactosyl epitope antibody, both present in human sera. Transgene expression is also reduced by inflammatory interferons, specifically IFN-alpha and IFN-gamma, acting on viral LTRs. As the retroviral genome integrates into the host genome, it is most likely that the viral LTR promoters are being inactivated; therefore, one approach has been to use promoters for host cell genes, such as tyrosine. Clearly, this is an area where continued research is needed.

9.7.1.1.2 Lentivirus Vectors

Lentiviruses are a subclass of retroviruses that are able to infect both proliferating and non-proliferating cells. They are considerably more complicated than simple retroviruses, containing an additional six proteins: tat, rev, vpr, vpu, nef, and vif. Current packaging cell lines have separate plasmids for a pseudotype env gene, a transgene construct, and a packaging construct supplying the structural and regulatory genes in *trans*. Early results using marker genes have been promising, showing prolonged *in vivo* expression in muscle, liver, and neuronal tissue. Interestingly, the transgenes are driven by an internally engineered cytomegalovirus promoter which, unlike in Mo-MLV vectors, is not inactivated. This may be due to the limited inflammatory response to the vector injection, which is equal in magnitude to the saline control. The lentiviral vectors used are derived from the human immunodeficiency virus (HIV) and are being evaluated for safety, with a view to removing some of the nonessential regulatory genes. Mutants of vpr and vif are able to infect neurons with reduced efficiency, but not muscle or liver cells.

9.7.1.1.3 Adenovirus Vectors

Adenoviruses are non-enveloped viruses containing a linear double-stranded DNA genome. While there are over 40 serotype strains of adenovirus, most of which cause benign respiratory tract infections in humans, subgroup C serotypes 2 or 5 are predominantly used as vectors. The life cycle does not normally involve integration into the host genome; rather, they replicate as episomal elements in the nucleus of the host cell, and consequently there is no risk of insertional mutagenesis. The wild type adenovirus genome is approximately 35 kb, of which up to 30 kb can be replaced with foreign DNA. There are four early transcriptional units (E1, E2, E3, and E4), which have regulatory functions, and a late transcript, which codes for structural proteins. Progenitor vectors have either the E1 or E3 gene inactivated, with the missing gene being supplied in *trans*, either by a helper virus or plasmid, or integrated into a helper cell genome (human fetal kidney cells, line 293). Second-generation vectors additionally use an E2a temperature-sensitive mutant or an E4 deletion. The most recent "gutless" vectors contain only the inverted terminal repeats (ITRs) and a packaging sequence around the transgene, all the necessary viral genes being provided in *trans* by a helper virus.

9.7.1.1.4 Adeno-Associated Viruses

AAVs are nonpathogenic human parvoviruses, dependent on a helper virus, usually an adenovirus, to proliferate. They are capable of infecting both dividing and nondividing cells and, in the absence of a helper virus, integrate into a specific point of the host genome (19q 13-qter) at a high frequency. The wild type genome is a single-stranded DNA molecule, consisting of two genes: *rep*, coding for proteins which control viral replication, structural gene expression, and integration into the host genome, and *cap*, which codes for capsid structural proteins. At either end of the genome is a 145 bp terminal repeat (TR) containing a promoter.

When used as a vector, the rep and cap genes are replaced by the transgene and its associated regulatory sequences. The total length of the insert cannot greatly exceed 4.7 kb, the length of the wild type genome. Production of the recombinant vector requires that rep and cap are provided in *trans*, along with helper virus gene products (E1a, E1b, E2a, E4, and VA RNA from the adenovirus genome). The conventional method is to co-transfect two plasmids, one for the vector and another for rep and cap, into 293 cells infected with adenovirus. This method, however, is cumbersome, low yielding ($<10^4$ particles/mL), and prone to contamination with adenovirus and wild type AAV. One of the reasons for the low yield is the inhibitory effect of the rep gene product on adenovirus replication. More recent protocols remove all adenoviral structural genes and use rep-resistant plasmids or conjugate a rep expression plasmid to the mature virus prior to infection.

In the absence of rep, the AAV vector will only integrate at random, as a single provirus or head-to-tail concatamers, once the TRs have been slightly degraded. Interest in AAV vectors has been due to their integration into the host genome, allowing prolonged transgene expression. Gene transfer into vascular epithelial cells, striated muscle, and hepatic cells has been reported, with prolonged expression when the transgene is not derived from a different species. Neutralizing antibody to the AAV capsid may be detectable, but does not prevent re-administration of the vector or shut down promoter activity. It is possibly due to the simplicity of the viral capsid that the immune response is so muted. As AAV antibodies will be present in the human population, this will require further investigation. There has been no attempt to target particular cell types other than by localized vector delivery.

9.7.1.1.5 Herpes Simplex Virus

HSV-1 is a human neurotropic virus. Consequently, interest has largely focused on using HSV-1 as a vector for gene transfer to the nervous system. The wild type HSV-1 virus is able to infect neurones and either proceed into a lytic life cycle or persist as an intranuclear episome in a latent state. Latently infected neurons work normally and are not rejected by the immune system. Though the latent virus is transcriptionally almost silent, it does possess neurone-specific promoters that are capable of functioning during latency. Antibodies to HSV-1 are common in the human population. However, there may be complications due to herpes infection, such as encephalitis, although these are very rare. The viral genome is a linear double-stranded DNA molecule of 152 kb. There are two unique regions, long and short (termed UL and US), which are linked in either orientation by internal repeat sequences (IRL and IRS). At the non-linker end of the unique regions are TRs (TRL and TRS). There are up to 81 genes, of which approximately half are not essential for growth in cell culture. Once these nonessential genes have been deleted, 40–50 kb of foreign DNA can be accommodated within the virus. Three main classes of HSV-1 genes have been identified: the immediate-early (IE or alpha) genes, early (E or beta) genes, and late (L or gamma) genes. Following infection in susceptible cells, lytic replication is regulated by a temporally co-coordinated sequence of gene transcription. *Vmw65* (a tegument structural protein) activates the IE genes (IP0, ICP4, ICP22, ICP27, and ICP477), which are transactivating factors allowing the production of E genes. The E genes encode genes for nucleotide metabolism and DNA replication. L genes are activated by the E genes and code for structural proteins. The entire cycle takes less than 10 h and invariably results in cell death.

9.7.1.2 Nonviral Methods of DNA Transfer

Viral vectors all induce an immunological response to some degree and may have safety risks such as insertional mutagenesis and toxicity problems. Furthermore, their capacity is limited and large-scale production may be difficult to achieve. Nonviral methods of DNA transfer require only a small number of proteins, have a virtually infinite capacity, have no infectious or mutagenic capability, and large-scale production is possible using pharmaceutical techniques.

9.7.1.2.1 Naked DNA

This is the simplest method of nonviral transfection. Clinical trials of an intramuscular injection of a naked DNA plasmid have been carried out with some success. However, the expression has been very low in comparison to other methods of transfection. In addition to trials with plasmids, there have been trials with a naked PCR product, which have had similar or greater success. This success, however, does not compare to that of the other methods, leading to research into more efficient methods for delivery of the naked DNA, such as electroporation, sonoporation, and the use of a "gene gun," which shoots DNA-coated gold particles into the cell using high-pressure gas.

9.7.1.2.2 Oligonucleotides

Oligonucleotides are being used to inactivate the genes involved in the disease process. There are several methods by which this is achieved. One strategy uses antisense specific to the target gene to disrupt the transcription of the faulty gene. Another uses small molecules of RNA called *siRNA* to signal the cell to cleave specific unique sequences in the messenger RNA (mRNA) transcript of the faulty gene, disrupting translation of the faulty mRNA, and therefore expression of the gene. A further strategy uses double-stranded oligodeoxynucleotides as a decoy for the transcription factors that are required to activate the transcription of the target gene. The transcription factors bind to the decoys instead of the promoter of the faulty gene, which reduces the transcription of the target gene, lowering expression. Additionally, single-stranded DNA oligonucleotides have been used to direct a single base change within a mutant gene. The oligonucleotide is designed to anneal with complementarity to the target gene (with the exception of a central base), the target base, which serves as the template base for repair. This technique is referred to as *oligonucleotide-mediated gene repair*, *targeted gene repair*, or *targeted nucleotide alteration*.

9.7.1.2.3 Anti-DNA Degradation

Gene therapy is based on the delivery of genes. To improve the delivery of a gene into the host cell, the DNA must be protected from damage, and its entry into the cell must be facilitated. To these ends, new molecules such as lipoplexes and polyplexes have been created, which have the ability to protect the DNA from undesirable degradation during the transfection process. Plasmid DNA can be covered with lipids in an organized structure like a micelle or a liposome. When the organized structure is complexed with DNA, it is called a lipoplex. There are three types of lipids: anionic (negatively charged), neutral, and cationic (positively charged). Initially, anionic and neutral lipids were used for the construction of lipoplexes for synthetic vectors. However, despite the facts that there is little toxicity associated with them, that they are compatible with body fluids, and that there was a possibility of adapting them to be tissue-specific, they are complicated and time consuming to produce, so attention was turned to the cationic versions. Cationic lipids, due to their positive charge, were first used to condense negatively charged DNA molecules so as to facilitate the encapsulation of DNA into liposomes. Later, it was found that the use of cationic lipids significantly enhanced the stability of lipoplexes. Also, as a result of their charge, cationic liposomes interact with the cell membrane; endocytosis was widely believed to be the major route by which cells uptake lipoplexes. Endosomes are formed as the results of endocytosis; however, if genes cannot be released into cytoplasm by breaking the membrane of endosome, they will be sent to lysosomes, where all DNA will be destroyed before they can achieve their functions.

The most common use of lipoplexes has been for gene transfer into cancer cells, where the supplied genes activate tumor suppressor control genes in the cell and decrease the activity of oncogenes. Recent studies have shown that lipoplexes are useful in transfecting respiratory epithelial cells so they may be used for treatment of genetic respiratory diseases such as cystic fibrosis. Complexes of polymers with DNA are called *polyplexes*. Most polyplexes consist of cationic polymers, and their production is regulated by ionic interactions. One large difference between the methods of action of polyplexes and lipoplexes is that polyplexes cannot release their DNA load into the cytoplasm; to this end, co-transfection with endosome-lytic agents (to lyse the endosome that is made during endocytosis, the process by which the polyplex enters the cell) such as inactivated adenovirus must occur. However, this is not always the case; polymers such as polyethylenimine have their own method of endosome disruption, as do chitosan and trimethylchitosan.

9.7.1.2.4 Hybrid Methods

Almost all methods of gene transfer have shortcomings, and there has been constant effort toward developing hybrid technology by combining two or more techniques. Virosomes are one example. They combine liposomes with an inactivated HIV or influenza virus. This has been shown to have more efficient gene transfer in respiratory epithelial cells than either viral or liposomal methods alone. Other methods involve mixing other viral vectors with cationic lipids or hybridizing viruses.

9.7.1.2.5 Dendrimers

A *dendrimer* is a highly branched macromolecule with a spherical shape. The surface of the particle may be functionalized in many ways, and many of the properties of the resulting construct are determined by its surface. In particular, it is possible to construct a cationic dendrimer (i.e., one with a positive surface charge). When in the presence of genetic material such as DNA or RNA, charge complementarity leads to a temporary association of the nucleic acid with the cationic dendrimer. On reaching its destination, the dendrimer–nucleic acid complex is then taken into the cell via endocytosis. In recent years, the benchmark for transfection agents has been cationic lipids. Limitations of these competing reagents have been reported to include the lack of ability to transfect a number of cell types, the lack of robust active targeting capabilities, incompatibility with animal models, and toxicity. Dendrimers offer robust covalent construction and extreme control over molecule structure, and therefore over size. Together, these give compelling advantages compared to existing approaches. Producing dendrimers has historically been a slow and expensive process consisting of numerous slow reactions, an obstacle that severely curtailed their commercial development. The Michigan-based company Dendritic Nanotechnologies discovered a method to produce dendrimers using kinetically driven chemistry, a process that not only reduced costs by a magnitude of three, but also cut reaction time from over a month to several days. These new "Priostar" dendrimers can be specifically constructed to carry a DNA or RNA payload that transfects cells at a high efficiency, with little or no toxicity.

9.7.2 *In Vivo* Gene Therapy

In vivo gene therapy means to repair or correct genes inside the body. In this section, we will study how *in vivo* gene therapies are being used to treat various genetic diseases. In 2003, a University of California, Los Angeles research team inserted genes into the brain using liposomes coated in a polymer called PEG. The transfer of genes into the brain is a significant achievement, because viral vectors are too big to get across the "blood–brain barrier." This method has potential for treating Parkinson's disease.

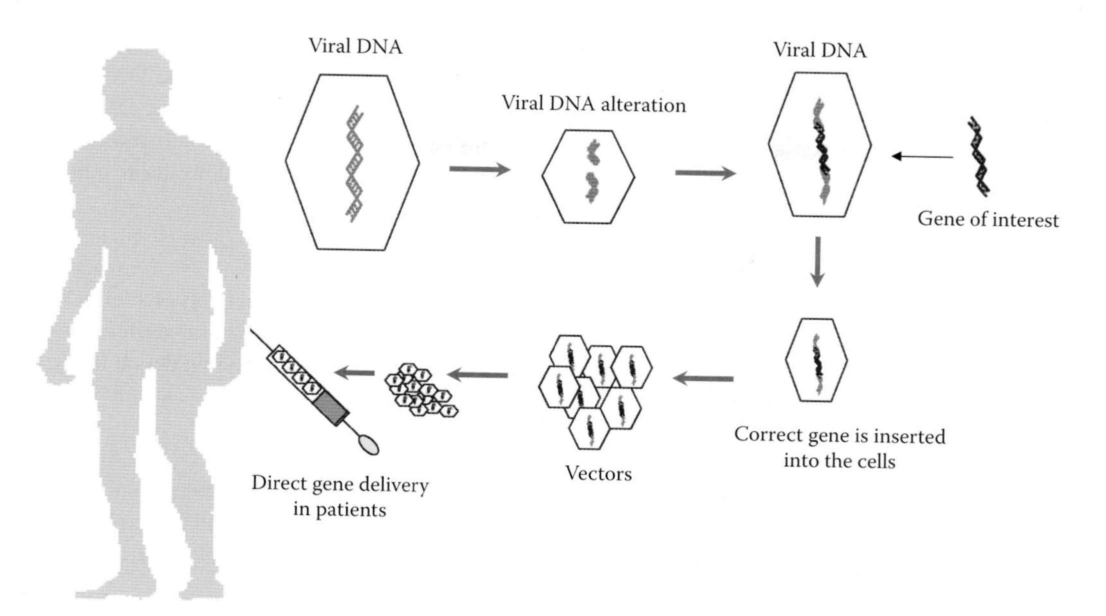

FIGURE 9.9 *In vivo* gene therapy.

RNA interference or gene silencing may be a new way to treat Huntington's disease. Short pieces of double-stranded RNA (short interfering RNAs or siRNAs) are used by cells to degrade RNA of a particular sequence. If a siRNA is designed to match the RNA copied from a faulty gene, then the abnormal protein product of that gene will not be produced (Figure 9.9).

In 2006, scientists at the National Institutes of Health (Bethesda, Maryland) successfully treated metastatic melanoma in two patients using killer T-cells genetically retargeted to attack the cancer cells. This study constitutes the first demonstration that gene therapy can be effective in treating cancer. In March 2006, an international group of scientists announced the successful use of gene therapy to treat two adult patients for a disease affecting myeloid cells. The study, published in *Nature Medicine*, is believed to be the first to show that gene therapy can cure diseases of the myeloid system.

In May 2006, a team of scientists led by Dr. Luigi Naldini and Dr. Brian Brown from the San Raffaele Telethon Institute for Gene Therapy (HSR-TIGET) in Milan, Italy, reported a breakthrough for gene therapy in which they developed a way to prevent the immune system from rejecting a newly delivered gene. Similar to organ transplantation, gene therapy has been plagued by the problem of immune rejection. So far, delivery of the "normal" gene has been difficult, because the immune system recognizes the new gene as foreign and rejects the cells carrying it. To overcome this problem, the HSR-TIGET group utilized a newly uncovered network of genes regulated by molecules known as microRNAs. Dr. Naldini's group reasoned that they could use this natural function of microRNAs to selectively turn off the identity of their therapeutic gene in cells of the immune system and prevent the gene from being found and destroyed. The researchers injected mice with the gene containing an immune-cell microRNA target sequence and, spectacularly, the mice did not reject the gene, as had occurred previously when vectors without the microRNA target sequence were used. This work will have important implications for the treatment of hemophilia and other genetic diseases by gene therapy.

9.7.3 Problems with Gene Therapy

There are technical or scientific issues associated with gene therapy that need to be resolved before using it in humans.

9.7.3.1 Short-Lived Nature of Gene Therapy

Before gene therapy can become a permanent cure for any condition, the therapeutic DNA introduced into target cells must remain functional, and the cells containing the therapeutic DNA must be long-lived and stable. Problems with integrating therapeutic DNA into the genome and the rapidly dividing nature of many cells prevent gene therapy from achieving any long-term benefits. Patients will have to undergo multiple rounds of gene therapy.

9.7.3.2 Immune Response

Anytime a foreign object is introduced into human tissues, the immune system evolves to attack the invader. The risk of stimulating the immune system in a way that reduces gene therapy effectiveness is always a possibility. Furthermore, the immune system's enhanced response to invaders it has seen before makes it difficult for gene therapy to be repeated in patients.

9.7.3.3 Viral-Induced Toxicity

The viruses that act as carrier vehicles in gene delivery may cause potential problems in patients, which include toxicity, immune and inflammatory responses, and gene control and targeting issues. In addition, there is always the fear that the viral vector, once inside the patient, may recover its ability to cause disease.

9.7.3.4 Not for Multi-Gene Disorders

Single gene mutations may cause genetic disorders, which are, in fact, considered the best candidates for gene therapy. Unfortunately, some of the most commonly occurring disorders, such as heart disease, high blood pressure, Alzheimer's disease, arthritis, and diabetes, are caused by the combined effects of variations in many genes. Multigene or multifactorial disorders such as these would be especially difficult to treat effectively using gene therapy.

9.7.3.5 Induced Mutagenesis

During gene delivery, there is a high possibility that DNA is delivered to nontargeted site in the genome, thus causing severe health problems in patients; for example, in a tumor suppressor gene, it could induce a tumor. This has occurred in clinical trials for X-linked severe combined immunodeficiency (X-SCID) patients, in which HSCs were transduced with a corrective transgene using a retrovirus, and this led to the development of T-cell leukemia in 3 of 20 patients.

9.7.4 Genetic Counseling

Genetic counseling for couples who believe that there may be a risk of producing a child with a congenital disease has now become a routine aspect of medical practice, particularly in the developed countries. These parents may either voluntarily abstain from having any children or may undergo selective abortions on suspicion or after ascertaining the occurrence of congenital disease through antenatal diagnosis, as discussed earlier. A genetic counselor should first be able to identify the disease and should therefore be first a clinician and then a geneticist. Testing for hereditary or developmental defects is done through amniocentesis. At approximately the 15th week of pregnancy, a sample of amniotic fluid containing fetal cells is removed and cultured *in vitro*. The cultured cells are then used for karyotype analysis and biochemical enzyme tests. The simplest cases asking for genetic counseling will be those with a family history of disease, and the parents may like to know the chances of having a child free of that disease. In the simplest case, a couple may have one abnormal child (carrying disease) and would like to know the chances of having a normal child with the next pregnancy.

In such a case, if the defect is controlled by single recessive gene and both parents are normal, the chance of having a normal child is three in four, although even a normal child will have a two-thirds chance of being a carrier. The parents may like to give birth to such a child who may be a carrier, because the chance of his or her spouse also being a carrier will be remote. However, in such cases, even the possibility of having abnormal grandchildren can be worked out if the frequency of heterozygotes in the population available for marriage of the child is known. In the United Kingdom, for instance, in case of fibrocystic disease (cystic fibrosis) of the pancreas, the frequency of heterozygotes in the general population is 1/22. The chance of the normal child being a carrier being 1/22, the chance of a grandchild being abnormal should be 2/3 × 1/22 × 1/4 = 1/132 (since 114 is the chance that a child born to both heterozygote parents will be defective). A risk of 1/132 may or may not be worth taking, depending on the parents' temperament and the circumstances.

When a heterozygote, though phenotypically normal, produces a particular enzyme activity intermediate to those found in two homozygotes, the heterozygote can be identified. In such cases, if a biochemical laboratory is available, deficiencies such as hypoxanthine-guanine phosphoribosyltransferase (HGPRT, causing Lesch–Nyhan syndrome) in a heterozygous condition can be detected from a blood sample or some skin cells. When the mutant produces an altered form of gene product, the heterozygote may produce two different forms of protein that can be separated by electrophoresis to enable the identification of a heterozygote. If the defect is associated with a chromosome structure, then such an abnormality in the heterozygous condition can be identified with the availability of a cytogenetic laboratory.

9.8 MOLECULAR DIAGNOSTICS

Molecular techniques are more sensitive and more specific than classical techniques. Over the last few years, molecular diagnostics have opened up a whole spectrum of new tools and approaches to the challenging task of diagnosing animal disease. Due to recent outbreaks of major diseases, rapid and accurate diagnosis is a fundamental requirement in the prevention and management of major epizootic outbreaks. In this section, we will discuss some methods or techniques employed in clinical or diagnostic investigations.

9.8.1 DNA Fingerprinting

DNA profiling—such as DNA testing, DNA typing, or genetic fingerprinting—is generally employed by forensic scientists to assist in the identification of individuals on the basis of their respective DNA profiles. DNA profiles are encrypted sets of numbers that reflect a person's DNA makeup, which can also be used as the person's identifier. DNA profiling should not be confused with full genome sequencing. Although 99.9% of human DNA sequences are the same in every person, enough of the DNA is different to distinguish one individual from another. DNA profiling uses repetitive ("repeat") sequences that are highly variable, called *variable number tandem repeats* (VNTR). VNTRs loci are very similar between closely related humans, but so variable that unrelated individuals are extremely unlikely to have the same VNTRs. The DNA profiling technique was first reported in 1985 by Sir Alec Jeffreys at the University of Leicester in England, and is now the basis of several national DNA databases. *DNA fingerprinting* is a technique to determine the likelihood that genetic material came from a particular individual or group. In the case of grapes, scientists compared the similarities between different species and were able to piece together parent subspecies that could have contributed to the present prize-winning varieties.

Recall that DNA alphabet is made up of four building blocks: A, C, T, and G, called *base pairs*, which are linked together in long chains to spell out the genetic words (i.e., genes) which tell our cells what to do. The order in which these four DNA letters are used determines the meaning (function) of the words (or genes) that they spell. However, not our entire DNA contains useful information. In fact, a large amount is said to be "noncoding" or "junk" DNA, which is not translated into useful proteins. Changes often crop up within these regions of junk DNA, because they make no contribution to the health or survival of the organism. However, compare the situation if a change occurs within an essential gene, preventing it from working properly. The organism will be strongly disadvantaged and will probably not survive, effectively removing that altered gene from the population. For this reason, random variations crop up in the noncoding (junk) DNA sequences as often as once in every 200 DNA letters. DNA fingerprinting takes advantage of these changes and creates a visible pattern of the differences to assess similarity.

There are many relevant applications of DNA fingerprinting technology in the modern world, and these fall into three main categories. We can address certain issues such as (1) to know where we came from, (2) to discover what we are doing at present, (3) and to predict where we are going. In terms of where we came from, DNA fingerprinting is commonly used to probe our heredity. Since people inherit the arrangement of their base pairs from their parents, comparing the banding patterns of a child and the alleged parent generates a probability of relatedness. If the two patterns are similar enough (taking into account that only half the DNA is inherited from each parent), then they are probably family. However, DNA fingerprinting cannot discriminate between identical twins, since their banding patterns are the same. In paternity suits involving identical twins—and yes, there have been such cases—if neither brother has an alibi to prove that he could not have impregnated the mother, the courts have been known to force them to split childcare costs. Thankfully, there are other less "Jerry Springer-esque" applications that teach us about our origins. When used alongside more traditional sociological methodologies, DNA fingerprinting can be used to analyze patterns of migration and claims of ethnicity.

DNA fingerprinting can also tell us about present-day situations. Perhaps best known is the use of DNA fingerprinting in forensic medicine. DNA samples gathered at a crime scene can be compared with the DNA of a suspect to prove whether or not he or she was present in the scene. Databases of DNA fingerprints are only available from known offenders, so it is not yet possible to fingerprint the DNA from a crime scene and then pull out names of probable matches from the general public. However, this may happen in future if DNA fingerprints replace more traditional and forgeable forms of identification. In a real case, trading standards agents found that 25% of caviar is bulked up with roe from different categories, the high-class equivalent of cheating the consumer by not filling the metaphorical pint glass all the way up to the top. DNA fingerprinting confirmed that the "suspect" (inferior) caviar was present at the crime scene.

Finally, genetic fingerprinting can help us to predict our future health. DNA fingerprinting is often used to track down the genetic basis of inherited diseases. If a particular pattern turns up time and time again in different patients, scientists can narrow down which gene(s), or at least which stretch(es) of DNA, might be involved. Since knowing the genes involved in disease susceptibility gives clues about the underlying physiology of the disorder, genetic fingerprinting aids in developing therapies.

Prenatally, genetic fingerprinting can also be used to screen parents and fetuses for the presence of inherited abnormalities, such as Huntington's disease or muscular dystrophy. The process begins with a sample of an individual's DNA (typically called a "reference sample"). The most desirable method of collecting a reference sample is the use of a buccal swab, as this reduces the possibility of contamination. When this is not available (such as because a court order may be needed and not obtainable), other methods may need to be used to collect a sample of blood, saliva, semen, or other appropriate fluid or tissue from personal items (such as a toothbrush or razor) or from stored samples (such as banked sperm or biopsy tissue). Samples obtained from blood relatives (biological relatives) can provide an indication of an individual's profile, as could human remains that have been previously profiled.

9.8.2 Techniques of DNA Profiling

In order to profile the DNA of a sample, a reference sample is analyzed to create the individual's DNA profile using one of a number of techniques discussed below. The DNA profile is then compared against another sample to determine whether there is a genetic match.

9.8.2.1 Restriction Fragment Length Polymorphism Analysis

The first methods for finding out genetics used for DNA profiling involved restriction enzyme digestion, followed by Southern blot analysis. Although polymorphisms can exist in the restriction enzyme cleavage sites, more commonly the enzymes and DNA probes were used to analyze VNTR loci. However, the Southern blot technique is laborious and requires large amounts of undegraded sample DNA. In addition, Karl Brown's original technique looked at many minisatellite loci at the same time, increasing the observed variability, but making it hard to discern individual alleles (and thereby precluding parental testing). These early techniques have been supplanted by polymerase chain reaction (PCR)-based assays.

9.8.2.2 PCR Analysis

With the invention of the PCR technique, DNA profiling took huge strides forward in both discriminating power and the ability to recover information from very small (or degraded) starting samples. PCR greatly amplifies the amounts of a specific region of DNA, using oligonucleotide primers and a thermostable DNA polymerase. Early assays, such as the human leukocyte antigen DQ subregion (HLA-DQ) alpha reverse dot blot strips, grew to be very popular due to their ease of use and the speed with which a result could be obtained. However, they were not as discriminating as restriction fragment length polymorphism (RFLP). It was also difficult to determine a DNA profile for mixed samples, such as a vaginal swab from a sexual assault victim. Fortunately, the PCR method is readily adaptable for analyzing VNTR loci. In the United States, the FBI has standardized a set of 13 VNTR assays for DNA typing and has organized the Combined DNA Index System (CODIS) database for forensic identification in criminal cases. Similar assays and databases have been set up in other countries. Also, commercial kits are available that analyze single nucleotide polymorphisms (SNPs). These kits use PCR to amplify specific regions with known variations and hybridize them to probes anchored on cards, which results in a colored spot corresponding to the particular sequence variation.

9.8.2.3 Short Tandem Repeats Analysis

The method of DNA profiling used today is based on PCR and uses short tandem repeats (STR). This method uses highly polymorphic regions that have short repeated sequences of DNA (the most common is four bases repeated, but there are other lengths in use, including three and five bases). Because different unrelated people have different numbers of repeat units, these regions of DNA can be used to discriminate between unrelated individuals. These STR loci (locations) are targeted with sequence-specific primers and are amplified using PCR. The DNA fragments are then separated and are detected using the electrophoresis technique. There are two common methods of separation and detection: CE and gel electrophoresis. The polymorphisms displayed at each STR region are very common by themselves; typically, each polymorphism will be shared by around 5%–20% of individuals. When looking at multiple loci, it is the unique combination of these polymorphisms in an individual that makes this method discriminating as an identification tool. The more STR regions that are tested in an individual, the more discriminating the test becomes. Different STR-based DNA profiling systems are in use from country to country. In North America, systems which amplify the CODIS 13 core loci are almost universal, while in the United Kingdom, the Second Generation Multiplex Plus (SGM+) DNA profiling system, which is compatible with the National DNA Database, is in use. Whichever system is used, many of the STR regions under test are the same. These DNA profiling systems are based around multiplex reactions, whereby many STR regions will be under test at the same time. CE works by electrokinetically (movement through the application of an electric field) injecting the DNA fragments into a thin glass tube (the capillary) filled with polymer. The DNA is pulled through the tube by the application of an electric field, separating the fragments so that the smaller fragments travel faster through the capillary. The fragments are then detected using fluorescent dyes that are attached to the primers used in PCR. This allows multiple fragments to be amplified and run simultaneously, a procedure known as *multiplexing*. Sizes are assigned using labeled DNA size standards that are added to each sample, and the number of repeats is determined by comparing the size to an allelic ladder, a sample that contains all of the common possible repeat sizes. Although this method is expensive, larger capacity machines with higher throughput are being used to lower the cost per sample and reduce the backlogs that exist in many government crime facilities.

The true power of STR analysis is in its statistical power of discrimination. In the United States, there are 13 core loci (DNA locations) that are currently used for discrimination in CODIS. Because these loci are independently assorted (having a certain number of repeats at one locus does not change the likelihood of having any number of repeats at any other locus), the product rule for probabilities can be applied. This means that if someone has the DNA type of ABC, where the three loci were independent, we can say that the probability of having that DNA type is the probability of having type A times the probability of having type B times the probability of having type C. This has resulted in the ability to generate match probabilities of one in a quintillion (1 with 18 zeros after it) or more. However, since there are approximately 12 million monozygotic twins on earth, that theoretical probability is useless. For example, the actual probability that two random persons have the same DNA is only 1 in 3 trillion.

9.8.2.4 Amplified Fragment Length Polymorphism Analysis

The amplified fragment length polymorphism (AmFLP) technique is found to be faster than RFLP analysis and uses PCR to amplify DNA samples. It relies on VNTR polymorphisms to distinguish various alleles, which are separated on a polyacrylamide gel using an allelic ladder (as opposed to a molecular weight ladder). Bands can be visualized by silver staining the gel. One popular locus for fingerprinting was the D1S80 locus. As with all PCR-based methods, highly degraded DNA or very small amounts of DNA may cause allelic dropout (causing a mistake in thinking a heterozygote is a homozygote) or other stochastic effects. In addition, because the analysis is done on a gel, very high number repeats may bunch together at the top of the gel, making it difficult to resolve. AmpFLP analysis can be highly automated and allows for easy creation of phylogenetic trees. Due to its relatively low cost and ease of setup and operation, AmpFLP remains popular in lower income countries.

9.8.2.5 Y-Chromosome Analysis

Recent innovations have included the creation of primers targeting polymorphic regions on the Y chromosome (Y-STR), which allows resolution of a mixed DNA sample from a male and female and/or cases in which a differential extraction is not possible. Y chromosomes are paternally inherited, so Y-STR analysis can help in the identification of paternally related males. Y-STR analysis was performed in the Sally Hemings controversy to determine if Thomas Jefferson had sired a son with one of his slaves.

9.8.2.6 Mitochondrial DNA Analysis

It is sometimes impossible to get a complete profile of the 13 CODIS STRs in highly degraded samples. In these situations, mitochondrial DNA (mtDNA) is sometimes mapped due to the large number of mtDNA copies, while there may only be 1–2 copies of the nuclear DNA. Forensic scientists amplify the HV1 and HV2 regions of the mtDNA, and then sequence each region and compare single nucleotide differences to a reference. Because mtDNA is maternally inherited, directly linked maternal relatives can be used as match references, such as one's maternal grandmother's sister's son. A difference of two or more nucleotides is generally considered to be exclusion. Heteroplasmy and poly-C differences may throw off straight sequence comparisons, so some expertise on the part of the analyst is required. mtDNA is useful in determining unclear identities, such as those of missing persons, when a maternally linked relative can be found. mtDNA testing was used in determining that Anna Anderson was not, Anastasia Romanov, the Russian princess she had claimed to be. mtDNA can be obtained from hair shafts and old bones/teeth of the body.

9.8.2.7 Confirmation of Genetic Relationship

It is also possible to use DNA profiling as evidence of genetic relationship, but test results are not entirely reliable. While almost all individuals have a single and distinct set of genes, rare individuals, known as "chimeras," have at least two different sets of genes. There have been several cases of DNA profiling that falsely "proved" that a mother was unrelated to her children.

9.8.2.8 Fake DNA Evidence

There are challenges in DNA profiling. For instance, DNA samples at crime scenes have been found to be faked. In one case, a criminal even planted fake DNA evidence in his own body: Dr. John Schneeberger raped one of his sedated patients in 1992 and left semen on her underwear. Police drew Schneeberger's blood and compared its DNA against the crime scene semen DNA on three occasions, never showing a match. It turned out that he had surgically inserted a Penrose drain into his arm and filled it with foreign blood and anticoagulants.

Familial searching is the use of family members' DNA to identify a closely related suspect in jurisdictions where large DNA databases exist, but no exact match has been found. The first successful use of the practice was in a United Kingdom case where a man was convicted of manslaughter when he threw a brick stained with his own blood into a moving car. Police could not get an exact match from the U.K.'s DNA database because the man had no criminal convictions, but police implicated him using a close relative's DNA.

9.8.2.9 Criminal DNA Data

Police forces may collect DNA samples without the suspect's knowledge and use them as evidence. The legality of this mode of proceeding has been questioned in Australia. In the United States, it has been accepted, with courts often claiming that there was no expectation of privacy, citing California versus Greenwood (1985), during which the Supreme Court held that the Fourth Amendment does not prohibit the warrantless search and seizure of garbage left for collection outside the curtilage of a home. Critics of this practice underline the fact that this analogy ignores that "most people have no idea that they risk surrendering their genetic identity to the police by, for instance, failing to destroy a used coffee cup. Moreover, even if they do realize it, there is no way to avoid abandoning one's DNA in public." In the United Kingdom, the Human Tissue Act of 2004 prohibited private individuals from covertly collecting biological samples (hair, fingernails, etc.) for DNA analysis, but excluded medical and criminal investigations from the offense.

9.9 ARTIFICIAL BLOOD

Blood is very important for survival of human life, and due to shortage of different types of blood in the world, efforts have been directed to generate blood through artificial means. In this section, we will closely examine how blood is generated outside the body, and what its advantages and disadvantages are. In brief, *blood* is a special type of connective tissue that is composed of white cells, red cells, platelets, and plasma. It has a variety of functions in the body. *Plasma* is the extracellular material made up of water, salts, and various proteins that, along with platelets, encourage blood to clot. Proteins in the plasma react with air and harden to prevent further bleeding. The white blood cells are responsible for the immune defense. They seek out invading organisms or materials and minimize their effect in the body. The red cells in blood create its bright red color. As little as two drops of blood contain approximately one billion red blood cells. These cells are responsible for the transportation of oxygen and carbon dioxide throughout the body, and are also responsible for the "typing" phenomena. Currently, artificial blood products are only designed to replace the function of red blood cells. It may even be better to call the products being developed now oxygen carriers instead of artificial blood.

There has been a need for blood replacements for as long as patients have been bleeding to death because of a serious injury. According to medical folklore, the ancient Incas were responsible for the first recorded blood transfusions. No real progress was made in the development of a blood substitute until 1616, when William Harvey described how blood circulates throughout the body. In the years to follow, medical practitioners tried numerous substances such as beer, urine, milk, plant resins, and sheep blood as a substitute for blood. They had hoped that changing a person's blood could have different beneficial effects such as curing diseases or even changing a personality. The first successful human blood transfusions were done in 1667. Unfortunately, the practice was halted because patients who received subsequent transfusions died. Of the different materials that were tried as blood substitutes over the years, only a few met with minimal success. Milk was one of the first of these materials. Other materials that were tried during the eighteenth century include hemoglobin and animal plasma. In 1868, researchers found that solutions containing hemoglobin isolated from red blood cells could be used as blood replacements. In 1871, they also examined the use of animal plasma and blood as a substitute for human blood. Both these approaches were hampered by significant technological problems. First, scientists found it difficult to isolate a large volume of hemoglobin. Second, animal products contained many materials that were toxic to humans. Removing these toxins was a challenge during the nineteenth century.

A significant breakthrough in the development of artificial blood came in 1883, with the creation of Ringer's solution—a solution composed of sodium, potassium, and calcium salts. In research using part of a frog's heart, scientists found that the heart could be kept beating by applying the solution. This eventually led to finding that the reduction in blood pressure caused by a loss of blood volume could be restored by using Ringer's solution. This product evolved into a human product when lactate was added. While it is still used today as a blood volume expander, Ringer's solution does not replace the action of red blood cells and it is therefore not a true blood substitute.

In 1966, experiments with mice suggested a new type of blood substitute, *perfluorochemicals* (PFC). These are long-chain polymers similar to Teflon. It was found that mice could survive even after being immersed in PFC. This finding gave scientists the idea to use PFC as a blood thinner. In 1968, the idea was tested on rats. The rat's blood was completely removed and replaced with a PFC emulsion. The animals lived for a few hours and recovered fully after their blood was replaced. The ideal artificial blood product has a number of characteristics. First, it must be safe to use and compatible within the human body. This means that different blood types should not matter when an artificial blood is used. It also means that artificial blood can be processed to remove all disease-causing agents such as viruses and microorganisms. Second, it must be able to transport oxygen throughout the body and release it where it is needed. Third, it must be shelf stable. Unlike donated blood, artificial blood can be stored for over a year or more. This is in contrast to natural blood, which can only be stored for one month before it breaks down. There are two significantly different products that are under development as blood substitutes. They differ primarily in the way that they carry oxygen. One is based on PFC, while the other is a hemoglobin-based product. The production of artificial blood can be done in a variety of ways. For hemoglobin-based products, this involves isolation of hemoglobin, molecular modification, and then reconstitution in an artificial blood formula.

9.10 ORGAN TRANSPLANT

In general terms, *organ transplant* is the moving of an organ from one body to another for the purpose of replacing the recipient's damaged or failing organ with a working one from the donor. Organ donors can be living or deceased. In the human body, both organs (such as the heart, kidneys, liver, lungs, pancreas, and intestines) and tissues (which include bones, tendons, cornea, heart valves, veins, arms, and skin) can be transplanted. Transplantation medicine is one of the most challenging and complex areas of modern medicine.

9.10.1 History of Organ Transplant

The first successful corneal allograft transplant was performed in 1837 on a gazelle model. The first successful human corneal transplant, a keratoplastic operation, was performed by Eduard Zirm in Olomouc, Czech Republic, in 1905. Pioneering work in the surgical technique of transplantation was done in the early 1900s with the transplantation of arteries or veins by the French surgeon, Alexis Carrel, with Charles Guthrie. Their skillful anastomosis operations and new suturing techniques laid the groundwork for subsequent transplant surgery and won Carrel the 1912

Nobel Prize in Physiology or Medicine. From 1902, Carrel performed transplant experiments on dogs. Surgically successful in moving kidneys, hearts, and spleens, he was one of the first to identify the problem of rejection, which remained insurmountable for decades. Major steps in skin transplantation occurred during World War I, notably in the work of Harold Gillies at Aldershot. Among his advances was the tubed pedicle graft, maintaining a flesh connection from the donor site until the graft established its own blood flow. Gillies' assistant, Archibald McIndoe, carried on the work into World War II as reconstructive surgery. In 1962, the first successful replantation surgery was performed, reattaching a severed limb and restoring (limited) function and feeling.

Transplant of a single gonad (testis) from a living donor was carried out in early July 1926 in Zaječar, Serbia, by a Russian emigré surgeon, Dr. Peter Vasil'evič Kolesnikov. The donor was a convicted murderer, Ilija Krajan, whose death sentence was commuted to 20 years imprisonment; he was led to believe that it was done because he had donated his testis to an elderly medical doctor. Both the donor and the receiver survived, but charges were brought in a court of law by the public prosecutor against Dr. Kolesnikov—not for performing the operation, but for lying to the donor. The first attempted human deceased-donor transplant was performed by the Ukrainian surgeon Yu Voronoy in the 1930s, where rejection resulted in failure. Joseph Murray performed the first successful transplant, a kidney transplant between identical twins, in 1954, successful because no immunosuppression was necessary in genetically identical twins.

In 1951, Peter Medawar improved the understanding of organ rejection by identifying the immune reactions. He suggested that immunosuppressive drugs could be used to minimize organ rejection. The discovery of cyclosporine in 1970 revolutionized organ transplantation techniques with greater success. Dr. Murray's success with the kidney led to attempts to transplant other organs. There was a successful deceased-donor lung transplant into a lung cancer sufferer in June 1963 by James Hardy in Jackson, Mississippi. The patient survived for 18 days before dying of kidney failure. Thomas Starzl of Denver attempted a liver transplant in the same year, but was not successful until 1967. The heart was a major prize for transplant surgeons. However, as well as rejection issues, the heart deteriorates within minutes of death, so any operation would have to be performed at great speed. The development of the heart–lung machine was also needed. Lung pioneer James Hardy attempted a human heart transplant in 1964, but a premature failure of the recipient's heart caught Hardy with no human donor, so he used a chimpanzee heart, which failed very quickly. The first success was achieved on December 3, 1967, by Christiaan Barnard in Cape Town, South Africa. Louis Washkansky, the recipient, survived for 18 days amid what many saw as a distasteful publicity circus. The media interest prompted a spate of heart transplants. Over a hundred were performed in 1968–1969, but almost all the patients died within 60 days. Barnard's second patient, Philip Blaiberg, lived for 19 months.

It was the advent of cyclosporine that altered transplants from research surgery to life-saving treatment. In 1968, surgical pioneer Denton Cooley performed 17 transplants, including the first heart–lung transplant. Fourteen of his patients were dead within 6 months. By 1984, two-thirds of all heart transplant patients survived for 5 years or more. With organ transplants becoming commonplace, limited only by donors, surgeons moved onto more risky fields—multiple organ transplants on humans and whole-body transplant research on animals. On March 9, 1981, the first successful heart–lung transplant took place at Stanford University Hospital. The head surgeon, Bruce Reitz, credited the patient's recovery to cyclosporine-A. As the rising success rate of transplants and modern immunosuppressants make transplants more common, the need for more organs has become critical. Advances in living related-donor transplants have made that increasingly common. Additionally, there is substantive research into xenotransplantation or transgenic organs. Although these forms of transplant are not yet being used in humans, clinical trials involving the use of specific cell types, such as using porcine islets of Langerhans to treat type 1 diabetes, have been conducted with promising results. However, there are still many problems that need to be solved before these would be feasible options in patients requiring transplants.

Recently, researchers have been looking into steroid-free immunosuppressants. This type of immunosuppression is being pioneered on a large scale at Northwestern University in Evanston, Illinois, and other smaller institutions, while steroid minimization is being employed at the University of Wisconsin, Madison, and other smaller institutions. This would avoid the side effects of steroids. While short-term outcomes appear promising, long-term outcomes are still unknown. In addition, calcineurin-inhibitor-free immunosuppression is currently undergoing extensive trialing, the result of which would be to allow sufficient immunosuppression, without the nephrotoxicity associated with standard regimens, which include calcineurin inhibitors. Positive results have yet to be demonstrated in any trial. An FDA-approved immune function test from Cylex has shown effectiveness in minimizing the risk of infection and rejection in post-transplant patients by enabling doctors to tailor immunosuppressant drug regimens. By keeping a patient's immune function within a certain window, doctors can adjust drug levels to prevent organ rejection, while avoiding infection. Such information could help physicians reduce the use of immunosuppressive drugs, lower drug therapy expenses while reducing the morbidity associated with liver biopsies, improve the daily life of transplant patients, and could prolong the life of the transplanted organ. There is minimal evidence that this monitoring can be used with clinical benefit to patients. In most countries, there is a shortage of suitable organs for transplantation. Countries often have formal systems in place to manage the allocation and reduce the risk of rejection. Some countries are associated with international organizations like Eurotransplant in order to increase the supply of appropriate donor organs and allocate organs to recipients. Transplantation also raises a number of bioethical issues, including the definition of death, when and how consent should be given for an organ to be transplanted, and payment for organs for transplantation.

9.10.2 Types of Transplants

9.10.2.1 Autograft

Tissue transplanted from one part of the body to another in the same individual is called an *autograft* or *autotransplant.* A common example is when a piece of bone (usually from the hip) is removed and ground into a paste when reconstructing another portion of bone. This is sometimes done with surplus tissue, or tissue that can regenerate, or tissues more desperately needed elsewhere (such as skin grafts). In orthopedic medicine, a bone graft can be sourced from a patient's own bone in order to fill space and produce an osteogenic response in a bone defect. However, due to the donor-site morbidity associated with autografts, other methods such as bone allograft and bone morphogenetic proteins and synthetic graft materials are often used as alternatives. Autografts have long been considered the "gold standard" in oral surgery and implant dentistry, because they offered the best regeneration results. Lately, the introduction of morphogen-enhanced bone graft substitutes have shown similar success rates and quality of regeneration. However, their price is still very high. Sometimes, an autograft is done to remove the tissue and then treat it or the person before returning it (examples include stem cell autograft and storing blood in advance of surgery).

9.10.2.2 Allograft

An *allograft* or *allotransplantation* is the transplantation of cells, tissues, or organs, sourced from a genetically nonidentical member of the same species as the recipient. The transplant is called an allograft or allogeneic transplant, or a homograft. Most human tissue and organ transplants are allografts. In bone marrow transplantation, the term for a genetically identical graft is *syngeneic*, whereas the equivalent of an autograft is termed *autologous transplantation.* When a host mounts an immune response against an allograft or xenograft, the process is termed *rejection.* For any number of reasons, before death, a person may decide to donate tissue from his or her body for the purpose of transplant to another who is in need. Additionally, consent for donation may also be given by the donor's family if the donor did not specify his or her wishes before death. After consent is obtained, potential donors are thoroughly screened for risk factors and medical conditions that would rule out donation. This screening includes interviews with family members, evaluation of medical and hospital records, and a physical assessment of the donor. Recovery of the tissue is performed with respect for the donor, using surgical techniques.

Personnel from tissue recovery agencies remove the tissue from the donor. These agencies are under the regulation of the FDA and must abide by the Current Good Tissue Practices (cGTP) rule. Once the tissue is removed, it is sent to tissue banks for processing and distribution. Each year, tissue banks accredited by the American Association of Tissue Banks (AATB) distribute 1.5 million bone and tissue allografts. These banks are also regulated by the FDA to ensure the quality of the tissue being distributed. An allogenic bone marrow transplant can result in an immune attack against the recipient, called *graft-versus-host disease*. Due to the genetic differences between the organ and the recipient, the recipient's immune system will identify the organ as foreign and attempt to destroy it, causing transplant rejection. To prevent this, the organ recipient must take immunosuppressants. This dramatically affects the entire immune system, making the body vulnerable to pathogens.

9.10.2.3 Isograft

An *isograft* is a graft of tissue between two individuals who are genetically identical (monozygotic twins). Transplant rejection between two such individuals virtually never occurs. As monozygotic twins have the same MHC, there is very rarely any rejection of transplanted tissue by the adaptive immune system. Furthermore, there is virtually no incidence of graft-versus-host disease. This forms the basis for why a monozygotic twin will be the preferred choice of a physician considering an organ donor.

9.10.2.4 Xenograft

A xenograft is a surgical graft of cells, tissues, or organs from one species to an unlike species, such as from pigs to humans. Human xenotransplantation offers a potential treatment for end-stage organ failure, a significant health problem in parts of the industrialized world. It also raises many novel medical, legal, and ethical issues. A continuing concern is that pigs have a shorter lifespan than humans and their tissues age at a different rate. Disease transmission (xenozoonosis) and permanent alteration to the genetic code of animals are also causes for concern. There are few published cases of successful xenotransplantation. Because there is a worldwide shortage of organs for clinical implantation, approximately 60% of patients awaiting replacement organs die on the waiting list. Recent advances in understanding the mechanisms of transplant rejection have brought science to a stage where it is reasonable to consider that organs from other species, probably pigs, may soon be engineered to minimize the risk of serious rejection and used as an alternative to human tissues, possibly ending organ shortages. Other procedures, some of which are being investigated in early clinical trials, aim to use cells or tissues from other species to treat life-threatening and debilitating illnesses such as cancer, diabetes, liver failure, and Parkinson's disease. If vitrification can be perfected, it could allow for long-term storage of xenogenic cells, tissues, and organs so that they would be more readily available for transplant. Xenotransplants could save thousands of patients waiting for donated organs. The animal organ, probably from a pig or baboon, could be genetically altered with human genes to trick a patient's immune system into accepting it as a part of his own body. Xenotransplants have re-emerged because of the lack of organs available and the constant battle to keep immune systems from rejecting allotransplants.

9.11 CLONING

The word "cloning," which literally means the exact copy of an individual, became a household name ever since researcher Ian Wilmut created a cloned sheep named Dolly. In biology, *cloning* is the process of producing populations of genetically-identical individuals and occurs in nature when organisms such as bacteria, insects, or plants reproduce asexually. In this section, we will discuss cloning and its significance, and how cloning has been used in medical biotechnology to produce an array of products.

9.11.1 Molecular Cloning

Molecular cloning refers to the process of making multiple copies of a defined DNA sequence. Cloning is frequently used to amplify DNA fragments containing whole genes, but it can also be used to amplify any DNA sequence, such as promoters, noncoding sequences, and randomly fragmented DNA. It is used in a wide array of biological experiments and practical applications, ranging from genetic fingerprinting to large-scale protein production. Occasionally, the term "cloning" is misleadingly used to refer to the identification of the chromosomal location of a gene associated with a particular phenotype of interest, such as in positional cloning. In practice, localization of the gene to a chromosome or genomic region does not necessarily enable one to isolate or amplify the relevant genomic sequence. In order to amplify any DNA sequence in a living organism, that sequence must be linked to an origin of replication, which is a sequence of DNA capable of directing the propagation of itself and any linked sequence. However, a number of other features are needed, and a variety of specialized cloning vectors exist, which allow protein expression, tagging, single-stranded RNA and DNA production, and a host of other manipulations. Cloning of any DNA fragment essentially involves four steps: (1) *fragmentation*—breaking apart a strand of DNA; (2) *ligation*—gluing together pieces of DNA in a desired sequence; (3) *transfection*—inserting the newly formed pieces of DNA into cells; and (4) *screening/selection*—selecting out the cells that were successfully transfected with the new DNA.

Although these steps are invariable among cloning procedures, a number of alternative routes can be selected. These are summarized as a "cloning strategy." Initially, the DNA of interest needs to be isolated to provide a DNA segment of suitable size. Subsequently, a ligation procedure is used, where the amplified fragment is inserted into a vector (piece of DNA). The vector (which is frequently circular) is linearized using restriction enzymes, and incubated with the fragment of interest under appropriate conditions with an enzyme called DNA ligase. Following ligation, the vector with the insert of interest is transfected into cells. A number of alternative techniques are available, such as chemical sensitization of cells, electroporation, and biolistics. Finally, the transfected cells are cultured. As the aforementioned procedures are of particularly low efficiency, there is a need to identify the cells that have been successfully transfected with the vector construct containing the desired insertion sequence in the required orientation. Modern cloning vectors include selectable antibiotic resistance markers, which allow only cells in which the vector has been transfected to grow. Additionally, the cloning vectors may contain color selection markers that provide blue/white screening (α-factor complementation) on X-gal medium.

9.11.2 Clonal Cell Technology

In recent times, enormous efforts have been made to derive a population of cells from a single cell, which is called *clonal cell technology*. In the case of unicellular organisms such as bacteria and yeast, this process is remarkably simple and essentially only requires the inoculation of the appropriate medium. However, cell cloning is an arduous task in the case of cell cultures from multicellular organisms, as these cells will not readily grow in standard media. A useful tissue culture technique used to clone distinct lineages of cell lines involves the use of cloning rings. According to this technique, a single-cell suspension of cells which have been exposed to a mutagenic agent or drug used to drive selection is plated at high dilution to create isolated colonies, each arising from a single and potentially clonally distinct cell. At an early growth stage when colonies consist of only a few cells, sterile polystyrene rings, also known as *cloning rings*, that have been dipped in grease are placed over an individual colony, and a small amount of trypsin is added. Cloned cells are collected from inside the ring and transferred to a new vessel for further growth. These clonal cells can be used for both therapeutic and diagnostic applications.

9.11.3 Clonal Embryo

With the strict limitation of getting human embryos for ES cells, scientist explored other means to develop human embryos, and they used *somatic cell nuclear transfer* (SCNT) technology to create a clonal embryo. The most likely purpose for this is to produce embryos for use in research, particularly stem cell research. This process is also called *therapeutic cloning*. The goal is not to create cloned human beings, but rather to harvest stem cells that can be used to study human development and to potentially treat disease.

9.11.4 Reproductive Cloning

Reproductive cloning uses SCNT to create animals that are genetically identical. This process entails the transfer of a nucleus from a donor adult cell to an egg that has no nucleus. If the egg begins to divide normally, it is transferred into the uterus of the surrogate mother. Such clones are not strictly identical, since the somatic cells may contain mutations in their nuclear DNA. Additionally, the mitochondria in the cytoplasm also contains DNA; during SCNT, this DNA is wholly from the donor egg, and the mitochondrial genome is therefore not the same as that of the nucleus donor cell from which it was produced. This may have important implications for cross-species nuclear transfer in which nuclear–mitochondrial incompatibilities may lead to death. Modern cloning techniques involving nuclear transfer have been successfully performed on several species, such as tadpole (1952), carp (1963), mice (1986), Dolly the sheep (1997), gaur (2001), cattle (2001), cat (late 2001), water buffalo (2009), and camel (2009). One of the setbacks to animal cloning is that most cloned animals died due to unknown health complications (Figure 9.10).

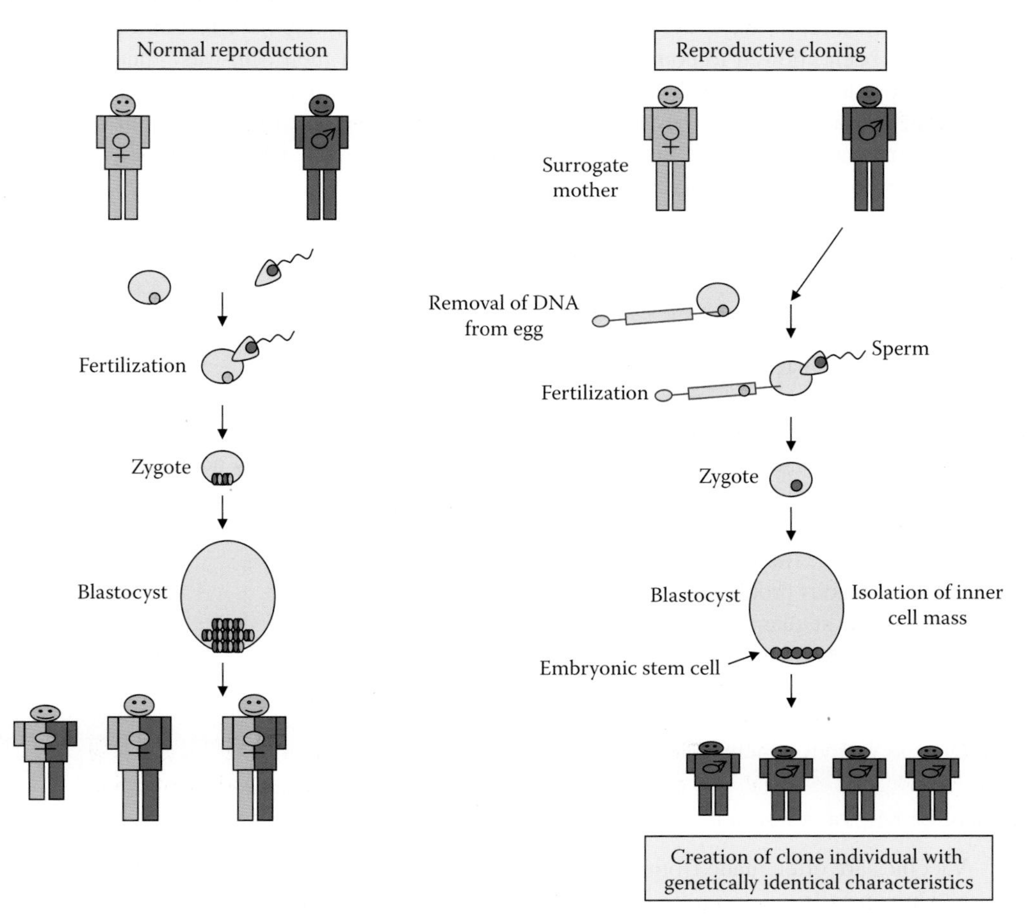

FIGURE 9.10 Reproductive cloning.

9.11.5 Human Cloning

After cloning so many species, the next big step would be cloning humans. *Human cloning* is the creation of a genetically identical copy of an existing or previously existing human. There are two commonly discussed types of human cloning: *therapeutic cloning* and *reproductive cloning*. Therapeutic cloning involves cloning cells from an adult for use in medicine and is an active area of research, while reproductive cloning involves making cloned human beings. Such reproductive cloning has not been performed and is illegal in many countries. A third type of cloning called *replacement cloning* is a theoretical possibility, and would be a combination of therapeutic and reproductive cloning. Replacement cloning would entail the replacement of an extensively damaged or a failed or failing body through cloning, followed by whole or partial brain transplant. The various forms of human cloning are controversial. There have been numerous demands for all progress in the human cloning field to be halted. Some people and groups oppose therapeutic cloning, but most scientific, governmental, and religious organizations oppose reproductive cloning. The American Association for the Advancement of Science (AAAS) and other scientific organizations have made public statements suggesting that human reproductive cloning be banned until safety issues are resolved. Serious ethical concerns have been raised by the idea that it might be possible in the future to harvest organs from clones. Some people have considered the idea of growing organs separately from a human organism. In doing this, a new organ supply could be established without the moral implications of harvesting them from humans.

The first hybrid human clone was created in November 1998 by American Cell Technologies (ACT). It was created from a man's leg cell, and a cow's egg with the DNA removed. However, it was destroyed after 12 days. While making an embryo, which may have resulted in a complete human had it been allowed to come to term, ACT claimed that their aim was "therapeutic cloning," not "reproductive cloning." In January 2008, Stemagen Corporation Lab, in La Jolla in the United States, announced that they had successfully created the first five mature human embryos using DNA from adult skin cells, aiming to provide a source of viable ES cells; however, the cloned embryos were destroyed for ethical reasons.

QUESTIONS

Section A: Descriptive Type

Q1. What are the different types of vaccines?
Q2. Describe various applications of MABs.
Q3. Explain the significance of adult stem cells.
Q4. Is it possible to isolate stem cells from the human body? Why or why not?
Q5. Explain the issue related to contamination of human ES cells.
Q6. What is reproductive cloning?
Q7. How does gene therapy work?
Q8. Explain various problems associated with gene therapy.
Q9. How is gene sequencing done?
Q10. What is DNA fingerprinting?
Q11. What is the difference between RFLP and STR analysis? Explain.
Q12. Describe allograft transplantation.
Q13. Discuss various benefits of bioengineered skin.

Section B: Multiple Choice

Q1. Vaccines are often made from weakened or killed forms of ...
 a. Human cells
 b. Fungi

c. Microbes
d. Chicken

Q2. Which of the following has no available vaccine to date?
a. Diphtheria
b. Measles
c. Malaria
d. Rubella

Q3. Who first proposed the idea of "magic bullet"?
a. Alexander Fleming
b. Paul Ehrlich
c. Kary Mullis
d. Georges Kohler

Q4. Hybridoma technology deals with …
a. Vaccine production
b. Antibiotic production
c. Polyclonal antibodies production
d. MABs production

Q5. Recombinant antibody engineering involves the use of viruses or yeast to create antibodies, rather than mice. True/False

Q6. Therapeutic proteins are also known as …
a. Biologicals
b. Biomaterials
c. Biopharmaceuticals
d. Pharmaceuticals

Q7. Mesenchymal stem cells are …
a. Unipotent stem cells
b. Multipotent stem cells
c. Pluripotent stem cells
d. Totipotent stem cells

Q8. From which embryonic stage are embryonic stem cells derived?
a. Blastocyst
b. Gastrula
c. Morula
d. Germ layer

Q9. The U.S. FDA approved the first clinical trial using embryonic stem cells in 2009. True/False

Q10. *Ex vivo* gene therapy means …
a. Externally modified cells transplanted back to same patient
b. Externally modified cells transplanted back to another patient
c. Internally modified cells

Q11. Xenograft involves a transplant from …
a. Mouse to mouse
b. Mouse to human
c. Human to human
d. None of the above

Section C: Critical Thinking

Q1. Do you think it would be better if recombinant therapeutic proteins are produced using human cells rather than using microbes or animals? If no, explain. If yes, what would be your strategy?

Q2. Will embryonic stem cells ever be a successful cell-based therapy? Why or why not?

Q3. Why does gene therapy fail in many aspects?

ASSIGNMENT

Read more about human genome project and submit a report on genetic variations and mutations found in the human genome.

REFERENCES AND FURTHER READING

Alexeyev, M.F., LeDoux, S.P., and Wilson, G.L. Mitochondrial DNA and aging. *Clin. Sci.* 107: 355–364, 2004.

Andrews, P., Matin, M., Bahrami, A., Damjanov, I., Gokhale, P., and Draper, J. Embryonic stem (ES) cells and embryonal carcinoma (EC) cells: Opposite sides of the same coin. *Biochem. Soc. Trans.* 33: 1526–1530, 2005.

Bammler, T., Beyer, R.P., Bhattacharya, S., Boorman, G.A., Boyles, A., Bradford, B.U., Bumgarner, R.E., Bushel, P.R., Chaturvedi, K., Choi, D., Cunningham, M.L. et al. Standardizing global gene expression analysis between laboratories and across platforms. *Nat. Methods* 2: 351–356, 2005.

Barrilleaux, B., Phinney, D.G., Prockop, D.G., and O'Connor, K.C. Review: Ex vivo engineering of living tissues with adult stem cells. *Tissue Eng.* 12: 3007–3019, 2006.

Beachy, P.A., Karhadkar, S.S., and Berman, D.M. Tissue repair and stem cell renewal in carcinogenesis. *Nature* 432: 324–331, 2004.

Bhattacharya, S. Killer convicted thanks to relative's DNA. *New Sci.* 2004.

Black, L.L., Gaynor, J., Adams, C., Dhupa, S., Sams, A.E., Taylor, R., Harman, S., Gingerich, D.A., and Harman, R. et al. Effect of intraarticular injection of autologous adipose-derived mesenchymal stem and regenerative cells on clinical signs of chronic osteoarthritis of the elbow joint in dogs. *Vet. Therap.* 9: 192–200, 2008.

Burt, R.K., Loh, Y., Pearce, W. et al. Clinical applications of blood-derived and marrow-derived stem cells for nonmalignant diseases. *JAMA* 299: 925–936, 2008.

Carson, H.L. Chromosomal tracers of evolution. *Science* 168: 1414–1418, 1970.

Carson, H.L. Chromosomal sequences and interisland colonizations in Hawaiian Drosophila. *Genetics* 103: 465–482, 1983.

Chaudhary, P.M. and Roninson, I.B. Expression and activity of P-glycoprotein, a multidrug efflux pump, in human hematopoietic stem cells. *Cell* 66: 85–94, 1991.

Churchill, G.A. Fundamentals of experimental design for cDNA microarrays. *Nat. Genet. Suppl.* 32: 490, 2002.

Clark, D.P. and Russell, L.D. *Molecular Biology Made Simple and Fun*, 2nd edn. Cache River Press, Vienna, IL, 2000.

Comai, L. The advantages and disadvantages of being polyploid. *Nat. Rev. Genet.* 6: 836–846, 2005.

Cutler, C. and Antin, J.H. Peripheral blood stem cells for allogeneic transplantation: A review. *Stem Cells* 19: 108–117, 2001.

Dulbecco, R. and Ginsberg, H.S. *Virology*, 2nd edn. J.B. Lippincott Company, Philadelphia, PA, 1988.

Edgar, B.A. and Orr-Weaver, T.L. Endoreduplication cell cycles: More for less. *Cell* 105: 297–306, 2001.

Evans, M. and Kaufman, M. Establishment in culture of pluripotent cells from mouse embryos. *Nature* 292: 154–156, 1981.

Giarratana, M.C., Kobari, L., Lapillonne, H. et al. Ex vivo generation of fully mature human red blood cells from hematopoietic stem cells. *Nat. Biotechnol.* 23: 69–74, 2005.

Gilbert, S.F. *Developmental Biology*, 8th edn., Chapter 9, Sinauer Associates, Stamford, CT, 2006.

Hacia, J.G., Fan, J.B., Ryder, O. et al. Determination of ancestral alleles for human single-nucleotide polymorphisms using high-density oligonucleotide arrays. *Nat. Genet.* 22: 164–167, 1999.

Haupt, Y., Bath, M.L., Harris, A.W., and Adams, J.M. Bmi-1 transgene induces lymphomas and collaborates with myc in tumorigenesis. *Oncogene* 8: 3161–3164, 1993.

Hwang, W.S., Roh, S.I., Lee, B.C. et al. Patient-specific embryonic stem cells derived from human SCNT blastocysts. *Science* 308: 1777–1783, 2005.

Keirstead, H.S., Nistor, G., Bernal, G. et al. Human embryonic stem cell-derived oligodendrocyte progenitor cell transplants remyelinate and restore locomotion after spinal cord injury. *J. Neurosci.* 25: 4694–4705, 2005.

Kennedy, D. Editorial retraction. *Science* 311(5759): 335, 2006.

Kikyo, N. and Wolffe, A.P. Reprogramming nuclei: Insights from cloning, nuclear transfer and heterokaryons. *J. Cell Sci.* 113: 11–20, 2000.

Klimanskaya, I., Chung, Y., Becker, S., Lu, S.J., and Lanza, R. Human embryonic stem cells derived without feeder cells. *Lancet* 365: 1636–1641, 2006.

Kohler, G. and Milstein, C. Continuous cultures of fused cells secreting antibody of predefined specificity. *Nature* 256: 495–497, 1975.

Kottler, M. Cytological technique, preconception and the counting of the human chromosomes. *Bull. Hist. Med.* 48: 465–502, 1974.

Kurinczuk, J.J. Safety issues in assisted reproduction technology. From theory to reality—Just what are the data telling us about ICSI offspring health and future fertility and should we be concerned. *Hum. Reprod.* 18: 925–931, 2003.

Langer, R. and Vacanti, J.P. Tissue engineering. *Science* 260: 920–926, 1993.

Lashkari, D.A., DeRisi, J.L., McCusker, J.H. et al. Yeast microarrays for genome wide parallel genetic and gene expression analysis. *Proc. Natl. Acad. Sci. USA* 94: 13057–13062, 1997.

Martin, G. Isolation of a pluripotent cell line from early mouse embryos cultured in medium conditioned by teratocarcinoma stem cells. *Proc. Natl. Acad. Sci. USA* 78: 7634–7638, 1981.

Martin, M.J., Muotri, A., Gage F., and Varki, A. Human embryonic stem cells express an immunogenic nonhuman sialic acid. *Nat. Med.* 11: 228–232, 2005.

McCarthy, M. Bio-engineered tissues move towards the clinic. *Lancet* 348(9025): 466, 1996.

McFarland, D. Preparation of pure cell cultures by cloning. *Methods Cell Sci.* 22: 63–66, 2000.

McLaren, A. Cloning: Pathways to a pluripotent future. *Science*. 288: 1775–1780, 2000.

Michael, W., Horn, M., and Holger, D. Fluorescence in situ hybridisation for the identification and characterization of prokaryotes. *Curr. Opin. Microbiol.* 6: 302–309, 2003.

Olson, W.P. *Separations Technology. Pharmaceutical and Biotechnology Applications*. Interpharm Press, Inc., Buffalo Grove, IL, 1995.

Painter, T.S. The spermatogenesis of man. *Anat. Res.* 23: 129, 1922.

Painter, T.S. Studies in mammalian spermatogenesis II. The spermatogenesis of humans. *J. Exp. Zoology* 37: 291–336, 1923.

Rundle, R.L. Cells 'Tricked' to make skin for burn cases. *Wall Street J.* 1994.

Yen, A.H. and Sharpe, P.T. Stem cells and tooth tissue engineering. *Cell Tissue Res.* 331: 359–372, 2008.

10 Nanobiotechnology

LEARNING OBJECTIVES

- Define nanotechnology
- Explain the significance of nanobiotechnology
- Discuss various applications of nanobiotechnology, including nanomedicine
- Discuss nanoparticles and nanomaterials
- Explain the application of nanobiotechnology in food testing and water pollution
- Discuss current research trends in nanobiotechnology

10.1 INTRODUCTION

Nanobiotechnology is the convergence of engineering and molecular biology that is leading to a new class of multifunctional devices and systems for biological and chemical analysis with better sensitivity and specificity and a higher rate of recognition. *Nanobiotechnology* is the branch of nanotechnology with biological and biochemical applications or uses. It often studies existing elements of nature in order to fabricate new devices. The term "bionanotechnology" is often used interchangeably with nanobiotechnology, though a distinction is sometimes drawn between the two; nanobiotechnology usually refers to the use of nanotechnology to further the goals of biotechnology, while bionanotechnology might refer to any overlap between biology and nanotechnology, including the use of biomolecules as part of or as an inspiration for nanotechnological devices. In the following two sections, we will examine both aspects in order to understand the significance of nanobiotechnology.

10.2 NANOTECHNOLOGY

Before discussing nanobiotechnology, we will first look at the field of nanotechnology and how it has shaped the development of nanobiotechnology. The concept of nanotechnology was first envisioned as early as 1959 by the renowned physicist Richard Feynman, who said, "I want to build a billion tiny factories, models of each other, which are manufacturing simultaneously … The principles of physics, as far as I can see, do not speak against the possibility of maneuvering things atom by atom. It is not an attempt to violate any laws; it is something, in principle, that can be done; but in practice, it has not been done because we are too big." The term "nanotechnology" was defined by Tokyo Science University's Professor Norio Taniguchi in a 1974 paper as follows: "Nanotechnology mainly consists of the processing of, separation, consolidation, and deformation of materials by one atom or by one molecule." However, it was K. Eric Drexler who popularized the word "nanotechnology" in the 1980s by publishing a research paper in the journal *Proceeding of National Academy of Sciences* in the United States. In order to prove his passion for nanotechnology, Drexler spent many years in his lab describing and analyzing nanodevices based on the molecular scale. Later on, as nanotechnology became an accepted concept, the meaning of the word shifted to encompass the simpler kinds of nanometer-scale technology.

The U.S. National Nanotechnology Initiative (NNI) was created to fund nanotechnology research, and their definition includes anything smaller than 100 nanometers (nm) with novel properties. Nanotechnology is extremely diverse, ranging from novel extensions of conventional device physics, to completely new approaches based upon molecular self-assembly, to developing new materials

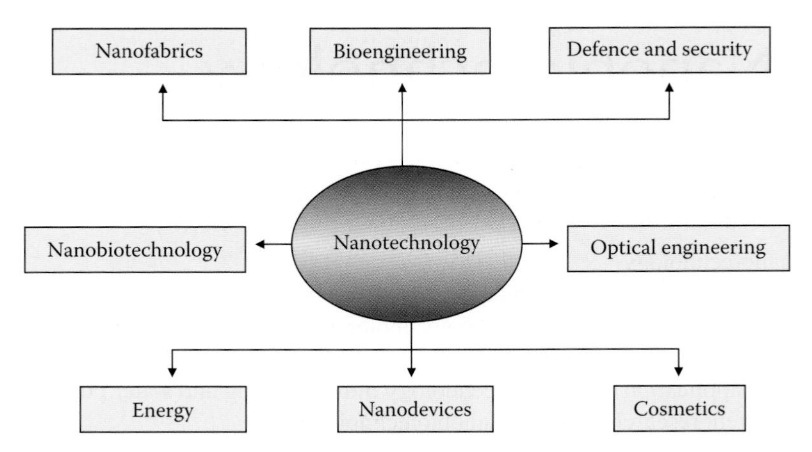

FIGURE 10.1 Applications of nanotechnology.

with dimensions on the nanoscale—even to speculation on whether we can directly control matter on the atomic scale. Nanotechnology and nanoscience originated in the early 1980s with two major developments: the birth of cluster science and the invention of the scanning tunneling microscope (STM). This development led to the discovery of fullerenes in 1985 and carbon nanotubes a few years later. In another development, the synthesis and properties of semiconductor nanocrystals was studied. This led to a fast increasing number of metal and metal oxide nanoparticles and quantum dots (qdots). The atomic force microscope was invented 6 years after the STM was invented. In 2000, the NNI was founded to coordinate federal nanotechnology research and development (R&D). The various applications of nanotechnology are illustrated in the Figure 10.1.

10.3 NANOBIOTECHNOLOGY

Nanobiotechnology—one of the relatively recent, promising (and yet, largely untapped) fields of science—has its origins in the nanotechnological advances made in the last four decades. It is a marriage between the fields of microelectronics and biological sciences, and combines the complementary strengths of biological molecules (to selectively bind with other molecules) and nanoelectronics (detection of the smallest electrical change caused by such binding). The integration of biological science with engineering is one of the most challenging and fastest growing sectors of nanobiotechnology. Nanotechnology is the creation and utilization of materials, devices, and systems through the control of matter on the nanometer-length scale (a nanometer is one billionth of a meter).

Nanobiotechnology is that branch of nanotechnology that deals with the study and application of biological and biochemical activities from elements of nature to fabricate new devices like biosensors. Nanobiotechnology is often used to describe the overlapping multidisciplinary activities associated with biosensors, particularly where photonics, chemistry, biology, biophysics, nanomedicine, and engineering converge. Measurement in biology using waveguide techniques such as dual polarization interferometer is another example. One example of current nanobiotechnological research involves nanospheres coated with fluorescent polymers. Researchers are seeking to design polymers whose fluorescence is quenched when they encounter specific molecules. Different polymers would detect different metabolites. The polymer-coated spheres could become part of new biological assays, and the technology might someday lead to particles that could be introduced into the human body to track down metabolites associated with tumors and other health problems.

Nanobiotechnology—an integration of physical sciences, molecular engineering, biology, chemistry, and biotechnology—holds considerable promise of advances in pharmaceuticals and healthcare.

Initial work is being done to improve the various techniques and materials that are relevant to nanobiotechnology. It includes some of the physical forms of energy, such as nanolasers. Some of the technologies are scaling down, such as microfluidics to nanofluidic biochips. Some of the earliest applications are in molecular diagnostics. Nanoparticles, particularly qdots, are playing important roles. Various nanodiagnostics that have been reviewed will improve the sensitivity and extend the present limits of molecular diagnostics. An increasing use of nanobiotechnology by the pharmaceutical and biotechnology industries is anticipated. Nanotechnology can be applied at all stages of drug development, from formulations to diagnostic applications in clinical trials. Many of the assays based on nanobiotechnology will enable high-throughput screening of drug candidates. Some nanostructures such as fullerenes are themselves drug candidates, as they allow precise grafting of active chemical groups in three-dimensional orientations. Apart from offering a solution to solubility problems, nanobiotechnology provides intracellular delivery possibilities. Skin penetration is improved in transdermal drug delivery. A particularly effective application is on nonviral gene therapy vectors.

Nanotechnology has the potential to provide controlled release devices with autonomous operation guided by need. Nanomedicine is now within the realm of reality, starting with nanodiagnostics and drug delivery facilitated by nanobiotechnology. Miniature devices such as nanorobots could carry out integrated diagnosis and therapy by refined and minimally invasive procedures (nanosurgery) as an alternative to crude surgery. Nanotechnology has markedly improved the implant and tissue engineering approaches as well. There is some concern about the safety of nanoparticles introduced in the human body and released into the environment, but research is underway to address these issues. As yet, there are no directives by the U.S. Food and Drug Administration (FDA) to regulate nanobiotechnology, but these are expected to be in place as products are ready to enter market.

10.4 APPLICATIONS OF NANOBIOTECHNOLOGY

10.4.1 Nanomedicine

One of the major applications of nanotechnology is to develop diagnostic tools for better understanding of human diseases. The application of nanotechnology in medicine is called *nanomedine.* The applications of nanomedicine range from the medical use of nanomaterials, to nanoelectronic biosensors, and even possible applications of molecular nanotechnology. Current challenges in the field of nanomedicine involve understanding the issues related to toxicity and the environmental impact of nanoscale materials. The application of nanomaterials has revolutionized drug discovery research. The journal *Nature Materials* showed that an estimated 130 nanotech-based drugs and delivery systems were being developed worldwide. The NNI expects new commercial applications in the pharmaceutical industry that may include advanced drug delivery systems, new therapies, and *in vivo* imaging. Neuroelectronic interfaces and other nanoelectronics-based sensors are another active goal of research. The speculative field of molecular nanotechnology believes that further down the line cell repair machines could revolutionize medicine and the medical field. Nanomedicine is a large industry, with nanomedicine sales reaching 6.8 billion dollars in 2004, and with over 200 companies and 38 products worldwide, a minimum of 3.8 billion dollars in nanotechnology R&D is being invested every year. As the nanomedicine industry continues to grow, it is expected to have a significant impact on the economy.

Buckyballs may be used to trap free radicals generated during an allergic reaction and block the inflammation that results from an allergic reaction. Nanoshells may be used to concentrate the heat from infrared light to destroy cancer cells with minimal damage to surrounding healthy cells. Nanospectra Biosciences has developed such a treatment using nanoshells illuminated by an infrared laser that has been approved for a pilot trial with human patients. Nanoparticles, when activated by x-rays that generate electrons cause the destruction of cancer cells to which they attach.

TABLE 10.1
Nanotechnology in Medicine: Company Directory

Company	Product
BioDelivery Sciences	Oral drug delivery of drugs encapsulated in a nanocrystalline structure, called a *cochleate*
CytImmune	Gold nanoparticles for targeted delivery of drugs to tumors
Invitrogen	Qdots for medical imaging
Nucryst	Antimicrobial wound dressings using silver nanocrystals
Luna Innovations	Buckyballs to block inflammation by trapping free radicals
NanoBio	Nanoemulsions for nasal delivery to fight viruses (such as influenza and colds) or through the skin to fight bacteria
NanoBioMagnetics	Magnetically responsive nanoparticles for targeted drug delivery and other applications
Nanobiotix	Nanoparticles that target tumor cells; when irradiated by x-rays, the nanoparticles generate electrons which cause localized destruction of the tumor cells
Nanospectra	AuroShell particles (nanoshells) for thermal destruction of cancer tissue
Nanosphere	Diagnostic testing using gold nanoparticles to detect low levels of proteins, indicating particular diseases
Nanotherapeutics	Nanoparticles for improving the performance of drug delivery by oral or nasal methods
Oxonica	Diagnostic testing using gold nanoparticles (biomarkers)
T2 Biosystems	Diagnostic testing using magnetic nanoparticles
Z-Medica	Medical gauze containing aluminosilicate nanoparticles that help blood clot faster in open wounds
Sirnaomics	Nanoparticle-enhanced techniques for delivery of siRNA

This is intended to be used in place of radiation therapy, with much less damage to healthy tissue. Nanobiotix has released preclinical results for this technique. Aluminosilicate nanoparticles can more quickly reduce bleeding in trauma patients by absorbing water, causing blood in a wound to clot quickly. Z-Medica is producing medical gauze that uses aluminosilicate nanoparticles. Nanofibers can stimulate the production of cartilage in damaged joints. The list company which makes the healthcare products is shown in the Table 10.1.

10.4.1.1 Drug Delivery

In recent times, drug delivery has become an important field to improve the effectiveness of drug therapy, and efforts have been made to improve the drug delivery system to avoid pains and to enhance target-specific delivery with minimum side effects. We will first find out what kinds of drug delivery methods are available. *Drug delivery* is the method or process of administering a drug to achieve a therapeutic effect in humans. The most common methods of drug delivery include the preferred noninvasive peroral (through the mouth), topical (skin), transmucosal (nasal, buccal/sublingual, vaginal, ocular, and rectal) and inhalation routes. Many medications, such as peptides and proteins, antibodies, vaccines and gene-based drugs in general, may not be delivered using these routes, because they might be susceptible to enzymatic degradation or cannot be absorbed into the systemic circulation efficiently enough to be therapeutically effective due to molecular size and charge issues For this reason, many protein and peptide drugs have to be delivered by injection. For example, many immunizations are based on the delivery of protein drugs and are often administered by injection (Figure 10.2).

One of the main issues with the current drug delivery system is nontargeted drug delivery, which has number of limitations that include not reaching the target site and drug action on normal cells or tissues. Constant efforts have been made in the area of drug delivery, which include the development of targeted delivery in the target area of the body (e.g., in cancerous tissues) and sustained release formulations in which the drug is released over a period of time in a controlled manner. Types of sustained release formulations include liposomes, drug-loaded biodegradable microspheres, and drug polymer conjugates. Nanomedical approaches to drug delivery

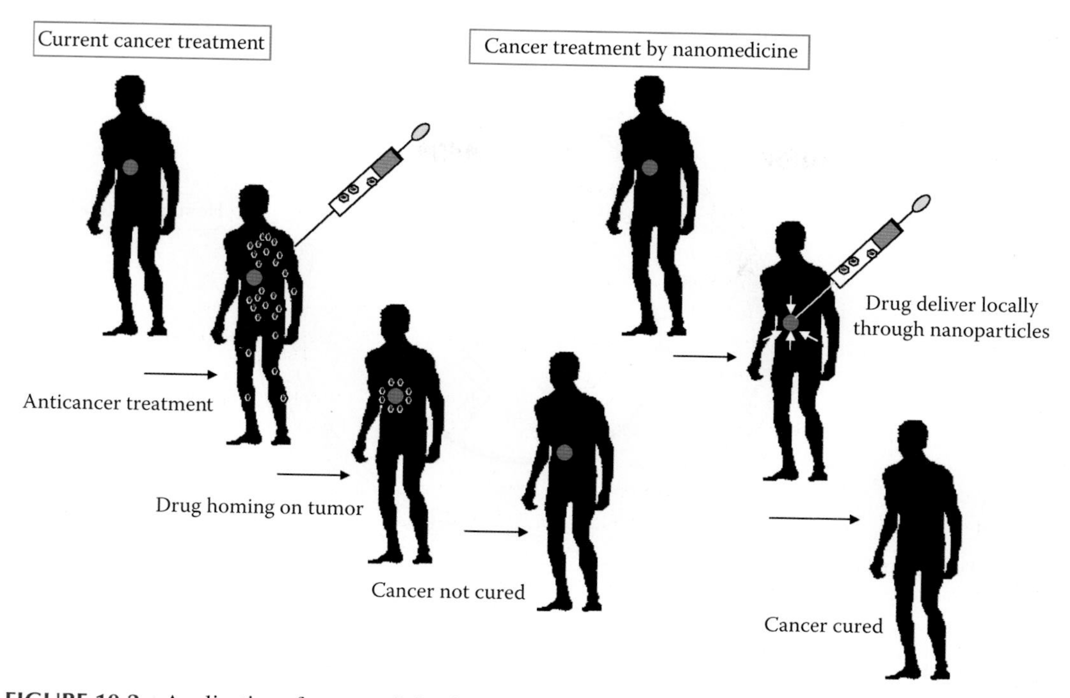

FIGURE 10.2 Application of nanomedicine in cancer treatment.

center on developing nanoscale particles or molecules to improve the bioavailability of a drug. *Bioavailability* refers to the presence of drug molecules where they are needed in the body and where they will do the most good. Drug delivery focuses on maximizing bioavailability, both at specific places in the body and over a period of time. This will be achieved by molecular targeting by nanoengineered devices. It is all about targeting molecules and delivering drugs with cell precision. More than $65 billion are wasted each year due to poor bioavailability of drug molecules. *In vivo* imaging is another area where tools and devices are being developed. Using nanoparticle contrast agents, images such as ultrasound and magnetic resonance imaging (MRI) have a favorable distribution and improved contrast. The new methods of nanoengineered materials that are being developed might be effective in treating illnesses and diseases such as cancer. What nanoscientists will be able to achieve in the future is beyond current imagination. This will be accomplished by self-assembled biocompatible nanodevices that will detect, evaluate, treat, and report to the clinical doctor automatically (Figure 10.3).

Drug delivery systems, lipid- or polymer-based nanoparticles, can be designed to improve the pharmacological and therapeutic properties of drugs (Figure 10.4). The strength of drug delivery systems is their ability to alter the pharmacokinetics and biodistribution of the drug. Nanoparticles have unusual properties that can be used to improve drug delivery. Where larger particles would have been cleared from the body, cells take up these nanoparticles because of their size. Complex drug delivery mechanisms are being developed, including the ability to get drugs through cell membranes and into cell cytoplasm. Efficiency is important, because many diseases depend upon processes within the cells and can only be impeded by drugs that make their way into the cell. A triggered response is one way for drug molecules to be used more efficiently. Drugs are placed in the body and only activate on encountering a particular signal. For example, a drug with poor solubility will be replaced by a drug delivery system where both hydrophilic and hydrophobic environments exist, improving the solubility. In addition, a drug may cause tissue damage, but regulated drug release with drug delivery systems can eliminate the problem. If a drug is cleared too quickly from the body, this could force a patient to use high doses, but clearance can be reduced by altering the pharmacokinetics of the drug with drug delivery systems. Poor biodistribution is a problem that

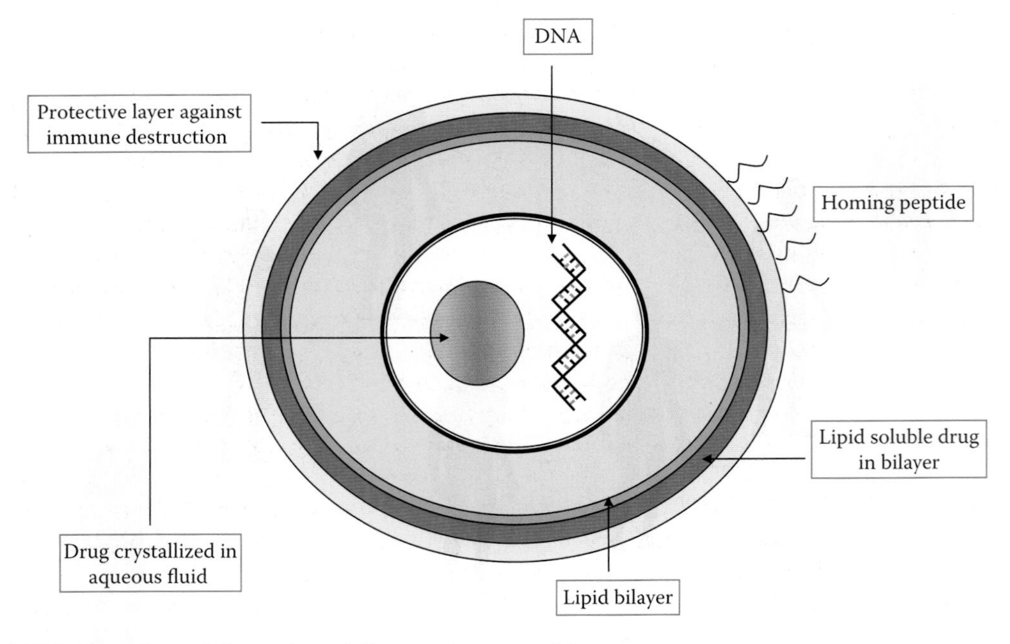

FIGURE 10.3 Drug delivery through liposomal nanoparticles.

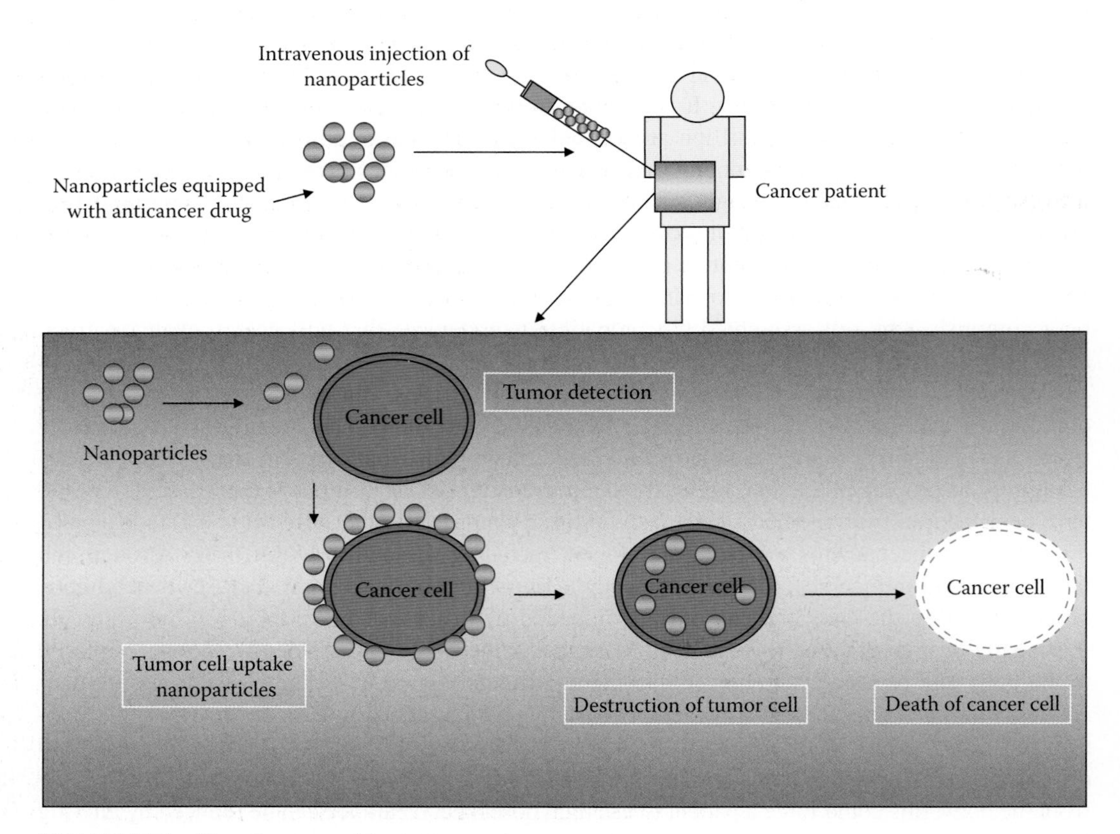

FIGURE 10.4 Use of nanoparticles in cancer treatment.

can affect normal tissues through widespread distribution, but the particulates from drug delivery systems lower the volume of distribution and reduce the effect on nontarget tissue. Potential nanodrugs will work by very specific and well-understood mechanisms. One of the major impacts of nanotechnology and nanoscience will be in leading development of completely new drugs with more useful performance and fewer side effects. The basic point in the use of drug delivery is based upon three facts: (a) efficient encapsulation of the drugs, (b) successful delivery of these drugs to the targeted region of the body, and (c) successful release of these drug in the targeted region.

Targeting ligands (such as monoclonal antibodies and sugar residues) can also be attached to nanoparticles to achieve active targeting of these particles. The hepatocytes of the liver can be an important target in case of hepatitis and also in other cases of gene therapy, when the administered gene needs to be expressed in these cells of the liver. The nanoparticles having ligands for active targeting should not only escape capture by the kupffer cells of the liver, but also need to be small enough (>50 nm, or even >20 nm in size) to be able to enter the hepatocytes with the help of specific receptors on these surfaces and through ferestrations that are 100–150 nm in diameter.

10.4.1.2 Cancer Diagnostics

The success of cancer treatment is primarily based on how quickly we detect the progression of disease so that effective treatment modalities can be implemented in patients. Over the past few years, scientists have shown that nanoparticles are very useful diagnostic tools in oncology, particularly in imaging. Qdots (nanoparticles with quantum confinement properties, such as size-tunable light emission), when used in conjunction with MRI, can produce exceptionally good images of tumor sites. These nanoparticles are much brighter than organic dyes and only need one light source for excitation. This means that the use of fluorescent qdots could produce a higher contrast image at a lower cost than today's organic dyes used as contrast media. The downside, however, is that qdots are usually made of quite toxic elements (Figure 10.5).

Another property of nanoparticles is that they have a high surface-area-to-volume ratio, allowing many functional groups to be attached, which can be used to detect tumor cells. Additionally, the small size of nanoparticles (10–100 nm), allows them to preferentially accumulate at tumor sites, because tumors lack an effective lymphatic drainage system. A very exciting research question is how to make these imaging nanoparticles do more things for cancer. For instance, is it possible to manufacture multifunctional nanoparticles that would detect, image, and then proceed to treat a tumor? This question is under vigorous investigation, and the answer could shape the future of cancer treatment.

A promising new cancer treatment that may one day replace radiation and chemotherapy is edging closer to human trials. Sensor test chips containing thousands of nanowires, able to detect

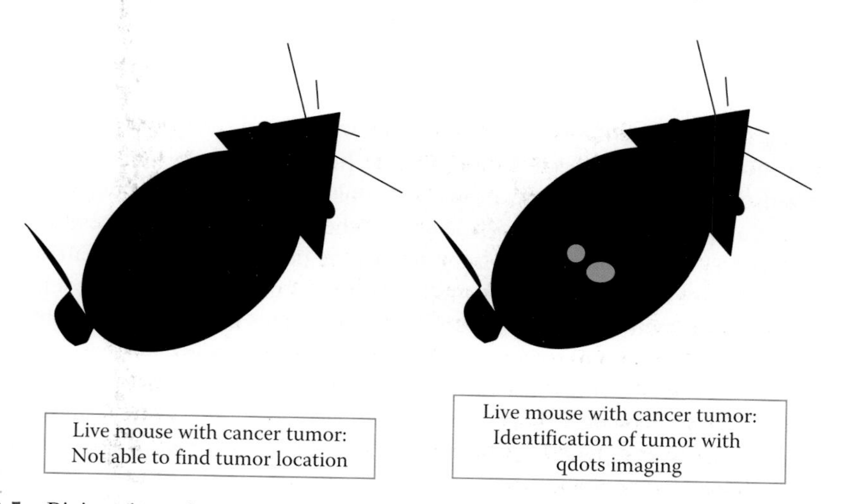

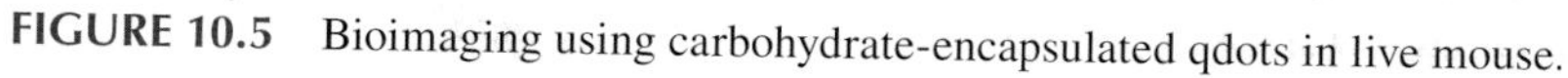

FIGURE 10.5 Bioimaging using carbohydrate-encapsulated qdots in live mouse.

proteins and other biomarkers left behind by cancer cells, could enable the detection and diagnosis of cancer in the early stages from a few drops of a patient's blood. Headed by Prof. Jennifer West, researchers at Rice University have demonstrated the use of 120 nm diameter nanoshells coated with gold to kill cancer tumors in mice. The nanoshells can be targeted to bond to cancerous cells by conjugating antibodies or peptides to the nanoshell surface. By irradiating the area of the tumor with an infrared laser, which passes through flesh without heating it, the gold is heated sufficiently to cause death to the cancer cells. Additionally, John Kanzius has invented a radio machine that uses a combination of radio waves and carbon or gold nanoparticles to destroy cancer cells. Nanoparticles of cadmium selenide (qdots) glow when exposed to ultraviolet (UV) light. When injected, they seep into cancer tumors. The surgeon can see the glowing tumor and use it as a guide for more accurate tumor removal.

One scientist, University of Michigan's James Baker, believes he has discovered a highly efficient and successful way of delivering cancer-treatment drugs that is less harmful to the surrounding body. Baker has developed a nanotechnology that can locate and then eliminate cancerous cells. He looks at a molecule called a dendrimer. This molecule has more than 100 hooks on it that allow it to attach to cells in the body for a variety of purposes. Baker then attaches folic acid to a few of the hooks (folic acid, being a vitamin, is received by cells in the body). Cancer cells have more vitamin receptors than normal cells, so Baker's vitamin-laden dendrimer will be absorbed by the cancer cell. To the rest of the hooks on the dendrimer, Baker places anticancer drugs that will be absorbed with the dendrimer into the cancer cell, thereby delivering the cancer drug to the cancer cell and nowhere else.

In photodynamic therapy, a particle is placed within the body and is illuminated with light from the outside. The light gets absorbed by the particle; if the particle is metal, energy from the light will heat the particle and surrounding tissue. Light may also be used to produce high-energy oxygen molecules that will chemically react with and destroy most organic molecules that are next to them (such as tumors). This therapy is appealing for many reasons. It does not leave a "toxic trail" of reactive molecules throughout the body (as in chemotherapy), because it is directed where only the light is made to shine and the particles exist. Photodynamic therapy has potential as a noninvasive procedure for dealing with diseases, growths, and tumors.

10.4.1.3 *In Vivo* Drug Imaging

Tracking the movement of the drug molecules can help determine how well drugs are being distributed or how substances are metabolized in the body. It is difficult to track a small group of cells throughout the body, so scientists previously dyed the cells. These dyes needed to be excited by light of a certain wavelength in order for them to light up. One of the main problems with this dye method was that it needs multiple light sources for visualization. This problem has been solved by using qdot nanoparticles that can attach to proteins that penetrate cell membranes. The dots can be random in size, can be made of bio-inert material, and they demonstrate the nanoscale property that color is size-dependent. As a result, sizes are selected so that the frequency of light used to make a group of qdots fluorescence is an even multiple of the frequency required to make another group incandesce; then both groups can be lit with a single light source. It has been widely observed that nanoparticles are promising tools for the advancement of drug delivery, medical imaging, and as diagnostic sensors. However, the biodistribution of these nanoparticles is mostly unknown, due to the difficulty in targeting specific organs in the body. Current research in the excretory systems of mice, however, shows the ability of gold composites to selectively target certain organs based on their size and charge. These composites are encapsulated by a dendrimer and assigned a specific charge and size. Positively-charged gold nanoparticles were found to enter the kidneys, while negatively-charged gold nanoparticles remained in the liver and spleen. It is suggested that the positive surface charge of the nanoparticle decreases the rate of osponization of nanoparticles in the liver, thus affecting the excretory pathway. Even at a relatively small size of 5 nm, though, these particles can become compartmentalized in the peripheral tissues, and will therefore accumulate in the body over time.

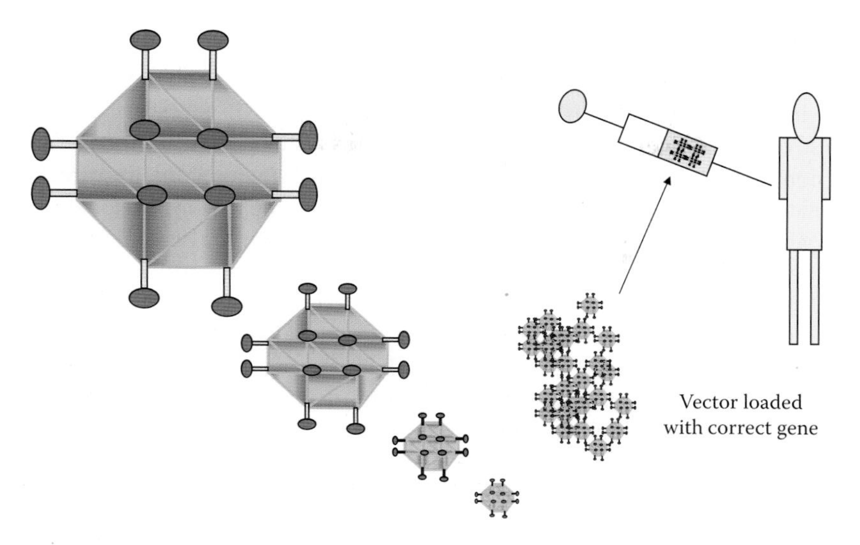

FIGURE 10.6 Polymer vector for gene therapy.

While advancement of research proves that targeting and distribution can be augmented by nanoparticles, the dangers of nanotoxicity become an important next step in further understanding of their medical uses. Qdots may be used in the future for locating cancer tumors in patients and in the near term for performing diagnostic tests in samples. Invitrogen's website provides information about qdots that are available for both uses, although at this time, the use of "*in vivo*" is limited to experiments with laboratory animals. Iron oxide nanoparticles can be used to improve MRI images of cancer tumors. The nanoparticles are coated with a peptide that binds to a cancer tumor. Once the nanoparticles are attached to the tumor, the magnetic property of the iron oxide enhances the images from the MRI scan (Figure 10.6).

Nanoparticles can attach to proteins or other molecules, allowing detection of disease indicators in a laboratory sample at a very early stage. There are several efforts to develop nanoparticle disease detection systems, and one system being developed by *Nanosphere, Inc.* uses gold nanoparticles. Nanosphere has clinical study results with their Verigene system, involving its ability to detect four different nucleic acids. Another system, developed by T2 Biosystems, uses magnetic nanoparticles to identify specimens such as proteins, nucleic acids, and other materials.

10.4.1.4 Nanonephrology

Nanonephrology is a branch of nanomedicine and nanotechnology that deals with (1) the study of kidney protein structures at the atomic level; (2) nano-imaging approaches to study cellular processes in kidney cells; and (3) nanomedical treatments that utilize nanoparticles to treat various kidney diseases. The creation and use of materials and devices at the molecular and atomic levels that can be used for the diagnosis and therapy of renal diseases is also a part of nanonephrology that will play a role in the management of patients with kidney disease in the future. Advances in nanonephrology will be based on discoveries in the above areas that can provide nanoscale information on the cellular molecular machinery involved in normal kidney processes and in pathological states. By understanding the physical and chemical properties of proteins and other macromolecules at the atomic level in various cells in the kidney, novel therapeutic approaches can be designed to combat major renal diseases. The nanoscale artificial kidney is a goal that many physicians dream of. Nanoscale engineering advances will permit programmable and controllable nanoscale robots to execute curative and reconstructive procedures in the human kidney at the cellular and molecular levels. Designing nanostructures compatible with the kidney cells and that can safely operate *in vivo* is also a future goal. The ability to direct events in a controlled fashion at the cellular nano-level has the potential to significantly improve the lives of patients with kidney diseases.

10.4.1.5 Gene Therapy Using Nanotechnology

Nanoparticles have also found use in the field of nanotechnology. A pilot experiment involving gene therapy in human subjects has already been conducted for diseases like cystic fibrosis and muscular dystrophy. Three main types of gene-delivery systems have been described: (1) viral vectors; (2) nonviral vectors (particles and polymers); and (3) gene guns for direct injection of the genetic material into the target tissue. Since viral vectors pose some serious problems, nonviral vectors are the gene-delivery system of choice. In this nonviral vector system, negatively charged plasmid DNA is condensed into nanoparticles that are 50–200 nm in size. The use of cationic lipids and cationic polymers gives a compact structure due to interaction between cationic material and anionic DNA. These compact structures also provide increased stability and uptake by the target cells. Some of the nonviral vectors for gene therapy based on nanoparticles are listed in Table 10.2. The targets for gene therapy using nanoparticles include liver hepatocytes, endothelial cells, the spleen, and lymph nodes, where some success has already been achieved (Figure 10.7).

TABLE 10.2
Gene Therapy Using Nanotechnology

System	Animal	Route	Fate
Naked DNA	Mouse	Intravenous	Liver (kupffer cells)
Cationic lipid	Mouse, human	Aerosol, nasal spray	Lung deposition
Poly lysine conjugate	Mouse, human	Intravenous	Liver parenchyma
Liposome	Rat	Intravenous	Liver (kupffer cells)
Poly vinyl derivative	Rat	Intramuscular	Muscle expression
Poly cyanoacrylate	—	Cell culture	—
Cationic polymers	—	Cell culture	—

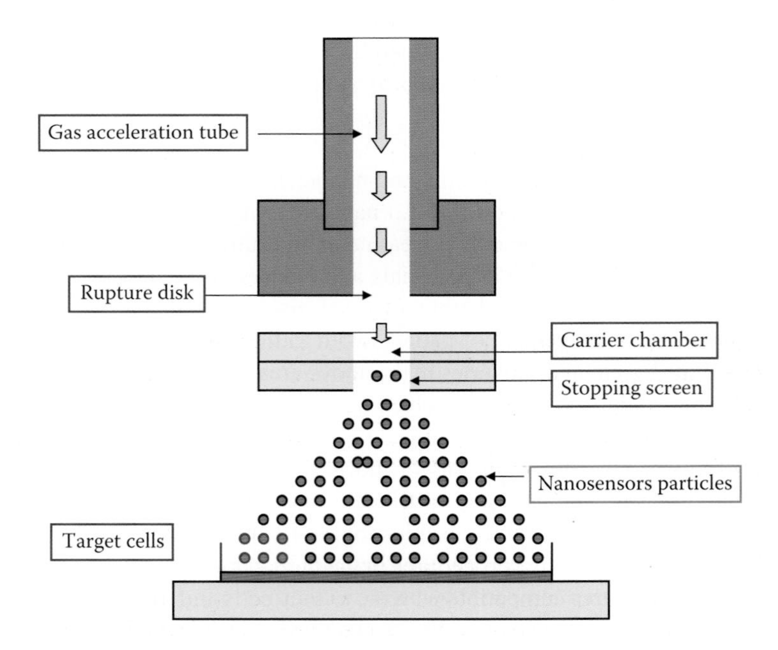

FIGURE 10.7 Bioimaging by nanosensor particles.

10.4.1.6 Antimicrobial Techniques

One of the earliest nanomedicine applications was the use of nanocrystalline silver, which is an antimicrobial agent for the treatment of wounds. A nanoparticle cream has been shown to fight staph infections. The nanoparticles contain nitric oxide gas, which is known to kill bacteria. Studies on mice have shown that using the nanoparticle cream to release nitric oxide gas at the site of staph abscesses significantly reduced the infection. A welcome idea in the early study stages is the elimination of bacterial infections in a patient within minutes, instead of delivering treatment with antibiotics over a period of weeks.

10.4.2 Neuro-Electronic Interfaces

Neuro-electronic interfaces are a visionary goal dealing with the construction of nanodevices that will permit computers to be joined and linked to the nervous system. This technology will allow a computer to control neural activity in patients suffering from various neural disorders. This idea requires the building of a molecular structure that will permit control and detection of nerve impulses by an external computer. The computer will be able to interpret, register, and respond to signals the body gives off when it feels sensations. The demand for such structures is huge, because many diseases such as amyotrophic lateral sclerosis (ALS, also referred to as Lou Gehrig's Disease) and multiple sclerosis involve the decay of the nervous system. Also, many injuries and accidents may impair the nervous system, resulting in dysfunctional systems and paraplegia. If computers could control the nervous system through neuro-electronic interface, problems that impair the system could be controlled so as to overcome the effects of diseases and injuries. Two approaches must be considered when selecting the power source for such applications: refuelable and non-refuelable strategies. A refuelable strategy implies that energy is refilled continuously or periodically with external sonic, chemical, tethered, magnetic, or electrical sources. A non-refuelable strategy implies that all power is drawn from an internal energy store, which would stop when all energy is drained. One limitation to this innovation is the fact that electrical interference is a possibility. Electric fields, electromagnetic pulses (EMPs), and stray fields from other *in vivo* electrical devices can all cause interference. Also, thick insulators are required to prevent electron leakage, and if high conductivity of the *in vivo* medium occurs there is a risk of sudden power loss and "shorting out." Finally, thick wires are also needed to conduct substantial power levels without overheating. Little practical progress has been made, even though research is ongoing. The wiring of the structure is extremely difficult, because the wires must be positioned precisely in the nervous system so that they are able to monitor and respond to nervous signals. The structures that will provide the interface must also be compatible with the body's immune system so that they will remain unaffected in the body for a long time. In addition, the structures must also sense ionic currents and be able to cause currents to flow backward. While the potential for these structures is amazing, there is no timetable for when they will be available.

10.4.3 Molecular Nanotechnology

Molecular nanotechnology is a speculative subfield of nanotechnology regarding the possibility of engineering molecular assemblers, machines which could reorder matter on a molecular or atomic scale. Molecular nanotechnology is highly theoretical, seeking to anticipate what inventions nanotechnology might yield and to propose an agenda for future inquiry. The proposed elements of molecular nanotechnology, such as molecular assemblers and nanorobots, are far beyond current capabilities.

10.4.4 Nanorobots

It sounds like Hollywood science fiction movie; yes, in the near future, there is a possibility of creating nanorobots that will totally change the world of medicine once it is realized. According to Robert Freitas of the Institute for Molecular Manufacturing, a typical blood-borne medical nanorobot would be between 0.5 and 3 μm in size, because that is the maximum size possible due to capillary passage requirements. Carbon would be the primary element used to build these nanorobots due to the inherent strength and other characteristics of some forms of carbon (diamond/fullerene composites), and nanorobots would be fabricated in desktop nanofactories specialized for this purpose. Nanodevices could be observed at work inside the body using MRI, especially if their components were manufactured using mostly 13C atoms rather than the natural 12C isotope of carbon, since 13C has a nonzero nuclear magnetic moment. Medical nanodevices would first be injected into a human body and would then go to work in a specific organ or tissue mass. The doctor would monitor the progress and make certain that the nanodevices have got to the correct target treatment region. The doctor would also be able to scan a section of the body and actually see the nanodevices congregated neatly around their target (such as a tumor mass) so that he or she can be sure that the procedure was successful.

10.4.5 Cell Repair Machines

If there is a problem in cells, we use drug molecules as a model to treat the problem, but it may be possible to treat dysfunctional cell with molecular machines in the future. A cell repair machine would be a device having a set of minuscule arms and tools controlled by a nanocomputer. The whole system would be much smaller than a cell. A repair machine could work like a tiny surgeon, reaching into a cell, sensing damaged parts, repairing them, closing up the cell, and moving on. By repairing and rearranging cells and surrounding structures, cell repair machines could restore tissues to health. Cells build and repair themselves using molecular machines, and cell repair machines will use the same principle. The main challenge will be to orchestrate these operations properly, once assemblers are able to build suitable tools. With molecular machines, there will be more direct repairs, as it would be possible to enter into a diseased cell with great accuracy and precision. Not only are molecular machines capable of entering the cell, all specific biochemical interactions show that molecular systems can recognize other molecules by touch, build or rebuild every molecule in a cell, and disassemble damaged molecules. Finally, cells that replicate prove that molecular systems can assemble every system found in a cell. Therefore, since nature has demonstrated the basic operations needed to perform molecular-level cell repair, in the future, nanomachine-based systems will be built that are able to enter cells, sense differences from healthy ones, and make modifications to the structure. The possibilities of these cell repair machines are impressive. Comparable to the size of viruses or bacteria, their compact parts would allow them to be more complex. The early machines will be specialized.

10.4.6 Nanosensors

Nanosensors are any biological, chemical, or surgery sensory points used to convey information about nanoparticles to the macroscopic world. Their use is mainly for various medicinal purposes and as gateways to building other nanoproducts, such as computer chips that work at the nanoscale and nanorobots. Presently, there are several ways proposed to make nanosensors, including top-down lithography, bottom-up assembly, and molecular self-assembly. Medicinal uses of nanosensors mainly revolve around the potential of nanosensors to accurately identify particular cells or places in the body in need. By measuring changes in volume, concentration, displacement, and velocity, and changes in gravitational, electrical, and magnetic forces, pressure, or temperature of cells in a body, nanosensors may be able to distinguish between and recognize certain cells (most notably, cancer cells) at the molecular level in order to deliver medicine to or monitor development in specific areas of the body.

In addition, nanosensors may be able to detect macroscopic variations from outside the body and communicate these changes to other nanoproducts working within the body. One example of nanosensors involves using the fluorescence properties of cadmium selenide qdots as sensors to uncover tumors within the body. By injecting a body with these qdots, a doctor could see where a tumor or cancer cell was by finding the injected qdots—an easy process because of their fluorescence. Developed nanosensor qdots would be specifically constructed to find only the particular cell for which the body was at risk. A downside to the cadmium selenide qdots, however, is that they are highly toxic to the body. As a result, researchers are working on developing alternate qdots made out of a different, less toxic material, while still retaining some of the fluorescence properties. In particular, they have been investigating the specific benefits of zinc sulfide qdots, which, though not quite as fluorescent as cadmium selenide qdots, can be augmented with other metals, including manganese and various lanthanide elements.

In addition, these newer qdots become more fluorescent when they bond to their target cells. Potential predicted functions may also include sensors used to detect specific DNA in order to recognize explicit genetic defects, especially for individuals at high risk, and implanted sensors that can automatically detect glucose levels for diabetic subjects more simply than current detectors. DNA can also serve as sacrificial layer for manufacturing a complementary metal-oxide semiconductor integrated circuit (CMOS IC), integrating a nanodevice with sensing capabilities. Therefore, using proteomic patterns and new hybrid materials, nanobiosensors can also be used to enable components configured into a hybrid semiconductor substrate as part of the circuit assembly. The development and miniaturization of nanobiosensors should provide interesting new opportunities. Other projected products most commonly involve using nanosensors to build smaller integrated circuits, as well as incorporating them into various other commodities made using other forms of nanotechnology for use in a variety of situations, including transportation, communication, improvements in structural integrity, and robotics. Nanosensors may also eventually be valuable as more accurate monitors of material states for use in systems where size and weight are constrained, such as in satellites and other aeronautic machines.

There are currently several hypothesized ways to produce nanosensors. Top-down lithography is the manner in which most integrated circuits are now made. It involves starting out with a larger block of some material and carving out the desired form. These carved-out devices, notably put to use in specific micro-electromechanical systems used as microsensors, generally only reach the microsize, but the most recent of these have begun to incorporate nanosized components. Another way to produce nanosensors is through the bottom-up method, which involves assembling the sensors out of even more minuscule components, most likely individual atoms or molecules. This would involve moving atoms of a particular substance one by one into particular positions which—though it has been achieved in laboratory tests using tools such as atomic force microscopes—is still significantly difficult, especially when moving atoms en masse, both for logistic and economic reasons. Most likely, this process would be mainly used to build starter molecules for self-assembling sensors. The third way, which promises far faster results, involves self-assembly, or "growing" particular nanostructures to be used as sensors. This most often entails one of two types of assembly. The first involves using a piece of some previously created or naturally formed nanostructure and immersing it in free atoms of its own kind. After a given period, the structure, having an irregular surface that would make it prone to attracting more molecules as a continuation of its current pattern, would capture some of the free atoms and continue to form more of itself to make larger components of nanosensors.

10.4.7 Nanoparticles

In nanotechnology, a particle is defined as a small object that behaves as a whole unit in terms of its transport and properties. It is further classified according to size. In terms of diameter, fine particles cover a range between 100 and 2500 nm, while ultrafine particles are sized between 1 and 100 nm. Similar to ultrafine particles, nanoparticles are sized between 1 and 100 nm, though the size limitation can be restricted to two dimensions. Nanoparticles may or may not exhibit size-related properties that differ significantly from those observed in fine particles or bulk materials. Nanoclusters have at least one dimension between 1 and 10 nm and a narrow size distribution. Nanopowders are agglomerates of ultrafine particles, nanoparticles, or nanoclusters. Nanometer-sized single crystals or single-domain ultrafine particles are often referred to as *nanocrystals*. The term "nanocrystal" is a registered trademark of Elan Pharma International (EPIL) used in relation to EPIL's proprietary milling process and nanoparticulate drug formulations. Nanoparticle research is currently an area of intense scientific research due to a wide variety of potential applications in the biomedical, optical, and electronic fields. The NNI has led to generous public funding for nanoparticle research in the United States.

Although nanoparticles are generally considered an invention of modern science, they actually have a very long history. Specifically, nanoparticles were used by artisans as far back as the ninth century in Mesopotamia to generate a glittering effect on the surface of pots. Even these days, pottery from the Middle Ages often retains a distinct gold- or copper-colored metallic glitter. This so-called luster is caused by a metallic film that was applied to the transparent surface of the glazing. The luster may still be visible if the film has resisted atmospheric oxidation and other weathering. The luster originates within the film itself, which contains silver and copper nanoparticles dispersed homogeneously in the glassy matrix of the ceramic glaze. These nanoparticles were created by the artisans by applying copper and silver salts and oxides together with vinegar, ochre, and clay on the surface of previously-glazed pottery. The object was then placed to a kiln and heated to approximately 600°C in a reducing atmosphere. The glaze would soften in the heat, causing the copper and silver ions to migrate into the outer layers of the glaze. There the reducing atmosphere reduced the ions back to metals, which then came together forming the nanoparticles that give the color and optical effects. Luster technique shows that craftsmen had a rather sophisticated empirical knowledge of materials. The technique originates in the Islamic world, as Muslims were not allowed to use gold in artistic representations; they had to find a way to create a similar effect without using real gold. The solution they found was using luster.

10.4.7.1 Classification of Nanoparticles

At the small end of the size range, nanoparticles are often referred to as clusters. Nanospheres, nanorods, nanofibers, and nanocups are just a few of the shapes that have been grown. Metal, dielectric, and semiconductor nanoparticles as well as hybrid structures (e.g., core-shell nanoparticles) have been formed. Nanoparticles made of semiconducting material may also be labeled qdots if they are small enough (typically <10 nm) for quantization of electronic energy levels to occur. Such nanoscale particles are used in biomedical applications as drug carriers or imaging agents. Semisolid and soft nanoparticles have been manufactured. A liposome is a prototype nanoparticle of semisolid nature. Various types of liposome nanoparticles are currently used clinically as delivery systems for anticancer drugs and vaccines.

10.4.7.2 Characterization of Nanoparticles

The characterization of nanoparticles is necessary to establish understanding and control of nanoparticle synthesis and applications. Characterization is done by using a variety of different techniques, which include electron microscopy, atomic force microscopy, dynamic light scattering, x-ray photoelectron spectroscopy, powder x-ray diffractometry, Fourier transform infrared spectroscopy (FTIR), matrix-assisted laser desorption/ionization time-of-flight mass spectrometry (MALDI TOF MS), and UV-visible spectroscopy.

10.4.7.3 Nanoparticles and Safety Issues

Besides having so many beneficial applications, nanoparticles do cause health and environmental hazards. Most of these are due to the high surface-to-volume ratio, which can make the particles very reactive or catalytic. They can also pass through cell membranes in organisms, and their interactions with biological systems are relatively unknown. According to the *San Francisco Chronicle*, "Animal studies have shown that some nanoparticles can penetrate cells and tissues, move through the body and brain, and cause biochemical damage. They also have been shown to cause a risk factor in men for testicular cancer. But whether cosmetics and sunscreens containing nanomaterials pose health risks remains largely unknown, pending completion of long-range studies recently begun by the FDA and other agencies." Diesel nanoparticles have been found to damage the cardiovascular system in a mouse model. In October 2008 the Department of Toxic Substances Control within the California Environmental Protection Agency, announced its intent to request information regarding "analytical test methods, fate and transport" in the environment, and other relevant information from manufacturers of carbon nanotubes. The purpose of this information request was to identify information gaps and to develop information about carbon nanotubes, an important emerging nanomaterial. The law places the responsibility to provide this information to the Department on those who manufacture and those who import the chemicals.

10.5 NANOTECHNOLOGY IN THE FOOD INDUSTRY

Nanotechnology is having an impact on several aspects of food science, from how food is grown to how it is packaged. Companies are developing nanomaterials that will make a difference not only in the taste of food, but also in food safety and the health benefits that food delivers. Nanoparticles are being developed that will deliver vitamins or other nutrients in food and beverages without affecting the taste or appearance. These nanoparticles actually encapsulate the nutrients and carry them through the stomach into the bloodstream. Researchers are using silicate nanoparticles to provide a barrier to gases (e.g., oxygen) or to moisture in the plastic film used for packaging. This could reduce the possibly of food spoiling or drying out. Zinc oxide nanoparticles can be incorporated into plastic packaging to block UV rays and provide antibacterial protection, while improving the strength and stability of the plastic film.

Nanosensors are being developed that can detect bacteria and other contaminates (such as salmonella) at packaging plants. This will allow for frequent testing at a much lower cost than sending samples to a lab for analysis. This point-of-packaging testing, if conducted properly, has the potential to dramatically reduce the chance of contaminated food reaching grocery store shelves. Research is also being conducted to develop nanocapsules containing nutrients that would be released when nanosensors detect a vitamin deficiency in your body. Basically, this research could result in a super vitamin storage system in your body that delivers the nutrients you need, when you need them. "Interactive" foods are being developed by Kraft that would allow you to choose the desired flavor and color. Nanocapsules that contain flavor or color enhancers are embedded in the food, remaining inert until a hungry consumer triggers them. The method has not been published, so it will be interesting to see how this particular trick is accomplished. Researchers are also working on pesticides encapsulated in nanoparticles that only release the pesticide within an insect's stomach, minimizing the contamination of plants. Another development being pursued is a network of nanosensors and dispensers used throughout a farm field. The sensors recognize when a plant needs nutrients or water, before there is any sign of their deficiency in the plant. The dispensers then release fertilizer, nutrients, or water as needed, optimizing the growth of each plant in the field.

10.6 WATER POLLUTION AND NANOTECHNOLOGY

Nanotechnology is being used to develop solutions to three very different problems in water quality. One challenge is the removal of industrial water pollution from ground water. Nanoparticles can be used to convert the contaminating chemical through a chemical reaction to make it harmless. Studies have shown that this method can be used successfully to reach contaminates dispersed in underground ponds and at much lower cost than methods which require pumping the water out of the ground for treatment. The challenge is the removal of salt or metals from water. A deionization method using electrodes composed of nano-sized fibers shows promise for reducing the cost and energy requirements of turning salt water into drinking water. The third problem concerns the fact that standard filters do not work on virus cells. A filter only a few nanometers in diameter is currently being developed that should be capable of removing virus cells from water.

10.7 RESEARCH TRENDS

With a rapid increase in technological breakthroughs, it would be possible to create new cure and new method of disease detection. One of the major issues with regard to human diseases such as cancer, Parkinson's disease, and Alzheimer's disease is how early you can detect the disease. Nanotechnology has the potential to overcome the limitations of current approaches and thereby advances the diagnosis and treatment of life-threatening diseases. To make rapid, accurate, real-time detection possible, a number of advances will be required. The molecular signatures of diseases, also called *biomarkers*, are often present at concentrations that are too low to be measured by current technology, so new devices will be needed to improve sensitivity. One example of a promising new medical approach that uses nanobiotechnology is the bio-barcode assay for the detection of disease-related proteins or DNA in tissue samples. In its first application, the bio-barcode assay uses gold nanoparticles to amplify and detect amyloid beta-derived diffusible ligands (ADDLs), a molecular signature for early-stage Alzheimer's disease. This method is a million times more sensitive than the current diagnostic tests.

A key challenge for nanobiotechnology is the fabrication and assembly of nanoscale materials, devices, and systems. Naturally occurring nano-structured organisms such as diatoms have been converted from silica into a range of ceramic nanomaterials with new properties. Cells and viruses can also be engineered to manufacture or assemble nanomaterials. Certain bacteria that naturally produce magnetic nanoparticles have been engineered to produce nanoparticles coated with specific proteins. Researchers have used biological methods to discover viruses that can be used as scaffolds for selective attachment and growth of semiconductor nanoparticles to produce the first virus-assembled battery. Scientists are also learning to create complex inorganic nanostructures, including those with unique chemical, optical, and mechanical properties, and efforts are being made to create nanoparticles to quickly and accurately measure drug toxicity.

R&D is an important factor for the creation of novel products. Nanotechnology has the potential to profoundly change our economy and to improve our standard of living in a manner not unlike the impact made by advances in information technology over the last two decades. While some commercial products are beginning to come to market, many major applications for nanotechnology are still 5–10 years away. Private investors look for shorter-term returns on investment, generally in the range of 1–3 years. Consequently, government support for basic R&D in its early stages needs to maintain a competitive position in the worldwide nanotechnology marketplace in order to realize nanotechnology's full potential. In the United States, federal funding for nanotechnology has increased from approximately $464 million in 2001 to nearly $1.5 billion for the fiscal year 2009. According to estimates, private industry is investing at least as much as the government. Besides the United States, the European Union and Japan have invested approximately $1.05 billion and $950 million, respectively, in nanotechnology. Korea, China, and Taiwan have invested $300 million, $250 million, and $110 million, respectively, in nanotechnology R&D.

PROBLEMS

Section A: Descriptive Type

Q1. What is nanobiotechnology?
Q2. How is nanobiotechnology used in medicine? Describe it by using an example.
Q3. How is qdots technology used in diagnosing cancer?
Q4. What is *in vivo* drug imaging?
Q5. Is it possible to repair cells using a machine?
Q6. How are nanoparticles characterized?
Q7. Describe the future trends of nanobiotechnology.

Section B: Multiple Choice

Q1. As per the U.S. NNI, what is the minimum size of a nanoparticle?
  a. 100 nm
  b. 100 pm
  c. 100 mm
  d. 100 cm

Q2. There are no U.S. FDA directives to regulate nanobiotechnology. True/False

Q3. Bioavailability of a drug refers to the ...
  a. Presence of the drug inside of the body
  b. Presence of the drug outside of the body
  c. Excretion of the drug
  d. Absorption of the drug

Q4. Nanoparticles have a high surface area, thus allowing many functional groups to attach. True/False

Q5. Dendrimer is a nanoparticle with ...
  a. 100 hooks
  b. >100 hooks
  c. >500 hooks
  d. None of the above

Q6. Nanoparticles cause health and environmental hazards. True/False

Q7. ADDL is related to detection of ...
  a. Parkinson's disease
  b. Huntington's disease
  c. Alzheimer's disease
  d. Mad-cow disease

Section C: Critical Thinking

Q1. Do you agree that nanoparticles have the capability to diagnose early onset of disease? Explain.

Q2. What would happen to the human body if nanorobots worked as an artificial defense shield against all infections?

REFERENCES AND FURTHER READING

Allen, T.M. and Cullis, P.R. Drug delivery systems: Entering the mainstream. *Science* 303: 1818–1822, 2004.

Allman III, R.M. *Structural Variations in Colloidal Crystals*, MS thesis, UCLA (1983). See Ref. 14 in Mangels, J.A. and Messing, G.L. (eds.) Forming of ceramics, microstructural control through colloidal consolidation, I.A. Aksay, *Adv. Ceram.* 9: 94, Proc. Amer. Ceramic Soc.

Books, B.R., Bruccoleri, R.E., Olafson, B.D. et al. CHARMM: A program for macromolecular energy, minimization, and dynamics calculations. *J. Comput. Chem.* 4: 187–217, 1983.

Buffat, P.H. and Burrel, J.P. Size effect on the melting temperature of gold particles. *Phys. Rev. A* 13: 2287–2298, 1976.

Cavalcanti, A., Shirinzadeh, B., Freitas, R.A. Jr., and Hogg, T. Nanorobot architecture for medical target identification. *Nanotechnology* 19: 15, 2008.

Cavalcanti, A., Shirinzadeh, B., Freitas, R.A. Jr., and Kretly, L.C. Medical nanorobot architecture based on nanobioelectronics. *Recent Pat. Nanotechnol.* 1: 1–10, 2007.

Chemical Information Call-In web page. Department of Toxic Substances Control. 2008. http://www.dtsc.ca.gov/PollutionPrevention/Chemical_Call_In.cfm

Choy, J.H., Jang, E.S., Won, J.H. et al. Hydrothermal route to ZnO nanocoral reefs and nanofibers. *Appl. Phys. Lett.* 84: 287–289, 2004.

Davidson, K. FDA urged to limit nanoparticle use in cosmetics and sunscreens. *San Francisco Chronicle.* http://www.sfgate.com/cgi-bin/article.cgi?file=/c/a/2006/05/17/MNGFHIT1161.DTL. Retrieved on April 20, 2007.

Department of Toxic Substances Control [Report] 2008. http://www.dtsc.ca.gov/TechnologyDevelopment/Nanotechnology/index.cfm

Fahlman, B.D. *Materials Chemistry*. Springer, Mount Pleasant, MI, Vol. 1, pp. 282–283, 2007.

Faraday, M. Experimental relations of gold (and other metals) to light. *Phil. Trans. Roy. Soc. London* 147: 145–181, 1857.

Freitas, R.A. *Nanomedicine, Volume I: Basic Capabilities*. Landes Bioscience, Georgetown, TX, 1999.

Freitas, R.A. Jr. What is nanomedicine? *Nanomedicine: Nanotech. Biol. Med.* 1: 2–9, 2005.

Hench L.L. and West, J.K. The Sol-Gel process. *Chem. Rev*. 90: 33–72, 1990.

LaVan, D.A., McGuire, T., and Langer, R. Small-scale systems for in vivo drug delivery. *Nat. Biotechnol*. 21: 1184–1191, 2003.

Loo, C., Lin, A., Hirsch, L. et al. Nanoshell-enabled photonics-based imaging and therapy of cancer. *Technol. Cancer Res. Treat*. 3: 33–40, 2004.

Minchin, R. Sizing up targets with nanoparticles. *Nat. Nanotechnol*. 3: 12–13, 2008.

Mnyusiwalla, A., Daar, A.S., and Singer, P.A. *Nanotechnology* 14: R9–R13, 2003.

Nanomedicine: A matter of rhetoric? [Editorial] *Nat. Mater*. 5: 243, 2006.

National Cancer Institute Alliance for Nanotechnology in Cancer. http://nano.cancer.gov/

Onoda, G.Y. Jr. and Hench, L.L. *Ceramic Processing before Firing*. Wiley & Sons, New York, 1979.

Roco, M.C. and Bainbridge, W.S. *Converging Technologies for Improving Human Performance: Nanotechnology, Biotechnology, Information Technology and Cognitive Science*. Kluwer Academic Publishers (Springer), Dordrecht, Boston, MA, 2003.

Shi, X., Wang, S., Meshinchi, S. et al. Dendrimer-entrapped gold nanoparticles as a platform for cancer-cell targeting and imaging. *Small* 3: 1245–1252, 2007.

Wagner, V., Dullaart, A., Bock, A.K., and Zweck, A. The emerging nanomedicine landscape. *Nat. Biotechnol.* 24(10): 1211–1217, 2006.

Whitesides, G.M. et al. Molecular self-assembly and nanochemistry: A chemical strategy for the synthesis of nanostructures. *Science* 254: 1312, 1991.

Ying, J. *Nanostructured Materials*. Academic Press, New York, 2001.

Zheng, G., Patolsky, F., Cui, Y., Wang, W.U., and Lieber, C.M. Multiplexed electrical detection of cancer markers with nanowire sensor arrays. *Nat. Biotechnol*. 23: 1294–1301, 2005.

11 Product Development in Biotechnology

LEARNING OBJECTIVES

- Define the product development process in the biotechnology industry
- Discuss scientific inventions and their commercialization
- Discuss various components of the biotechnology industry
- Explain the various phases of biotechnology product development
- Explain how a biotechnology company can be started
- Discuss the role of investment and financial implications in the biotechnology industry
- Discuss intellectual property rights in biotechnology
- Discuss the roles of Good Manufacturing Practices (GMP), Good Laboratory Practice (GLP), the World Health Organization (WHO), and the U.S. Food and Drug Administration (FDA) in biotechnology product development

11.1 INTRODUCTION

The biotechnology product development process is based on the kind of product you intend to develop or manufacture. For example, the product development process for medical applications is different from that for agricultural and food products. Likewise, the product development process for agricultural products is different from that for industrial products. In this chapter, we discuss product development pathways for all types of biotechnology products. We also describe in detail the development pathways for all products derived from the medical, agricultural, environmental, industrial, and nanobiotechnology sectors (Figure 11.1).

Biotechnology product development normally starts with a novel idea. A novel idea is basically a new way of thinking that can offer better products for human welfare. Let us find out the historical aspects of scientific inventions by using the scientific method. The *scientific method* refers to a body of techniques for investigating phenomena, acquiring new knowledge, or correcting and integrating previous knowledge. To be termed scientific, a method of inquiry must be based on gathering observable, empirical, and measurable evidence subject to specific principles of reasoning. A scientific method consists of the collection of data through observation and experimentation, and the formulation and testing of hypotheses. Although procedures vary from one field of inquiry to another, identifiable features distinguish scientific inquiry from other methodologies of knowledge.

Scientific researchers propose hypotheses as explanations of phenomena and then design experimental studies to test these hypotheses. These steps must be repeatable in order to dependably predict any future results. Theories that encompass wider domains of inquiry may bind many independently derived hypotheses together in a coherent, supportive structure. This in turn may help form new hypotheses or place groups of hypotheses into context. Since Ibn al-Haytham (Alhazen, 965–1039), one of the key figures in the development of the scientific method, the emphasis has been on seeking truth. Ibn al-Haytham discovered by experimentation that light travels in a straight line. Ibn al-Haytham's discovery was based on scientific experimentation, proved to be a milestone in scientific discovery and continues to be of scientific relevance.

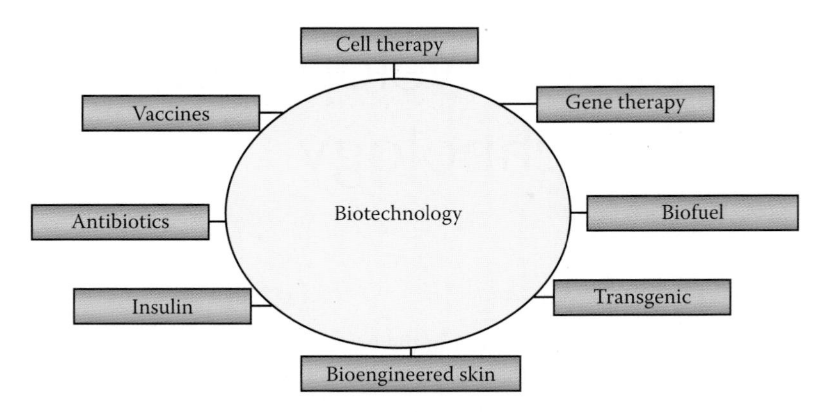

FIGURE 11.1 Products of biotechnology.

11.2 METHODS OF SCIENTIFIC ENQUIRY

There are different ways of outlining the basic method used for scientific inquiry. The scientific community and philosophers of science generally agree on a general classification of the method's components. These methodological elements and organization of procedures tend to be more characteristic of natural sciences than social sciences. Nonetheless, the cycle of formulating hypotheses, testing and analyzing the results, and formulating new hypotheses, will resemble the sequence described below. This is also the expected format and outline of a scientific report.

The scientific method of investigation is based on four essential elements:

1. *Characterizations*—observations, definitions, and measurements of the subject of inquiry
2. *Hypotheses*—theoretical, hypothetical explanations of observations and measurements of the subject
3. *Predictions*—reasoning (including logical deduction) from the hypothesis or theory, or the identification of distinct and (ideally) mutually exclusive possible discernible outcomes
4. *Experiments*—tests of characterizations, hypotheses, and predictions

Each element of a scientific method is subject to peer review for possible mistakes. These activities do not describe all that scientists do, but apply mostly to experimental sciences such as physics and chemistry. The elements above are often taught in the educational system (Figure 11.2).

The scientific method is not a recipe; it requires intelligence, imagination, and creativity. It is also an ongoing cycle, constantly developing more useful, accurate, and comprehensive models and methods. For example, when Einstein developed the special and general theories of relativity, he did

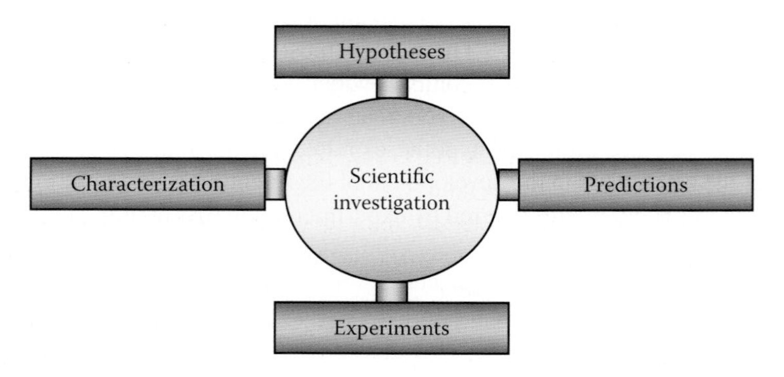

FIGURE 11.2 Methods of scientific investigation.

not in any way refute or discount Newton's *Principia*. On the contrary, if the astronomically large, the vanishingly small, and the extremely fast are reduced out from Einstein's theories—all phenomena that Newton could not have observed—Newton's equations remain. Einstein's theories are expansions and refinements of Newton's theories, and thus increase our confidence in Newton's work.

11.2.1 Characterizations of Scientific Investigation

The scientific method depends upon increasingly more sophisticated characterizations of the subjects of investigation (the *subjects* can also be called *unsolved problems* or the *unknowns*). The history of the discovery of the structure of DNA is a classic example of the elements of the scientific method. In 1950, it was known that genetic inheritance had a mathematical description, starting with the studies of Gregor Mendel, but the mechanism of the gene was unclear. Researchers in Bragg's laboratory at Cambridge University made x-ray diffraction pictures of various molecules, starting with crystals of salt and proceeding to more complicated substances. Using clues which were painstakingly assembled over the course of decades, beginning with the chemical composition of DNA, it was determined that it should be possible to characterize the physical structure of DNA, and x-ray images would be the vehicle. Linus Pauling proposed that DNA might be a triple helix. This hypothesis was also considered by Francis Crick and James Watson, but discarded. When Watson and Crick learned of Pauling's hypothesis, they understood from existing data that Pauling was wrong and that Pauling would soon admit his difficulties with that structure. Therefore, the race was on to figure out the correct structure (although, at the time, Pauling did not realize that he was in a race). James Watson, Francis Crick, and others hypothesized that DNA had a helical structure. This implied that DNA's x-ray diffraction pattern would be "x-shaped." This prediction followed from the work of Cochran, Crick, and Vand (and independently from the work by Stokes). The Cochran–Crick–Vand–Stokes theorem provided a mathematical explanation for the empirical observation that diffraction from helical structures produces x-shaped patterns. Also in their first paper, Watson and Crick predicted that the double helix structure provided a simple mechanism for DNA replication, documenting "It has not escaped our notice that the specific pairing we have postulated immediately suggests a possible copying mechanism for the genetic material."

11.2.2 Scientific Inventions

The popular image of an inventor is someone toiling away for months or years in a back shed and finally emerging with some wondrous gizmo that no one else had ever thought of. Another image is of a person in a white coat running out of their laboratory shouting "*I've done it*!" Sometimes, an invention is the result of the work of a team of people, particularly in the case of academic research or corporate research and development (R&D). Sometimes, Person A discovers a substance or process, while Person B later invents a way that it can be used or harnessed. Thus, while penicillin was "discovered" by Alexander Fleming, it was later isolated by Howard Florey and Ernst Chain at Cambridge, and it was Florey and his team who turned it into the practical medication that was used to save millions of lives. There were numbers of important people in the "invention" of penicillin as we know it, and Fleming, Chain, and Florey shared the 1945 Nobel Prize.

A patent application also includes one or more claims, although it is not always a requirement to submit these when first filing the application. The claims set out what the applicant is seeking to protect, in that they define what the patent owner has a right to exclude others from making, using, or selling, as the case may be. For a patent to be granted (i.e., to take legal effect in a particular country), the patent application must meet the patentability requirements of that country. Most patent organizations examine the application for compliance with these requirements. If the application does not comply, objections are communicated to the applicant or their patent agent or attorney, and one or more opportunities to respond to the objections to bring the application into compliance are usually provided. After rigorous technical evaluation, the patent agencies grant patent rights to the inventor, which in fact allow the inventor to manufacture the product. You may now wonder how to make a biotechnology product in this scenario.

11.3 COMMERCIALIZATION OF SCIENTIFIC DISCOVERY

Once scientific data is tested and proved by various techniques, the next step is to use the scientific information to create a product that may not be scientifically relevant but is economically viable. Most commercialization of a health product is largely dependent on the basic scientific data generated and published by the researchers working in the universities or federal research centers. Note that no company would like to invest in a project with no sound scientific basis, and at the same time they do not want to invest in noncommercial scientific projects. Let us take, for example, recombinant DNA (rDNA) insulin, which is basically synthetically manufactured insulin, a product of rDNA technology and used by most diabetic patients on a regular basis. One of the greatest breakthroughs of rDNA technology is the biosynthetic "human" insulin, the first medicine made via rDNA technology ever to be approved by the U.S. Food and Drug Administration (U.S. FDA). Due to the efforts of various scientists, rDNA technology has become one of the most important technologies in the modern times for generating various therapeutic proteins that are used in treating several diseases. Big companies had started investing billions of dollars to take advantage of this technology and to successfully produce large quantities of recombinant products.

11.4 BUSINESS PLAN

One of the basic requirements of any biotechnology product is to make a business plan, which includes work on the novel and economically feasible project. To ensure that the project is novel and economically feasible, a company consults with scientific and technical experts. The experts then review the project and examine it to ensure that the product is easy to use, safe for all users, and economical to manufacture. The design process is a cyclic, ongoing developmental process of generating ideas, testing these ideas, and selecting the ideas that work best. This developmental cycle begins with a range of possibilities and general ideas (conceptual designs) and gradually reaches a point where particular and fine details of the final product (final specifications) have been decided.

11.4.1 Project Feasibility

Before investing into a project, a company must know a lot of things to ensure its feasibility. The following are some of the questions that a company must know the answer to before investing into a project: Will the project work? Can it be done in time? Can it be done within budget? One of the biggest problems faced by any biotechnology company is how to make a biotechnology product in large quantities. It may be possible to produce a product in small amounts or numbers, perhaps in a lab setting, but is it feasible to upscale this process for mass production? This is very important to know before a company can think of investing money into a biotechnology project. Biotechnology companies ask industry experts to prepare a feasibility report on the proposed project, which generally consists of the following components.

11.4.1.1 Market Research

The first thing that any biotechnology or healthcare company does is to carry out extensive research on existing products that could be potential competitors in the future. This kind of survey is conducted to identify the problems of existing products in terms of effectiveness of the treatment, side effects, cost, and—most important—the market demand for the product, such as how big a market there is for the product in terms of sale. Market research can provide information about customer preferences, identify gaps in the market, get customer feedback, and identify any social or cultural issues that may be involved in the production and sale of a product.

11.4.1.2 Significance of a Project

The proposed project must have a clear understanding of both scientific and technical content, with each and every component mentioned in great detail, and how the proposed project will be unique in terms of the quality of the products and treatment modalities in comparison with existing product lines.

11.4.1.3 Technical Outline

An outline specification lists key features and performance of a product. It lists what the product is expected to do and what materials, properties, approximate amounts, processes, equipments, and expertise are needed.

11.4.1.4 Time Plan

It is very important to know the deadlines for product R&D, manufacturing, formulation, and the preclinical and clinical testing phases. These deadlines help both owner and company management to successfully launch the product on time.

11.4.1.5 Project Cost

The monetary requirement is very critical, and any company must know how much they have to invest to manufacture the product. This cost estimate also helps a company to generate funds through various sources. To make a biotechnology product, approximately 500–700 million U.S. dollars is primarily used for R&D, patent application, and preclinical and clinical trial costs. The money is either funded by the biotechnology company (or corporate investor) itself, or by a group of investors (venture investor). The venture capitalist can contribute at various phases by signing an agreement with the owner of the biotechnology product.

11.4.1.6 Legal and Regulatory Issues

Besides having scientific and technical information about a product, a company must know the intellectual property (IP), regulatory controls, environmental issues, and health and safety issues associated with the proposed product. This is to protect the company's ideas or to make sure that other companies do not transgress others' IP rights.

11.4.1.7 Quality of Product

In product development terms, quality is a relative term that is defined by how well the product meets or exceeds what it was designed to do. A product's quality or fitness for purpose is judged by the degree to which it matches the desired specifications such as (1) measuring against specifications and (2) meeting customer requirements. Measuring against specifications requires a quality assurance system.

11.5 BIOTECHNOLOGY PRODUCT DEVELOPMENT

The field of biotechnology keeps expanding with new knowledge and advanced technology. Various components of biotechnology product development are illustrated in Figure 11.3. Even though such products may be very different and may use different biological processes for production, most biotechnology companies operate in a similar way. Industrial and environmental biotechnology makes manufacturing processes cleaner and more efficient; creates new materials, food ingredients, and other products; unlocks cleaner and greener sources of energy; and reduces industrial waste. For example, biotechnology-based enzymes are used in such wide-ranging products as cheese, detergents, environmentally friendly plastics, and renewable fuels like cellulosic ethanol. We will now look at the pharmaceutical industry process of the major functions involved in the discovery, development, and marketing of a new biotechnology product (Figure 11.4). Some of these processes

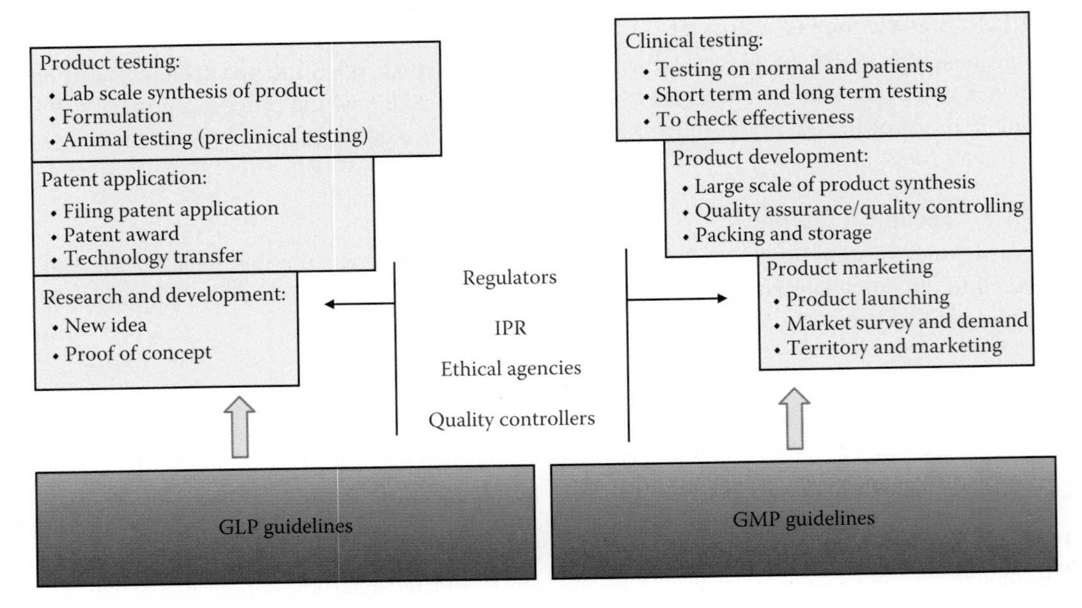

FIGURE 11.3 Components of biotechnology product development.

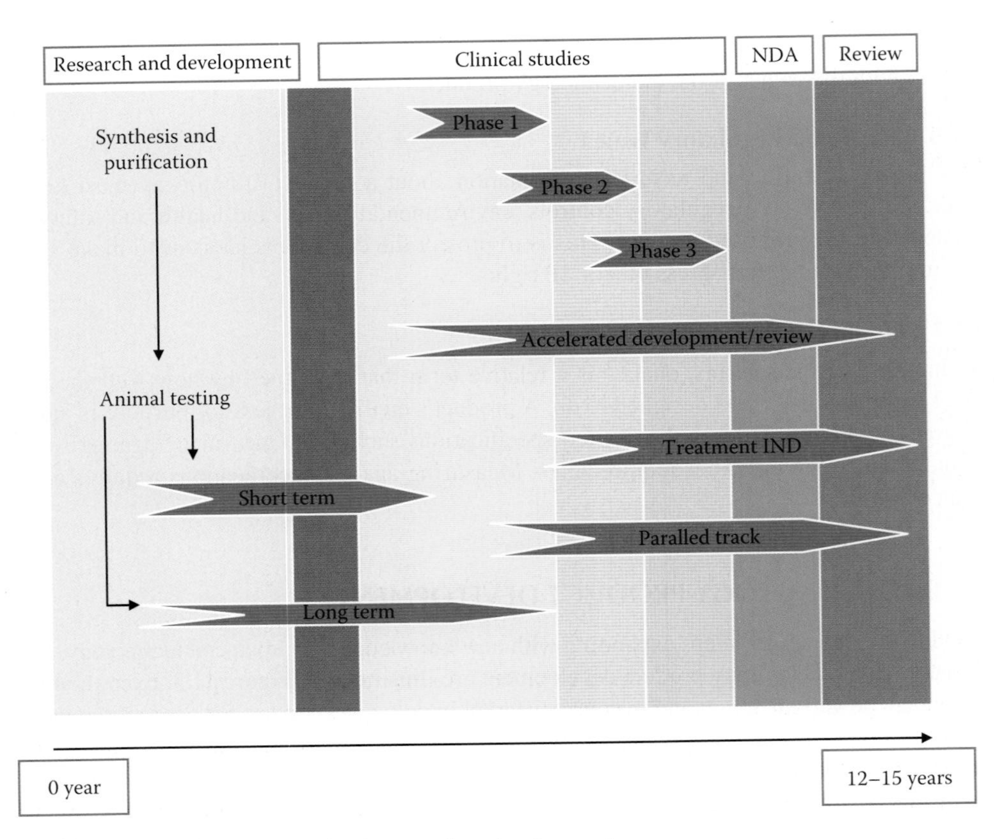

FIGURE 11.4 Timelines for biotechnology product development.

occur in lab settings, some in the field, and some in manufacturing facilities. In order to prove that intended research work is novel and reproducible, the researchers must work in the approved laboratories or centers, and carryout experiments to generate sufficient data under strict good laboratory practice (GLP) guidelines.

11.5.1 Infrastructure Requirements

One of the most important components of biotechnology product development is to set up the infrastructure required to manufacture the biotechnology products based on the GLP or good manufacturing practice (GMP) requirements and as per FDA guidelines. The infrastructure is primarily based on the kind of product required to be manufactured, but generally, any biotechnology manufacturing company has an R&D lab, animal testing lab, manufacturing area, quality assurance and quality control lab, and packaging and storage areas. This briefly describes the various components of the any standard biotechnology manufacturing company.

11.5.1.1 Research and Development Facility

One of the first things any company does is to create a research facility and hire expert researchers and scientists to work on the project. The construction of research labs should be as per GLP requirements. In this section, we will discuss GLP for conducting research on biotechnology products.

A new product begins in the research laboratory, where scientists and laboratory technicians use biotechnology tools to learn about the causes of disease. Their discoveries lead to new ideas about how to combat diseases. For example, an antibody might be a cure for a particular disease. Many different antibodies are then tested to see which one works best. The laboratory facilities consist of a main working lab where researchers work on various research activities from DNA isolation to protein assays in well-defined designated areas. For example, all DNA- and RNA-related activities are mainly performed under sterile conditions, such as in the biosafety hood or laminar flow, to avoid contamination. To avoid contamination, nonclinical labs (which handle microbes and virus) are separated from clinical labs (which handle human cells and DNA) by a proper barrier. The instruments (such as a PCR thermal cycler, spectrophotometer, laminar flow, centrifuge, and micropipettes) used in the labs should also be well calibrated, as per GLP requirements.

11.5.1.2 Animal Testing Facility or Preclinical Lab

Virtually every available medical treatment today has, to some degree, involved animal testing. The animals themselves may be bred specifically for testing or they may be captured in the wild. There are also commercial establishments that sell animals specifically for use in animal testing facilities. Animals are considered to be similar to humans in terms of assessing safety, which means that there are strict requirements for testing on animals with regards to new drugs. In the United Kingdom, for instance, a new drug is required to have been tested on two different species of live mammals. However, those who oppose animal testing and view it as an unnecessary infliction of suffering cite that the stress that an animal experiences will impact the accuracy of the results, rendering them useless. For now, however, animal testing is required before drugs and some other products are available for consumer use. Animal testing is a phrase that most people have heard at some point in their lives, but they are perhaps still unsure of exactly what is involved in animal testing. On top of that, there is still a great deal of subjectivity around the meaning behind animal testing. In addition, how animal testing is interpreted is partly related to the person's personal views of animal testing. Whether it is called animal testing, animal experimentation, or animal research, it refers to the experimental use of nonhuman animals. This type of experimentation is not done directly for healing purposes, although the end result may involve medications used for healing both humans and other animals. Instead, healing an animal would be akin to veterinary medicine, which is entirely different from animal testing. It is also important to note that those who oppose animal testing may believe it to mean the torture and suffering of animals, with no room for any additional definition.

Animal testing is used for countless products and applications. Everything from items in your home to products you use and medications you take have likely been tested on animals at some point prior to their distribution. Some of the products that commonly involve animal testing are cosmetics, drugs, food additives, food supplements, pesticides, and industrial chemicals. An animal facility has three major components: the breeding room, the surgical room, and the testing room. All these rooms are properly separated from each other to avoid contamination and infections. Aside from this, the temperature, light-dark cycle, and humidity are fully regulated as per international animal standards.

Preclinical testing, also known as *animal testing*, is a stage of research that begins before clinical trials (testing in humans) can begin. The main goal of preclinical studies is to determine a product's ultimate safety profile. Products may include new, iterated, or like-kind medical devices, drugs, or gene therapy solutions. Each class of product may undergo different types of preclinical research. For instance, drugs may undergo pharmacodynamics (PD); pharmacokinetics (PK); absorption, distribution, metabolism, and excretion (ADME); and toxicity testing through animal testing. This data allows researchers to allometrically estimate a safe starting dose of the drug for clinical trials in humans. Medical devices that do not have a drug attached will not undergo these additional tests and may go directly to GLP testing for safety of the device and its components. Some medical devices will also undergo biocompatibility testing, which helps to show whether a component of the device or all its components are sustainable in a living model. Most preclinical studies must adhere to GLP in guidelines issued by the International Conference on Harmonization (ICH guidelines) to be acceptable for submission to regulatory agencies such as the U.S. FDA. Typically, both *in vitro* and *in vivo* tests will be performed. Studies of a drug's toxicity include which organs are targeted by that drug and whether there are any long-term carcinogenic effects or toxic effects on mammalian reproduction.

The information collected from these studies is vital before safe human testing can begin. In drug development studies, animal testing typically involves two species. The most commonly used models are murine and canine, although primate and porcine are also used. The choice of species is based on which will give the best correlation to human trials. Differences in the gut, enzyme activity, circulatory system, or other considerations make certain models more appropriate, based on the dosage form, site of activity, or noxious metabolites. For example, canines may not be good models for solid oral dosage forms because the characteristic carnivore intestine is underdeveloped compared to that of omnivores and gastric emptying rates are increased. In addition, rodents cannot act as models for antibiotic drugs because the resulting alteration to their intestinal flora causes significant adverse effects. Depending on a drug's functional group, it may be metabolized in similar or different ways between species, which will affect both efficacy and toxicology. Medical device studies also use this basic premise. Most studies are performed on larger species such as dogs, pigs, and sheep, which allow for testing in a similar sized model as a human. In addition, some species are used for similarity in specific organs or organ system physiology (swine for dermatological and coronary stent studies, goats for mammary implant studies, dogs for gastric studies, etc.).

Animal testing in the research-based biotechnology industry has been reduced in recent years, both for ethical and cost reasons. However, most research will still involve animal-based testing due to the need for similarity in anatomy and physiology that is required for diverse product development. An animal facility is generally equipped with veterinary support, husbandry, and administrative services, all of which give researchers a variety of resources for accomplishment of their particular protocols. The centralized facility permits the most efficient and up-to-date environmental control for sanitation and animal health monitoring.

11.5.1.3 Bioequivalence Lab

The U.S. FDA has defined *bioequivalence* as "the absence of a significant difference in the rate and extent to which the active ingredient in pharmaceutical equivalents or pharmaceutical alternatives becomes available at the site of drug action when administered at the same molar dose under similar conditions in an appropriately designed study." In determining bioequivalence—for example, between two products such as a commercially-available brand product and a potential to-be-marketed generic product—PK studies are conducted, whereby each of the preparations is administered in a crossover study to volunteer subjects (generally, to healthy individuals, but occasionally to patients). Serum/plasma samples are obtained at regular intervals and assayed for parent drug (or occasionally metabolite) concentration. Occasionally, if blood-concentration levels are neither feasible nor possible to compare the two products (such as inhaled corticosteroids), then PD endpoints rather than PK endpoints are used for comparison. For a PK comparison, the plasma concentration data are used to assess key PK parameters such as area under the curve (AUC), peak concentration (Cmax), time to peak concentration (Tmax), and absorption lag time (Tlag). Testing should be conducted at several different doses, especially when the drug displays nonlinear PK. Bioequivalence cannot be claimed based only on *in vitro* testing or only on the basis of animal studies. Bioequivalence of human drugs must be determined in humans via established measures of bioavailability. By the same token, animal drugs must be tested for bioequivalence in the animal species for which the drug is intended.

11.5.1.4 Clinical Trial Center

Once a drug has completed preclinical trials with a great success rate, the drug molecule is now ready for test in humans, starting with normal and healthy individuals, to check the toxicity or side effects of the drug molecules. If the drug does not show any apparent toxicity or side effects, it may now be tested in patients suffering from specific disease conditions. There are four phases of clinical trials: Phase I, Phase II, Phase III, and Phase IV. All these phases are conducted in specialized and FDA-approved clinical centers and hospitals. In healthcare, clinical trials are conducted to allow safety and efficacy data to be collected for new drugs or devices. These trials can only take place once satisfactory information has been gathered on the quality of the product and its nonclinical safety. Health authority or ethics committee approval is granted in the country where the trial is taking place. Depending on the type of product and the stage of its development, investigators enroll healthy volunteers and/or patients in small pilot studies initially, followed by larger-scale studies in patients that often compare the new product with the currently prescribed treatment. As positive safety and efficacy data are gathered, the number of patients is typically increased. Clinical trials can vary in size from a single center in one country to multicenter trials in multiple countries. Due to the sizable cost a full series of clinical trials may incur, the burden of paying all the necessary people and services is usually borne by the sponsor, who may be the pharmaceutical or biotechnology company that developed the agent under study. Since the diversity of roles may exceed the sponsor's resources, a clinical trial is often managed by an outsourced partner such as a contract research organization (CRO).

11.5.1.5 Manufacturing Plant

During the clinical phase, a company builds a manufacturing plant to synthesize drug in great capacity. A manufacturing plant normally consists of several divisions, such as a raw material room, a bioreactor room, a formulation room, a packaging room, a quality assurance and quality control lab, and storage room. Unlike a research laboratory, a manufacturing plant works round-the-clock, and most personnel work in shifts, with assigned duties and responsibilities. A manufacturing plant strictly follows the GMP guidelines as prescribed by the FDA and other competent organizations (Figure 11.5).

11.5.1.6 Formulation Lab

After successful testing in laboratories, the desired product is synthesized at smaller levels to check its efficacy and toxicity in animals. The formulation lab is basically used to synthesize molecules in the final form where it can be delivered in tablet form, syrup form, injection form, or ointment form. Researchers and formulation teams discuss and decide which formulation will be effective

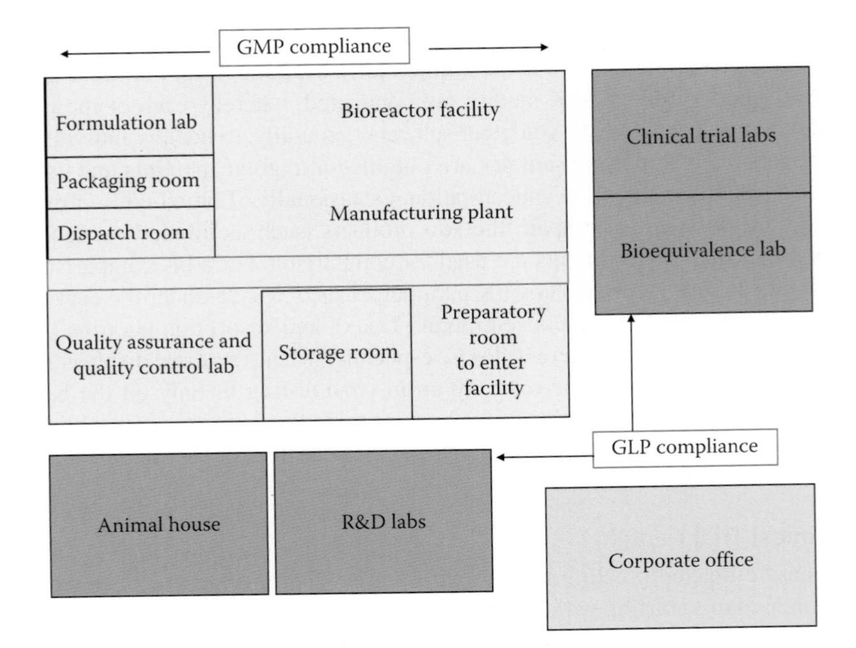

FIGURE 11.5 Biotechnology manufactory company: Plant layout.

in patients. The formulation team mainly consists of biochemists and chemists. *Formulation* in biotechnology is the process in which different chemical substances, including the active drug, are combined to produce a final medicinal product. Formulation studies involve developing a preparation of the drug that is not only effective and stable, but is also palatable to the patient. For oral drugs, this usually involves incorporating the drug into tablet, capsule, or syrup form. It is important to appreciate that a tablet contains a variety of other substances apart from the drug itself, and studies have to be carried out to ensure that the drug is compatible with these other substances. Before using it for formulation, the chemical's physical, chemical, and mechanical properties must be completely characterized in order to choose what other ingredients should be used in the preparation. In dealing with protein preformulation, the important aspect is to understand the solution behavior of a given protein under a variety of stress conditions such as freeze/thaw, temperature, and shear stress, among others, to identify mechanisms of degradation and therefore their mitigation. Formulation studies then consider other factors such as particle size, polymorphism, pH, and solubility, all of which can influence bioavailability and hence the activity of a drug. The drug must be combined with inactive additives by a method which ensures that the quantity of drug present is consistent in each dosage unit, such as in each tablet. The dosage should have a uniform appearance, with an acceptable taste, tablet hardness, or capsule disintegration. It is unlikely that formulation studies will be complete by the time clinical trials commence. This means that simple preparations are developed initially for use in Phase I clinical trials. These typically consist of hand-filled capsules containing a small amount of the drug and a diluent. Proof of the long-term stability of these formulations is not required, as they will be tested in a matter of days. Consideration has to be given to what is called the *drug load*, the ratio of the active drug to the total content of the dose. A low drug load may cause homogeneity problems. A high drug load may pose flow problems or require large capsules if the compound has a low bulk density.

By the time Phase III clinical trials are reached, the formulation of the drug should have been developed to be close to the preparation that will ultimately be used in the market. Knowledge of stability is essential by this stage, and conditions must have been developed to ensure that the drug is stable during its preparation. If the drug proves unstable, it will invalidate the results from clinical trials, since it would be impossible to know what the administered dose actually was. Stability studies are carried out to test whether temperature, humidity, oxidation, or photolysis (ultraviolet light or visible light) have any effect, and the preparation is analyzed to see if any degradation products have been formed. It is also important to check whether there are any unwanted interactions between the preparation and the container. If a plastic container is used, tests are carried out to see whether any of the ingredients become adsorbed on to the plastic and whether any plasticizers, lubricants, pigments, or stabilizers leach out of the plastic into the preparation. Even the adhesives for the container label need to be tested to ensure they do not leach through the plastic container into the preparation.

11.5.1.7 Quality Assurance and Quality Control Lab

One of the important aspects of biotechnology product development is to ensure the quality of a product at all given times. Companies cannot afford to lose market share due to poor quality, hence the companies themselves create an internal quality assurance and quality control system. The people who works as quality controllers are basically trained biotechnology researchers whose role is to make sure that the company produces high-quality and safe products at all times.

11.6 PHASES OF BIOTECHNOLOGY PRODUCT DEVELOPMENT

Currently, most clinical trial programs follow ICH guidelines, aimed at "ensuring that good quality, safe and effective medicines are developed and registered in the most efficient and cost-effective manner." Clinical trials involving new drugs are commonly classified into four phases. Each phase of the drug approval process is treated as a separate clinical trial. The drug-development process will normally proceed through all four phases over many years. Drugs that successfully pass through Phases I, II, III, and IV are approved by the national regulatory authority for use by the general population. Phase IV covers "post-approval" studies. Before pharmaceutical companies start clinical trials on a drug, they conduct extensive preclinical studies (Figure 11.6).

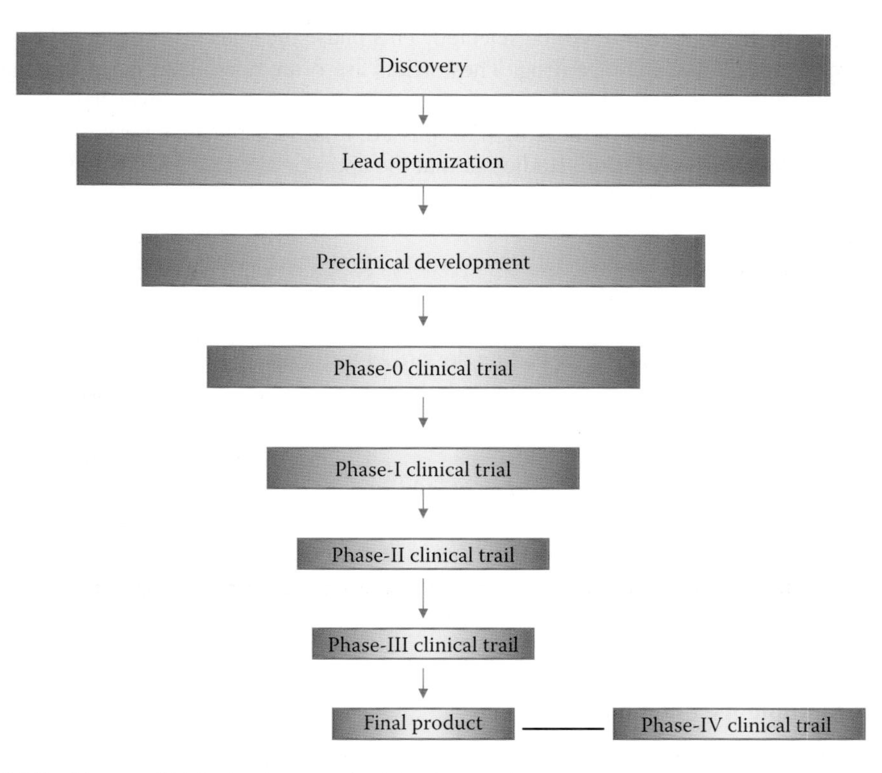

FIGURE 11.6 Phases of biotechnology product development.

11.6.1 Preclinical Studies

Preclinical studies involve *in vitro* (test tube) and *in vivo* (animal) experiments using a wide range of doses of the drug under study to obtain preliminary efficacy, toxicity, and PK information. Such tests assist pharmaceutical companies to decide whether a drug candidate has scientific merit for further development as an investigational new drug.

11.6.2 Phase 0

Phase 0 is a recent designation for exploratory, first-in-human trials conducted in accordance with the U.S. FDA's 2006 Guidance on Exploratory Investigational New Drug (IND) Studies. Phase 0 trials are also known as *human microdosing studies* and are designed to speed up the development of promising drugs or imaging agents by establishing very early on whether the drug or agent behaves in human subjects as was expected from preclinical studies. Distinctive features of Phase 0 trials include the administration of single subtherapeutic doses of the drug under study to a small number of subjects (usually ranging from 10 to 15) to gather preliminary data on the agent's PK (how the body processes the drug) and PD (how the drug works in the body). A Phase 0 study gives no data on safety or efficacy, being by definition too low a dose to cause any therapeutic effect. Drug development companies carry out Phase 0 studies to rank drug candidates so as to decide which has the best PK parameters in humans to take forward into further development. They enable go/no-go decisions to be based on relevant human models instead of relying on sometimes inconsistent animal data. Questions have been raised by experts about whether Phase 0 trials are useful, ethically acceptable, feasible, speed up the drug development process, or save money, and whether there is room for improvement.

11.6.3 Phase I

Phase I trials are the first stage of testing in human subjects. Normally, a small group of healthy volunteers (usually from 20 to 80) will be selected. This phase includes trials designed to assess the safety, tolerability, PK, and PD of a drug. These trials are often conducted in an inpatient clinic, where the subject can be observed by a full-time staff. The subject who receives the drug is usually observed until several half-lives of the drug have passed. Phase I trials also normally include dose-ranging (also called *dose escalation*) studies so that the appropriate dose for therapeutic use can be found. The tested range of doses will usually be a fraction of the dose that causes harm in animal testing. Phase I trials most often include healthy volunteers; however, there are some circumstances when real patients are used, such as patients who have end-stage disease and lack other treatment options. This exception to the rule most often occurs in oncology (cancer) and HIV drug trials. Volunteers are paid an inconvenience fee for their time spent in the volunteer center. There are different kinds of Phase I trials: single ascending dose (SAD) studies, multiple ascending dose (MAD) studies, and food effect studies.

11.6.3.1 Single Ascending Dose

SAD studies are those in which small groups of subjects are given a single dose of the drug while they are observed and tested for a period of time. If they do not exhibit any adverse side effects and if the PK data is roughly in line with predicted safe values, the dose is escalated, and a new group of subjects is then given a higher dose. This is continued until either precalculated PK safety levels are reached or intolerable side effects start showing up, at which point the drug is said to have reached the maximum tolerated dose (MTD).

11.6.3.2 Multiple Ascending Doses

MAD studies are conducted to better understand the PK and PD of multiple doses of the drug. In these studies, a group of patients receives multiple low doses of the drug, while samples of blood and other fluids are collected at various time points and analyzed to understand how the drug is processed within the body. The dose is subsequently escalated for further groups, up to a predetermined level.

11.6.3.3 Food Effect

Food effect studies are short trials designed to investigate any differences in absorption of a drug by the body caused by eating before the drug is given. These studies are usually run as a crossover study, with volunteers being given two identical doses of the drug on two different occasions: one after having fasted, and one after being fed.

11.6.4 Phase II

Once the safety of a drug under study has been confirmed in Phase I trials, Phase II trials are performed on larger groups (usually ranging from 20 to 300 participants). These trials are designed to assess how well the drug works and to continue Phase I safety assessments in a larger group of volunteers and patients.

Phase II studies are sometimes divided into Phases IIA and IIB. *Phase IIA* is specifically designed to assess dosing requirements or how much drug should be given. *Phase IIB* is specifically designed to study efficacy or how well the drug works at the prescribed dose(s). Some trials combine Phase I and Phase II, and test both efficacy and toxicity. Some Phase II trials are designed as case series, demonstrating a drug's safety and activity in a selected group of patients. Other Phase II trials are designed as randomized clinical trials, where some patients receive the drug/device and others receive placebo/standard treatment. Randomized Phase II trials have far fewer patients than randomized Phase III trials.

11.6.5 Phase III

Phase III studies are randomized controlled multicenter trials on large patient groups (usually ranging from 300 to 3000 or more, depending on the disease/medical condition studied) and are aimed at being the definitive assessment of how effective a drug is in comparison with the current "gold standard" treatment. Because of their size and comparatively long duration, Phase III trials are the most expensive, time-consuming, and difficult trials to design and run, especially in therapies for chronic medical conditions. It is common practice that certain Phase III trials will continue while the regulatory submission is pending at an appropriate regulatory agency. This allows patients to continue to receive possibly lifesaving drugs until the drug can be obtained by purchase. Other reasons for performing trials at this stage include attempts by the sponsor at "label expansion" (to show that the drug works for additional types of patients/diseases beyond the original use for which the drug was approved for marketing), to obtain additional safety data, or to support marketing claims for the drug. Studies in this phase are categorized as Phase IIIB studies by some companies.

While not required in all cases, it is typically expected that there will be at least two successful Phase III trials demonstrating a drug's safety and efficacy in order to obtain approval from the appropriate regulatory agencies such as the FDA in the United States, the Therapeutic Goods Administration (TGA) in Australia, the European Medicines Agency (EMA) in the European Union, or the Central Drug Standard Control Organization (CDSCO) or the Indian Council of Medical Research (ICMR) in India. Once a drug has proved satisfactory after Phase III trials, the trial results are usually combined into a large document containing a comprehensive description of the methods and results of human and animal studies, manufacturing procedures, formulation details, and shelf life. This collection of information makes up the "regulatory submission" that is provided for review to the appropriate regulatory authorities in different countries. They will review the submission, and it is hoped that the sponsor be given the approval to market the drug.

Most drugs undergoing Phase III clinical trials can be marketed under FDA norms with proper recommendations and guidelines. In case any adverse effects are reported anywhere, the drugs need to be recalled immediately from the market. While most pharmaceutical companies refrain from this practice, it is not abnormal to see many drugs undergoing Phase III clinical trials in the market.

11.6.6 Phase IV

A Phase IV trial is also known as *post-marketing surveillance trial*. Phase IV trials involve the safety surveillance and ongoing technical support of a drug after it receives permission to be sold. Phase IV studies may be required by regulatory authorities or may be undertaken by the sponsoring company for competitive (finding a new market for the drug) or other reasons (e.g., the drug may not have been tested for interactions with other drugs, or on certain population groups such as pregnant women, who are unlikely to subject themselves to trials). The safety surveillance is designed to detect any rare or long-term adverse effects over a much larger patient population and longer time period than was possible during the Phase I–III clinical trials. Harmful effects discovered by Phase IV trials may result in a drug being no longer sold or being restricted to certain uses. Recent examples involve cerivastatin (brand names Baycol and Lipobay), troglitazone (Rezulin), and rofecoxib (Vioxx).

11.7 BIOTECHNOLOGY ENTREPRENEURSHIP

During the last two decades of the twentieth century, a large number of biotechnology companies came into existence. According to U.S. Biotechnology Industry Organization (BIO), there were 1379 biotechnology companies in the United States in 2001. Similarly, according to Biotechnology Information Databank (BID) maintained at the University of Siena (Italy), there were 2104 biotechnology companies in Europe in the same year. Due to an anticipated growth in the biotechnology industry, biotechnology companies in both the United States and Europe also attracted huge investments in the stock market. While some of these companies achieved success, others met with failure and closure.

In the past, many scientists who started using biotechnology companies for commercial gains had no background or familiarity with business world. Similarly, experts in the world of business were sometimes unfamiliar with the tools of biotechnology research that have an important bearing on the biotechnology business. Consequently, those who were in the biotechnology business learned through their mistakes. Furthermore, young entrepreneurs in this area often repeat these mistakes, since the biotechnology business is young and the mistakes committed by others do not become quickly and widely known. In view of this, one would like to have knowledge of factors that influence the success of a biotechnology company. In this chapter, therefore, an effort is being made to familiarize readers with various aspects of biotechnology that have a bearing on the success of biotechnology businesses.

11.7.1 Starting a Biotechnology Company

To develop a successful biotechnology product, a company must have a sound economic plan. There are different ways a company can generate funds to develop a product. One way is to fund from the company's own source to maintain market monopoly. This kind of investment is called *corporate investment*. If it is difficult for a company to find the required funds, they usually find investors, present their business proposals, and influence them to invest in the project. There are several options to obtain funding for a project.

11.7.1.1 Grants

Academics still in the research stages might qualify for government grants for equipment costs and staff (graduate students and technicians) salaries. There are grants available for academic collaborations with industry to facilitate invention commercialization—for example, in Canada, the Industrial Research Assistance Program (IRAP) funds a large number of collaborative biotechnology projects. In the United States, funding from the National Institutes of Health (NIH) comes with certain data sharing policies that must be followed. Universities that have recognized the potential of their research programs have organizations to help commercialize the discoveries of their scientists.

11.7.1.2 Private Investors

Many startup companies rely on funding from private investors that have an interest and belief in a biotechnology product. This may come from friends and family or from acquaintances with money. These people might be the easiest to convince that your product is a viable investment, and they typically demand the least control over your company. However, if the company folds, you have the most to lose in terms of your relationships with them.

11.7.1.3 Angel Investors

Friends and families do not often provide substantial amounts of funding, but you might score bigger with "angel investors." These are individuals with money or capital who invest privately in new businesses. A typical angel investor will demand a larger share of the company than friends or family, and thus more control over the company. However, you may actually benefit from their experience and advice. Angels know what they are up against, and you risk less in terms of personal relationships by taking this route.

11.7.1.4 Venture Capitalists

Like angel investors, venture capitalists will also demand a fair amount of control over your operations and decision making. However, the role of a venture capitalist is also to rally around the business, help with the management, promote the business, and provide contacts to protect their investment.

11.7.1.5 Bank Loans

Look into loans for new businesses and be sure to have a thorough business plan. It is usually easier to get a small business loan if you already have paying customers. If this is not an option, you can try for a personal loan (if it is enough to get you started). The downside of doing so is that, if the business fails, you still have to repay the loan. Although the amount of funding you gain may be less than with investors as listed above, by starting with "debt" financing (loans, lines of credit, and credit cards), you demonstrate to investors that you have faith in the company and are willing to take risks to make it work.

11.8 BIOTECHNOLOGY INDUSTRY: FACTS AND FIGURES

The biotechnology industry emerged in the 1970s, based largely on a new rDNA technique whose details were published in 1973 by Stanley Cohen of Stanford University and Herbert Boyer of the University of California, San Francisco. Boyer went on to co-found Genentech, which is today's largest biotechnology company by market capitalization. Let us take a brief look at some significant biotechnology applications and economic statistics.

The biotechnology industry in the United States is regulated by the FDA, the Environmental Protection Agency (EPA), and the Department of Agriculture. BIO was founded in 1993 to represent biotechnology companies at the local, state, federal, and international levels. BIO comprises more than 1200 members, including biotechnology companies, academic centers, state and local associations, and related enterprises.

Biotechnology has created more than 200 new therapies and vaccines, including products to treat cancer, diabetes, HIV/AIDS, and autoimmune disorders. There are more than 400 biotechnology drug products and vaccines currently in clinical trials targeting more than 200 diseases, including various cancers, Alzheimer's disease, heart disease, diabetes, multiple sclerosis, AIDS, and arthritis. In 1982, recombinant human insulin became the first biotechnology therapy to earn FDA approval. The product was developed by Genentech and Eli Lilly and Co. Since then, biotechnology is responsible for hundreds of medical diagnostic tests that keep blood supply safe from HIV and detect other conditions early enough to be successfully treated. Home pregnancy tests are also biotechnology diagnostic products. Agricultural biotechnology benefits farmers, consumers, and the environment by increasing yields and farm income, decreasing pesticide applications and improving soil and water quality, and providing healthful foods to consumers. Environmental biotechnology products make it possible to clean up hazardous waste more efficiently by harnessing pollution-eating microbes. Industrial biotechnology applications have led to cleaner processes that produce less waste and use less energy and water in such industrial sectors as chemicals, pulp and paper, textiles, food, energy, and metals and minerals. For example, most laundry detergents produced in the United States contain biotechnology-based enzymes. DNA fingerprinting, a biotechnology process, has dramatically improved criminal investigation and forensic medicine. It has also led to significant advances in anthropology and wildlife management.

Noteworthy statistics include the following:

- The biotechnology industry has mushroomed since 1992, with U.S. healthcare biotechnology revenues from publicly traded companies rising from $8 billion in 1992 to $58.8 billion in 2006.
- As of December 31, 2006, there were 1452 biotechnology companies in the United States, of which 336 were publicly held, and approximately 180,000 employees. The average annual wage of U.S. bioscience workers was $71,000 in 2006, more than $29,000 higher than the average annual wage in the private sector.
- The biosciences, including all life-sciences activities, employed 1.3 million people in the United States in 2006 and generated an additional 7.5 million related jobs.
- Biotechnology is one of the most research-intensive industries in the world. U.S. publicly traded biotechnology companies spent $27.1 billion on R&D in 2006.
- The top five biotechnology companies invested an average of $170,000 per employee in R&D in 2007.
- Most biotechnology companies are young companies developing their first products, and depend on investor capital for survival. According to BioWorld, biotechnology attracted more than $24.8 billion in financing in 2007 and raised more than $100 billion in the 5-year span between 2003 and 2007.
- Corporate partnering has been critical to the success of biotechnology. According to BioWorld, in 2007, biotechnology companies struck 417 new partnerships with pharmaceutical companies and 473 deals with fellow biotechnology companies. The industry also saw 126 mergers and acquisitions.
- Market capitalization, the total value of publicly traded U.S. biotechnology companies at market prices, was $360 billion as of late April 2008 (based on stocks tracked by BioWorld).

11.8.1 Capital Investment in Biotechnology

The first biotechnology company in the United States to make an initial public offering (IPO) in the capital market was Genentech, which made an offer of $35 million at the rate of $35 per share. This share soared to $88 within the first 20 min and closed at the end of the day at $56 per share, thus giving the company a valuation of $400 million. Similarly, in March 1981, Cetus's gross IPO was $120 million, giving a valuation of approximately $500 million to this company. However, this trend in the United States did not continue later in the 1980s and 1990s. The average IPO raised by an individual company during 1980–2000 did not exceed $20–30 million/year, although the total capital raised due to biotechnology business improved significantly due to an increase in the number of biotechnology companies.

In the year 1986 (which was the best year in the biotechnology business in the 1980s), the U.S. biotechnology industry raised $900 million (all companies together), which steadily improved over the years, reaching a level of several billion dollars per year during the 1990s. However, in general, the biotechnology industry did not attract investors very much during the 1990s, except towards the end of the twentieth century. For instance, in the year 1999, the share price of Tularik (a premier biotechnology company) improved from $11 to 13 per share in October 1999 to $90 in February 2000, and the company valuation improved from $500 million to $4 billion during the same period. Other companies, impressed by the performance of Tularik, suddenly began filing for IPOs, so that the year 2000 was a record year for biotechnology financing, with 63 IPOs completed and $5.4 billion (with an average IPO proceeds of an individual company rising from $30 million to $85 million) raised for the biotechnology industry. As many as 37 new biotechnology companies were floated in the area of *genomics research* alone, although most of them may not be able to sustain growth and may therefore merge with or be acquired by other successful companies. The revival of the biotechnology industry was also witnessed in Europe, as suggested by several European biotechnological companies such as Neurotech, Transgene, NicOx, and Cytomix, which raised a total of $194 million in May 2001 alone.

11.8.2 Mergers and Acquisitions of Biotechnology Companies

Another feature witnessed in the biotechnology industry during 1990–2000 was the merger and/or acquisitions of biotechnology companies. In most cases, this was due to the inability of several companies to retain their independent existence. For instance, Celera Genomics announced its acquisition of Axys Pharmaceuticals, while Lexicon Genetics announced its purchase of Coelacanth. Many more mergers, acquisitions, and collaborations took place in the first decade of the twenty-first century. In 2001, Sequinom Biotech Company merged with Gemini Genomics. Both companies specialized in information mining for the Human Genome Project. While Sequinom has the mass spectrometry-based genetic analysis system, Gemini Genomics specialized in clinical population genomics. The purpose of the merger was to combine human genetic resources with the rapid analytical system to generate a more powerful data-generating machine. This merger is seen as a synergistic effort, where two youthful and vigorous companies, which raised $250 million in IPO, decided to merge—not due to poverty, but due to a cleverly formulated strategy.

11.9 FORMATION OF A NEW BIOTECHNOLOGY COMPANY

Transformation of an important discovery or a brilliant idea into a commercial opportunity is not always easy, but some young biotechnology companies have done it successfully. To ensure success, a new company has to announce or display its focus; for instance, Amgen (California, United States) is a 20-year-old pharmaceutical company that succeeded due to its superb early product selection, thus eventually succeeding in getting its product to the market. Other successful companies include Sugen (United States), Enzon (United States), Morphosys (Germany), and Vertex (Cambridge, United Kingdom). But such a focus can also meet failure, when the selected product fails in clinical trials; one such example is British Biotechnology (United Kingdom), whose matrix metalloproteinase inhibitor program failed.

Young companies should also be careful in focusing their attention on product development and/or R&D. In the field of drug discovery and development, the company may have skill for drug discovery through R&D, but may not be equipped for drug development. A small, young company may enter into drug discovery or clinical trials without a realistic estimate of cost and the risk involved. For instance, even the simplest Phase I clinical study may cost approximately $250,000, which is much more than a single experiment on the bench. One should also assess the treatment for which a drug is being developed. For example, a single Phase III study of a drug for heart stroke will cost approximately $30 million, but a treatment for migraine will cost only about one-tenth of this amount. Planning, therefore, is essential to avoid surprises that may prove fatal for the company.

11.10 SUCCESSFUL BIOENTREPRENEURSHIP

A variety of skills is also needed to be a successful entrepreneur. Although research experience, degrees, and publications may help, a successful bioentrepreneur should possess the following intangible attributes for coping with many complex and ambiguous situations: adaptability, intelligence, confidence, and the ability to interact with others, ambition, persistence, risk-taking capacity, humility, flexibility, and patience. A successful bioentrepreneur also needs to have great communication skills. He or she should be able to write and speak clearly and confidently, and should be persuasive and creative. Scientists who aspire to become bioentrepreneurs can learn these attributes by experience, by attending biotechnology conferences, and/or by taking one or more of the several bioentrepreneurial training and educational courses offered by the university systems in several countries.

11.11 BIOTECHNOLOGY PRODUCTS AND INTELLECTUAL PROPERTY RIGHTS

IP is the term used to describe the branch of law that protects the application of thoughts, ideas, and information which are of commercial value. It thus covers the law relating to patents, copyrights, trademarks, trade secrets, and other similar rights. IP protection for biotechnology is currently in a state of flux. While it used to be the case that living organisms were largely excluded from protection, attitudes are now changing, and biotechnology is increasingly receiving some form of protection. These changes have largely taken place in the United States and other industrialized countries, but other countries that wish to compete in the new biotechnological markets are likely to change their national laws in order to protect and encourage investment in biotechnology.

At the moment, there is no clear international consensus on how biotechnology should be treated. Although bodies such as the World Intellectual Property Organization (WIPO), the permanent body of the United Nations (UN) that is primarily responsible for international cooperation in IP, and the Organization for Economic Cooperation and Development (OECD) have conducted separate studies and produced various reports, these initiatives have only sought to make governments more aware of the potential problems and to offer some suggested solutions. In view of the highly controversial nature of providing IP protection for biotechnology, it is likely that in the short term, developments will be at a national and regional level.

11.11.1 Patenting, Licensing, and Partnership in the Biotechnology Industry

The success of a biotechnology company also depends on its patent portfolio and licensing revenue; therefore, from the very beginning, a biotechnology company needs to carefully design its patent portfolio. Patents for biological inventions were not allowed till 1980, but doing so became necessary only with the advent of biotechnology. Consequently, patents are now allowed for biological inventions involving microorganisms, vectors, DNA/RNA, proteins, monoclonal antibodies and hybridoma, isolated antigens, vaccine compositions, and transgenic animals and plants. Patents have also been allowed for methods involving isolation or purification of biological material; gene cloning and production of proteins; diagnosis, treatment, and use of a product; and screening methods.

The development of the biotechnology industry partly depends on its licensing revenue, which stems from a good patent portfolio. Similarly, the success of a pharmaceutical company partly depends upon its revenue that is generated due to marketing of compounds that are licensed by a biotechnology company. The revenue generated by pharmaceutical companies due to licensed compounds is estimated to have increased from 24% in 1992 to 35% in 2002. Similarly, the licensing revenue in the biotechnology industry rose from $5.7 billion in 1998 to more than $6.4 billion in 2000, which was expected to rise to $7.8 billion mark in the year 2003. Biogen, for example, made about $600 million from its licensing activity during 1991–1995, which was a prerequisite for licensing its first product, *Avonex*, in the year 1996. In 2000, the licensing revenue of Biogen comprised approximately 18% of its total revenue. This example suggests that the biotechnology industry depends heavily on long-term deals with their pharmaceutical partners, which not only determines the total revenue, but also the company's share price in the stock market. For instance, the very news that Curagen (United States) has entered into a deal for a drug development alliance with Bayer (Germany) sent Curagen's share price up by 35% to approximately $36. This illustrates that survival of biotechnology companies sometimes depends on successful deals with pharmaceutical and other companies, as does the success of deals. The success of deals also depends on finding the right partner and the post-deal governance.

Selection of a right partner in the biotechnology industry involves three steps:

1. Designing of search criteria, which should include both science (skills and R&D competence) and business (geographic proximity of partners and the market).
2. Large and diverse database on biotechnology companies that meet the search criteria and therefore could be prospective partners.
3. A screening process that involves negotiation on the terms of the deal.

Once a deal has materialized following the above three steps, it is necessary to nurture and improve the relationship between partners, since there are examples of bad relationships affecting business adversely and also those of good relationship boosting the business. For instance, Johnson and Johnson and Amgen entered into a deal to market *erythropoietin* (EPO) for anemia and had serious problems affecting the business. In contrast, the productive relationship between Pfizer and Neurogen, which involved licensing the drug *NGD-91* for treatment of Alzheimer's disease, led to further fruitful collaboration, since both partners invested heavily in building trust in their partnership.

11.11.2 Intellectual Property Protection

There are currently two main systems of protection for biotechnology: rights in plant varieties and patents. Both systems provide exclusive, time-limited rights of commercial exploitation. Keeping biotechnology a "secret" is a valuable form of protection. National treatment of trade secrets is diverse, and all attempts to harmonize trade secret laws in Europe, for example, have failed. Most jurisdictions do provide some form of protection against those who steal or use others' trade secrets unfairly. However, the problem with this form of protection is that the secret generally becomes public once the biotechnology is used commercially and thus the protection is lost. It is conceivable that copyright laws could afford some protection for biotechnology. Lines of genetic code are analogous to some extent with computer program code, which has now been incorporated into the copyright systems of most industrialized countries. However, this route to protection is fraught with practical and conceptual difficulties and is generally thought to be unsuitable. There is as yet no recorded case of biotechnologists claiming copyright for their inventions. Trademarks are also unlikely to be of much use in protecting biotechnology, though they may of course prove important later with regard to marketing products, processes, or services. An attempt to register the name of a plant or an animal as a trademark is unlikely to be successful, as public policy would prevent it. For example, in England, registrations for names of varieties of roses have been removed from the Trademark Register for lack of distinctiveness and because of the likelihood of confusion.

11.11.3 Intellectual Property Rights for Plants

Prior to the mid-1960s, only a few countries (such as Germany and the United States) gave any IP protection to plant varieties. Because of pressure from their plant breeding industries, 10 western European countries entered into a diplomatic process in the early 1960s which eventually culminated in the formation of International Union for the Protection of New Varieties of Plants (UPOV) and the signing of a convention (the 1961 UPOV Convention). Since that time, a number of other countries have become parties to the UPOV Convention. Amendments were made to the UPOV Convention in 1978, principally to facilitate the entry of the United States. The UPOV Convention requires that, in accordance with the provisions of the Convention, each member country must adopt national legislation to give at least 24 genera or species protection within 8 years of signing. A plant variety is protectable under the UPOV system if it is distinct, uniform, stable (DUS), and satisfies a novelty requirement. Novelty and distinctiveness equate broadly to novelty under patent law, but are more leniently applied in comparison to the patent rule. Also, an important requirement is that the variety must be maintained throughout the duration of protection. A country may apply the system to all genera or species, but there is no obligation to do so, and thus the system has been extended only gradually. In addition, the UPOV Convention allows national legislation to discriminate against foreigners (including nationals of a UPOV Convention country) under the principle of reciprocity. Thus, there is still some disparity in protection among UPOV members.

Duration of protection depends on national legislation and on the plant species to which the variety belongs, but is generally for 20–30 years. Grant of plant variety rights confers certain exclusive rights on the holder, including the exclusive right to sell the reproductive material (e.g., seed, cuttings, whole plants) of the protected variety. However, the rights do not extend to consumption material such as fruit or wheat seed grown for milling flour. Essentially, the exclusive rights define what others may or may not do in relation to the protected varieties. Plant breeders were for some time dissatisfied with the protection provided by the UPOV system. This eventually resulted in a major diplomatic conference in March 1991, at which the UPOV Convention was substantially revised. The new 1991 text provided greater protection, most notably by requiring that all member countries apply the convention to all genera and species by extending the exclusive rights to include harvested material and, most controversially, by allowing enforcement against farm-saved seed (where a farmer produces further seed of the protected variety from the previous year's crop). However, until the national governments ratify the new convention, the system will continue to be based on the 1978 text. There will be considerable national opposition to the strengthening of plant variety rights, and thus these changes may take years before they are implemented and, in the meantime, may even be superseded by greater availability of patent protection.

11.11.4 Patents and Biotechnology Products

A *patent* is a grant of exclusive rights for a limited time in respect of a new and useful invention. The exact requirements for grant of a patent, the scope of protection it provides, and its duration differs, depending on national legislation. However, generally the invention must be of patentable subject matter, novel (new), nonobvious (inventive), of industrial application, and sufficiently disclosed. A patent will provide a wide range of legal rights, including the right to possess, use, transfer by sale or gift, and to exclude others from similar rights. The duration will be for approximately 20 years (although for only 17 years in the United States). These rights are generally restricted to the territorial jurisdiction of the country granting the patent, and thus an inventor wishing to protect his or her invention in a number of countries will need to seek separate patents in each of those countries. Whilst the majority of countries provide some form of patent protection, only a few provide patent protection for biotechnology. These include Australia, Bulgaria, Canada, Czechoslovakia, Hungary, Japan, Romania, the Soviet Union, and the parties to the European Patent Convention (EPC). The reasons for this may differ, but generally it has been because biotechnology has been thought inappropriate for patent protection, either because the system was originally designed for mechanical inventions, for technical or practical reasons, or for one or more ethical, religious, or social concerns. In all the national patent offices where patents are granted for biotechnology, there is a considerable backlog of pending applications. Even in those countries where patent protection is provided, the type and extent of that protection is different in nearly every national system.

It has largely been the United States that has broken new ground in providing the possibility of patent protection. The patents have been granted for plants, under the Plant Patent Act, since 1930. However, prior to 1980, the U.S. Patent Office would not grant utility patents (separate from the Plant Patent Act) for living matter, because it deemed products of nature not to be within the terms of the utility patent statute. That was until the landmark decision of the U.S. Supreme Court in *Diamond v. Chakrabarty* which held that a particular genetically engineered bacterium was statutory subject matter for a utility patent. This decision has been the basis upon which patents have been granted for higher life forms. Subsequently, it has been held that a utility patent may be granted for plants.

Elsewhere, the treatment of applications for patents for living matter is far from certain. While patents are granted in many countries for plants and microorganisms, it has been the issue of patents for animals which has been the most controversial. Whilst it is not possible to succinctly summarize the position in the rest of the world, it is possible to describe the present approach of those countries that are party to the EPC. The EPC is a regional arrangement agreed by14 European countries to make multiple patent applications and to introduce a common system for patent protection. An application under the EPC is for a European patent (or Europatent). If a Europatent is granted by the European Patent Office (EPO), it has the same effect and is subject to the same conditions as a national patent in each of the member countries designated in the application—in other words, a bundle of national patents can be obtained through a single application. The EPC provides that "plant or animal varieties or essentially biological processes for the production of plants or animals" are excluded from patent protection (although the exclusion is expressly stated not to apply to microbiological processes and products). These exclusions would appear to place unequivocal prohibition on Europatents for macrobiotechnology. However, the EPO has been taking an increasingly narrow view of these exclusions, and has held that they do not exclude all plants and animals per se, but only claims for varieties of plants or animals, and that a process is not "essentially biological" if there has been substantial interference by man. It is also important to note that there is currently before the European Parliament of the European Community (EC) a proposal for a Council Directive for harmonization of the legal protection provided for biotechnology in the EC. This does not propose to amend the EPC, but the present draft proposal would make even more opportunities available for patenting biotechnology and thus make the EC more attractive in terms of investment in biotechnology research.

11.11.5 International Treaties

There are three international IP treaties which are of particular importance for the protection of biotechnology: The Paris Convention for the Protection of Industrial Property (the Paris Convention), the Budapest Treaty on the International Recognition of the Deposit of Microorganisms for the Purposes of Patent Procedure (the Deposit Treaty), and the Patent Cooperation Treaty (PCT).

The Paris Convention was originally signed in 1883 by just 11 countries, but now the majority of countries who have any form of IP law are parties to it. The keystone of the convention is the principle of national treatment: An applicant from one convention country shall have the same rights in a second convention country as a national of that second country. The convention covers patents and defines them so broadly that it permits application to any of the forms of industrial patents granted under the laws of the convention countries. The most important practical result of the convention is that it is possible to claim priority from an application made in a convention country for all subsequent convention countries within 12 months of the original filing.

The Deposit Treaty, as the full title suggests, is concerned with the deposit of samples of microorganisms for the purposes of patent applications. Applications for biotechnology patents often face considerable difficulties in describing the nature of the invention sufficiently. The Deposit Treaty is a vehicle for solving these problems, primarily through the setting up of a series of International Depository Authorities (IDA) and through the recognition by all member countries of a deposit in a single IDA.

The PCT simplifies the process of simultaneously filing patent applications in a number of countries. Under the PCT, a single application may be filed in one of the official receiving offices (ROs), designating any number of PCT member countries, which can eventually result in a national patent (and/or a Europatent) being granted in each of the designated states. Unfortunately, the eventual outcome is not a "world patent," and there is no harmonization patent law under the PCT apart from the procedural aspects. The PCT is discussed in greater detail next.

11.11.5.1 Patent Cooperation Treaty

The PCT is an international patent law treaty, concluded in 1970. It provides a unified procedure for filing patent applications to protect inventions in each of its contracting states. A patent application filed under the PCT is called an *international patent application*, or *PCT application*. A single filing of an international application is made with a RO in one language. It then results in a search performed by the relevant International Searching Authority (ISA), accompanied by a written opinion regarding the patentability of the invention that is the subject of the application. It is optionally followed by a preliminary examination, performed by the relevant International Preliminary Examining Authority (IPEA). Finally, the relevant national or regional authorities administer matters related to the examination of application (if provided by national law) and issuance of the patent. The PCT does not provide for the grant of an "international patent," as such a multinational patent does not exist, and the grant of patent is a prerogative of each national or regional authority.

The contracting states (the states that are parties to the PCT) constitute the International Patent Cooperation Union. The main advantages of the PCT procedure, also referred to as the *international procedure*, are the possibility to delay the national or regional procedures as much as possible, along with the respective fees and translation costs, and the unified filing procedure. An international patent application has two phases. The first phase is the international phase, in which patent protection is pending under a single patent application filed with the patent office of a contracting state of the PCT. The second phase is the national and regional phase, which follows the international phase, in which rights are continued by filing the necessary documents with the patent offices of separate contracting states of the PCT. A PCT application, as such, is not an actual request that a patent be granted, and it is not converted into one until it enters the "national phase." At the end of year 2004, one million patent applications have been filed. Eighteen months after the filing date or the priority date, the international application is published by the International Bureau at WIPO (based in Geneva, Switzerland) in one of the ten "languages of publication": Arabic, Chinese, English, French, German, Japanese, Korean, Portuguese, Russian, and Spanish. There is an exception to this rule, however. If 18 months after the priority date, the international application designates only the United States, the application is not automatically published. Finally, at 30 months from the filing date of the international application or from the earliest priority date of the application if a priority is claimed, the international phase ends, and the international application enters the national and regional phase. However, any national law may fix time limits, which expire after 30 months. For instance, it is possible to enter the European regional phase at 31 months from the earliest priority date. National and regional phases can also be started earlier on the express request of the applicant. If the entry into the national or regional phase is not performed within the prescribed time limit, the international application generally ceases to have the effect of a national or regional application.

11.11.5.2 World Intellectual Property Organization

The WIPO is one of the 16 specialized agencies of the UN. WIPO was created in 1967 "to encourage creative activity, to promote the protection of IP throughout the world." WIPO currently has 184 member states, and is headquartered in Geneva, Switzerland. The current Director-General of WIPO is Francis Gurry, who took office on October 1, 2008. Almost all UN Members as well as the Holy See are Members of WIPO (nonmembers are the states of Kiribati, Marshall Islands, Micronesia, Nauru, Palau, Solomon Islands, Timor-Leste, Tuvalu and Vanuatu, as well as the entities of the Palestinian Authority, the Sahrawi Republic, and Taiwan).

11.11.5.3 Agreement on Trade-Related Aspects of Intellectual Property Rights

The Agreement on Trade-Related Aspects of Intellectual Property Rights (TRIPS) is an international agreement administered by the World Trade Organization (WTO) that sets down minimum standards for many forms of IP regulation as applied to nationals of other WTO members. It was negotiated at the end of the Uruguay Round of the General Agreement on Tariffs and Trade (GATT) in 1994. Specifically, TRIPS contains requirements that nations' laws must meet for copyright rights, including the rights of performers, producers of sound recordings, and broadcasting organizations; geographical indications, including appellations of origin; industrial designs; integrated circuit layout-designs; patents; monopolies for the developers of new plant varieties; trademarks; trade dress; and undisclosed or confidential information. TRIPS also specifies enforcement procedures, remedies, and dispute resolution procedures. Protection and enforcement of all IP rights must meet the objectives to contribute to the promotion of technological innovation and to the transfer and dissemination of technology, to the mutual advantage of producers and users of technological knowledge, and in a manner conducive to social and economic welfare and to a balance of rights and obligations.

The TRIPS agreement introduced IP law into the international trading system for the first time and remains the most comprehensive international agreement on IP to date. In 2001, developing countries, concerned that developed countries were insisting on an overly narrow reading of TRIPS, initiated a round of talks that resulted in the Doha Declaration. The Doha declaration is a WTO statement that clarifies the scope of TRIPS, stating, for example, that TRIPS can and should be interpreted in light of the goal "to promote access to medicines for all." TRIPS has been criticized by the alter-globalization movement. Members of the movement object, for example, to its consequences with regards to the AIDS pandemic in Africa. The most visible conflict has been over AIDS drugs in Africa. Despite the role that patents have played in maintaining higher drug costs for public health programs across Africa, this controversy has not led to a revision of TRIPS. Instead, an interpretive statement, the Doha Declaration, was issued in November 2001, which indicated that TRIPS should not prevent states from dealing with public health crises. After Doha, the Pharmaceutical Research and Manufacturers of America (PhRMA) and the United States, and to a lesser extent other developed nations, began working to minimize the effect of the declaration. A 2003 agreement loosened the domestic market requirement and allows developing countries to export to other countries where there is a national health problem, as long as drugs exported are not part of a commercial or industrial policy. Drugs exported under such a regime may be packaged or colored differently to prevent them from prejudicing markets in the developed world. In 2003, the Bush administration also changed its position, concluding that generic treatments might, in fact, be a component of an effective strategy to combat HIV. Bush created the President's Emergency Plan for AIDS Relief (PEPFAR) program, which received $15 billion from 2003 to 2007, and was reauthorized in 2007 for $30 billion over the next 5 years. Despite wavering on the issue of compulsory licensing, PEPFAR began to distribute generic drugs in 2004–2005.

11.11.5.4 Issues with Biotechnology Patents

TRIPS covers everything from pharmaceuticals to information technology software and human gene sequences, and is emerging as a major issue dividing the North and the South. The TRIPS agreement is controversial in at least two areas. First, it threatens the right of poor countries to manufacture or to import cheap generic versions of patented drugs. The AIDS epidemic and other diseases are killing millions every year because people in poor countries cannot afford the exorbitant prices the pharmaceutical giants are charging for the patented drugs. The existing TRIPS agreement also forces all countries to accept a medley of new biotechnology patents covering genes, cell lines, organisms, and living processes that turn living matter into commodities. Governments all over the world have been persuaded into accepting these "patents on life" before anyone understood the scientific and ethical implications. The patenting of life forms and living processes is covered under Article 27.3(b) of the TRIPS agreement.

11.11.5.5 Patent Infringements

Farmers may face legal liability issues in agricultural biotechnology and in legal disputes involving IP. More particularly, companies that create transgenic crops have IP rights, usually patents, in those crops and take legal action against farmers who grow the transgenic crops without the companies' permission. However, one should not forget that seed companies also protect their patents in non-transgenic seeds and plants. Patent infringement cases in agriculture are not unique to transgenic seeds and plants. Four cases for patent infringement regarding transgenic crops have resulted in written opinions by courts in Canada and the United States: *Monsanto Canada, Inc. v. Schmeiser* (2001); *Monsanto Company v. Trantham* (2001); *Monsanto Company v. McFarling* (2002); and *Monsanto Company v. Swann* (2003).

The Schmeiser Case: The Canadian courts found the following to be factually true. Mr. Schmeiser sprayed three or four acres of his canola in 1997 with Roundup (Monsanto Corp., St. Louis) herbicide because he thought his canola field contained Roundup Ready canola. Sixty percent of the sprayed area survived the herbicide, thereby showing herbicide tolerance. Mr. Schmeiser separately harvested and stored the canola seed from the sprayed acres. In 1998, Mr. Schmeiser decided to use the canola from the sprayed acres as his seed canola for the 1998 crop. When Monsanto Canada pursued Mr. Schmeiser for these actions in a lawsuit, the grow-out and DNA tests of Schmeiser's 1998 crop showed 95%–98% Roundup Ready (Monsanto Corp., St. Louis) canola from tests conducted by Monsanto, 95%–98% Roundup Ready canola on the 1998 crop from a Canadian laboratory, 63%–70% Roundup Ready canola from grow-out tests by Mr. Schmeiser himself on 1997 and 1998 crops, and 0%–98% on various samples from the 1997 and 1998 crops submitted by Mr. Schmeiser to the University of Manitoba, with the 1997 saved seed specifically testing in the 95%–98% range.

The Canadian courts held that Mr. Schmeiser infringed the Monsanto patents after consideration of two defenses. First, the courts ruled that the fact that Mr. Schmeiser did not use Roundup herbicide was irrelevant to the patent violation. The court stated that the Monsanto patent claims related to the gene and cells of the canola plants, and this particular patent had nothing to do with the herbicide. In light of the patent relevant to this lawsuit, Mr. Schmeiser could only grow these patented seeds if he had permission to do so. Mr. Schmeiser admitted that he had not signed a technology use agreement with Monsanto. Second, the court ruled that regardless of the origin of the 1997 seeds—for example pollen drift, seed spills from bags, or wind-blown from truck beds—Mr. Schmeiser infringed the patent when he knew or should have known that he was planting Roundup Ready canola in 1998. The court rejected Mr. Schmeiser's claim as an innocent grower, because his actions demonstrated that he was not innocent—that is, he knew or should have known that he was growing patented seeds. The court left undecided whether Monsanto would have an infringement claim against a truly innocent grower.

The Trantham, Swann, and McFarling Cases: In these three U.S. cases, the farmers defended against Monsanto's infringement claims relating to saved seed by arguing that U.S. antitrust law blocked Monsanto from pursuing legal remedies against farmers who saved seed from one crop year to the next. The arguments can be conflated into three main points about the interplay between antitrust law and patent law. First, the U.S. courts ruled that the patent law allows Monsanto to condition its permissive use of patented seeds through technology use agreements prohibiting the saving of seeds and authorizing the growing of seeds for one crop year only for commercial sale as a commodity crop. In other words, Monsanto does not create an illegal monopoly solely because Monsanto grants permission through technology use agreements that impose use restrictions. Second, the courts held that the farmers were free to buy or to not buy Monsanto patented seeds each crop year. The courts stated that the fact that the farmers desired to use Monsanto's patented seeds in future years signified only superior performance of patented technology, not an illegal tying arrangement under antitrust law. Finally, the courts held that Monsanto does not violate U.S. antitrust law on the basis of the fact that Monsanto markets its Roundup Ready seeds differently in the United States than it does in Argentina. The court noted that Monsanto is responding to different market circumstances in the United States, where patents on seeds exist, than in Argentina, where patents on seeds do not exist. Under U.S. law, Monsanto is entitled to enforce its U.S. patents.

11.12 BIOTECHNOLOGY STOCK INVESTMENT: PROS AND CONS

The biotechnology sector is an innovative and growing industry. Back in 1982, the sector had virtually no products, but now biotechnology firms have successfully commercialized a number of products. While biotechnology companies have introduced many new drug products, several biotechnology firms are still in the developmental stage, with their fortunes largely determined by investor perceptions of the relative merits of their R&D pipelines. In addition, the endeavors of biotechnology research have extended beyond the field of medicine to other fields, such as agriculture, energy, and environmental protection. How does an individual investor analyze a company with no revenues and no products? How is one to access the quality of research in such a technical and specialized field? One of the most striking aspects of the biotechnology industry is its scarcity of earnings. Most biotechnology companies carry losses for years, even after they have launched a new product. As a result, valuing biotechnology is a precarious undertaking at best. The details of a biotechnology firm's business, its research methods, its test results—even the products themselves—can be highly complex. The success or failure of a drug during clinical trials is difficult and often impossible to predict.

Investing in biotechnology stocks is somewhat unlike investing in other stocks, because in valuing biotechnology stocks, it has always been difficult to use traditional net present value and discounted cash flow approaches, particularly for the clinical and preclinical-stage companies. Predicting the probability of a single product's success in the clinic depends on many variables such as clinical trial design, difficulty of indication, and quality of Phase II data. In addition, the company's financial well-being and corporate partnerships may further complicate the valuation analysis. The large cap and profitable biotechnology companies have had the broadest appeal to investors, but that is only a handful of companies. Investors in biotechnology stocks take a long-term approach to investing. There are stocks that can significantly appreciate in value overnight if a trial is successful; conversely, they can also drop by 30%–70% in value with disappointing results. Biotechnology companies' stocks tend to be heavily influenced by favorable or unfavorable news regarding the development or testing of a product. In this section, we discuss some of the critical factors that need to be considered before investing in biotechnology companies.

11.12.1 Products in the Pipeline

Look for companies with at least two drugs in clinical trials, because if for some reason the product proves to lack efficacy, then at least the company has something to fall back on. Another approach is to look for companies diversified around a specific disease class or those that have a niche technology that can be used as a platform for a range of different drugs.

11.12.2 Collaboration and Merger

Companies that fail to link up with a corporate or academic partner can have trouble surviving. To ensure survival or lower risk, biotechnology companies will engineer several collaborative agreements with various pharmaceutical companies for research or marketing. Look for substantial milestone payments and cash commitments when the deal is announced, not just "talk" about a research alliance. For example, Abgenix has made a number of deals with biotechnology and pharmaceuticals for licensing its XenoMouse technology for making humanized monoclonal antibodies that are worth hundreds of millions of dollars.

11.12.3 Experienced Management

Early-phase companies may or may not have someone in senior management with a proven track record of taking a drug through the regulatory hurdles and/or to the marketplace; even so, look at their financials and go to the section on management to see who is working there and what they have accomplished in the past.

11.12.4 Cash Flow

For many biotechnology companies, the release of a commercial product is often many years away and requires millions of dollars. Thus, a company's burning of cash in ongoing R&D, or "burn rate," is a critical measure of a company's longevity. Look for companies that have a minimum of 2 years' cash reserves.

11.13 MARKETING TRENDS IN BIOTECHNOLOGY

A favored alternative to pharmaceuticals due to higher growth and lower generic risk, biotechnology is considered "an industry that cannot be judged with a rearview mirror, because it is moving forward so fast," particularly focusing on cancer and other hard-to-treat diseases. Between impending couples of years of expiring drug patents, Merck & Co. found itself on the defensive against a number of Vioxx-safety-related lawsuits; the traditional pharmaceutical industry, which has historically used chemical techniques to discover its products, had a rough year in 2005.

On the other hand, the biotechnology industry, which uses genetic techniques, is fairly new in its product offerings, so patent expiration is still decades away. As the year 2005 ended, shares of the nation's largest biotechnology companies were trading at or near record levels, and the *Washington Post* acknowledged, "a payoff for investors in companies that have been putting intensive focus on cancer and other hard-to-treat diseases over the last few years." In fact, life-sciences company Burrill & Company prognosticated that the biotechnology industry would raise more than $100 billion in 2010. According to Thomson/Baseline, these S&P 500 companies gained 16.2% in 2005, while traditional pharmaceutical companies fell about 7%.

11.14 ROLE OF REGULATORS IN BIOTECHNOLOGY PRODUCT DEVELOPMENT

11.14.1 World Health Organization

The World Health Organization (WHO) is a specialized agency of the UN that acts as a coordinating authority on international public health. Established on April 7, 1948, and headquartered in Geneva, Switzerland, the agency inherited the mandate and resources of its predecessor, the Health Organization, which had been an agency of the League of Nations. As well as coordinating international efforts to monitor outbreaks of infectious diseases such as SARS, malaria, and AIDS, the WHO also sponsors programs to prevent and treat such diseases. The WHO supports the development and distribution of safe and effective vaccines, pharmaceutical diagnostics, and drugs. After over two decades of fighting smallpox, the WHO declared in 1980 that the disease had been eradicated—the first disease in history to be eliminated by human effort. The WHO aims to eradicate polio within the next few years. The organization has already endorsed the world's first official HIV/AIDS Toolkit for Zimbabwe (from October 3, 2006), making it an international standard. In addition to its work in eradicating disease, the WHO also carries out various health-related campaigns—for example, to boost the consumption of fruits and vegetables worldwide and to discourage tobacco use. Experts met at the WHO headquarters in Geneva in February 2007 and reported that their work on pandemic influenza vaccine development had achieved encouraging progress. More than 40 clinical trials have been completed or are ongoing. Most have focused on healthy adults. Some companies, after completing safety analyses in adults, have initiated clinical trials in the elderly and in children. So far, all vaccines appear to be safe and well-tolerated in all the age groups tested.

The WHO also conducts research on, for instance, whether the electromagnetic field surrounding cell phones has a negative influence on health. Some of this work can be controversial, as illustrated by the April 2003 joint WHO/Food and Agricultural Organization (FAO) report, which recommended that sugar should form no more than 10% of a healthy diet. This report led to the sugar industry lobbying against the recommendation, to which the WHO/FAO responded by including in the report the statement "The Consultation recognized that a population goal for free sugars of less than 10% of total energy is controversial," but also stood by its recommendation based upon its own analysis of scientific studies.

11.14.2 International Conference on Harmonization

In much of the industrialized world, the history of medicinal product registration has followed a similar pattern, which could be described as initiation, acceleration, rationalization, and harmonization. The realization that it was important to have an independent evaluation of medicinal products before they are allowed on the market was reached at different times in different regions. In the United States, a tragic mistake in the formulation of a children's syrup in the 1930s was the trigger for setting up the product authorization system under the FDA. In Japan, government regulations requiring all medicinal products to be registered for sale started in the 1950s. In many countries in Europe, the trigger was the thalidomide tragedy of the 1960s, which revealed that the new generation of synthetic drugs that were revolutionizing medicine at the time had the potential to harm as well as heal.

For most countries, whether or not they had initiated product registration controls earlier, the 1960s and 1970s saw a rapid increase in laws, regulations, and guidelines for reporting and evaluating the data on safety, quality, and efficacy of new medicinal products. At the same time, the industry was becoming more international and seeking new global markets, but the registration of medicines remained a national responsibility. Although different regulatory systems were based on the same fundamental obligations to evaluate the quality, safety, and efficacy, the detailed technical requirements had diverged over time to such an extent that the industry found it necessary to duplicate many time-consuming and expensive test procedures in order to market new products internationally. The urgent need to rationalize and harmonize regulation was impelled by concerns over rising costs of healthcare, escalation of R&D costs, and the need to meet public expectations that there should be a minimum of delay in making safe and efficacious new treatments available to patients in need.

The birth of the ICH took place in April 1990 at a meeting hosted by the European Federation of Pharmaceutical Industries and Associations (EFPIA) in Brussels. Representatives of the regulatory agencies and industry associations of Europe, Japan, and the United States met primarily to plan an international conference, but the meeting also discussed the ICH's wider implications and terms of reference. The ICH Steering Committee, which was established at that meeting, has since met at least twice a year, with the location rotating between the three regions.

The ICH of Technical Requirements for Registration of Pharmaceuticals for Human Use is a unique project that brings together the regulatory authorities of Europe, Japan, and the United States and experts from the pharmaceutical industry in the three regions to discuss scientific and technical aspects of product registration. The purpose is to make recommendations on ways to achieve greater harmonization in the interpretation and application of technical guidelines and requirements for product registration, in order to reduce or obviate the need to duplicate the testing carried out during the R&D of new medicines. The objective of such harmonization is a more economical use of human, animal, and material resources, and the elimination of unnecessary delay in the global development and availability of new medicines, while maintaining safeguards on quality, safety, and efficacy, and regulatory obligations to protect public health.

11.14.3 United States Food and Drug Administration

The U.S. FDA is an agency within the U.S. Public Health Service, which is a part of the Department of Health and Human Services. The FDA regulates over $1 trillion worth of products, which account for 25 cents of every dollar spent annually by American consumers and touch the lives of virtually every American every day. It is the FDA's job to see that food is safe and wholesome, cosmetics will not hurt people, medicines and medical devices are safe and effective, and that radiation-emitting products such as microwave ovens will not do harm. Feed and drugs for pets and farm animals also come under FDA scrutiny. The FDA ensures that all of these products are labeled truthfully with the information that people need to use them properly. The FDA requires that both prescription and over-the-counter drugs be proven safe and effective. In deciding whether to approve new drugs, the FDA does not itself conduct research, but rather examines the results of studies done by the manufacturer. The FDA must determine that the new drug produces the benefits it is supposed to, without causing side effects that would outweigh those benefits.

The FDA tests food samples to see if any substances, such as pesticide residues, are present in unacceptable amounts. If contaminants are identified, the FDA takes corrective action. The FDA also sets labeling standards to help consumers know what is in the foods they buy. The FDA also ensures that medicated feeds and other drugs given to pets and animals raised for food are not threatening to health. FDA investigators examine blood bank operations, from record keeping to testing for contaminants. The FDA also ensures the purity and effectiveness of biological products (medical preparations made from living organisms and their products) such as insulin and vaccines. These are classified and regulated by the FDA according to their degree of risk to the public. Devices that are life-supporting, life-sustaining, or implanted (such as pacemakers) must receive agency approval before they can be marketed. The FDA can have unsafe cosmetics removed from the market. Dyes and other additives used in drugs, foods, and cosmetics are also subject to FDA scrutiny. The agency must review and approve these chemicals before they can be used.

FDA investigators and inspectors collect domestic and imported product samples for scientific examination and for label checks. If a company is found violating a law that the FDA enforces, the FDA can encourage the firm to voluntarily correct the problem or to recall a faulty product from the market. When a company cannot (or will not) correct a public health problem with one of its products voluntarily, the FDA has legal sanctions it can bring to bear. The FDA can go to court to force a company to stop selling a product and to have items already produced seized and destroyed. When warranted, criminal penalties—including prison sentences—are sought against manufacturers and distributors. About 3000 products a year are found to be unfit for consumers and are withdrawn from the marketplace, either by voluntary recall or by court-ordered seizure. In addition, approximately 30,000 import shipments a year are detained at the port of entry because the goods appear unacceptable. Evidence to back up FDA legal cases is prepared by FDA laboratory scientists. Some analyze samples to see, for instance, if products are contaminated with illegal substances. Other scientists review test results submitted by companies seeking agency approval for drugs, vaccines, food additives, coloring agents, and medical devices.

11.14.4 Good Laboratory Practice

GLP generally refers to a quality system of management controls for laboratories and research organizations that regulate how nonclinical safety studies are planned, performed, monitored, recorded, reported, and archived. It ensures the consistency and reliability of results for submissions to the USFDA, the OECD, and other national organizations. GLP practices are intended to promote the quality and validity of test data. GLP is a regulation that goes beyond good analytical practice. Good analytical practice is important, but it is not enough. For example, the laboratory must have a specific organizational structure and procedures to perform and document lab work. The objective is not only quality of data, but also traceability and integrity of data. However, the biggest difference between GLP and non-GLP work is the type and amount of documentation. A GLP inspector normally examines the documentation process to find out who has done a study, how the experiment was carried out, which procedures have been used, whether there has been any problem, and if so, how it has been solved.

11.14.4.1 GLP Requirements

The role and responsibilities of each activity must be defined by the company, with names of individuals who are involved in the project. All routine lab work should follow written standard operating procedures (SOPs). Facilities such as laboratories should be large enough and have the right construction to ensure the integrity of a study (e.g., to avoid cross-contamination). Test and control articles should have the right quality, and instruments should be calibrated and well maintained. People should be trained or otherwise qualified for the job. Raw data and other data should be acquired, processed, and archived to ensure integrity. Unlike a company's research labs, a university's research labs do not follow the GLP norms, as they adhere to only institutional laboratory norms and unfortunately most laboratories are in situations where they have had to interpret the regulations. Procedures have been developed on an ad hoc basis, in isolation, in response to inspections by both the company's quality assurance unit and regulatory bodies. Under such duress, many scientists in the industry have developed procedures to validate their instrumentation, even though the same approach will already have been applied at the instrument manufacturer's site. SOPs written to accompany such validation efforts often duplicate extracts from operation manuals. This being the case, why don't the manufacturers provide the SOPs directly? When it comes to validating the instrument's application software, the person responsible has to take the manufacturer's word for it that the software has been validated and hope that supporting documents, such as test results and source code are available to regulatory agencies upon request.

11.14.4.2 National Legislation

In view of the importance of GLP, various pieces of legislation have been established in the United States to ensure the safety and efficacy of human and veterinary drugs (and devices) and the safety of food and color additives on the sponsor (manufacturer) of the regulated product. Public agencies (such as the U.S. FDA) are responsible for reviewing the sponsor's test results and whether or not they can demonstrate the product's safety and efficacy. Only when the agencies are satisfied that safety and efficacy have been established adequately is the marketing of the product permitted. Until the mid-1970s, the underlying assumption at the FDA was that the reports submitted by the sponsors to the agency accurately described study conduct and precisely reported the study data. A suspicion that this assumption was mistaken was raised during a review of studies submitted by a major pharmaceutical manufacturer in support of new drug applications (NDAs) for two important therapeutic products. Data inconsistencies and evidence of unacceptable laboratory practices came to light. The FDA requested a "for cause" inspection—which is initiated at the request of an agency when there are grounds for doubt surrounding an FDA-regulated product—of the manufacturer's laboratories to determine the cause and the extent of the discrepancies, which revealed defects in design, conduct, and reporting of the studies. Further inspections at several other sites found similar problems.

GLP regulations were finally proposed on November 19, 1976, to assure a study's validity. The proposed regulations were designated as a new part, 3.e, of Chapter 21 of the Code of Federal Regulations (CFR). The final regulations were codified as Part 58 (21 CFR). The U.S. EPA issued almost identical regulations in 1983 to cover the required health and safety testing of agricultural and industrial chemicals under the Federal Insecticide, Fungicide and Rodenticide Act (FIFRA) and the Toxic Substances Control Act (TSCA). The GLPs were promulgated in response to problems encountered with the reliability of submitted studies. Some of the studies were so poorly conducted that "the resulting data could not be relied upon for the EPA's regulatory decision-making process." The EPA regulations were extensively amended in 1989 and now cover essentially all testing required to be submitted to EPA under either Act. Both GLP regulations are of a similar format and have, with few exceptions, the same wording.

11.14.4.3 Facilities

All GLP regulations also have requirements for facilities; for example, animal care facilities are listed along with animal supply facilities, facilities for handling test and control articles, and laboratories and storage facilities. The main purpose of this requirement is to ensure the integrity of the study and of study data. Three main requirements for facilities are: limited access to buildings and rooms, adequate size, and adequate construction. For example, if a testing facility is too small to handle the specified volume of work, there may be a risk of mixing incompatible functions; if the air conditioning system is wrongly designed, there may be cross-contamination between different areas.

11.14.4.4 GLP Inspection and Enforcement

It is FDA's responsibility to enforce the federal Food, Drug, and Cosmetic Act to ensure safety and effectiveness of drugs and medical devices. This is enforced through regulations, guidance documents, and FDA inspections. The FDA has the responsibility to inspect GLP studies related to products that are marketed in the United States—it does not matter where the products are developed or manufactured. The FDA has developed an inspection program with two types of inspections: routine inspections and "for cause" inspections. Routine inspections should be conducted at least every second year. It is an ongoing evaluation of a laboratory's compliance with GLP regulation. "For cause" inspections are less frequent, and they constitute only about 20% of all GLP inspections. Reasons for such inspections could be a follow-up of an inspection with serious deficiencies or when the FDA suspects noncompliance when investigating an NDA. It also may happen that the FDA gets some hints from external sources about noncompliance in laboratories. Typically, the FDA does not announce GLP inspections. If a laboratory refuses to accept FDA inspections, either in full or in part, the FDA will not accept studies in support of NDAs. Deviations from GLP requirements are documented in different ways. If the inspection team finds deviations, they write them in a specific form, Form 483. The deviations are discussed during the exit meeting and the laboratory can respond. Next, the lead inspector writes a full inspection report called *establishment inspection report*, which may be up to 20 or 30 pages in length. Depending on the deviations, the inspector will or will not to write a warning letter. This letter is sent to the company's management. The company should respond with a corrective action plan within 14 days.

11.14.5 Good Manufacturing Practice

GMP regulations promulgated by the U.S. FDA under the authority of the Federal Food, Drug, and Cosmetic Act have the force of law, and require that manufacturers, processors, and packagers of drugs, medical devices, food, and blood take proactive steps to ensure that their products are safe, pure, and effective. GMP regulations require a quality approach to manufacturing, enabling companies to minimize or eliminate instances of contamination, mix-ups, and errors. This in turn protects the consumer from purchasing a product that is not effective or may even be dangerous. Failure of firms to comply with GMP regulations can result in very serious consequences, including recall, seizure, fines, and jail time. GMP regulations address issues including record keeping, personnel qualifications, sanitation, cleanliness, equipment verification, process validation, and complaint handling. Most GMP requirements are very general and open-ended, allowing each manufacturer to decide individually how to best implement the necessary controls. This provides much flexibility, but also requires that the manufacturer interpret the requirements in a manner that makes sense for each individual business. GMP is also sometimes referred to as "cGMP." The "c" stands for "current," reminding manufacturers that they must employ technologies and systems that are up-to-date in order to comply with the regulation. In the United States, the phrase "current GMP" appears in 501(B) of the 1938 Food, Drug, and Cosmetic Act (21 USC 351). U.S. courts may theoretically hold that a drug product is adulterated even if there is no specific regulatory requirement that was violated, as long as the process was not performed according to industry standards.

11.14.5.1 Cleanroom Facility

Unlike a research laboratory, manufacturing of healthcare products takes place in a highly contaminant-free environment called a *cleanroom facility*, which is strictly regulated and controlled. Typically used in manufacturing or scientific research, a cleanroom is an environment that has a low level of environmental pollutants such as dust, airborne microbes, aerosol particles, and chemical vapors. More accurately, a cleanroom has a *controlled* level of contamination that is specified by the number of particles per cubic meter at a specified particle size. To provide perspective, the ambient air outside in a typical urban environment might contain as many as 35,000,000 particles per m^3, 0.5 μm and larger in diameter, corresponding to an ISO 9 cleanroom.

11.14.5.2 Cleanroom Classifications

Cleanrooms are classified according to the number and size of particles permitted per volume of air. Large numbers like "class 100" or "class 1000" refer to U.S. Federal Standard (FED STD or FS) 209E, and denote the number of particles of size 0.5 μm or larger permitted per cubic foot of air. The standard also allows interpolation, so it is possible to describe, e.g., "class 2000." Small numbers refer to ISO 14644-1 standards, which specify the decimal logarithm of the number of particles of size 0.1 μm or larger permitted per cubic meter of air. Therefore, for example, an ISO class 5 cleanroom has at most 10^5 or 100,000 particles per cubic meter. Both FS 209E and ISO 14644-1 assume log–log relationships between particle size and particle concentration; consequently, there is no such thing as a "zero" particle concentration. The table locations without entries are not applicable (N/A) combinations of particle sizes and cleanliness classes and should not be read as zero. Because 1 m^3 is approximately 35 ft^3, the two standards are mostly equivalent when measuring 0.5 μm size particles, although the testing standards differ. Ordinary room air is approximately class 1,000,000 or ISO 9. BS 5295 *Class* 1 also requires that the greatest particle present in any sample does not exceed 5 μm.

11.14.5.3 GMP Enforcement

GMP regulations are enforced in the United States by the FDA. Within the European Union, GMP inspections are performed by the national regulatory agencies, such as the Medicines and Healthcare Products Regulatory Agency (MHRA) in the United Kingdom. In the Republic of Korea (South Korea), the enforcement is done by the Korea Food and Drug Administration (KFDA), while the TGA enforces GMP in Australia. GMP enforcement in South Africa is done by the Medicines Control Council (MCC), and in Brazil by the Agência Nacional de Vigilância Sanitária (ANVISA, the National Health Surveillance Agency). In case of Iran, India, and Pakistan, GMP enforcement is done by the Ministry of Health, and by similar national organizations worldwide. Each of the inspectorates carries out routine GMP inspections to ensure that drug products are produced safely and correctly. Additionally, many countries perform a preapproval inspection (PAI) for GMP compliance prior to the approval of a new drug for marketing. Regulatory agencies (including the FDA and regulatory agencies in many European nations) are authorized to conduct unannounced inspections, though some are scheduled. FDA routine domestic inspections are usually unannounced, but must be conducted according to 704(A) of the Food, Drugs, and Cosmetics Act (21 USC 374), which requires that they are performed at a "reasonable time." Courts have held that any time the firm is open for business is a reasonable time for an inspection.

11.15 CERTIFICATION AND ACCREDITATION

The quality of a product or diagnostic assay is based on how good a product is manufactured or how well the diagnostic assay is conducted. A company that makes biotechnology products applies to international organizations that specialize in biotechnology product assessment and obtains the license and complies with guidelines laid down by the regulatory agencies. There are several international organizations that rate a biotechnology lab product or a diagnostic procedure. We next discuss some of the most renowned organizations that deal with biotechnology lab and product development.

11.15.1 International Operating Procedures

ISO 9000 is a family of standards for quality management systems maintained by the International Organization for Standardization (ISO) and administered by accreditation and certification bodies. The rules are updated, and the time and changes in the requirements for quality motivate change. More recently, ISO 9000 made a few changes for the requirement of getting ISO 9000 certification, some which are as follows:

- A set of procedures that cover all key processes in the business
- Monitoring all processes
- Keeping adequate records
- Checking output for defects with an appropriate and corrective action plan
- Regularly reviewing individual processes for effectiveness
- Facilitating continual improvement

A company or organization that has been independently audited and certified to be in conformance with ISO 9001 may publicly state that it is "ISO 9001 certified" or "ISO 9001 registered." Certification to an ISO 9001 standard does not guarantee any quality of end products and services; rather, it certifies that formalized business processes are being applied. Although the standards originated in manufacturing, they are now employed across several types of organizations. A "product," in ISO vocabulary, can mean a physical object, services, or software.

11.15.1.1 ISO 14644-1

ISO 14644-1 was the first ISO 14644 International Standard prepared by ISO Technical Committee 209 (ISO/TC 209). The document was submitted as an American National Standard and has been adopted as ANSI/IEST/ISO 14644-1: 1999 in the United States, following the cancellation of FED STD 209E.

11.15.1.2 ISO 14698-1

ISO 14698-1 was first written in 2003. ISO 14698-1 describes the principles and basic methodology for a formal system to assess and control biocontamination where clean room technology is applied, in order that biocontamination in zones at risk can be monitored in a reproducible way and appropriate control measures can be selected. In zones of low or negligible risk, this standard may be used as a source of information.

11.15.1.3 ISO 14698-2

ISO 14698-2 became available to the public in October 2003. ISO 14698-2 gives guidance on basic principles and methodological requirements for all microbiological data evaluation and the estimation of biocontamination data obtained from sampling for viable particles in zones at risk, as specified by the system selected. This standard is not intended to test the performance of microbiological counting techniques of determining viable units.

11.15.2 British Standard 5295

Clean rooms are classified by the cleanliness of their air. The method most easily understood and most universally applied is the one suggested in the earlier versions (A to D) of Federal Standard 209 of the USA. In this old standard the number of particles equal to and greater than 0.5 mm is measured in one cubic foot of air and this count used to classify the room. The most recent 209E version has also accepted a metric nomenclature. In the UK the British Standard 5295, published in 1989, is also used to classify clean rooms. This standard is about to be superseded by BS EN ISO 14644-1.

PROBLEMS

Section A: Descriptive Type

Q1. What are the different methods of scientific enquiry?
Q2. Describe a scientific invention, and provide an example.
Q3. How do companies commercialize scientific discovery? Explain.
Q4. Explain the significance of a bioequivalence lab in biotechnology product testing.
Q5. Why are quality assurance and quality control essential for biotechnology companies?
Q6. What is a PCT?
Q7. What are some of the challenges in filing patents in biotechnology?
Q8. Explain how a biotechnology company can be started.

Section B: Multiple Choice

Q1. Ibn al-Haytham (Alhazen) is known for the …
  a. Development of the scientific method
  b. Development in the field of medicine
  c. Development in the field of chemistry

Q2. GLP deals with …
  a. Clinical testing
  b. Research laboratories
  c. Manufacturing plants
  d. None of the above

Q3. A biosafety hood or laminar hood is used to …
  a. Process tissues
  b. Avoid contamination
  c. Prepare reagents

Q4. To avoid contamination, nonclinical labs are separated from clinical labs by a proper barrier. True/False

Q5. Bioequivalence tests deal with …
  a. Animal testing
  b. *In vitro* testing
  c. Human testing
  d. None of the above

Q6. Preclinical testing is done after clinical testing. True/False

Q7. Phase IV trials are also known as post-marketing surveillance trials. True/False

Q8. Which of the following is the first biotechnology company in the United States to make an IPO?
  a. Genzyme
  b. Biogen
  c. Genentech
  d. Immunogen

Q9. The PCT is a regional patent law treaty. True/False

Q10. WIPO is affiliated with …

a. WHO
b. UNESCO
c. UN
d. USFDA

Q11. Both U.S. FDA and ICH are drug regulatory bodies. True/False

Q12. ________ is a systematic preventive approach to food safety and pharmaceutical safety.

a. U.S. FDA
b. ICH
c. HACCP

Section C: Critical Thinking

Q1. Why are market research and analysis important in biotechnology product development?

Q2. Why cannot biotechnology products be sold at "discount rates" like other products?

FIELD TRIP

Organize a field trip to any pharmaceutical or biotechnology manufacturing bases in your city. Learn how its biotechnology product is manufactured and submit your observations in the form of field report to your class instructor.

REFERENCES AND FURTHER READING

Angell, M. *The Truth about Drug Companies*. Random House Trade Paperbacks, New York, 2004. p. 30.

Angell, M. Drug companies & doctors: A story of corruption. *New York Rev. Books* 56(1): 8–12, 2009.

Arcangelo, V.P. and Peterson, M.A. *Pharmacotherapeutics for Advanced Practice: A Practical Approach*. Lippincott Williams & Wilkins, Philadelphia, PA, 2005. ISBN 0781757843.

Brater, C.D. and Daly, W.J. Clinical pharmacology in the middle ages: Principles that presage the 21st century. *Clin. Pharmacol. Ther*. 67: 447–450, 2000.

Cancellation of FED-STD-209E—Institute of Environmental Sciences and Technology Nov. 2001. http://www.iest.org/i4a/pages/index.cfm?pageid=3480 (accessed on 2007-12-03).

Chronicle of the World Health Organization. World Health Organization, 1948, p. 54. http://whqlibdoc.who.int/hist/chronicles/chronicle_1948.pdf. Retrieved on July 18, 2007.

Cleanroom Classification/Particle Count/FS209E/ISO TC209/. http://cleanroom.byu.edu/particlecount.phtml

David, W.T. Arab roots of European medicine. *Heart Views* 4: 2, 2003.

Demain, A.L. and Davies, J.E. *Manual of Industrial Microbiology and Biotechnology*. ASM Press, Washington, DC, 1999.

FAO/WHO. FAO/WHO guidance to governments on the application of HACCP in small and/or less-developed food businesses (PDF).

Food Safety Research Information Office. A focus on hazard analysis and critical control points. Created June 2003, Updated March 2008.

Glossary of Clinical Trial Terms. NIH Clinicaltrials.gov/ct/info/glossary (accessed on 2008-18-03).

Guidance for Industry 2006. Investigators, and Reviewers Exploratory IND Studies. Food and Drug Administration. http://www.fda.gov/cder/guidance/7086fnl.htm. Retrieved on 2007-05-01.

ICH Guideline for Good Clinical Practice: Consolidated Guidance (www.vichsec.org/en/GL48-st4.doc).

International Conference on Harmonization of Technical Requirements for Registration of Pharmaceuticals for Human Use. http://www.ich.org

International Conference on Harmonization. http://www.ich.org/cache/compo/276-254-1.html

International HACCP Alliance. International HACCP Alliance (PDF). http://haccpalliance.org/alliance/HACCPall.pdf. Retrieved on October 12, 2007.

Iriye, A. *Global Community: The Role of International Organizations in the Making of the Contemporary World*. University of California Press, Berkeley, CA, 2002.

ISO 14644-1 Scope 2007. http://www.iest.org/i4a/pages/index.cfm?pageid=3322#ISO_14644-1

Patents: Frequently Asked Questions, World Intellectual Property Organization, Retrieved on February 22, 2009. http://www.wipo.int

Sox, H.C. and Rennie, D. Seeding trials: Just say "no." *Ann. Intern. Med.* 149: 279–280, 2008.

Toby, E.H. *The Rise of Early Modern Science*. Islam, China, and the West, p. 218. Cambridge University Press, Cambridge, New York, 2003.

What is informed consent? US National Institutes of Health. www.Clinicaltrials.gov;en.wikipedia.org/wiki/Informed_consent (Accessed on 2011-18-05).

World Health Organization. Workers' health: Global plan of action. *Sixtieth World Health Assembly*, May 23, 2007. Retrieved September 15, 2008.

12 Industrial Biotechnology

LEARNING OBJECTIVES

- Discuss the significance of industrial biotechnology
- Explain the large-scale production of biomolecules using bioreactors
- Discuss types of fermenters and their applications
- Discuss upstream and downstream processes for biotechnology products
- Explain biosafety issues and automation of industrial plants
- Discuss product validation and regulation of biotechnology products
- Discuss industrial applications of microbes, plants, and mammalian cells in biotechnology product development

12.1 INTRODUCTION

With the rapid increase in human population that created an increased demand for healthcare products, manufacturing large quantities of healthcare products (such as vaccines and antibiotics) became essential. This could only be possible by using large-sized bioreactors or fermenters. When we talk about industrial biotechnology, the first thing that comes to mind is large-scale production. In the broadest sense, industrial biotechnology is concerned with all aspects of business that relate to microbiology. In a more restricted sense, industrial microbiology is concerned with employing microorganisms to produce a desired product and preventing microbes from diminishing the economic value of various products. Some of the commercial products made by microbes are antibiotics, steroids, human protein, vaccines, vitamins, organic acids, amino acids, enzymes, alcohols, organic solvents, and synthetic fuels. The potential of microbes was also realized in the recovery of metals from ores through bioleaching, recovery of petrol, and single-protein production. The term fermentation in industrial microbiology is used in a wider sense to include any chemical transformation of organic compounds carried out by using microbes and their enzymes. Production methods in industrial microbiology bring together raw materials (substrates), microorganisms (specific strains or microbial enzymes), and a controlled favorable environment (created in a fermenter) to produce the desired substance (Figure 12.1).

The essence of an industrial process is the combination of the right organism, an inexpensive substrate, and the proper environment to produce high yields of a desired product. Microbes possess a wealth of metabolic equipment that brings about diverse chemical transformations. This characteristic of microbes could be exploited to obtain some valuable products for daily use. The cheap raw materials available in nature as waste may be converted into useful commercial products by the activity of microbes. Microbes thus serve a dual purpose: first, they are good agents of waste disposal and, second, the resultant end products of their breakdown are useful commercial products. Cheap raw materials are converted to valuable products through the metabolism of microbes. For this purpose, microbes could be exploited in different ways, such as synthesis of fermentation products as acids, alcohols, or other organic compounds; transformation of one compound into another desired type; production of enzymes, antibiotics, or insecticides; or use of microbes themselves as food.

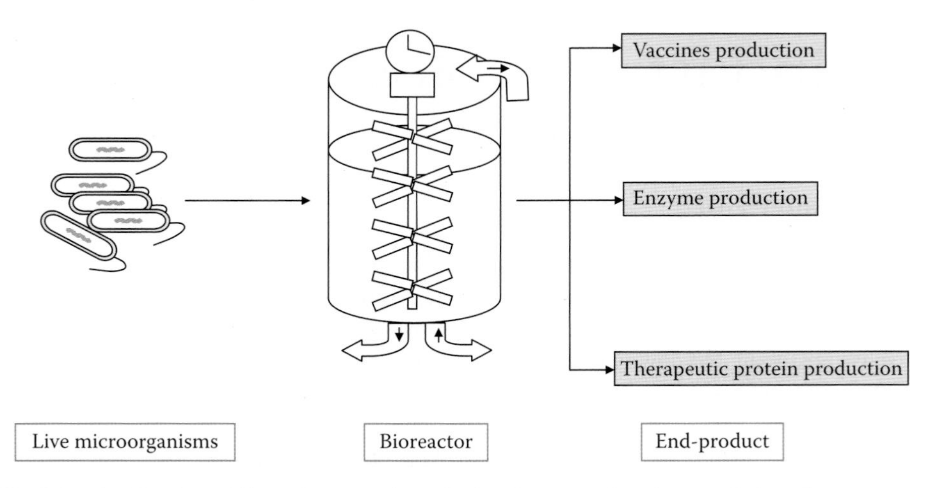

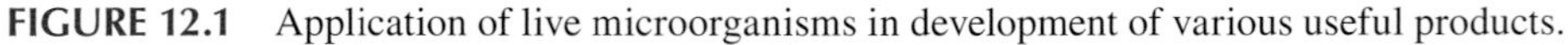

FIGURE 12.1 Application of live microorganisms in development of various useful products.

12.2 FERMENTER OR BIOREACTOR

A *bioreactor* is a device in which a substrate of low value is utilized by living cells or enzymes to generate a product of higher value. Bioreactors are extensively used for food processing, fermentation, waste treatment, etc. On the basis of the agent used, bioreactors are grouped into two broad classes: those based on living cells and those employing enzymes. All bioreactors deal with heterogeneous systems having two or more phases (liquid, gas, solid); therefore, optimal conditions for fermentation necessitate efficient transfer of mass, heat, and momentum from one phase to the other. A bioreactor normally has the following functions: (1) agitation, for mixing of cells and medium; (2) aeration, such as aerobic fermenters for oxygen supply; (3) regulation of factors such as temperature, pH, pressure, aeration, nutrient feeding, and liquid level; (4) sterilization and maintenance of sterility; and (5) withdrawal of cells or medium, such as for continuous fermenters. Modem fermenters are usually integrated with computers for efficient process monitoring and data acquisition. However, in terms of process requirements, bioreactors can be classified into four categories: aerobic, anaerobic, solid-state, and immobilized cell bioreactors. The size of fermenters ranges from 1.0 to 500,000L or occasionally even more; fermenters of up to 1.2 million liters have been used. Generally, 20%–25% of fermenter volume is left unfilled with medium as "head space" to allow for splashing, foaming, and aeration. The design of fermenters varies greatly, depending on the type of fermentation for which it is used (Figure 12.2).

12.3 PRINCIPLE OF FERMENTATION

In a general sense, *fermentation* is the conversion of a carbohydrate such as sugar into an acid or an alcohol. More specifically, fermentation can refer to the use of yeast to change sugar into alcohol or the use of bacteria to create lactic acid in certain foods. Given the right conditions, fermentation occurs naturally in many different foods, and humans have intentionally made use of it for thousands of years. The earliest uses of fermentation were to create alcoholic beverages such as mead, wine, and beer. These beverages have been created as far back as 7000 BCE in parts of the Middle East. Fermentation of foods such as milk and various vegetables took place some thousands of years later in both the Middle East and China. While the general principle of fermentation is the same across all drinks and foods, the precise methods of achieving it and the end results differ. Pickling foods such as cucumbers can be accomplished by submerging the vegetable one wants to pickle in a saltwater solution with vinegar added. Over time, bacteria create the lactic acid that gives the food its distinctive flavor and helps to preserve it. Other foods can be pickled simply by packing them in

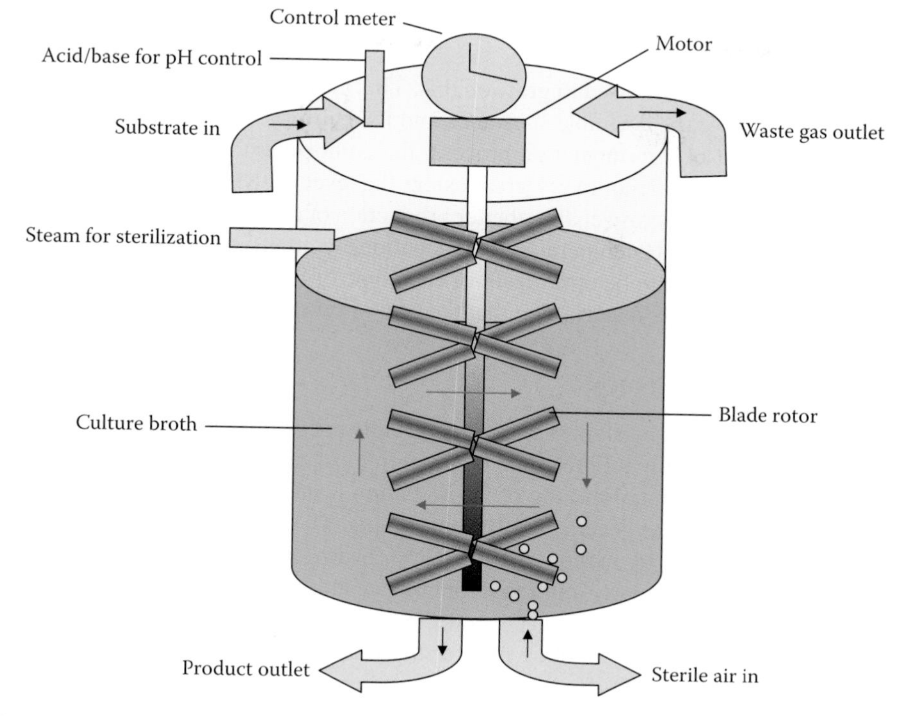

FIGURE 12.2 Typical bioreactor or fermenter.

dry salt and allowing the natural fermentation process to occur. There are two types of fermentation process: *aerobic fermentation* and *anaerobic fermentation*.

12.3.1 Aerobic Fermentation

The main feature of aerobic fermentation is the provision for adequate aeration. In some cases, the amount of air needed per hour is about 60 times the medium volume; therefore, bioreactors used for aerobic fermentation have a provision for adequate supply of sterile air, which is generally sparged into the medium. In addition, these fermenters may have a mechanism for stirring and mixing of the medium and cells. Aerobic fermenters may either be of the stirred-tank type, in which mechanical motor-driven stirrers are provided, or of the air-lift type, in which no mechanical stirrers are used and the agitation is achieved by the air bubbles generated by the air supply. Generally, these bioreactors are of closed or batch type, but continuous flow reactors are also used. Such reactors provide a continuous source of cells and are also suitable for product generation when the product is released into the medium. There are two major types of aerobic fermentation processes: *submerged culture method* and *semisolid or solid-state method*.

12.3.1.1 Submerged Culture Method

In this process, an organism is grown in a liquid medium that is vigorously aerated and agitated in large tanks (fermenters). The fermenter could be either an open tank or a closed tank, may be a batch type or a continuous type, and is generally made of a noncorrosive type of metal or is glass lined or is made of wood. In batch fermentation, the organism is grown in a known amount of culture medium for a defined period of time, after which the cell mass is separated from the liquid before further processing in the continuous culture. The culture medium is withdrawn depending on the rate of product formation and the inflow of fresh medium. Most fermentation industries today use the submerged process for the production of microbial products.

12.3.1.2 Semisolid/Solid-State Methods

In this method, the culture medium is impregnated in a carrier such as bagasse, wheat bran, potato pulp, etc. and the organism is allowed to grow on this. This method allows a greater surface area for growth. The production of the desirable substance and its recovery is generally easier and satisfactory. In the development of a fermentation process, the composition of the culture medium plays a major role and will determine to a very great extent the level of the end product. For example, a culture medium containing sucrose enables better production of citric acid by *Aspergillus niger* than any other carbohydrate. The pH, temperature of incubation, aeration, etc. are all important factors in fermentations, and these have to be optimized for each type of fermentation. Emphasis is generally placed on the use of cheap raw materials so that the cost of production is low.

12.3.2 Anaerobic Fermentation

In anaerobic fermentation, a provision for aeration is usually not needed. Nevertheless, in some cases, aeration may be needed initially for inoculum buildup. In most cases, a mixing device is also unnecessary, while in some cases initial mixing of the inoculum is necessary. Once fermentation begins, the gas produced in the process generates sufficient mixing. The air present in the headspace of the fermenter should be replaced by carbon dioxide (CO_2), hydrogen (H_2), nitrogen (N_2), or a suitable mixture of these. This is particularly important for obligate anaerobes like *Clostridium*. The fermentation usually liberates CO_2 and H_2, which are collected and used. For example, CO_2 is used for making dry ice and methanol and for bubbling into freshly inoculated fermenters. In case of acetogens and other gas-utilizing bacteria, O_2-free sterile CO_2 or other gases are bubbled through the medium.

Acetogens have been cultured in 400 L fermenters by bubbling sterile CO_2 and 3 kg cells could be harvested in each run. Recovery of products from anaerobic fermenters does not require anaerobic conditions. However, many enzymes of such organisms are highly O_2 sensitive; therefore, when recovery of such enzymes is the objective, cells must be harvested under strictly anaerobic conditions. Under anaerobic conditions, the pyruvate molecule can follow other anaerobic pathways to regenerate the nicotinamide adenine dinucleotide (NAD^+) necessary for glycolysis to continue. These include alcoholic fermentation and lactate fermentation (Figure 12.3). In the absence of oxygen, the further reduction or addition of H_2 ions and electrons to the pyruvate molecules that were produced during glycolysis is termed fermentation. This process recycles the reduced or hydrogenated NAD (NADH), to the free NAD^+ coenzyme, which once again serves as the H_2 acceptor, enabling glycolysis to continue. Alcoholic fermentation, which is characteristic of some plants and many microorganisms, yields alcohol and CO_2 as its products. Yeast is used by the biotechnology industry to generate CO_2 gas necessary for bread making and in the fermentation of hops and grapes to produce alcoholic beverages. Reduction of pyruvate by NADH to release the NAD^+ necessary for the glycolytic pathway can also result in lactate fermentation, which takes place in some animal tissues and in some microorganisms. Lactic acid-producing bacterial cells are responsible for the souring of milk to produce yogurt. In working animal muscle cells, lactate fermentation follows the exhaustion of the adenosine triphosphate (ATP) stores. Fast-twitch muscle fibers store little energy and rely on quick spurts of anaerobic activity, but the lactic acid that accumulates within the cells eventually leads to muscle fatigue and cramp. The fermentation unit in industrial biotechnology is analogous to a chemical plant in the chemical industry. A fermentation process is a biological process and therefore has requirements of sterility and use of cellular enzymatic reactions instead of chemical reactions aided by inanimate catalysts, sometimes operating at elevated temperature and pressure. Industrial fermentation processes may be divided into two main types, with various combinations and modifications: *batch fermentation* and *continuous fermentation*.

12.3.2.1 Batch Fermentation Process

The tank of a fermenter is filled with the prepared mash of raw materials to be fermented. The temperature and pH for microbial fermentation are properly adjusted and nutritive supplements

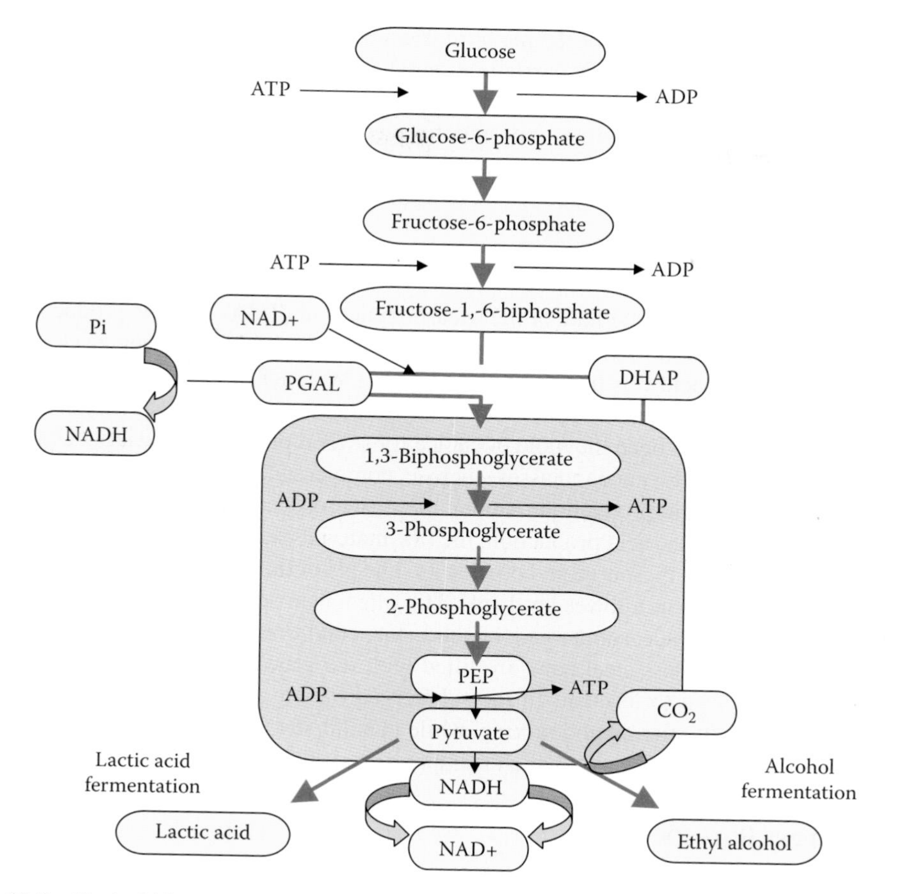

FIGURE 12.3 Typical bioreactor for producing lactic acid and ethyl alcohol.

are occasionally added to the prepared mash. The mash is steam sterilized in a pure culture process. The inoculum of a pure culture is added to the fermenter from a separate pure culture vessel. Fermentation proceeds, and, after the proper time, the contents of the fermenter are taken out for further processing. The fermenter is cleaned and the process is repeated.

12.3.2.2 Continuous Fermentation Process

Growth of microorganisms during batch fermentation confirms to the characteristic growth curve, with a lag phase followed by a logarithmic phase. This, in turn, is terminated by progressive decrements in the rate of growth until the stationary phase is reached. This is because of the limitation of one or more of the essential nutrients. In continuous fermentation, the substrate is added to the fermenter continuously at a fixed rate, which maintains the organisms in the logarithmic growth phase. The fermentation products are taken out continuously. The design and arrangements for continuous fermentation are somewhat complex.

12.4 PRODUCTION OF BIOMOLECULES USING FERMENTER TECHNOLOGY

12.4.1 Gluconic Acid

Gluconic acid used in pharmaceutical industries is produced by the fermentation of glucose by strains of *A. niger*, *Penicillium* sp., or selected bacteria. In the commercial process, a nutrient solution containing 24%–38% glucose, corn steep liquor, a N_2 source and salts, with pH 4.5 is used to culture a selected strain of fungus in shallow pans or in submerged culture conditions to convert glucose into gluconic acid. The pH of the medium is controlled by the addition of a strong solution of sodium hydroxide. Fermentation is carried out at 33°C or 34°C. The medium composition and fermentation conditions determine the production of acids other than gluconic acid (such as citric acid and oxalic acid), and it is therefore important to select a mold strain and the fermentation conditions that will avoid the formation of unwanted organic acid. After the fermentation, the cell-free broth is centrifuged and processed to recover gluconic acid.

12.4.2 Citric Acid

Citric acid, which is a key intermediate of the tricarboxylic acid (TCA) cycle, is produced by fungi, yeast, and bacteria as an overflow product due to a faulty operation of the citric acid cycle. The ability of fungi to produce citric acid was first discovered by Carl Wehmer in 1893, and today all the citric acid commercially produced comes from mold fermentation. Among the organisms used for citric acid production, *A. niger* has been the mold of choice for several decades. A variety of carbohydrate sources such as beet molasses, cane molasses, sucrose, commercial glucose, and starch hydrolysates have been used for citric acid production. Among these, sucrose cane and beet molasses have been found to be the best. For citric acid production, the raw material is diluted to a 20%–25% sugar concentration and mixed with a N_2 source and other salts. The pH of the medium is maintained around 5 when molasses is used and at a lower level (pH 3.0) when sucrose is used. The fermentations are carried out either under surface, submerged, or solid-state conditions. In the surface culture method, shallow aluminum or stainless steel pans are filled with the growth medium, inoculated with the fungal spores, and allowed to ferment. In the submerged culture method, the mold is cultured in fermenters under vigorous stirring and mixing, while in solid-state fermentation, the mold is grown over carrier material, such as bagasse, which is impregnated with the fermentation medium.

12.4.3 Acetone Butanol

Acetone butanol fermentation is one of the oldest fermentation methods known. The fermentation is based on culturing various strains of *Clostridia* in a carbohydrate-rich media under anaerobic conditions to yield butanol and acetone. *Clostridium acetobutylicum* is the organism of choice in the production of these organic solvents. Until very recently, these fermentation methods were out of favor because of the availability of acetone and butanol from the petroleum industry. Today, there is considerable interest in these fermentations. However, the concentration of the end products in these fermentations is quite small, and the fermentations are a type of mixed fermentation yielding a mixture of compounds such as butyric acid, butanol, and acetone. Attempts to increase yields by use of genetically altered strains or changes in fermentation conditions have been partially successful.

12.4.4 Itaconic Acid

Itaconic acid is used as a resin in detergents. The transformation of citric acid by *Aspergillus terreus* can be used for commercial production of itaconic acid. The fermentation process involves a well-aerated molasses–mineral salts medium at a pH below 2.2. At higher pH, this microbe degrades into itaconic acid. Like citric acid, low levels of trace metals must be used to achieve acceptable product yields.

12.4.5 Gibberellic Acid

Gibberellic acid and related gibberellins are important growth regulators of plants. Commercial production of these acids helps in boosting agriculture. This acid is formed by the fungus *Gibberella fujikuroi* (imperfect state, *Fusarium moniliforme*) and can be produced commercially using aerated submerged cultures. A glucose–mineral salt medium, incubation at 25°C, and slightly acidic pH are used for fermentation. This process normally takes 2–3 days.

12.4.6 Lactic Acid

Lactic acid is produced from various carbohydrates such as cornstarch, potato starch, molasses, and whey. Starchy materials when used are first hydrolyzed to simple sugars. The medium is then supplemented with a N_2 source and calcium carbonate. Arid fermentation is carried out by the inoculation with homofermentative lactobacilli such as *Lactobacillus bulgaricus* or *Lactobacillus delbruckli*. During fermentation, the temperature is controlled at 43°C–50°C, depending on the organism, and the medium is kept in constant agitation to keep the calcium carbonate in suspension. After completion of the fermentation (in 4–6 days), the fermented liquor is heated to 82°C and filtered. The filtrate containing calcium lactate is spray dried after treating with sodium sulfide. To obtain lactic acid, the calcium lactate is treated with sulfuric acid and the lactic acid thus obtained is further purified (Figure 12.3).

12.4.7 Amino Acids

In recent years, there has been rapid development in the production of particular amino acids by fermentation. Microorganisms can synthesize amino acids from inorganic nitrogen compounds. The rate and the amount of synthesis of some amino acids may exceed the cells needed for protein synthesis, whereupon the amino acids are excreted into the medium. Some microorganisms are capable of producing sufficient amounts of certain amino acids to justify their commercial production. The amino acids can be obtained from hydrolyzing protein or from chemical synthesis, but in several instances the microbial process is more economical. In addition, the microbiological method yields the naturally occurring L-amino acids. The demand for amino acids for use in foods, feeds, and in the pharmaceutical industry is expanding. When production costs decrease, new usage is anticipated as raw material for amino acid polymers. Microbial production of amino acids shows outstanding features which are not usually encountered in the development of other microbiological processes; one is the importance of auxotrophic microorganisms (Figure 12.4).

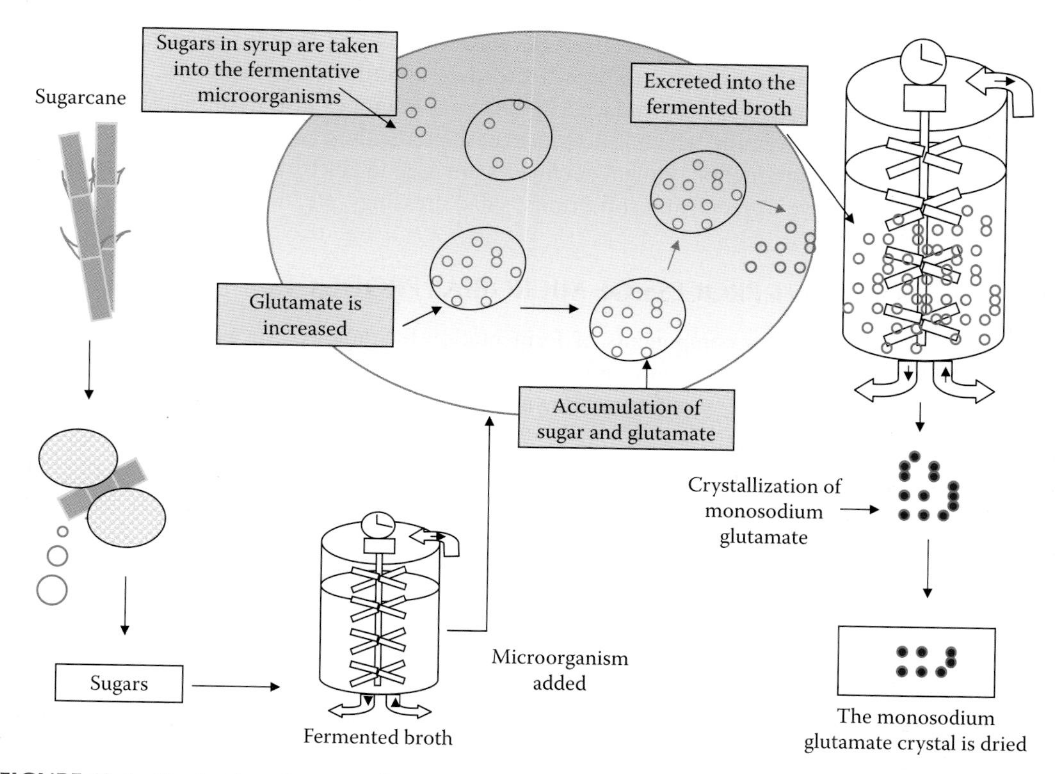

FIGURE 12.4 Production of monosodium glutamate by fermentation.

12.4.8 Enzymes

Enzymes have important applications in industry and these are produced by different microbes, mostly by fungi. Some enzymes produced for industry are proteases, amylases, alpha-amylase, glucamylase, glucose isomerases, glucose oxidases, rennin, pectinases, and lipases. Of these, proteases, glucamylase, alpha-amylase, and glucose isomerase are produced extensively.

12.4.9 Proteases

Microbial proteases have been extensively used in modern detergent formulations and also in laundries, where they are used to remove stains caused by milk, eggs, or blood. Proteases are heat stable and are largely produced by *Bacillus licheniformis*. Other alkaline proteases are being developed using recombinant DNA (rDNA) technology, to function over a wide range of pH and temperature. One such recombinant strain is *Bacillus* sp. *Gx 6644*, active for milk casein. Another strain *Bacillus* sp. *Gx 6638*, produces several alkaline proteases. Through this technology, a recombinant strain of a bacterium has also been developed. It produces the enzyme kerazyme, which is used for dissolving hair and opening hair clogged drains. These enzymes also have important applications in the baking industry. Microbial proteases reduce mixing time and improve the quality of bread loaves. Fungal proteases are mainly obtained from *Aspergillus* spp., while bacterial proteases are obtained from *Bacillus* spp. These enzymes are used as meat tenderizers and for the bating of hides in the leather industry.

12.4.10 Amylases

Amylases are used for preparation of sizing agents in the textile industry; for preparation of starch sizing pastes for use in the paper industry; in the production of bread, chocolate syrup, and corn syrup; and for stain removal in laundries. There are various types of amylases: α-amylases, β-amylases, and glucamylases. *Aspergillus oryzae*, *A. Niger*, *Bacillus subtilis*, and *Bacillus diastaticus* are principally used. The conversion of starch to high-fructose syrup utilizes amylases (i.e., in producing sweeteners). Various other enzymes produced by different microbes also have industrial applications. Some examples are rennin, which is used in cheese production, and *Mucor pussilus*, which is used for its commercial production. Fungal pectinases are used to clarify fruit juices, while glucose oxidase is used to remove oxygen from soft drinks, salad dressings, etc.

12.5 DEVELOPMENT PROCESS OF MICROBIAL PRODUCTS

After describing the various components of fermentation technology, we will next take a brief look at the development process of microbial products using fermentation technology in brief so that readers have a better understanding of the various phases and stages of product development (Table 12.1).

12.5.1 Isolation and Screening of Microorganisms

The success of an industrial fermentation process chiefly depends on the strain of the microorganism used. An ideal producer or economically important strain should have the following characteristics:

- It should be pure and free from phage.
- It should be genetically stable but amenable to genetic modification.
- It should produce both vegetative cells and spores; species producing only mycelium are rarely used.
- It should grow vigorously after inoculation in seed stage vessels.

TABLE 12.1
Microbial Products with Potential Importance

Product/Activity	Examples
Products	
1. Amino acids	L-glutamic acid, L-lysine
2. Antibiotics	Streptomycin, penicillin, tetracyclines, polymyxin
3. Beverages	Wine, beer, distilled beverages
4. Biodegradable plastic	β-Polyhydroxybutyrate
5. Enzymes	Amylase, proteases, cellulase
6. Flavoring agents	Monosodium glutamate, nucleotides
7. Foods	Cheese, pickles, yoghurt, bread, vinegar
8. Gases	CO_2, H_2, CH_4
9. Organic acids	Lactic, citric, acetic, butyric, fumaric
10. Organic solvents	Acetone, ethanol, butanol, amyl alcohol
11. Others	Glycerol, fats, steroids, gibberellins
11(a) Vitamins	B12, riboflavin, A
12. Recombinant proteins	Insulin, interferon, subunit vaccines
13. Substrates	A wide range of compounds used for chemical syntheses of valuable products
Cells/biomass	
14. Biomass	Food and feed yeast, other organisms used as single cell protein (SCP)
15. Cells	Biofertilizers, biocontrol agents, bacterial insecticides, mycorrhizae
16. Vaccines	A variety of viral and bacterial vaccines
Activities	
Biotransformation	Steroids, antibiotics D-sorbitol
Degradation	Disposal of biological and industrial wastes, detoxification of toxic compounds, petroleum
Solubilization/accumulation	Improved recovery of oil and metals, discovery of new oil reserves, removal of toxic metals

- It should produce a single valuable product and no toxic by-products.
- Product should be produced in a short time.
- It should be amenable to long-term conservation.
- The risk of contamination should be minimal under optimum performance conditions.

12.5.1.1 Isolation of Microorganisms

The first step in developing a producer strain is the isolation of the relevant microorganisms from their natural habitats. Alternatively, microorganisms can be obtained as pure cultures from organizations that maintain culture collections, such as the American Type Culture Collection (ATCC) in Rockville, Maryland, United States; the Commonwealth Mycological Institute (CMI) in Kew, Surrey, England; the Fermentation Research Institute (FERM) in Tokyo, Japan; and the Research Institute for Antibiotics, in Moscow, Soviet Union. Microorganisms of industrial importance are usually bacteria, actinomycetes, fungi, and algae. These organisms occur virtually everywhere, such as in air, water, soil, surfaces of plants and animals, and in plant and animal tissues. However, the most common sources of industrial microorganisms are soils and lake and river mud. Often, the ecological habitat from which a desired microorganism is more likely to be isolated will depend on the characteristics of the product desired from it and on the development process; for example, if the objective is to isolate a source of enzymes that can withstand high temperatures, the obvious place to look will be hot springs.

A variety of complex isolation procedures have been developed, but no single method can reveal all the microorganisms present in a sample. Many different microorganisms can be isolated by using specialized enrichment techniques such as soil treatment (ultraviolet irradiation, air drying or heating at 70°C–120°C, filtration or continuous percolation, washings from root systems, treatment with detergents or alcohols, preinoculation with toxic agents), selective inhibitors (antimetabolites, antibiotics, etc.), nutritional (specific C and N) sources, variations in pH, temperature, aeration, and so on. The enrichment techniques are designed for selective multiplication of only some of the microorganisms present in a sample. These approaches, however, take a long time (about 20–40 days) and require considerable labor and money. The main isolation methods used routinely for isolation from soil samples are sponging (soil directly), dilution, gradient plate, aerosol dilution, flotation, and differential centrifugation. Often, these methods are used in conjunction with an enrichment technique. Table 12.1 lists some microbial products with examples.

12.5.1.2 Screening of Microorganisms

After the isolation of microorganisms, the next step is their screening. A set of highly selective procedures, which allows the detection and isolation of microorganisms producing the desired metabolite, constitutes primary screening. Ideally, primary screening should be rapid, inexpensive, predictive, and specific, but effective for a broad range of compounds and applicable on a large scale. Primary screening is time consuming and labor intensive, since a large number of isolates have to be screened to identify a few potential ones. However, this is possibly the most critical step, since it eliminates the large bulk of unwanted useless isolates, which are either nonproducers or producers of known compounds. Computer-based databases play an important role by instantaneously providing detailed information about the already known microbial antibiotic compounds.

Rapid and effective screening techniques have been devised for a variety of microbial products, which utilize either a property of the product or that of its biosynthetic pathway for detection of desirable isolates. Some of the screening techniques are relatively simple, such as that for extracellular enzymes and enzyme inhibitors. However, for most microbial products of high value, the screening is usually complex and tedious, and often may involve two or more steps, such as for antimicrobials. In some cases, it may be desirable to concentrate on a group of organisms expected to yield new products. For example, the search for new antibiotics now focuses on rare actinomycetes, (i.e., actinomycetes other than those belonging to the genus *Streptomyces*). Suitably designed specialized screening techniques may be used to detect compounds having various pharmacological activities other than antibiotics.

12.5.2 Inoculum Development

The preparation of a population of microorganisms from a dormant stock culture to an active state of growth that is suitable for inoculation in the final production stage is called *inoculum development*. As a first step in inoculum development, the inoculum is taken from a working stock culture to initiate growth in a suitable liquid medium. Bacterial vegetative cells and spores are suspended, usually in sterile tap water, which is then added to the broth. In case of non-sporulating fungi and actinomycetes, the hyphae are fragmented and then transferred to the broth. Inoculum development is generally done using flask cultures. Flasks of 50 mL to 12 L may be used and their number can be increased as per need. In some cases, small fermenters may be used. Inoculum development is usually done in a stepwise sequence to increase the volume to the desired level. At each step, the inoculum is used at 0.5%–5% of the medium volume, which allows a 20–200-fold increase in inoculum volume at each step. Typically, the inoculum used for the production stage is about 5% of the medium volume.

12.5.3 Culture Media

Inoculum preparation media are quite different from production media. These media are designed for rapid microbial growth, and little or no product accumulation will normally occur. Many production processes depend on inducible enzymes. In all such cases, the appropriate inducers must be included either in all the stages or at least in the final stages of inoculum development. This will ensure the presence of the concerned inducible enzymes at high levels for the production to start immediately after inoculation.

12.5.4 Contamination

The inoculum used for production tanks must be free from contamination. However, as the risk of contamination is always present during inoculum development, every effort must be made to detect as well as prevent contamination. The most common contaminants of different industrial processes are considerably different. Some examples are as follows:

- In the canning industry, *Clostridium butylicum* is the chief concern. This obligate anaerobe can grow in sealed cans, producing heat-resistant spores and a deadly toxin. However, this is not a problem for ketchup (too acidic), jam and jellies (too high a sugar concentration), and milk (stored at too low a temperature).
- Organisms like lactobacillus are a problem in wine production.
- In the antibiotic industry, some potential contaminants are molds, yeast, and bacteria such as *Bacillus*.
- The most dreaded contaminants in the fermentation industry are phages. The only effective protection against phages is to develop resistant strains.

12.5.5 Sterilization

Sterilization is the process of inactivating or removing all living organisms from a substance or surface. In theory, it is regarded as absolute, in that all living cells must be inactivated or removed, usually in a single step at the given time; in practice, however, the success of sterilization procedures is only a probability. Therefore, the probability of a cell escaping inactivation/filtration does exist, although it is usually very small. When a closed system is sterilized once, it remains so indefinitely, since it has no openings for the entry of microorganisms. Most fermentation vessels are open systems. Such systems are initially sterilized and must be kept sterilized by ensuring the removal of living cells at their entry points, such as the cotton plug of a culture flask. Some methods of sterilization are discussed in the following.

12.5.5.1 Heating

Heat is the most commonly used and the least expensive sterilizing agent. Dry heat is used in ovens and is suitable for sterilization of solids which can withstand the high temperatures needed for sterilization. Steam, which is moist or wet heat, is used for sterilization of media and fermenter vessels. An autoclave uses steam for sterilization (at temperature 121°C and pressure 15 psi). The period of time at this temperature and pressure depends on medium volume; for example, 12–15 min for 200 mL, 17–22 min for 500 mL, 20–25 min for 1 L, and 30–35 min for 2 L. However, sterilization of oils will require a few hours, and concentrated media (10%–20% solid) must be agitated for effective sterilization.

Autoclaves can also be used to sterilize laboratory vessels, small volumes of media, and even small fermenters. Large fermenters are sterilized either by a direct injection of steam or by indirect heating by passing steam through heat exchange coils or a jacket. The steam should always be saturated. Media sterilization may be achieved in a continuous flow sterilization system, either by direct steam injection or by indirect steam heating, and then filled in a sterile fermenter. Alternatively, the medium may be filled in the fermenter and steam-sterilized with the latter. Heat killing is largely due to protein inactivation. In general, moist heat is far superior to dry heat. Bacterial spores are the most heat resistant. For example, spores of thermophilic bacteria can survive steam at 30 psi at 134°C for 1–10 min and dry heat at 180°C for up to 15 min.

12.5.5.2 Radiation

High-energy x-rays are used for sterilization of a variety of labware and food. In general, vegetative cells are much more susceptible than bacterial spores (*Clostridium* spores can resist nearly 0.5 M rad). However, *Deinococcus radiodurans* vegetative cells can survive 6 M rad. Viruses are usually similar to bacterial spores, but some viruses (such as the encephalitis virus) require up to 4.5 M rad. In practice, 2.5 M rad is used for sterilizing pharmaceutical and medical products. X-rays cause inactivation by inducing single- and double-strand DNA breaks, and by producing free radicals and peroxides, to which SH enzymes are particularly susceptible.

12.5.5.3 Chemicals

The chemicals used for sterilization, such as formaldehyde, hydrogen peroxide (H_2O_2), ethylene oxide, and propylene oxide, cause inactivation by oxidation or alkylation. H_2O_2 (10%–25% w/v) is being increasingly used in the sterilization of milk and of containers for food products.

H_2O_2 is a powerful oxidizing agent, kills both vegetative cells and spores, and is very safe. Ethylene oxide is used to sterilize equipment that is likely to be damaged by heat and is very effective, but highly toxic and violently explosive if mixed with air.

12.5.5.4 Filtration

Aerobic fermentation requires a very high rate of air supply, often equaling one volume of air (equal to medium volume) every minute. Air contains both fungal spores and bacteria, which are ordinarily removed by filtration using either a depth filter or a screen filter. Depth filters are made from fibrous or powdered materials pressed or bonded together in a relatively thick layer. Some materials used are fiberglass, cotton, mineral wool, and cellulose fibers in the form of mats, wads, or cylinders.

12.5.6 Strain Improvement

After an organism producing a valuable product is identified, it becomes necessary to increase the product yield from fermentation to minimize production costs. Product yields can be increased by developing a suitable medium for fermentation, refining the fermentation process, and by improving the productivity of the strain. Generally, major improvements arise from the last approach, and all fermentation enterprises therefore place considerable emphasis on this activity. The techniques and approaches used to genetically modify strains and to increase the production of the desired product are called *strain improvement* or *strain development*. Strain improvement can be done by using mutant selection, recombination, and rDNA technology. Table 12.2 shows different approaches to improve strains.

12.5.6.1 Mutant Selection

The large-scale mutant selection program begins when favorable reports of clinical trials are obtained. In the early stages, selection of spontaneous mutants may be helpful, but induced mutations are the most common sources of improvements. Many mutations bring about marked changes in the biochemical character of practical interest and are called *major mutations*. Some major mutations can be useful in strain improvement. For example, a mutant strain (S-604) of *Streptomyces aureofaciens* produces 6-demethyl tetracycline in place of tetracycline. This demethylated form

TABLE 12.2
Mutation and Genetic Recombination in Strain Improvement

Approach	Chief Feature	Example/Remark
(A) Mutant selection: types		The main approach to strain improvement; produces new alleles of existing genes
1. Spontaneous mutations	Occur without any treatment with a mutagen	Used in the initial stages of strain improvement; also for maintenance of improved strains
2. Induced mutations	Induced by chemical (mainly) or physical mutagens	Mutagenesis followed by selection; several cycles employed
3. Major mutations	Affect the pattern of metabolite production	Production of 6-demethyl tetracycline in place of tetracycline by *S. aureofaciens*
4. Minor mutations	Affect the rate metabolite production	Small gains in each cycle of selection; substantial improvement after several cycles
(B) Mutant selection: strategies		
1. Auxotrophic mutants	Defective biosynthesis of a biochemical	Enhanced production of an amino acid, e.g., phe mutants accumulate tyrosine
2. Analogue-resistant mutants	Feedback insensitive enzymes	Overproduction of metabolites, e.g., amino acids by *C. glutamicus*
3. Revertants of nonproducing mutants		Some mutants are high producers, e.g., chlortetracycline by *S. viridifaciens*
4. Revertants of auxotrophic mutants		Some are high producers, e.g., chlortetracycline by *S. viridifaciens*
5. Resistance to the antibiotic produced by the organism itself		Increased production, e.g., chlortetracycline by *S. aureofaciens*
(C) Recombination		Produces new combinations of existing alleles
1. Sexual reproduction	Conjugation; fusion of gametes	Some bacteria and actinomycetes; fungi and yeast
2. Heterokaryosis	Nuclear fusion followed, by mitotic recombination and mitotic reduction	Fungi
3. Protoplast fusion	Protoplasts produced by lytic enzymes fusion by PEG, recombinant recovery	Bacteria, actinomycetes, fungi

of tetracycline is the major commercial form of tetracycline. In contrast, most improvements in biochemical production have been due to the stepwise accumulation of the so-called minor genes. These genes lead to small increases (or decreases) in antibiotic or other biochemical production, and selection may be expected to result in a 10%–15% increase in yield.

The selected strains are usually subjected to successive cycles of mutagenesis and selection. After several cycles, a large increase in yield is likely to be obtained. Mutants of *Penicillium chrysogenum* were selected for increased penicillin production. Each cycle of selection was preceded by mutagen treatment and resulted in only small changes in penicillin yield. Refer to Table 12.2 for different approaches to select mutants.

12.5.6.2 Selective Isolation of Mutants

The majority of desirable mutants showing increased production, especially the minor gene mutants, are isolated by screening a large number of clones surviving the mutagen treatment; this is called *secondary screening*. However, as this approach requires a large amount of work, efforts have increasingly focused on developing techniques for the isolation of particular classes of mutants that are likely to be overproducers.

- Isolation of auxotrophic mutants from the bacterium *Corynebacterium glutamicus* is the basis for commercial amino acid production in Japan. For example, phe-mutants of *C glutamicus* accumulate tyrosine.
- Many analogue-resistant mutants have feedback-insensitive enzymes of the biosynthetic pathway, the analogue of whose product was used for the selection of such cells. Such mutants tend to overproduce the end product of the concerned pathway.
- Revertants from nonproducing mutants of a strain are sometimes high producers. One such reversion mutant of *Streptomyces viridifaciens* showed an over sixfold increase in chlortetracycline production over the original strain from which the nonproducing mutant was obtained. Reversion mutants of appropriate auxotrophs may often be high producers.
- In some cases, selection for resistance to the antibiotic produced by the organism itself may lead to increased yields.
- Mutants with altered cell membrane permeability sometimes show high production of some metabolites.
- Mutants have been selected to produce altered metabolites, especially in case of aminoglycoside antibiotics. For example, *Pseudomonas aureofaciens* produces the antibiotic pyrrolnitrin; a mutant of this fungus yields 4′-fluoropyrrolnitrin.

Mutant selection has been the most successful approach for strain improvement, but major advances are being made in the exploitation of other strategies (i.e., by recombination and rDNA technology).

12.6 UPSTREAM BIOPROCESS

The upstream part of a bioprocess refers to the first step in which biomolecules are grown, usually by bacterial or mammalian cell lines in bioreactors. When they reach the desired density (for batch and fed-batch cultures), they are harvested and moved to the downstream section of the bioprocess. The following discusses the different kinds of culture methods in biotechnology.

12.6.1 Industrial Microbial Culture

The use of microbes to produce natural products and processes that benefit and improve human healthcare has been a part of human history since early civilization. The production of secondary metabolites (such as antibiotics) often occurs during the stationary growth phase (or idiophase) of a bacterial culture. Secondary metabolites are nongrowth related and seem to play no obvious role in cell maintenance. The production of secondary metabolites largely concerns the first growth phase of a bacterial culture. This production is considered a two-stage process, with a trophophase (growth phase) and an idiophase (production phase), which usually occurs when culture growth has slowed or ceased. The major aim for mass production of secondary metabolites is to maximize the biomass in a short trophophase, while optimizing the conditions for high, sustained idiophase production. The growth of a bacterial culture can be represented by a curve that consists of the following four stages or phases:

1. *Lag phase*—growth and reproduction are just beginning
2. *Log phase*—reproduction is occurring at an exponential rate
3. *Stationary phase*—the environmental surroundings and food supply cannot support any more exponential growth
4. *Death phase*—when all the nutrients have been exhausted and the population dies off

A variety of parameters is important to influence the production of desired secondary metabolites in bioreactor cultures, which include substrate, slow or fast utilization, N_2 requirements, amino acid supplements, temperature, pH, oxygen concentration, and pressure.

12.6.2 Mammalian Cell Culture

Cell culture is the process by which cells are grown under controlled conditions. In practice, the term "cell culture" has come to refer to the culturing of cells derived from multicellular eukaryotes, especially animal cells. The historical development and methods of cell culture are closely interrelated to those of tissue culture and organ culture. Animal cell culture became a common laboratory technique in the mid-1900s, but the concept of maintaining live cell lines separated from their original tissue source was discovered in the nineteenth century. Cells are grown and maintained at an appropriate temperature and gas mixture (typically, 37°C, 5% CO_2 for mammalian cells) in a cell incubator. Culture conditions vary widely for each cell type, and variation of conditions for a particular cell type can result in different phenotypes being expressed. Aside from temperature and gas mixture, the most commonly varied factor in culture systems is the growth medium. Recipes for growth media can vary in pH, glucose concentration, growth factors, and the presence of other nutrients. The growth factors used to supplement media are often derived from animal blood, such as calf serum. One complication of these blood-derived ingredients is the potential for contamination of the culture with viruses or prions, particularly in biotechnology medical applications. Current practice is to minimize or eliminate the use of these ingredients wherever possible, but this cannot always be accomplished.

Cells can be grown in *suspension* or as *adherent* cultures. Some cells naturally live in suspension without being attached to a surface, such as cells that exist in the bloodstream. There are also cell lines that have been modified to be able to survive in suspension cultures so that they can be grown to a higher density than adherent conditions would allow. Adherent cells require a surface, such as tissue culture plastic or microcarrier, which may be coated with extracellular matrix components to increase adhesion properties and provide other signals needed for growth and differentiation. Most cells derived from solid tissues are adherent. Another type of adherent culture is the *organotypic culture,* which involves growing cells in a three-dimensional environment as opposed to two-dimensional culture dishes. This three-dimensional culture system is biochemically and physiologically more similar to *in vivo* tissue, but is technically challenging to maintain because of many factors (e.g., diffusion).

12.6.2.1 Manipulation of Cultured Cells

Generally, as cells continue to divide in a culture, they grow to fill the available area or volume. This can generate several issues, which include nutrient depletion in the growth media; accumulation of apoptotic/necrotic (dead) cells; cell-to-cell contact that can stimulate cell cycle arrest, causing cells to stop dividing, known as *contact inhibition* or *senescence*; and cell-to-cell contact that can stimulate cellular differentiation. Among the common manipulations carried out on culture cells are media changes, passaging cells, and transfecting cells. These are generally performed using tissue culture methods that rely on sterile technique. Sterile technique aims to avoid contamination with bacteria, yeast, or other cell lines. Manipulations are typically carried out in a biosafety hood or laminar flow cabinet to exclude contaminating microorganisms. Antibiotics (e.g., penicillin and streptomycin) and antifungals (e.g., amphotericin B) can also be added to the growth media. As cells undergo metabolic processes, acid is produced and the pH decreases. Often, a pH indicator is added to the medium in order to measure nutrient depletion.

12.6.2.2 Generation of Hybridomas

It is possible to fuse normal cells with an immortalized cell line. This method is used to produce monoclonal antibodies. In brief, lymphocytes isolated from the spleen (or possibly from the blood) of an immunized animal are combined with an immortal myeloma cell line (B cell lineage) to produce a hybridoma which has the antibody specificity of the primary lymphocyte and the immortality of the myeloma. Selective growth hypoxanthine-aminopterin-thymidine medium (HAT medium) is used to select against unfused myeloma cells. Primary lymphocytes die quickly in culture and only the fused cells survive. To start with, these are screened for production of the required antibody, generally in pools, and then single cloning is done afterwards.

12.6.3 Nonmammalian Culture Methods

12.6.3.1 Industrial Plant Cell Culture

Plant secondary metabolites are important for plant interactions with their environments. These metabolites give plants important properties such as antibiotic, antifungal, antiviral, ultraviolet protection, and pest deterrence. These chemicals have also been used as medicines by humans for thousands of years. Today, plants are responsible for the production of over 30,000 types of chemicals, including pharmaceuticals, aromas, pigments, cosmetics, nutraceutics, and other fine chemicals. Early on, it was discovered that metabolites could be accumulated in higher quantities using plant cell cultures as compared with whole plant cultivation. However, since plant cell cultures have low yield compared to their bacterial counterparts, they are primarily used only for high-value metabolites. The biosynthetic routes and mechanisms describing the production of plant secondary metabolites are often very complex and can involve dozens of enzymes. Through the study of plant metabolism, a new discipline called *metabolic engineering* has emerged. The goal of metabolic engineering is to improve cellular activity by manipulating enzymatic, transport, and regulatory functions of a cell. This is accomplished through the use of rDNA technology. Plant cell cultures are typically grown as cell suspension cultures in liquid medium or as callus cultures on solid medium. The culturing of undifferentiated plant cells and calli requires the proper balance of the plant growth hormones auxin and cytokinin.

12.6.3.2 Bacterial/Yeast Culture Methods

For bacteria and yeast, small quantities of cells are usually grown on a solid support that contains nutrients embedded in it, usually a gel such as agar, while large-scale cultures are grown with the cells suspended in a nutrient broth.

12.6.3.3 Viral Culture Methods

The culture of viruses requires the culture of cells of mammalian, plant, fungal, or bacterial origin as hosts for the growth and replication of the virus. Whole wild type viruses, recombinant viruses, or viral products may be generated in cell types other than their natural hosts under the right conditions. Depending on the species of the virus, infection and viral replication may result in host cell lysis and formation of a viral plaque.

12.7 DOWNSTREAM BIOPROCESS

The various processes used for the actual recovery of useful products from fermentation or any other industrial process are called *downstream processing* (DSP). As the cost of DSP is often more than 50% of the manufacturing cost, and there is product loss at each step, DSP should be efficient and cost-effective and should involve as few steps as possible to avoid product loss. DSP refers to the recovery and purification of biosynthetic products, particularly pharmaceuticals, from natural sources such as animal or plant tissue or fermentation broth, including the recycling of salvageable components and the proper treatment and disposal of waste. It is an essential step in the manufacture of pharmaceuticals such as antibiotics, hormones (e.g., insulin and human growth hormone), antibodies (e.g., infliximab and abciximab), and vaccines; antibodies and enzymes used in diagnostics; industrial enzymes; and natural fragrance and flavor compounds. DSP is usually considered a specialized field in biochemical engineering, itself a specialization within chemical engineering, though many of the key technologies were developed by chemists and biologists for laboratory-scale separation of biological products. DSP and analytical bioseparation both refer to the separation or purification of biological products, but at different scales of operation and for different purposes. DSP implies manufacture of a purified product fit for a specific use, generally in marketable quantities, while analytical bioseparation refers to purification for the sole purpose of measuring a component or components of a mixture and may deal with sample sizes as small as a single cell.

12.7.1 Stages in DSP

A widely recognized heuristic for categorizing DSP operations divides them into four groups which are applied in order to bring a product from its natural state as a component of a tissue, cell, or fermentation broth through progressive improvements in purity and concentration.

12.7.1.1 Removal of Insolubles

This first step involves the capture of the product as a solute in a particulate-free liquid, for example the separation of cells, cell debris, or other particulate matter from fermentation broth containing an antibiotic. Typical operations to achieve this are filtration, centrifugation, sedimentation, flocculation, electroprecipitation, and gravity settling. Additional operations such as grinding, homogenization, or leaching, which are required to recover products from solid sources such as plant and animal tissues, are usually included in this group.

12.7.1.2 Product Isolation

Product isolation is the removal of those components whose properties vary markedly from that of the desired product. For most products, water is the chief impurity and isolation steps are designed to remove most of it, reducing the volume of material to be handled, and concentrating the product. Solvent extraction, adsorption, ultrafiltration, and precipitation are some of the unit operations involved.

12.7.1.3 Product Purification

Product purification is done to separate those contaminants that resemble the product very closely in physical and chemical properties. Steps in this stage are expensive to carry out and require sensitive and sophisticated equipment. This stage contributes a significant fraction of the entire DSP expenditure. Examples of operations include affinity, size exclusion, reversed phase chromatography, crystallization, and fractional precipitation.

12.7.1.4 Product Polishing

Product polishing describes the final processing steps which end with packaging of the product in a form that is stable, easily transportable, and convenient. Crystallization, desiccation, lyophilization, and spray drying are typical unit operations. Depending on the product and its intended use, polishing may also include operations to sterilize the product and remove or deactivate trace contaminants which might compromise product safety. Such operations might include the removal of viruses or depyrogenation. A few product recovery methods may be considered to combine two or more stages; for example, expanded bed adsorption accomplishes removal of insoluble and product isolation in a single step. Affinity chromatography often isolates and purifies in a single step (Figure 12.5).

12.8 BIOPROCESS AUTOMATION

Maximizing profits by operating the most efficient process is the primary goal of all industrial companies. To help create efficient operations, companies use process simulation. This involves the use of a range of software tools to analyze complete processes, not just single unit operations. Process engineers and scientists use simulation models to investigate complex and integrated biochemical operations without the need for extensive experimentation. Simulation tools can be used at any stage of process development, from initial concept, through design, to final plant operation. These tools tackle a range of tasks, such as creating process flow diagrams, generating material and energy

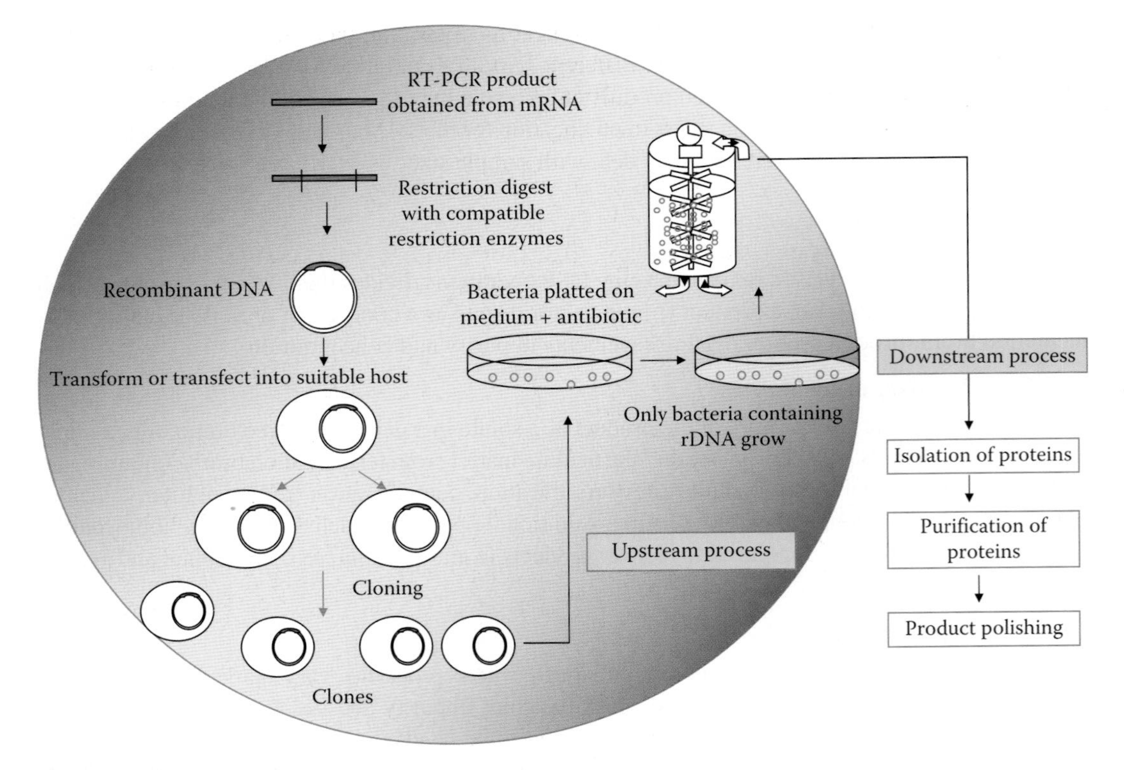

FIGURE 12.5 Production of biotechnology product: upstream and downstream processes.

balances, determining equipment sizing, and estimating capital and operating costs. Although the application of bioprocess simulation is not yet widespread, large-scale bioprocesses have been studied for many years, and their viability and usefulness is well established in some areas. For example, National Renewable Energy Laboratories (NREL) in Denver, Colorado, has been developing comprehensive process models to study different processes for conversion of biomass to ethanol for over 15 years.

A recent development in bioprocess modeling is the use of computational fluid dynamics (CFD). Commercial software packages, such as FLUENT (Fluent Inc., Lebanon, New Hampshire), and Star-CD (CD-Adapco, New York) can model mixing effects by incorporating physical properties of fluids and aeration patterns with detailed information of vessel internals, such as impeller geometry and baffle location. CFD programs compute the velocity profiles within the fluid to model gas dispersion, calculate residence time distributions, and visualize the mixing process. Results from these models highlight such phenomena as areas of poor mixing or areas of high fluid shear. The recent interest in CFD can be attributed to two reasons. First, computationally intensive CFD models are more readily accessible, as they now run on inexpensive workstations and even laptops. Second, companies are operating at larger scales where flow phenomena, such as degree of mixing, are more important.

12.8.1 Modeling Individual Unit Operations

Performance of a process is sometimes highly dependent on the operation of a few key unit operations. If the design and performance characteristics of these units can be described by a mathematical representation, then potential problems can be investigated in silico, reducing the need for expensive experimentation. Popular tools for writing such models include Excel (Microsoft), MATLAB® (Mathworks, Natick, Massachusetts) and MathCAD (Mathsoft, Cambridge, Massachusetts). Mathematical models can also be written in native software languages (FORTRAN, C++, etc.), but there is the issue of maintaining and documenting these approaches, which makes it more convenient to use a commercial software package. In addition to MATLAB and MathCAD, there are more targeted process simulation tools. Simulation packages usually contain a library of common unit operation models. They also support the development of customized models. Some examples of commercial software that support this development are Aspen Engineering Suite (Aspen Technology; Cambridge, Massachusetts) and gPROMS (PSE; London, U.K.). Both packages have the ability to solve complex systems of differential and algebraic equations, and can be used for complete flow sheet simulations or for rigorous modeling of single unit operations.

The level of detail in any model depends on the questions posed. Typically, you start from a simple model and add layers of complexity until the model meets its requirements.

12.8.2 Simple Mass Balance

In the simplest case, the process of crystallization is described as a mass conversion or reaction. The "reaction" is mass-based, with a percentage reaction extent used to determine the formation of crystals. If the reaction extent were 40%, then every 100 kg product would produce 40 kg crystals. This approach ensures that a consistent material balance is generated. Any stream entering and leaving the operation is identified with a list of all the components in the process. This level of detail is insufficient when considering a single unit operation, but is perfectly acceptable when describing a flow sheet and is a starting point for any model.

12.8.3 Mass and Heat Balance

Typically, a crystallizer evaporates a solvent to produce a supersaturated solution of the product. Assume that 10 kg of water is evaporated for every kilogram of crystal formed. In addition to the heat of vaporization of the solvent, there is the heat of crystallization. Both these properties are included as heat sinks. In reality, heat is also required to increase the temperature of the liquor to boiling point, and a more comprehensive heat balance would be covered in a detailed model description.

12.8.4 Rate-Based Model

One reason to increase the level of model complexity for a crystallizer model is to study factors affecting the particle size distribution (PSD). Including the amount of seed, the rates of nucleation, crystal growth, and crystal breakage allows for the calculation of the complete PSD as a function of time. Controlling the PSD improves product quality, which directly affects product cost, which further explains the increasing interest in this type of approach.

12.8.5 Batch Processing

Most bioprocess simulation is targeted at batch processing, since most bioprocesses are batch operations. There are generic software packages for solving these procedural or batch process problems, such as Extend (Imagine That; San Jose, California). Extend is a discrete event modeling environment that can be customized for bioprocess applications and is the basis for some proprietary solutions. SuperProDesigner (Intelligen; Scott's Plain, New Jersey) combines drawing, calculation, and scheduling features in a moderately priced package. While this simulator can be used to study both batch and continuous processes, it is particularly well suited to batch bioprocesses.

Aspen Technology offers a more expensive range of products. For bioprocessing, the principal product is Batch Plus, a recipe-driven modeling environment for batch processes. Process Systems Enterprise Limited (PSE) offers Model Builder, which is another environment for modeling batch operations.

12.8.6 Continuous Processes

Large-volume processes, such as the production of ethanol from corn, require the ability to model both batch and continuous processes. For continuous processes, it is possible to use the conventional Aspen Plus simulator, but this requires customization for many bioprocess operations. gPROMS is a moderately expensive package that offers a library of commonly used operations, but this also requires customization for bioprocess applications.

12.8.7 Dynamic Simulation

For batch processes, the productivity focus is the overall throughput and time-averaging the operations is usually adequate. Typically, process parameters include the duration of an operation and the mass of the reactants and products. For a continuous operation, the overall process is studied as operating at a steady-state condition. However, if more information is needed—such as the study of process transients or upset conditions—then dynamic simulation is used. *Dynamic simulation* is the study of process variations with time. Dynamic simulations are much more complex to construct and run and require considerable expertise, but can yield insights that are not available by other methods, making the investment worthwhile.

12.8.8 Water Consumption

Flow sheet simulation can be used to tackle a wide range of process problems, but here is an example of some specific issues. All bioprocess operations consume significant quantities of water, both as part of the process and for cleaning purposes, such as with clean-in-place (CIP) systems. Many operations use several different qualities of water such as purified water (USP), water for injection (WFI), and steam for sterilization, all of which require varying degrees of processing. The general approach today is to reduce, recycle, and reuse as much water as possible.

12.8.9 Waste Recycling

Any final product will contain little water; therefore, most of the water used in the process requires disposal or treatment. If the water streams contain significant amounts of bioburden, then this contributes to a high biological oxygen demand (BOD) value for the stream and is a pollution hazard. Design of on-site treatment plants may be required. Additionally, it may be desirable to segregate water waste streams and recycle the cleaner streams with minimal purification. Simulation models can compare the impact of each scenario to determine the most cost-effective solution.

12.9 INDUSTRIAL APPLICATION OF MICROBES

12.9.1 *Corynebacterium*

Corynebacterium is a genus of Gram-positive rod-shaped bacteria. They are widely distributed in nature and are mostly innocuous. Some, such as *C. glutamicum*, are useful in industrial settings; others, such as *C. diphtheriae* (the pathogen responsible for diphtheria) cause human disease. Nonpathogenic species of *Corynebacterium* are used for very important industrial applications, such as the production of amino acids, nucleotides, and other nutritional factors; bioconversion of steroids; degradation of hydrocarbons; cheese aging; and production of enzymes. Some species produce metabolites similar to antibiotics, such as bacteriocins of the corynecin–linocin type and anti-tumor agents. One of the most studied species is *C. glutamicum*, whose name refers to its capacity to produce glutamic acid in aerobic conditions. It is used in the food industry as monosodium glutamate in the production of soy sauce and yogurt. Species of *Corynebacterium* have been used in the mass production of various amino acids such as L-glutamic acid, a popular food additive that is produced by *Corynebacterium* at a rate of 1.5 million tons/year. The metabolic pathways of *Corynebacterium* have been further manipulated to produce L-lysine and L-threonine.

12.9.2 *Bacillus*

Bacillus is a genus of rod-shaped bacteria and a member of the division Firmicutes. *Bacillus* species are obligate aerobes, and test positive for the enzyme catalase. Ubiquitous in nature, *Bacillus* includes both free-living and pathogenic species. Under stressful environmental conditions, the cells produce oval endospores that can stay dormant for extended periods. These characteristics originally defined the genus, but not all such species are closely related, and many have been moved to other genera. Many *Bacillus* species are able to secrete large quantities of enzymes. *Bacillus amyloliquefaciens* is a species of *Bacillus* that is the source of a natural antibiotic protein barnase (a ribonuclease), alpha amylase used in starch hydrolysis, the protease subtilisin used with detergents, and the BamH1 restriction enzyme used in DNA research. A portion of the *Bacillus thuringiensis* genome was incorporated into corn and cotton crops. The resulting GMOs are therefore resistant to some insect pests.

12.9.3 *Saccharomyces cerevisiae*

Saccharomyces cerevisiae is a species of budding yeast. It is perhaps the most useful yeast, owing to its use since ancient times in baking and brewing. It is believed that it was originally isolated from the skins of grapes (one can see the yeast as a component of the thin white film on the skins of some dark-colored fruits such as plums; it exists among the waxes of the cuticle). It is one of the most intensively studied eukaryotic model organisms in molecular and cell biology, much like *Escherichia coli* as the model prokaryote. It is the microorganism behind the most common type of fermentation. *S. cerevisiae* cells are round to ovoid and 5–10 μm in diameter. It reproduces by a division process known as budding. Many proteins important in human biology were first discovered by studying their homologs in yeast. These proteins include cell cycle proteins, signaling proteins, and protein-processing enzymes. The petite mutation in *S. cerevisiae* is of particular interest.

Antibodies against *S. cerevisiae* are found in 60%–70% of patients with Crohn's disease and 10%–15% of patients with ulcerative colitis. *S. cerevisiae* is known as a top-fermenting yeast, so called because during the fermentation process its hydrophobic surface causes the flocs to adhere to CO_2 and rise to the top of the fermentation vessel. It is one of the major types of yeast used in the brewing of ale, along with *Saccharomyces pastorianus*, which is used in the brewing of lager. Top-fermenting yeasts are fermented at higher temperatures than lager yeasts, and the resulting ales have a different flavor than the same beverage fermented with lager yeast. "Fruity esters" may be formed if the ale yeast undergoes temperatures near 21°C (or 70°F) or if the fermentation temperature of the beverage fluctuates during the process. Lager yeast normally ferments at a temperature of approximately 5°C (or 40°F), where ale yeast becomes dormant. Lager yeast can be fermented at a higher temperature normally used for ale yeast, and this application is often used in a beer style known as "steam beer."

12.9.4 *Pseudomonas*

Pseudomonas is a genus of gamma proteobacteria, belonging to the larger family of pseudomonads. Recently, 16S rRNA sequence analysis has redefined the taxonomy of many bacterial species. As a result, the genus *Pseudomonas* includes strains formerly classified in the genera *Chryseomonas* and *Flavimonas*. Other strains previously classified in the genus *Pseudomonas* are now classified in the genera *Burkholderia* and *Ralstonia*. Since the mid-1980s, certain members of the *Pseudomonas* genus have been applied to cereal seeds or applied directly to soils as a way of preventing the growth or establishment of crop pathogens. This practice is generically referred to as *biocontrol*. Currently, the biocontrol properties of *P. fluorescens* strains (CHA0 or Pf-5, for example) are best understood, although it is not clear exactly how the plant growth promoting properties of *P. fluorescens* are achieved. Theories include that the bacteria might induce systemic resistance in the host plant, so it can better resist attack by a true pathogen. The bacteria might compete with other (pathogenic) soil microbes (e.g., by siderophores giving a competitive advantage at scavenging for iron); the bacteria might produce compounds antagonistic to other soil microbes, such as phenazine-type antibiotics or hydrogen cyanide. There is experimental evidence to support all of these theories, in certain conditions. A good review of the topic is written by Haas and Defago. Other notable *Pseudomonas* species with biocontrol properties include *P. chlororaphis*, which produces a phenazine-type antibiotic (an active agent against certain fungal plant pathogens) and the closely related species *P. aurantiaca*, which produces di-2,4-diacetylfluoroglucylmethan (a compound antibiotically active against Gram-positive organisms). Some members of the genus *Pseudomonas* are able to metabolize chemical pollutants in the environment and can therefore be used for bioremediation.

12.9.5 *Clostridium*

Clostridium is a genus of Gram-positive bacteria, belonging to the Firmicutes. These are obligate anaerobes capable of producing endospores. Individual cells are rod-shaped, which gives them their name, from the Greek *kloster* or spindle. These characteristics traditionally defined the genus; however, many species originally classified as *Clostridium* have been reclassified in other genera. *Clostridium thermocellum* can utilize lignocellulosic waste and generate ethanol, thus making it a possible candidate for use in ethanol production. It also has no oxygen requirement and is thermophilic, thus reducing cooling cost. *C. acetobutylicum*, also known as the *Weizmann organism*, was first used by Chaim Weizmann in 1916 to produce acetone and biobutanol from starch for the production of gunpowder and TNT. The anaerobic bacterium *Clostridium ljungdahlii*, recently discovered in commercial chicken wastes, can produce ethanol from single-carbon sources including synthesis gas, a mixture of carbon monoxide and H_2 that can be generated from the partial combustion of either fossil fuels or biomass. Use of these bacteria to produce ethanol from synthesis gas has progressed to the pilot plant stage at the BRI Energy facility in Fayetteville, Arkansas. Genes from *C. thermocellum* have been inserted into transgenic mice to allow the production of endoglucanase. The experiment was intended to learn more about how the digestive capacity of monogastric animals could be improved.

12.9.6 Thermophiles

Thermophiles are microorganisms that live and grow in extremely hot environments that would kill most other microorganisms. Thermophiles are grouped into either prokaryotes or eukaryotes, and these two groups of extremophiles are classified in the group of Archaea. They grow best in temperatures that are between 50°C/120°F and 70°C/158°F. They will not grow if the temperature reaches 20°C/68°F. Thermophiles are not easy to study because of the extreme conditions that they need to survive. Thermophiles either live in geothermal habitats or in environments that themselves create heat. A pile of compost and garbage landfills are two examples of environments that produce heat on their own. Some thermophiles are fungi such as *Chaetomium thermophile*, *Humicola insolens*, *Humicola* (Thermomyces) *lanuginosus*, *Thermoascus aurantiacus* (a Paecilomyces-like fungus), and *Aspergillus fumigates*. *Thermus aquaticus* and *Thermococcus litoralis* are two thermophiles that are used as an enzyme in DNA fingerprinting in criminal cases or in identification of parents or siblings. *Bacillus stearothermophilus* is another thermophile used as an enzyme in biological detergents. Thermophiles in self-heating environments must have a supply of organic matter like food scraps in order to grow. These kinds of thermophiles turn this organic matter into a rich source of nutrients for living microorganisms and plants to use as food.

12.10 INDUSTRIAL PRODUCTION OF HEALTHCARE PRODUCTS

12.10.1 Antibiotic Manufacturing

The manufacturers of antibiotics maintain pilot plant facilities whose purpose is to upgrade yields and to bring about improvements in processing procedures. Studies are continually being made on strain improvement, inoculum conditions, fermentation conditions, and various combinations of these factors. For example, improved mutant strains almost always require adjustments in fermentation conditions in order to achieve the high yields in fermenters that are obtainable in shaken flasks. Since all of the antibiotics are made by aerobic fermentations, there are a number of similarities in the processes used for their production. Many of the specific details regarding the plant scale production of a given antibiotic remain trade secrets, with certain exceptions. However, the general outline of these methods is fairly well known to those connected with the industry.

12.10.1.1 Penicillin Production

The mold from which Alexander Fleming isolated penicillin was later identified as *Penicillium notatum*. A variety of molds belonging to other species and genera were later found to yield greater amounts of the antibiotic and a series of closely related penicillins. The naturally occurring penicillins differ from each other in the side chain (R group). Penicillin was produced by a surface culture method early in World War II. Submerged culture methods were introduced by 1943 and are now almost exclusively employed. Penicillin production needs strict aseptic conditions. Contamination by other microorganisms reduces the yield of penicillin. This is caused by the widespread occurrence of penicillinase-producing bacteria that inactivate the antibiotic. Penicillin production also needs tremendous amount of air. In all methods, deep tanks with a capacity of several thousand gallons are filled with a culture medium. The medium consists of corn steep liquor, lactose, glucose, nutrient, salts, phenylacetic acid or a derivative, and calcium carbonate as buffer. The medium is inoculated with a suspension of conidia of *P. chrysogenum*. The medium is constantly aerated and agitated, and the mold grows throughout as pellets. After about seven days, growth is complete, the pH rises to 8.0 or above, and penicillin production ceases. When the fermentation is complete, the masses of mold growth are separated from the culture medium by centrifugation and filtration.

12.10.1.2 Cephalosporins Production

Similar semisynthetic approaches can be used for manufacture of other antibiotics. Cephalosporin C is made as the fermentation product of *Cephalosporium acremonium*. However, this form is not potent for clinical use. Its molecule can be transformed by removal of an aminoadipic acid side chain to form 7-α aminocephalosporanic acid (7-ACA), which can be further modified by adding side chains to form clinically useful broad-spectrum antimicrobials. Various side chains can be added to as well as removed from both 6-APA and 7-ACA to produce antibiotics with varying spectra of activities and varying degrees of resistance to inactivation by enzymes produced by pathogenic microbes. Thus, we have entered into the so-called third-generation cephalosporins, such as *moxalactam*, developed for control of bacteria that produce enzymes capable of degrading penicillins and cephalosporins.

12.10.1.3 Streptomycin Production

Streptomycin and other various antibiotics are produced using strains of *Streptomyces griseus*. Spores of this actinomycete are inoculated into a medium to establish a culture with a high mycelial biomass for introduction into an inoculum tank, with subsequent use of the mycelial inoculum to initiate the fermentation process in the production tank. The medium contains soybean meal (N-source), glucose (C-source) and NaCl. The process is carried out at 28°C and the maximum production is achieved at pH range of 7.6–8.0. High agitation and aeration are needed. The process lasts for about 10 days. The classic fermentation process involves three phases. During the first phase, there is a rapid growth of the microbe with production of mycelial biomass. Proteolytic activity of the microbe releases ammonia (NH_3) to the medium from the soybean meal, causing a rise in pH. During this initial fermentation phase, there is little production of streptomycin. During second phase, there is little additional production of mycelium, but the secondary metabolite, streptomycin, accumulates in the medium. The glucose and NH_3 released are consumed during this phase. The pH remains fairly constant—between 7.6 and 8.0. In the third and final phase, when carbohydrates become depleted, streptomycin production ceases and the microbial cells begin to lyse pH increases, and process normally ends by this time. After the process is complete, mycelium is separated from the broth by filtration and the antibiotic is recovered. In one method of recovery and purification, streptomycin is adsorbed onto activated charcoal and eluted with acid alcohol. It is then precipitated with acetone and further purified by use of column chromatography.

Manufacturing process of antibiotics

Step 1: Before fermentation can begin, the desired antibiotic-producing organism must be isolated and its numbers must be increased by many times. To do this, a starter culture from a sample of previously isolated, cold-stored organisms is created in the lab. In order to grow the initial culture, a sample of the organism is transferred to an agar-containing plate. The initial culture is then put into shake flasks along with food and other nutrients necessary for growth. This creates a suspension, which can be transferred to seed tanks for further growth.

Step 2: The seed tanks are steel tanks designed to provide an ideal environment for growing microorganisms. They are filled with all things a specific microorganism would need to survive and thrive, including warm water and carbohydrates like lactose or glucose sugars. Additionally, they contain other necessary carbon sources such as acetic acid, alcohols, or hydrocarbons, and N_2 sources like ammonia salts. Growth factors like vitamins, amino acids, and minor nutrients round out the composition of the seed tank contents. The seed tanks are equipped with mixers, which keep the growth medium moving, and a pump to deliver sterilized, filtered air. After approximately 24–28 h, the material in the seed tanks is transferred to the primary fermentation tanks.

Step 3: The fermentation tank is essentially a larger version of the steel seed tank, which is able to hold about 30,000 gal. The fermentation tank is filled with growth media that provide an environment suitable for growth. Here, the microorganisms are allowed to grow and multiply. During this process, they excrete large quantities of the desired antibiotic. The tank is cooled to keep the temperature between 73°F and 81°F (23°C and 27.2°C). It is constantly agitated, and a continuous stream of sterilized air is pumped into it. For this reason, anti-foaming agents are periodically added. Since pH control is vital for optimal growth, acids or bases are added to the tank as necessary.

Step 4: After 3–5 days, the maximum amount of antibiotic will have been produced and the isolation process can begin. Depending on the specific antibiotic produced, the fermentation broth is processed by various purification methods. For example, for antibiotic compounds that are water soluble, an ion-exchange method may be used for purification. In this method, the compound is first separated from the waste organic materials in the broth and then sent through equipment, which separates the other water-soluble compounds from the desired one. To isolate an oil-soluble antibiotic such as penicillin, a solvent extraction method is used. In this method, the broth is treated with organic solvents such as butyl acetate or methyl isobutyl ketone, which can specifically dissolve the antibiotic. The dissolved antibiotic is then recovered using various organic chemical means. At the end of this step, the manufacturer is typically left with a purified powdered form of the antibiotic, which can be further refined into different product types.

Step 5: Antibiotic products can take on many different forms. They can be sold in solutions for intravenous bags or syringes, in pill or gel capsule form, or as powders that are incorporated into topical ointments. Depending on the final form of the antibiotic, various refining steps may be taken after the initial isolation. For intravenous bags, the crystalline antibiotic can be dissolved in a solution and put in the bag, which is then hermetically sealed. For gel capsules, the powdered antibiotic is physically filled into the bottom half of a capsule, after which the top half is mechanically put in place. When used in topical ointments, the antibiotic is mixed into the ointment.

Step 6: From this point, the antibiotic product is transported to the final packaging stations where the products are stacked and put in boxes. They are loaded up on trucks and transported to various distributors, hospitals, and pharmacies. The entire process of fermentation, recovery, and processing can take anywhere from 5 to 8 days.

Step 7: Quality control is of utmost importance in the production of antibiotics. Since it involves a fermentation process, steps must be taken to ensure that absolutely no contamination is introduced at any point during production. To this end, the medium and all of the processing equipment are thoroughly steam sterilized. During manufacturing, the quality of all the compounds is checked on a regular basis. Of particular importance are frequent checks of the condition of the microorganism culture during fermentation. These checks are accomplished using various chromatography techniques. Also, various physical and chemical properties of the finished product are checked, such as pH, melting point, and moisture content. In the United States, antibiotic production is highly regulated by the FDA. Depending on the application and type of antibiotic, more or less testing must be completed. For example, the FDA requires that for certain antibiotics each batch must be checked by them for effectiveness and purity; only after they have certified the batch can it be sold for general consumption.

12.10.2 Steroids Production

Microbial biotransformation of steroids is very important in the pharmaceutical industry. Steroids are used in treatment of various disorders and also involved in regulation of sexuality. Chemical synthesis of steroids is very complex because of the requirement to achieve the necessary precision of substitute location. For example, cortisone can be synthesized chemically from deoxycholic acid, but the process requires 37 steps, many of which must be carried out under extreme conditions of temperature and pressure, with the resulting product costing over $200 per g. The most difficult is the introduction of an oxygen atom at number 11 position of the steroid ring, but this can be accomplished by some microorganisms. The fungus, *Rhizopus nigricans*, for example, hydroxylates progesterone, forming another steroid with the introduction of oxygen at the number 11 position. Similarly, the fungus *Cunninghamella blakesleeana* can hydroxylate the steroid cortexolone to form hydrocortisone with the introduction of oxygen at number 11 position. Other transformations of the steroid nucleus carried out by microbes include hydrogenations, dehydrogenations, epoxidations, removal, and addition of the side chains.

12.10.3 Textile Production

There are two principal aspects to the microbiology of textiles. One is the use of microorganisms in preparing fibers such as flax and hemp; the other is deterioration of textiles, including cordage and ropes, and the preservation of such materials. The long fibers of flax, hemp, jute, and sisal are loosened from the plant stem by felting. The fiber bundles are just held within the outer layers of cells and outside the central pithy and woody layers by an intercellular cement of pectin. A number of bacteria and molds can digest pectin. This permits the fiber bundles to be separated mechanically from the stems and from each other. Fibers can then be collected and woven into linen, or used in the form of ropes and packaging. There are two basic process of retting. In the first process, the flax or hemp is spread out on the ground, particularly under somewhat acid conditions such as would occur in peat moss, and is exposed to the air and rain for some period. Under these conditions, the molds and aerobic bacteria that hydrolyzed pectin can grow.

Retting is achieved, but the process is somewhat uncertain and subject to continuous temperature and moisture changes. Frequently, certain organisms are observed that produce pigment or some growth on cellulose fibers, and thus color or weaken it. The fibers obtained by this method are frequently of poor quality and the yield is small. In the second process, the plant stalks are immersed in flowing stream, ponds, or tanks of water and are allowed to remain there for several days. During this period, the anaerobic organisms, especially *Clostridium felsineum*, digest the pectin. The process, when properly operated, yields a nicer fiber that can be made into linen of quite good quality. Jute and sisal are tropical plants that are subjected to retting, generally under higher temperature condition and under water in bays and lagoons, or sometimes in flowing streams. The product resulting from these operations is relatively crude, but relatively cheap. Much of the rope fibers and packaging cloth used throughout the world originates from these plants.

12.10.4 Microbial Synthesis of Vitamin B12

It seems probable that the only primary source of vitamin B12 in nature is the metabolic activity of the microorganisms. It is synthesized by a wide range of bacteria and *Streptomycetes*, though not to any extent by yeasts and fungi. While over 100 fermentation processes have been described for the production of vitamin Bl2, only half a dozen have apparently been used on a commercial scale. Some of these are as follows: (1) recovery of vitamin B12 as by-product of the streptomycin and aureomycin antibiotic fermentations, and (2) fermentation processes using *Bacillus megatherium*, *Streptomyces olivaceus*, *Propionibacterium freudemeichii*, and *Propionibacterium shermanii*. The processes using the *Propionibacterium* species are the most productive and are now widely used commercially. Both batch and continuous processes have been described.

It is important to select microbial species that make the 5, 6-dimethyl benzimidazolylcobamid exclusively. Several manufacturers have been led astray by organisms that gave high yields of the related cobamides including pseudo-vitamin (adeninylcobamide). The natural form of the vitamin is Barker's coenzyme, where a deoxyadenosyl residue replaces the cyano group found in the commercial vitamin.

Practically, all of the cobamides formed in the fermentation are retained in the cells, and the first step is the separation of the cells from the fermentation medium. Large high-speed centrifuges are used to concentrate the bacteria to a cream, while filters are used to remove *Streptomycete* cells. The vitamin B12 activity is released from the cells by acid, heating, cyanide or other treatments. Addition of cyanide solutions decomposes the coenzyme form of the vitamin and results in the formation of the cyanocobalamin. The cyanocobalamin is adsorbed on ion exchange resin IRC-50 or charcoal and is eluted. It is then purified further by partition between phenolic solvents and water. The vitamin is finally crystallized from aqueous-acetone solutions. The crystalline product often contains some water of crystallization.

PROBLEMS

Section A: Descriptive Type

Q1. What is a fermenter? Describe its significance.
Q2. Differentiate between aerobic and anaerobic fermentation.
Q3. Describe the screening method for microorganisms.
Q4. What is an inoculum?
Q5. What is an upstream bioprocess?
Q6. Describe the various stages in DSP.
Q7. Write a short note on *Corynebacterium*.
Q8. Describe the process for microbial synthesis of vitamin B12.
Q9. Describe the significance of contract manufacturing in biotechnology.

Section B: Multiple Choice

Q1. A ________ is a device in which a substrate of low value is utilized by living cells to generate a product of higher value.
  a. PCR
  b. Bioreactor
  c. Chemical reactor
  d. Thermal reactor

Q2. Fermentation is the conversion of carbohydrates such as sugar into an acid or an alcohol. True/False

Q3. ________ is the organism of choice for acetone butanol fermentation.
  a. *Lactobacillus*
  b. *Nocardia*
  c. *Clostridium*
  d. *Aspergillus*

Q4. Inoculum development refers to the ...
  a. Dormant stage of microbial culture
  b. Active stage of microbial culture

Q5. Sterilization is the process of inactivating or removing all living organisms from a substance or surface. True/False

Q6. An autoclave uses a temperature of ________ for steam sterilization.
  a. 100°C
  b. 150°C
  c. 121°C

Q7. Mutant strain S-604 of *S. aureofaciens* produces 6-demethyl tetracycline in place of tetracycline. True/False

Q8. The upstream part of a bioprocess refers to ...
  a. Growth of bacterial or mammalian cells
  b. Synthesis of protein
  c. Recombination DNA technology
  d. None of the above

Q9. The ideal temperature for mammalian cells to grow in a bioreactor is ...
  a. 37°C
  b. 38°C
  c. 30°C
  d. 40°C

Q10. The actual recovery of useful products from fermentation is known as ...
  a. Upstream bioprocess
  b. Downstream bioprocess

Q11. Cloning of *E. coli* K-12 was exempted from NIH guidelines. True/False

Section C: Critical Thinking

Q1. Why do mammalian cells fail to manufacture rDNA products as microorganisms do?

Q2. What would you do to improve the efficiency of existing bioreactors?

LABORATORY ASSIGNMENT

Write a protocol to make a cheese in the laboratory and discuss it with your course instructor. Then make a cheese using microbes. You may use different methods of making cheese.

REFERENCES AND FURTHER READING

American cheese and cheeses. *Consumer Rep.* 728–732, November 1990.

Brown, B. *The Complete Book of Cheese*. Gramercy Publishing, New York, 1995.

Cambrosio, A. and Keating, P. Between fact and technique: The beginnings of hybridoma technology. *J. Hist. Biol.* 25(2): 175–230, 1992.

Carr, S. *The Simon and Schuster Pocket Guide to Cheese*. Simon and Schuster, New York, 1981.

Crueger, W. *Biotechnology: A Textbook of Industrial Microbiology*. Sinauer Associates, Inc., Sunderland, MA, 1989.

Dinsmoor, R.S. Insulin: A never-ending evolution. *Countdown* (Spring 2001), 2001.

Demain, A.L. and Davies, J.E. *Manual of Industrial Microbiology and Biotechnology*, 2nd edn. ASM Press, Washington, DC, 1999.

Dulbecco, R. and Ginsberg, H.S. *Virology*, 2nd edn. J.B. Lippincott Company, Philadelphia, PA, 1988.

Eli Lilly Corporation. *Humulin and Humalog Development*. CD-ROM, 2001. *Eli Lilly Diabetes Web Page*, November 16, 2001. <http://www.lillydiabetes.com>

Kosikowski, F. *Cheese and Fermented Milk Foods*. Cornell University, Ithaca, NY, 1966.

Kirk Othmer Encyclopedia of Chemical Technology. John Wiley & Sons, New York, 1992.

Kohler, G. and Milstein, C. Continuous cultures of fused cells secreting antibody of predefined specificity. *Nature* 1256: 495–497, 1975.

Ladisch, M.R. *Bioseparations Engineering: Principles, Practice, and Economics*. John Wiley & Sons, Inc., Hoboken, NJ, 2001.

Mills, S. *The World Guide to Cheese*. Gallery Books, 1988. http://www.madehow.com/Volume-1/Cheese.html;http://www.enotes.com/how-products-encyclopedia/cheese

Morell, V. Antibiotic resistance: Road of no return. *Science* 278: 575–576, 1997.

Novo Nordisk Diabetes Web Page, November 15, 2001. <http://www.novonet.co.nz>

Olson, W.P. *Separations Technology: Pharmaceutical and Biotechnology Applications*. Interpharm Press, Inc., Buffalo Grove, IL, 1995.

PhRMA Reports Identifies More Than 400 Biotech Drugs in Development. Pharmaceutical Technology, August 24, 2006. http://pharmtech.findpharma.com/pharmtech/News/article/detail/367489. Retrieved 2006-09-04.

Rang, H.P. *Pharmacology*. Churchill Livingstone, Edinburgh, U.K., p. 241, for the examples infliximab, basiliximab, abciximab, daclizumab, palivusamab, palivusamab, gemtuzumab, alemtuzumab, etanercept and rituximab, and mechanism and mode. ISBN 0-443-07145-4, 2003.

Riechmann, L., Clark, M., Waldmann, H., and Winter, G. Reshaping human antibodies for therapy. *Nature* 332: 323–327, 1988.

Siegel, D.L. Recombinant monoclonal antibody technology. *Transfus. Clin. Biol.* 9: 15–22, 2002.

Schmitz, U., Versmold, A., Kaufmann, P., and Frank, H.G. Phage display: A molecular tool for the generation of antibodies-a review. *Placenta* 21(Suppl A): S106–S112, 2000.

Schwaber, J. and Cohen, E.P. Human x mouse somatic cell hybrid clones secreting immunoglobulins of both parental types. *Nature* 244(5416): 444–447, 1973.

Shepherd, P. and Christopher, D. *Monoclonal Antibodies*. Oxford University Press, New York, 2000.

Stinson, S. Drug firms restock antibacterial arsenal. *Chem. Eng. News* 75–100, 1997.

13 Ethics in Biotechnology

LEARNING OBJECTIVES

- Discuss the significance of ethics in biotechnology
- Discuss ethical issues related to genetically modified (GM) food, plants, and animals
- Discuss ethical issues associated with animal testing
- Discuss ethical issues associated with xenotransplantation, human cloning, genetic screening, DNA fingerprinting, biometrics, organ transplantation, and embryonic stem cell research

13.1 INTRODUCTION

Bioethicists are people who raise ethical questions about the use of any living organism for experimentation purposes. Bioethicists are not against the development of new drugs or new therapies; they just wanted judicious use of animals and human subjects in experimentation. The whole objective of bioethicists is to minimize trauma in living organisms during experimentation. Human experimentation was one of the first areas addressed by modern bioethicists. The National Commission for the Protection of Human Subjects of Biomedical and Behavioral Research was initially established in 1974 to identify the basic ethical principles that should underlie the conduct of biomedical and behavioral research involving human subjects. However, the fundamental principles announced in the Belmont Report (1979), namely, autonomy, beneficence, and justice, have influenced the thinking of bioethicists across a wide range of issues. Others have added non-malfeasance, human dignity, and the sanctity of life to this list of cardinal values.

Although bioethical issues have been debated since ancient times and public attention briefly focused on the role of human subjects in biomedical experiments following the revelation of Nazi experiments conducted during World War II, the modern field of bioethics first emerged as an academic discipline only in the 1960s. Technological advances in such diverse areas as organ transplantation and end-of-life care, including the development of kidney dialysis and respirators, posed novel questions regarding when and how care might be withdrawn. These questions often fell upon philosophers and religious scholars, but, by the 1970s, bioethical think tanks and academic bioethics programs emerged. Among the earliest such institutions were the Hastings Center (originally known as the Institute of Society, Ethics, and the Life Sciences), founded in 1970 by philosopher Daniel Callahan and psychiatrist Willard Gaylin, and the Kennedy Institute of Ethics, established at Georgetown University in 1971.

During the subsequent three decades, bioethical issues gained widespread attention through the court cases surrounding the deaths of Karen Ann Quinlan, Nancy Cruzan, and Terri Schiavo. The field developed its own cadre of widely known advocates, such as Al Jonsen at the University of Washington; John Fletcher at the University of Virginia, Minnesota; Glenn McGee at the University at Albany, State University of New York; Jacob M. Appel at Brown University; Ruth Faden at Johns Hopkins University; and Arthur Caplan at the University of Pennsylvania, Philadelphia. In 1995, President Bill Clinton established the President's Council on Bioethics, a sign that the field had finally reached an unprecedented level of maturity and acceptance. President George W. Bush also relied upon a Council on Bioethics in rendering decisions in areas such as the public funding of embryonic stem cell research. In this chapter, we discuss the various ethical issues concerning biotechnology products with suitable examples so that you can understand the importance of ethical concerns and analyze the claims of both scientists and bioethicists for yourself.

13.2 GENETICALLY MODIFIED FOODS AND PLANTS

Millions of people worldwide have consumed foods derived from genetically modified (GM) plants (mainly maize, soybean, and oilseed rape) and no adverse effects have been observed so far. The lack of evidence of negative effects, however, does not mean that new GM foods are without risk. The possibility of long-term effects from GM plants cannot be excluded and must be examined on a case-by-case basis. The question of the safety of GM foods has been reviewed by the International Council of Science (ICSU), which based its opinion on 50 authoritative independent scientific assessments from around the world. Currently available GM crops and foods derived from them have been judged safe to eat, and the methods used to test them have been deemed appropriate (Figure 13.1).

Allergens and toxins occur in some traditional foods and can adversely affect some people, leading to concerns that GM plant-derived foods may contain elevated levels of allergens and toxins. Extensive testing of GM foods currently on the market has not confirmed these concerns. The use of genes from plants with known allergens is discouraged, and if a transformed product is found to pose an increased risk of allergies it should be discontinued. All new foods, including those derived from GM crops, should be assessed with caution. One concern about food safety is the potential transfer of genes from consumed food into human cells or into microorganisms within the body.

Many GM crops were created using antibiotic-resistant genes as markers. Therefore, in addition to having the desired characteristics, these GM crops contain antibiotic-resistant genes. If these genes were to transfer in the digestive tract from a food product into human cells or to bacteria, this could lead to the development of antibiotic-resistant strains of bacteria. Although scientists believe the probability of such a transfer is extremely low, the use of antibiotic-resistant genes has been discouraged. Methods are now being developed whereby only the strict minimum of transgenic DNA is present in GM plants. Some of these techniques involve the complete elimination of the genetic marker once the selection process has been made. Scientists generally agree that genetic engineering can offer some health benefits to consumers. Direct benefits can come from improving the nutritional quality of food and from reducing the presence of toxic compounds and allergens in certain foods. Indirect health benefits can come from diminished pesticide use, less insect or disease damage to plants, increased availability of affordable food, and the removal of toxic compounds from soil. These direct and indirect benefits need to be better documented (Figure 13.2).

How do we identify a GM food or crop? The draft guidelines for the labeling of foods obtained through certain techniques of genetic modification/genetic engineering (FAO/WHO, 2003c) are still

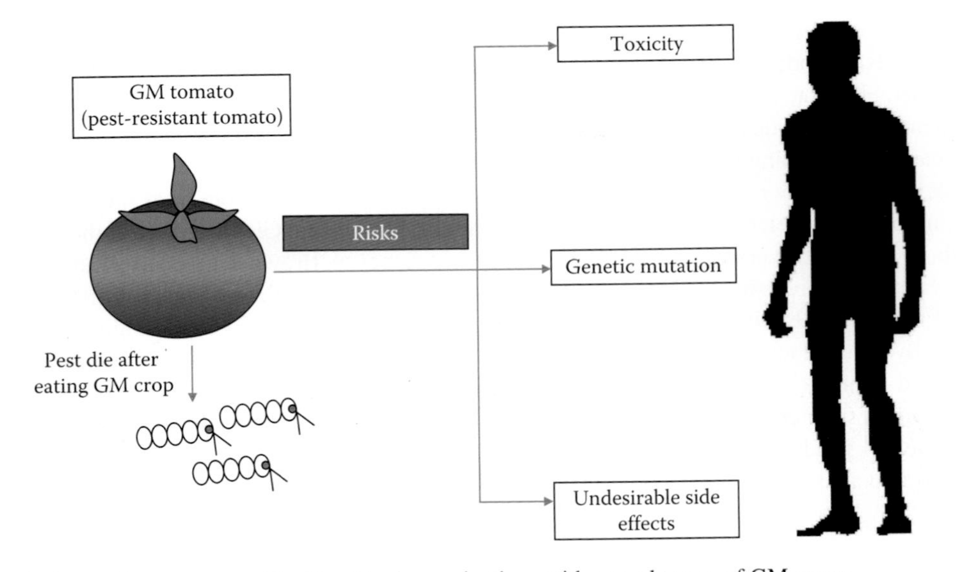

FIGURE 13.1 Concerns raised by bioethical organizations with regard to use of GM crops.

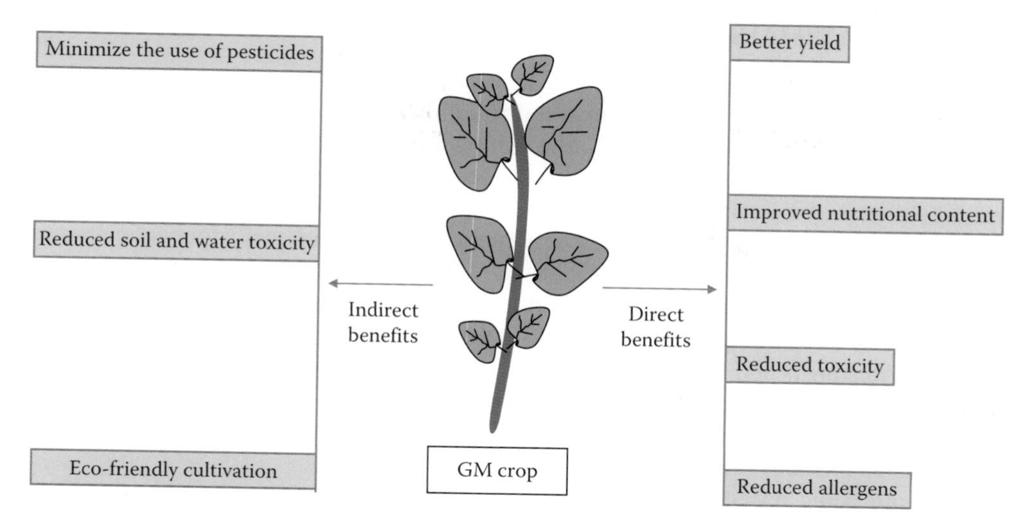

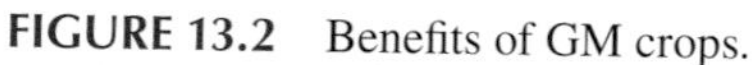

FIGURE 13.2 Benefits of GM crops.

at an early stage of discussion and many sections are bracketed, indicating that the language has not yet been agreed upon. The guideline is proposed to apply to labeling of foods and food ingredients in three situations, when foods are (1) significantly different from their conventional counterparts; (2) composed of or contain GM and/or genetically engineered (GE) organisms or contain protein or DNA resulting from gene technology; and (3) produced from but do not contain GM/GE organisms, protein, or DNA from gene technology. According to the ICSU, scientists do not fully agree about the appropriate role of labeling. Although mandatory labeling is traditionally used to help consumers identify foods that may contain allergens or other potentially harmful substances, labels are also used to help consumers who wish to select certain foods on the basis of their mode of production, on environmental (e.g., organic), ethical (e.g., fair trade), or religious (e.g., kosher or halal) grounds. Countries differ in the type of labeling information that is mandatory or permitted. According to the ICSU, "labeling of foods as GM or non-GM may enable consumer choice as to the process by which the food is produced [but] it conveys no information as to the content of the foods, and whether there are any risks and/or benefits associated with particular foods." The ICSU suggests that more informative food labeling that explains the type of transformation and any resulting compositional changes could enable consumers to assess the risks and benefits of particular foods (Figure 13.3).

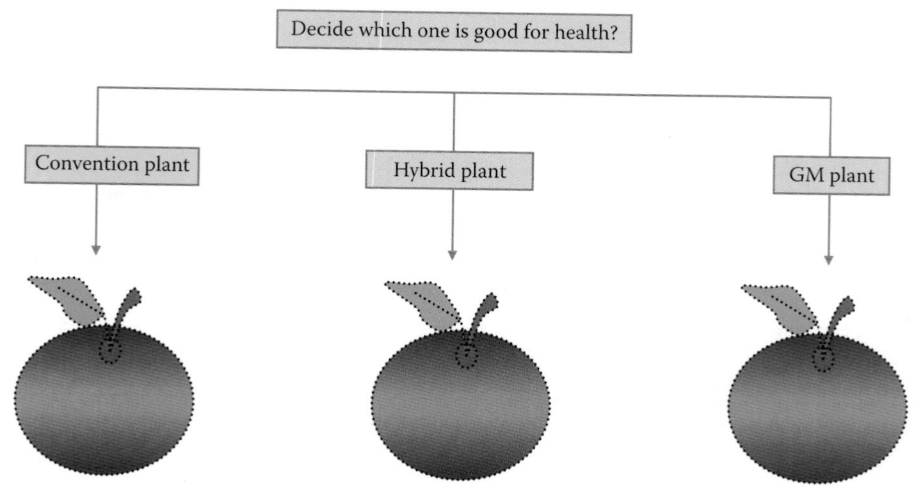

FIGURE 13.3 Can you decide which one is good for health?

13.3 USE OF ANIMALS AS EXPERIMENTAL MODELS

Animal testing is a phrase that most people have heard at some point in their lives, but they are perhaps still unsure of exactly what is involved in animal testing. Additionally, there is still a great deal of subjectivity around the meaning behind animal testing. Also, how animal testing is interpreted is partly related to the person's personal views on animal testing. Whether it is called animal testing, animal experimentation, or animal research, it refers to the experimental use of non-human animals. This type of experimentation is not done directly for healing purposes, although the end result may involve medications used for healing both humans and other animals. Instead, healing an animal would be akin to veterinary medicine, which is entirely different from animal testing. It is also important to note that those who are opposed to animal testing may believe animal testing is nothing but torture and suffering of animals. Animal testing is conducted virtually everywhere and its uses are broad. Global standards are quite strict and rigorous with regard to animal testing and monitoring. Animal testing only occurs if there is no other viable alternative to the methods (Figure 13.4).

Animals are being tested at various centers located in universities, medical schools, pharmaceutical and biotechnology companies, and military defense establishments. Universities are often at the forefront of research that uses animal testing. Research students led by experienced scientists use the animals to better understand the anatomy and physiology of the body's function as well as to study the effects of pharmaceutical and drug treatments. Universities are quite often in the limelight, which means their treatment of animals is well scrutinized. Despite this, offenses have occurred at universities, such as previous primate experiments at the University of Cambridge. Medical research is considered integral to medical understanding and medical progress. For this reason, medical schools are generally avid supporters of animal testing for improving knowledge and health outcomes in disease prevention, treatment, and overall management. At the same time, medical schools are also strong supporters of animal testing alternatives whenever appropriate. In fact, new methods are being used as an alternative to animal testing even in the training of medical students. Some of the leading, highly respected universities in the United States and Britain are using innovative, interactive clinical teaching methods to replace the more old-fashioned types of laboratories.

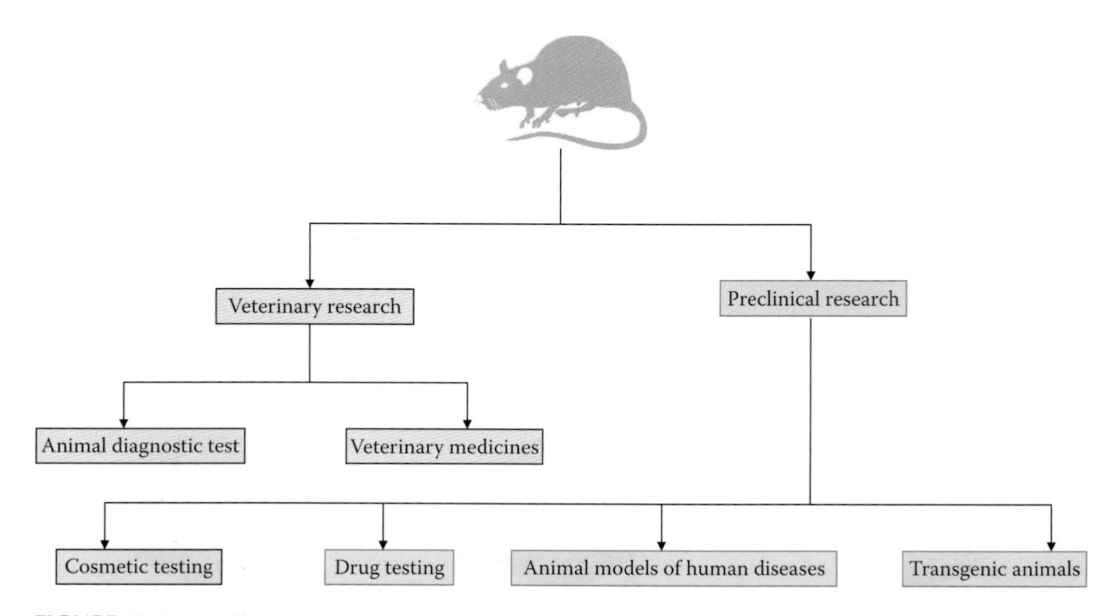

FIGURE 13.4 Application animals in the development and testing of veterinary and therapeutic products.

Pharmaceuticals comprise a billion dollar industry, with enormous competition to find the next breakthrough drug. The new drug requires rigorous pharmaceutical testing to ensure its safety. Prior to performing human clinical trials, drugs require preliminary testing on animals. Many drugs will never make it beyond this stage, which is precisely why they are first tested on animals rather than immediately on humans. This practice allows researchers to understand how the drugs are metabolized and how they affect the various body systems, and also allows researchers to determine their efficacy. Efficacy, however, must not carry an unacceptable level of risk, which means that a drug working for the intended purpose in an animal is not cause enough to move forward to human trials if the drug causes major side effects, harm, or death.

Many are not aware that animal testing occurs quite regularly in military establishments. One important reason for their use is to protect the military and the general public from weapons and tactics that may be used by other countries. The military performs animal testing to determine how various biological weapons (such as viruses) will affect the intended recipients. By using animals, they can gauge the effect and develop the means to prevent and successfully handle the consequences. Vaccines are produced to prevent the disease development, whereas the objective of medicine is to treat or cure disease, particularly on a wide scale, where an enormous number of people could be affected. Other testing uses include wound testing, where the military examines how various weapons create wounds and how those wounds heal.

13.3.1 Disadvantages of Animal Testing

In animal testing, countless animals are experimented on and then killed after their use. Others are injured and will still live the remainder of their lives in captivity. The unfortunate aspect is that many of these animals have undergone tests for substances that will never actually see approval for public consumption and use. It is this aspect of animal testing that many view as a major negative against the practice. This aspect seems to focus on the idea that the animal died in vain, because no direct benefit to humans occurred from the animal tests. Another disadvantage to the issue of animal testing is the sheer cost. Animal testing generally costs an enormous amount of money. Animals must be fed, housed, cared for, and treated with drugs or a similar experimental substance. The controlled environment is important, but it comes at a high cost. In addition, animal testing may occur more than once and over the course of months, which means that additional costs are incurred. The price of animals themselves must also be factored into the equation. There are companies who breed animals specifically for testing, and animals can be purchased through them. There is also the argument that the reaction of a drug in an animal's body is quite different from the reaction in humans. The main criticism here is that some believe animal testing is unreliable. Following on that criticism is the premise that because animals are in an unnatural environment, they will be under stress and therefore will not react to the drugs in the same way, compared to their potential reaction in a natural environment. This argument further weakens the validity of animal experimentation.

13.3.2 Attacks on Researchers

The increase in groups against animal testing led to a backlash against researchers who conduct these tests. There was less resistance in the earlier part of the twentieth century, but as experimental use of animals increased, the bonding of groups against animal testing began to take shape. The Internet also added glue to the network of those against animal testing, because it allowed them to more easily reach people around the world who would support their belief that animals should not be used in experimental testing. People for the Ethical Treatment of Animals (PETA) is one group that has led the way in animal rights campaigns as well as spurring on the creation of many other groups. PETA campaigns are not only directed at animal testing but also at areas such as promoting vegetarianism and ending the killing of animals for fur. Their campaigns regarding animal testing have, however, been well publicized, garnering a great deal of media attention and reaching people around the world. Threats to researchers and their families left some of the more prominent ones requiring bodyguards for protection. Arson attacks and violence became a reality for those who performed animal testing, particularly those who used primates. In fact, in 2006, a primate researcher at a university in the United States received numerous threats for his research. After regular demonstrations and threats to his well-being, he halted the research. In a similar incident in 2007, a bomb was actually placed under a researcher's car, but the bomb fortunately had a faulty fuse and nobody was injured. In the United Kingdom, millions in estimated damages are caused each year to property by animal rights activists. Millions are also spent each year on policing and preventing damage.

13.3.3 Regulations for Animal Testing in the United States

A large part of the regulation process is governed by the Animal Welfare Act, which is supported and maintained through several sectioned agencies within the United States Department of Agriculture. The goals are to ensure that animals used for research receive a high standard of care and are treated humanely and with respect. Suffering should be minimized, but the stance is that it should not be minimized at the expense of the research goals and aims. This means that animal suffering is seen as secondary to the goals of research, and is only minimized if it does not interfere with the success of the experiment. The guidelines and regulations themselves are also only relevant to mammals, excluding birds, rats and mice, whether caught in the wild or purpose-bred for the experiment. Given the enormous numbers of rats and mice used in the United States each year, it is clear that a great deal of legislation does not cover these animals, much to the criticism of animal welfare groups. The actual care of the animals and their subsequent use in experiments, however, is mostly monitored and handled by Institutional Animal Care and Use Committees.

13.3.4 Role of Animal Welfare Groups

It is perhaps in the United States that animal welfare groups are the most prominent and play the largest roles. In particular, members of PETA sometimes attempt to obtain "hidden" videos of abuse within government, private, or university laboratories in the hopes of exposing cruelty to animals and making an example of the offending research organization and scientists. PETA is also known to target researchers by threatening them or actively holding protests in front of their property. Indeed, in some instances, their actions have resulted in researchers withdrawing from their research. Although PETA targets animal testing across numerous species, they do tend to focus on the use of non-human primates as well as animals such as rabbits that are used for cosmetics and toxicology testing. The United States is competitive in terms of scientific advancement, which means that research into drugs and medical knowledge will continue to rise. The role of animal testing is a large one, but with aggressive groups like PETA that constantly challenge the laws and regulations within the country, it is anticipated that pressure to develop better alternatives to animal testing will prevail.

13.3.5 Future of Animal Testing

It is likely that animal testing will decrease as alternatives continue to be developed and regulations similarly continue to moderate and hold accountable those who conduct animal testing. At the same time, issues such as outsourcing to countries where animal testing is poorly regulated may continue to be a problem—perhaps even increasing, until appropriate measures are created to hold companies responsible. Animal rights activists are also not, in all likelihood, going to stop their quest to ban animal testing or increase the use of alternative methods. In an ideal world, there would be a perfect substitute for animals in experimental procedures, but for now, the scientific community overwhelmingly supports its use and the debate over animal testing will quite probably persist for many years.

13.3.6 Decide for Yourself

While there are numerous pros and cons of animal testing, the ethical aspect overshadows both of them, which means that emotion may be the ultimate determining factor in whether a person believes the benefits of animal testing outweigh the problems associated with the practice. Animal testing occurs frequently in the United States, where millions of animals are used each year for experimentation. Regulations in the United States, as in the United Kingdom, focus on monitoring and regulating the use of vertebrates rather than invertebrates. Animal testing is riddled with debate and misconceptions, both on the side of animal testing supporters and on the side of those who are against it. Aim to learn more about the facts of animal testing so that you can make an informed decision regarding its benefit and validity in your life.

13.4 USE OF HUMANS AS EXPERIMENTAL MODELS

Biotechnology or pharmaceutical companies test their drug molecules in both animals and humans. The testing phase in animals is called the *preclinical phase,* whereas testing in humans is called the *clinical phase*. Clinical trials were first introduced in Avicenna's *The Canon of Medicine* in AD 1025, in which he laid down rules for the experimental use and testing of drugs and wrote a precise guide for practical experimentation in the process of discovering and proving the effectiveness of medical drugs and substances. Avicenna laid out several rules and principles for testing the effectiveness of new drugs and medications, which still form the basis of modern clinical trials. One of the most famous clinical trials was James Lind's demonstration in 1747 that citrus fruits cure scurvy. He compared the effects of various different acidic substances, ranging from vinegar to cider, on groups of afflicted sailors, and found that the group who were given oranges and lemons had largely recovered from scurvy after 6 days. Frederick Akbar Mahomed (d. 1884), who worked at Guy's Hospital in London, made substantial contributions to the process of clinical trials during his detailed clinical studies, where he "separated chronic nephritis with secondary hypertension from what we now term essential hypertension." He also founded the Collective Investigation Record for the British Medical Association. This organization collected data from physicians practicing outside the hospital setting and was the precursor of modern collaborative clinical trials.

In planning a clinical trial, the sponsor or investigator first identifies the medication or device to be tested. Usually, one or more pilot experiments are conducted to gain insights for the design of the clinical trial to follow. In the United States, the elderly comprise only 14% of the population, but they consume over one-third of pharmaceutical drugs that are manufactured. Despite this, they are often excluded from trials because their more frequent health issues and drug use produces more messy data. Women, children, and people with common medical conditions are also frequently excluded. In coordination with a panel of expert investigators, the sponsor decides what to compare the new agent with (one or more existing treatments or a placebo) and what kind of patients might benefit from the medication/device. If the sponsor cannot obtain enough patients with this specific disease or condition at one location, then investigators at other locations who can obtain the same kind of patients to receive the treatment would be recruited into the study. During the clinical trial, the investigators recruit patients with the predetermined characteristics, administer the treatment(s), and collect data on the patients' health for a defined time period. These data include measurements like vital signs, amount of the study drug in the blood, and whether or not the patient's health improves. The researchers send the data to the trial sponsor, who then analyzes the pooled data using statistical tests.

Note that while most clinical trials compare two medications or devices, some trials compare three or four medications, doses of medications, or devices against each other.

Except for very small trials limited to a single location, the clinical trial design and objectives are written into a document called a *clinical trial protocol*. The protocol is the "operating manual" for the clinical trial, and ensures that researchers in different locations all perform the trial in the same way on patients with the same characteristics. (This uniformity is designed to allow the data to be pooled.) A protocol is always used in multicenter trials. Because the clinical trial is designed to test hypotheses and rigorously monitor and assess what happens, clinical trials can be seen as the application of the scientific method to understanding human or animal biology.

Beginning in the 1980s, harmonization of clinical trial protocols was shown as feasible across countries of the European Union. At the same time, coordination between Europe, Japan, and the United States led to a joint regulatory–industry initiative on international harmonization, named after 1990 as the International Conference on Harmonization (ICH). Currently, most clinical trial programs follow ICH guidelines, aimed at "ensuring that good quality, safe and effective medicines are developed and registered in the most efficient and cost-effective manner. These activities are pursued in the interest of the consumer and public health, to prevent unnecessary duplication of clinical trials in humans, and to minimize the use of animal testing without compromising the regulatory obligations of safety and effectiveness."

13.4.1 Bioethical Implications

Clinical trials are closely supervised by the appropriate regulatory authorities. All studies that involve a medical or therapeutic intervention on patients must be approved by a supervising ethics committee before permission is granted to run the trial. The local ethics committee has discretion on how it will supervise non-interventional studies (observational studies or those using already collected data). In the United States, this body is called the Institutional Review Board (IRB). Most IRBs are located at the local investigator's hospital or institution, but some sponsors allow the use of a central (independent/for profit) IRB for investigators who work at smaller institutions. To be ethical, researchers must obtain the full and informed consent of participating human subjects. (One of the IRB's main functions is ensuring that potential patients are adequately informed about the clinical trial.) If the patient is unable to give consent for himself or herself, researchers can seek consent from the patient's legally authorized representative. In California, the state has prioritized the individuals who can serve as the legally authorized representative. In some U.S. locations, the local IRB must certify researchers and their staff before they can conduct clinical trials. They must understand the federal patient privacy law as promulgated by the Health Insurance Portability and Accountability Act (HIPAA) and the principles of good clinical practice. The International Conference of Harmonization Guidelines for Good Clinical Practice (ICH GCP) is a set of standards used internationally for the conduct of clinical trials. The guidelines aim to ensure that the "rights, safety and well being of trial subjects are protected." The notion of informed consent by participating human subjects exists in many countries all over the world, but its precise definition may still vary. Informed consent is clearly a necessary condition for ethical conduct, but does not ensure ethical conduct. The final objective is to serve the community of patients or future patients in the best possible and most responsible way.

13.5 XENOTRANSPLANTATION

The controversial concept of cross-species transplantation, also known as *xenotransplantation*, has existed in myths and science fiction stories for a long time. Today, the lack of human organ donors has prompted an intense research effort throughout the medical community in the potential for animal organ transplants. Building on the overwhelming success of human-to-human transplantation, xenotransplantation aims to reduce the demand–supply gap for organs. In this section, we explore whether or not an individual who might benefit from xenotransplantation should be denied these benefits due to the ethical implications that accompany this practice on a mass scale. There are four components of xenotransplantation: (1) the individual receiving the xenograft, (2) the society which must incur the benefits and damages of decisions regarding xenotransplantation, (3) the long-term effect on humanity, and (4) the concern for animals used for humans' benefits.

There is no positive flip side to this danger of xenotransplantation, according to the experts in the field. Recently, a world-renowned primate virologist named Jonathan Allan from the Southwest Foundation for Biomedical Research in San Antonio addressed the possibility of viruses being able to jump the species gap and transfer to human beings. He declared that "primates are known to carry undiscovered viruses. In 1994, eight or nine baboons at Southwest were infected with a mysterious nervous-system virus which was later identified as a new virus." Adopting a precautionary approach could turn out to be ethical in the worst-case scenario. The consequentialist examination of society's recourse in the event that a new virus is discovered probes the issue of our morality if we isolate the infected human recipients and treat them as specimens until the mode of transmission of the diseases is understood. The widespread objection to xenotransplantation is issued from utilitarian ethical theory, which was embraced by the Nuttfield Council on Bioethics. They reemphasized that it is not ethical for an individual to affect negatively the whole human population as a result of a singular decision: "the risks associated with the possible transmission of infectious diseases as a consequence of xenotransplantation have not been adequately dealt with. It would be unethical, therefore, to use xenografts involving human beings." While new epidemics remain, for the present time, only caveats to animal organ use, a very real consideration is the macro-allocation of resources in healthcare which will affect the society considering human-to-animal transplantation. The latter is just a small part of the medical realm, and yet it is certainly a very expensive field due to the complex manipulations necessary for a successful xenotransplantation. The ethical dilemma arises because our global economic system is built on limited resources juxtaposed with unlimited wants. The idea of depriving an individual suffering from organ malfunction of care because of society's interests in treating other diseases is not unique to xenotransplantation. Having said that, the appropriateness and ethicality of society's decision is extremely relevant to whether or not an individual who may survive a xenograft is going to have this option in a hospital. If we do decide to spend money on this field, then we should be quick to examine the benefits lost through the investment into other domains (e.g., cancer treatment) in order to make an ethical decision. Even if we attempt to disregard the economic burden of xenotransplantation as well as the potential for radiation of new diseases, human society will have to come to terms with what it means to be human in the long term.

Many will object to this characterization of humanity because they view appearance as an integral part of being human. Opponents of the perseverance in this domain often use the expression "playing God" to refer to the unethical use of animals to save human beings. Their most meaningful objection is based on the "unnaturalness" of such practices. It is indeed very difficult to define what is natural and to be able to distinguish it from what is unnatural without running into some difficulties. An article in the *Journal of Medical Ethics* states that "on one account, everything that humans do is by definition unnatural, because it constitutes an interference with the non-human natural order. On another account, nothing that humans do is unnatural, since humans are themselves a part of nature" (Hugues 1998). Another relevant point which deters from the "naturalness" argument is the fact that any alterations made to the patient are somatic (i.e., they will not be passed on to any offspring) and temporary, since the transplanted organ can at any time be removed or replaced with a human organ. Looking at xenotransplantation as a temporary transition phase that will help many patients survive until a better solution is uncovered is a rational method of evaluating the ethical dimensions of cross-species transplantation without becoming overwhelmed with horrifying images originating from science fiction.

On the other hand, our reckless use of animals may cause mayhem in the "natural order" in its effect on the animals, and this should lead us to question the morality of doing so. A hastened assessment will direct us to a false conclusion that may be phrased in the following way: since it is ethical to use animals for consumption, then their use as organ donors should be even more acceptable. Indeed, using animals for organs may well be looked upon as a more efficient use of animals, which will lead to an increase in their social value and worth. In addition, justifying the use of non-humans for organs as being ethical simply because of our membership into the human species is misguided and superficial. In order to make animal use for saving a human being's life ethical, one has to justify this use based on the superiority of humans over animals in a certain aspect. Superior brain and mental capacities are obvious candidates, but that excludes mentally retarded individuals from the batch.

First, in order to prevent disease acquisition in the donor animals, increased suffering on the part of the animal of choice (e.g., pigs) in the form of isolated development and rigorous viral and bacterial tests is inevitable. More important, these specialized farms will probably make use of the complex genetic knowledge that had been acquired in the last decade to modify these animals so that the compatibility of their organs to those of humans will become greater. The increased resemblance of the animals' organs to those of humans will make it more and more unethical to use them simply for organs. Even more problematic is the increased compatibility of the genetic makeup of humans and genetically altered animals, which will transform the boundary between humans and non-humans into a chaotic blur.

13.6 GENETIC SCREENING

Despite having already succeeded in mapping and identifying the whole genetic sequence of humans, scientists do not yet know the exact function of most genes. A gene contains information on hereditary characteristics such as hair color, eye color, and height, as well as susceptibility to certain diseases. The term *embryo* refers to the first 6 weeks of development after a sperm fertilizes an egg. This can happen naturally, but as one in six Irish couples experience fertility problems, some seek medical assistance in the form of *in vitro* fertilization (IVF), during which fertilization occurs outside of the mother's body, with the embryos then being placed in her womb. While it is unknown to what extent genes can influence somebody's skills, psychology, or overall health, it is possible to carry out tests on embryos created during IVF (before placement in the womb) in order to detect severe genetic defects. This technique is known as *preimplantation genetic diagnosis* (PGD). Concerns have been raised that this technique will result in a modern form of eugenics where people with certain disabilities will be screened out of existence, while others argue that PGD is a way of reducing pain and suffering. PGD is also used to help treat ill children. For example, in the United States, the parents of a girl named Molly Nash, who was born with a rare incurable disease called Fanconi's anemia, underwent a controversial procedure to conceive another child who could help cure Molly. Embryos created using IVF were tested for signs of the gene responsible for the disease. Embryos that did not carry the disease underwent further testing to find those that would be a tissue match for Molly. As a result, a boy, Adam, was born, who was free of the disease and who could donate cells from his umbilical cord to treat his sister. Many people argue that there are a number of ethical problems with this "savoir sibling" technology. The question that arises is whether Adam would have been born at all had Molly been a healthy child or whether he was born only to save her life. Adam and Molly's parents refute this and say that they would have had another child anyway, but that this technology ensured Adam would be healthy and that Molly's life would be saved. Critics also express concern over whether children like Adam will constantly be called upon to donate tissue or maybe even organs (e.g., kidneys) to their sick siblings. Concerns are also raised about the possibility of "savior siblings" viewing their birth merely as a means to an end and seeing themselves as an instrument for curing disease. Another argument against selecting "savior siblings" is that a number of healthy embryos which are not exact tissue matches will be discarded.

13.7 BIOMETRICS

Biometrics refers to the technique of identifying people using unique physical characteristics such as fingerprints, the iris of the eye, voice pattern, or facial pattern. Biometric data can serve many purposes, including civil and criminal identification, surveillance and screening, healthcare, e-commerce, and e-government. Given the post-9/11 political and social anxieties, biometric data is being used for security purposes as a means of identifying citizens. Indeed, biometrics is increasingly being encountered in air travel as passengers are required to undergo fingerprint and iris scans before being allowed to travel to certain destinations. Under the U.S. Visa Waiver Program, participating countries have to begin using biometric passports. In 2005, Irish citizens made approximately 500,000 visits to the United States. Therefore, the department of Foreign Affairs launched the new e-passports, which have an inbuilt microchip containing biographical information as well as a digital photograph of the holder. While the use of biometric data promises to augment security and reduce the occurrence of specific crimes such as identity theft and the abuse of social funds, it evokes a range of social, legal, and ethical concerns. In particular, it raises questions about the effect it will have on the concept of identity and whether its use will result in discrimination against particular individuals or communities. As with DNA, there are fears that personal information contained in e-passports might be accessed by organizations not involved in security, such as insurance companies and employers, which might lead to discrimination against individuals. There are also concerns about the reliability of biometrics and whether people will be at risk of having their personal information copied. At a recent hacker's conference, a German security consultant showed he could clone data contained in e-passports similar to those introduced. There are also concerns that biometrics will allow governments to intrude into citizens' privacy and that people might be monitored without their knowledge or consent.

13.8 DNA FINGERPRINTING

The discovery of the polymerase chain reaction (PCR) technique made it possible to map the entire human genome and has tremendous diagnostic potentials, especially in the fields of forensic science and individual identification. Over the past decades, DNA fingerprinting has made an invaluable contribution to crime detection and crime prevention. The UK DNA database provides many positive statistics. For instance, it is estimated that in a typical month, the database links suspects to 15 murders, 31 rapes, and 770 car crimes. However, numerous ethical questions arise with respect to who should have samples taken, for what crimes samples should be taken, how long samples should be retained on file, and who shall have access to stored genetic information. Under what circumstances should people have their DNA taken is unclear—that is, should a database include samples from people who are cautioned, arrested, or charged but acquitted, or only those who are convicted of a crime? Should samples be taken only in the case of serious crimes such as murder or assault, or should samples also be taken for lesser crimes such as driving and public order offenses? There are those who argue that only people guilty of criminal behavior would fear having their profile stored on a database and that everybody should be willing to provide a sample. However, 32% of black men in England and Wales are DNA profiled, as opposed to only 8% of white men, which has led to concerns about discrimination. Concerns are also raised about how long DNA should be stored in a forensic database. There is no doubt that scene-of-crime samples should be retained, in case an individual convicted of an offense alleges that a miscarriage of justice has occurred. In fact, DNA evidence was used to vindicate a number of convicted individuals who were on death row in the United States. However, what of an individual's sample? For instance, should it be destroyed once a convicted person has completed his or her sentence or should it be retained indefinitely? Some say the retention of DNA samples will greatly benefit society, while others fear that even those who have repaid their debt to society would be continually under suspicion and discriminated against. DNA can provide sensitive information about individuals and their families, such as their susceptibility to a genetic disorder like Huntington's disease. This increases the potential for genetic discrimination by government, insurance companies, employers, financial institutions, and other organizations. For instance, were genetic information about a person to be given to an insurance company, that person might be prohibited from taking out life cover and in turn be prevented from acquiring a mortgage.

13.9 ORGAN DONATION AND TRANSPLANTATION

There is currently a severe global shortage in the number of donor organs available for transplantation. For example, despite Ireland having one of the highest rates of organ donation (per million population) in the world, the numbers of people on transplant waiting lists there are increasing. Presently, Ireland operates an opt-in system where people are asked to sign an organ donor card. However, in order to reduce transplant waiting lists, other options are being explored, each with their own social, ethical, and economic implications. Traditionally, human organs used in transplants were acquired from brain dead donors (i.e., where permanent stoppage of all brain activity occurs). Now, the expansion of posthumous donation criteria to include cardiac death (i.e., where stoppage of heart and lung function occurs) is being contemplated, as is the establishment of a live donor program where organs such as kidneys are donated by a relative. Another system being considered is that of "presumed consent," whereby if someone does not expressly opt-out (e.g., via their driving license form), they are presumed to be willing donors. Some international commentators have gone so far as to call for automatic posthumous donation, similar to other mandatory civil obligations like jury service or compulsory military service, where individual consent is not required. Due to global shortage of donors, people have been using more sinister methods to procure organs for donation. In recent years, a black market in organs has arisen, where people from developed nations in desperate need of a transplant travel to developing countries and pay tens of thousands of dollars to receive a life-saving transplant. There are grave concerns that the commercialization of organ donation has led to the coercion and exploitation of the economically disadvantaged. People have also raised concerns regarding the practice in the Philippines, where prisoners are offered the chance to have their death sentences commuted to life imprisonment in return for an organ, and in China, where organs are procured from executed prisoners.

13.10 EUTHANASIA

Euthanasia is the intentional and painless taking of the life of another person, by act or omission, for compassionate motives. The word *euthanasia* is derived from the ancient Greek language and can be literally interpreted as "good death." Despite its etymology, the question of whether or not euthanasia is, in fact, a "good death" is highly controversial. Correct terminology in debates about euthanasia is crucial. Euthanasia may be performed by act or omission, either by administering a legal drug or by withdrawing basic healthcare which normally sustains life (such as air, food, water, or antibiotics). The term euthanasia mostly refers to the taking of the life of a person on his or her request, a *voluntary euthanasia*. On the other hand, the taking of a person's life against that person's expressed wish/direction is called *involuntary euthanasia*.

Central to the discussion on euthanasia is the notion of intention. While death may be caused by an action or omission of medical staff during treatment in hospital, euthanasia only occurs if death was intended. For example, if a doctor provides a dying patient extra morphine with the intention of relieving pain but knowing that his actions may hasten death, he has not performed euthanasia unless his intention was to cause death (the principle of double effect). Euthanasia may be distinguished from a practice called *physician-assisted suicide*, which occurs when death is brought about by the person's own hand by means provided to him or her by another person. All practices of euthanasia and physician-assisted suicide are illegal in Australia.

13.11 NEUROETHICS

Neuroethics is a field of inquiry that is very broad in scope and is closely related to both cognitive neuroscience and bioethics, though it is now formally recognized as a discipline in its own right. Neuroethics can be roughly divided into two streams. One stream concerns the more direct or proximal implications of cognitive neuroscience, which can be referred to as the "ethics of neuroscience." It deals with the ethical implications of neuroscientific knowledge and technology, such as enhancing neurological function through novel neuropharmacological, neurostimulation, and neurogenetic engineering techniques. The implications of brain imaging technology, which is now commonly used in both research and medical practice, raises issues concerning mental privacy, diagnostics, and predicting behavior. Furthermore, knowledge gained through neuroscience, along with brain imaging technology, may one day allow us to probe the human mind to even observe a person's thoughts and predilections. The second stream of neuroethics can be referred to as the "neuroscience of ethics." This stream of neuroethics lies at the border between philosophy, metaethics, and normative ethics; one of the central issues concerns moral agency. How do we impute moral responsibility given that cognitive neuroscience may shed new light on the way humans make their decisions, as well as the nature of our underlying motivations to act in certain ways? How can we trust our moral beliefs if it turns out that one's belief was not the product of rational contemplation but a post hoc rationalization of an emotive judgment, an attitude of disapprobation, or a pre-reflective moral intuition that is distinct, impenetrable, and encapsulated from rational contemplation?

13.12 ASSISTED REPRODUCTIVE TECHNOLOGY

Assisted reproductive technology (ART) is a medical intervention developed to improve an "infertile" couple's chance of pregnancy. "Infertility" is clinically accepted as the inability to conceive after 12 months of actively trying to conceive. The means of ART involves separating procreation from sexual intercourse. The importance of this association is addressed in bioethics. Some techniques used in clinical ART include artificial insemination, IVF, gamete intrafallopian transfer (GIFT), gestational surrogate mothering, gamete donation, sex selection, and preimplantation genetic diagnosis. Issues addressed in bioethics are the appropriate use of these technologies and the techniques employed to carry out procedures for quality and ethical reviews. Assisted reproductive technology (ART) and its use directly impact the foundational unit of society—the family. ART enables children to be conceived who have no genetic relationship to one or both of their parents. Children can also be conceived who will never have a social relationship with one or both of their genetic parents, such as a child conceived using donor sperm. Non-infertile people in today's society—including both male and female homosexual couples, single men and women, and postmenopausal women—are seeking the assistance of ART. Concerns in all situations include the child and his or her welfare, including the right to have one biological mother and father. The fragmented family created by ART can disconnect genetic, gestational, and social child–parent relationships which have typically been one and the same in the traditional nuclear family. Other important bioethical issues include the appropriate use of PGD screening, use, and storage; destruction of excess IVF embryos; and research involving embryos. ART research requires human participants, donors and donated embryos, oocytes and sperm. Ethical issues that arise in ART research surround the creation and destruction of embryos. One approach in bioethics involves preserving justice, beneficence, non-malfeasance, and the autonomous interests of all involved. Bioethicists contribute to ethical guidelines and moral evaluations of new technologies and techniques in ART as well as to public discourse that lead to the development of national regulations and restrictions of unacceptable practices.

13.13 EMBRYONIC STEM CELL RESEARCH

Stem cell research is a rapidly evolving technology and has the potential to offer a wide range of medical benefits to patients. However, stem cell research—in particular, research involving embryos, which results in the destruction of the embryos—raises many ethical issues. The wide spectrum of opinions on the moral status of the embryo ranges from viewing it as a ball of cells, as a developing human being, or as a person with the same rights as someone who has already been born. There are numerous examples of how divisive the debate on the status of the embryo is, including cases where couples have disputed what should happen to their frozen embryos and where widows have asked to have embryos which were created while their husbands were alive placed in their wombs. Supernumerary embryos (embryos that are not placed in the mother's womb during IVF) are usually either frozen for future use or destroyed about 5 years after creation. Some people argue that harvesting stem cells from these embryos is permissible, given that they will be destroyed regardless. Indeed, they argue that there is a moral imperative to use these supernumerary embryos for research into fatal and debilitating diseases suffered by those already born. However, those who afford higher moral status to the embryo reject this argument and compare research on supernumerary embryos to performing research on terminally ill people. They also argue that allowing embryos to die is not equivalent to actively terminating them. The creation of embryos specifically for research may be problematic even for people in favor of stem cell research involving supernumerary embryos. Those who confer significant moral status on the embryo argue that their creation specifically for research represents disrespect for human life, because creating an embryo that is never intended for transfer to the womb treats the embryo merely as a means to an end. Others fear that allowing embryos to be created for research would set science on a "slippery slope" to reproductive cloning. The creation of embryos for research purposes would also require the donation of a large number of eggs, and there are fears that vulnerable women might be induced into donating their eggs for financial gain.

Ever since embryonic stem cell research came into existence, it has not only generated tremendous hopes for millions of patients who are suffering from cancer, Parkinson's diseases, and diabetes, but at the same time it has raised serious issues pertaining to how embryonic stem cells are being collected after killing human embryos. Debates over the ethics of embryonic stem cell research continue to divide scientists, politicians, and religious groups. However, promising developments in other areas of stem cell research might lead to solutions that bypass these ethical issues. These new developments tilt the pros and cons scale toward the former, since the main concern of those against this research seems to be the destruction of human blastocysts, considered by pro-life advocates to be an immoral act of disregard for human life. It all started in November 1998, with the publication of a research paper which reported that stem cells could be isolated from human living embryos. Subsequent research led to the ability to maintain undifferentiated stem cell lines (pluripotent cells) and techniques for differentiating them into cells specific to various tissues and organs. The debates over the ethics of stem cell research began almost immediately, in 1999, despite reports that stem cells cannot grow into complete organisms. In 2000–2001, governments worldwide were beginning to draft proposals and guidelines in an effort to control stem cell research and the handling of embryonic tissues, and to reach universal policies to prevent "brain-drain" (emigration of top scientists in the field) between countries. The Canadian Institute of Health Research (CIHR) drafted a list of recommendations for stem cell research in 2001. The Clinton administration drafted guidelines for stem cell research in 2000, but Clinton left office prior to their release. The Bush government had to deal with the issue throughout his administration. Australia, Germany, the United Kingdom, and other countries have also formulated their policies on this matter.

13.13.1 In Favor of Embryonic Stem Cell Research

The excitement about stem cell research is primarily due to the medical benefits in areas of regenerative medicine and therapeutic cloning. Stem cells provide huge potential for finding treatments and cures for a vast array of diseases, including different cancers, diabetes, spinal cord injuries, Alzheimer's disease, multiple sclerosis, Huntington's disease, Parkinson's disease, and more. There is endless potential for scientists to learn about human growth and cell development from studying stem cells. Use of adult-derived stem cells from blood, skin, and other tissues has been demonstrated to be effective for treating different diseases in animal models. Umbilical cord-derived stem cells have also been isolated and utilized for various experimental treatments. Although these cell lines derived from adult stem cells trigger no ethical issues, there are some disadvantages or shortcomings compared to embryonic cell lines, because they are difficult to isolate, poor in quantity, and unsuitable for long-term expansion.

13.13.2 Against Embryonic Stem Cell Research

The use of embryonic stem cells involves the destruction of human blastocysts. Many people believe that it is in the blastocyst that life begins, and they thus have the notion that it is a human life, which it is unacceptable and immoral to destroy. This seems to be the only controversial issue standing in the way of stem cell research in North America and elsewhere. In the summer of 2006, U.S. President George W. Bush stood his ground on the issue of stem cell research and vetoed a bill passed by the Senate that would have expanded federal funding of embryonic stem cell research. Currently, American federal funding can only extend to research on stem cells from existing (already destroyed) embryos. Similarly, in Canada, as of 2002, scientists cannot create or clone embryos for research. They can only use existing embryos discarded by couples. On the other hand, the United Kingdom allows embryonic stem cell cloning. The use of stem cell lines from alternative non-embryonic sources has received more attention in recent years and has already been demonstrated as a successful option for treatment of certain diseases. For example, adult stem cells can be used to replace blood-cell-forming cells killed during chemotherapy in bone marrow transplant patients. Biotech companies such as Revivicor and ACT are researching techniques for cellular reprogramming of adult cells, use of amniotic fluid, or stem cell extraction techniques that do not damage the embryo, and that also provide alternatives for obtaining viable stem cell lines. Of necessity, the research on these alternatives is catching up with embryonic stem cell research and, with sufficient funding, other solutions might be found that are acceptable to everyone.

13.13.3 Current Status

On March 9, 2009, U.S. President Barack Obama overturned Bush's ruling, allowing U.S. federal funding to go to embryonic stem cell research. Despite the progress being made in other areas of stem cell research, such as using pluripotent cells from other sources, many American scientists were putting pressure on the government to allow their participation and compete with the Europeans. However, many people are still strongly opposed.

13.14 HUMAN CLONING

Cloning is a technology to create genetically identical organisms (such as bacteria, insects or plants) asexually. In biotechnology, cloning refers to processes used to create copies of DNA fragments (molecular cloning), cells (cell cloning), or organisms. The ethics of cloning refers to a variety of ethical positions regarding the practice and possibilities of cloning, especially human cloning. While many of these views are religious in origin, the questions raised by cloning are faced with secular perspectives as well. As the science of cloning continues to advance, governments have dealt with ethical questions through legislation.

Advocates of human therapeutic cloning believe the practice could provide genetically identical cells for regenerative medicine, and tissues and organs for transplantation. Such cells, tissues, and organs would neither trigger an immune response nor require the use of immunosuppressive drugs. Both basic research and therapeutic development for serious diseases such as cancer, heart disease, and diabetes, as well as improvements in burn treatment and reconstructive and cosmetic surgery, are areas that might benefit from such new technology. One bioethicist, Jacob M. Appel of New York University, has gone so far as to argue that "children cloned for therapeutic purposes" such as "to donate bone marrow to a sibling with leukemia" may someday be viewed as heroes. Proponents claim that human reproductive cloning also would produce benefits. Severino Antinori and Panos Zavos hope to create a fertility treatment that allows parents who are both infertile to have children with at least some of their DNA in their offspring. Some scientists, including Dr. Richard Seed, suggest that human cloning might obviate the human aging process. How this could work is not entirely clear, since the brain or identity would have to be transferred to a cloned body. Dr. Preston Estep has suggested the terms "replacement cloning" to describe the generation of a clone of a previously living person and "persistence cloning" to describe the production of a cloned body for the purpose of obviating aging, although he maintains that such procedures currently should be considered science fiction and current cloning techniques risk producing a prematurely aged child. In Aubrey de Grey's proposed Strategies for Engineered Negligible Senescence (SENS), one of the considered options to repair cell depletion related to cellular senescence is to grow replacement tissues from stem cells harvested from a cloned embryo.

At present, the main non-religious objection to human cloning is that cloned individuals are often biologically damaged due to the inherent unreliability of their origin. For example, researchers are currently unable to safely and reliably clone non-human primates. Bioethicist Thomas Murray of the Hastings Center argues that "it is absolutely inevitable that groups are going to try to clone a human being. But they are going to create a lot of dead and dying babies along the way." UNESCO's Universal Declaration on Human Genome and Human Rights asserts that cloning contradicts human nature and dignity. Cloning is an asexual reproductive mode that could distort generation lines and family relationships and limit genetic differentiation, which ensures that human life is largely unique. Cloning can also imply an instrumental attitude toward humans, which risks turning them into manufactured objects and interferes with evolution, the implications of which we lack the insight or prescience to predict. Furthermore, proponents of animal rights argue that non-human animals possess certain moral rights as living entities and should therefore be afforded the same ethical considerations as human beings. This would negate the exploitation of animals in scientific research on cloning, cloning used in food production, or as other resources for human use or consumption. Rudolph Jaenisch, a professor at Harvard, has pointed out that we have become more efficient at producing clones which are still defective. Other arguments against cloning come from various religious orders (believing cloning violates God's will or the natural order of life) and a general discomfort some have with the idea of "meddling" with the creation and basic function of life. This unease often manifests itself in contemporary novels, movies, and popular culture, as it did with numerous prior scientific discoveries and inventions. Various fictional scenarios portray clones being unhappy, soulless, or unable to integrate into society. Furthermore, clones are often depicted not as unique individuals but as "spare parts," providing organs for the clone's original (or any non-clone that requires replacement organs). Other ethical (and legal) concerns surround the concept of "identity." Since both the "original" and the "copy" are genetically the same person, which one is legally the "real" individual? Moreover, since, for all practical purposes, they are both "the same person," how are criminal actions prosecuted when one individual is indistinguishable from another? On December 28, 2006, the U.S. Food and Drug Administration (FDA) approved eating meat from cloned animals; it was said to be virtually indistinguishable from that of non-cloned animals. Furthermore, companies would not be required to provide labels informing the consumer that the meat comes from a cloned animal. In the year 2007, the FDA announced plans for a national database system to track all cloned animals as they move through the food chain, so that food sources can be clearly labeled.

QUESTIONS

Section A: Descriptive Type

Q1. Define bioethics.
Q2. Describe ethical issues associated with GM foods.
Q3. Explain why animal testing is important for research.
Q4. Discuss some major advantages of animal testing.

Q5. What are the different kinds of ethical issues associated with human clinical trials?
Q6. What is xenotransplantation? Discuss the ethical issues associated with it.
Q7. Define biometrics and cite suitable examples.
Q8. Describe organ donation and transplantation, with examples.

Section B: Multiple Choice

Q1. The safety of GM foods is supervised by the ...
   a. International Council of Science
   b. National Science Foundation
   c. National Institute of Health
Q2. Clinical trials were first introduced by ...
   a. Avicenna (Ibne Sina)
   b. Alexander Fleming
   c. Robert Koch
Q3. Which of the following does not come under the ethical scanner?
   a. Human embryonic stem cell
   b. Human cloning
   c. Xenotransplantation
   d. Antibiotics
Q4. Biometrics refers to the technique of identifying people using unique physical properties. True/False

Section C: Critical Thinking

Q1. How does DNA fingerprinting violate human rights?
Q2. Do you think it is ethical to create a human embryo by ART? Why or why not?

ASSIGNMENTS

The following lists some questions raised, based on bioethics. Discuss these issues with your course instructor and submit your answers in the form of assignment. The information may be accessed through the Internet.

- *Research issues*: Should scientists be held to some standard of integrity and honesty? Who should enforce this? Why is peer review so important? Should scientists be held responsible for creating (or discovering) technology that can be used to harm others or have unforeseen side effects (chemical, nuclear warfare)?
- *Reproductive technologies*: *In vitro* fertilization, surrogacy, RU-486, preimplantation embryo screening, cloning: Is there a significant difference between cloning sheep for pharmaceutical production and cloning humans?
- *Human genome project*: Should employers be able to screen job applicants for specific genetic conditions? Who should have access to this information: Family members, lawyers, insurance agencies?
- *Gene therapy*: What are the potential ramifications of somatic and germ-line gene therapy? Should genes be tinkered with? If so, what limits should be placed on this type of technology?
- *Fetal rights*: Does a fetus have rights? If so, what rights does a fetus have and who is responsible for representing its interests? Does a fetus have rights that supersede the mother's? Can the government step in to ensure the health of a fetus if the mother is not willing to do so? What about embryos?

- *Euthanasia*: What is the right to die? How does withdrawing or withholding treatment differ from physician-assisted suicide? Who has the right to decide when and how a person dies? Should doctors be held legally responsible if they assist a patient's death? What laws should be passed to protect doctors and patients?
- *Environmental issues*: How do we decide between conservation and economic interests? How much land should be allocated to other species and to parks? Should industries be responsible for damage done to the environment (pollution) by them?
- *Animal rights issues*: Is animal testing acceptable when it benefits humans? Which animals should be tested for these tests and which should not? Is animal research justified by its benefits to mankind?
- *Population control*: Who has the right to decide who should have children (and how many)? What measures should be taken to control the world population?
- *Human research*: Should humans be used for medical and psychological studies? What guidelines should be instated to protect subjects?
- *Minors and medicine*: What medical procedures do minors have available to them without parental consent? Do doctors have an obligation to inform parents of conditions a teen has (such as pregnancy and AIDS) even if the teen does not wish it?
- *GM crops*: What rights do consumers have? What rights do farmers have to grow GM crops? Who decides whether a food is safe?

REFERENCES AND FURTHER READING

A dozen questions (and answers) on human cloning. World Health Organization. http:/www.who.int/ethics/topics/cloning/en/ (retrieved on February 27, 2008).

Andre, J. *Bioethics as Practice*. University of North Carolina Press, Chapel Hill, NC, 2002.

Aulisio, M., Robert, A., and Stuart, Y. *Ethics Consultation: From Theory to Practice*. Johns Hopkins University Press, Baltimore, MD, 2003.

Beauchamp, T. and Childress, J. *Principles of Biomedical Ethics*. Oxford University Press, Oxford, 2001.

Bioethical issues. http://www.bioethics.ie/uploads/docs/IntroBioethics.pdf

Callahan, D. Bioethics as a discipline. *Hastings Cent. Stud.* 1: 66–73, 1973.

Caplan, A.L. *Am I My Brother's Keeper? The Ethical Frontiers of Biomedicine*. Indiana University Press, Bloomington, IN, 1997.

Emanuel, E., Crouch, R., Arras, J. et al. *Ethical and Regulatory Aspects of Clinical Research*. Johns Hopkins University Press, Baltimore, MD, 2003.

FDA Probes Sect's Human Cloning. *Wired News*, 2002. http://www.wired.com

Food Agriculture Organization World Health of the United Nations Organization (FAO/WHO). 2003c. ftp://ftp.fao.org/es/esn/jecfa/jecfa61sc.pdf

Friend, T. The real face of cloning. *USA Today*, 2003.

Glad, J. *Future Human Evolution: Eugenics in the Twenty-First Century*. Hermitage Press, Ewing, NJ, 2008.

Hugues, J. Xenografting: Ethical issues. *J. Med. Ethics* 24: 18–24, 1998.

Jonathan, B. *Against Bioethics*. The MIT Press, Cambridge, MA, 2006.

Jonsen, A.R. *The Birth of Bioethics*. Oxford University Press, Oxford, 1998.

Jonsen, A., Veatch, R., and Walters, L. *Source Book in Bioethics*, Georgetown University Press, Washington, DC, 1998.

Kaufmann, R.E. Clinical trials in children: Problems and pitfalls. *Pediatr. Drugs* 2: 411–418, 2000.

Khushf, T. *Handbook of Bioethics: Taking Stock of the Field from a Philosophical Perspective*, Kluwer Academic Publishers, Dordrecht, the Netherlands, 2004.

Korthals, M. and Bogers, R.J. *Ethics for Life Scientists*. Springer, Dordrecht, the Netherlands, 2004.

Kuczewski, M.G. and Polansky, R. *Bioethics: Ancient Themes in Contemporary Issues*. The MIT Press, Cambridge, MA, 2002.

McGee, G. *Pragmatic Bioethics*. Massachusetts Institute of Technology Press, Cambridge, MA, 2003.

Murphy, T. *Case Studies in Biomedical Research Ethics*. The MIT Press, Cambridge, MA, 2004.

Nuttfield Council on Bioethics. Animal-to-human transplants: The ethics of xenotransplantation. Nuttfield Council on Bioethics, London, U.K., 27, 1996.

Politis, P. Transition from the carrot to the stick: The evolution of pharmaceutical regulations concerning pediatric drug testing. *Widener Law Rev.* 12: 271, 2005.
Reich, W.T. The word "bioethics": Its birth and the legacies of those who shaped its meaning. *Kennedy Inst. Ethics J.* 4: 319–336, 1994.
Rhoden, N.K. Treating baby Doe: The ethics of uncertainty. *Hastings Cent. Rep.* 16: 34–43, 1986.
Rothman, D.J. *Strangers at the Bedside: A History of How Law and Bioethics Transformed Medical Decision Making*, 2nd edn. Aldine de Gruyter, New York, 2003.
Singer, P.A. and Viens, A.M. *Cambridge Textbook of Bioethics.* Cambridge University Press, Cambridge, U.K., 2008.
Stevens, M.L.T. *Bioethics in America: Origins and Cultural Politics.* Johns Hopkins University Press, Baltimore, MD, 2000.
Sugarman, J. and Sulmasy, D. *Methods in Medical Ethics.* Georgetown University Press, Washington, DC, 2001.
Universal Declaration on the Human Genome and Human Rights. UNESCO. 1997. http://portal.unesco.org
Wikler, D. Ethics and rationing: Whether, how, or how much. *J. Am. Geriatr. Soc.* 40: 398–403, 1992.
Aetna Recommends Guidelines for Access to Genetic Testing—Company advocates education and information privacy. www.aetna.com
American Academy of Actuaries—Issue Brief: Genetic Information and Voluntary Life Insurance. www.actuary.org
American Academy of Actuaries—Issue Brief: The Use of Genetic Information in Bridging the Gap between Life Insurer and Consumer in the Genetic Testing Era—Article from the *Indiana Law Journal.* www.ornl.gov
Cancer Genetics—A collection of resources on genetic testing for cancer. From the National Cancer Institute. www.nci.nih.gov
Disability Income and Long-Term Care Insurance. www.actuary.org
Ethical, Legal, and Social Issues of the Human Genome Project. www.ornl.gov
Evaluating Gene Testing—Information on evaluating test quality, the potential usefulness of test results, available preventive or treatment options, and social issues. From the Human Genome Project Information Website. www.ornl.gov
Genes, Dreams, and Reality: The Promises and Risks of the New Genetics—Article from Judicature. www.ornl.gov
Gene Tests—Find reviews of genetic disorders, genetics glossary, clinics, and laboratories conducting gene testing. www.genetests.org
Gene Tests Website. www.genetests.org
Genetic Testing—A collection of news, overviews, and other Web resources from MEDLINEplus. www.nlm.nih.gov
Genetic Testing—An introduction to genetic testing and the different types of tests available. From the Genetics and Public Policy Center. www.dnapolicy.org
Genetic Testing for Inherited Predisposition to Breast Cancer—An article from the National Breast Cancer Coalition website—April 2003. www.natlbcc.org
Genetic Testing of Newborn Infants—An activity for considering government policies to guide a program for genetic screening of newborns. From the Genetic Science Learning Center. http://gslc.genetics.utah.edu
Human Gene Testing 1996—An article on gene testing and science behind genetic disorders from the National Academy Sciences Beyond Discovery series that traces the origins of important recent technological and medical advances. www.beyonddiscovery.com
Human Genetic Testing and DNA Analysis—A collection of Websites and recorded broadcasts of radio programs. From the DNA Files provided by National Public Radio (NPR). www.dnafiles.org
If Genetic Tests Were Available for Diseases Which Could Be Treated or Prevented, Many People Would Have Them. A report on the results of a poll conducted in 2002 by Harris Interactive. www.harrisinteractive.com
Prenatal Testing: A Modern Eugenics?—From the DNA Files provided by National Public Radio (NPR). www.ornl.gov
Resources from the National Conference of State Legislatures.
Understanding Gene Testing. A basic introduction to genes and genetic testing from the U.S. Department of Health and Human Services. www.accessexcellence.org
Understanding Gene Testing. A tutorial that illustrates what genes are, explains how mutations occur and are identified within genes, and discusses the benefits and limitations of gene testing for cancer and other disorders. From the National Cancer Institute. www.cancer.gov
What Can the New Gene Tests Tell Us? Article from the Judges'. *Journal of the American Bar Association.* www.ornl.gov

14 Careers in Biotechnology

LEARNING OBJECTIVES

- Discuss various careers in biotechnology
- Explain how one can get a suitable position in the biotechnology industry
- Discuss the status of job openings in the academic and industrial sectors involving biotechnology
- Explain how jobs in academic institutes are different than those in corporates

14.1 INTRODUCTION

Biotechnology is an emerging field that keeps on expanding with arrays of opportunities and challenges, especially in terms of finding out new methods and better treatment modalities for various human ailments, protecting environmental degradation, reestablishment of extinct species, enhancing crops and food products, and finding out alternative and eco-friendly energy sources. One can imagine the impact of biotechnology on everyday life. It has now become obvious to think about how we can shape our career in various areas of biotechnology.

The biotechnology industry consists of six segments: (1) research and development (R&D); (2) laboratory-level synthesis and testing; (3) industrial-level synthesis/manufacturing; (4) quality control and quality assurance; (5) sales, marketing, and finance; and (6) legal and intellectual property rights (IPRs). In these fields, people with diverse educational qualifications, backgrounds, and skill sets are in great demand. We will discuss each of these six components in greater detail to know about challenging job opportunities.

14.2 EDUCATION AND INVESTMENT IN BIOTECHNOLOGY

There is a dramatic increase in the number of institutes and universities around the world that offer graduate, postgraduate, doctoral, and postdoctoral research in various specialized fields of biotechnology. They provide challenging training and research opportunities in cutting-edge fields such as gene therapy or stem cell technology. In recent years, there has been a great demand for specialized training programs, especially in the field of stem cell biology and its therapeutic potential. We will now list some of the major institutes and universities that offer various degrees and research programs in the field of biotechnology.

North America, which includes the United States and Canada, has the largest number of biotechnology institutes and universities in the world, where thousands of students are pursuing their education in various fields of biotechnology. Universities in Canada offer bachelor's, master's, and doctorate (PhD) programs in biotechnology. In Alberta, the University of Lethbridge offers an undergraduate degree in Agricultural Biotechnology, while the British Columbia University offers associate certificate and Bachelor of Science (BS) degrees in biotechnology. Universities in Ontario, such as York University, the University of Windsor, Carleton University, and Brock University, offer bachelor's, master's, and doctoral degrees. McGill University and Concordia University in Quebec offer a graduate degree in biotechnology, while universities located in Saskatchewan region offers an undergraduate degree in biotechnology management.

European countries have also shown great interest in biotechnology education and research. Some European countries such as Germany, the United Kingdom, and Switzerland have the largest number of biotechnology institutes and universities. Over the past decade, there has been a dramatic increase in the number of biotechnology institutes and universities in Asia (such as in China and India) that have shown interest in both biotechnology education and research activities. In China, most biotechnology education and research is concentrated on agricultural aspects, as life sciences and biotechnology are of great importance to China. The Chinese government pays a great deal of attention to the development of modern life sciences. During the period 1986–1990, the Chinese government implemented a couple of programs that prioritize the development of biotechnology. The programs covered a wide spectrum of basic research programs, high-tech R&D programs, key science and technology problem-solving programs, key science infrastructure programs, key industrial pilot programs, as well as the establishment of key laboratories and engineering centers.

In the Middle East, the field of biotechnology is not growing as rapidly as it is in other Asian countries such as China, India, and Singapore. However, the federal governments of United Arab Emirates, Saudi Arabia, and Qatar have indicated interest in the development of the biotechnology industry. The government of the United Arab Emirates has invested $150 million to create the Dubai Biotechnology and Research Park to attract biotechnology companies and to promote biotechnology and biopharmaceutical product manufacturing and marketing. Efforts have been made to create similar kinds of research parks in Saudi Arabia (Biocity Jeddah) and Qatar (Biocity Bahrain). Until 2003, there was not a single university in the entire Middle East to offer biotechnology education, research, and training. In 2004, Manipal University, Dubai (a branch campus of India's reputed Manipal University), became the first university in the Middle East to offer bachelor's, master's, and doctoral programs in various specialized fields of biotechnology.

The funding from the government of China for R&D has been increasing continuously during the last 20 years. Many R&D centers and bases on biotechnology have been set up, and an increasing number of scientists with PhD degrees are returning home after receiving training in western countries. There are more than 400 universities, research institutes, and companies in China, and a total of over 20,000 scientists and researchers involved in biotechnology. The number of research papers being published internationally is increasing rapidly. The total sales of biotechnological products in China have increased 50-fold during the past 10 years, surging from approximately US$ 1.6 billion in 1997 to more than US$ 2.5 billion in 2000. A total of 18 biopharmaceutical products have been commercialized, including recombinant medicines and vaccines, while thirty more are at the stage of clinical trials. Crop breeding techniques are generating huge economic value in China. The yield of super hybrid rice is as high as 100 kg/ha/day during the rice growing season. China is one of the earliest countries to carry out gene therapy for patients, and Chinese scientists have identified several disease genes, including the neural deafness gene.

With the development of life sciences and bioindustry, the biosafety issue becomes inevitable. Statistics suggest that there are more than 44 million hectares of transgenic plants in the world. The transgenic technique has brought huge benefits to many countries. However, governments, scientists, and the general public of the countries also face the challenge of biosafety and bioethics issues. The Chinese government actively supports R&D in biotechnology and transgenic techniques. They passed the first Bio-Engineering Safety Regulations in 1983. China has been adopting policies to protect patents and IPRs, and has gradually established various laws and regulations regarding patents, trademarks, and copyrights. The Chinese government formally accepted the international consensus on patenting and IPR protection. They have been working hard to create an environment that recognizes patents and protects the interests of patent owners. As a result, the number of patent applications in China has increased significantly during the past few years.

India is not far behind China. Like China, India has an agricultural-based economy and also concentrates on agricultural-based education and research. The success of the biotechnology industry has led to sustained economic growth in developed nations like the United States, Canada, and several European countries. India is also fast emerging as a key player in biotechnology-related activities and investments. Currently, although the biotechnology market in India contributes a mere 2% share in the global market, it is on the threshold of colossal growth in the coming decade.

In 1986, the Department of Biotechnology (DBT) was set up under the Ministry of Science and Technology. This gave a new impetus to the development of biotechnology in India. The DBT has set up many centers of excellence in the country, which are responsible for generating skilled manpower as well as supporting the R&D efforts of private industries. This has promoted interactions between academia and industry that have resulted in several industry houses from the "old industry" and entrepreneur initiatives taking root and developing biotechnology in India. The Indian government has evolved biosafety guidelines and has helped to lay down patent rules. It has also participated in technology transfers and international collaborations. The growth of biotechnology in India has led to an increase in budgetary allocations, which have gone up by tremendously, from Indian Rupees (INR) 404 million in 1987–1988 to INR 1138 million in 1997–1998 and to INR 2356 million in 2002–2003. The centers are also planning to introduce additional venture capital funds in line with the Technology Development Fund (TDF) to promote small and medium biotechnology enterprises.

The Indian government has laid down a decent regulatory framework to approve genetically modified (GM) crops and recombinant DNA (rDNA) products for human consumption. A proactive government policy allows stem cell research in the country, while sound ethical guidelines have been established. The second amendment of the Indian Patents Bill that was recently cleared by the Parliament includes a 20 year patent term, emergency provisions, and commencement of R&D immediately after the filing of patents. The bill is compatible with the provisions of the World Trade Organization (WTO) and the Agreement on Trade-Related Aspects of Intellectual Property Rights (TRIPS). Over the next 5 years, biotechnology can offer opportunities for fresh investment of US$200–300 million in India. This fresh investment could result in a turnover of US$400–500 million during the next 5–7 years. The total allocation for biotechnology is divided into three major areas: agricultural biotechnology (60%), diagnostics and therapeutics (25%), and vaccines (15%). The Indian pharmaceutical market is growing exponentially. In 1997, its value was US$3 billion, and this is expected to rise dramatically in the future. In 2001, the vaccine market in India was US$100 million and growing at 20%, while the diagnostic market was worth US$200 million. Vaccines form the major percentage of products, of which recombinant Hepatitis B vaccine dominates the medical biotechnology segment; others include granulocyte colony-stimulating factor (GCSF), erythropoietin (EPO), and Interferon Alpha 2b. Multinational companies like Monsanto, Pfizer, Unilever, Dupont, Bayer, Eli-Lilly, Ranbaxy, Hoechst, and Chiron–Boerhringer have also set up their businesses in India.

India is the second largest food producer after China, and thus offers a huge market for biotechnology products such as transgenic rice, brassica, moonbean, pigeon pea, cotton, and tomato. Some vegetables such as cabbage and cauliflower are already into field trials. The nutraceuticals market is valued at US$532–638 million. Due to the rising costs of R&D, companies from the United States and Europe are looking at contract research in India. India also offers a suitable population for clinical trials because of its diverse gene pools covering a large number of diseases.

14.3 RESEARCH AND DEVELOPMENT IN BIOTECHNOLOGY

R&D is a critical component of the biotechnology industry and is, in fact, critical for all industrial revolutions. Without research work, one cannot make a new product or make a better product. Research often refers to laboratory-based activity to discover something new, whereas development refers to the exploitation of discoveries. Research involves identification of possible chemical compounds or theoretical mechanisms. The people who usually work in the R&D labs are mostly PhD researchers or scientists, supported by technical staff who may be BS and Master of Science (MS) degree holders. The R&D labs located in universities usually conduct basic research of least commercial relevance, whereas R&D labs in industries mainly concentrate on applied research with mostly commercial relevance. Basic research, also called *fundamental research* (or *pure research*), is carried out to increase the understanding of fundamental principles. Many times, the end results

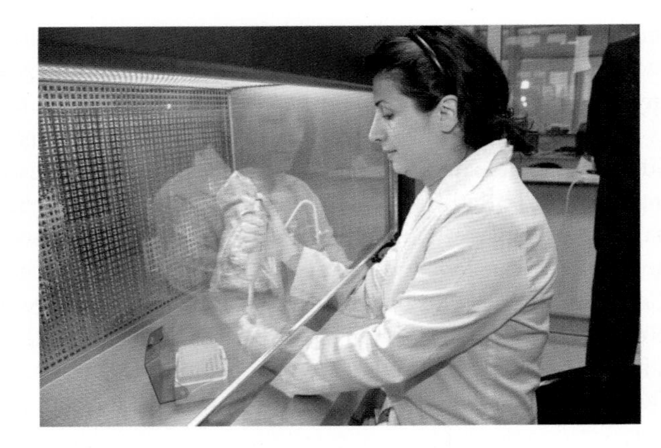

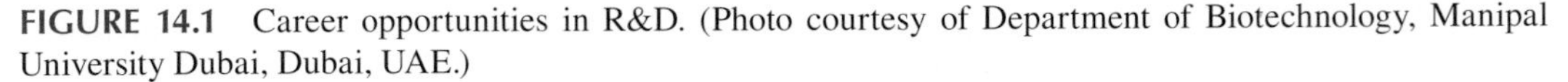

FIGURE 14.1 Career opportunities in R&D. (Photo courtesy of Department of Biotechnology, Manipal University Dubai, Dubai, UAE.)

of basic research have no direct or immediate commercial benefits, but in the long term these results can become the basis for many commercial products and applied research. Basic research is normally conducted by university professors or scientists in their respective fields. Their primary emphasis is to generate data of scientific relevance and publish the result in scientific journals. There are funding agencies available in each country to carry out basic research, and some of the well-known funding agencies are the National Science Foundation (United States), the National Institute of Health (United States), the World Health Organization (WHO, based in Switzerland), the Welcome Trust (United Kingdom), the Department of Biotechnology (India), and some non-profit organizations such as the Bill Gates Foundation and the Rockefeller Foundation in the United States (Figure 14.1).

R&D has a special economic significance apart from its conventional association with scientific and technological development. R&D investment generally reflects a government's or organization's willingness to forgo current operations or profit to improve future performance or returns and its ability to conduct R&D. In 2006, the world's four largest spenders on R&D were the United States (US$343 billion), the European Union (US$231 billion), China (US$136 billion), and Japan (US$130 billion). In terms of percentage of GDP, the order of these spenders for 2006 was China (US$115 billion of US$2,668 billion GDP), Japan, the United States, and the European Union, with approximate percentages of 4.3, 3.2, 2.6, and 1.8, respectively. The top 10 spenders in terms of percentage of GDP were Israel, China, Sweden, Finland, Japan, South Korea, Switzerland, Iceland, the United States, and Germany.

In general, R&D activities are conducted by specialized units or centers belonging to companies, universities, and state agencies. In the context of commerce, "R&D" normally refers to future-oriented, longer-term activities in science or technology, using similar techniques to scientific research, without predetermined outcomes and with broad forecasts of commercial yield. Statistics on organizations devoted to R&D may express the state of an industry, the degree of competition, or the lure of progress. Some common measures include budgets, numbers of patents, or rates of peer-reviewed publications. In the United States, a typical ratio of R&D for an industrial company is approximately 3.5% of revenues. A high-tech company such as a computer manufacturer may spend 7%. Although Allergan, a biotechnology company, tops the spending table at 43.4% investment, anything over 15% is remarkable and usually gains the organization a reputation for being a high-tech company. Companies in this category include pharmaceutical companies such as Merck & Co. (14.1%) and Novartis (15.1%), and engineering companies such as Ericsson (24.9%).

Universities are usually the main source of early discoveries or inventions. Commercial companies buy licenses from universities after paying hefty money to inventors. Universities and commercial companies then proceed with developing a biotechnology product by hiring scientists and technical staff. During the drug development process, efforts are being made to test the proof-of-concept and safeness and to determine ideal levels and delivery mechanisms. Development often occurs in phases that are defined by drug safety regulators in the country of interest. In the United States, the development phase can cost between $10 and $200 million. Approximately one in ten compounds identified by basic research passes all development phases and reaches the market. New product design and development is most often a crucial factor in the survival of a company. In an industry that is fast changing, companies must continually revise their design and range of products. This is necessary due to continuous technology change and development as well as increasing competitors in the market. In a challenging and competitive world, companies must conduct market research to know about product demand and economic feasibility. If the development is technology-driven, then it is a matter of selling what it is possible to make. The product range is developed so that production processes are as efficient as possible and the products are technically superior, hence possessing a natural advantage in the market place.

14.4 BIOTECHNOLOGY INDUSTRY AND PRODUCTS

In this section, we will look at various facts and figures of biotechnology industries worldwide. As per estimate, biotechnology is a $30 billion a year industry that has so far produced approximately 160 drugs and vaccines. There are more than 370 biotechnology drug products and vaccines currently in clinical trials, targeting more than 200 diseases, including various cancers, Alzheimer's disease, heart disease, diabetes, multiple sclerosis, AIDS, and arthritis. Biotechnology is responsible for hundreds of medical diagnostic tests that keep blood supply safe from the AIDS virus and detect other conditions early enough to be successfully treated. Home pregnancy tests are also biotechnology diagnostic products. Genetic engineering is sweeping the world's farms, and approximately 7 million farmers in 18 countries grew genetically engineered crops on about 16.72 million acres. Consumers are already enjoying biotechnology foods such as papaya, soybeans, and corn. Hundreds of biopesticides and other agricultural products are also being used to improve our food supply and to reduce our dependence on conventional chemical pesticides. Environmental biotechnology products make it possible to clean up hazardous waste more efficiently by harnessing pollution-eating microbes without the use of caustic chemicals. Industrial biotechnology applications have led to cleaner processes that produce less waste and use less energy and water in such industrial sectors as chemicals, pulp and paper, textiles, food, energy, and metals and minerals. For example, most laundry detergents produced in the United States contain biotechnology-based enzymes. DNA fingerprinting, a biotechnology process, has dramatically improved criminal investigation and forensic medicine, as well as afforded significant advances in anthropology and wildlife management.

14.5 BIOTECHNOLOGY STATUS IN THE UNITED STATES

The United States is currently the world leader in the research, development, and commercialization of biotechnology products. These advances have brought life-saving healthcare products and microbial pesticides to the market, and will soon offer healthier foods, disease- and insect-resistant crops, additional energy resources, environmental clean-up techniques, and more. There are 1473 biotechnology companies in the United States, of which 314 are publicly held. Market capitalization, the total value of publicly traded biotechnology companies at market prices, was $311 billion as of mid-March 2004. Biotechnology is one of the most research-intensive industries in the world. The U.S. biotechnology industry spent $17.9 billion on R&D in 2003. The top eight biotechnology companies spent an average of $104,000 per employee on R&D in 2003. The biotechnology industry in the United States is regulated by the Food and Drug Administration (FDA), the Environmental Protection Agency (EPA), and the Department of Agriculture.

14.6 CAREER OPPORTUNITIES IN BIOTECHNOLOGY

Biotechnology is a very exciting sector with tremendous job possibilities and career development potential. The biotechnology field offers promising jobs in universities, R&D centers, and corporate organizations such as drug manufacturing and drug marketing companies. The basic qualification for getting a job in any organized sector of biotechnology is a BS degree in biotechnology or any related field. However, a mere degree will not fetch jobs; candidates should also have at least 6 months to 1 year of work experience. Now, you may well ask, "How can one get work experience?" Obviously, biotechnology companies do not provide jobs just to give work experience to fresh graduates. So, one option is to apply to university research labs and work as a trainee to get work experience. Of course, as a trainee, you would not earn much money, but it would positively help you to get ready for the next job. In case you do not want a job after completing a BS degree, then you can proceed for higher education by getting a MS degree in any of the various specialized domains of biotechnology, such as medical, agricultural, industrial, microbial and environmental biotechnology, nanobiotechnology, biotechnology business and marketing, bioinformatics, or IPRs and biotechnology regulations. An MS degree in biotechnology with specialization will open up tremendous job opportunities and you can choose to work in any organized sector of biotechnology. Even if you do not want a job after completing an MS degree, you can do doctoral research by doing a PhD in any specialized or super-specialized field of biotechnology such as stem cell research, cloning, or gene therapy. During PhD research work, you have an opportunity to conduct research by employing various laboratory techniques independently, and if the research finding is novel, it may lead to a very commendable research paper publication in scientific journals. A PhD degree in biotechnology with work experience and decent publication tracks brings tremendous job opportunities, both in academic institutes and in corporate organizations. If you do not want a job in R&D labs, you can either go in for an master's degree in business administration (MBA) or do short courses in bioinformatics, IPRs, and healthcare management (Figure 14.2).

There is a tremendous job satisfaction working in the R&D sector of biotechnology. By publishing or patenting their research, scientists establish a name as well as large fortunes as fruits of their research. Researchers (mostly PhDs) not only perform experiments in labs, but also devote much of their time to attending seminars, conferences, teaching university students, and writing papers. A researcher has a very creative role working on the forefront of scientific discovery and can decide what to study and how to study, thus learning new things. Laboratory assistants (mostly BS degree holders) and lab technicians (mostly MS degree holders) work on a variety of sophisticated equipment, prepare chemicals, and maintain experimental microbial, plant, or tissue cultures

FIGURE 14.2 Training opportunities in biotechnology. (Photo courtesy of Department of Biotechnology, Manipal University Dubai, Dubai, UAE.)

in research labs. Lab technicians also help research scientists in their scientific discoveries. After doing an MBA, lab technicians can get management positions in biotechnology firms and can be involved in marketing biotechnology industry products to doctors, farmers, and other industries, or sell and supply scientific equipment to biotechnology researchers. After a BS or MS degree, one can do courses in scientific instrumentation and can work as a quality assurance technician to test biotechnology products in labs or production facilities, to ensure that they are safe and meet the required standards. Bioinformatics personnel work on designing and maintaining scientific data and analyzing the results of various research trials.

As the biotechnology industry continues to grow and commercialize, the number of job vacancies continues to increase. Companies are staffing more and more sales force to push their products on the market. They are also employing trained people for clinical research so as to get through the clinical approval process. This has increased the demand for clinical trial managers, clinical research professionals, clinical research associates, and outsourcing managers. Due to commercialization of the industry, the demand for finance and administration personnel is also increasing. Thus, biotechnology companies are recruiting people at all levels. A biotechnologist may also work in both the public and private sectors. Today, agriculture, dairy, horticulture, and drug companies are also employing large numbers of biotechnologists.

Compensation in biotechnology companies is competitive and includes incentives, such as stock option plans, 401K plans (in the United States), company-wide stock purchase plans, and cash bonus plans. The salary for any career in biotechnology varies from individual to individual and from organization to organization. Indeed, compensation is a big factor in choosing any career in biotechnology as a profession. There are some disparities in compensation between commercial biotechnology companies and academic or noncommercial organizations. Usually, the compensation package in commercial companies is higher than those in academic institutions. In addition, there is a greater flexibility in negotiating compensation packages in commercial companies than in academic institutions.

14.7 ENTRY-LEVEL JOB POSITIONS IN BIOTECHNOLOGY

This section discusses some typical entry-level biotechnology positions with their corresponding job descriptions.

14.7.1 Research and Development Division

14.7.1.1 Glass Washer

A glass washer is responsible for washing and drying glassware and distributing it to appropriate locations within the laboratories. He or she maintains the glass washing facility, keeps it clean, and picks up dirty glassware. He or she also sterilizes glassware and other items using an autoclave. A glass washer performs routine maintenance of glass washing equipment and performs other related duties as required. This position requires a high school diploma or equivalent and a minimum of 0–2 years of laboratory experience.

14.7.1.2 Laboratory Assistant

A laboratory assistant is responsible for performing a wide variety of research laboratory tasks and experiments, making detailed observations, analyzing data, and interpreting results. He or she maintains laboratory equipment and inventory levels for laboratory supplies. He or she may also write reports, summaries, and protocols regarding experiments. A laboratory assistant also performs limited troubleshooting and calibration of instruments. An entry-level laboratory assistant position requires a basic associate degree in science and 0–2 years of laboratory experience.

14.7.1.3 Research Associate

A research associate is responsible for R&D in collaboration with the other staff involved in a project. He or she makes detailed observations, analyzes data, and interprets results. Research associates prepare technical reports, summaries, protocols, and quantitative analyses. An incumbent maintains familiarity with current scientific literature and contributes to the process of a project within his or her scientific discipline, along with investigating, creating, and developing new methods and technologies for project advancement. He or she may also be responsible for identifying patentable inventions and acting as principal investigator in conducting his or her own experiments. A research associate may also be asked to participate in scientific conferences and contribute to scientific journals.

14.7.1.4 Research Assistant

The job description for a research assistant is similar to that of a research associate. At the entry level, the job can require a BS or an advanced degree in a scientific discipline or an equivalent qualification, with little or no related experience.

14.7.1.5 Postdoctoral Fellow

A postdoctoral fellow has a PhD, but little or no job experience, and joins the staff for a maximum of 2–3 years to gain the necessary experience before moving onto a more senior scientist's position.

14.7.1.6 Media Preparation Technician

A media preparation technician, or media prep technician, is responsible for media preparation in the R&D area. He or she performs experiments as required and outlined, and develops and maintains record keeping for procedures and experiments performed. An entry-level media prep technician position requires a basic associate degree in science and 0–2 years of related experience and/or completion of a company's on-the-job training program.

14.7.1.7 Greenhouse Assistant

A greenhouse assistant performs a variety of greenhouse research tasks and experiments. He or she may be required to make detailed observations, detecting horticultural or pest problems and instituting corrective action. Greenhouse assistants determine optimal cultural requirements and perform tasks related to disease and pest prevention. They are often required to collect, record, and analyze data, and also to interpret results. In addition, a greenhouse assistant may perform troubleshooting and equipment maintenance. An entry-level greenhouse assistant position requires a high school diploma, an associate degree or equivalent, and a minimum of 0–2 years of relevant greenhouse/plant experience.

14.7.1.8 Plant Breeder

A plant breeder is responsible for the design, development, execution, and implementation of plant breeding research projects in collaboration with a larger research team. He or she may be responsible for project planning and personnel management within the project. Plant breeders may use exotic germ plasm, work with various mating systems, and integrate them with biotechnology, as needed, to enhance selection methods and accelerate product development. A plant breeder's diverse responsibilities can include contributing to and developing good public relations with scientific and other professional communities. He or she may participate in the development of patents or proposals and assist with the management and development of a plant breeding group. An entry-level plant breeder position requires a BS degree or equivalent. A minimum of 0–2 years of plant breeding or agronomical experience and/or training in plant breeding or plant science is also required.

14.7.2 Quality Control

14.7.2.1 Quality Control Analyst

A quality control analyst is responsible for conducting routine and nonroutine analysis of raw materials. He or she compiles data for documentation of test procedures and reports abnormalities. A quality control analyst also reviews data obtained for compliance with specifications and reports abnormalities. He or she revises and updates standard operating procedures and may perform special projects on analytical and instrumental problem solving. An entry-level quality control analyst position requires a BS degree in a scientific discipline or equivalent and a minimum of 0–2 years of experience in quality control systems.

14.7.2.2 Quality Control Engineer

A quality control engineer is responsible for developing, applying, revising, and maintaining quality standards for processing materials into partially finished or finished products. He or she designs and implements methods and procedures for inspecting, testing, and evaluating the precision and accuracy of products and prepares documentation for inspection testing procedures. Depending on the job level, a quality control engineer is responsible for ensuring conformance to in-house specifications and good manufacturing practices and may conduct training programs. He or she may also be responsible for supervising the development and efforts of a quality control engineering group. An entry-level job requires a BS degree or equivalent and a minimum of 0–2 years of experience in quality control systems.

14.7.2.3 Environmental Health and Safety Specialist

An environmental health and safety specialist is responsible for developing, implementing, and monitoring industrial safety programs within the company. He or she inspects plant areas to ensure compliance with Occupational Safety and Health Administration (OSHA) regulations. He or she evaluates new equipment and raw materials for safety, and monitors employee exposure to chemicals and other toxic substances. Depending on the job level, a safety specialist may also conduct training programs in hazardous waste collection, disposal, and radiation safety regulations. An entry-level safety specialist job requires a BS degree or equivalent and a minimum of 0–2 years of related experience.

14.7.2.4 Quality Assurance Auditor

A quality assurance auditor is responsible for performing audits of production and quality control. He or she ensures compliance with in-house specifications, standards, and good manufacturing practices. The job requires a BS degree in a scientific discipline or equivalent and a minimum of 0–2 years of experience in biological or pharmaceutical manufacturing.

14.7.2.5 Validation Engineer

A validation engineer is responsible for the calibration and validation of equipment and systems and for assisting in the selection, specification, and negotiation of competitive pricing of equipment. He or she maintains all of the documentation pertaining to qualification and validation, and serves as an information resource for validation technicians, contractors, and vendors. At the entry level, the job requires a BS degree or equivalent, with 0–2 years of related experience.

14.7.2.6 Validation Technician

An entry-level validation technician would be responsible for developing, preparing the installation of, and revising test validation procedures/protocols to ensure that a product is manufactured in accordance with appropriate regulatory agency validation requirements, internal company standards, and current industry practices. A validation technician compiles and analyzes validation data, prepares reports, and makes recommendations for changes and/or improvements. He or she may also investigate and troubleshoot problems and determine solutions. He or she maintains appropriate validation documentation and files. An entry-level validation technician requires a high school diploma or equivalent and a minimum of 0–2 years of related experience.

14.7.3 Clinical Research

14.7.3.1 Clinical Research Administrator

A clinical research administrator is responsible for clinical data entry and validation to ensure legibility, completeness, and consistency of data. He or she assists users with requests for clinical documents and is responsible for working with physicians and/or their staff to clarify any questionable information. He or she may be responsible for auditing internal patient files and studies and for assisting with the development and evaluation of clinical record documents. At the entry level, a clinical research administrator typically develops an internal record-keeping system, including maintaining, auditing data, and providing status and activity reports as required. The job requires a high school diploma or equivalent, with a minimum of 0–2 years of related experience.

14.7.3.2 Clinical Coordinator

A clinical coordinator must be familiar with the scientific/investigative process. Expertise may be limited to a specific functional area. A clinical coordinator must have good communications skills, both written and oral. He or she must also have project team experience and a familiarity with standard computer applications. Responsibilities include coordinating the clinical development plan as outlined by the company or clinical department and defining objectives, strategies, and studies. The clinical coordinator must provide support for planning, including detailed effort estimates, scheduling, and critical path analysis. He or she must monitor clinical activities to identify issues, variances, and conflicts, and analyze and recommend solutions. The clinical coordinator is also responsible for project staffing requirements and tracking drug supply to outside vendors, as well as providing ongoing objective updates on progress and problems with projects, tracking and following up on action items. The position requires a BS or a Bachelor of Arts (BA) degree in Health Science, Information Technology, or Business, and 3–5 years of experience in the healthcare industry.

14.7.3.3 Clinical Programmer

A novice clinical programmer is responsible for coordinating and monitoring the flow of clinical data into the computer database. He or she analyzes and evaluates clinical data, recognizes inconsistencies, and initiates the resolution of data problems. He or she implements data management plans designed to meet project and protocol deadlines, and consults in the design and development of clinical trials, protocols, and case report forms. A clinical programmer also acts as a liaison between clinical management and subcommittees and project teams on an as-needed basis. An entry-level position as a clinical programmer requires a BS degree or equivalent, although an MS degree is often preferred. A minimum of 0–2 years of experience in pharmaceutical programming in the clinical research area is also required.

14.7.3.4 Biostatistician

A biostatistician works with others to define and perform analyses of databases for publications, presentations to investigator meetings, and for meetings of professional societies. A position as a biostatistician requires at least an MS degree in biostatistics and 1–4 years of related experience.

14.7.3.5 Clinical Data Specialist

A clinical data specialist is responsible for collaborating with various departments on the design, documentation, testing, and implementation of clinical data studies. He or she develops systems for organizing data to analyze, identify, and report trends. A clinical data specialist also analyzes the interrelationships of data and defines logical aspects of data sets. A starting position as a clinical data specialist requires a BS degree or equivalent and a minimum of 0–2 years of related experience.

14.7.3.6 Drug Experience Coordinator

The major responsibility of a drug experience coordinator is to handle drug experience activities for marketed products. The candidate will also provide drug information on the products. He or she will oversee day-to-day processing of adverse event information for marketed products and will coordinate the receipt, classification, investigation, and processing of adverse experience reports. A position as a drug experience coordinator requires a Doctor of Pharmacy (Pharm D) degree or equivalent clinical training and 0–2 years of related experience.

14.7.3.7 Clinical Research Associate

A clinical research associate is responsible for the design, planning, implementation, and overall direction of clinical research projects. He or she evaluates and analyzes clinical data and coordinates activities of associates to ensure compliance with protocol and overall clinical objectives. He or she may also travel to field sites to supervise and coordinate clinical studies. An entry-level clinical research associate position typically requires a BS degree, a registered nurse degree, or an equivalent qualification, and a minimum of 0–2 years of clinical experience in medical research, nursing, or the pharmaceutical industry. Knowledge of FDA regulatory requirements is preferred.

14.7.3.8 Animal Handler

An animal handler is responsible for the daily care of research animals for experimental purposes. He or she cleans animal cages and racks, maintains records to comply with regulatory requirements and standard operating procedures, and performs preventive maintenance on facility equipment. The incumbent may also perform animal observation, grooming, and minor clinical tasks. An animal handler position requires a high school diploma or equivalent experience with a scientific background. A minimum of 0–2 years of relevant laboratory experience is also expected.

14.7.3.9 Animal Technician

An animal technician is responsible for the daily care of research animals for experimental purposes. He or she also coordinate with vendors and supervisors on operational, administrative, and technical responsibilities. The animal technician performs some surgery and postoperative care as directed, and is responsible for overseeing procurement of animals and supplies, preventive maintenance of facility equipment, cleaning of animal cages and racks, daily rounds, and observation to check the animals' health status. He or she develops standard operating procedures and maintains records to comply with regulatory requirements. An entry-level animal technician job requires a high school diploma or equivalent experience with a scientific background and a minimum of 0–2 years of related laboratory experience.

14.7.3.10 Technical Writer

An entry-level technical writer is responsible for writing and editing standard operating procedures, clinical study protocols, laboratory procedure manuals, and other related documents. He or she edits and/or rewrites various sources of information into a uniform style and language for regulatory compliance, and assists in developing documentation for instructional, descriptive, reference, and/or informational purposes. An entry-level position requires a BS degree or equivalent and a minimum of 0–2 years of experience in technical documentation.

14.7.4 Product and Development

14.7.4.1 Product Development Engineer

An entry-level product development engineer is responsible for the design, development, modification, and enhancement of existing products and processes. The incumbent is involved in new product scale-up, process optimization, technology transfer, and process validation. He or she ensures that processes and design implementations are consistent with good labor and manufacturing practices.

A product development engineer may also be responsible for contact with outside vendors and for the administration of contracts to accomplish goals. He or she works on problems of moderate scope, where analysis of a situation or data requires a review of identifiable factors. The job requires a BS degree or equivalent, and 0–2 years of related experience.

14.7.4.2 Production Planner/Scheduler

A production planner/scheduler is responsible for planning, scheduling, and coordinating the final approval of products through the production cycle. He or she coordinates production plans to ensure that materials are provided according to schedules to maintain production and provides input to management. When necessary, a production planner/scheduler works with the customer service, marketing, production, quality control, and sales departments to review backorder status, prioritize production orders, and deal with other potential schedule interruptions or rescheduling. An entry-level production planner/scheduler requires a bachelor's degree or equivalent and a minimum of 0–2 years of related experience.

14.7.4.3 Manufacturing Technician

Manufacturing technicians are responsible for the manufacture and packaging of potential and existing products. They operate and maintain small production equipment. They also weigh, measure, and check raw materials and ensure that manufactured batches contain the proper ingredients and quantities. They maintain records and clean production areas to comply with regulatory requirements, good manufacturing practices, and standard operating procedures. A manufacturing technician may also assist with in-process testing to make sure that batches meet product specifications. An entry-level position requires an associate degree in science and a minimum of 0–2 years of related experience in a manufacturing environment.

14.7.4.4 Packaging Operator

A packaging operator uses manual and/or automated packaging systems to label, inspect, and package final container products. He or she also enters data and imprints computer-generated labels, maintains records, and maintains the manufacturing/production area to comply with regulatory requirements, good manufacturing practices, and standard operating procedures. A packaging operator may also perform initial checks of completed documents for completeness and accuracy. The position requires a high school diploma or equivalent and a minimum of 0–2 years of experience in a manufacturing environment.

14.7.4.5 Manufacturing Research Associate

A manufacturing research associate is responsible for the implementation of production procedures to optimize manufacturing processes and regulatory requirements, and has responsibilities in packaging and distribution processes. He or she may also help maintain production equipment. The position requires a BS degree in a scientific discipline or equivalent and a minimum of 0–2 years of experience in a manufacturing environment.

14.7.4.6 Instrument/Calibration Technician

An entry-level instrument/calibration technician is responsible for performing maintenance, testing, troubleshooting, calibration, and repair on a variety of circuits, components, analytical equipment, and instrumentations. He or she also calibrates instrumentation, performs validation studies, and specifies and requests purchase of components. He or she analyzes results, may develop test specifications and electrical schematics, and maintains logs and required documentation. An instrument/calibration technician also maintains spare parts inventories and may prepare technical reports with recommendations for solutions to technical problems. An associate degree in electronics technology or equivalent is required, as is a minimum of 0–2 years of related experience.

14.7.4.7 Biochemical Development Engineer

A biochemical development engineer is responsible for the design and scale-up of processes, instruments, and equipment from the laboratory through the pilot plant and manufacturing process. He or she assists the manufacturing operations in problem solving with regards to equipment and systems, and participates in the design and start-up of new manufacturing facilities and equipment. He or she develops and recommends new process formulas and technologies to achieve cost effectiveness and product quality. A biochemical development engineer also establishes operating equipment specifications and improves manufacturing techniques. He or she is involved in new product scale-up, process improvement, technology transfer, and process-validation activities. He or she works with various departments to ensure that processes and designs are compatible for new product technology transfer and to establish future process and equipment automation technology. The position requires a BS degree in biological, chemical, or pharmaceutical engineering, or a related discipline. A minimum of 0–2 years of experience is also required, preferably in the areas of pharmaceutical processes or research product development.

14.7.4.8 Process Development Associate

A process development associate is responsible for the implementation of production procedures to optimize manufacturing processes and regulatory requirements. He or she may also assist in process development, in creating scalable processes with improved product yield and reduced manufacturing systems costs. An entry-level process development associate may also be involved in packaging and distribution processes and in the maintenance of production equipment. He or she may research and implement new methods and technologies to enhance operations. The position requires a BS degree in a scientific discipline or equivalent and a minimum of 0–2 years of experience.

14.7.4.9 Assay Analyst

An assay analyst is responsible for doing cell cultures and performing assays and tests on tissue and cell cultures, following standard protocols. He or she prepares glassware, reagents, and media for cell culture use. He or she also performs, prepares and maintains tissues and cell cultures and maintains records required by good manufacturing procedures. An assay analyst also participates in the modification of assay procedures for routine implementation. The position requires a high school diploma or equivalent and a minimum of 0–2 years of related experience.

14.7.4.10 Manufacturing Engineer

A manufacturing engineer is responsible for developing, implementing, and maintaining methods, operation sequences, and processes in manufacturing. He or she works with the engineering department to coordinate the release of new products. He or she estimates manufacturing costs, determines time standards, and makes recommendations for process requirements of new or existing product lines. A manufacturing engineer also maintains records and reporting systems as required for the coordination of manufacturing operations. An entry-level job as a manufacturing engineer requires a BS degree in a scientific discipline or equivalent and a minimum of 0–2 years of related experience, preferably in research product development or in a manufacturing environment.

14.7.5 Regulatory Affairs

14.7.5.1 Regulatory Affairs Specialists

An entry-level regulatory affairs specialist coordinates and prepares document packages for submission to regulatory agencies, internal audits, and inspections. He or she compiles all materials required for submissions, license renewals, and annual registrations. An incumbent monitors and improves tracking and control systems and keeps abreast of regulatory procedures and changes.

He or she may work with regulatory agencies and recommend strategies for earliest possible approvals of clinical trial applications. An entry-level position requires a BS degree or equivalent and a minimum of 0–2 years of related experience.

14.7.5.2 Documentation Coordinator

A documentation coordinator provides clerical and administrative support related to a company's documentation system requirements. He or she audits all documentation manuals to ensure that they are accurate, up-to-date, and available to appropriate personnel. A documentation coordinator also files and retrieves all master documents. The position requires a high school diploma or equivalent and a minimum of 0–2 years of related experience.

14.7.5.3 Documentation Specialist

An entry-level documentation specialist is responsible for coordinating all activities related to providing required documentation and implementing related documentation systems. He or she coordinates the review and revision of procedures, specifications, and forms. He or she also assists in compiling regulatory filing documents and in maintaining computerized files to support all documentation systems. The job requires a BS degree in a related field or equivalent and a minimum of 0–2 years of experience in documentation, quality assurance, technical writing, or an equivalent position.

14.7.6 Information Systems

14.7.6.1 Library Assistant

A library assistant maintains serial control and locates and orders journal articles and/or books on relevant subjects that are unavailable at local libraries. He or she performs special data-gathering projects as requested and is responsible for online computer searching of scientific databases. The position of entry-level library assistant requires an associate degree or equivalent and a minimum of 0–2 years of relevant library experience or the completion of an on-the-job training program.

14.7.6.2 Scientific Programmer/Analyst

A scientific programmer/analyst designs, develops, evaluates, and modifies computer programs for the solution of scientific or engineering problems and for the support of R&D efforts. He or she analyzes existing systems and formulates logic for new systems. A scientific programmer/analyst also devises logical procedures, prepares flowcharts, performs coding, tests, and debugs programs. He or she provides input for the documentation of new or existing programs and determines system specifications, input/output processes, and working parameters for hardware/software compatibility. An analyst also contributes to decisions on policies, procedures, expansion strategies, and product evaluations. An entry-level scientific programmer/analyst position requires a BS degree in a related discipline or equivalent and a minimum of 0–2 years of experience.

14.7.7 Marketing and Sales

14.7.7.1 Market Research Analyst

A market research analyst is responsible for researching and analyzing the company's markets, competition, and product mix. He or she performs literature research, analyzes and summarizes data, and makes presentations on new market and technical areas. He or she also analyzes the competitive environment, as well as future marketing trends, and makes appropriate recommendations. He or she conducts market surveys, summarizes results, and assists in the preparation, presentation, and follow-up of research proposals. An entry-level position as a market research analyst requires a bachelor's degree or equivalent and a minimum of 0–2 years of experience in market research, competitive analysis, and product planning, as well as excellent writing skills.

14.7.7.2 Systems Analyst

A systems analyst is responsible for system-level software and for maintenance of the operating system(s), various layered products, system tuning, and various levels of user assistance. The systems analyst is responsible for the operation of all system software, performing upgrades, and maintaining related system and user documentation. He or she assists in the implementation of system validation and documentation, in system capacity planning, and system configuration. In addition, the systems analyst troubleshoots system-related problems and interacts with vendors. An entry-level position as a systems analyst requires a degree in data processing and a minimum of 0–2 years of experience.

14.7.7.3 Sales Representative

A sales representative is responsible for direct sales of company products or services. He or she calls on prospective customers, provides product information and/or demonstrations, and quotes appropriate customer prices. A sales representative is also responsible for new account development and growth of existing accounts within an established geographic territory. A sales representative must meet assigned sales quotas and may handle key company accounts or act as an account manager for national or major accounts. Depending on the level of the position, an experienced incumbent may also assist in the training of other sales representatives. An entry-level sales representative position requires a bachelor's degree or equivalent, a minimum of 0–2 years of related sales experience, and some knowledge of the company's products.

14.7.7.4 Customer Service Representative

Customer service representatives are responsible for ensuring product delivery in accordance with customer requirements and manufacturing capabilities and for responding to customer product inquiries and satisfaction issues. He or she answers telephones, takes product orders, and inputs sales order data into consumer data systems. He or she also investigates problems related to the shipment of products, credits, and new orders. He or she may also be responsible for sale order administration and/or inside sales. The entry-level position requires a college education or equivalent, with preference toward a BS degree in a technical or scientific field. A minimum of 0–2 years of related experience in diagnosing and troubleshooting products in the pharmaceutical industry may also be required.

14.7.7.5 Technical Service Representative

A technical service representative provides technical direction and support to customers on the operation and maintenance of company products. He or she also serves as a contact for customers regarding technical and service-related problems. A technical service representative also demonstrates the uses and advantages of products. An entry-level technical service representative position requires an associate's degree or equivalent and a minimum of 0–2 years of experience.

14.8 ADMINISTRATION

14.8.1 Technical Recruiter

A technical recruiter is responsible for recruiting, interviewing, and screening applicants for technical exempt and nonexempt positions. He or she coordinates preemployment physicals, travel, reporting dates, security clearances, and employment processing for new hires. He or she also conducts employee advertising and reviews employment agency placements. In addition, a technical recruiter maintains college recruiting, affirmative action, and career development programs. An entry-level position as a technical recruiter requires a BS degree or equivalent and a minimum of 0–2 years of experience.

14.8.2 Human Resource Representative

A human resources representative is responsible for a variety of activities in personnel administration, including employment, compensation and benefits, employee relations, equal employment opportunities, and training programs. He or she conducts job interviews, counsels employees, maintains records, and conducts research and analyzes data on assigned projects. An entry-level position as a human resource representative requires a BS degree or equivalent and a minimum of 0–2 years of related experience.

14.8.3 Patent Administrator

A patent administrator is responsible for preparing and coordinating all procedural documentation for patent filings and applications. He or she tracks in-house research studies and recommends the need for and timing of patent filings. A patent administrator also assists attorneys with the drafting and editing of patent applications, collects, and evaluates supporting data. This position requires the maintenance of a tracking system to comply with trademark regulations. He or she may also be called upon to assist with determining the necessity of and approach to contracts to ensure protection of the company's proprietary technology. A patent administrator is also typically responsible for tracking and paying legal fees. An entry-level patent administrator position requires a BS degree or equivalent and a minimum of 0–2 years of experience.

14.8.4 Patent Agent

A patent agent is responsible for preparing, filing, and processing patent applications for the company. He or she negotiates and drafts patent licenses and other agreements. A patent agent also conducts state-of-the-art searches and may assist with appeal and interference proceedings. A patent agent also performs other duties as required. The entry-level patent agent position requires a BS degree or equivalent with 0–2 years of related experience. It also requires registration to practice before the U.S. Patent and Trademark Office.

14.9 WHICH JOB IS GOOD FOR ME?

Most jobs in academic institutes are usually available from BS graduates onward, starting with laboratory technicians or assistants to the laboratory head/director, who should have a PhD degree with 10–15 years of laboratory experience. After getting a BS degree in biotechnology, students can either join any university or pursue an MS/MSc in a specialized branch of biotechnology, such as agricultural biotechnology, microbial biotechnology, industrial biotechnology, animal biotechnology, environmental biotechnology, or genetic and forensic fields. Although a laboratory assistant position may not be very lucrative, it will definitely give hands-on experience. Job prospects are high after 2–3 years of laboratory experience. After obtaining BS/MS degrees, candidates can also do specialized courses in patents and IPRs, healthcare management, and they can also do MBA courses. A BS degree with an MBA qualification from a reputed university/institute may also help candidates to enter into sales and marketing jobs in pharmaceuticals, biopharmaceuticals, biotechnology, or any other healthcare industries. Jobs in sales and marketing are very challenging and require numerous field visits, but they are also equally lucrative. Students who have obtained an MS/MSc degree but do not want to work in labs or in research projects can also choose to do an MBA and enter into the sales and marketing segment of biotechnology industries. Students with an MS/MSc degree in biotechnology can enroll for a PhD program in their respective field of interest and may pursue their careers as research scientists in various positions, from junior scientist to the president level in R&D. R&D labs offer the most challenging job opportunities in the biotechnology industry, because these are the people who bring about new discoveries and new inventions.

If scientists work in universities and discover new inventions that cause the development of new fields, they may reap rewards in the form of a Nobel Prize in Physiology and Medicine, with a great deal of public recognition and appreciation. On the other hand, if scientists who work in industries discover new inventions, they do not get much public accolade and appreciation, but they get an exuberantly high salary from their company. It is up to candidates to decide which field he or she would like to take his or her career.

Another frequently asked question about careers in biotechnology is that which field of biotechnology is the most challenging or career oriented. The answer would be that all fields of biotechnology provide challenging career opportunities. For example, researchers who work in agriculture, biology, genetics, and medicine are at the forefront of new biotechnology discoveries. These men and women are working to unravel the genetic codes that govern the biological processes of different forms of life so that they can be understood and, when appropriate, modified. Life sciences researchers may work in an academic environment such as a university, or for a company or a government agency. They can focus their work on animals, bacteria, humans, plants, viruses, or any other life forms in which they have a special interest. The discoveries made in government, university, or corporate laboratories are the first steps toward genetically engineered products or processes such as new vaccines, drugs, or plant varieties. There are many jobs related to biotechnology that are held by people without extensive science or engineering expertise. These individuals must understand the science of biotechnology, but their primary talent may be in communications or in some other area. Regulatory officials develop guidelines for biotechnology research and the development of new products and processes. They work with company, government, or university researchers to review proposed research plans and assess the safety of resulting products. Regulatory officials must approve biotechnology research plans before they can be executed and biotechnology products before they can be marketed.

Individuals involved with the regulation of biotechnology research and product development generally work for a federal or state government agency. Public relations personnel provide comprehensible information to the general public about new biotechnology products and processes. They translate complex scientific information about new discoveries for nonscientists. Sales people work with the dealers and distributors of biotechnology products. They have expertise in marketing skills and are knowledgeable about the products. Patent lawyers who specialize in biotechnology help scientists, companies, or universities protect their legal rights to new discoveries. They file patent applications for their clients and interact with the U.S. Patent Office.

14.10 WHY ARE R&D JOBS THE MOST CHALLENGING?

Among all phases of biotechnology, the R&D phase is considered to be the most challenging one, as it allows scientists to discover something new, such as the discovery of a new drug molecule, better treatment modalities, or discovery of a new diagnostic test for the early detection of diseases. Challenging career opportunities are available in specialized fields of biotechnology where candidates can work in the domain of drug discovery to find new treatments for various dreaded diseases. We present a list of some of the major fields where one can work and be part of a research team.

14.10.1 Medical and Diagnostics Sectors

14.10.1.1 Detecting and Treating Hereditary Diseases

Many diseases—including some types of anemia, cystic fibrosis, Huntington's disease, and some blood disorders—are the result of a defective gene that parents pass to their children. Biotechnologists are working to identify and locate where defects occur in genes that are related to hereditary diseases. Once the correct genetic code is known, healthcare professionals hope that in the future they will be able to replace the missing or defective genes to make the individual healthy. Currently, prospective parents can be screened for such genetic defects and counseled about the likelihood of their children being affected. Fetuses are being screened for genetic disorders before they are born and genetic counselors play an important role in informing parents concerning the test results. Genetic counselors prepare parents for the birth and early medical treatment of a child with a genetic disorder.

14.10.1.2 Heart Disease

Heart attacks occur when a blood clot enters one of the coronary arteries and cuts off blood flow to a portion of the heart. If the artery is not reopened quickly, severe damage to the heart can occur. Doctors can now prescribe a genetically engineered drug called *tissue plasminogen activator* (TPA) that travels to the blood clot and breaks it up within minutes, restoring blood flow to the heart and lessening the chance of permanent damage.

14.10.1.3 Cancer

Medical doctors are using biotechnology to treat cancer in several ways. Genetically engineered proteins called lymphokines seem to work with the body's immune system to attack cancer cells and growth inhibitor proteins seem to slow the reproduction of cancer cells. Highly specific and purified antibodies can be loaded with poisons that locate and destroy cancer cells.

14.10.1.4 AIDS

Genetic engineering has produced several substances that show promise in the treatment of AIDS. These substances stimulate the body's own immune system to fight the disease.

14.10.1.5 Veterinary Medicine

Veterinarians and professionals in animal science are using biotechnology discoveries to improve animal health and production. Genetically engineered vaccines, monoclonal antibody technology, and growth hormones are three developments that are making this possible. Questions concerning food safety, economic impacts, and animal health issues have been raised by those opposing the use of growth hormones and have made their use controversial.

14.10.1.6 Vaccines

Most vaccines are made from viruses or bacteria that have been weakened or killed. However, since live viruses or bacteria are often included in these vaccines, they are not without side effects. An animal could become sick from the vaccine. rDNA technology allows the production of synthetic vaccines that do not have this risk. rDNA vaccines have been developed for swine and cattle diarrhea, and research on other vaccines is continuing.

14.10.1.7 Monoclonal Antibodies

Antibodies are produced naturally by animals when invaded by a disease-causing organism. Each type of antibody is very specific in that it recognizes and attacks only one particular disease organism. Monoclonal antibody technology allows biotechnologists to produce large amounts of purified antibodies for use in the development of vaccines. Antibodies can also be used to diagnose illnesses and can detect drugs, viral and bacterial products, and other substances. For example, home pregnancy test kits use antibodies to detect the presence of a certain hormone in the urine.

14.10.1.8 Growth Hormones

Several biotechnology companies are seeking approval by the federal government of genetically engineered proteins that improve meat and milk production in cattle or pigs. Bovine somatotropin for cattle and porcine somatotropin for pigs could impact the life cycle of farm animals by increasing their rate of growth and milk production and producing leaner carcasses.

14.10.2 Agricultural Sector

Farmers and other agricultural professionals are being faced with decisions about the use of biotechnology products in their operations. In addition to animal health products and growth hormones that are available for livestock production, a host of crop-production products is (or will soon be) on the market. Scientists are exploring the genetic modification of food crops to achieve desirable characteristics such as high yield, increased protein or oil production, disease resistance, or pest resistance.

14.10.2.1 Crop Yield

Crop yields are controlled not by one gene, but by many genes acting together. Scientists are working to identify these genes and their contribution to yield so that crops can be genetically modified for increased production.

14.10.2.2 Protein and Oil Content of Seeds

By modifying the genes that control the accumulation of protein and oil in seeds such as corn and soybeans, biotechnology researchers hope to develop more nutritious crops or crops that produce modified oils for food or industrial uses. For example, by changing the kinds or amounts of fatty acids stored in soybeans, new oils can be developed.

14.10.2.3 Environmental Conditions

Most crops do not grow well in dry, salty, or alkaline soils. Most cannot withstand heavy frosts or extreme temperature changes. Biotechnologists are trying to genetically engineer crops that will grow well in the poorest food-producing areas of the world, where these conditions are often present.

14.10.2.4 Disease and Pest Resistance

Genes for disease or pest resistance have been identified for several crops. If crops can be genetically modified to include a resistant gene that makes them undesirable to pests, the amount of chemical pesticide needed could be reduced, a less expensive and more environmentally friendly option. Crops that "tolerate" herbicides to which they are normally sensitive are now on the market. When used properly, the insertion of an herbicide-resistant gene into crops can give a farmer more choices in selecting an herbicide. There is also the opportunity to develop crops that are tolerant of herbicides and are less damaging to the environment.

Many other diseases can be treated with genetically engineered products. Doctors can use a genetically engineered vaccine to treat human hepatitis B or a growth hormone to help children with dwarfism. Other treatments developed through genetic engineering techniques include a protein to control blood clotting in hemophiliacs, a hormone that stimulates red blood cell production to fight anemia, and antibodies that discourage organ rejection by transplant patients.

14.10.3 Law Enforcement Sector

Biotechnology has provided law enforcement professionals with another way of placing a suspect at the scene of a crime. This area of study, called forensic biotechnology, uses a method called DNA fingerprinting. This method is based on the fact that each individual's DNA is highly unlikely to be identical to any other person's DNA (unless he or she has an identical twin). By examining traces of tissue, hair, tooth pulp, blood, or other body fluids left at the scene of a crime, a suspect can be linked to a crime location with great accuracy. Many states now accept DNA fingerprinting results as admissible evidence in criminal and civil trials.

14.10.4 Product Manufacturing Sector

After a biotechnology product has been approved for use, engineers with a BS, Bachelor of Technology (Btech), and/or Bachelor of Engineering (BE) degree in biotechnology engineering are needed to manufacture it. The engineers are trained to handle bioreactors to produce biotechnology products in large quantities while maintaining the quality as per FDA guidelines.

A biotechnology company offers many of the same career opportunities as any other manufacturing business. However, in addition to specific skills in engineering, scale-up, quality control, and other manufacturing processes, individuals employed in the biotechnology industry will need a solid background in the biological sciences. Industrial chemists find that many natural biological products such as amino acids, enzymes, and vitamins can be manufactured more efficiently using biotechnology. The gene or genes that produce the natural biological product can be transferred to an organism, perhaps a bacterium that now starts producing it. Microorganisms are capable of producing many common organic chemicals, such as ethanol. They can also produce proteins for vaccines and for other uses through a fermentation process.

14.10.5 Microbial Engineering Sector

As the world's population grows, so does the problem of waste disposal. Biotechnology is helping waste management experts in several ways. Microorganisms such as bacteria and microbes can easily adapt to different environments and live off their surroundings. Biotechnologists have found bacteria in solid waste sites that can break down various kinds of waste for their own use. rDNA techniques can enhance these capabilities, and new strains of waste degraders could be developed. Biotechnology can also be used to improve microorganisms used in the treatment of wastewater to make the process cheaper and more efficient. Those involved in energy industries are finding that living organisms modified by rDNA technologies can improve energy production and use. Ore-containing rock is often mixed with minerals that must be separated from the rock by a heat process called smelting. Some microorganisms can dissolve and absorb the minerals, lessening the need for smelting. Other bacteria can force oil out of rocks where conventional drilling is not possible. The production of energy from biomass, especially waste plant materials like wood chips or corn stalks, also benefits from biotechnology. Microorganisms can produce enzymes that degrade the plant materials, making them useful in energy production.

14.11 SALARY AND INCENTIVES IN BIOTECHNOLOGY

Biotechnology as a subject has grown rapidly and, as far as employment is concerned, it has become one of the fast growing sectors. Employment records show that biotechnology has great scope in the future. Biotechnologists can find careers with pharmaceutical companies, chemical, agriculture, and allied industries. They can be employed in the areas of planning, production, and management of bioprocessing industries. There is large-scale employment in research laboratories run by the government as well as those in the corporate sector. Biotechnology students in India may find work in government-based entities such as state universities and research institutes, or at private centers as research scientists/assistants. Alternatively, they may find employment in specialized biotechnology companies or biotechnology-related companies such as pharmaceutical firms, food manufacturers, or aquaculture and agricultural companies. Companies that are engaged in businesses related to life sciences (ranging from equipment, chemicals, pharmaceuticals, diagnostics, etc.) also consider a biotechnology degree relevant to their field. The work scope can range from research, sales, marketing, administration, quality control, animal breeders, and technical support. Armed with this powerful combination of fundamental cell and molecular biology and applied science, graduates are well placed to take up careers in plant, animal, or microbial biotechnology laboratories or in horticulture, food science, commerce, and teaching.

Biotechnology is a highly remunerative career option. Biotechnologists employed in public sector industries and research institutes have pay packages according to their qualifications and experiences. Those working in private sector, corporate houses, and industries have very lucrative remuneration. Those employed in teaching have better pay packages, which are even very lucrative in private colleges and institutions. The average annual salary in the industrial sector is listed below for your information, but note that a compensation package can always be negotiated with the employer.

- Lab technician (BS level): $15,000–20,000
- Lab manager (BS level plus experience): $20,000–30,000
- Research assistant (MS level plus experience): $30,000–50,000
- Postdoctoral fellow (PhD level): $35,000–50,000
- Research scientist (PhD level plus experience): $50,000–80,000
- Research director/vice president (PhD level plus experience): $80,000–120,000
- Company CEO/president: $150,000 or more

However, in academic institutions, the compensation package for all positions may vary depending on the university and location. An opportunity to negotiate a salary package is not a common phenomenon because most positions have a fixed compensation package.

PROBLEMS

Q1. Discuss how different countries invest money in the progress of biotechnology.
Q2. What career opportunities are available in R&D?
Q3. Describe the job situation in the U.S. biotechnology sector.

ASSIGNMENTS

Make a chart showing various job opportunities available in the biotechnology field and discuss these with your course instructor and classmates. Next, organize a quiz contest based on who is who in biotechnology industries.

FIELD VISIT

With the help of your course instructor, organize a field visit to any biotechnology-based industry located in your own city or town. The field experience will help you to get real-time exposure to all segments of biotechnology—from R&D labs, manufacturing plants, animal testing facilities, to quality assurance/quality control labs—and may also help you to decide on the kind of field you might be keen to undertake.

REFERENCES AND FURTHER READING

Dineen, J.K. and Leuty, R. Amgen slows its Bay Area expansion. *San Francisco Business Times* Web site. http://www.bizjournals.com (Retrieved August 14, 2007).

Annual Report 2008. UCB. http://www.ucb.com/_up/ucb_com_investors/documents/UCB_Annual_Report_2008_EN.pdf (retrieved on May 25, 2009).

Biogen Idec. Product pipeline, in *2007 Annual Report*, 2008a.

Biogen Idec. SEC Form 10-K, pp. 1, 11, 2008b. http://edgar.sec.gov/Archives/edgar/data/875045/000095013508001029/b68103bie10vk.htm (retrieved on June 7, 2008).

Calabro, S. Genzyme put patients first, and grew to become a multi-billion-dollar company. But empries don't survive on altruism. Pharmaceutical Executive, [March 1, 2006]. http://www.pharmexec.com/pharmexec/article/articleDetail.jsp?id=310976&&pageID=7&searchString=genzyme

Coy, P. The search for tomorrow. *Business Week*, October 11, 2001. http://www.businessweek.com

Eugene R. Special report: The birth of biotechnology. *Nature*. 421: 456–457, 2003. http://www.nature.com

Food and Drug Administration. FDA approves new orphan drug for treatment of pulmonary arterial hypertension. Press release, 2007. http://www.fda.gov/bbs/topics/NEWS/2007/NEW01653.html (retrieved on June 22, 2007).

Forbes (12/2007). Henri Teemer profile. http://www.forbes.com/finance/mktguideapps/personinfo/FromPersonIdPersonTearsheet.jhtml?passedPersonId=222120 (Retrieved 2008-07-10).

Genentech. Corporate overview. http://www.gene.com

Gilead Sciences. Donald H. Rumsfeld named chairman of Gilead Sciences. Press release, 1997. http://www.gilead.com/wt/sec/pr_933190157/ (Retrieved on June 3, 2007).

Gilead Sciences. Gilead invests $25 million in Corus Pharma; establishes equity position in company with late-stage product candidate for Cystic Fibrosis. Press release, 2006a. http://www.gilead.com/pr_842319 (retrieved on August 15, 2007).

Gilead Sciences. Gilead Sciences completes acquisition of Raylo Chemicals Inc. Press release, 2006b. http://www.gilead.com/wt/sec/pr_926645 (retrieved on June 7, 2007).

Gilead Sciences. Gilead Sciences to Acquire Myogen, Inc. for $2.5 Billion. Press release, 2006c. http://www.gilead.com/pr_910704 (retrieved on August 15, 2007).

Gilead Sciences. Parion Sciences and Gilead Sciences sign agreement to advance drug candidates for pulmonary disease. Press release, 2007a. http://www.gilead.com/pr_1040975 (retrieved on August 15, 2007).

Gilead Sciences. U.S. Food and Drug Administration approves Gilead's Letairis treatment of pulmonary arterial hypertension. Press release, 2007b. http://www.gilead.com/wt/sec/pr_1016053 (retrieved on June 16, 2007).

Gilead Sciences and Bristol-Myers Squibb. U.S. Food And Drug Administration (FDA) approves Atripla. Press release, 2006. http://www.gilead.com/pr_881419 (retrieved on December 15, 2007).

Langer-Gould, A., Atlas, S.W., Green, A.J., Bollen, A.W., and Pelletier, D. Progressive multifocal leukoencephalopathy in a patient treated with natalizumab. *N. Engl. J. Med.* 353: 375–381, 2005.

Pennsylvania Bio—Member listings. *Pennsylvania Bio* web site. http://www.pennsylvaniabio.org

Pollack, A. F.D.A. backs AIDS pill to be taken once a day. *New York Times*, 2006. http://www.nytimes.com/2006/07/13/business/13drug.html (retrieved on September 20, 2007).

Pollack, A. Gilead's drug is approved to treat a rare disease. *New York Times*, 2007. http://www.nytimes.com/2007/06/16/business/16gilead.html (retrieved on June 16, 2007).

Schouten, E. Genzyme's lifelong commitments. NRC Handelsblad, 2005. http://www.gaucher.org.uk

Schwartz, N.D. Rumsfeld's growing stake in Tamiflu. *CNN*, 2005. http://money.cnn.com/2005/10/31/news/newsmakers/fortune_rumsfeld/?cnn=yes (retrieved on June 3, 2007).

Staff writers. Roche makes $43.7B bid for Genentech. *Genetic Engineering and Biotechnology News*, ISSN 1935472X, July 21, 2008. http://www.genengnews.com

The National Medal of Technology Recipients 2005 Laureates. The National Medal of Technology and Innovation. United States Patent and Trademark Office. http://www.uspto.gov/nmti/recipients_05.html (retrieved on February 26, 2009). Who made America? Herbert Boyer. PBS. http://www.pbs.org

15 Laboratory Tutorials

Laboratory training is a very critical component of biotechnology research and product development. It is necessary to know basic laboratory techniques in order to shape a career in biotechnology. Keeping this in mind, we have decided to include some of the basic and advanced techniques employed in biotechnology. Since it is not possible to include all the laboratory techniques employed in biotechnology, we have made sure to include fundamental techniques. One of the important aspects of biotechnology laboratory experiments is to understand good laboratory practice (GLP), which has been described in the previous chapter in great detail. However, it will be helpful to shed some light on the significance of GLP in biotechnology research. GLP deals with the organization, process, and conditions under which laboratory studies are planned, performed, monitored, recorded, and reported. GLP is intended to promote the quality and validity of test data. One of the main criteria of GLP-based laboratory experiments is ensuring that all routine work follows written standard operating procedures (SOPs). Facilities such as laboratories should be large enough and should have the right construction to ensure the integrity of a study—for example, to avoid cross-contamination. In addition, instruments employed in the study should be regularly calibrated and maintained. Moreover, raw data should be acquired, processed, and archived to ensure the integrity of the data. Although GLP regulations may not be directly applicable to basic research conducted at academic institutes or universities, there is no harm in conducting laboratory experiments as per GPL guidelines.

15.1 LABORATORY EXPERIMENTS

In scientific inquiry, an *experiment* is a method of investigating a hypothesis by generating laboratory data. An experiment can be used to help solve practical problems and to support or negate theoretical assumptions. In biology, the experiments are conducted to identify something new or to compare one thing with others. In this section, we discuss the different kinds of experiments which are conducted in any standard biotechnology laboratory.

15.1.1 Controlled Experiment

A controlled experiment generally compares the results obtained from an experimental sample against a control sample, which is practically identical to the experimental sample except for the one aspect whose effect is being tested (the independent variable). A good example would be a drug trial. The sample or group receiving the drug would be the experimental group, and the one receiving the placebo would be the control group. In many laboratory experiments, it is good practice to have several replicate samples for the test being performed and have both a positive control and a negative control. The results from replicate samples can often be averaged, or if one of the replicates is obviously inconsistent with the results from the other samples, it can be discarded as being the result of an experimental error (some step of the test procedure may have been mistakenly omitted for that sample). Most often, tests are done in duplicate or triplicate. A positive control is a procedure which is known from previous experience to give a positive result. A negative control is known to give a negative result. The positive control confirms that the basic conditions of the experiment were able to produce a positive result, even if none of the actual experimental samples produces a positive result. The negative control demonstrates the baseline result obtained when a test does not produce a measurable positive result. Often, the value of the negative control is treated as a "background" value to be subtracted from the test sample results. Sometimes, the positive control takes the quadrant of a standard curve.

An example that is often used in teaching laboratories is a controlled protein assay. Students might be given a fluid sample containing an unknown (to the student) amount of protein. It is their job to correctly perform a controlled experiment in which they determine the concentration of protein in the fluid sample (usually called the "unknown sample"). The teaching lab would be equipped with a protein standard solution with a known protein concentration. Students could make several positive control samples containing various dilutions of the protein standard. Negative control samples would contain all of the reagents for the protein assay, but no protein. In this example, all samples are performed in duplicate. The assay is a colorimetric assay in which a spectrophotometer can measure the amount of protein in samples by detecting a colored complex formed by the interaction of protein molecules and molecules of an added dye. The results for the diluted test samples can be compared to the results of the standard curve in order to determine an estimate of the amount of protein in the unknown sample.

Controlled experiments can be performed when it is difficult to precisely control all the conditions in an experiment. In this case, the experiment begins by creating two or more sample groups that are probabilistically equivalent, which means that measurements of traits should be similar among the groups and that the groups should respond in the same manner if given the same treatment. This equivalency is determined by statistical methods that take into account the amount of variation between individuals and the number of individuals in each group. In fields such as microbiology and chemistry, where there is very little variation between individuals and the group size is easily in the millions, these statistical methods are often bypassed, and simply splitting a solution into equal parts is assumed to produce identical sample groups.

Once equivalent groups have been formed, the experimenter tries to treat them identically, except for the one variable that he or she wishes to isolate. Human experimentation requires special safeguards against outside variables, such as the placebo effect. Such experiments are generally double blind, meaning that neither the volunteer nor the researcher knows which individuals are in the control group or the experimental group until after all of the data have been collected. This ensures that any effects on the volunteer are due to the treatment itself and are not a response to the knowledge that he or she is being treated. In human experiments, a subject (person) may be given a stimulus to which he or she should respond. The goal of the experiment is to measure the response to a given stimulus by a test method.

15.1.2 Observational Experiment

Observational studies are very much like controlled experiments, except that they lack probabilistic equivalency between groups. These types of experiments often arise in the area of medicine where, for ethical reasons, it is not possible to create a truly controlled group. For example, one would not want to deny all forms of treatment for a life-threatening disease to one group of patients to evaluate the effectiveness of another treatment on a different group of patients. The results of observational studies are considered much less convincing than those of designed experiments, as they are much more prone to selection bias. Researchers attempt to compensate for this with complicated statistical methods such as propensity score-matching methods.

15.2 LABORATORY SAFETY

Laboratory safety is very critical for all individuals who work in the lab most of the time, and it is very important to know the dos and don'ts while using lab facilities. We have listed some useful guidelines to ensure the safety of both individuals and the facility.

1. Conduct yourself in a responsible manner at all times in the laboratory.
2. Follow all written and verbal instructions carefully. If you do not understand a direction or part of a procedure, ask your teacher.
3. Never work alone in the laboratory. No student may work in the science classroom without the presence of a teacher.
4. When first entering a science room, do not touch any equipment, chemicals, or other materials in the laboratory area until you are instructed to do so.
5. Perform only those experiments authorized by your teacher. Carefully follow all instructions, both written and oral. Unauthorized experiments are not allowed.
6. Do not eat food, drink beverages, or chew gum in the laboratory. Do not use laboratory glassware as containers for food or beverages.
7. Be prepared for your work in the laboratory. Read all procedures thoroughly before entering the laboratory. Never fool around in the laboratory. Horseplay, practical jokes, and pranks are dangerous and prohibited.
8. Always work in a well-ventilated area.
9. Observe good housekeeping practices. Work areas should be kept clean and tidy at all times.
10. Be alert and proceed with caution at all times in the laboratory. Notify the teacher immediately about any unsafe conditions you observe.
11. Dispose of all chemical waste properly. Never mix chemicals in sink drains. Sinks are to be used only for water. Check with your teacher regarding the disposal of chemicals and solutions.
12. Labels and equipment instructions must be read carefully before use. Set up and use the equipment as directed by your teacher.
13. Keep your hands away from your face, eyes, mouth, and body while using chemicals or lab equipment. Wash your hands with soap and water after performing an experiment.
14. Experiments must be personally monitored at all times. Do not wander around the room, distract other students, startle other students, or interfere with the laboratory experiments of others.
15. Know the locations of and operating procedures for all safety equipment such as first aid kits and fire extinguishers. Know where the fire alarm and the exits are located.
16. Know what to do if there is a fire drill during a laboratory period; containers must be closed and any electrical equipment turned off.
17. Take only as much chemical as you need from the chemical container.
18. Always wear safety goggles when chemicals, heat, or glassware are used.
19. Do not wear contact lenses in the laboratory.
20. Dress properly during laboratory activities. Long hair, dangling jewelry, and loose or baggy clothing are hazards in the laboratory. Long hair must be tied back, and dangling jewelry and baggy clothing must be secured. Shoes must completely cover the foot. No sandals will be allowed on lab days.
21. A lab coat or smock should be worn during laboratory experiments.
22. Report any accident (spill, breakage, etc.) or injury (cut, burn, etc.) to the teacher immediately, no matter how trivial it seems. Do not panic.
23. If you or your lab partner is hurt, immediately (and loudly) yell out the teacher's name to get the teacher's attention. Do not panic.
24. If a chemical should splash in your eye(s) or on your skin, immediately flush with running water for at least 20 min. Immediately (and loudly) yell out the teacher's name to get the teacher's attention.
25. All chemicals in the laboratory are to be considered dangerous. Avoid handling chemicals with fingers. Always use tweezers. When making an observation, keep at least one foot away from the specimen. Do not taste or smell any chemicals.

26. Check the label on all chemical bottles twice before removing any of the contents. Take only as much chemical as you need.
27. Never return unused chemicals to their original container.
28. Never remove chemicals or other materials from the laboratory area.
29. Never handle broken glass with your bare hands. Use a brush and dustpan to clean up broken glass. Place broken glass in the designated glass disposal container.
30. Examine glassware before each use. Never use chipped, cracked, or dirty glassware.
31. If you do not understand how to use a piece of equipment, ask your teacher.
32. Do not immerse hot glassware in cold water. The glassware may shatter.
33. Do not operate a hot plate by yourself. Take care that hair, clothing, and hands are at a safe distance from the hot plate at all times. Use of a hot plate is only allowed in the presence of the teacher.
34. Heated glassware remains very hot for a long time. It should be set aside in a designated place to cool, and picked up with caution. Use tongs or heat protective gloves if necessary.
35. Never look into a container that is being heated.
36. Do not place hot apparatus directly on the laboratory desk. Always use an insulated pad. Allow plenty of time for hot apparatus to cool before touching it.

15.3 GOOD LABORATORY PRACTICES FOR BIOTECHNOLOGY LABS

1. All food and drinks should remain outside the lab or in your backpack. This is nonnegotiable.
2. Wash your hands before leaving the lab.
3. Clean your lab table benchtop with soap and water (or disinfectant, when working with bacteria) before and after your work.
4. Always prepare fresh media or solution and label it with your name and the date of preparation.
5. Nothing should be casually thrown in the trash or down the sink.
6. Check with your lab instructor or lab manager for proper disposal of waste materials, including chemical solutions, biological wastes (biohazard bags may be required), broken glass, sharps, and other supplies.
7. If the use of a lab coat is required, wear the lab coat only in the lab and not outside of the lab area.
8. Do not use aluminum foil or metal in the microwave.
9. Balance tubes before using centrifuges.
10. Workers handling human material must be vaccinated against Hepatitis B and tuberculosis (TB).
11. All blood or other human tissue should be treated as a potential Hepatitis/human immunodeficiency virus (HIV)/TB risk.
12. Wear personal protective equipment at all times.
13. Clinical samples should be collected by hospital staff and not by the laboratory staff.
14. All clinical samples have to be collected wearing an N95 mask.
15. Clinical samples must be properly labeled and stored.
16. Use latex disposable gloves.
17. Clinical samples should be processed only in a designated laboratory having the appropriate containment facilities.
18. All technical procedures should be performed in the hood to minimize the formation of aerosols and droplets.
19. Adequate and conveniently located biohazard waste containers must be used for disposal of contaminated materials. Work surfaces must be decontaminated after any spill of potentially dangerous material and at the end of the working day. Generally, a 5% bleach solution is appropriate for dealing with biohazard spillage.
20. Personnel must wash their hands often, especially after handling infectious materials, before leaving the laboratory working areas, and before eating.

21. Personal protective wear must be removed before leaving the laboratory.
22. Familiarize yourself with the rules of clinical sample processing, and make sure that relevant consent forms are filled out before any work is started.
23. Never use your own blood for any experiments such as culturing/transforming cells or DNA extraction.
24. Always use disposable plastic containers and tubes.
25. When using centrifuges, use sealed buckets. Should a breakage occur during centrifugation, leave the machine for 30 min after it stops before opening the lid. Clean any spillages immediately and thoroughly by absorbing onto paper tissues and discarding it as clinical waste, then clean with a disinfectant such as 5% Hycolin or diluted bleach.
26. To avoid contamination, do not enter into a cell or tissue culture lab with symptoms of a cold, flu, or sore throat, even if you are on medications.
27. Use only sterilized glassware or plasticware for DNA or RNA isolation and processing.
28. To avoid harming your eyes, do not look at the polymerase chain reaction (PCR) product on the casting gel without using UV-protected glasses or goggles.

15.4 GENERAL LABORATORY TECHNIQUES

15.4.1 Pipetting Technique

15.4.1.1 Plastic Pipette

- Pipettes are used as tools for transferring fluids. Pipettes can be graduated for measurement of small volumes of material.
- Squeeze the bulb before inserting the tip of the pipette into the liquid of choice.
- Insert the pipette into the liquid to be moved.
- Release your grip on the bulb. The liquid will be sucked into the pipette.
- Move your pipette to the next container and squeeze the bulb. Its contents will be pushed into the container.
- Be cautious about sucking particulate matter into the shaft of the pipette. The tip of the pipette can be enlarged, if required, by cutting the end.

15.4.1.2 Plastic Pipette Pump

- Insert the cotton-filled end of a sterile pipette into the end of a plastic pipette pump.
- Both pipettes and pumps come in different sizes. Be sure to use a pump that can "suck" the quantity of liquid you want (e.g., 10 mL pump with a 5 or 10 mL sterile pipette).
- Using the wheel at the top of the pump, raise the white top of the pump approximately 1 cm BEFORE inserting the pipette into the liquid. This step is necessary in order to create additional air pressure in the pipette as you expel the liquid.
- Next, place the tapered tip of the pipette into the liquid. The tip should be within the liquid during the pipetting procedure.
- Hold the pipette pump with one hand—your thumb should be placed on the wheel.
- Use your thumb to rotate the wheel downward. This will cause the liquid to rise into the pipette. Do this carefully, and watch the meniscus of the liquid rise to your desired level.
- *Remember*: Measure the level of the liquid at the BOTTOM of the meniscus.
- Next, take the tip of the pipette out of the liquid and move the entire apparatus to the place where you desire to put the measured liquid.
- Use your thumb to rotate the wheel upward. This will cause the liquid to be dispensed from the pipette. Lower the white top of the pump all the way to the pump shaft. Because you created the extra space before pipetting, the entire contents of the pipette should be now be dispensed.

15.4.1.3 Pipette with a Bulb

- There are three areas on the bulb that can be squeezed: A (air), S (suck), and E (empty). Identify these areas and give them a quick squeeze, just to check.
- Put a pipette into the bulb under area S. Insert the cotton-filled end carefully.
- Always hold the pipette with the open end down to prevent solutions from entering the bulb.
- Be aware that there are different graduations on different sizes of pipettes. Make sure you use the correct size!
- Squeeze the bulb while holding area A. This creates a vacuum.
- Put the end of the pipette tube into the liquid, and squeeze area S. Watch carefully, as the liquid rises quickly. Fill the pipette to a level just higher than you need for the experiment. Release area S.
- Put the pipette tube into the container in which you want to dispense the liquid. Squeeze area E to release the fluid. The harder you squeeze, the faster the liquid comes out.
- Dispose of the pipette properly when finished.

15.4.1.4 Micropipette

A micropipette is used to measure a very small and accurate volume of fluid. Micropipettes can be found in varieties that have set, unalterable fluid volumes or in varieties with various adjustable fluid ranges.

Select the appropriate pipette for the desired volume allocation.

- The range of the pipette can be found at the top of the pipette plunger.
- Use the correct tip for the specific micropipette size.
- Adjust the volume of the micropipette by turning the dial to the desired volume needed.
- Depress the plunger to the first level for the intake of the fluid.
- Place the tip of the micropipette in the desired fluid and slowly release pressure, allowing the plunger to rise, thus allowing for the intake of the fluid.
- Transfer the fluid to its desired location by depressing the plunger to its first position. Click the plunger a second time to allow for the full transfer of the fluid.
- Remove the tip from the pipette, with the plunger down, by pressing on the eject button.
- Always hold the point of the barrel down at all times, to prevent fluids from entering the pump mechanism.
- Always dispense used tips between each transfer.
- Do not stick micropipette tips in ears, eyes, nose, or mouth or other openings in the body.

15.4.2 Centrifugation Technique

1. Before running a centrifuge, check the classification decal on the centrifuge to ensure that the rotor is safe to use in the centrifuge at hand.
2. Never use an alkali detergent on a rotor (most are highly alkaline—be sure to check before use).
3. Always clean and completely dry the rotor after every use. Any spilled materials, especially salts and corrosive solvents, must be removed immediately with running water. Fixed-angle rotors are stored upside down to drain after thorough cleaning and rinsing. Always keep swinging buckets clean, dry, and store in inverted position with the caps removed. NEVER immerse the rotor portion of a swinging-bucket rotor. The linkage pins will inevitably rust, as it is virtually impossible to remove all fluids from them.
4. Be especially careful not to scratch the surface of a rotor or bucket. Use plastic brushes only. Normal wire brushes will scratch the anodized surface of aluminum rotors, which will increase the likelihood of corrosion. The anodized layer is extremely thin and is the main defense against corrosion of an aluminum rotor.

5. Always use the proper centrifuge tube. Glass tubes are used in clinical centrifuges only. High-speed Corex tubes can be used up to 15,000 rpm (in an SS34 rotor) if there are no scratches or imperfections in the glass and if the proper rubber or plastic adapter is employed. All ultracentrifugation use employs nitrocellulose or polyallomer tubes. Nitrocellulose ages and collapses in a strong centrifugal field if old.
6. Always fill the centrifuge tubes to the proper level (usually filled to within 1/2 in. of the top).
7. Always balance the rotor properly. Use a precision scale for most work. Always balance the tube with a medium that is identical to that being centrifuged—i.e., do not balance an alcohol solution with water or a dense sucrose solution with water only, as the distribution of the densities will be incorrect. For swinging buckets, be sure the buckets are weighed with their caps in place, that the seals are intact, and that the caps are secure. Be careful with the placement of tubes within the rotor to ensure proper balance—check the manufacturer's guide for complex rotors that hold multiple tubes.
8. Ensure that the rotor is properly seated within the centrifuge. For swinging buckets, ensure that they are hanging properly. For preparative rotors, be sure the rotor cover is in place and properly screwed down, where appropriate. Never use a rotor without its lid when one is supplied—the screw actually holds the rotor to the motor shaft.
9. Check that the centrifuge chamber is clean and defrosted, that all membranes or measuring devices are intact and functional, and that the lid is securely closed.
10. Adjust acceleration rates, deceleration rates, temperature, and rpm controls, as appropriate. Set brake on or off, as appropriate, and check the vacuum level, where appropriate.
11. Start the centrifuge and set the timer. Do not attempt to open the centrifuge until the rotor has come to a complete stop.
12. Before opening the centrifuge, record the appropriate information in the centrifuge log.
13. If properly balanced and used, the rotor should accelerate smoothly and with a constant change in the pitch of the motor sound. Any vibrations or unusual sounds should cause the operator to immediately stop the operation. Never leave the centrifuge until you are certain that it has reached its operating speed and is functioning properly. All rotors go through a minor vibration phase when they first start. There will be a minor flutter when the rotor reaches this vibration point; do not confuse this with a serious vibration caused by imbalance. If in doubt, halt the centrifuge and get assistance.

15.4.3 Spectrophotometer Technique

Spectrophotometers and colorimeters make use of the transmission of light through a solution to determine the concentration of a solute within the solution. A spectrophotometer differs from a colorimeter in the manner in which light is separated into its component wavelengths. A spectrophotometer uses a prism to separate light, while a colorimeter uses filters. Both are based on a simple design of passing light of a known wavelength through a sample and measuring the amount of light energy that is transmitted. This is accomplished by placing a photocell on the other side of the sample. All molecules absorb radiant energy at one wavelength or another. Those that absorb energy from within the visible spectrum are known as pigments. Proteins and nucleic acids absorb light in the ultraviolet (UV) range. The design of the single beam spectrophotometer involves a light source, a prism, a sample holder, and a photocell. Connected to each are the appropriate electrical or mechanical systems to control the illuminating intensity, the wavelength, and for conversion of energy received at the photocell into a voltage fluctuation. The voltage fluctuation is then displayed on a meter scale, is displayed digitally, or is recorded via connection to a computer for later investigation.

Spectrophotometers are useful because of the relation of intensity of color in a sample and its relation to the amount of solute within the sample. For example, if you use a solution of red food coloring in water and measure the amount of blue light absorbed when it passes through the solution, a measurable voltage fluctuation can be induced in a photocell on the opposite side. If the solution of red dye is now diluted to half its original concentration by the addition of water, the color will be approximately half as intense and the voltage generated on the photocell will be approximately half as great. Thus, there is a relationship between the voltage and the amount of dye in the sample.

Given the geometry of a spectrophotometer, what is actually measured at the photocell is the amount of light energy which arrives at the cell. The voltage meter reads the amount of light transmitted to the photocell. Light transmission is not a linear function, but an exponential function. That is why the solution was approximately half as intense when viewed in its diluted form. We can, however, monitor the transmission level and convert it to a percentage of the amount transmitted when no dye is present. The transmission is the relative percent of light passed through the sample, if half the light is transmitted; we can say that the solution has a 50% transmittance. Note that transmittance is always relative to a solution containing no dye. What makes all of this easy to use, however, is the conversion of that information from a percent transmittance to an inverse log function, known as *absorbance* (*or optical density*).

15.4.4 Aseptic Techniques

Aseptic techniques are used by cell culture specialists in handling products such as mammalian cells.

Materials required: Laminar flow or biological safety hood, as appropriate to the hazardous nature of the project, lint-free wipes, disinfectant solution, 70% alcohol, sterile pipettes of appropriate size, and biohazard waste container

Protocol:

1. Carry out all culturing operation in a laminar flow hood.
2. Disinfect all surfaces with a disinfectant solution prior to use.
3. Swab down the working surface liberally with 70% alcohol.
4. Periodically spread a solution of 70% alcohol over the exterior of gloves to minimize contamination. Replace them if torn.
5. In case of any spillage, spread a solution of 70% alcohol and swab immediately with non-linting wipes.
6. Discard gloves after use and do not wear them when entering any other lab area.
7. Bring into the work area only those items needed for a particular procedure.
8. Leave a wide, clear space in the center of the hood (not just the front edge) to work on. To prevent blockage of proper airflow and to minimize turbulence, do not clutter the area.
9. Swab all glassware (medium bottles, beakers, etc.) with 70% alcohol before placing them inside the hood.
10. Arrange the work area to have easy access to all of it without having to reach over one item to get at another (especially over an open bottle or flask).
11. Use sterile wrapped pipettes and discard them after use into a biohazard waste container.
12. Check that the wrapping of the sterile pipette is not broken or damaged.
13. Inspect the vessels to be used.
14. Discard any biohazardous or contaminated material immediately.
15. Never perform mouth pipetting. A pipettor must be used.
16. Make sure not to touch the tip of the pipette to the rim of any flask or sterile bottle.
17. When finished, clean the work area by wiping it with 70% alcohol.

15.5 GENERAL PRINCIPLES OF ANIMAL HANDLING

1. All animals must be allowed to acclimate to the facility for 3 days. During this time, they may not be experimentally manipulated. Acclimation periods of up to 1 week are recommended for all animals.
2. If a study will involve significant handling of animals, it is recommended that the animals be acclimated to the handling. Prior to experimental manipulation, handle the animal on a regular basis in a nonthreatening situation such as weighing, petting, and giving food treats. Most animals, even rodents, will respond positively to handling and will learn to recognize individuals.
3. Handle animals gently. Do not make loud noises or sudden movements that may startle them.
4. Handle animals firmly. The animal will struggle more if it sees a chance to escape.
5. Use an assistant whenever possible.
6. Chemical restraint should be considered for any prolonged or potentially painful procedure.

15.6 ANIMAL ANESTHESIA

An animal can be made unconscious by using anesthesia. The three components of anesthesia are analgesia (pain relief), amnesia (loss of memory), and immobilization. The drugs used to achieve anesthesia usually have varying effects in each of these areas. Some drugs may be used individually to achieve all three; others have only analgesic or sedative properties and may be used individually for these purposes or in combination with other drugs to achieve full anesthesia. Analgesia is the relief of pain. *Pain* is normally defined as an unpleasant sensory and emotional experience associated with potential or actual tissue damage. Pain is difficult to assess in animals because of the animal's inability to communicate directly about what it is experiencing. Instead, indirect signs of pain are often noted. Because of the difficulty of determining when an animal is in pain, animal welfare regulations require that analgesia be provided whenever a procedure is being performed or a condition is present that is likely to cause pain. In the absence of evidence to the contrary, it is assumed that something that is painful to a human will also be painful to an animal. It is best if analgesia can be provided to animals preemptively, or prior to the painful procedure, rather than waiting until after clinical signs of pain are observed. Analgesia is normally provided using one of several types of pharmaceutical preparations.

15.6.1 Anesthesia by Gas

Isoflurane gas anesthesia is recommended because isoflurane is minimally metabolized by the liver and is therefore less toxic to the animal's metabolism as compared to injectable anesthetics.

Materials: Gas anesthesia instrument equipped with a vaporizer, charcoal scavenger filters, induction chambers, and nose cones, isoflurane anesthesia gas

Method:

1. Weigh scavenger filter and record weight (saturation weight is indicated).
2. Dial vaporizer to "0"/OFF and fill with isoflurane.
3. Turn on O_2 tank at 55 psi.
4. Set O_2 flowmeter to 1 L/min.
5. Place the animal in the induction chamber.
6. Induce anesthesia by dialing vaporizer to 3%. The animal will be sedated in 1–2 min.
7. Reduce anesthesia to 1.5%–2% (imaging is non-traumatic; only a low dose of anesthetic is required to immobilize the animal; consistency of anesthesia dosing procedures is recommended in order to reduce variability in anesthetic effects).

8. Cover eyes with eye lubricant in order to prevent corneal dehydration (isoflurane inhibits blinking reflex).
9. Transfer the animal to imaging chamber and place in front of individual nose cones.
10. Maintain warmth in the animal during anesthesia in order to prevent hypothermia.

15.6.2 Injectable Anesthesia

Anesthetic induction using injectable anesthetics is fairly simple. It involves administration of the drug and monitoring the depth of anesthesia. Supportive care may be needed. Maintenance of injectable anesthesia can be through repeated bolus doses of the drug or through a constant infusion. Infusion rates are calculated based on the clearance time of the drug. Bolus dosing is simpler. Typically, half the original dose is given for repeat doses. Injectable anesthetics can be administered by various routes, depending upon the specific compound. The most frequently used routes of administration in laboratory animals are intravenous (IV), intramuscular (IM), intraperitoneal (IP), and subcutaneous (SQ). Less frequently used routes are intrathoracic, oral, and rectal, among others. The techniques are described below.

15.6.2.1 Intravenous Method

An appropriate vein must be selected. For large animals, the saphenous, cephalic, or jugular veins are best. For rodents, the tail veins are best. For rabbits and swine, ear veins may be used. The vein is held off proximal to the venipuncture site. The vessel may be stroked with a finger to stimulate blood flow into it. The needle is inserted into the vessel at an angle of 30°–45°. The needle is then lowered to align with the longitudinal axis of the vessel and advanced slightly, after which it should be drawn back. If blood appears in the hub of the needle, the drug may be injected. If not, try redirecting the needle (before you pull it out of the skin) and repeat. You may need to try several times while learning. Using a new, sharp needle for each stick, even if it is the same animal, will improve your chances of success. Once the needle is withdrawn, it is necessary to put pressure on the vessel to prevent bleeding.

Advantages: Rapid delivery of drug, ability to titrate dose, irritating substances may be given.

Disadvantages: Small veins are hard to access (especially in small animals), restraint is critical, developing skill in venipuncture takes experience.

15.6.2.2 Intramuscular Method

Insert the needle into a large muscle mass and draw back slightly. If blood is aspirated, you are in a blood vessel. Redirect the needle. When the needle is placed correctly, inject the drug. The best muscle masses to use for small animals are the caudal thigh muscles. For larger animals, the lateral dorsal spinal muscles or the cranial or caudal thigh muscles may be used. When administering a drug into thigh muscles, inject from the lateral aspect or, if from the caudal aspect, direct the needle slightly lateral. This will help avoid injecting into the sciatic nerve.

Advantages: Fairly rapid absorption, the technique is simple.

Disadvantages: IM injections are painful, small volumes are necessary, the animal may try to bite or escape.

15.6.2.3 Intraperitoneal Method

The animal is usually restrained in dorsal recumbency. The drug may be injected anywhere in the caudal two-thirds of the abdomen. However, it is best to try to avoid the left side in rodents and rabbits, because of the presence of the cecum. After the needle is inserted, draw it back. If anything is aspirated, you have likely hit the viscera. Withdraw and get a new needle before trying again. If the needle is placed correctly, the drug may be injected.

Advantages: Relatively large volumes may be injected (0.5 mL in mice, 2 mL in rats, etc.).

Disadvantages: The technique is more difficult than IM injections, drug may be administered into the viscera, resulting in no effect or in a complication.

15.6.2.4 Subcutaneous Method

Pinch an area of loose skin. Inject into the center of the "tent" created by pinching the skin.

Advantages: This technique is the simplest of all, and large volumes may be administered (basically as much as the tent of skin will hold that does not cause discomfort to the animal).

Disadvantages: Irritating substances cannot be given this way, absorption is slow.

15.6.3 Animal Euthanasia

The objective of this experiment is learning how to sacrifice animals after completing animal experimentation. Animals are normally euthanized at the end of a study for the purpose of sample collection or postmortem examination. Animals may be euthanized because they are experiencing pain or distress. Euthanasia is defined as a pain-free or stress-free death. The Institutional Animal Care and Use Committee (IACUC) has approved certain methods for humanely killing animals that meet the definition for euthanasia. The appropriateness of the method may vary from species to species. These guidelines are adapted from the report of the American Veterinary Medical Association Panel on Euthanasia.

15.6.3.1 Criteria for Euthanasia

Euthanasia of animals is expected if animals demonstrate the conditions listed below, whether the animal has been manipulated or not. Additional criteria may be specified on the Animal Usage Form. Fulfillment of one criterion can constitute grounds for euthanasia. Exceptions are permitted only if approved by the IACUC as part of the protocol review process (i.e., the clinical signs listed below are expected as part of the experiment and appropriate measures are taken to minimize pain or discomfort in the animals).

- *Weight loss*: Loss of 20%–25% (depending on attitude, weight recorded at time of arrival, and age: growing animals may not lose weight, but may not gain normally) or, if not measured, characterized by cachexia and muscle wasting
- *Impedance*: Complete anorexia for 24 h in small rodents, up to 5 days in large animals; partial anorexia (less than 50% of caloric requirement) for 3 days in small rodents, 7 days in large animals
- *Inability to obtain feed or water*: Animals unable to stand on its own, which condition persists for 24 h
- *Moribund state*: Depression coupled with body temperature falling below 99°F or nonresponsive to stimulation
- *Infection*: Infection involving any organ system (either overt or indicated by increased body temperature or white blood count parameters), which fails to respond to antibiotic therapy within an appropriate time and is accompanied by systemic signs of illness
- Signs of severe organ system dysfunction, nonresponsive to treatment, or with a poor prognosis as determined by a veterinarian

15.6.3.2 Surgical Operation

Surgical operations may be performed on animals under anesthesia, as detailed later, to avoid pain to animals. Chloroform is not acceptable for either anesthesia or euthanasia as it is very toxic to many species of mice. Additionally, this compound has been shown to be carcinogenic. Ether is irritating, flammable, and explosive, and should not be used in animal rooms. In addition, animals euthanized with ether must be left in a fume hood for several hours so that the carcasses are not explosive when disposed of. Precautions on ether use are available from the Department of Health and Safety (DEHS). Chloral hydrate used as a sole agent is not adequate to reliably achieve euthanasia.

15.6.3.3 Use of CO_2 Chamber

Whenever possible, euthanize animals in their home cage rather than transferring them to a new cage or chamber for euthanasia. Do not prefill the cage or chamber with CO_2. Open the tank and adjust the regulator to read no higher than 5 psi. Slow filling will minimize the nasal/ocular irritation and aversion to CO_2. Wait approximately 3–5 min for the animal to stop moving or breathing. Eyes should be fixed and dilated.

15.7 ANIMAL HISTOLOGY

Objective: The objective of this experiment is to learn to prepare animal tissue slides for histological staining.

General principles: *Histology* is the study of the microscopic anatomy of cells and tissues of plants and animals. It is performed by examining a thin slice (section) of tissue under a light microscope or electron microscope. Animal histology is a rigorous procedure of tissue and organ microanatomy and related structural changes, which deals with four basic tissues (epithelium, connective tissue, muscle, and nerves), and the organ systems (integument, digestive, respiratory, urinary, male and female reproductive, endocrine, and sensory systems), stressing human histology. The process involves animal dissection, tissue preparation, tissue sectioning, and tissue staining and analysis.

Tissue processing: Fixatives are used to preserve tissue from degradation and to maintain the structure of the cells, inclusive of subcellular components such as cell organelles (nucleus, endoplasmic reticulum, mitochondria). The most common fixative for light microscopy is 10% neutral buffered formalin, which is 4% formaldehyde in phosphate buffered saline (PBS). For electron microscopy, the most commonly used fixative is glutaraldehyde, usually as a 2.5% solution in PBS. These fixatives preserve tissues or cells mainly by irreversibly cross-linking proteins. The main action of these aldehyde fixatives is to cross-link amino groups in proteins through the formation of methylene (CH_2) linkage in the case of formaldehyde, or by trimethylethylene (C_5H_{10}) cross-links in the case of glutaraldehyde. This process, while preserving the structural integrity of the cells and tissue, can damage the biological functionality of proteins, particularly enzymes, and can also denature them to a certain extent. This can be detrimental to certain histological techniques. Further fixatives such as osmium tetroxide or uranyl acetate are often used for electron microscopy.

Frozen section fixation: A frozen section is a rapid way to fix and mount histology sections. It is used in surgical removal of tumors and allows rapid determination of margin (that the tumor has been completely removed). It is done using a refrigeration device called a *cryostat*. The frozen tissue is sliced using a microtome, and the frozen slices are mounted on a glass slide and stained the same way as in other methods. It is a necessary way to fix tissue for certain stains such as antibody-linked immunofluorescence staining.

Dehydration and infiltration: Biological tissue must be supported in a hard matrix to allow sufficiently thin sections to be cut, typically 5 μm thick for light microscopy and 80–100 nm thick for electron microscopy. For light microscopy, paraffin wax is most frequently used. Since it is immiscible with water, the main constituent of biological tissue, water must first be removed by the process of dehydration. Samples are transferred through baths of progressively more concentrated ethanol to remove the water, followed by a clearing agent, usually xylene, to remove the alcohol, and finally molten paraffin wax, which replaces the xylene. Paraffin wax does not provide a sufficiently hard matrix for cutting very thin sections for electron microscopy. Instead, resins are used. Epoxy resins are the most commonly employed embedding media, but acrylic resins are also used, particularly where immunohistochemistry is required. Thicker sections (0.35–5 μm) of resin-embedded tissue can also be cut for light microscopy. Again, the immiscibility of most epoxy and acrylic resins with water necessitates the use of dehydration, usually with ethanol.

Embedding: After the tissues have been dehydrated and infiltrated with the embedding material, they are ready for embedding. During this process, the tissue samples are placed into molds along with liquid embedding material, which is then hardened. This is achieved by cooling in the case of paraffin wax or heating in the case of the epoxy resins (curing). The acrylic resins are polymerized by heat, UV light, or chemical catalysts. The hardened blocks containing the tissue samples are then ready to be sectioned. Formalin-fixed paraffin-embedded (FFPE) tissues may be stored indefinitely at room temperature, and nucleic acids (both DNA and RNA) may be recovered from them decades after fixation, making FFPE tissues an important resource for historical studies in medicine. Embedding can also be accomplished using frozen, non-fixed tissue in a water-based medium. Prefrozen tissues are placed into molds with the liquid embedding material, usually a water-based glycol or resin, which is then frozen to form hardened blocks.

Sectioning: The tissue is cut using a glass knife mounted on a microtome or, for transmission electron microscopy, a diamond knife mounted on an ultramicrotome is used to cut 50-nm-thick tissue sections which are mounted on a 3-mL-diameter copper grid. The mounted sections then are treated with the appropriate stain. Frozen tissue embedded in a freezing medium is cut on a microtome in a cryostat.

Staining: Hematoxylin and eosin (H&E) is the most commonly used light microscopical stain in histology and histopathology. Hematoxylin stains nuclei blue; eosin stains the cytoplasm pink. Uranyl acetate and lead citrate are commonly used to impart contrast to tissue in the electron microscope. There are hundreds of various other techniques that have been used to selectively stain cells and cellular components. Other compounds used to color tissue sections include safranin, oil red o, Congo red, fast green fungal culture filtrate (FCF), silver salts, and numerous natural and artificial dyes that largely originated from the development dyes for the textile industry. *Histochemistry* refers to the science of using chemical reactions between laboratory chemicals and components within tissue. A commonly performed histochemical technique is the Perls Prussian blue reaction, used to demonstrate iron deposits in diseases like hemochromatosis.

Histology samples have often been examined by radioactive techniques. In historadiography, a slide (sometimes stained histochemically) is x-rayed. More commonly, autoradiography is used to visualize the locations to which a radioactive substance has been transported within the body, such as cells in S phase (undergoing DNA replication) which incorporate tritiated thymidine or sites to which radiolabeled nucleic acid probes bind in *in situ* hybridization. For autoradiography on a microscopic level, the slide is typically dipped into liquid nuclear tract emulsion, which dries to form the exposure film. Individual silver grains in the film are visualized with dark field microscopy. Recently, antibodies have been used to specifically visualize proteins, carbohydrates, and lipids, which is called *immunohistochemistry*; when the stain is a fluorescent molecule, it is called *immunofluorescence*. This technique has greatly increased the ability to identify categories of cells under a microscope. Other advanced techniques, such as nonradioactive *in situ* hybridization, can be combined with immunochemistry to identify specific DNA or RNA molecules with fluorescent probes or tags that can be used for immunofluorescence and enzyme-linked fluorescence amplification (especially alkaline phosphatase and tyramide signal amplification). Fluorescence microscopy and confocal microscopy are used to detect fluorescent signals with good intracellular detail. Digital cameras are increasingly used to capture histological and histopathological images. The Nissl method and Golgi's method are useful in identifying neurons.

15.8 BLOOD COLLECTION IN ANIMALS

The amount of blood needed and other factors will govern the method and sites for the collection of blood. Proper insertion of the needle into a vein or other part of the vascular system is normally the most difficult part of the procedure. Animal ethical guidelines must be followed in order to minimize pain in animals. Sometimes, veins may be expected to roll, collapse, or shift, making the entrance difficult. A precise, careful introduction of the needle is best, and several attempts may be required. The needle is inserted parallel to the vein and the tip directed into the lumen along the longitudinal axis. While withdrawing blood from a vein, aspiration should be reduced so the vessel does not collapse.

Site preparation: The area of injection or incision should be cleaned with alcohol. Some procedures will require sedation or anesthesia; others may be carried out without anesthesia, provided suitable restraint is used. In order to better visualize veins, dilation can be accomplished by immersing the tail in warm water for 5–10 s or by warming the animal with a low-wattage light bulb for 5–15 min prior to venipuncture. This also aids by providing additional light.

Tail vein venipuncture in mice: A scalpel blade and a 25–30 gauge needle are required. The mouse is restrained using a mechanical device. The veins may be seen laterally near the base of the tail, but good illumination and dilation will normally be required. A small blood sample may be collected by capillary action using a microhematocrit tube inserted into the hub of a small needle previously placed into the tail vein. This technique normally recovers a few drops of blood, adequate for hemoglobin, microhematocrit, and cell counts. Larger blood samples can be obtained by making a small incision over the vessels, 0.5–2 cm from the tail base, using a scalpel blade. 0.5–1 mL of blood can be withdrawn using this method. Anesthesia or sedation should be used.

Toe clipping or tail clipping to obtain blood samples: Clipping toes is an unacceptable method of blood collection. Likewise, tail clipping is not a preferable method for blood collection.

Cardiac puncture: A 0.90–0.50 mm needle is required. Cardiac puncture represents an accepted method of blood collection from mice when more than a few drops are required. However, this method also carries considerable risk to the animal, and occasionally deaths occur. It is not recommended as a repetitive blood sampling procedure. Animals must be anesthetized and restrained in dorsal recumbency. The needle is inserted under the xyphoid cartilage slightly to the left of midline. The needle is advanced at a 20°–30° angle from the horizontal axis to the sternum to enter the heart. Aspirate lightly while advancing. Blood should be withdrawn slowly, and the amount must be limited (up to 1 mL) unless euthanasia is planned.

Orbital sinus venipuncture in mice: Papillary tubes are required. Blood collection from the orbital sinus of mice is frequently used. One-quarter milliliter of blood can be repeatedly collected from mice at weekly intervals from alternate sides. Bleeding requires the tube be directed into the orbital sinus which surrounds the globe. In the mouse, the tube is inserted into the medial canthus of the eye and directed caudally and slightly dorsally. Knowledge of the location of the venous structures of the orbit of the mouse aids in establishing a successful periorbital bleeding technique. Pressure should be applied after blood collection to prevent hematomas. Anesthesia is required for all periorbital bleeding procedures.

Axillary bleed: A scalpel blade and a 3–5 cc syringe are required. Blood can be collected from the axillary region in a terminal exsanguination. Exsanguination of the mouse can be achieved by incising the right or left axillary region of an anesthetized mouse in dorsal recumbency. One to two milliliters of blood can be harvested in this manner.

15.9 HISTOLOGY AND MICROSCOPY

15.9.1 Immunocytochemistry Technique

Materials required: PHEM buffer: (25 mM HEPES, 10 mM EGTA, 60 mM PIPES, 2 mM $MgCl_2$ pH = 6.9). Antifade: 1 mL (1 mg p-phenylene diamine hydrochloride, Dissolve in 0.1 ml 10× PBS (20 min at RT). Add 0.9 mL 100% glycerol. Keep covered at all times and no vortexing. If it turns brown, it is no good. Aliquot and store at −70°C.

Protocol:

1. To sterilize glass coverslips, dip in ethanol and flame.
2. Use 22 × 22 × 1 mm^3 coverslips and put them in six well plates.
3. Seed 100,000 cells per well overnight and fix the next day.
4. Remove the media and rinse once with PBS.
5. Remove the PBS and immediately add −20°C methanol. (Do not allow the cells to dry.)
6. Put the plate in a −20°C freezer for 5 min.
7. Remove the methanol and add PHEM buffer. Fixed cells are kept at 4°C in PHEM.
8. Block with appropriate sera (2.5%–5%) in PHEM buffer for 1 h with gentle rocking.
9. Add primary antibody to the blocking buffer and incubate for 1 h with gentle rocking.
10. Remove and wash 4 × 10 min with PHEM buffer.
11. Add secondary antibody in PHEM buffer with sera and incubate for 30 min with gentle rocking.
12. Remove and wash 4 × 10 min with PHEM buffer.
13. Pick up coverslip with forceps and drain away excess buffer (can gently aspirate if desired).
14. Put ~20 μL "antifade" on slide and gently lay coverslip on top.
15. After removing excess antifade, either by blotting with Kimwipe or aspirating, seal with Sally Hansen clear nail polish. (This brand supposedly works better than others do.)
16. Store in freezer at −20°C.

15.9.1.1 Direct Immunofluorescence

Staining cells with antibodies directly linked to fluorochromes is known as *direct immunofluorescence* (DIF). DIF lends itself to multicolor experiments where a cell suspension is simultaneously stained with two, three, or four antibodies, each tagged with a different fluorescent dye. When designing multicolor experiments, it is important to use dyes that are compatible with each other. The inclusion of sodium azide prevents "capping" and so enables shorter incubations at room temperature.

Controls: All multicolor experiments require compensation controls, where the same cells are stained with each of the fluorochromes separately. The correct negative control for all immunofluorescent experiments are cells treated in the same way, but incubated with isotype-matched control antibodies with no known specificity, tagged with the same fluorochromes as the test antibodies. Sometimes, when dealing with suspensions where there will be negative and positive cells, such as peripheral T-cells, it is permissible to consider the unstained cells in the sample the negatives and exclude the negative control.

Equipment and reagents: Tissue culture medium with 10 mM sodium azide (TCM-N3), wash buffer: DPBS with 10 mM sodium azide (DPBS-N3), fixative, 1% paraformaldehyde in PBS (PFA).

Method:

1. Harvest cells in the usual way. Do not use trypsin to detach adherent cells. Rinse cells with cation-free PBS, and then use ethylenediaminetetraacetic acid (EDTA) in PBS. Wash cells in TCM-N3 by centrifuging in 50 mL conical tubes for 10 min at 2000 rpm. Aspirate supernatant and resuspend in 1 mL (TCM-N3). Count the cells using a hemocytometer. The ideal number of cells is approximately 5 × 10^5 per tube, and less than 5 × 10^4 is too few. Resuspend in n × mL of TCM-N3, where *n* is the number of tubes you will have.
2. Pipette 1 mL of the cell suspension into each of your 12 × 75 mm tubes (having labeled them first). Centrifuge for 5 min at 1500 rpm. Aspirate all but approximately 10 μL of the supernatant. Resuspend cells by vortexing or flicking the tube.
3. Dispense the antibodies into microcentrifuge tubes; with multicolur experiments, all the antibodies for one tube can be premixed before addition to cells. Centrifuging briefly in a microcentrifuge removes aggregates and reduces background staining. Add the appropriate amount of antibody to each tube and incubate for 15 min at room temperature.

4. Resuspend cells by flicking the tube or vortexing gently and add 1 mL of DPBS-N3. Centrifuge tubes at 1500 rpm for 5 min. Aspirate supernatant; resuspend cells by flicking or vortexing. Slowly add 0.5 mL of 1% formaldehyde in PBS while vortexing gently. Samples should be left for at least 1 h in the formaldehyde solution before running on the cytometer if they are virus infected or possibly virus infected.
5. *Paraformaldehyde* is the solid form of polymerized formaldehyde. Formaldehyde solution should be made fresh and not kept for more than a week. The most convenient form is vials of 20% formaldehyde. Old formaldehyde can spoil your data by increasing nonspecific background fluorescence. If you have to make it up from the powdered paraformaldehyde, mix the powder with PBS and leave in a water bath for a day or two. Heating on a hotplate is not recommended, as the fumes are dangerous.
6. A method of removing adherent cells from flasks: Make up PBS with 0.53 mM EDTA. It is important that there are no divalent cations in the PBS, for example, Ca2+ or Mg2+. Wash the monolayer free of tissue culture medium with PBS-EDTA. Decant off fluid and add more PBS-EDTA. Incubate for 15 min at 35°C. Rocking the flask may help. For really stubborn cells, you may need to give the flask a few hard raps to detach them.

15.9.1.2 Indirect Immunofluorescence

Staining cells with antibodies that are not directly conjugated to fluorochromes and then using a second labeled reagent to bind to your primary antibody is known as *indirect immunofluorescence* (IIF). This method was pretty standard earlier, so many antibodies became available directly conjugated to fluorochromes. Mostly with unlabelled antibodies, the second-step reagent will be another antibody with specificity for the primary antibody allotype (or sometimes, isotype). The typical second-step reagent would be goat anti-mouse immunoglobulin G (IgG) conjugated to fluorescein isothiocyanate (FITC). However, care should be taken to ensure that your second-step reagent will recognize the primary antibody. For instance, if your primary antibody is mouse IgG2a and your second step reagent is anti-mouse IgG1, then it will not work. In addition, some primary antibodies are made in hamster or rat, so beware.

Some primary antibodies are sold conjugated to biotin. In this case, the second-step reagent will be a fluorochrome-labeled streptavidin or avidin. The affinity of streptavidin for biotin is awesome, so short incubations can be used. Multicolor experiments are much more difficult with IIF, but are possible. The following protocol assumes it is a single-color experiment.

Controls: The correct negative control for all immunofluorescent experiments are cells treated in the same way, but incubated with isotype-matched control antibodies with no known specificity, tagged with the same fluorochromes as the test antibodies. Sometimes, when dealing with suspensions where there will be negative and positive cells, such as peripheral T-cells, it is permissible to consider the unstained cells in the sample as the negatives and exclude the negative control.

Method:

1. Harvest cells in the usual way. Do not use trypsin to detach adherent cells. Rinse cells with cation-free phosphate buffered slaine (PBS), then use ethylenediaminetetraacetic acid (EDTA) in PBS. Wash cells in sodium azide (TCM-N3) by centrifuging in 50 mL conical tubes for 10 min at 2000 rpm. Aspirate supernatant and resuspend in 1 mL (TCM-N3). Count the cells using a hemocytometer. The ideal number of cells is approximately 5×10^5 per tube, and less than 5×10^4 is too few. Resuspend in n × mL of TCM-N3, where n is the number of tubes you will have.
2. Pipette 1 mL of the cell suspension into each of your 12 × 75 mm tubes, having labeled them first. Centrifuge for 5 min at 1500 rpm. Aspirate all but about 10 μL of the supernatant. Resuspend cells by vortexing or flicking the tube.

3. Dispense the antibodies into microcentrifuge tubes. Centrifuging briefly in a microcentrifuge removes aggregates and reduces background staining. Add the appropriate amount of antibody to each tube (read the manufacturer's suggestions) and incubate for 15 min at room temperature.
4. Resuspend cells by flicking the tube or vortexing gently and add 1 mL of DPBS-N3; centrifuge tubes at 1500 rpm for 5 min. Aspirate supernatant.
5. Repeat step 4, leaving about 10 μL of fluid in the tube. Add the second-step reagent; incubate for 15 min at room temperature. Incubations can probably be reduced with streptavidin.
6. Resuspend cells by flicking the tube or vortexing gently and add 1 mL of DPBS-N3, centrifuge tubes at 1500 rpm for 5 min. Aspirate supernatant, resuspend cells by flicking or vortexing. Slowly add 0.5 mL of 1% formaldehyde in PBS while vortexing gently. Samples should be left for at least 1 h in the formaldehyde solution before running on the cytometer if they are virus infected or could be virus infected.

15.10 SEPARATION TECHNIQUES

15.10.1 Agarose Gel Electrophoresis (Basic Method)

Agarose gel electrophoresis is the easiest and commonest way of separating and analyzing DNA. The purpose of the gel might be to look at the DNA, to quantify it, or to isolate a particular band. The DNA is visualized in the gel by addition of ethidium bromide (EtBr). This binds strongly to DNA by intercalating between the bases and is fluorescent, meaning that it absorbs invisible UV light and transmits the energy as visible orange light.

15.10.2 How Much Percentage Gel Will Be Made?

Most agarose gels are made between 0.7% and 2%. A 0.7% gel will show good separation (resolution) of large DNA fragments (5–10 kb) and a 2% gel will show good resolution for small fragments (0.2–1 kb). Some people go as high as 3% for separating very tiny fragments, but a vertical polyacrylamide gel (PAGE) is more appropriate in this case. Low-percentage gels are very weak and may break when you try to lift them. High-percentage gels are often brittle and do not set evenly.

15.10.3 Which Gel Tank to Use?

Small 8 × 10 cm gels (minigels) are very popular and make good photographs. Larger gels are used for applications such as Southern and northern blotting. The volume of agarose required for a minigel is approximately 30–50 mL, while a larger gel may require 250 mL agarose.

15.10.4 How Much DNA Should Be Loaded?

You may be preparing an analytical gel to just look at your DNA. Alternatively, you may be preparing a preparative gel to separate a DNA fragment before cutting it out of the gel for further treatment. Either way, you want to be able to see the DNA bands under UV light in an ethidium bromide-stained gel. Typically, a band is easily visible if it contains approximately 20 ng of DNA.

Example: Suppose you are digesting a plasmid that comprises 3 kb of vector and 2 kb of insert. You are using EcoRI (a common restriction enzyme) and you expect to see three bands: the linearized vector (3 kb), the 5′ end of the insert (0.5 kb), and the 3′ end of the insert (1.5 kb). In order to see the smallest band (0.5 kb), you want it to contain at least 20 ng of DNA. The smallest band is one-tenth the size of the uncut plasmid. Therefore, you need to cut 10 × 20 ng, that is, 200 ng of DNA (0.2 μg). Then, your three bands will contain 120 ng, 20 ng, and 60 ng of DNA, respectively. All three bands will be clearly visible on the gel and the biggest band will be six times brighter than the smallest band.

Now, imagine cutting the same plasmid with BamHI (another popular restriction enzyme) and that BamHI only cuts the plasmid once, to linearize it. If you digest 200ng of DNA in this case, then the band will contain 200ng of DNA, will be very bright, and will probably be overloaded. Too much DNA loaded onto a gel is a bad thing. The band appears to run fast (implying that it is smaller than it really is) and, in extreme cases, can mess up the electrical field for the other bands, also making them appear the wrong size. Too little DNA is only a problem in that you will not be able to see the smallest bands because they are too faint.

15.10.5 Which Comb?

This depends on the volume of DNA you are loading and the number of samples. Combs with many tiny teeth may hold 10μL. This is no good if you want to load 20μL of restriction digest plus 5μL of loading buffer. When deciding whether a comb has enough teeth, remember that you need to load at least one marker lane, preferably two.

For making 1% gel in 50mL volume, measure out 0.5g of agarose into a 250mL conical flask. Add 50mL of 0.5× TBE and swirl to mix. It is good to use a large container, as long as it fits in the microwave, because the agarose boils over easily.

The agarose solution can boil over very easily, so keep checking it. It is good to stop it after 45s and give it a swirl. It can become superheated and NOT boil until you take it out, whereupon it boils out all over your hands, so wear gloves and hold it at arm's length. You can use a bunsen burner instead of a microwave—just remember to keep watching the agarose solution. Cool it down to about 60°C (just too hot to keep holding in your bare hands). If you had to boil it for a long time to dissolve the agarose, then you may have lost some water to water vapor. You can weigh the flask before and after heating and add in a little distilled water to make up this lost volume. While the agarose is cooling, prepare the gel tank on a level surface.

Add 1μL of EtBr (10mg/mL) and swirl to mix. The reason for allowing the agarose to cool a little before this step is to minimize production of EtBr vapor. EtBr is mutagenic and should be handled with extreme caution. Dispose of the contaminated tip into a dedicated EtBr waste container. A 10mg/mL EtBr solution is made up using tablets (to avoid weighing out powder) and is stored in the dark at 4°C, with a "TOXIC" label on it. Pour the gel slowly into the tank. Push any bubbles away to the side using a disposable tip. Insert the comb and double-check that it is correctly positioned.

The benefit of pouring slowly is that most bubbles stay up in the flask. Rinse out the flask immediately. Leave to set for at least 30min, preferably 1h, with the lid on if possible. The gel may look set much sooner, but running DNA into a gel too soon can give results that look terrible, with smeary, diffuse bands. Pour 0.5× TBE buffer into the gel tank to submerge the gel to a 2–5mm depth. This is the running buffer. You must use the same buffer at this stage as you used to make the gel—that is, if you used 0.6× TBE in the gel, then use 0.6× TBE for the running buffer. Remember to remove the metal gel formers if your gel tank uses them.

Prepare the samples: Transfer an appropriate amount of each sample to a fresh microfuge tube. It may be 10μL of a 50μL PCR reaction or 5μL of a 20μL restriction enzyme digestion. If you are loading the entire 20μL of a 20μL PCR reaction or enzyme digestion, then there is no need to use fresh tubes; just add in the loading buffer into the PCR tubes. Write down the physical order of the tubes in your lab book so you can identify the lanes on the gel photograph. Add an appropriate amount of loading buffer into each tube and leave the tip in the tube. Add 0.2 volume of loading buffer, e.g., 2μL into a 10μL sample. The tip will be used again to load the gel.

Load the first well with marker: Store the markers ready-mixed with loading buffer at 4°C. Load 2μL and avoid using the end wells, if possible. For example, if you have 12 samples and two markers, then you will use 14 lanes in total. If your comb formed 18 wells, then you will not be using four wells. It is best to not use the outer wells because they are the most likely to run aberrantly.

Continue loading the samples and finish off with a final lane of marker: Load the gels from right to left, with the wells facing you. This is because, by convention, gels are published as if the wells were at the top and the DNA had run down the page. If this seems confusing, then you can load left to right, with the wells facing away from you.

Close the gel tank, switch on the power-source, and run the gel at 5 V/cm: For example, if the electrodes are 10 cm apart, then run the gel at 50 V. It is fine to run the gel slower than this, but do not run it any faster. Above 5 V/cm, the agarose may heat up and begin to melt, with disastrous effects on your gel's resolution. Some people run the gel slowly at first (e.g., 2 V/cm for 10 min) to allow the DNA to move into the gel slowly and evenly, and then speed up the gel later. This may give better resolution. It is also fine to run gels overnight at very low voltages, e.g., 0.25–0.5 V/cm.

Check if electrical current is flowing: You can check this on the power source. The milliamps should be in the same ballpark as the voltage, but the best way is to look at the electrodes and check that they are evolving gas (i.e., bubbles). If not, then check the connections and make sure that the power-source is plugged in. This has been known to happen if people use water instead of running buffer.

Monitor the progress of the gel by reference to the marker dye: Stop the gel when the bromophenol blue has run three-fourths the length of the gel. Switch off and unplug the gel tank, and carry the gel (in its holder, if possible) to the dark room to look at on the UV light box. Some gel holders are not UV-transparent, so you have to carefully place the gel onto the glass surface of the light box. UV light is carcinogenic and must not be allowed to shine on naked skin or eyes, so wear face protection, gloves, and long sleeves.

Loading buffers: The loading buffer gives color and density to the sample to make it easy to load into the wells. In addition, the dyes are negatively charged in neutral buffers and thus move in the same direction as the DNA during electrophoresis. This allows you to monitor the progress of the gel. The most common dyes are bromophenol blue (Sigma B8026) and xylene cyanol (Sigma X4126). Density is provided by glycerol or sucrose.

Typical recipe:

- 25 mg bromophenol blue or xylene cyanol
- 4 g sucrose
- H_2O to 10 mL

The exact amount of dye is not important. Store at 4°C to avoid mold growing in the sucrose. Ten milliliters of loading buffer will last for years. Bromophenol blue migrates at a rate equivalent to 200–400 bp DNA. If you want to see fragments anywhere near this size (i.e., anything smaller than 600 bp), then use the other dye, because the bromophenol blue will obscure the visibility of the small fragments. Xylene cyanol migrates at approximately 4 kb equivalence, so do not use this if you want to visualize fragments of 4 kb.

Size markers: There are lots of different kinds of DNA size markers. In the old days, the cheapest defined DNA was from bacteriophages, so a lot of markers are phage DNA cut with restriction enzymes. Many of these, such as lambda HindIII, lambda PstI, and PhiX174 HaeIII, are still very popular. These give bands with known sizes, but the sizes are arbitrary. Choose a marker with good resolution for the fragment size you expect to see in your sample lanes. For example, for tiny PCR products, you might choose PhiX174 HaeIII, but for 6 kb fragments, you would choose lambda HindIII. More recently, companies have started producing ladder markers with bands at defined intervals such as 0.5, 1, 1.5, 2, 2.5 kb, and so on, up to 10 kb. If you know the total amount of DNA loaded into a marker lane and the sizes of all the bands, you can calculate the amount of DNA in each band visible on the gel. This can be very useful for quantifying the amount of DNA in your sample bands by comparison with the marker bands. It is good to load two markers lanes, flanking the samples. Many companies sell DNA size markers. It pays to shop around for the cheapest. Often, the local kitchen-sink biotech company sells excellent markers.

15.11 MICROBIOLOGY TECHNIQUES

15.11.1 Isolation of Pure Culture

A pure culture theoretically contains a single bacterial species. There are a number of procedures available for the isolation of pure cultures from mixed populations. A pure culture may be isolated by the use of special media with specific chemical or physical agents that allow the enrichment or selection of one organism over another. Simpler methods for isolation of a pure culture include (i) spread plating on solid agar medium with a glass spreader and (ii) streak plating with a loop. The purpose of spread plating and streak plating is to isolate individual bacterial cells (colony-forming units) on a nutrient medium. Both procedures (spread plating and streak plating) require understanding of the aseptic technique. *Asepsis* can be defined as the absence of infectious microorganisms, although the term is usually applied to any technique designed to keep unwanted microorganisms from contaminating sterile materials.

Materials required: Seven 9 mL dilution tubes of sterile saline, seven nutrient agar plates, 1.0 and 0.1 mL pipettes, glass spreader or hockey stick, 95% ethyl alcohol in glass beaker, broth culture of *Staphylococcus aureus* and *Serratia marcescens,* mixed overnight

Procedure:

A. Spread plate technique: In this technique, the number of bacteria per unit volume of sample is reduced by serial dilution before the sample is spread on the surface of an agar plate.

1. Prepare serial dilutions of the broth culture as shown below. Be sure to mix the nutrient broth tubes before each serial transfer. Transfer 0.1 mL of the final three dilutions (10^{-5}, 10^{-6}, and 10^{-7}) to each of three nutrient agar plates and label the plates.
2. Position the beaker of alcohol containing the glass spreader away from the flame. Remove the spreader and very carefully pass it over the flame just once (lab instructor will demonstrate). This will ignite the excess alcohol on the spreader and effectively sterilize it.
3. Spread the 0.1 mL inoculum evenly over the entire surface of one of the nutrient agar plates until the medium no longer appears moist. Return the spreader to the alcohol.
4. Repeat the flaming and spreading for each of the remaining two plates.
5. Invert the three plates and incubate at room temperature until the next lab period.

B. Streak plate technique: The streak plating technique isolates individual bacterial cells (colony-forming units) on the surface of an agar plate using a wire loop. Once again, the idea is to obtain isolated colonies after incubation of the plate.

1. Label two nutrient agar plates as "No. 1" and "No. 2."
2. Prepare two streak plates by following two of the three streaking methods. Use the 10^{-1} dilution as inoculum.
3. Invert the plates and incubate at room temperature until the next lab period.

C. Exposure plates: Exposure of sterile media to the environment will demonstrate the importance of the aseptic technique.

1. Label two nutrient agar plates as "Exposure I" and "Exposure II."
2. Uncover the plate marked "Exposure I" and allow it to remain exposed in the lab for about 5 min.
3. Expose the plate marked "Exposure II" to a source of possible contaminants.
4. Use your imagination: cough or sneeze, place your fingers on the surface of the agar, etc.
5. Invert the plates and incubate at room temperature until the next lab period.

15.11.2 Streaking Bacteria for Single Colonies

1. For the Initial inoculum obtained from solid media, touch isolated colonies with a sterile applicator or toothpick.
2. Draw the inoculated end of the applicator across an agar media plate, keeping toward the edge of the plate.
3. Select a fresh applicator and make a second single streak across the first one.
4. If unskilled at the technique, selected a fresh applicator; if skilled, rotate the applicator 180° and use the sterile edge.
5. Streak across the second streak, avoiding the first streak, and continue to make dense streaks until the designated area has been filled.
6. Single colonies will form where a single bacterium was deposited on the medium. It is easy to streak four strains for single colonies on a single 100 mm plate. Twelve or more strains can be done by skilled workers.
7. Invert plate so that the media face points down. This prevents water droplets from dripping onto the plate surface. Incubate at the appropriate temperature.

15.11.3 Gram-Staining Procedure

Gram-staining is a four-part procedure which uses certain dyes to make a bacterial cell stand out against its background. The specimen should be mounted and fixed on a slide before you proceed to stain it.

Reagents required: crystal violet (the primary stain), iodine solution (the mordant), decolorizer (ethanol is a good choice), safranin (the counterstain), water (preferably in a squirt bottle)

Step 1: Place your slide on a slide holder or a rack. Flood (cover completely) the entire slide with crystal violet. Let the crystal violet stand for about 60 s. When the time has elapsed, wash your slide with water for 5 s. The specimen should appear blue-violet when observed with the naked eye.

Step 2: Now, flood your slide with the iodine solution and let it stand for approximately a minute. When time has expired, rinse the slide with water for 5 s and immediately proceed to step 3. At this point, the specimen should still be blue-violet.

Step 3: This step involves the addition of the decolorizer, which is ethanol. This step is also somewhat subjective, because using too much decolorizer could result in a false Gram-negative result. Likewise, not using enough decolorizer may yield a false Gram-positive result. To be safe, add the ethanol dropwise until the blue-violet color is no longer emitted from your specimen. As in the previous steps, rinse with water for 5 s.

Step 4: The final step involves applying the counterstain, safranin. Flood the slide with the dye as you did in steps 1 and 2. Let this stand for about a minute, to allow the bacteria to incorporate the safranin. Gram-positive cells will incorporate little or no counterstain and will remain blue-violet in appearance. Gram-negative bacteria, however, take on a pink color and are easily distinguishable from the Gram positives. Again, rinse with water for 5 s to remove any excess dye.

After you have completed steps 1 through 4, you should blot the slide gently with bibulous paper or allow it to air-dry before viewing it under the microscope.

15.12 BIOCHEMISTRY TECHNIQUES

15.12.1 Estimation of Fat in Milk Samples

Milk fat measurement is a common task in the dairy industry, because milk fat content is one factor that determines milk price and it is necessary to know for casein/fat ratio normalization. It is also important for a dairyman to know the exact milk fat content, because discrepancies in the results of milk fat tests (usually performed in the dairy industrial plant) have economic relevance. In addition, a low milk fat content could indicate the existence of animal health deficiencies.

Principle: The fat can be separated from fat-containing milk/milk powder through the addition of sulfuric acid. The separation is made by using amyl alcohol and centrifugation. The fat content is read directly on a special calibrated butyrometer after centrifugation.

Apparatus: Good quality balance, special butyrometer, caoutchouc stoppers, pipettes (1 and 10 mL), centrifuge (1200 rpm) equipped with heating element

Reagents: 90% sulfuric acid, 100% amyl alcohol

Procedure:

1. Mix the milk sample (at a temperature of approximately 20°C) thoroughly, taking care to minimize incorporation of air. Allow the sample to stand for a few minutes to discharge any air bubbles. Mix again gently before pipetting.
2. Pipette sequentially into the butyrometer:
   a. 10 mL of 90% sulfuric acid
   b. 10.75 mL milk (must not be mixed with the acid)
   (*Note*: Pipette the required volume of milk into the butyrometer. Care must be taken to avoid charring of the milk, by ensuring that the milk flows gently down the inside of the butyrometer. It then rests on top of the acid.)
   a. 1 mL amyl alcohol
   b. Some drops of distilled water
3. Clean the neck of the butyrometer with a tissue or dry cloth.
4. Close the butyrometer with a caoutchouc stopper and shake until the milk has dissolved. The butyrometer is turned upside down five or more times.
5. Spin in the Gerber centrifuge for 15 min.
6. The fat column is adjusted by using the stopper so that it will be in the graduated part of the butyrometer. The fat percentage is read directly.
   Note: Measurements should be carried out in duplicate.

Hazards: Sulfuric acid is toxic, highly corrosive, and will cause severe burning if it comes in contact with the skin or eyes. When mixing the butyrometer contents, considerable heat is generated. If the stopper is slightly loose, leakage may occur during mixing, centrifuging, or holding in the water bath.

Precautions: Wear protective eye goggles. Avoid all spillage and dropping of sulfuric acid from acid dispensers. When mixing, hold the butyrometer stopper firmly to ensure that it cannot slip. Use a cloth or gloves to protect the hands when mixing.

Do not point the butyrometer at anyone when mixing.

15.12.2 Protein Quantification by Bradford Assay

The *Bradford dye-binding assay* is a colorimetric assay for measuring total protein concentration. It involves the binding of Coomassie brilliant blue to protein. There is no interference from cations or anions from carbohydrates such as sucrose. However, detergents such as sodium dodecyl sulfate (SDS) and triton x-100, as well as strongly alkaline solutions, can interfere with the assay.

Procedure:

1. Dilute 5× Bradford reagent (BioRad) 4:1 with water and run through a gravity filter.
2. Get a 96 well enzyme-linked immunosorbent assay (ELISA) plate (Limbro microtitration plate).

3. Turn on machine to allow the bulb to warm up (approximately 10 min before use).
4. Prepare standards.
5. Plate standards in triplicate (10 μL per well). Leave column 1 blank and begin plating in column 2.
6. Plate samples in triplicate (10 μL per well). Since your standard curve is from 0–1 mg/mL (for bovine serum albumin, BSA), your samples should be in this range. Taking an optical density (OD) reading on your sample and generalizing that 1 OD = 1 mg/mL is a good way to get a ballpark value. Dilute your samples appropriately, and if you do not know the concentration of protein, make several samples of various dilutions.
7. Add 200 μL of diluted Bradford reagent to each well and let it stand for 5 min.
8. Insert the 595 nm filter into the machine. Set the plate on the reader. Press "blank," and after the machine gives a printout of the blank reading, press "start." The reading is complete when the plate slides back out from underneath the reader.
9. Use the results to graph the standard curve (axes are commonly labeled as y = A, 600 nm and x = mg/mL). Use the curve and data from the Bradford values to determine unknown protein concentration.

Preparation of standards: A set of standards is created from a stock of protein whose concentration is known. The Bradford values obtained for the standard are then used to construct a standard curve to which the unknown values obtained can be compared to determine their concentration. As your standard, use a protein that most closely resembles the protein you are assaying. BSA and IgG are typical standards used to construct the curve. For BSA, use 0–1 mg/mL as your standard curve concentration; for IgG, use 0–1.6 mg/mL. Examples of typical standards created from BSA stock (1 mg/mL) to give a standard curve from 0–1 mg/mL:

- Sample #1 (0.0 mg/mL): 0 μL BSA + 30 μL buffer
- Sample #2 (0.2 mg/mL): 6 μL BSA + 24 μL buffer
- Sample #3 (0.4 mg/mL): 12 μL BSA + 18 μL buffer
- Sample #4 (0.6 mg/mL): 18 μL BSA + 12 μL buffer
- Sample #5 (0.8 mg/mL): 24 μL BSA + 6 μL buffer
- Sample #6 (1.0 mg/mL): 30 μL BSA + 0 μL buffer

15.12.3 Indirect ELISA

The indirect ELISA is used primarily to determine the strength and/or amount of antibody response in a sample, whether it is from the serum of an immunized animal or the cell supernatant from growing hybridoma clones.

Procedure:

1. All incubations are done as follows: Cover with plate with sealer tape or place in a sealed box containing a wet paper towel and incubate for 2 h at room temperature.
2. For the four controls: Two are negative controls using pre-bleeds of the same concentration as your starting dilution of primary antibody; the third is a negative control where no primary antibody is added, just blocking buffer at this step; and the fourth is a positive control, either from a previously positive bleed or cell supernatant, or you can lay down primary antibody as antigen.
3. In a 96 well ELISA plate (Nunc MaxiSorp is best), add 100 ng of antigen in 50 μL in each well you will be using for your test, as well as four control wells. The perimeter wells on the plate are generally not used, as they tend to give poor results. Incubate.
4. Dump out the antigen solution and add 100 μL of blocking buffer (1% BSA, 0.1 M KPi, 0.1% Tween-20, 0.02% thimerisol, pH 7). If your carrier protein for injection was BSA,

then substitute 1% nonfat dry milk for the BSA. Incubate. The blocking step incubation can be also done at 4°C overnight.

5. Dump out the blocking buffer and bang the plate upside down on some paper towels to remove all the liquid. Wash three times with wash buffer (0.1 M KPi, 0.05% Tween-20, pH 7), shaking the wash out vigorously each time. Again bang out the residual wash buffer.
6. Add 50 μL/well of your 1° antibody. For screening hybridomas, this will be the cell supernatant. For obtaining a titer on serum from an immunized animal, you will need to perform serial dilutions of the serum in blocking buffer in the plate. 1:1 serial dilutions are done by placing 100 μL of your starting dilution of serum in the first column of your plate (1:499 in blocking buffer is usually a good starting point).
7. Place 50 μL/well of blocking buffer down all the wells remaining in the rows you are using. Pipette out 50 μL from the first well (with your starting dilution) and place in the next well in the row. Mix by pipetting the solution up and down, and then transfer 50 μL of this solution to the next well and mix again. Continue these dilutions down the row until the last well, where you remove 50 μL and throw it away. Incubate.
8. Wash three times, as before.
9. Add 50 μL/well of a 1:1999 dilution in blocking buffer of horseradish peroxidase (HRP)-labeled 2° antibody that is directed against the species of your primary antibody (anti-mouse for monoclonal and mouse serum, anti-rabbit for polyclonal antibodies raised in rabbits). Incubate.
10. Wash three times, as before.
11. Add 100 μL/well of ABTS HRP substrate. Incubate at room temperature for 5–20 min, depending on the rate of color development. Keep the time identical for subsequent comparisons of titer, and for hybridoma screening go the full 20 min.
12. Add 100 μL/well stop solution (0.5 M oxalic acid).
13. Read absorbance at 414 nm in an ELISA reader.

15.12.4 Sandwich ELISA

The sandwich ELISA measures the amount of antigen between two layers of antibodies. The antigens to be measured must contain at least two antigenic sites, capable of binding to the antibodies, since at least two antibodies act in the sandwich. Therefore, sandwich assays are restricted to the quantitation of multivalent antigens such as proteins or polysaccharides. Sandwich ELISA for quantitation of antigens is especially valuable when the concentration of antigens is low and/or they are contained in high concentrations of contaminating protein.

Procedure:

1. A capture antibody is first diluted in 0.1 M bicarbonate buffer, pH 9.2, and then 50 μL is added to each well of the microtiter plate.
2. The antibody-coated plate is covered with paraffin and incubated in the cold room overnight in a moist box containing a wet paper towel or at room temperature and humidity for 2 h.
3. The plate is emptied and the unoccupied sites are blocked with 100 μL of blocking buffer containing 100 mM phosphate buffer, pH 7.2, 1% BSA, 0.5% Tween-20, and 0.02% Thimerosol for 30 min at room temperature.
4. The plate is emptied and washed three times with wash buffer (100 mM phosphate buffer, 150 mM sodium chloride (NaCl), 0.2% BSA, and 0.05% Tween-20).
5. The antigen solution is first diluted in antigen buffer (100 mM phosphate buffer, 150 mM NaCl) and then added to the plate in a volume of 50 μL per well. The plate is incubated at room temperature for 45 min to 1 h.
6. The plate is emptied again and washed three times with wash buffer.

7. The enzyme-labeled antibody against antigen is diluted appropriately in 0.1 M bicarbonate buffer, pH 9.2, after which 50 μL is added to each well and incubated at room temperature for 30 min.
8. The plate is emptied again and washed three times with wash buffer.
9. The color development system is added and the color intensities are measured.

15.12.5 Sonication of Bacteria

This procedure is used to break cells that contain a protein to be purified. This procedure is valid only for the use on Appling's lab sonicator.

Preparation:

1. Place your sample in a 15 mL elongated plastic tube such that the sample tube wall interface area is maximized.
2. Fill a small container with ice and NaCl and then add a small amount of water (the NaCl lowers the freezing temperature).
3. Place the tube with the sample in the ice container, making sure that the ice bath completely covers the walls of the tube.

Sonication:

1. Turn on the sonicator and calibrate by placing the 20/40 switch in the 20 position and the control power switch in the 60 position. Start the apparatus by pressing the foot pedal and move the calibration knob such that the output signal is at a minimum.
2. Move the power switch to the 100 position and repeat the above calibration procedure.
3. Clean the probe with ethanol, let dry, and place it inside the tube. Make sure that the cold tube is inside the ice container. Move the 20/40 switch to the 40 position and the power switch to 70. Leave the sonicator on for 30 s and then release the foot pedal.
4. Wait about 1 or 2 min until the tube cools down again, and then gently shake the tube. Place it on ice and repeat the experiment.
5. Move the power switch to 90 and repeat the experiment. The sequence of sonication is 2× at 70 for 30 s and 1× at 90 for 30 s, waiting 1–2 min on the ice bath in between.
6. Some proteins resist sonication more than others, so different times may be needed. The initial cell sample is viscous and, as the sonication processes, the viscosity of the sample decreases due to macromolecular shearing.
7. Another way to monitor the amount of sonication needed to break open the bacteria is to monitor the release of DNA and RNA by measuring the absorbance at 260 nm of your sample after every sonication step. The A260 will increase until it reaches a plateau—this means all the bacteria are lysed and no more nucleic acids can be liberated. Always remember to cool down your sample in between and after sonication to prevent heat denaturing your proteins.

15.13 MOLECULAR TECHNIQUES

15.13.1 Isolation of Genomic DNA from Blood

In order to isolate genomic DNA (gDNA), blood samples which are stored at −70°C in EDTA vacutainer tubes are thawed, standard citrate buffer is added and mixed, and the tubes are centrifuged. The top portion of the supernatant is discarded, additional buffer is added and mixed, and the tube is centrifuged again. After the supernatant is discarded, the pellet is resuspended in a solution of SDS detergent and proteinase K, and the mixture is incubated at 55°C for 1 h. The sample is then phenol-extracted once with a phenol/chloroform/isoamyl alcohol solution and, after centrifugation the aqueous layer, is removed to a fresh microcentrifuge tube. The DNA is ethanol precipitated, resuspended in buffer, and then ethanol precipitated the second time. Once the pellet is dried, buffer is added, and the DNA is resuspended by incubation at 55°C overnight. Then the gDNA solution is assayed by PCR. Please refer to Table 15.1 for preparation of molecular reagents.

Protocol:

1. Blood samples typically were obtained as 1 mL of whole blood stored in EDTA vacutainer tubes frozen at −70°C.
2. Thaw the frozen samples, and to each 1 mL sample add 0.8 mL 1× saline–sodium citrate (SSC) buffer, and mix. Centrifuge for 1 min at 12,000 rpm in a microcentrifuge.
3. Remove 1 mL of the supernatant and discard into disinfectant.
4. Add 1 mL of 1× SSC buffer, vortex, and centrifuge as above for 1 min, and remove all of the supernatant.
5. Add 375 L of 0.2 M sodium acetate (NaOAc) to each pellet and vortex briefly. Then add 25 μL of 10% SDS and 5 μL of proteinase K (20 mg/mL H_2O) (Sigma P-0390), vortex briefly, and incubate for 1 h at 55°C.
6. Add 120 μL phenol/chloroform/isoamyl alcohol and vortex for 30 s. Centrifuge the sample for 2 min at 12,000 rpm in a microcentrifuge tube.
7. Carefully remove the aqueous layer to a new 1.5 mL microcentrifuge tube, add 1 mL of cold 100% ethanol, mix, and incubate for 15 min at −20°C.
8. Centrifuge for 2 min at 12,000 rpm in a microcentrifuge. Decant the supernatant and drain.
9. Add 180 μL 10:1 Tris–EDTA (TE) buffer, vortex, and incubate at 55°C for 10 min.
10. Add 20 μL 2 M NaOAc and mix. Add 500 μL of cold 100% ethanol, mix, and centrifuge for 1 min at 12,000 rpm in a microcentrifuge.
11. Decant the supernatant and rinse the pellet with 1 mL of 80% ethanol. Centrifuge for 1 min at 12,000 rpm in a microcentrifuge.
12. Decant the supernatant and dry the pellet in a Speed-Vac for 10 min (or until dry).
13. Resuspend the pellet by adding 200 μL of 10:1 TE buffer. Incubate overnight at 55°C, vortexing periodically to dissolve the gDNA. Store the samples at −20°C.

15.13.2 Isolation of DNA from Fresh or Frozen Tissue

Reagents required: chloroform, EDTA, 0.5 M ethanol, absolute isoamyl alcohol, phenol, PBS, 1×, proteinase K, RNase A, SDS solution, 10% digene diagnostics

Preparation of DNA buffer TE: 1 M Tris pH 8.0 20 mL, 0.5 M EDTA 20 mL, sterile water 100 mL, proteinase K (10 mg/mL). Dissolve 100 mg proteinase K in 10 mL TE for 30 min at room temperature. Aliquot and store at −20°C.

RNase A (20 mg/mL): Dissolve 200 mg RNase A in 10 mL sterile water, boil for 15 min, and cool to room temperature. Aliquot and store at −20°C.

Procedure:

1. Place 60–80 mg of tissue in a petri dish with culture media, and divide the tissue into two pieces.
2. Put the tissue into two sterile 15 mL tubes and centrifuge for 2 min at 4°C at 1500 rpm.
3. Remove the supernatant, and wash twice with 1 mL 1× PBS or DNA buffer.
4. (It is possible to store the pellet at −80°C. In that case, add 1 mL 1× PBS and resuspend the pellet. Use a cryotube, and centrifuge at 1500 rpm for 2 min at 4°C. Remove the supernatant and freeze the pellet.)

TABLE 15.1
Conversion Table for Molecular Biology Solutions

3 M NaOAc (5.2)
NaOAc $3H_2O$ 40.81 g
Double-distilled water (DD-H_2O) 80 mL
Adjust pH to 5.2 with glacial HOAc Adjust volume to 100 mL with DD-H_2O, and autoclave to sterilize

Gel-Loading Buffer (6×)
Bromophenol Blue 0.025 g
Xylene Cyanole 0.025 g
Ficoll 400 1.5 g
DD-H_2O to 10 mL
May be stored at room temperature or at 4°C

1 M Tris (per 500 mL)
Tris base 60.55 g
DD-H_2O 400 mL
Dissolve and adjust to the desired pH by adding concentrated hydrochloric acid (HCl)
Adjust volume to 500 mL with DD-H_2O
Autoclave to sterilize

0.5 M EDTA (8.0) (per liter)
EDTA.$2H_2O$ (disodium salt) 186.1 g
DD-H_2O 800 mL
Stir vigorously and adjust pH to 8.0 with concentrated NaOH in order to dissolve the EDTA salt
Adjust the volume to 1 L and autoclave to sterilize

20% SDS (per 500 mL)
In a fume hood, carefully add 1 100 g bottle of SDS to 450 mL DD-H_2O
Heat to assist dissolution
Adjust pH to 7.2 using concentrated HCl
Adjust volume to 500 mL with DD-H_2O
Sterilize by filtration

EtBr (10 mg/mL) (per 100 mL)
In an amber bottle, add 1 g of EtBr to 100 mL of DD-H_2O
Stir for several hours and store at 4°C
Use gloves

$CaCl_2$ Solution (per 500 mL) for preparing competent cells
$CaCl_2$ $2H_2O$ 4.41 g
Glycerol 75 mL
1 M Tris 7.5 5 mL
DD-H_2O 400 mL
Adjust volume to 500 mL with DD-H_2O and filter sterilize. Store at 4°C

TAE(50×) (per liter)
Tris base 242 g
Glacial HOAc 57.1 mL
0.5 M EDTA 8.0 100 mL
Add DD-H_2O to 500 mL and autoclave to sterilize

GTE (per 100 mL)
Glucose 0.90 g
0.5 M EDTA 2 mL
1 M Tris 8.0 2.5 mL
Add DD-H_2O to 100 mL, filter sterilize and store at 4°C

TBE(20×) (per liter)
Tris Base 216 g
Boric acid 110 g
0.5 M EDTA 8.0 80 mL
Add DD-H_2O to 500 mL and autoclave to sterilize

KOAc Solution (per 100 mL)
5 M KOAc 60 mL
Glacial HOAc 11.5 mL
DD-H_2O 28.5 mL
Filter sterilize and store at 4°C

TE(10×) (per 500 mL)
1 M Tris 8.0 50 mL
0.5 M EDTA 8.0 5 mL
DD-H_2O 445 mL
Autoclave to sterilize

Prehybridization Solution (per 500 mL)
NaCl 29.22 g
Ficoll 400 1 g
PVP 360 1 g
BSA 1 g
Sodium Pyrophosphate 0.5 g
1 M Tris 7.4 25 mL
Formamide 250 mL
Combine all ingredients in a 1 L flask by stirring
Add DD-H_2O to just under 500 mL
Stir for several hours or overnight
Then add 2.5 mL 20% SDS
Before use, heat sonicated sperm DNA to 95°C for a minute and add to Prehyb

5 M NaCl (per liter)
292.2 g NaCl
800 mL DD-H_2O
Dissolve, adjust volume to1 L, and autoclave to sterilize

5 M ammonium acetate (NH_4OAc) (per 100 mL)
NH_4OAc 38.5 g
DD-H_2O 80 mL
Adjust volume to 100 mL with DD-H_2O and sterilize by filtration

5. Remove supernatant and resuspend the pellet in 2.06 mL DNA-buffer.
6. Add 100 μL proteinase K (10 mg/mL) and 240 μL 10% SDS, shake gently, and incubate overnight at 45°C in a water bath.
7. If there are still some tissue pieces visible, add proteinase K again, shake gently, and incubate for another 5 h at 45°C.
8. Add 2.4 mL of phenol, shake by hand for 5–10 min, and centrifuge at 3000 rpm for 5 min at 10°C.
9. Pipette the supernatant into a new tube, add 1.2 mL phenol and 1.2 mL chloroform/isoamyl alcohol (24:1), shake by hand for 5–10 min, and centrifuge at 3000 rpm for 5 min at 10°C.
10. Pipette the supernatant into a new tube, add 2.4 mL chloroform/isoamyl alcohol (24:1), shake by hand for 5–10 min, and centrifuge at 3000 rpm for 5 min at 10°C.
11. Pipette the supernatant into a new tube, add 25 μL 3 M NaOAc (pH 5.2) and 5 mL ethanol, and shake gently until the DNA precipitates.
12. Take a glass pipette, heat it over a gas burner, and bend the end to a hook. Fish the DNA thread out of the solution using the hook, and transfer DNA to a new tube.
13. Wash the DNA in 70% ethanol, and dry it in the Speed-Vac.
14. Dissolve the DNA in 0.5–1 mL sterile water overnight (or longer if necessary) at 4°C on a rotating shaker.
15. Measure the DNA concentration in a spectrophotometer and run 200 ng on a 1% agarose gel.

15.13.3 Preparation of Genomic DNA from Bacteria

1. Grow *E. coli* culture overnight in rich broth.
2. Transfer 2 mL to a 2 mL microcentrifuge tube and spin for 2 min. Decant the supernatant. Drain well onto a Kimwipe. Resuspend the pellet in 467 μL TE buffer by repeated pipetting.
3. Add 30 μL of 10% SDS and 3 μL of 20 mg/mL proteinase K, mix, and incubate 1 h at 37°C. Add an equal volume of phenol/chloroform and mix well but very gently, to avoid shearing the DNA, by inverting the tube until the phases are completely mixed.
4. Carefully transfer the DNA/phenol mixture into a Phase Lock Gel™ tube (green) and spin at 12,000 rpm for 10 min.
5. Transfer the upper aqueous phase to a new tube, and add an equal volume of phenol/chloroform. Again, mix well and transfer to a new Phase Lock Gel™ tube, and spin for 10 min.
6. Transfer the upper aqueous phase to a new tube. Add one-tenth volume of NaOAc. Mix. Add 0.6 vol of isopropanol, and mix gently until the DNA precipitates. Spool DNA onto a glass rod (or Pasteur pipette with a heat-sealed end).
7. Wash the DNA by dipping the end of the rod into 1 mL of 70% ethanol for 30 s. Resuspend the DNA in at least 200 μL TE buffer. Complete resuspension may take several days.
8. Store the DNA at 4°C (short term), −20°C or −80°C (long term). After the DNA has dissolved, determine the concentration by measuring the absorbance at 260 nm.

15.13.4 DNA Isolation Procedure

1. Grow cells overnight in 500 mL broth medium.
2. Pellet cells by centrifugation, and resuspend in 5 mL 50 mM Tris (pH 8.0), 50 mM EDTA.
3. Freeze cell suspension at −20°C.
4. Add 0.5 mL 250 mM Tris (pH 8.0), 10 mg/mL lysozyme to frozen suspension, and let thaw at room temperature. When thawed, place on ice for 45 min.
5. Add 1 mL 0.5% SDS, 50 mM Tris (pH 7.5), 0.4 M EDTA, 1 mg/mL proteinase K. Place in 50°C water bath for 60 min.
6. Extract with 6 mL Tris-equilibrated phenol, and centrifuge at 10,000× g for 15 min. Transfer top layer to new tube (avoid interface). Re-do this step if necessary.

7. Add 0.1 vol 3 M NaOAc, mix gently, then add 2 vol 95% ethanol (mix by inverting).
8. Spool out DNA and transfer to 5 mL 50 mM Tris (pH 7.5), 1 mM EDTA, 200 g/ml RNase. Dissolve overnight by rocking at 4°C.
9. Extract with equal volume chloroform, mix by inverting, and centrifuge at 10,000× g for 5 min. Transfer top layer to a new tube.
10. Add 0.1 vol 3 M NaOAc, mix gently, then add 2 vol 95% ethanol (mix by inverting).
11. Spool out DNA and dissolve in 2 mL 50 mM Tris (pH 7.5), 1 mM EDTA.
12. Check purity of DNA by electrophoresis and spectrophotometer analysis.

15.13.5 Polymerase Chain Reaction

Objective: In a PCR, a thermostable DNA polymerase amplifies DNA that is flanked by known sequences. The known sequences correspond to those on synthetic oligonucleotide primers which are used to initiate the reaction. PCR can be used in many complex ways to achieve different results.

General Issues to Consider: Because a PCR is very sensitive, product contamination by undesired sequences is problematic. Every effort must be made to keep the template free of contamination, especially previously amplified sequences. Leave the DNA in the refrigerator until all the other reagents are mixed and aliquoted in the reaction tubes. Wear gloves when handling the DNA template and the PCR reaction mixtures. It is usually easier to make a master mix of everything (including the enzyme, which is thermostable) except the template. Aliquot the mixture into tubes, and add the template just prior to putting them in a thermal cycler. Other precautions include the use of cotton-plugged pipette tips to reduce aerosol contaminants of amplified products and separating initial templates from amplified DNA in separate rooms.

Negative template controls should be included in all amplifications. One of the most critical factors for successful amplification of DNA is the magnesium ion (Mg^{++}) concentration. Too much magnesium chloride ($MgCl_2$) will cause high levels of nonspecific amplification, while too little will inhibit the reaction. This protocol uses 1.5 mM Mg^{++} as the final concentration for Taq DNA polymerase. Use this as a starting point when using previously untested primers and templates. Optimal concentrations may vary from 0.5 to 6 mM and are determined by titration with $MgCl_2$. Because of this variation, 10× stock solutions frequently have no magnesium in them and the $MgCl_2$ is added separately. Generally, the deoxyribonucleotide triphosphate (dNTP) and $MgCl_2$ concentrations are adjusted simultaneously. Annealing temperatures are critical and may also require experimental optimization. Low-temperature annealing increases nonspecific amplification; high temperatures inhibit annealing but may increase specificity. Typical reactions are performed in a range of 55°C–65°C. The optimum temperature for Taq DNA polymerase is 72°C.

Choice of primers: A number of considerations go into designing primers. Primers should not be self-complementary or complementary to each other, especially at their 3′ ends, to avoid primer dimers from forming. Keep the guanine and cytosine (G+C) content between 40% and 60%. Avoid long stretches of G + C, since they may form secondary structures. Addition of restriction enzyme sites to the 5′ end of the primers serve as useful vehicles for subsequent subcloning. Primer concentrations should be in excess of the template throughout the cycling. Typically, the primers are used over a 0.1–1.0 μM range. Lower concentrations may reduce artifacts and formation of primer dimers.

Choice of templates: A variety of different templates have been successfully used in amplification reactions. Most human DNA preparations are from fresh peripheral blood leukocytes or cell lines. The cells are lysed, treated with proteanase-K, RNAase-treated, and phenol-extracted. The final A260:280 ratio is approximately 1.8. For amplification of portions of plasmids, cesium chloride (CsCl)-banded DNA, quick-prep DNA, and heat-denatured, and transformed bacteria have been used. In the last case, a single bacterial colony (or as small a portion of a frozen stock as possible) was added to 0.5 mL of TE, vortexed, and then centrifuged to pellet the bacteria. The pellet was resuspended or washed once in 100 μL of TE, centrifuged again, and resuspended in 10 μL of TE. The entire sample was added as the DNA template to a total of 50 μL reaction volume.

Choice of DNA polymerases: Taq DNA Polymerase—Taq has 5′ to 3′ exonuclease activity, but neither 3′–5′ exonuclease nor any endonuclease activity. Its half-life at 95°C is 35–40 min and is 10 min at 97.5°C. Molecular weight is equal to 94,000 by SDS-PAGE. GeneAmp 10× PCR buffer II" from Perkin-Elmer Cetus is 500 mM potassium chloride (KCl) and 100 mM Tris–HCl (pH 8.3). Buffer I had 15 mM $MgCl_2$ added to it. We dilute the 10× buffer in H_2O; Perkin–Elmer recommends 0.15% NP-40, 0.15% Tween-20, 0.1 mM EDTA and 25 mM Tris–HCl, pH 8.3 for AmpliTaq buffer.

Protocol for Competitive Reverse Transcriptase PCR (RT-PCR): For quantifying mRNA, we use a competitive RT-PCR protocol with internal standard RNAs. These are added in a defined quantity to the RNA sample prior to the reverse transcriptase (RT) reaction. The resulting standard complementary DNA (cDNA) is co-amplified with the same primers as the endogenous target sequence. Its PCR product is approximately 50 nucleotides smaller. This method allows measurement of small differences (as low as factor 2) in messenger RNA (mRNA) amount between RNA samples. RNA standards have the big advantage that the variation of the RT efficiency as well as the variation of the PCR efficiency are irrelevant.

Making internal standard RNAs: The sequence to be amplified should ideally span an intron so that gDNA contamination does not play a crucial role. Otherwise, the RNA has to be treated with DNase I (RNase-free), and the success of this treatment has to be controlled by a PCR without prior RT. It is, however, always recommendable to treat the RNA samples with DNase so that no gDNA competes for the PCR components. To make a standard, first a PCR with a conventional downstream primer and a modified upstream primer (40 nucleotides in length) is performed, and cDNA is used as template for the PCRs. The PCR products are isolated from 1.5% agarose gel and cloned into a pGEM 3Z vector that contains a T7 RNA promoter sequence. The *in vitro* transcription of the cloned fragments is performed using T7 RNA polymerase (e.g., Gibco BRL). The internal standard RNA is then treated with RNase-free DNase I (e.g., Gibco BRL) to remove the plasmid DNA (success also checked by PCR without prior RT) and finally quantified by measurement of the OD at 260 nm and stored at −70°C. The main problem with RNA standards is their instability. We found that thawing and refreezing especially damages them. Therefore, it is best to store the standards in small aliquots in different dilutions and discard them if thawed too often. In addition, this problem means that no absolute amounts can be measured, because there is no way of knowing how much standard is already degraded in the aliquot used. However, this method is very reliable and accurate for comparison of different samples if the same standard aliquots are used for measuring their mRNA amount.

Quantitation of mRNA: For the quantitation of the mRNA of one RNA preparation, 4 RT reactions are prepared with 1 μg total RNA each and different amounts of standard RNA (if more than one mRNA is to be quantitated, the different standards can be mixed together). We use the SuperScript Preamplification System from Gibco BRL for our RTs. For the first measurements of an mRNA, it is best to add standard amounts which differ by factor 10 (i.e., 100 fg, 1 pg, 10 pg, 100 pg, 1 ng) to determine the range in which the transcript amount is to be found. If that is known, factor 2–2.5 between the standard amounts gives more accurate results (i.e., 25 pg, 50 pg, 100 pg, 250 pg). In the following PCR, 3–5 μL of cDNA, 1.5 units Taq DNA polymerase (Pharmacia), 200 μM of each dNTP, 250 nM of each primer and one-tenth volume of a 10× PCR standard buffer (15 mM $MgCl_2$; 100 mM Tris/HCl, pH 8.3; 500 mM KCl) are added to a total volume of 50 μL. The PCR is run in the thermal cycler GeneAmp 9600 (Perkin Elmer). The PCR products are then separated on a 1.5% agarose gel, stained with EtBr, SYBR-Green or SYBR-Gold (molecular probes, higher sensitivity, but also lighter sensitive) and scanned by a CCD camera. The amount of cDNA used for a PCR, the number of cycles, and the nucleic acid stain used depend on how abundant the transcript is that is being measured. If heteroduplices appear, one possibility is to run less PCR cycles.

15.13.6 Semiquantitative RT-PCR

The RT-PCR method can be used not only to detect specific mRNAs, but also to semi-quantitate their levels. Thus, one can compare levels of transcripts in different samples. This can be done in two different ways. One is to quantify against levels of transcripts from a control, housekeeping gene, such as actin and glyceraldehyde 3-phosphate dehydrogenase (GAPDH). (Transcription of housekeeping genes is believed to be unaffected by almost all experimental conditions.) The second method is to add an exogenous, primer-specific PCR template during PCR.

15.13.6.1 Role of Housekeeping Gene Transcript

The method involves reverse transcription using an oligodeoxythymidylic (oligo-dT) or random hexamer primer. The resulting cDNA thus represents both housekeeping gene transcripts as well as specific transcripts one is quantitating. The RT reaction is then amplified in a pair of PCR series—one series is to amplify the housekeeping gene cDNA (using GAPDH-specific or other specific primers) and the other is for the specific cDNA of interest (in a separate PCR, using gene-specific primers). The different PCR tubes within each series are set up such that they either vary in the amount of template (amount of RT reaction product [cDNA]) or undergo PCR for a different number of amplification cycles. This is because PCR amplification, though theoretically logarithmic, is not so at a low or high number of amplification cycles. The logarithmic or exponential amplification usually occurs only during the middle cycles, and this depends on the concentration of the target template. Comparison can therefore be done only during this phase. After PCR, the same volume of reaction products are electrophoresed on an agarose gel (preferably on the same gel). Images of stained DNA (PCR products) are then obtained and analyzed by visual comparison or with computer software.

1. A recommended practice is to first conduct pilot experiments to make sure that only single, specific products are obtained with the PCR primers and to get an idea of the range of cycles or RT reaction dilutions to use.
2. One should start with equal amounts of RNA for the RT reaction.
3. All reactions should be carried out at the same time, using reagents that are prepared as a supermix to minimize any variation.
4. PCR products should be run on the same gel (again, to minimize any variations).
5. When quantitating relatively rare transcripts, it may be better to use for controls those housekeeping genes whose transcripts are present at a low level. For example, transcripts of aldolase A are expressed at a low level compared to those of GAPDH gene.
6. When doing PCR for different numbers of cycles, set up PCR reactions together, removing them one after another during the middle of the extension phase and during the desired number of PCR cycles. Next, place them in a water bath at 65°C–75°C for a few minutes to ensure that extension is complete.
7. When doing PCR with different RT reaction volumes (usually a series of twofold or threefold dilutions), dilute the RT reaction using an RT reaction that had no RNA template so the different PCR reactions have the same *volume* of RT reaction (but different amounts of cDNA). This is necessary because components other than cDNA in the RT reaction, such as RT, diothiothreitol (DTT), etc., can affect PCR efficiency.

15.13.6.2 Isolation of Total RNA

We suggest the following procedure for total RNA isolation, based on years of experience working with complicated samples and small amounts of starting material. The procedure is suitable for all types of tissues from a wide variety of animal species. The method is based on the well-known protocol of Chomczynski and Sacchi (Chomczynski and Sacchi, 1987), except that all steps are performed at neutral pH instead of an acidic pH, as was originally suggested. In addition, we precipitate the RNA with lithium chloride (LiCl) for increased stability of the RNA preparation and improvement of cDNA synthesis. As an alternative, the popular Trizol method (GIBCO/Life Technologies) may be used, although it does not work on some nonstandard species such as jellyfish. Kits for RNA isolation that utilize columns (such as Qiagen's RNeasy kit) are generally not recommended for nonstandard samples. The following protocol is designed for large tissue samples (tissue volume 10–100 L) which normally yield approximately 10–100 µg of total RNA. Smaller amounts of starting material (expected to yield approximately 1 µg RNA or less) should be prepared in the same way, with the exclusion of the second phenol–chloroform extraction (step 4) and LiCl precipitation (step 6). Additionally, the final "pellet" should be dissolved in 5 µL instead of 40 µL of water and transferred directly to cDNA synthesis, omitting the agarose gel analysis.

Materials for total RNA isolation: Simple precautions such as wearing gloves, use of aerosol-barrier tips, and fresh sterile water for all solutions are sufficient to obtain stable RNA preparations. All organic liquids (phenol, chloroform, and ethanol) can be considered essentially RNAse-free by definition, as is the dispersion buffer containing 4 M guanidine thiocyanate. We do not recommend using DEPC-treated aqueous solutions, as such treatment often leads to RNA preparations that are very stable but completely unsuitable for cDNA synthesis. Commonly, gDNA contamination does not affect cDNA synthesis. DNase treatment to degrade gDNA is not recommended. In some cases, excess of gDNA can be removed by LiCl precipitation or by phenol–chloroform extraction.

- Dispersion buffer ("buffer D"): 4 M guanidine thiocyanate
- 30 mM disodium citrate
- 30 mM β-mercaptoethanol, pH 7.0–7.5
- chloroform–isoamyl alcohol mix (24:1)
- 96% ethanol
- 80% ethanol
- 12 M LiCl
- Fresh sterile water (e.g., milliQ-purified)
- Agarose gel (1%) containing EtBr

Total RNA Isolation Protocol

1. Dissolve the tissue sample in buffer D. The volume of tissue should not exceed one-fifth of the buffer D volume. To avoid RNA degradation, tissue dispersion should be carried out as quickly and completely as possible, ensuring that cells do not die slowly on their own. To adequately disperse a piece of tissue (usually takes 2–3 min of triturating using a pipette), take all or nearly all the volume of the buffer into the tip each time. The piece being dissolved must go up and down the tip, so it is sometimes helpful to cut the tip to increase the diameter of the opening for larger tissue pieces. Tissue dispersion can be performed at room temperature. The tissue dispersed in buffer D produces a highly viscous solution. The viscosity is usually due to gDNA. This normally has no effect on the RNA isolation (except for dictating longer periods of spinning at the phenol–chloroform extraction steps), unless the amount of dissolved tissue was indeed too great.

   In some cases (e.g., with freshwater planarians or mushroom anemones), mucus produced by the animal contributes to viscosity. This substance tends to copurify with RNA, making it very difficult to collect the aqueous phase at the phenol–chloroform extraction step. It likewise lowers the efficiency of cDNA synthesis. The RNA sample contaminated with such mucus, although completely dissolved in water, does not enter the agarose gel during electrophoresis. The EtBr-stained material stays in the well, probably because the mucus adsorbs RNA. We have found that including cysteine in buffer D can diminish the mucus problem. To buffer D, add 0.1 volume of solution containing 20% cysteine chloride and 50 mM tricine–KOH, pH 7 (takes a lot of titration!). The cysteine solution should be freshly prepared. After dissolving the tissue, incubate the sample for 2 h at +4°C, and then proceed with the above protocol.

2. Spin the sample at 1500 rpm for 5 min at room temperature to remove debris.
3. Transfer the supernatant to a new tube.
4. Put the tube on ice. Add an equal volume of buffer-saturated phenol and mix. There will be no phase separation at this time. Add one-fifth volume of chloroform–isoamyl alcohol (24:1) and vortex the sample. Two distinct phases will separate.
5. Vortex three to four more times at approximately 1 min intervals between steps. Incubate the tube on ice between steps.
6. Spin at 1500 rpm for 30 min at +4°C. Remove and save the upper, aqueous phase. Avoid warming the tube with your fingers or the interphase may become invisible.
7. Repeat step 3.
8. Add 1 μL of co-precipitant, and then add one volume of 96% ethanol and mix. Spin immediately at maximum speed on a table microcentrifuge at room temperature for 10 min. The precipitate may not form a pellet, instead spreading over the back wall of the tube and thus being almost invisible, even with co-precipitant added. Wash the pellet once with 0.5 mL 80% ethanol. Dry the pellet briefly until no liquid is seen in the tube (do not over-dry).
9. Dissolve the pellet in 100 μL fresh milliQ water. If the pellet cannot be dissolved completely, remove the debris by spinning the sample at maximum speed on a table microcentrifuge for 3 min at room temperature. Transfer the supernatant to a new tube, then add an equal volume of 12 M LiCl and chill the solution at −20°C for 30 min. Spin at maximum 1500 rpm for 15 min at room temperature. Wash the pellet once with 0.5 mL 80% ethanol, and dry as previously done. The precipitated RNA is usually invisible, since co-precipitant does not precipitate in LiCl.
10. Dissolve the pellet in 40 μL fresh sterile water.
11. After RNA isolation, we recommend RNA quality estimation using gel electrophoresis. Denaturing formaldehyde/agarose gel electrophoresis should be performed as described (Sambrook et al., 1989). Alternatively, standard agarose/EtBr gel electrophoresis can be used to quickly estimate RNA quality.
12. To store the isolated RNA, add 0.1 volumes of 3 M NaOAc and 2.5 volumes 96% ethanol to the RNA in water, and mix thoroughly. The sample may be stored for several years at −20°C.

15.14 GENETIC TECHNIQUES

15.14.1 Preparation of Human Metaphase Chromosomes

Materials required: RPMI 1640 medium; fetal calf serum (FCS); 20% colcemid (e.g., Boehringer Mannheim cell biology reagents, Best.-Nr. 295892); cell culture flask; Phythemagglutinin, PHA-L (Seromed, M 5030); CO_2 cell culture incubator; 50 mL Nunc/Falcon tubes, 15 mL Nunc/Falcon tubes; KCl (0.075 M); fixative (methanol/acetic acid 3:1); glass microscopy slides

Solution preparation: Making an amount per 5 mL blood requires 40 mL RPMI 1640 medium, 10 mL FCS (20%), 5 mL peripheral blood (anticoagulation by heparin) 1.5 mL PHA, 1 cell culture flask (e.g., Falcon 250 mL flask). Prepare up to 10 flasks (one flask will yield approximately 50 slides).

Protocol:

1. Incubate culture for 72 h in CO_2 cell culture incubator. Mix flask 1–2 times per day.
2. Add colcemid (approximately 45 min before harvesting).
3. Make 2 aliquots and transfer cell into 50 mL Falcon tubes.
4. Incubate in cell culture incubator or 37°C water bath for additional 45 min.
5. Centrifuge for 10 min at 1000 rpm.
6. Remove supernatant (e.g., with a cell culture pipette) until 5 mL remain.
7. Gently add 40 mL KCl (0.075 M, 37°C), adding the first 5 mL drop by drop (hypotonic treatment).

8. Incubate for 25 min in 37°C water bath.
9. Centrifuge 10 min at 1000 rpm.
10. Remove supernatant. Leave approximately 5 mL; resuspend pellet.
11. Add 2 mL fixative, and then mix well.
12. Add fixative until 40 mL, mix meanwhile.
13. Repeat steps 9–12 until the pellet is white (at least four times).
14. After removal and resuspension of the pellet, transfer cells into 15 mL Falcon tube.
15. Repeat steps 9–12, add just 10 mL fixative.
16. Remove fixative until approximately 2 mL final volume remains.
17. Resuspend pellet and apply suspension on slides.
18. Cool slides to −20°C (e.g., put approximately 10 slides in a cuvette in the freezer at −20°C and keep the cuvette on ice while preparing the metaphase slides).
19. Take one slide and moisten it by breathing on it from very close. Either drop 50–100 μL of the suspension on the slide or apply the same volume to the inclined slide (the fast draining and drying of the fluid is usually an indication for good spreading).
20. Let the suspension begin to dry (the fluid film starts to retract), then put the slide briefly in 70% acetic acid.
21. Air-dry the chromosome slide, check for chromosome spreading and cytoplasm debris in a phase contrast lab microscope. Adjust volume of fixative so that the density of nuclei/metaphases is appropriate.
22. If conditions are favorable, prepare a batch of metaphase spreads.
23. Keep slide in a box at room temperature (up to approximately 1–2 months). Metaphase spreads may be kept longer at −80°C or in 70% ethanol at 4°C.
24. Keep fixative with lymphocytes at −20°C until the preparation of new slides. Add new fixative and wash cells before the preparation of new metaphase spreads.

15.14.2 Structural Analysis of Human Chromosomes by Karyotype

Materials required: fresh venous blood, heparinized syringes, Eagle's spinner modified media with PHA, culture flasks, tissue culture-grade incubator at 37°C, 10 μg/mL colcemid, clinical centrifuge and tubes, 0.075 M KCl, absolute methanol and glacial HOAc (3:1 mixture, prepared fresh), dry ice, slides, cover slips and permount, Alkaline solution for G-banding, SSC for G-banding, ethanol (70% and 95% (v/v)), and Giemsa stain.

Protocol:

1. Draw 5 mL of venous blood into a sterile syringe containing 0.5 mL of sodium heparin (1000 units/mL). The blood may be collected in a heparinized vacutainer and transferred to a syringe.
2. Bend a clean, covered 18 gauge needle to a 45° angle and place on the syringe. Invert the syringe (needle pointing up, plunger down), and stand it on end for 1.5– 2 h at room temperature.

   During this time, the erythrocytes settle by gravity, leaving approximately 4 mL of leukocyte-rich plasma on the top, and a white buffy coat of leukocytes in the middle.
3. Carefully tip the syringe (do not invert) and slowly expel the leukocyte-rich plasma and the buffy coat into a sterile tissue culture flask containing 8 mL of Eagle's spinner modified media supplemented with 0.1 mL of PHA.
4. Be extremely careful not to disturb the red blood cells in the bottom of the syringe because red blood cells will inhibit growth of the leukocytes.
5. Incubate the culture for 66–72 h at 37°C. Gently agitate the culture once or twice daily during the incubation period.
6. Add 0.1 mL of colcemid (10 μg/mL) to the culture flasks and incubate for an additional 2 h.
7. Transfer the colcemid-treated cells to a 15 mL centrifuge tube and centrifuge at 350 rpm for 10 min.

8. Aspirate and discard all but 0.5 mL of the supernatant. Gently tap the bottom of the centrifuge tube to resuspend the cells in the remaining 0.5 mL of culture media.
9. Add 10 mL of 0.075 M KCl to the centrifuge tube, dropwise at first and then with gentle agitation. Gently mix with each drop. (Start timing the next immediately with the first drop of KCL.)
10. Let the cells stand exactly 6 min in the hypotonic KCl. The hypotonic solution should not be in contact with the cells in excess of 15 min from the time it is added.
11. Centrifuge the cells at 350 rpm for 6 min. Aspirate the KCl and discard all but 0.5 mL of the supernatant. Gently resuspend the cells in this small volume of fluid.
12. Add 10 mL freshly prepared fixative, dropwise at first and then with gentle agitation. Gentle and continuous agitation is important at this step to prevent clumping of the cells. If the cells were not properly resuspended in step 10, the cells will clump beyond any further use.
13. Allow the cells to stand in fixative at room temperature for 30 min.
14. Centrifuge at 300 rpm for 5 min and remove all but 0.5 mL of the supernatant. Resuspend the cells in fresh fixative.
15. Wash the cells twice more in 10 mL volumes of fixative. Add the fixative slowly, recentrifuge, and aspirate the fixative as previously directed. The fixed and pelleted cells may be stored for several weeks at 4°C.
16. Resuspend the pellet of cells in just enough fixative to give a slightly turbid appearance.

    Prop a piece of dry ice against the side of a styrofoam container and lace a clean slide onto the dry ice to chill the slide. Use a siliconized Pasteur pipette to draw up a few drops of the suspended cells, and drop the cells onto the surface of the chilled slide. Spreading of the chromosomes may be enhanced by dropping the cell suspension from a height of at least 12 in. As soon as the cells strike the slide, blow hard on the slide to rapidly spread the cells.
17. Remove the slides from the dry ice and allow them to air-dry. Perform the desired banding and/or staining procedures.
18. Preparation of chromosomes for karyotype analysis can be performed in a number of ways, and each will yield differing pieces of information.
19. The chromosomes may be stained with aceto–orcein, feulgen, or a basophilic dye such as toluidine blue or methylene blue if only the general morphology is desired.
20. The chromosomes can be treated with various enzymes in combination with stains to yield banding patterns on each chromosome. These techniques have become commonplace and will yield far more diagnostic information than a Giemsa stain alone (the most commonly used process).
21. A band is an area of a chromosome which is clearly distinct from its neighboring area, but may be lighter or darker than its neighboring region. The standard methods of banding are the Q-, G-, R-, and C-banding techniques.
22. Q-banding requires quinacrine stain and fluorescence microscopy.
23. G-banding requires Giemsa stain, additional conditions (heat hydrolysis, trypsin treatment, Giemsa at pH 9.0).
24. R-banding requires Giemsa or acridine orange, negative bands of Q and Greversed; heat hydrolysis in buffered salt.
25. C-banding requires Giemsa stain and pretreatment with BaOH or NaOH, followed by application of heat and salt.

Protocol for G-banding:

- Treat fixed and flamed slides in alkaline solution in room temperature for 30 s.
- Rinse in SSC solution; three changes for 5–10 min each.
- Incubate in SSC solution, 65°C for 60–72 h.
- Treat with three changes of 70% ethanol and three changes of 95% ethanol (3 min) each.
- Air-dry.

- Stain in buffered Giemsa for 5 min.
- Rinse briefly in distilled water.
- Air-dry and mount.

26. Photograph appropriate spreads and produce 8 × 10 high-contrast photographs of your chromosome spreads.
27. Cut each chromosome from the photograph and arrange the chromosomes according to size and position of the centromere.
28. Tape or glue each chromosome to the form supplied for this purpose.

15.14.3 DNA Amplification Fingerprinting Protocol

PCR amplification: The PCR reaction mix (10 μL) contains template DNA (2 ng/μL), primer (0.3 μM), Taq DNA polymerase, Stoffel Fragment (Perkin Elmer; 5 U), Mg++ (2.5 mM), buffer (1×), overlaid with a drop of mineral oil. Amplifications are performed using the 96 well plates in a MJ Research thermal cycler for 35 cycles after an initial denaturation at 94°C for 5 min and a final extension at 72°C for 5 min. For arbitrary and mini-hairpin primers, each cycle consists of 5 s at 94°C, 20 s at either 35°C or 45°C (depending on the primer) and 30 s at 72°C. For simple sequence repeat (SSR) primers, each cycle is 1 min at 94°C, 1 min at 55°C and 2 min at 72°C.

Gel electrophoresis: DNA fragments are separated in a vertical electrophoresis system using a polyacrylamide-based vinyl polymer (GeneAmp; Perkin Elmer, Norwalk, CT) (He et al., 1994).

Protocol:

1. Add 3.5 mL deionized water to a clean beaker (10 mL).
2. Add 1.25 mL of GeneAmp Detection Gel solution and 0.25 mL of 10× TBE buffer (1 M Tris–HCl, 0.83 M boric acid, 10 mM Na2 EDTA, pH 8.3). Swirl to mix.
3. Add 60 μL of 10% (w/v) ammonium persulfate and 5 μL of N,N,N′,N′-tetramethylenediamine (TEMED). Mix thoroughly.
4. Immediately pour the gel mixture into the gel cassette (Mini-Protean II, BioRad Co, Richmond, CA) (0.75 mm thick; 8 × 10 cm).
5. Insert the comb at the top of the gel. Let the gel solidify for 20 min.
6. Add 1 μL of the loading buffer to 2.5 μL of the final amplified reaction mix.
7. Load this sample into the gel and conduct electrophoresis at 200 V.
8. Stop the electrophoresis when the front of the dye migrates to the bottom of the gel.

Silver Staining for DNA visualization: Gels were silver stained using a modified procedure of Bassam et al. (1991).

Protocol:

1. Gently shake the gel in 7.5% (v/v) glacial HOAc for 10 min at room temperature.
2. Rinse the gel in deionized water twice for approximately 2 min each.
3. Incubate the gel in 10% oxidizer solution (Bio-Rad #161-0444) for 5–10 min.
4. Rinse the gel in water three times for approximately 5 min each. Use fresh deionized water each time.
5. Immerse the gel in silver staining solution (100 mg silver nitrate and 150 μL formaldehyde in 100 mL water) for 20 min.
6. Pour out the silver staining solution, and wash the gel quickly with deionized water.
7. Immerse the gel in an ice-cold developer solution (8°C) (3 g sodium carbonate, 300 μL formaldehyde, and 200 μg sodium thiosulfate in 100 mL water) until optimal image intensity is obtained.
8. Stop the developing process by immersing the gel in 7.5% ice-cold glacial acetic acid.
9. Air-dry the gel and back it with a GelBond plastic film (FMC BioProducts, Rockland, ME).

15.14.4 Single-Strand Conformation Polymorphism Technique

Single-strand conformation polymorphism (SSCP) technique is a simple and efficient means to detect any small alteration in PCR-amplified products. It is based on the assumption that subtle nucleic acid changes affect the migration of single-stranded DNA fragments and, therefore, results in visible mobility shifts across a non-denaturing PAGE. PAGEs were used for analysis of DNA with specialized buffer systems and without urea. In non-denaturing PAGE, the components used to synthesize matrix were acrylamide monomers, *N,N*-methylenebisacrylamide (Bis), ammonium persulphate (APS) and TEMED. APS when dissolved in water generates free radicals, which activate acrylamide monomers, inducing them to react with other acrylamide molecules, forming long chains. These chains cross-linked with Bis. TEMED acts as a catalyst for gel formation because of its ability to exist in free radical form. The acrylamide and bisacrylamide was used in a 49:1 ratio, adding autoclaved HPLC water to make 100 mL volume. This 49:1 acrylamide–bisacrylamide solution was dissolved completely using a magnetic stirrer and kept in refrigerator until required.

Materials required: Fifty percent acrylamide–bisacrylamide (24 mL), 10× TBE (5 mL), glycerol (10 mL), autoclaved HPLC water (61 mL), make total volume to 100 mL.

Preparation of SSCP gel: The percentage of SSCP gels used varied from 8% to 20%, but most of the primers were optimized with good results in 12% PAGE solution. The gel mixture was kept dissolved completely and stored at −20°C till it was used.

Protocol:

1. The PCR–SSCP procedure included the following steps: PCR amplification of the gene fragments, resolution in non-denaturing PAGE, and visualization using silver staining. PCR conditions were optimized for PCR–SSCP by testing a number of variables such as concentration of DNA, Taq polymerase, dNTPs, $MgCl_2$ and temperature profiles. The PCR amplification protocol for all the SSCP primers used was the same, except the annealing temperature, which varied between primers.
2. The single-strand conformation polymorphism analysis of amplified gene fragments was carried out using a Bio-Rad Protein II xi Cell vertical gel electrophoresis unit (Bio-Rad Laboratories).
3. The two glass plates were washed thoroughly using tap water and detergent and rinsed under running tap water till no remains of detergent were left. The plates were wiped two times: first with tissue paper soaked in distilled water and then with 70% alcohol and air-dried. The similar thorough cleaning treatment was given to spacers and the comb to ensure proper alignment of the 20 cm glass plates.
4. The gel sandwich was assembled on a clean surface, laying down the long rectangular plate first, then two spacers of equal thickness along the long edges of plate, and the short plate was placed on the rectangular plate. The two glass plates with spacers between them were fitted well, with proper alignment, by tightening the screws of two sandwich clamps. The cleaned comb (20 wells) was inserted from the top side of the gel sandwich and the clamps were immediately applied over the plates containing the comb to create sharp wells.
5. The bottom side of the gel sandwich was sealed using 10 mL of 12% gel mix. The gel sandwich was kept in a slanting position and the solution mixed with 50 μL APS, and 20 μL TEMED was injected between the two glass plates using a syringe with a fine tip and allowed to polymerize for 10 min.
6. After polymerization, the assembled gel sandwich was placed in the alignment slot of casting stand. The 12% native PAGE gel mix (25 mL) was prepared by adding APS (100 μL) and TEMED (40 μL) one at a time and mixing well. This gel mix was filled

from the upper side of the gel sandwich using a syringe, smoothly, without any bubbles, and clamps were immediately applied over the comb to ensure sharp wells. The gel was kept undisturbed at least 45 min for polymerization.

7. After polymerization, the comb was removed and the wells were flushed with 0.5× buffer. The gel sandwich was placed in the electrophoresis tank with the notched plate facing toward the buffer reservoir. The reservoir of the electrophoresis tank was filled with 0.5× TBE and the gel was given a pre-run at 200 V at constant temperature for a minimum of 45 min. Ice-cooled water was circulated with an electric pump applied to a central cooling core of assembly to maintain constant temperature.
8. Approximately 4 μL PCR product and 12 μL of a formamide dye were prepared in the PCR tube and denatured at 95°C for 10 min in the Biometra PCR machine. After denaturation, the samples were immediately placed in an ice-chilled box and kept at −20°C in a deep freezer for 10 min.
9. After completion of the pre-run, the wells were flushed again using buffer. The samples were loaded on a non-denaturing 12%–20% acrylamide–bisacrylamide (49:1) gel with a gel loading tip and electrophoresis was immediately performed in 0.5× Tris borate (pH 8.3)–EDTA buffer at 10–12.5 V/cm for 2–24 h at room temperature, depending on the optimized conditions for each primer.
10. After completion of the electrophoresis for the required time, the glass plates were removed from the assembly.

15.15 AGRICULTURAL BIOTECHNOLOGY

15.15.1 Plant DNA Isolation

Materials required: 750 μL/sample EB buffer: 100 mM Tris pH 8.0, 50 mM EDTA, pH 8.0, 500 mM NaCl; 10 mM beta-mercaptoethanol (beta-merc); 50 μL/sample—20% SDS, 250 μL/sample—5 M KOAc (potassium acetate), 700 μL/sample—50 mM Tris, 10 mM EDTA pH 8.0, 1 mL/sample, isopropanol, 75 μL/sample 3 M NaOAc, 700 μL/sample—80% EtOH 100 μL/sample, 10 mM Tris, 0.5 mM EDTA pH 8.0. Blue plastic pestles, −20°C freezer, water bath, refrigerator, ice, spectrophotometer, mira cloth.

Protocol:

1. Weigh 0.1–0.2 g tissue (~4 full trifoliate leaves) in 1.5 mL microfuge tube. Quick-freeze and grind to powder with blue pestle—do not let thaw.
2. Add 0.75 mL EB buffer. EB buffer (100 mM Tris pH 8.0, 50 mM EDTA pH 8.0, 500 mM NaCl, 10 mM beta-merc).
3. Add beta-merc fresh every time. Continue mixing until mixed well and thawed.
4. Add 0.05 mL 20% SDS, shake well and incubate tubes at 65°C for 10 min.
5. Add 0.25 mL 5 M KOAc, shake well and incubate tubes on ice (0°C) 20 min. (Keep on ice until all samples are lysed).
6. Spin tubes at 21,000 rpm for 20 min.
7. Remove supernatant w/pipetman (~1 mL) and put through mira cloth into new 1.5 mL tube.
8. Add 500 μL isopropanol, invert to mix well, and incubate at −20°C for 30–60 min.
9. Centrifuge DNA at 21,000 rpm for 15 min, pour off supernatant, and dry pellets by inverting tube on paper towel for 5 min.
10. Redissolve pellets with 0.7 mL TE (50 mM Tris, 10 mM EDTA pH 8) by agitating, making sure the pellet is resuspended.
11. Spin tube at max speed for 10 min to remove insoluble debris.
12. Transfer supernatant to a new tube and add 75 μL 3 M NaOAc and 500 μL isopropanol.
13. Mix well by inverting, and spin to pellet DNA for 30 s.

14. Wash pellet with 700 μL 80% EtOH—spin and remove EtOH, dry ~5 min inverted on paper towel.
15. Redissolve in 100 μL 10 mM Tris, 0.5 mM EDTA pH 8.0.
16. Check OD 260 nm and dilute for PCR (~1/100 or 5 ng/μL).

15.15.2 Plant Regeneration by Protoplast Fusion

Phaseolineae seeds were surface sterilized in 12% calcium hypochlorite for 10 min, and then immersed in 70% ethanol for 30 s and rinsed three times in sterile deionised water. Seed scarification, humidification, and pre-germination were carried out in sterile Petri dishes for 10 days. Germinated seeds were first transferred into standard bottles containing 100 mL vermiculite and a standard half-strength medium solidified with 2 g L^{-1} phytagel until lateral buds developed. Plantlets were then transferred into a new standard bottle onto a solidified Murashige–Skoog (MS) medium containing 20 g L^{-1} sucrose and 5 g L^{-1} agar. Growing conditions were 24°C/21°C day/night temperatures with a 16 h photoperiod.

Standard protoplast isolation: Protoplasts of plant were isolated from green leaves. Fresh green leaves were more difficult to obtain with PV and PP genotypes when grown *in vitro*. Therefore, protoplasts were isolated from 10-day-old hypocotyl explants after pre-germination in Petri dishes for PV (NI637 and NI638) and PP (NI1015) accessions. Material was finely chopped and plasmolysed for 1 h in 10 cm^3 CPW medium with 10 mM calcium chloride ($CaCl_2$), 13% mannitol, and adjusted to pH 5.5 (CPW 13 M). Tissues of all accessions (PV, PC, PP) were digested overnight on a continuous rotary shaker (60 T min^{-1}) with an enzyme mixture of 3% Macerozyme R10, 4% cellulase Onozuka RS, and 0.2% Pectolyase Y-23 (described as 3402RS by Ochatt et al., 2000). For PC accessions, we compared the use of cellulase Onozuka YC (described as 3402YC by Ochatt et al., 2000) versus cellulase Onozuka RS in the enzyme mixture. Onozuka YC was tested regarding the difference in source tissues (leaf explants versus hypocotyl).

Isolation of protoplasts for fusion: Protoplasts were sieved (40 μ for PV and PP and 50 μ for PC) and centrifuged successively at 35 g (5 min, 10°C) and 70 g (5 min, 10°C). Each pellet was resuspended in 200 mm^3 CPW 13 M. Pellets were mixed together and labeled with five drops (approximately 150 mm^3) of fluorescein diacetate (green, described as FDA) for PV accessions, while rhodamine B isothiocyanate (red, described as RBi) was used for PP and PC accessions. Stock solutions of fluorochromes were made from 5 mg for FDA or 30 mg for RBi per cm^3 acetone solution. Pellets were finally layered on top of 6 cm^3 of CPW solution containing 21% sucrose (CPW 21S) and spun at 80 g (10 min, 10°C, maximum acceleration). Under UV light, protoplasts with FDA staining gave a yellow-green fluorescence, allowing density and viability evaluation, while those with RBi gave a red fluorescence (Durieu et al., 2000). Density was determined using a Bürker cell (Marienfeld, Germany). Optimum plating density is between 5×10^4 and 1×10^6, maximizing wall regeneration and concomitant daughter cell formation. Viability expressed as a percentage is determined as the number of protoplasts that fluoresced yellow-green under UV light out of the total number of isolated protoplasts observed in the same microscopic field under normal light.

Protoplast fusion: Regarding the low density of protoplasts obtained for NI637 as PV accession, electrofusion could not be easily realized. Therefore, chemical fusion was conducted to perform protoplast fusion between NI637 with all PP and PC accessions. The efficiency of protoplast fusion with the two tested methods was evaluated under UV light, as fluorochromes are linked to different parental protoplasts, whereby heterokaryons can be observed and counted through their double fluorescence, green and red.

Culture: Protoplasts were cultured at 10^5 cm^{-3} on a medium based on KM (Kao et al., 1975) with 0.1 mg L^{-1} 2,4-D, 0.2 mg L^{-1} zeatin and 1 mg L^{-1} NAA (described as KP). After 1 week, a dilution was performed with the same medium, and as soon as the majority of cells had regenerated their wall, weekly dilutions (weekly adding 1 mL media per mL initial protoplast culture) were carried out with the culture medium containing 20 g L^{-1} sucrose and 10 g L^{-1} glucose.

15.15.3 Simplified Arabidopsis Transformation

With this method, you should be able to achieve transformation rates above 1% (one transformant for every 100 seeds harvested from Agrobacterium-treated plants).

Protocol:

1. Grow healthy Arabidopsis plants until they are flowering. Grow under long days in pots, in soil covered with bridal veil, window screen, or cheesecloth.
2. (*optional*) Clip first bolts to encourage proliferation of many secondary bolts. Plants will be ready roughly 4–6 days after clipping. Clipping can be repeated to delay plants. Optimal plants have many immature flower clusters and not many fertilized siliques, although a range of plant stages can be successfully transformed.
3. Prepare *Agrobacterium tumefaciens* strain carrying the gene of interest on a binary vector. Grow a large liquid culture at 28°C in lysogeny broth (LB) with antibiotics to select for the binary plasmid or grow in other media. You can use mid-log cells or a recently stationary culture.
4. Spin down Agrobacterium, resuspend to $OD_{600} = 0.8$ (can be higher or lower) in 5% sucrose solution (if made fresh, no need to autoclave). You will need 100–200 mL for each two or three small pots to be dipped or 400–500 mL for each two or three 3.5″ (9 cm) pots.
5. Before dipping, add Silwet L-77 to a concentration of 0.05% (500 µL/L) and mix well. If there are problems with L-77 toxicity, use 0.02% or as low as 0.005%.
6. Dip above-ground parts of plant in Agrobacterium solution for 2–3 s, with gentle agitation. You should then see a film of liquid coating the plant. Some investigators dip the inflorescence only, while others also dip the rosette to hit the shorter axillary inflorescences.
7. Place dipped plants under a dome or cover for 16–24 h to maintain high humidity (plants can be laid on their side, if necessary). Do not expose to excessive sunlight (air under dome can get hot).
8. Water and grow plants normally, tying up loose bolts with wax paper, tape, stakes, twist-ties, or other means. Stop watering as seeds become mature.
9. Harvest dry seeds. Transformants are usually all independent, but are guaranteed to be independent if they come off separate plants.
10. Select for transformants using antibiotic or herbicide selectable marker. For example, vapor-phase sterilize and plate 40 mg = 2000 seed (resuspended in 4 mL 0.1% agarose) on 0.5× MS/0.8% tissue culture Agar plates with 50 ug/mL Kanamycin, cold treat for 2 days, and grow under continuous light (50–100 µE) for 7–10 days.
11. Transplant putative transformants to soil. Grow, test, and use.
12. For higher rates of transformation, plants may be dipped two or three times at 7-day intervals. We suggest one dip 2 days after clipping and a second dip 1 week later. Do not dip less than 6 days apart.

15.15.4 Agrobacterium-Mediated Gene Transfer via Hypocotyls

1. Grow cotton seedlings aseptically in culture tubes.
2. Using strictly aseptic practices, transfer hypocotyls to aluminum foil cutting pads and cut into approximately 2 in. sections.
3. Transfer hypocotyls to Petri dishes containing *Agrobacterium spp.* plus Murashigee–Skoog nonhormone (MSNH) solution. Further section tissue into approximately 5 mm explants sections. Tissue is wounded for generating *Agrobacterium spp.* infection sites.

4. Remove excess *Agrobacterium spp.* plus MSNH by blotting tissue on sterile filter paper.
5. Transfer explants sections to Petri dishes containing T2 medium. Twenty explants per dish works well (four rows of five).
6. Incubate Petri dishes for 3 days at 28°C for cocultivation.
7. Remove explants, rinse in MSNH, blot on sterile filter paper, and transfer to MS2NK plus antibiotics. Incubate Petri dishes at 30°C in growth room for 3 weeks for callus initiation.

15.15.5 Isolation of DNA from Onion

Materials required: Fresh onions, graduated cylinders (10 and 100 mL), knife, 15 mL test tube, blender, test tube rack, or 250 mL beaker, strainer, glass stirring rod, coffee filters, noniodized salt, Adolph's natural meat tenderizer, Palmolive detergent soap, beaker, distilled water, ice-cold 95% ethanol

Preparation of solution: Prepare detergent/salt solution by mixing 20 mL detergent, 20 g noniodized salt in 180 mL distilled water. Prepare 5% meat tenderizer solution by mixing 5 g meat tenderizer in 95 mL distilled water.

Protocol:

1. Cut an inch square out of the center of three medium onions. Chop and place in a blender.
2. Add 100 mL of detergent/salt solution.
3. Blend on high 30 s to 1 min.
4. Strain the mixture into a beaker using a strainer with a coffee filter.
5. Add 20–30 mL meat tenderizer and stir to mix.
6. Place 6 mL filters in a test tube.
7. Pour 6 mL ice-cold ethanol carefully down the side of the tube to form a layer.
8. Let the mixture sit undisturbed 2–3 min until bubbling stops.
9. The DNA will float in the alcohol. Swirl a glass stirring rod at the interface of the two layers to see the small threads of DNA.

15.15.6 Isolation of DNA from Wheat Germ

Materials required: 250 mL beaker, baking soda, hot plate, Adolph's natural meat tenderizer, non-roasted wheat germ, ice-cold 95% ethanol, thermometer, 15 mL test tube, pH meter, glass stirring rod, Palmolive detergent, distilled water, test tube rack, graduated cylinders (10 and 100 mL)

Preparation of solution: Prepare baking solution by mixing baking soda to distilled water until a pH of approximately 8.0 is reached.

Protocol:

1. Add 100 mL distilled water to a beaker and heat to 50°C–60°C.
2. Add 1.5 g wheat germ and mix until dissolved.
3. Add 5 mL of detergent. Maintain 50°C–60°C temperature and stir for 5 min.
4. Add 3 g meat tenderizer.
5. Add baking soda solution to bring the pH to approximately 8.0.
6. Maintain the 50°C–60°C temperature and stir for 10 min.
7. Remove from heat.
8. Add 6 mL of the solution to a test tube and cool to room temperature.
9. Pour 6 mL ice-cold ethanol carefully down the side of the tube to form a layer.
10. Let the mixture sit undisturbed 2–3 min until bubbling stops.
11. The DNA will float in the alcohol. Swirl a glass stirring rod at the interface of the two layers to see the small threads of DNA.

15.16 MICROBIAL BIOTECHNOLOGY

15.16.1 Gram Positive/Negative Staining

Materials required: Colonies of bacteria from toothpicks, crystal violet, Gram's iodine, 95% ethanol, safranin, oil immersion microscope

Procedure:

1. Before staining the individual colonies, you should first practice the technique by observation of the Gram-positive microorganisms normally found in the gum linings of your mouth.
2. Use a clean toothpick to rub along the gingival crevices (area between tooth surface and gums) of your mouth.
3. Mix the scrapings with a drop of water previously placed on a clean slide, spread in a thin film over the center of the slide, and allow to air-dry.
4. Fix the smear to the slide by passing the slide (smear side up) quickly through a flame three times. If the slide is held directly in the flame, it will heat up too rapidly and break. The trick is to gently dry the smear without overheating the slide.
5. Place the slide on a staining rack. Apply the stains on the fixed smear as follows:
   a. Flood the slide with crystal violet for 30 s.
   b. Rinse with water.
   c. Flood with Gram's iodine for 60 s.
   d. Rinse with water.
   e. Decolorize with 95% ethanol.
   f. Rinse with water.
   g. Counterstain with safranin for 60 s.
   h. Rinse with water and blot dry (no rubbing!)
   i. Examine under oil immersion objective lens.
6. Gram-positive bacteria retain crystal violet after washing with 95% ethanol, while Gram-negative bacteria lose the purple dye after washing with 95% ethanol. The positive or negative reaction is a measure of the presence or absence of specific polysaccharide components of their cell walls. Safranin is used as a pink counterstain so that Gram-negative cells can be visualized. In practice, then, the distinction is made between purple cells (Gram-positive) and pink cells (Gram-negative).
7. Determine the basic cell shape of the bacteria.

REFERENCES

Bassam B.J., Caetano-Anolles, G., and Gresshoff, P.M. Fast and sensitive silver staining of DNA in polyacrylamide gels. *Anal. Biochem.* 196: 80–83, 1991.

Chomczynski, P. and Sacchi, N. Single-step method of RNA isolation by acid guanidinium thiocyanate-phenol chloroform extraction. *Anal. Biochem.* 162:156–159, 1987.

Durieu P. and Ochatt S.J., 2000. Efficient intergeneric fusion of pea (*Pisum sativum* L.) and grass pea (Lathyrus sativus L.) protoplasts. *J. Exp. Bot.* 51(348): 1237–1242, 2000.

Ochatt, S.J., Mousset-Déclas, C., and Rancillac, M. Fertile pea plants from protoplasts when calluses have not undergone endoreduplication. *Plant Sci.* 156, 177–183, 2000.

Sambrook, J., Fritsch, E.F., and Maniatis, T. (1989) *Molecular Cloning: A Laboratory Manual* 2nd/Ed. CSH Laboratory Press, Cold Spring Harbor, New York.

Glossary

Abiotic stress: The negative impact of nonliving factors on the living organisms in a specific environment.

Acclimatization; acclimation: The process of an individual organism adjusting to a gradual change in its environment (such as a change in temperature, humidity, photoperiod, or pH) allowing it to maintain performance across a range of environmental conditions.

Acellular: Containing no cells; not made of cells.

Acentric chromosome: A segment of a chromosome that lacks a centromere.

Acetyl co-enzyme A; acetyl CoA: An important molecule in metabolism, used in many biochemical reactions. Its main function is to convey the carbon atoms within the acetyl group to the citric acid cycle to be oxidized for energy production.

Acrocentric: A chromosome (one of the microscopically visible carriers of the genetic material DNA) with its centromere (the "waist" of the chromosome) located quite near one end of the chromosome.

Acrylamide gels: A polyacrylamide gel is a separation matrix used in electrophoresis of biomolecules, such as proteins or DNA fragments. Traditional DNA sequencing techniques such as Maxam–Gilbert or Sanger methods used polyacrylamide gels to separate DNA fragments differing by a single base pair in length so the sequence could be read. Most modern DNA separation methods now use agarose gels, except for particularly small DNA fragments.

Actin: A globular, roughly 42 kDa moonlighting protein found in all eukaryotic cells (the only known exception being nematode sperm) where it may be present at concentrations of over 100 μM. It is also one of the most highly conserved proteins, differing by no more than 20% in species as diverse as algae and humans.

Activated charcoal; activated charcoal; activated coal: A form of carbon that has been processed to make it extremely porous and thus to have a very large surface area available for adsorption or chemical reactions.

Activated macrophage: Macrophages are components of the monocyte–macrophage system. Macrophages are usually immobile but become actively mobile when stimulated by inflammation, immune cytokines, and microbial products. They are an important class of antigen presenting cells (APCs).

Activated sludge system: A process for treating sewage and industrial wastewaters using air and a biological floc composed of bacteria and protozoans.

Activator: A DNA-binding protein that regulates one or more genes by increasing the rate of transcription.

Active site: In molecular biology, the active site is part of an enzyme where substrates bind and undergo a chemical reaction. The majority of enzymes are proteins, but RNA enzymes called ribozymes also exist. The active site of an enzyme is usually found in a cleft or pocket that is lined by amino acid residues (or nucleotides in ribozymes) that participate in recognition of the substrate. Residues that directly participate in the catalytic reaction mechanism are called active site residues.

Adaptation: Evolutionary process whereby a population becomes better suited to its habitat. This process takes place over many generations and is one of the basic phenomena of biology.

Adaptor: In genetic engineering, a short, chemically synthesized, double-stranded DNA molecule that is used to link the ends of two other DNA molecules. It may be used to add sticky ends to cDNA allowing it to be ligated into the plasmid much more efficiently.

Additive gene effects: When the combined effects of alleles at different loci are equal to the sum of their individual effects. In this type of inheritance, there is no sharp distinction between genotypes, but there are many gradations between the two extremes. An example of how additive genes express themselves may be illustrated by imagining a large glass cylinder of clear water on a desk top. The water would represent genes on the chromosomes that are neutral (have no expression). If a red pill is added to the water, it begins to turn pink. If another pill is added, it turns light red, add another pill and it turns red. The point is that the neutral genes on the chromosomes are replaced with genes that have additive expression and the phenotype of the individual changes. Thus, each pill added changes the water color in a linear manner. The same would be true when replacing neutral alleles on the chromosomes with additive alleles.

Adenine (A, Ade): A nucleobase (a purine derivative) with a variety of roles in biochemistry including cellular respiration, in the form of both the energy-rich adenosine triphosphate (ATP) and the cofactors nicotinamide adenine dinucleotide (NAD) and flavin adenine dinucleotide (FAD), and protein synthesis, as a chemical component of DNA and RNA. The shape of adenine is complementary to either thymine in DNA or uracil in RNA.

Adenosine disphosphate (ADP): A nucleotide. It is an ester of pyrophosphoric acid with the nucleoside adenosine. ADP consists of the pyrophosphate group, the pentose sugar ribose, and the nucleobase adenine.

Adenosine triphosphate: A nucleotide of fundamental importance as a carrier of chemical energy in all living organisms. It consists of adenosine with three phosphate groups, linked together linearly. ATP is regenerated by rephosphorylation of AMP and ADP, using chemical energy derived from the oxidation of food.

Adenovirus: Medium-sized (90–100 nm), nonenveloped (without an outer lipid bilayer) icosahedral viruses composed of a nucleocapsid and a double-stranded linear DNA genome. There are 55 described serotypes in humans, which are responsible for 5%–10% of upper respiratory infections in children and many infections in adults as well.

Adenylate cyclase: A lyase enzyme. It is a part of the cAMP-dependent pathway.

Adhesion: Any attraction process between dissimilar molecular species that can potentially bring them in "direct contact." By contrast, cohesion takes place between similar molecules.

A-DNA: A-DNA is one of the many possible double helical structures of DNA. A-DNA is thought to be one of three biologically active double helical structures along with B- and Z-DNA. It is a right-handed double helix fairly similar to the more common and well-known B-DNA form, but with a shorter more compact helical structure.

Adsorption: The adhesion of atoms, ions, biomolecules, or molecules of gas, liquid, or dissolved solids to a surface. This process creates a film of the adsorbate (the molecules or atoms being accumulated) on the surface of the adsorbent. It differs from absorption, in which a fluid permeates or is dissolved by a liquid or solid.

Aerobic: Adjective that means "requiring air," where "air" usually means oxygen.

Aerobic bacteria: Bacteria that require oxygen in order to grow and survive.

Aerobic respiration: Process that releases energy inside each of the body's cells.

Affinity chromatography: A method of separating biochemical mixtures and based on a highly specific biological interaction such as that between antigen and antibody, enzyme and substrate, or receptor and ligand. Affinity chromatography combines the size fractionation capability of gel permeation chromatography with the ability to design a chromatography that reversibly binds to a known subset of molecules. The method was discovered and developed by Pedro Cuatrecasas and Meir Wilchek for which the Wolf Prize in Medicine was awarded in 1987.

Aflatoxin: Naturally occurring mycotoxins that are produced by many species of *Aspergillus*, a fungus, most notably *Aspergillus flavus* and *Aspergillus parasiticus*. Aflatoxins are toxic and among the most carcinogenic substances known.

Agar; agar–agar: Gelatinous substance derived from a polysaccharide that accumulates in the cell walls of agarophyte red algae.

Agarose: Polysaccharide obtained from agar that is used for a variety of life science applications especially in gel electrophoresis. Agarose forms an inert matrix utilized in separation techniques.

Agrobacterium: Agrobacterium is a genus of Gram-negative bacteria established by H. J. Conn that uses horizontal gene transfer to cause tumors in plants. *Agrobacterium tumefaciens* is the most commonly studied species in this genus. Agrobacterium is well known for its ability to transfer DNA between itself and plants, and, for this reason, it has become an important tool for genetic engineering.

***Agrobacterium tumefaciens*:** *Agrobacterium tumefaciens* (updated scientific name: *Rhizobium radiobacter*) is the causal agent of crown gall disease (the formation of tumors) in over 140 species of dicot. It is a rod-shaped Gram-negative soil bacterium (Smith et al., 1907). Symptoms are caused by the insertion of a small segment of DNA (known as the T-DNA, for "transfer DNA") into the plant cell, which is incorporated at a semirandom location into the plant genome.

Albinism: Albinism is a congenital disorder characterized by the complete or partial absence of pigment in the skin, hair, and eyes due to absence or defect of an enzyme involved in the production of melanin.

Allele: An allele is one of two or more forms of a gene. Sometimes, different alleles can result in different traits, such as color. Other times, different alleles will have the same result in the expression of a gene.

Allele frequency: Allele frequency is the proportion of all copies of a gene that is made up of a particular gene variant (allele). In other words, it is the number of copies of a particular allele divided by the number of copies of all alleles at the genetic place (locus) in a population. It can be expressed for example as a percentage.

Allergen: An allergen is any substance that can cause an allergy. Technically, an allergen is a nonparasitic antigen capable of stimulating a type-I hypersensitivity reaction in atopic individuals.

Allogamy: Allogamy (cross-fertilization) is a term used in the field of biological reproduction describing the fertilization of an ovum from one individual with the spermatozoa of another. By contrast, autogamy is the term used for self-fertilization. In humans, the fertilization event is an instance of allogamy.

Allopolyploid: Allopolyploids are hybrids that have a chromosome number double that off their parents. Some of these are created via selective breeding to produce new varieties of plants from previously sterile species.

Allosteric control: Allosteric control, in enzymology, inhibition, or activation of an enzyme by a small regulatory molecule that interacts at a site (allosteric site) other than the active site (at which catalytic activity occurs). The interaction changes the shape of the enzyme so as to affect the formation at the active site of the usual complex between the enzyme and its substrate (the compound upon which it acts to form a product).

Allosteric enzyme: Allosteric enzymes are enzymes that change their conformation upon binding of an effector. An allosteric enzyme is an oligomer whose biological activity is affected by altering the conformation(s) of its quaternary structure. Allosteric enzymes tend to have several subunits. These subunits are referred to as protomers.

Allosteric regulation: In biochemistry, allosteric regulation is the regulation of an enzyme or other protein by binding an effector molecule at the protein's allosteric site (i.e., a site other than the protein's active site). Effectors that enhance the protein's activity are referred to as allosteric activators, whereas those that decrease the protein's activity are called allosteric inhibitors.

Alternative mRNA splicing: Alternative splicing (or differential splicing) is a process by which the exons of the RNA produced by transcription of a gene (a primary gene transcript or pre-mRNA) are reconnected in multiple ways during RNA splicing. The resulting different mRNAs may be translated into different protein isoforms; thus, a single gene may code for multiple proteins.

Ambient temperature: Ambient temperature simply means "the temperature of the surroundings" and will be the same as room temperature indoors.

Aminoacyl site: A nucleotide sequence near the 5′ terminus of mRNA required for binding of mRNA to the small ribosomal subunit.

Aminoacyl tRNA synthetase: An aminoacyl tRNA synthetase (aaRS) is an enzyme that catalyzes the esterification of a specific amino acid or its precursor to one of all its compatible cognate tRNAs to form an aminoacyl-tRNA. This is sometimes called "charging" the tRNA with the amino acid. Once the tRNA is charged, a ribosome can transfer the amino acid from the tRNA onto a growing peptide, according to the genetic code.

Amitosis: Direct division of the nucleus and cell, without the complicated changes in the nucleus that occur during the ordinary process of cell reproduction.

Amniocentesis: Amniocentesis (also referred to as amniotic fluid test or AFT) is a medical procedure used in prenatal diagnosis of chromosomal abnormalities and fetal infections, in which a small amount of amniotic fluid, which contains fetal tissues, is extracted from the amnion or amniotic sac surrounding a developing fetus, and the fetal DNA is examined for genetic abnormalities. A procedure for obtaining amniotic fluid from a pregnant mammal for the diagnosis of some diseases in the unborn fetus.

Amnion: The amnion is a membrane building the amniotic sac that surrounds and protects an embryo. It is developed in reptiles, birds, and mammals, which are hence called "Amniota"; but not in amphibians and fish that are consequently termed "Anamniota."

Amniotic fluid: Amniotic fluid is the nourishing and protecting liquid contained by the amniotic sac of a pregnant woman.

Amorph: A mutant gene that produces no detectable phenotypic effect.

Amphidiploid: An organism or individual having a diploid set of chromosomes derived from each parent.

Amphimixis: The union of the sperm and egg in sexual reproduction.

Ampicillin: Ampicillin is a beta-lactam antibiotic that has been used extensively to treat bacterial infections since 1961. Until the introduction of ampicillin by the British company Beecham, penicillin therapies had only been effective against Gram-positive organisms such as staphylococci and streptococci.

Amplification: A mechanism leading to multiple copies of a chromosomal region within a chromosome arm. The DNA amplification technique of the polymerase chain reaction (PCR) in molecular biology is a laboratory method for creating multiple copies of small segments of DNA.

Amplified fragment length polymorphism (AFLP): Amplified fragment length polymorphism PCR (or AFLP-PCR or just AFLP) is a PCR-based tool used in genetics research, DNA fingerprinting, and in the practice of genetic engineering. AFLP uses restriction enzymes to digest genomic DNA, followed by ligation of adaptors to the sticky ends of the restriction fragments. A subset of the restriction fragments is then selected to be amplified.

Amylase: Amylase is an enzyme that catalyzes the breakdown of starch into sugars. Amylase is present in human saliva, where it begins the chemical process of digestion.

Amylopectin: Amylopectin is a soluble polysaccharide and highly branched polymer of glucose found in plants. It is one of the two components of starch, the other being amylose.

Amylose: Amylose is a linear polymer made up of D-glucose units. This polysaccharide is one of the two components of starch, making up approximately 20%–30% of the structure. The other component is amylopectin, which makes up 70%–80% of the structure.

Anabolic pathway: The series of chemical reactions that construct or synthesize molecules from smaller units, usually requiring input of energy (ATP) in the process.

Anaerobic: Anaerobic is a technical word, which literally means without oxygen, as opposed to aerobic. In wastewater treatment, the absence of oxygen is indicated as anoxic, and anaerobic is used to indicate the absence of a common electron acceptor such as nitrate, sulfate, or oxygen. An anaerobic adhesive is a bonding agent that does not cure in the presence of air.

Anaerobic digestion: Anaerobic digestion is a series of processes in which microorganisms break down biodegradable material in the absence of oxygen, used for industrial or domestic purposes to manage waste and/or to release energy.

Anaerobic respiration: Anaerobic respiration is a form of respiration using electron acceptors other than oxygen. Although oxygen is not used as the final electron acceptor, the process still uses a respiratory electron transport chain; it is respiration without oxygen. In order for the electron transport chain to function, an exogenous final electron acceptor must be present to allow electrons to pass through the system.

Anaphase: Anaphase is the stage of mitosis or meiosis when chromosomes separate in an eukaryotic cell. Each chromatid moves to opposite poles of the cell, the opposite ends of the mitotic spindle, near the microtubule organizing centers. During this stage, anaphase lag could happen.

Anchor gene: A gene that has been positioned on both the physical map and the linkage map of a chromosome and thereby allows their mutual alignment.

Aneuploidy: Aneuploidy is an abnormal number of chromosomes and is a type of chromosome abnormality. An extra or missing chromosome is a common cause of genetic disorders (birth defects). Some cancer cells also have abnormal numbers of chromosomes. Aneuploidy occurs during cell division when the chromosomes do not separate properly between the two cells.

Animal cloning: Animal cloning is the process by which an entire organism is reproduced from a single cell taken from the parent organism and in a genetically identical manner. This means the cloned animal is an exact duplicate in every way of its parent; it has the same exact DNA.

Annealing temperature: The reaction temperature is lowered to 50°C–65°C for 20–40 s allowing annealing of the primers to the single-stranded DNA template. Typically, the annealing temperature is about 3°C–5°C below the Tm of the primers used. Stable DNA–DNA hydrogen bonds are only formed when the primer sequence very closely matches the template sequence. The polymerase binds to the primer-template hybrid and begins DNA synthesis.

Antagonism: When a substance binds to the same site, an agonist would bind to without causing activation of the receptor.

Antagonist: A chemical compound that reversed the effects of agonist is called as antagonist.

Anthocyanins: Anthocyanins are water-soluble vacuolar pigments that may appear red, purple, or blue according to pH. They belong to a parent class of molecules called flavonoids synthesized via the phenylpropanoid pathway; they are odorless and nearly flavorless, contributing to taste as a moderately astringent sensation. Anthocyanins occur in all tissues of higher plants, including leaves, stems, roots, flowers, and fruits.

Antiauxin: A substance that inhibits the growth-regulating function of an auxin.

Antibiotic: The term "antibiotic" was coined by Selman Waksman in 1942 to describe any substance produced by a microorganism that is antagonistic to the growth of other microorganisms in high dilution.

Antibiotic resistance: Antibiotic resistance is a type of drug resistance where a microorganism is able to survive exposure to an antibiotic. Genes can be transferred between bacteria in a horizontal fashion by conjugation, transduction, or transformation. Thus, a gene for antibiotic resistance that had evolved via natural selection may be shared. Evolutionary stress such as exposure to antibiotics then selects for the antibiotic-resistant trait.

Antibody: An antibody, also known as an immunoglobulin, is a large Y-shaped protein used by the immune system to identify and neutralize foreign objects like bacteria and viruses.

Anticoding strand: The DNA strand that forms the template for both the transcribed mRNA and the coding DNA strand.

Anticodon: A sequence of three adjacent nucleotides located on one end of transfer RNA. It bounds to the complementary coding triplet of nucleotides in messenger RNA during translation phase of protein synthesis.

Antigen: An antigen is a substance/molecule that, when introduced into the body triggers the production of an antibody by the immune system, which will then kills or neutralizes the antigen that is recognized as a foreign and potentially harmful invader.

Anti-idiotype vaccines: Anti-idiotypic vaccines comprise antibodies that have three-dimensional immunogenic regions, designated idiotopes that consist of protein sequences that bind to cell receptors. Idiotopes are aggregated into idiotypes specific of their target antigen.

Antimicrobial agent: An antimicrobial is a substance that kills or inhibits the growth of microorganisms such as bacteria, fungi, or protozoans. Antimicrobial drugs either kill microbes (microbiocidal) or prevent the growth of microbes (microbiostatic). Disinfectants are antimicrobial substances used on nonliving objects or outside the body.

Antioxidant: An antioxidant is a molecule capable of inhibiting the oxidation of other molecules. Oxidation is a chemical reaction that transfers electrons from a substance to an oxidizing agent.

Antiparallel orientation: Two strands of DNA arranged in opposite directions.

Antisense DNA: DNA normally has two strands, that is, the sense strand and the antisense strand. In double-stranded DNA, only one strand codes for the RNA that is translated into protein. This DNA strand is referred to as the antisense strand. The strand that does not code for RNA is called the sense strand.

Antisense gene: A gene that produces an mRNA complementary to the transcript of a normal gene (usually constructed by inverting the coding region relative to the promoter).

Antisense RNA: Antisense RNA is a single-stranded RNA that is complementary to a messenger RNA (mRNA) strand transcribed within a cell. Antisense RNA may be introduced into a cell to inhibit translation of a complementary mRNA by base pairing to it and physically obstructing the translation machinery.

Antisense therapy: Antisense therapy is a form of treatment for genetic disorders or infections. When the genetic sequence of a particular gene is known to be causative of a particular disease, it is possible to synthesize a strand of nucleic acid (DNA, RNA, or a chemical analogue) that will bind to the messenger RNA (mRNA) produced by that gene and inactivate it, effectively turning that gene "off." This is because mRNA has to be single stranded for it to be translated. Alternatively, the strand might be targeted to bind a splicing site on pre-mRNA and modify the exon content of an mRNA.

Antiseptic: Antiseptics are antimicrobial substances that are applied to living tissue/skin to reduce the possibility of infection, sepsis, or putrefaction. Antiseptics are generally distinguished from antibiotics by the latter's ability to be transported through the lymphatic system to destroy bacteria within the body, and from disinfectants, which destroy microorganisms found on nonliving objects.

Apoenzyme: The protein component of an enzyme to which the coenzyme attaches to form an active enzyme.

Apomixis: Reproduction without meiosis or formation of gametes.

Apoptosis: The process of cell death by disintegration of cells into membrane-bound particles that are then eliminated by phagocytosis or by shedding.

Arabidopsis: Genus of the mustard family having white or yellow or purplish flowers; closely related to genus Arabis.

Artificial insemination: Introduction of animal semen into the uterus without sexual contact.

Artificial selection: Artificial selection (or selective breeding) describes intentional breeding for certain traits or combination of traits. The term was utilized by Charles Darwin in contrast to natural selection, in which the differential reproduction of organisms with certain traits is attributed to improved survival or reproductive ability ("Darwinian fitness"). As opposed to artificial selection, in which humans favor specific traits, in natural selection, the environment acts as a sieve through which only certain variations can pass.

Aseptic: Free from the living germs of disease, fermentation, or putrefaction.

Asexual reproduction: Asexual reproduction is a mode of reproduction by which offspring arise from a single parent and inherit the genes of that parent only; it is the reproduction that does not involve meiosis, ploidy reduction, or fertilization.

Attenuated vaccine: An attenuated vaccine is a vaccine created by reducing the virulence of a pathogen, but still keeping it viable (or "live"). Attenuation takes an infectious agent and alters it, so that it becomes harmless or less virulent. These vaccines contrast to those produced by "killing" the virus (inactivated vaccine).

Autoclave: An autoclave is an instrument used to sterilize equipment and supplies by subjecting them to high pressure saturated steam at 121°C for around 15–20 min depending on the size of the load and the contents. It was invented by Charles Chamberland in 1879.

Autoimmune disease: Autoimmune diseases arise from an overactive immune response of the body against substances and tissues normally present in the body. In other words, the body actually attacks its own cells. The immune system mistakes some part of the body as a pathogen and attacks it.

Autologous cells: Cells taken from an individual, cultured (or stored), and, possibly, genetically manipulated before being transferred back into the original donor.

Autonomous replicating sequence: An autonomously replicating sequence (ARS) contains the origin of replication in the yeast genome. It contains four regions (A, B1, B2, and B3), named in order of their effect on plasmid stability; when these regions are mutated, replication does not initiate.

Autopolyploid: An individual or strain whose chromosome complement consists of more than two complete copies of the genome of a single ancestral species.

Autoradiography: An autoradiograph is an image on an x-ray film or nuclear emulsion produced by the pattern of decay emissions (e.g., beta particles or gamma rays) from a distribution of a radioactive substance.

Autosome: An autosome is a chromosome that is not a sex chromosome or allosome; that is, to say, there are an equal number of copies of the chromosome in males and females. For example, in humans, there are 22 pairs of autosomes. In addition to autosomes, there are sex chromosomes, to be specific: X and Y. So, humans have 23 pairs of chromosomes.

Autotrophic: An autotroph[α], or producer, is an organism that produces complex organic compounds (such as carbohydrates, fats, and proteins) from simple inorganic molecules using energy from light (by photosynthesis) or inorganic chemical reactions (chemosynthesis).

Auxotrophy: Auxotrophy is the inability of an organism to synthesize a particular organic compound required for its growth (as defined by IUPAC). An auxotroph is an organism that displays this characteristic; auxotrophic is the corresponding adjective. Auxotrophy is the opposite of prototrophy, which is characterized by the ability to synthesize all the compounds that the parent organism could.

B cells: B cells are lymphocytes that play a large role in the humoral immune response (as opposed to the cell-mediated immune response, which is governed by T cells).

B lymphocyte: B cells are lymphocytes that play a large role in the humoral immune response (as opposed to the cell-mediated immune response, which is governed by T cells).

***Bacillus thuringiensis*:** *Bacillus thuringiensis* (or Bt) is a Gram-positive, soil-dwelling bacterium, commonly used as a biological alternative to a pesticide; alternatively, the Cry toxin may be extracted and used as a pesticide. *B. thuringiensis* also occurs naturally in the gut of caterpillars of various types of moths and butterflies as well as on the dark surface of plants.

Back mutation: The process that causes reversion of mutation.

Bacterial artificial chromosome: Bacterial artificial chromosome (BAC) is a DNA construct, based on a functional fertility plasmid (or F-plasmid), used for transforming and cloning in bacteria, usually *E. coli*. F-plasmids play a crucial role, because they contain partition genes that promote the even distribution of plasmids after bacterial cell division. The bacterial artificial chromosome's usual insert size is 150–350 kbp, but can be greater than 700 kbp.

Bacterial toxin: Bacterial toxin is a type of toxin that is generated by bacteria.

Bactericide: A bactericide or bacteriocide is a substance that kills bacteria and, ideally, nothing else. Bactericides are disinfectants, antiseptics, or antibiotics.

Bacteriocin: Bacteriocins are proteinaceous toxins produced by bacteria to inhibit the growth of similar or closely related bacterial strain(s). They are typically considered to be narrow spectrum antibiotics, though this has been debated. They are phenomenologically analogous to yeast and paramecium killing factors and are structurally, functionally, and ecologically diverse.

Bacteriophage: A bacteriophage is any one of a number of viruses that infect bacteria. Bacteriophages are among the most common biological entities on Earth. The term is commonly used in its shortened form, phage.

Bacteriostat: Bacteriostat is a biological or chemical agent that causes bacteriostasis. It stops bacteria from reproducing, while not necessarily harming them otherwise. Upon removal of the bacteriostat, the bacteria usually start to grow again. Bacteriostats are often used in plastics to prevent growth of bacteria on the plastic surface. This is in contrast to bacteriocides that kill bacteria.

Balanced lethal system: An arrangement of alleles of two recessive lethal genes in repulsion phase that maintains a heterozygous chromosome combination whereas homozygotes for any lethal-bearing chromosome will be lethal.

Balanced polymorphism: Balanced polymorphism is an equilibrium mixture of homozygotes and heterozygotes maintained by natural selection against both homozygotes.

Barr body: Barr body (named after discoverer Murray Barr) is the inactive X chromosome in a female somatic cell, rendered inactive in a process called lyonization, in those species (including humans) in which sex is determined by the presence of the Y or W chromosome rather than the diploidy of the X or Z.

Basal body: A basal body (also called a basal granule or kinetosome) is an organelle formed from a centriole and a short cylindrical array of microtubules. It is found at the base of a eukaryotic undulipodium (cilium or flagellum) and serves as a nucleation site for the growth of the axoneme microtubules.

Base pair: In molecular biology and genetics, two nucleotides on opposite complementary DNA or RNA strands that are connected via hydrogen bonds are called a base pair (often abbreviated bp). In the canonical Watson–Crick DNA base pairing, adenine (A) forms a base pair with thymine (T) and guanine (G) forms a base pair with cytosine (C).

Basophil: Basophils are part of your immune system that normally protects your body from infection, but can also be partly responsible for your asthma symptoms. Basophils are a type of white blood cell that is involved in inflammatory reactions in your body, especially those related to allergies and asthma.

Batch culture: A large-scale closed system culture in which cells are grown in a fixed volume of nutrient culture medium under specific environmental conditions (e.g., nutrient type, temperature, pressure, and aeration) up to a certain density in a tank or airlift fermenter, harvested, and processed as a batch, especially before all nutrients are used up.

Batch fermentation: Fermentation of the anaerobic enzymatic conversion of organic compounds, especially carbohydrates, to simpler compounds, especially to ethyl alcohol, producing energy in the form of ATP.

Binary vector system: Binary vector systems include the most commonly used vectors devised for agrobacterium gene transfer to plants. In these systems, the T-DNA region containing a gene of interest is contained in one vector, and the vir region is located in a separate disarmed (without tumor-genes) Ti plasmid. The plasmids co-reside in agrobacterium and remain independent.

Bioassay: Bioassay (commonly used shorthand for biological assay) or biological standardization is a type of scientific experiment. Bioassays are typically conducted to measure the effects of a substance on a living organism and are essential in the development of new drugs and in monitoring environmental pollutants. Both are procedures by which the potency or the nature of a substance is estimated by studying its effects on living matter.

Bioaugmentation: Bioaugmentation is the introduction of a group of natural microbial strains or a genetically engineered variant to treat contaminated soil or water. Usually, the steps involve studying the indigenous varieties present in the location to determine if biostimulation is possible. If the indigenous variety do not have the metabolic capability to perform the remediation process, exogenous varieties with such sophisticated pathways are introduced.

Bioconversion: The term bioconversion, also known as biotransformation, refers to the use of live organisms often microorganisms to carry out a chemical reaction that is more costly or not feasible nonbiologically. These organisms convert a substance to a chemically modified form. An example is the industrial production of cortisone. One step is the bioconversion of progesterone to 11-alpha-hydroxyprogesterone by *Rhizopus nigricans*.

Biodegradation: Biodegradation or biotic degradation or biotic decomposition is the chemical dissolution of materials by bacteria or other biological means. The term is often used in relation to ecology, waste management, biomedicine, and the natural environment (bioremediation) and is now commonly associated with environmentally friendly products that are capable of decomposing back into natural elements.

Biodiversity: Biodiversity is the degree of variation of life forms within a given ecosystem, biome, or an entire planet. Biodiversity is a measure of the health of ecosystems. Greater biodiversity implies greater health. Biodiversity is in part a function of climate. In terrestrial habitats, tropical regions are typically rich, whereas Polar Regions support fewer species.

Bioenrichment: Adding nutrients or oxygen to increase microbial breakdown of pollutants.

Bioethics: Bioethics is the study of controversial ethics brought about by advances in biology and medicine. Bioethicists are concerned with the ethical questions that arise in the relationships among life sciences, biotechnology, medicine, politics, law, and philosophy.

Biofuel: Biofuel is a type of fuel that is in some way derived from biomass. The term covers solid biomass, liquid fuels, and various biogases. Biofuels are gaining increased public and scientific attention, driven by factors such as oil price spikes, the need for increased energy security, concern over greenhouse gas emissions from fossil fuels, and government subsidies.

Biogas: Biogas typically refers to a gas produced by the biological breakdown of organic matter in the absence of oxygen. Biogas originates from biogenic material and is a type of biofuel.

Bioinformatics: Bioinformatics is the application of statistics and computer science to the field of molecular biology. The primary goal of bioinformatics is to increase the understanding of biological processes.

Biolistics: Gene gun or a biolistic particle delivery system, originally designed for plant transformation, is a device for injecting cells with genetic information. The payload is an elemental particle of a heavy metal coated with plasmid DNA. This technique is often simply referred to as bioballistics or biolistics.

Biological containment: Biological containment (or biocontainment) describes measures aimed at preventing genetically modified organisms (GMOs) and their transgenes from spreading into the environment.

Biometric: Biometrics consists of methods for uniquely recognizing humans based upon one or more intrinsic physical or behavioral traits. In computer science, in particular, biometrics is used as a form of identity access management and access control. It is also used to identify individuals in groups that are under surveillance.

Biopesticide: Biopesticides are biochemical pesticides that are naturally occurring substances that control pests by nontoxic mechanisms.

Biopolymer: Biopolymers are polymers produced by living organisms. Since they are polymers, biopolymers contain monomeric units that are covalently bonded to form larger structures.

Bioprocess: A bioprocess is any process that uses complete living cells or their components (e.g., bacteria, enzymes, and chloroplasts) to obtain desired products.

Bioreactor: A bioreactor may refer to any manufactured or engineered device or system that supports a biologically active environment. In one case, a bioreactor is a vessel in which a chemical process is carried out, which involves organisms or biochemically active substances derived from such organisms. This process can either be aerobic or anaerobic. These bioreactors are commonly cylindrical, ranging in size from liters to cubic meters, and are often made of stainless steel.

Bioremediation: Bioremediation is the use of microorganism metabolism to remove pollutants. Technologies can be generally classified as *in situ* or *ex situ*. *In situ* bioremediation involves treating the contaminated material at the site, while ex situ involves the removal of the contaminated material to be treated elsewhere.

Biosensor: A biosensor is an analytical device for the detection of a chemical compound that combines a biological component with a physicochemical detector component.

Biosphere: Our biosphere is the global sum of all ecosystems. It can also be called the zone of life on Earth, a closed (apart from solar and cosmic radiation), and self-regulating system.

Biosynthesis: Biosynthesis (also called biogenesis) is an enzyme-catalyzed process in cells of living organisms by which substrates are converted to more complex products. The biosynthesis process often consists of several enzymatic steps in which the product of one step is used as substrate in the following step. Examples for such multistep biosynthetic pathways are those for the production of amino acids, fatty acids, and natural products. Biosynthesis plays a major role in all cells, and many dedicated metabolic routes combined constitute general metabolism.

Biotic factor: A factor created by a living thing or any living component within an environment in which the action of the organism affects the life of another organism, for example, a predator consuming its prey.

Biotic stress: Biotic stress is stress that occurs as a result of damage done to plants by other living organisms, such as bacteria, viruses, fungi, parasites, beneficial and harmful insects, weeds, and cultivated or native plants.

Biotin: Biotin is a water-soluble B-complex vitamin (vitamin B7). It was discovered by Bateman in 1916. It is composed of a ureido (tetrahydroimidizalone) ring fused with a tetrahydrothiophene ring. A valeric acid substituent is attached to one of the carbon atoms of the tetrahydrothiophene ring. Biotin is a coenzyme in the metabolism of fatty acids and leucine, and it plays a role in gluconeogenesis.

Biotoxin: A toxic substance produced by a living organism.

Biotransformation: Biotransformation is the chemical modification (or modifications) made by an organism on a chemical compound. If this modification ends in mineral compounds like CO_2, NH_4^+, or H_2O, the biotransformation is called mineralization.

Bivalent: A molecule formed from two or more atoms bound together as a single unit molecule.

Blastocyst: The blastocyst is a structure formed in the early embryogenesis of mammals, after the formation of the morula. It is a specific mammalian example of a blastula.

Blastomere: A blastomere is a type of cell produced by division of the egg after fertilization.

Blastula: The blastula is a solid sphere of cells formed during an early stage of embryonic development in animals. The blastula is created when the zygote undergoes the cell division process known as cleavage.

Blot: Blot (biology), method of transferring proteins, DNA, RNA, or a protein onto a carrier.

Blunt end: The end of a DNA fragment resulting from the breaking of DNA molecule in which there are no unpaired bases; hence, both strands are of the same length.

Blunt-end cut: To cut a double-stranded DNA with a restriction endonuclease that generates blunt ends.

Bovine spongiform encephalopathy: Bovine spongiform encephalopathy (BSE), commonly known as mad-cow disease is a fatal neurodegenerative disease in cattle that causes a spongy degeneration in the brain and spinal cord. BSE has a long incubation period, about 30 months to 8 years, usually affecting adult cattle at a peak age onset of 4–5 years, all breeds being equally susceptible.

Breed: A breed is a group of domestic animals or plants with a homogeneous appearance, behavior, and other characteristics that distinguish it from other animals of the same species.

Breeding: Breeding is the reproduction, that is, producing of offspring, usually animals or plants.

Brewing: Brewing is the production of beer through steeping a starch source (commonly cereal grains) in water and then fermenting with yeast. Brewing has taken place since around the sixth millennium BC, and archeological evidence suggests that this technique was used in ancient Egypt.

Bubble column fermenter: A bioreactor in which the cells or microorganisms are kept suspended in a tall cylinder by rising air, which is introduced at the base of the vessel.

Buffer: Solution that reduces the change of pH upon addition of small amounts of acid or base or upon dilution.

Buoyant density: A measure of the tendency of a substance to float in some other substance; large molecules are distinguished by their differing buoyant densities in some standard fluid.

CAAT box: In molecular biology, a CCAAT box (also sometimes abbreviated a CAAT box or CAT box) is a distinct pattern of nucleotides with GGCCAATCT consensus sequence that occur upstream by 75–80 bases to the initial transcription site. The CAAT box signals the binding site for the RNA transcription factor and is typically accompanied by a conserved consensus sequence.

Callus culture: The callus culture is a technique of tissue culture; it is usually carried out on solidified gel medium in the presence of growth regulators and initiated by inoculation of small explants or sections from established organ or other cultures.

Calorie: The calorie is a pre-SI metric unit of energy. It was first defined by Nicolas Clément in 1824 as a unit of heat, entering French and English dictionaries between 1841 and 1867. In most fields, its use is archaic, having been replaced by the SI unit of energy, the joule. However, in many countries, it remains in common use as a unit of food energy.

Cancer: Cancer is the uncontrolled growth of abnormal cells in the body. Cancerous cells are also called malignant cells.

Candidate gene: A candidate gene is a gene, located in a chromosome region suspected of being involved in the expression of a trait such as a disease, whose protein product suggests that it could be the gene in question. A candidate gene can also be identified by association with the phenotype and by linkage analysis to a region of the genome.

Cap: The structure found on the 5′-end of eukaryotic mRNA, consisting of an inverted, methylated guanosine residue, is called as Cap.

Capsid: The protein coat of a virus.

Cap site: Gene translation initiation site.

Carcinogen: A carcinogen is any substance, radionuclide, or radiation that is an agent directly involved in causing cancer. This may be due to the ability to damage the genome or to the disruption of cellular metabolic processes.

Carcinoma: Carcinoma is the medical term for the most common type of cancer occurring in humans.

Carotene: The term carotene is used for several related hydrocarbon substances having the formula $C_{40}H_x$, which are synthesized by plants but cannot be made by animals.

Carotenoids: Carotenoids are tetraterpenoid organic pigments that are naturally occurring in the chloroplasts and chromoplasts of plants and some other photosynthetic organisms like algae, some types of fungus some bacteria, and at least one species of aphid. Carotenoids are generally not manufactured by species in the animal kingdom, although one species of aphid is known to have acquired the genes for the synthesis of the carotenoid torulene from fungi, by the known phenomenon of horizontal gene transfer.

Carrier DNA: DNA of undefined sequence that is added to the transforming (plasmid) DNA used in physical DNA-transfer procedures. This additional DNA increases the efficiency of transformation in electroporation and chemically mediated DNA delivery systems.

Carrier molecule: A molecule that plays a role in transporting electrons through the electron transport chain. Carrier molecules are usually proteins bound to a nonprotein group; they can undergo oxidation and reduction relatively easily, thus allowing electrons to flow through the system.

Casein: Casein is the name for a family of related phosphor-protein proteins. These proteins are commonly found in mammalian milk, making up 80% of the proteins in cow milk and between 60% and 65% of the proteins in human milk.

CAT box: In molecular biology, a CCAAT box (also sometimes abbreviated a CAAT box or CAT box) is a distinct pattern of nucleotides with GGCCAATCT consensus sequence that occur upstream by 75–80 bases to the initial transcription site. The CAAT box signals the binding site for the RNA transcription factor and is typically accompanied by a conserved consensus sequence.

Catabolic pathway: A sequence of degradative chemical reactions that break down complex molecules into smaller units, usually releasing energy in the process.

Catabolism: Catabolism is the set of pathways that break down molecules into smaller units and release energy. In catabolism, large molecules such as polysaccharides, lipids, nucleic acids, and proteins are broken down into smaller units such as monosaccharides, fatty acids, nucleotides, and amino acids, respectively.

Catabolite repression: Carbon catabolite repression, or simply catabolite repression, is an important part of global control system of various bacteria and other microorganisms. Catabolite repression allows bacteria to adapt quickly to a preferred (rapidly metabolisable) carbon and energy source first.

Catalysis: Catalysis is the change in rate of a chemical reaction due to the participation of a substance called a catalyst. Unlike other reagents that participate in the chemical reaction, a catalyst is not consumed by the reaction itself. A catalyst may participate in multiple chemical transformations. Catalysts that speed the reaction are called positive catalysts.

Catalytic RNA: A ribozyme is an RNA molecule with a well-defined tertiary structure that enables it to catalyze a chemical reaction. Ribozyme means ribonucleic acid enzyme. It may also be called an RNA enzyme or catalytic RNA.

Cation: A positively charged ion.

cDNA library: A cDNA library is a combination of cloned cDNA (complementary DNA) fragments inserted into a collection of host cells, which together constitute some portion of the transcriptome of the organism. cDNA is produced from fully transcribed mRNA found in the nucleus and therefore contains only the expressed genes of an organism.

Cell culture: Cell culture is the complex process by which cells are grown under controlled conditions. In practice, the term "cell culture" has come to refer to the culturing of cells derived from multicellular eukaryotes, especially animal cells.

Cell cycle: The cell cycle, or cell-division cycle, is the series of events that takes place in a cell leading to its division and duplication (replication). In cells without a nucleus (prokaryotic), the cell cycle occurs via a process termed binary fission.

Cell fusion: Cell fusion is an important cellular process that occurs during differentiation of muscle, bone and trophoblast cells, during embryogenesis, and during morphogenesis. Cell fusion is a necessary event in the maturation of cells, so that they maintain their specific functions throughout growth.

Cell generation time: The doubling time is the period of time required for a quantity to double in size or value. It is applied to population growth, inflation, and resource extraction, consumption of goods, compound interest, the volume of malignant tumors, and many other things, which tend to grow over time. When the relative growth rate (not the absolute growth rate) is constant, the quantity undergoes exponential growth and has a constant doubling time or period that can be calculated directly from the growth rate.

Cell hybridization: Fusion of two or more dissimilar cells, leading to formation of a synkaryon.

Cell line: A cell line is a product of immortal cells that are used for biological research. Cells used for cell lines are immortal that happens if a cell is cancerous. The cells can perpetuate division indefinitely, which is unlike regular cells which can only divide approximately 50 times. These cells are "useful" for experimentation in labs as they are always available to researchers as a product and do not require what is known as "harvesting" (the acquiring of tissue from a host) every time cells are needed in the lab.

Cell membrane: The cell membrane is a biological membrane that separates the interior of all cells from the outside environment. The cell membrane is selectively permeable to ions and organic molecules and controls the movement of substances in and out of cells.

Cell sap: The liquid inside the large central vacuole of a plant cell that serves as storage of materials and provides mechanical support, especially in nonwoody plants. It also has a vital role in plant cell osmosis.

Cell strain: A cell strain is derived either from a primary culture or a cell line by the selection or cloning of cells having specific properties or markers. In describing a cell strain, its specific features must be defined. The terms finite or continuous are to be used as prefixes if the status of the culture is known. If not, the term strain will suffice. In any published description of a cell strain, one must make every attempt to publish the characterization or history of the strain. If such has already been published, a reference to the original publication must be made. In obtaining a culture from another laboratory, the proper designation of the culture, as originally named and described, must be maintained, and any deviations in cultivation from the original must be reported in any publication.

Cellular immune response: Cell-mediated immunity is an immune response that does not involve antibodies or complement but rather involves the activation of macrophages, natural killer cells (NK), antigen-specific cytotoxic T-lymphocytes, and the release of various cytokines in response to an antigen.

Cellulose: Cellulose is an organic compound with the formula $(C_6H_{10}O_5)_n$, a polysaccharide consisting of a linear chain of several hundred to over ten thousand $\beta(1 \rightarrow 4)$ linked D-glucose units.

Cellulose nitrate: Nitrocellulose (also, cellulose nitrate and flash paper) is a highly flammable compound formed by nitrating cellulose through exposure to nitric acid or another powerful nitrating agent.

Cellulosomes: Cellulosomes are complexes of cellulolytic enzymes created by bacteria such as clostridium and bacteroides. They consist of catalytic subunits such as glycoside hydrolases, polysaccharide lyases, and carboxyl esterases bound together by scaffoldins consisting of cohesins connected to other functional units such as the enzymes and carbohydrate binding modules via dockerins. They assist in digestion or degradation of plant cell wall materials, most notably cellulose.

Central dogma: The central dogma of molecular biology deals with the detailed residue-by-residue transfer of sequential information. It states that information cannot be transferred back from protein to either protein or nucleic acid.

Centrifugation: Centrifugation is a process that involves the use of the centrifugal force for the separation of mixtures with a centrifuge, used in industry and in laboratory settings. More-dense components of the mixture migrate away from the axis of the centrifuge, while less-dense components of the mixture migrate toward the axis.

Centromere: A centromere is a region of DNA typically found near the middle of a chromosome where two identical sister chromatids come closest in contact. It is involved in cell division as the point of mitotic spindle attachment. The sister chromatids are attached all along their length, but they are closest at the centromere.

Centrosome: In cell biology, the centrosome is an organelle that serves as the main microtubule organizing center (MTOC) of the animal cell as well as a regulator of cell-cycle progression.

Chain termination: Chain termination is any chemical reaction that ceases the formation of reactive intermediates in a chain propagation step in the course of a polymerization, effectively bringing it to a halt.

Charcoal: Charcoal is the dark gray residue consisting of impure carbon obtained by removing water and other volatile constituents from animal and vegetation substances. Charcoal is usually produced by slow pyrolysis, the heating of wood, or other substances in the absence of oxygen.

Chelate: A chemical compound in the form of a heterocyclic ring, containing a metal ion attached by coordinate bonds to at least two nonmetal ions.

Chemical mutagens: Chemical mutagens are defined as those compounds that increase the frequency of some types of mutations. They vary in their potency, since this term reflects their ability to enter the cell, their reactivity with DNA, their general toxicity, and the likelihood that the type of chemical change they introduce into the DNA will be corrected by a repair system. Most of the following mutagens are used *in vivo* treatments, but some of them can also be used *in vitro.*

Chemically defined medium: A chemically defined medium is a growth medium suitable for the *in vitro* cell culture of human or animal cells in which all of the chemical components are known. The term chemically defined medium was defined by Jayme and Smith (2000) as a "Basal formulation, which may also be protein-free and is comprised solely of biochemically defined low molecular weight constituents."

Chemiluminescence: Chemiluminescence (sometimes "chemoluminescence") is the emission of light with limited emission of heat (luminescence), as the result of a chemical reaction.

Chemostat: A chemostat (from Chemical environment is static) is a bioreactor to which fresh medium is continuously added, while culture liquid is continuously removed to keep the culture volume constant. By changing the rate with which medium is added to the bioreactor the growth rate of the microorganism can be easily controlled.

Chemotaxis: Chemotaxis is the phenomenon in which somatic cells, bacteria, and other single-cell or multicellular organisms direct their movements according to certain chemicals in their environment. This is important for bacteria to find food (e.g., glucose) by swimming toward the highest concentration of food molecules or to flee from poisons (for example, phenol).

Chemotherapy: Chemotherapy (sometimes cancer chemotherapy) is the treatment of cancer with an antineoplastic drug or with a combination of such drugs into a standardized treatment regimen.

Chimeric DNA: A molecule of DNA that has resulted from recombination or DNA from two sources being spliced together.

Chimeric gene: Chimeric genes form through the combination of portions of one or more coding sequences to produce new genes. These mutations are distinct from fusion genes, which merge whole gene sequences into a single reading frame and often retain their original functions.

Chimeric protein: A hybrid protein encoded by a nucleotide sequence spliced together from 2+ complete or partial genes produced by recombinant DNA technology.

Chimeric selectable marker gene: A gene that is constructed from parts of two or more different genes and allows the host cell to survive under conditions where it would otherwise die.

Chi-squared: A chi-square test (also chi-squared test or χ^2 test) is any statistical hypothesis test in which the sampling distribution of the test statistic is a chi-square distribution when the null hypothesis is true, or any in which this is asymptotically true, meaning that the sampling distribution (if the null hypothesis is true) can be made to approximate a chi-square distribution as closely as desired by making the sample size large enough.

Chitin: Chitin is a long-chain polymer of an *N*-acetylglucosamine, a derivative of glucose, and is found in many places throughout the natural world.

Chitinase: Chitinases are digestive enzymes that break down glycosidic bonds in chitin. Because chitin composes the cell walls of fungi and exoskeletal elements of some animals (including worms and arthropods), chitinases are generally found in organisms that either need to reshape their own chitin or to dissolve and digest the chitin of fungi or animals.

Chloramphenicol: Chloramphenicol (INN) is a bacteriostatic antimicrobial. It is considered a prototypical broad-spectrum antibiotic, alongside the tetracyclines.

Chlorenchyma: Plant tissue consisting of parenchyma cells that contain chloroplasts.

Chlorophyll: Chlorophyll is a green pigment found in almost all plants, algae, and cyanobacteria. Chlorophyll is an extremely important biomolecule, critical in photosynthesis, which allows plants to obtain energy from light.

Chloroplasts: Chloroplasts are organelles found in plant cells and other eukaryotic organisms that conduct photosynthesis. Chloroplasts capture light energy to conserve free energy in the form of ATP and reduce NADP to NADPH through a complex set of processes called photosynthesis.

Chromatid: A chromatid is one of the two identical copies of DNA making up a duplicated chromosome, which are joined at their centromeres, for the process of cell division (mitosis or meiosis). They are called sister chromatids as long as they are joined by the centromeres. When they separate (during anaphase of mitosis and anaphase 2 of meiosis), the strands are called daughter chromosomes.

Chromatin: Chromatin is the combination of DNA and other proteins that make up the contents of the nucleus. The primary functions of chromatin are to package DNA into a smaller volume to fit in the cell, to strengthen the DNA to allow mitosis and meiosis and prevent DNA damage, and to control gene expression and DNA replication.

Chromatography: Chromatography is the collective term for a set of laboratory techniques for the separation of mixtures. It involves passing a mixture dissolved in a "mobile phase" through a stationary phase, which separates the analyte to be measured from other molecules in the mixture based on differential partitioning between the mobile and stationary phases. Subtle differences in a compound's partition coefficient result in differential retention on the stationary phase and thus changing the separation.

Chromosomal aberration: A chromosome anomaly, abnormality, or aberration reflects an atypical number of chromosomes or a structural abnormality in one or more chromosomes. A karyotype refers to a full set of chromosomes from an individual that can be compared to a "normal" Karyotype for the species via genetic testing. A chromosome anomaly may be detected or confirmed in this manner. Chromosome anomalies usually occur when there is an error in cell division following meiosis or mitosis. There are many types of chromosome anomalies. They can be organized into two basic groups, numerical and structural anomalies.

Chromosomal polymorphism: In genetics, chromosomal polymorphism is a condition where one species contains members with varying chromosome counts or shapes. Polymorphism is a general concept in biology where more than one version of a trait is present in a population. In some cases of differing counts, the difference in chromosome counts is the result of a single chromosome is undergoing fission, where it splits into two smaller chromosomes, or two undergoing fusion, where two chromosomes join to form one.

Chromosome banding: The banding patterns lend each chromosome a distinctive appearance, and so the 22 pairs of human nonsex chromosomes and the X and Y chromosomes can be identified and distinguished without ambiguity. Banding also permits the recognition of chromosome deletions (lost segments), chromosome duplications (surplus segments), and other types of structural rearrangements of chromosomes.

Chromosome jumping: Chromosome jumping is a tool of molecular biology that is used in the physical mapping of genomes. It is related to several other tools used for the same purpose, including chromosome walking.

Chromosome mutations: Chromosomal mutations take place when the number of chromosomes changes or when structural changes occur in the chromosomes. This process occurs generally during the formation of a zygote where changes in the number of chromosomes may result in fission (two into one or one into two) or fusion (two into one).

Chromosome theory of inheritance: The theory that chromosomes are linear sequences of genes and genes is located in specific sites on chromosomes.

Chromosome walking: Chromosome walking is a technique to clone a gene (e.g., a disease gene) from its known closest markers. The closest linked marker (e.g., EST or a known gene) to the gene is used to probe a genomic library. A restriction fragment isolated from the end of the positive clones is used to reprobe the genomic library for overlapping clones. This process is repeated several times to walk across the chromosome and reach the gene of interest.

Chymosin: Chymosin or rennin is an enzyme found in rennet. It is produced by cows in the lining of the abomasum (the fourth and final, chamber of the stomach). Chymosin is produced by gastric chief cells in infants to curdle the milk they ingest, allowing a longer residence in the bowels and better absorption.

Cistron: A cistron is a term used to describe the locus responsible for generating a protein. It can also be defined as the segment of DNA that contains all the information for the production of single polypeptide.

Clonal propagation: Asexual propagation of many new plants (ramets) from an individual (ortet); all have the same genotype.

Clonal selection: The clonal selection hypothesis has become a widely accepted model for how the immune system responds to infection and how certain types of B and T lymphocytes are selected for destruction of specific antigens invading the body.

Cloning: Cloning in biology is the process of producing similar populations of genetically identical individuals that occurs in nature when organisms such as bacteria, insects, or plants reproduce asexually. Cloning in biotechnology refers to processes used to create copies of DNA fragments (molecular cloning), cells (cell cloning), or organisms. The term also refers to the production of multiple copies of a product such as digital media or software.

Cloning vector: A cloning vector is a small piece of DNA into which a foreign DNA fragment can be inserted. The insertion of the fragment into the cloning vector is carried out by treating the vehicle and the foreign DNA with a restriction enzyme that creates the same overhang and then ligating the fragments together. There are many types of cloning vectors. Genetically, engineered plasmids and bacteriophages (such as phage λ) are perhaps most commonly used for this purpose. Other types of cloning vectors include bacterial artificial chromosomes (BACs) and yeast artificial chromosomes (YACs).

Co-culture: In the study of cell interaction, co-culture is obviously necessary. In some cases, actual contact with cells grown in a mixture is desired while in others the mutual effect of cell types on one another is of interest while the cells themselves are kept apart. In the former case, the usual cell culture plastics are ideal, while, in the latter, the use of cell culture inserts may allow control both of the physical contact and also of the duration of that contact.

Coding sequence: The coding region of a gene is that portion of a gene's DNA or RNA, composed of exons, that codes for protein. The region is bounded nearer the 5′ end by a start codon and nearer the 3′ end with a stop codon. The coding region in mRNA is bounded by the five prime untranslated regions and the three prime untranslated regions, which are also parts of the exons.

Coding strand: When referring to DNA transcription, the coding strand is the DNA strand, which has the same base sequence as the RNA transcript produced (although with thymine replaced by uracil). It is this strand that contains codons, while the noncoding strand contains anticodons.

Codon: Codon a series of three adjacent bases in one polynucleotide chain of a DNA or RNA molecule, which codes for a specific amino acid.

Coenzyme: Coenzyme A (CoA, CoASH, or HSCoA) is a coenzyme, notable for its role in the synthesis and oxidation of fatty acids, and the oxidation of pyruvate in the citric acid cycle. All sequenced genomes encode enzymes that use coenzyme A as a substrate and around 4% of cellular enzymes use it (or a thioester, such as acetyl-CoA) as a substrate. It is adapted from cysteamine, pantothenate, and adenosine triphosphate.

Cofactor: A cofactor is a nonprotein chemical compound that is bound to a protein and is required for the protein's biological activity. These proteins are commonly enzymes, and cofactors can be considered "helper molecules" that assist in biochemical transformations.

Cohesive ends: DNA end or sticky end refers to the properties of the end of a molecule of DNA or a recombinant DNA molecule. The concept is important in molecular biology, especially in cloning or when subcloning inserts DNA into vector DNA. All the terms can also be used in reference to RNA. The sticky ends or cohesive ends form base pairs. Any two complementary cohesive ends can anneal, even those from two different organisms. This bondage is temporary however, and DNA ligase will eventually form a covalent bond between the sugar-phosphate residues of adjacent nucleotides to join the two molecules together.

Co-integrate vector: Called co-integrated vectors or hybrid Ti plasmids, and these vectors were among the first types of modified and engineered Ti plasmids devised for agrobacterium-mediated transformation, but are not widely used today. These vectors are constructed by homologous recombination of a bacterial plasmid with the T-DNA region of an endogenous Ti plasmid in agrobacterium. Integration of the two plasmids requires a region of homology present in both.

Colchicine: Colchicine is a medication used for gout. It is a toxic natural product and secondary metabolite, originally extracted from plants of the genus Colchicum (autumn crocus, *Colchicum autumnale*, also known as "meadow saffron").

Complementary DNA: In genetics, complementary DNA (cDNA) is DNA synthesized from a mature mRNA template in a reaction catalyzed by the enzyme reverse transcriptase and the enzyme DNA polymerase, and cDNA is often used to clone eukaryotic genes in prokaryotes.

Dalton: Dalton (symbol: Da) is a unit that is used for indicating mass on an atomic or molecular scale. It is defined as one-twelfth of the rest mass of an unbound atom of carbon-12 in its nuclear and electronic ground state and has a value of $1.660538291(73) \times 10^{-27}$ kg. One dalton is approximately equal to the mass of one proton or one neutron. The CIPM has categorized it as a "non-SI unit whose values in SI units must be obtained experimentally."

ddNTP: Dideoxy nucleoside triphosphates (ddNTPs).

***De novo* synthesis:** *De novo* synthesis refers to the synthesis of complex molecules from simple molecules such as sugars or amino acids, as opposed to their being recycled after partial degradation. For example, nucleotides are not needed in the diet as they can be constructed from small precursor molecules such as formate and aspartate. Methionine, on the other hand, is needed in the diet, because while it can be degraded to and then regenerated from homocysteine, it cannot be synthesized *de novo*.

Death phase: The final growth phase, during which nutrients have been depleted and cell number decreases.

Dedifferentiation: Regression of a specialized cell or tissue to a simpler, more embryonic, unspecialized form. Dedifferentiation may occur before the regeneration of appendages in plants and certain animals and in the development of some cancers.

Degeneration: Deterioration of a tissue or an organ in which its function is diminished or its structure is impaired.

Dehydrogenase: A dehydrogenase (also called DHO in the literature) is an enzyme that oxidizes a substrate by transferring one or more hydrides (H^-) to an acceptor, usually NAD^+/$NADP^+$ or a flavin coenzyme such as FAD or FMN.

Deionized: Purified water is water from any source that is physically processed to remove impurities. Distilled water and deionized (DI) water have been the most common forms of purified water, but water can also be purified by other processes including reverse osmosis, carbon filtration, microfiltration, ultrafiltration, ultraviolet oxidation, or electrodialysis.

Denaturated DNA: The denaturation of nucleic acids such as DNA due to high temperatures is the separation of a double strand into two single strands, which occurs when the hydrogen bonds between the strands are broken. This may occur during polymerase chain reaction. Nucleic acid strands realign when "normal" conditions are restored during annealing. If the conditions are restored too quickly, the nucleic acid strands may realign imperfectly.

Denitrification: Denitrification is a microbially facilitated process of nitrate reduction that may ultimately produce molecular nitrogen (N_2) through a series of intermediate gaseous nitrogen oxide products.

Deoxyribonuclease: A deoxyribonuclease (DNase) is any enzyme that catalyzes the hydrolytic cleavage of phosphodiester linkages in the DNA backbone. Thus, deoxyribonucleases are one type of nuclease. A wide variety of deoxyribonucleases are known, which differ in their substrate specificities, chemical mechanisms, and biological functions.

Deoxyribonucleic acid: Deoxyribonucleic acid or DNA is a nucleic acid that contains the genetic instructions used in the development and functioning of all known living organisms (with the exception of RNA viruses).

Derepression: In genetics and biochemistry, a repressor gene inhibits the activity of an operator gene. By inactivating the repressor, the operator gene becomes active again. This effect is called derepression.

Dessicator: Desiccators are sealable enclosures containing desiccants used for preserving moisture-sensitive items. A common use for desiccators is to protect chemicals that are hygroscopic or that react with water from humidity.

Detergent: A detergent is a surfactant or a mixture of surfactants having "cleaning properties in dilute solutions." Commonly, "detergent" refers to alkylbenzenesulfonates, a family of compounds that are similar to soap but are less affected by hard water. In most household contexts, the term detergent by itself refers specifically to laundry detergent or dish detergent, as opposed to hand soap or other types of cleaning agents. Detergents are commonly available as powders or concentrated solutions.

Determinate growth: In biology, determinate growth means not continuing to grow indefinitely. Determinate growth describe a more or less rapid growth to a mature conclusive size, with no growth thereafter like in the animals and leaves that stop growing at the reaching of the adult final condition.

Deviation: The difference between the value of an observation and the mean of the population in mathematics and statistics.

Dextrins: Dextrins are a group of low-molecular-weight carbohydrates produced by the hydrolysis of starch or glycogen. Dextrins are mixtures of polymers of D-glucose units linked by α-(1 $\rightarrow$ 4) or α-(1 $\rightarrow$ 6) glycosidic bonds.

Diabetes: Diabetes is a chronic (lifelong) disease marked by high levels of sugar in the blood.

Diakinesis: The final stage of the prophase in meiosis, characterized by shortening and thickening of the paired chromosomes, formation of the spindle fibers, disappearance of the nucleolus, and degeneration of the nuclear membrane.

Dicentric chromosome: Dicentric chromosome is an aberrant chromosome having two centromeres. Dicentric chromosomes form when two chromosome segments (from different chromosomes or from the two chromatids of a single one), each with a centromere, fuse end to end, with loss of their acentric fragments.

Dichogamy: The maturing of pistils and stamens at different times, preventing self-pollination.

Dideoxynucleotide: Dideoxynucleotides, or ddNTPs, are nucleotides lacking a 3′-hydroxyl (-OH) group on their deoxyribose sugar. Because deoxyribose already lacks a 2′-OH, dideoxyribose lacks hydroxyl groups at both its 2′ and 3′ carbons. The lack of this hydroxyl group means that, after being added by a DNA polymerase to a growing nucleotide chain, no further nucleotides can be added as no phosphodiester bond can be created based on the fact that deoxyribonucleoside triphosphates (which are the building blocks of DNA) allow DNA chain synthesis to occur through a condensation reaction between the 5′ phosphate (following the cleavage of pyrophospate) of the current nucleotide with the 3′ hydroxyl group of the previous nucleotide.

Differential centrifugation: Differential centrifugation is a common procedure in microbiology and cytology used to separate certain organelles from whole cells for further analysis of specific parts of cells. In the process, a tissue sample is first homogenized to break the cell membranes and mix up the cell contents.

Differentiation: In developmental biology, cellular differentiation is the process by which a less specialized cell becomes a more specialized cell type. Differentiation occurs numerous times during the development of a multicellular organism as the organism changes from a simple zygote to a complex system of tissues and cell types. Differentiation is a common process in adults as well: adult stem cells divide and create fully differentiated daughter cells during tissue repair and during normal cell turnover.

Diffusion: Diffusion describes the spread of particles through random motion from regions of higher concentration to regions of lower concentration.

Digest: A restriction digest is a procedure used in molecular biology to prepare DNA for analysis or other processing. It is sometimes termed as DNA fragmentation (this term is used for other procedures as well).

Dimer: A dimer is a chemical entity consisting of two structurally similar subunits called monomers joined by bonds that can be either strong or weak.

Dimethyl sulfoxide: Dimethyl sulfoxide (DMSO) is an organo-sulfur compound with the formula $(CH_3)_2SO$. This colorless liquid is an important polar aprotic solvent that dissolves both polar and nonpolar compounds and is miscible in a wide range of organic solvents as well as water.

Dimorphism: The existence of a part (as leaves of a plant) in two different forms.

Direct embryogenesis: The formation in culture, on the surface of zygotic or somatic embryos or on explant tissues (leaf section, root tip, etc.) of embryoids without an intervening callus phase.

Direct repeat: Direct repeats are nucleotide sequences present in multiple copies in the genome. There are several types of repeated sequences. Interspersed (or dispersed) DNA repeats (interspersed repetitive sequences) are copies of transposable elements interspersed throughout the genome. Flanking (or terminal) repeats (terminal repeat sequences) are sequences that are repeated on both ends of a sequence, for example, the long terminal repeats (LTRs) on retroviruses.

Directed mutagenesis: Site-directed mutagenesis, also called site-specific mutagenesis or oligonucleotide-directed mutagenesis, is a molecular biology technique in which a mutation is created at a defined site in a DNA molecule. In general, this form of mutagenesis requires that the wild-type gene sequence be known.

Directional cloning: DNA inserts and vector molecules are digested with two different restriction enzymes to create noncomplementary sticky ends at either end of each restriction fragment. This allows the insert to be ligated to the vector in a specific orientation and prevent the vector from self-ligation.

Disaccharide: A disaccharide or biose is the carbohydrate formed when two monosaccharides undergo a condensation reaction, which involves the elimination of a small molecule, such as water, from the functional groups only. Like monosaccharides, disaccharides also dissolve in water, taste sweet, and are called sugars.

Discontinuous variation: Discontinuous variation is variation within a population of a characteristic that falls into two or more discrete classes. Classic examples include such things as eye color in animals and the tall and short pea phenotypes used by Austrian botanist Gregor Johann Mendel. Characteristics that display discontinuous variation are present in one state or another; there is no blending or merging of the different forms possible. Unlike continuous variation, discontinuous variation is displayed by characteristics that are usually controlled by only one or two genes and that have little or no environmental component in their expression.

Dissecting microscope: The stereo or dissecting microscope is an optical microscope variant designed for low magnification observation or a sample using incident light illumination rather than trans-illumination. It uses two separate optical paths with two objectives and two eyepieces to provide slightly different viewing angles to the left and right eyes. In this way, it produces a three-dimensional visualization of the sample being examined.

Distillation: Distillation is a method of separating mixtures based on differences in their boiling points. Distillation is a unit operation or a physical separation process and not a chemical reaction.

Di-sulfide bond: In chemistry, a disulfide bond is a covalent bond, usually derived by the coupling of two thiol groups. The linkage is also called an SS-bond or disulfide bridge.

Diurnal: Occurring or active during the daytime rather than at night.

DNA amplification: The polymerase chain reaction (PCR) is a scientific technique in molecular biology to amplify a single or a few copies of a piece of DNA across several orders of magnitude, generating thousands to millions of copies of a particular DNA sequence.

DNAase: Deoxyribonuclease I (DNase I) is a single, glycosylated polypeptide that degrades unwanted single- and double-stranded DNA. The enzyme works by cleaving DNA into 5′ phosphodinucleotide and small oligonucleotide fragments. DNase I is commonly added to cell lysis reagents to remove the viscosity caused by the DNA content in bacterial cell lysates or to remove the DNA templates from RNAs produced by *in vitro* transcription.

DNA chip: DNA microarray (also commonly known as gene chip, DNA chip, or biochip) is a collection of microscopic DNA spots attached to a solid surface. Scientists use DNA microarrays to measure the expression levels of large numbers of genes simultaneously or to genotype multiple regions of a genome. Each DNA spot contains picomoles (10–12 mol) of a specific DNA sequence, known as probes (or reporters). These can be a short section of a gene or other DNA element that is used to hybridize a cDNA or cRNA sample (called target) under high-stringency conditions.

DNA cloning: DNA cloning is a technique to reproduce DNA fragments. It can be achieved by two different approaches: (1) cell based and (2) using polymerase chain reaction (PCR). In the cell-based approach, a vector is required to carry the DNA fragment of interest into the host cell.

DNA construct: A DNA construct (stress on first syllable) is an artificially constructed segment of nucleic acid that is going to be "transplanted" into a target tissue or cell. It often contains a DNA insert, which contains the gene sequence encoding a protein of interest that has been subcloned into a vector, which contains bacterial resistance genes for growth in bacteria, and promoters for expression in the organism.

DNA delivery system: Gene (DNA) delivery is the process of introducing foreign DNA into host cells. Gene delivery is, for example, one of the steps necessary for gene therapy and the genetic modification of crops. There are many different methods of gene delivery developed for a various types of cells and tissues, from bacterial to mammalian. Generally, the methods can be divided into two categories, viral and nonviral.

DNA fingerprint: DNA profiling (also called DNA testing, DNA typing, or genetic finger printing) is a technique employed by forensic scientists to assist in the identification of individuals by their respective DNA profiles. DNA profiles are encrypted sets of numbers that reflect a person's DNA makeup, which can also be used as the person's identifier. DNA profiling should not be confused with full genome sequencing. It is used in, for example, parental testing and criminal investigation.

DNA helicase: The role of helicases is to unwind the duplex DNA in order to provide a single-stranded DNA for replication, transcription, and recombination for instance.

DNA hybridization: DNA–DNA hybridization generally refers to a molecular biology technique that measures the degree of genetic similarity between pools of DNA sequences. It is usually used to determine the genetic distance between two species. When several species are compared that way, the similarity values allow the species to be arranged in a phylogenetic tree; it is therefore one possible approach to carrying out molecular systematics.

DNA ligase: In molecular biology, DNA ligase is a specific type of enzyme, a ligase that repairs single-stranded discontinuities in double-stranded DNA molecules, in simple words strands that have double-strand break (a break in both complementary strands of DNA).

DNA marker: A genetic marker is a gene or DNA sequence with a known location on a chromosome that can be used to identify cells, individuals, or species. It can be described as a variation (which may arise due to mutation or alteration in the genomic loci) that can be observed. A genetic marker may be a short DNA sequence, such as a sequence surrounding a single base-pair change (single nucleotide polymorphism, SNP), or a long one, like minisatellites.

DNA micro-array: DNA microarray (also commonly known as gene chip, DNA chip, or biochip) is a collection of microscopic DNA spots attached to a solid surface. Scientists use DNA microarrays to measure the expression levels of large numbers of genes simultaneously or to genotype multiple regions of a genome. Each DNA spot contains picomoles (10–12 mol) of a specific DNA sequence, known as probes (or reporters). These can be a short section of a gene or other DNA element that are used to hybridize a cDNA or cRNA sample (called target) under high-stringency conditions. Probe-target hybridization is usually detected and quantified by detection of fluorophore-, silver-, or chemiluminescence-labeled targets to determine relative abundance of nucleic acid sequences in the target.

DNA polymerase: DNA polymerase is an enzyme that helps catalyze in the polymerization of deoxyribonucleotides into a DNA strand. DNA polymerases are best known for their feedback role in DNA replication, in which the polymerase "reads" an intact DNA strand as a template and uses it to synthesize the new strand. This process copies a piece of DNA.

DNA polymorphism: One of two or more alternate forms (alleles) of a chromosomal locus that differ in nucleotide sequence or have variable numbers of repeated nucleotide units.

DNA primase: DNA primases are enzymes whose continual activity is required at the DNA replication fork. They catalyze the synthesis of short RNA molecules used as primers for DNA polymerases.

DNA probe: A labeled segment of DNA is used to find a specific sequence of nucleotides in a DNA molecule. Probes may be synthesized in the laboratory, with a sequence complementary to the target DNA sequence.

DNA repair: DNA repair refers to a collection of processes by which a cell identifies and corrects damage to the DNA molecules that encode its genome. In human cells, both normal metabolic activities and environmental factors such as UV light and radiation can cause DNA damage, resulting in as many as one million individual molecular lesions per cell per day.

DNA replication: DNA replication is a biological process that occurs in all living organisms and copies their DNA; it is the basis for biological inheritance. The process starts with one double-stranded DNA molecule and produces two identical copies of the molecule. Each strand of the original double-stranded DNA molecule serves as template for the production of the complementary strand.

DNA sequencing: DNA sequencing includes several methods and technologies that are used for determining the order of the nucleotide bases adenine, guanine, cytosine, and thymine in a molecule of DNA.

DNA transformation: In molecular biology, transformation is the genetic alteration of a cell resulting from the direct uptake, incorporation, and expression of exogenous genetic material (exogenous DNA) from its surrounding and taken up through the cell membrane(s). Transformation occurs most commonly in bacteria and in some species occurs naturally.

Dolly: Dolly was a female domestic sheep, and the first mammal to be cloned from an adult somatic cell, using the process of nuclear transfer. She was cloned by Ian Wilmut, Keith Campbell, and colleagues at the Roslin Institute near Edinburgh in Scotland. She was born on July 5, 1996, and she lived until the age of six.

Dominant gene: Gene that produces the same phenotype in the organism whether or not its allele identical; "the dominant gene for brown eyes."

Dominant marker selection: Selection of cells via a gene encoding a product that enables only the cells that carry the gene to grow under particular conditions. For example, plant and animal cells that express the introduced neo gene are resistant to neomycin and analogous antibiotics, while cells that do not carry neo are killed.

Dominant selectable marker gene: Selectable marker is a gene introduced into a cell, especially a bacterium or to cells in culture that confers a trait suitable for artificial selection. They are a type of reporter gene used in laboratory microbiology, molecular biology, and genetic engineering to indicate the success of a transfection or other procedure meant to introduce foreign DNA into a cell. Selectable markers are often antibiotic resistance genes; bacteria that have been subjected to a procedure to introduce foreign DNA are grown on a medium containing an antibiotic, and those bacterial colonies that can grow have successfully taken up and expressed the introduced genetic material.

Dormancy: Dormancy is a period in an organism's life cycle when growth, development, and (in animals) physical activity are temporarily stopped. This minimizes metabolic activity and therefore helps an organism to conserve energy. Dormancy tends to be closely associated with environmental conditions.

Double crossing-over: Chromosomal crossover (or crossing over) is an exchange of genetic material between homologous chromosomes. It is one of the final phases of genetic recombination, which occurs during prophase I of meiosis (pachytene) in a process called synapsis. Synapsis begins before the synaptonemal complex develops and is not completed until near the end of prophase I. Crossover usually occurs when matching regions on matching chromosomes break and then reconnect to the other chromosome.

Double helix: The term double helix refers to the structure formed by double-stranded molecules of nucleic acids such as DNA and RNA. The double helical structure of a nucleic acid complex arises as a consequence of its secondary structure and is a fundamental component in determining its tertiary structure.

Double recessive: A diploid individual homozygous for (containing two copies of) the same recessive allele of a gene, as indicated by the expression of the recessive allele in the phenotype.

Doubling time: The doubling time is the period of time required for a quantity to double in size or value. It is applied to population growth, inflation, and resource extraction, consumption of goods, compound interest, the volume of malignant tumors, and many other things which tend to grow over time. When the relative growth rate (not the absolute growth rate) is constant, the quantity undergoes exponential growth and has a constant doubling time or period which can be calculated directly from the growth rate.

Downstream processing: Downstream processing refers to the recovery and purification of biosynthetic products, particularly pharmaceuticals, from natural sources such as animal or plant tissue or fermentation broth, including the recycling of salvageable components and the proper treatment and disposal of waste. It is an essential step in the manufacture of pharmaceuticals such as antibiotics, hormones (e.g., insulin and human growth hormone), antibodies (e.g., infliximab and abciximab) and vaccines; antibodies and enzymes used in diagnostics; industrial enzymes; and natural fragrance and flavor compounds. Downstream processing is usually considered a specialized field in biochemical engineering, itself a specialization within chemical engineering, though many of the key technologies were developed by chemists and biologists for laboratory-scale separation of biological products.

Drug delivery: Drug delivery is the method or process of administering a pharmaceutical compound to achieve a therapeutic effect in humans or animals. Drug delivery technologies are patent protected formulation technologies that modify drug release profile, absorption, distribution, and elimination for the benefit of improving product efficacy and safety as well as patient convenience and compliance. Most common routes of administration include the preferred noninvasive peroral (through the mouth), topical (skin), transmucosal (nasal, buccal/sublingual, vaginal, ocular, and rectal), and inhalation routes.

dscDNA: Double-stranded complementary DNA.

dsDNA: Double-stranded DNA.

***E. coli*:** *Escherichia coli.*

Ecology: Ecology is the scientific study of the relations that living organisms have with respect to each other and their natural environment. Variables of interest to ecologists include the composition, distribution, amount (biomass), number, and changing states of organisms within and among ecosystems.

Ecosystem: An ecosystem is a biological environment consisting of all the organisms living in a particular area, as well as all the nonliving, physical components of the environment with which the organisms interact, such as air, soil, water, and sunlight. It is all the organisms in a given area, along with the nonliving (abiotic) factors with which they interact; a biological community and its physical environment.

EDTA: Ethylene-diamine tetra-acetic acid.

Effector cells: The muscle, gland, or organ cell capable of responding to a stimulus at the terminal end of an efferent neuron or motor neuron.

Effector molecule: An effector is a molecule (originally referring to small molecules but now encompassing any regulatory molecule, including proteins) that binds to a protein and thereby alters the activity of that protein. A modulator molecule binds to a regulatory site during allosteric modulation and allosterically modulates the shape of the protein. An effector can also be a protein that is secreted from a pathogen, which alters the host organism to enable infection, for example, by suppressing the host's immune system capabilities.

Electroblotting: Electroblotting is a method in molecular biology to transfer proteins or nucleic acids onto a membrane by using PVDF or nitrocellulose, after gel electrophoresis. The protein or nucleic acid can then be further analyzed using probes such as specific antibodies, ligands like lectins or stains. This method can be used with all polyacrylamide and agarose gels. An alternative technique for transferring proteins from a gel is capillary blotting.

Electron microscope: An electron microscope is a type of microscope that uses a particle beam of electrons to illuminate the specimen and produce a magnified image. Electron microscopes

(EM) have a greater resolving power than a light-powered optical microscope, because electrons have wavelengths about 100,000 times shorter than visible light (photons) and can achieve better than 50 pm resolution and magnifications of up to about 10,000,000X, whereas ordinary, nonconfocal light microscopes are limited by diffraction to about 200 nm resolution and useful magnifications below 2000X.

Electrophoresis: Electrophoresis is the motion of dispersed particles relative to a fluid under the influence of a spatially uniform electric field. This electrokinetic phenomenon was observed for the first time in 1807 by Reuss (Moscow State University), who noticed that the application of a constant electric field caused clay particles dispersed in water to migrate. It is ultimately caused by the presence of a charged interface between the particle surface and the surrounding fluid.

Electroporation: Electroporation, or electro-permeabilization, is a significant increase in the electrical conductivity and permeability of the cell plasma membrane caused by an externally applied electrical field. It is usually used in molecular biology as a way of introducing some substance into a cell, such as loading it with a molecular probe, a drug that can change the cell's function, or a piece of coding DNA.

ELISA: Enzyme-linked immunosorbent assay (ELISA), also known as an enzyme immunoassay (EIA), is a biochemical technique used mainly in immunology to detect the presence of an antibody or an antigen in a sample. The ELISA has been used as a diagnostic tool in medicine and plant pathology as well as a quality-control check in various industries. In simple terms, in ELISA, an unknown amount of antigen is affixed to a surface, and then a specific antibody is applied over the surface, so that it can bind to the antigen. This antibody is linked to an enzyme, and, in the final step, a substance is added that the enzyme can convert to some detectable signal, most commonly a color change in a chemical substrate.

Elongation factors: Elongation factors are a set of proteins that facilitate the events of translational elongation, the steps in protein synthesis from the formation of the first peptide bond to the formation of the last one.

Embryo cloning: Artificial embryo splitting or embryo twinning may also be used as a method of cloning, where an embryo is split in the maturation before embryo transfer.

Embryo culture: Embryo culture has been used to produce plants from embryos that would not normally develop within the fruit. This occurs in early ripening peaches and in some hybridization between species. Embryo culture can also be used to circumvent seed dormancy.

Embryogenesis: Embryogenesis is the process by which the embryo is formed and develops, until it develops into a fetus. Embryogenesis starts with the fertilization of the ovum (or egg) by sperm. The fertilized ovum is referred to as a zygote. The zygote undergoes rapid mitotic divisions with no significant growth (a process known as cleavage) and cellular differentiation, leading to development of an embryo.

Embryoid bodies: Embryoid bodies are aggregates of cells derived from embryonic stem cells and have been studied for years with mouse embryonic stem cells. Cell aggregation is imposed by hanging drop, plating upon nontissue culture treated plates or spinner flasks; either method prevents cells from adhering to a surface to form the typical colony growth. Upon aggregation, differentiation is initiated, and the cells begin to a limited extent to recapitulate embryonic development.

Embryonic stem cells: Embryonic stem cells (ES cells) are pluripotent stem cells derived from the inner cell mass of the blastocyst, an early-stage embryo. Human embryos reach the blastocyst stage 4–5 days postfertilization, at which time they consist of 50–150 cells. Isolating the embryoblast or inner cell mass (ICM) results in destruction of the fertilized human embryo, which raises ethical issues?

Encapsidation: Process by which a virus' nucleic acid is enclosed in a capsid.

Encapsulation: Molecular encapsulation in supramolecular chemistry is the confinement of a guest molecule inside the cavity of a supramolecular host molecule (molecular capsule, molecular container, or cage compounds). Examples of supramolecular host molecule include carcerands and endohedral fullerenes.

Encode: ENCODE (the ENCyclopedia Of DNA Elements) is a public research consortium launched by the US National Human Genome Research Institute (NHGRI) in September 2003. The goal is to find all functional elements in the human genome, one of the most critical projects by NHGRI after it completed the successful Human Genome Project. All data generated in the course of the project will be released rapidly into public databases.

5′ end: The 5′ cap is a specially altered nucleotide on the 5′ end of precursor messenger RNA and some other primary RNA transcripts as found in eukaryotes. The process of 5′ capping is vital to creating mature messenger RNA, which is then able to undergo translation. Capping ensures the messenger RNA's stability while it undergoes translation in the process of protein synthesis, and is a highly regulated process that occurs in the cell nucleus.

Endangered species: An endangered species is a population of organisms, which is at risk of becoming extinct, because it is either few in numbers, or threatened by changing environmental or predation parameters. The International Union for Conservation of Nature (IUCN) has calculated the percentage of endangered species as 40% of all organisms based on the sample of species that have been evaluated through 2006.

End-labeling: There are two ways to label a DNA molecular; by the ends or all along the molecule. End labeling can be performed at the 3′- or 5′-end. Labeling at the 3′ end is performed by filling 3′-end recessed ends with a mixture or labeled and unlabeled dNTPs using Klenow or T4 DNA polymerases.

Endocrine gland: Endocrine glands are glands of the endocrine system that secrete their products, hormones, directly into the blood rather than through a duct. The main endocrine glands include the pituitary gland, pancreas, ovaries, testes, thyroid gland, and adrenal glands. The hypothalamus is a neuroendocrine organ. Other organs that are not so well known for their endocrine activity include the stomach, which produces such hormones as ghrelin.

Endocytosis: Endocytosis is the process by which cells absorb molecules (such as proteins) by engulfing them. It is used by all cells of the body, because most substances important to them are large polar molecules that cannot pass through the hydrophobic plasma or cell membrane. The process opposite to endocytosis is exocytosis.

Endoderm: Endoderm is one of the germ layers formed during animal embryogenesis. Cells migrating inward along the archenteron form the inner layer of the gastrula, which develops into the endoderm. The endoderm consists at first of flattened cells, which subsequently become columnar. It forms the epithelial lining of multiple systems.

Endodermis: The innermost layer of the cortex that forms a sheath around the vascular tissue of roots and some stems. In the roots, the endodermis helps regulate the intake of water and minerals into the vascular tissues from the cortex.

Endogenous: Growing or developing from within; originating within the body or cell.

Endomitosis: Endomitosis is reproduction of nuclear elements not followed by chromosome movements and cytoplasmic division.

Endonuclease: Endonucleases are enzymes that cleave the phosphodiester bond within a polynucleotide chain, in contrast to exonucleases, which cleave phosphodiester bonds at the end of a polynucleotide chain. Typically, a restriction site will be a palindromic sequence four to six nucleotides long. Most restriction endonucleases cleave the DNA strand unevenly, leaving complementary single-stranded ends. These ends can reconnect through hybridization and are termed "sticky ends." Once paired, the phosphodiester bonds of the fragments can be joined by DNA ligase.

Endoplasmic reticulum: The endoplasmic reticulum (ER) is a eukaryotic organelle that forms an interconnected network of tubules, vesicles, and cisternae within cells. Rough endoplasmic reticula synthesize proteins, while smooth endoplasmic reticula synthesize lipids and steroids metabolize carbohydrates and steroids (but not lipids) and regulate calcium concentration, drug metabolism, and attachment of receptors on cell membrane proteins. Sarcoplasmic reticula solely regulate calcium levels.

Endopolyploidy: An increase in the number of chromosome sets caused by replication without cell division.

Endosperm: Endosperm is the tissue produced inside the seeds of most flowering plants around the time of fertilization. It surrounds the embryo and provides nutrition in the form of starch, though it can also contain oils and protein. This makes endosperm an important source of nutrition in human diet. For example, wheat endosperm is ground into flour for bread (the rest of the grain is included as well in whole wheat flour), while barley endosperm is the main source for beer production. Other examples of endosperm that forms the bulk of the edible portion are coconut "meat" and coconut "water", and corn, including popcorn. Some plants, like the orchid, lack endosperm in their seeds.

Endotoxin: Endotoxins are toxins associated with certain Gram-negative bacteria. An "endotoxin" is a toxin that is a structural molecule of the bacteria that is recognized by the immune system.

End-product inhibition: End product inhibition is negative feedback used to regulate the production of a given molecule.

Enhancer: In genetics, an enhancer is a short region of DNA that can be bound with proteins (namely, the trans-acting factors, much like a set of transcription factors) to enhance transcription levels of genes (hence the name) in a gene cluster. While enhancers are usually cis-acting, an enhancer does not need to be particularly close to the genes it acts on, and sometimes need not be located on the same chromosome.

Enterotoxin: An enterotoxin is a protein toxin released by a microorganism in the intestine. Enterotoxins are chromosomally encoded exotoxins that are produced and secreted from several bacterial organisms.

Enucleated ovum: Egg cell without nucleus.

Enzyme: Enzymes are proteins that catalyze (i.e., increase the rates of) chemical reactions. In enzymatic reactions, the molecules at the beginning of the process are called substrates, and they are converted into different molecules, called the products.

Enzyme Commission number: The Enzyme Commission number (EC number) is a numerical classification scheme for enzymes, based on the chemical reactions they catalyze. As a system of enzyme nomenclature, every EC number is associated with a recommended name for the respective enzyme.

Epicotyl: In plant physiology, the epicotyl is the embryonic shoot above the cotyledons. In most plants, the epicotyl will eventually develop into the leaves of the plant. In dicots, the hypocotyl is what appears to be the base stem under the spent withered cotyledons and the shoot just above that is the epicotyl. In monocot plants, the first shoot that emerges from the ground or from the seed is the epicotyl, from which the first shoots and leaves emerge.

Epidermis: The epidermis is the outer layer of the skin, which together with the dermis forms the cutis. The epidermis is a stratified squamous epithelium, composed of proliferating basal and differentiated suprabasal keratinocytes. The epidermis acts as the body's major barrier against an inhospitable environment.

Epigenesis: The unfolding development in an organism and in particular the development of a plant or animal from an egg or spore through a sequence of steps in which cells differentiates and organs form.

Epigenetic variation: In biology, and specifically genetics, epigenetics is the study of changes produced in gene expression caused by mechanisms other than changes in the underlying DNA sequence—hence the name epigenetics. Examples of such changes might be DNA methylation or histone deacetylation, both of which serve to suppress gene expression without altering the sequence of the silenced genes.

Epinasty: A downward bending of leaves or other plant parts, resulting from excessive growth of the upper side.

Epiphyte: An epiphyte (or air plants) is a plant that grows upon another plant (such as a tree) nonparasitically or sometimes upon some other object (such as a building or a telegraph wire), derives its moisture and nutrients from the air and rain and sometimes from debris accumulating around it, and is found in the temperate zone (as many mosses, liverworts, lichens, and algae) and in the tropics (as many ferns, cacti, orchids, and bromeliads).

Episome: A genetic element in bacteria that can replicate free in the cytoplasm (has a different number of copies) or can be inserted into the main bacterial chromosome and replicate with the chromosome. Plasmids are an example.

Epistasis: In genetics, epistasis is the phenomenon where the effects of one gene are modified by one or several other genes, which are sometimes called modifier genes. The gene whose phenotype is expressed is called epistatic, while the phenotype altered or suppressed is called hypostatic. Epistasis can be contrasted with dominance, which is an interaction between alleles at the same gene locus. Epistasis is often studied in relation to Quantitative Trait Loci (QTL) and polygenic inheritance.

Epitope: An epitope, also known as antigenic determinant, is the part of an antigen that is recognized by the immune system, specifically by antibodies, B cells, or T cells. The part of an antibody that recognizes the epitope is called a paratope. Although epitopes are usually thought to be derived from non-self-proteins, sequences derived from the host that can be recognized are also classified as epitopes.

Epizootic: In epizoology, an epizootic is a disease that appears as new cases in a given animal population, during a given period, at a rate that substantially exceeds what is "expected" based on recent experience (i.e., a sharp elevation in the incidence rate).

Equational division: The second meiotic division is an equational division, because it does not reduce chromosome numbers. A nuclear division that maintains the same ploidy level of the cell.

Equatorial plate: The plane located midway between the poles of a dividing cell during the metaphase stage of mitosis or meiosis. It is formed from the migration of the chromosomes to the center of the spindle.

Equilibrium: The state in which the concentrations of the reactants and products have no net change over time.

Equimolar: Having an equal number of moles.

Erlenmeyer flask: An Erlenmeyer, also known as a conical flask, is a widely used type of laboratory flask which features a flat bottom, a conical body, and a cylindrical neck. It is named after the German chemist Emil Erlenmeyer, who created it in 1861.

Essential amino acid: An essential amino acid or indispensable amino acid is an amino acid that cannot be synthesized *de novo* by the organism (usually referring to humans) and therefore must be supplied in the diet.

Established culture: Established cultures are those cultures that have been completely characterized and tested.

Estimated breeding value: An "estimated breeding value" (EBV) is a statistical numerical prediction of the relative genetic value of a particular dog (male or female) available for breeding. EBVs are used to rank breeding stock for selection, based upon the genetic risk of each dog with regard to one or more specified traits.

Estrogens: Estrogens are a group of compounds named for their importance in the estrous cycle of humans and other animals. They are the primary female sex hormones. Natural estrogens are steroid hormones, while some synthetic ones are nonsteroidal.

Ethanol: Ethanol, also called ethyl alcohol, pure alcohol, grain alcohol, or drinking alcohol, is a volatile, flammable, colorless liquid. It is a psychoactive drug and one of the oldest recreational drugs. Best known as the type of alcohol found in alcoholic beverages, it is also used in thermometers, as a solvent, and as a fuel. In common usage, it is often referred to simply as alcohol or spirits.

Ethidium bromide: Ethidium bromide is an intercalating agent commonly used as a fluorescent tag (nucleic acid stain) in molecular biology laboratories for techniques such as agarose gel electrophoresis. It is commonly abbreviated as "EtBr," which is also an abbreviation for bromoethane. When exposed to ultraviolet light, it will fluoresce with an orange color, intensifying almost 20-fold after binding to DNA.

Ethylenediaminetetraacetic acid: Ethylenediaminetetraacetic acid, widely abbreviated as EDTA is a polyamino carboxylic acid and a colorless, water-soluble solid. Its conjugate base is named ethylenediaminetetraacetate. It is widely used to dissolve limescale. Its usefulness arises because of its role as a hexadentate ("six-toothed") ligand and chelating agent, that is, its ability to "sequester" metal ions such as Ca^{2+} and Fe^{3+}. After being bound by EDTA, metal ions remain in solution, but exhibit diminished reactivity. EDTA is produced as several salts, notably disodium EDTA and calcium disodium EDTA.

Euchromatin: Euchromatin is a lightly packed form of chromatin (DNA, RNA, and protein) that is rich in gene concentration and is often (but not always) under active transcription. Unlike heterochromatin, it is found in both cells with nuclei (eukaryotes) and cells without nuclei (prokaryotes). Euchromatin comprises the most active portion of the genome within the cell nucleus.

Eugenics: Eugenics is the "applied science or the biosocial movement that advocates the use of practices aimed at improving the genetic composition of a population," usually referring to human populations.

Eukaryote: A eukaryote is an organism whose cells contain complex structures enclosed within membranes. Eukaryotes may more formally be referred to as the taxon Eukarya or Eukaryota.

Euploid: The normal number of chromosomes for a species. In humans, the euploid number of chromosomes is 46 with the notable exception of the unfertilized egg and sperm in which it is 23.

Evolution: Evolution (also known as biological or organic evolution) is the change over time in one or more inherited traits found in populations of organisms. Inherited traits are particular distinguishing characteristics, including anatomical, biochemical, or behavioral characteristics that are passed on from one generation to the next. Evolution may occur when there is variation of inherited traits within a population.

Ex novo: From the beginning.

Ex vitro: Grown in the natural condition, example *ex vitro* plant means field grown plants.

***Ex vivo* gene therapy:** *Ex vivo* means that which take place outside an organism. In science, *ex vivo* refers to experimentation or measurements done in or on tissue in an artificial environment outside the organism with the minimum alteration of natural conditions. *Ex vivo* conditions allow experimentation under more controlled conditions than possible in *in vivo* experiments (in the intact organism), at the expense of altering the "natural" environment.

Excinuclease: Excision endonuclease also known as "excinuclease" is a nuclease (enzyme), which excises a fragment of nucleotides during DNA repair. The excinuclease cuts out a fragment by hydrolyzing two phosphodiester bonds, one on either side of the lesion in the DNA. This process is part of "nucleotide excision repair," a mechanism that can fix specific damages to the DNA in the G1 phase of the eukaryotic cell cycle. Such damages can include the thymine dimers created by UV rays.

Excision repair: Excision repair mechanisms that remove the damaged nucleotide replacing it with an undamaged nucleotide complementary to the nucleotide in the undamaged DNA strand.

Excision: In surgery, the complete removal of an organ, tissue, or tumor from a body.

Exo III: Exonuclease III (ExoIII) is an enzyme that belongs to the exonuclease family. ExoIII catalyzes the stepwise removal of mononucleotides from 3′-hydroxyl termini of duplex DNA.

A limited number of nucleotides are removed during each binding event, resulting in coordinated progressive deletions within the population of DNA molecules.

Exocrine gland: Exocrine glands are glands that secrete their products (including hormones and other chemical messengers) into ducts (duct glands), which lead directly into the external environment. They are the counterparts to endocrine glands, which secrete their products (hormones) directly into the bloodstream (ductless glands) or release hormones (paracrines) that affect only target cells nearby the release site.

Exogenous: Produced by growth from superficial tissue.

Exon: An exon is a nucleic acid sequence that is represented in the mature form of an RNA molecule either after portions of a precursor RNA (introns) has been removed by cis-splicing or when two or more precursor RNA molecules have been ligated by trans-splicing. The mature RNA molecule can be a messenger RNA or a functional form of a noncoding RNA such as rRNA or tRNA. Depending on the context, exon can refer to the sequence in the DNA or its RNA transcript.

Exonuclease: Exonucleases are enzymes that work by cleaving nucleotides one at a time from the end (exo) of a polynucleotide chain. A hydrolyzing reaction that breaks phosphodiester bonds at either the 3′ or the 5′ end occurs. Its close relative is the endonuclease, which cleaves phosphodiester bonds in the middle (endo) of a polynucleotide chain.

Exonuclease III: Exonuclease III (ExoIII) is an enzyme that belongs to the exonuclease family. ExoIII catalyzes the stepwise removal of mononucleotides from 3′-hydroxyl termini of duplex DNA. A limited number of nucleotides are removed during each binding event, resulting in coordinated progressive deletions within the population of DNA molecules.

Exotoxin: An exotoxin is a toxin excreted by a microorganism, including bacteria, fungi, algae, and protozoa. An exotoxin can cause damage to the host by destroying cells or disrupting normal cellular metabolism. They are highly potent and can cause major damage to the host. Exotoxins may be secreted, or, similar to endotoxins, may be released during lysis of the cell.

Expected progeny differences: Expected progeny differences (EPDs) are the differences in performance to be expected from the future progeny of a sire, compared with that expected from the future progeny of the average sire in the same population. EPDs are generally expressed either as a plus or minus difference from the population average, reported in units of measure of the trait.

Explant: In biology, explant culture is a technique used for the isolation of cells from a piece or pieces of tissue. Tissue harvested in this manner is called an explant. It can be a portion of the shoot, leaves, or some cells from a plant or can be any part of the tissue from an animal.

Exponential phase: A growth phase. In the exponential (log) phase, cells divide as fast as possible according to the growth medium, the microorganism itself, and environmental conditions. This phase has a limited duration.

Expressed sequence tag: An expressed sequence tag or EST is a short subsequence of a cDNA sequence. They may be used to identify gene transcripts and are instrumental in gene discovery and gene sequence determination. The identification of ESTs has proceeded rapidly, with approximately 65.9 million ESTs now available in public databases (e.g., GenBank June 18, 2010, all species).

Expression library: Expression cloning is a technique in DNA cloning that uses expression vectors to generate a library of clones, with each clone expressing one protein. This expression library is then screened for the property of interest and clones of interest recovered for further analysis. An example would be using an expression library to isolate genes that could confer antibiotic resistance.

Expression system: Gene expression is the process by which information from a gene is used in the synthesis of a functional gene product. These products are often proteins, but in nonprotein-coding genes such as ribosomal RNA (rRNA), transfer RNA (tRNA), or small nuclear RNA (snRNA) genes, the product is a functional RNA.

Expression vector: An expression vector, otherwise known as an expression construct, is generally a plasmid that is used to introduce a specific gene into a target cell. Once the expression vector is inside the cell, the protein that is encoded by the gene is produced by the cellular-transcription and translation machinery ribosomal complexes. The plasmid is frequently engineered to contain regulatory sequences that act as enhancer and promoter regions and lead to efficient transcription of the gene carried on the expression vector.

Extrachromosomal: Extrachromosomal DNA (sometimes called extranuclear DNA or nonchromosomal DNA) is DNA located or maintained in a cell apart from the chromosomes.

Extrachromosomal inheritance: Inheritance of traits through DNA that is not connected with the chromosomes but rather to DNA from organelles in the cell and it is also called cytoplasmic inheritance.

F factor: A sequence of bacterial DNA.

F_1: The first filial generation.

F_2: The second filial generation.

FACS: Fluorescence-activated cell sorting (FACS) is a specialized type of flow cytometry. It provides a method for sorting a heterogeneous mixture of biological cells into two or more containers, one cell at a time, based upon the specific light scattering and fluorescent characteristics of each cell. It is a useful scientific instrument, as it provides fast, objective, and quantitative recording of fluorescent signals from individual cells as well as physical separation of cells of particular interest. The acronym FACS is trademarked and owned by Becton, Dickinson and Company.

Factorial mating: A mating scheme in which each male parent is mated with each female parent. It is made possible in animals by means of *in vitro* embryo production. Such a mating scheme substantially reduces the rate of inbreeding in a selection program.

False fruit: A fruit, as the apple, strawberry, or pineapple, that contains, in addition to a mature ovary and seeds, a significant amount of other tissue.

False-negative: A result that appears negative but fails to reveal a situation. An example of a false-negative: a particular test designed to detect cancer of the toenail is negative, but the person has toenail cancer.

False-positive: A result that is erroneously positive when a situation is normal. An example of a false-positive: a particular test designed to detect cancer of the toenail is positive, but the person does not have toenail cancer.

Fed-batch fermentation: A fed-batch is a biotechnological batch process that is based on feeding of a growth limiting nutrient substrate to a culture. The fed-batch strategy is typically used in bioindustrial processes to reach a high cell density in the bioreactor. Mostly, the feed solution is highly concentrated to avoid dilution of the bioreactor. The controlled addition of the nutrient directly affects the growth rate of the culture and allows to avoid overflow metabolism (formation of side metabolites, such as acetate for *Escherichia coli*, lactic acid in cell cultures, and ethanol in *Saccharomyces cerevisiae*), oxygen limitation (anaerobiosis).

Feedback inhibition: A cellular control mechanism in which an enzyme that catalyzes the production of a particular substance in the cell is inhibited when that substance has accumulated to a certain level, thereby balancing the amount provided with the amount needed.

Fermentation: In a general sense, fermentation is the conversion of a carbohydrate such as sugar into an acid or an alcohol. More specifically, fermentation can refer to the use of yeast to change sugar into alcohol or the use of bacteria to create lactic acid in certain foods. Fermentation occurs naturally in many different foods given the right conditions, and humans have intentionally made use of it for many thousands of years.

Fermentation substrates: Substrate for fermentation is usually glucose. But depending on the yeast type, it can be fructose or other monossaccharides too.

Fermenter: An apparatus for carrying out fermentation.

Fertilization: Fertilization (also known as conception, fecundation, and syngamy) is the fusion of gametes to produce a new organism. In animals, the process involves the fusion of an ovum with a sperm, which eventually leads to the development of an embryo. Depending on the animal species, the process can occur within the body of the female in internal fertilization or outside (external fertilization). The entire process of development of new individuals is called reproduction.

Fertilizer: Fertilizer is any organic or inorganic material of natural or synthetic origin (other than liming materials) that is added to a soil to supply one or more plant nutrients essential to the growth of plants. A recent assessment found that about 40%–60% of crop yields are attributable to commercial fertilizer use.

Feulgen's test: A test used to detect DNA in nuclei, especially during cell division. A section of tissue is first placed in dilute hydrochloric acid for 10 min at 60°C to hydrolyze DNA, removing the purine bases and exposing the aldehyde groups of deoxyribose. When the tissue is soaked in Schiff's reagent, the location of the DNA is shown by the development of a magenta color.

Filter bioreactor: A cell-culture system, in which cells are grown on a fine mesh of an inert material, that allows the culture medium to flow past it but retains the cells. This is similar in idea to membrane and hollow fiber reactors, but can be much easier to set up, being similar to conventional tower bioreactors, but with the mesh replacing the central reactor space.

Filtration: Filtration is commonly the mechanical or physical operation, which is used for the separation of solids from fluids (liquids or gases) by interposing a medium through which only the fluid can pass. Oversize solids in the fluid are retained, but the separation is not complete; solids will be contaminated with some fluid and filtrate will contain fine particles (depending on the pore size and filter thickness).

Fission: In biology, fission is the carp of a body, population, or species into parts and the regeneration of those parts into separate individuals. Binary fission, or prokaryotic fission, is a form of asexual reproduction and cell division used by all prokaryotes, some protozoa, and some organelles within eukaryotic organisms.

Flagellum: Flagellum is a tail-like projection that protrudes from the cell body of certain prokaryotic and eukaryotic cells and functions in locomotion. There are some notable differences between prokaryotic and eukaryotic flagella, such as protein composition, structure, and mechanism of propulsion.

Flanking region: A region of DNA that is adjacent to the 5′ end of the gene. The 5′ flanking region contains the promoter and may contain enhancers or other protein binding sites. It is the region of DNA that is not transcribed into RNA.

Flavin adenine dinucleotide: In biochemistry, flavin adenine dinucleotide (FAD) is a redox cofactor involved in several important reactions in metabolism. FAD can exist in two different redox states, which it converts between by accepting or donating electrons. The molecule consists of a riboflavin moiety (vitamin B2) bound to the phosphate group of an ADP molecule.

Floccule: A small, loosely held mass or aggregate of fine particles, resembling a tuft of wool and suspended in or precipitated from a solution.

Flow cytometry: Flow cytometry (abbreviated: FCM) is a technique for counting and examining microscopic particles, such as cells and chromosomes, by suspending them in a stream of fluid and passing them by an electronic detection apparatus. It allows simultaneous multiparametric analysis of the physical and/or chemical characteristics of up to thousands of particles per second. Flow cytometry is routinely used in the diagnosis of health disorders, especially blood cancers, but has many other applications in both research and clinical practice.

Fluorescence *in situ* hybridization: FISH (fluorescence *in situ* hybridization) is a cytogenetic technique developed by Christoph Lengauer that is used to detect and localize the presence or absence of specific DNA sequences on chromosomes. FISH uses fluorescent probes that bind to only those parts of the chromosome with which they show a high degree of sequence similarity.

Folded genome: The condensed state of the chromosomal DNA of a bacterium. The DNA is segregated into domains, and each domain is independently negatively supercoiled.

Follicle: A follicle is a small spherical or vase-like group of cells containing a cavity in which some other structure grows. Follicles are best known as the sockets from which hairs grow in humans and other mammals, but the bristles of annelid worms also grow from such sockets.

Follicle stimulating hormone: Follicle-stimulating hormone (FSH) is a hormone found in humans and other animals. It is synthesized and secreted by gonadotrophs of the anterior pituitary gland. FSH regulates the development, growth, pubertal maturation, and reproductive processes of the body. FSH and luteinizing hormone (LH) act synergistically in reproduction.

Forced cloning: The insertion of foreign DNA into a cloning vector in a predetermined orientation.

Functional gene cloning: Gene therapy involves supplying a functional gene to cells lacking that function, with the aim of correcting a genetic disorder or acquired disease. Gene therapy can be broadly divided into two categories. The first is alteration of germ cells, that is, sperm or eggs, which results in a permanent genetic change for the whole organism and subsequent generations. This "germ line gene therapy" is considered by many to be unethical in human beings.

Fungicide: Fungicides are chemical compounds or biological organisms used to kill or inhibit fungi or fungal spores. Fungi can cause serious damage in agriculture, resulting in critical losses of yield, quality, and profit. Fungicides are used both in agriculture and to fight fungal infections in animals. Chemicals used to control oomycetes, which are not fungi, are also referred to as fungicides as oomycetes use the same mechanisms as fungi to infect plants.

Fusion gene: A fusion gene is a hybrid gene formed from two previously separate genes. It can occur as the result of a translocation, interstitial deletion, or chromosomal inversion. Often, fusion genes are oncogenes. Most fusion genes are found from hematological cancers, sarcomas, and prostate cancer.

Fusion protein: Fusion proteins or chimeric proteins are proteins created through the joining of two or more genes, which originally coded for separate proteins. Translation of this fusion gene results in a single polypeptide with functional properties derived from each of the original proteins. Recombinant fusion proteins are created artificially by recombinant DNA technology for use in biological research or therapeutics.

Beta-galactosidase: β-galactosidase, also called beta-gal or β-gal, is a hydrolase enzyme that catalyzes the hydrolysis of β-galactosides into monosaccharides. Substrates of different β-galactosidases include ganglioside GM1, lactosylceramides, lactose, and various glycoproteins. Lactase is often confused as an alternative name for β-galactosidase, but it is actually simply a subclass of β-galactosidase.

Gamete: Gamete is a cell that fuses with another cell during fertilization (conception) in organisms that reproduce sexually.

Gastrula: An early metazoan embryo in which the ectoderm, mesoderm, and endoderm are established either by invagination of the blastula (as in fish and amphibians) to form a multilayered cellular cup with a blastopore opening into the archenteron or by differentiation of the blastodisc (as in reptiles, birds, and mammals) and inward cellular migration.

Gel electrophoresis: Gel electrophoresis refers to using a gel as an anticonvective medium and or sieving medium during electrophoresis. Gel electrophresis is most commonly used for the separation of biological macromolecules such as deoxyribonucleic acid (DNA), ribonucleic acid (RNA), or protein; however, gel electrophoresis can be used for separation of nanoparticles.

Gelatin: Gelatin is a translucent, colorless, brittle (when dry), tasteless solid substance, derived from the collagen inside animals' skin and bones. It is commonly used as a gelling agent in food, pharmaceuticals, photography, and cosmetic manufacturing.

Gene addition: Gene addition inserts a functioning copy of a misfunctioning or nonfunctional native gene. Viral-based gene addition involves the "domestication" of viral genomes as vectors.

Gene amplification: A cellular process characterized by the production of multiple copies of a particular gene or genes to amplify the phenotype that the gene confers on the cell. Drug resistance in cancer cells is linked to amplification of the gene that prevents absorption of the chemotherapeutic agent by the cell.

Gene conversion: Gene conversion is an event in DNA genetic recombination, which occurs at high frequencies during meiotic division but which also occurs in somatic cells.

Gene expression: Gene expression is the process by which information from a gene is used in the synthesis of a functional gene product. These products are often proteins, but in nonprotein-coding genes such as ribosomal RNA (rRNA), transfer RNA (tRNA), or small nuclear RNA (snRNA) genes, the product is a functional RNA.

Gene flow: In population genetics, gene flow (also known as gene migration) is the transfer of alleles of genes from one population to another.

Gene imprinting: Genomic imprinting is a genetic phenomenon by which certain genes are expressed in a parent-of-origin-specific manner. It is an inheritance process independent of the classical Mendelian inheritance. Imprinted alleles are silenced, such that the genes are either expressed only from the nonimprinted allele inherited from the mother (e.g., H19 or CDKN1C) or in other instances from the nonimprinted allele inherited from the father (e.g., IGF-2). Forms of genomic imprinting have been demonstrated in insects, mammals, and flowering plants.

Gene insertion: The process by which one or more genes from one organism are incorporated into the genetic makeup of a second organism.

Gene interaction: The collaboration of several different genes in the production of one phenotypic character (or related group of characters).

Gene library: A genomic library is a population of host bacteria, each of which carries a DNA molecule that was inserted into a cloning vector, such that the collection of cloned DNA molecules represents the entire genome of the source organism. This term also represents the collection of all of the vector molecules, each carrying a piece of the chromosomal DNA of the organism, before the insertion of these molecules into the host cells.

Gene pool: The total number of genes of every individual in an interbreeding population.

Gene probe: A gene probe is a specific segment of single-strand DNA that is complementary to a desired gene. For example, if the gene of interest contains the sequence AATGGCACA, then the probe will contain the complementary sequence TTACCGTGT. When added to the appropriate solution, the probe will match and then bind to the gene of interest. To facilitate locating the probe, scientists usually label it with a radioisotope or a fluorescent dye, so that it can be visualized and identified.

Gene recombination: Genetic recombination is a process by which a molecule of nucleic acid (usually DNA, but can also be RNA) is broken and then joined to a different one. Recombination can occur between similar molecules of DNA, as in homologous recombination, or dissimilar molecules, as in nonhomologous end joining. Recombination is a common method of DNA repair in both bacteria and eukaryotes. In eukaryotes, recombination also occurs in meiosis, where it facilitates chromosomal crossover.

Gene sequencing: Gene sequencing includes several methods and technologies that are used for determining the order of the nucleotide bases adenine, guanine, cytosine, and thymine in a molecule of DNA.

Gene splicing: The process in which fragments of DNA from one or more different organisms are combined to form recombinant DNA.

Gene therapy: Gene therapy is the insertion, alteration, or removal of genes within an individual's cells and biological tissues to treat disease. It is a technique for correcting defective genes that are responsible for disease development.

Gene tracking: Gene tracking is the method used to trace throughout a family the inheritance of a gene such as those causing cystic fibrosis or Huntington's chorea, in order to diagnose and predict genetic disorders.

Gene translocation: The movement of a gene fragment from one chromosomal location to another, which often alters or abolishes expression.

Genetic code: The genetic code is the set of rules by which information encoded in genetic material (DNA or mRNA sequences) is translated into proteins (amino acid sequences) by living cells.

Genetic complementation: In genetics, complementation refers to a relationship between two different strains of an organism, which both have homozygous recessive mutations that produce the same phenotype (e.g., a change in wing structure in flies).

Genetic disease: A genetic disease is an illness caused by abnormalities in genes or chromosomes. While some diseases, such as cancer, are due in part to genetic disorders, they can also be caused by environmental factors. Most disorders are quite rare and affect one person in every several thousands or millions. Some types of recessive gene disorders confer an advantage in the heterozygous state in certain environments.

Genetic diversity: Genetic diversity, the level of biodiversity, refers to the total number of genetic characteristics in the genetic makeup of a species. It is distinguished from genetic variability, which describes the tendency of genetic characteristics to vary.

Genetic drift: Genetic drift or allelic drift is the change in the frequency of a gene variant (allele) in a population due to random sampling. The alleles in the offspring are a sample of those in the parents, and chance has a role in determining whether a given individual survives and reproduces.

Genetic engineering: Genetic engineering, also called genetic modification, is the direct human manipulation of an organism's genome using modern DNA technology. It involves the introduction of foreign DNA or synthetic genes into the organism of interest. The introduction of new DNA does not require the use of classical genetic methods; however, traditional breeding methods are typically used for the propagation of recombinant organisms.

Genetic equilibrium: A genetic equilibrium is at hand for an allele in a gene pool when the frequency of that allele is not changing (i.e., when it is not evolving). For this to be the case, evolutionary forces acting upon the allele must be equal and opposite.

Genetic heterogeneity: The phenomenon that a single phenotype or genetic disorder may be caused by any one of a multiple number of alleles or nonallele (locus) mutations.

Genetic linkage: Genetic linkage is the tendency of certain loci or alleles to be inherited together. Genetic loci that are physically close to one another on the same chromosome tend to stay together during meiosis and are thus genetically linked.

Genetic mapping: Gene mapping, also called genome mapping, is the creation of a genetic map assigning DNA fragments to chromosomes. When a genome is first investigated, this map is nonexistent. The map improves with the scientific progress and is perfect when the genomic DNA sequencing of the species has been completed.

Genetic marker: A gene or DNA sequence having a known location on a chromosome and associated with a particular gene or trait. Genetic markers associated with certain diseases can be detected in the blood and used to determine whether an individual is at risk for developing a disease.

Genetic polymorphism: The existence together of many forms of DNA sequences at a locus within the population. Genetic polymorphism promotes diversity within a population. It often persists over many generations, because no single form has an overall advantage or disadvantage over the others regarding natural selection. A common example is the different allelic forms that give rise to different blood types in humans.

Genetic selection: The process of determining genetic attributes.

Genetic transformation: A process by which the genetic material carried by an individual cell is altered by the incorporation of foreign (exogenous) DNA into its genome.

Genetic variation: Genetic variation, variation in alleles of genes, occurs both within and among populations. Genetic variation is important, because it provides the "raw material" for natural selection. Genetic variation is brought about by mutation, a change in a chemical structure of a gene.

Genetically engineered organism: Genetically modified organism (GMO) or genetically engineered organism (GEO) is an organism whose genetic material has been altered using genetic engineering techniques. These techniques, generally known as recombinant DNA technology, use DNA molecules from different sources, which are combined into one molecule to create a new set of genes.

Genetically modified food: Genetically modified foods (or GM foods) are foods derived from genetically modified organisms. Genetically modified organisms have had specific changes introduced into their DNA by genetic engineering techniques.

Genome: In modern molecular biology and genetics, the genome is the entirety of an organism's hereditary information. It is encoded either in DNA or, for many types of virus, in RNA. The genome includes both the genes and the noncoding sequences of the DNA/RNA.

Genomic DNA library: A genomic library is a population of host bacteria, each of which carries a DNA molecule that was inserted into a cloning vector, such that the collection of cloned DNA molecules represents the entire genome of the source organism. This term also represents the collection of all of the vector molecules, each carrying a piece of the chromosomal DNA of the organism, before the insertion of these molecules into the host cells.

Genotype: The genetic makeup, as distinguished from the physical appearance, of an organism or a group of organisms.

Germ cell gene therapy: Germline gene therapy involves altering the genetic makeup of either an egg or sperm cell before fertilization or altering the genetic makeup of the blastomere when it is in a very early stage of division. The goal of germline gene therapy is to affect changes in the genetic code of an organism that will be passed on to future generations.

Germicide: An agent that kills germs, especially pathogenic microorganisms; a disinfectant.

Gestation: Gestation is the carrying of an embryo or fetus inside a female viviparous animal.

Glucocorticoid: Glucocorticoids (GC) are a class of steroid hormones that bind to the glucocorticoid receptor (GR), which is present in almost every vertebrate animal cell.

Glycolysis: Glycolysis is the metabolic pathway that converts glucose into pyruvate.

Glycosylation: Glycosylation is the enzymatic process that attaches glycans to proteins, lipids, or other organic molecules. This enzymatic process produces one of the fundamental biopolymers found in cells (along with DNA, RNA, and proteins).

Good laboratory practice: In the experimental research arena, the laboratory practice or GLP specifically refers to a quality system of management controls for research laboratories and organizations to try to ensure the uniformity, consistency, reliability, reproducibility, quality, and integrity of chemical (including pharmaceuticals) safety and efficacy tests.

Gram-negative bacteria: Gram-negative bacteria are bacteria that do not retain crystal violet dye in the Gram staining protocol. In a Gram stain test, a counterstain (commonly safranin) is added after the crystal violet, coloring all Gram-negative bacteria with a red or pink color. The test itself is useful in classifying two distinct types of bacteria based on the structural differences of their bacterial cell walls. Gram-positive bacteria will retain the crystal violet dye when washed in a decolorizing solution.

Gram-positive bacteria: Gram-positive bacteria are those that are stained dark blue or violet by Gram staining. This is in contrast to Gram-negative bacteria, which cannot retain the crystal violet stain, instead taking up the counterstain (safranin or fuchsine) and appearing red or pink. Gram-positive organisms are able to retain the crystal violet stain because of the high amount of peptidoglycan in the cell wall. Gram-positive cell walls typically lack the outer membrane found in Gram-negative bacteria.

Green Revolution: Green Revolution refers to a series of research, development, and technology transfer initiatives, occurring between the 1940s and the late-1970s, which increased agriculture production around the world, beginning most markedly in the late-1960s.

Growth curve: Growth curves are widely used in biology for quantities such as population size, body height, or biomass. Values for the measured property can be plotted on a graph as a function of time.

Growth factor: A growth factor is a naturally occurring substance capable of stimulating cellular growth proliferation, and cellular differentiation. Usually, it is a protein or a steroid hormone. Growth factors are important for regulating a variety of cellular processes.

Growth hormone: Growth hormone (GH) is a protein-based peptide hormone. It stimulates growth, cell reproduction, and regeneration in humans and other animals.

Guide RNA: Guide RNA (gRNA) is the RNAs that guide the insertion or deletion of uridine residues into mitochondrial mRNAs in kinetoplastid protists in a process known as RNA editing.

Guide sequence: An RNA molecule (or a part of it) that hybridizes with eukaryotic mRNA and aids in the splicing of intron sequences. Guide sequences may be either external (EGS) or internal (IGS) to the RNA being processed and may hybridize with either intron or exon sequences close to the splice junction.

Haploid cell: A haploid cell is a cell that contains one complete set of chromosomes. Gametes are haploid cells that are produced by meiosis.

Haplotype: Haplotype in genetics is a combination of alleles (DNA sequences) at different places (loci) on the chromosome that are transmitted together.

Hardy–Weinberg equilibrium: The Hardy–Weinberg equilibrium states that both allele and genotype frequencies in a population remain constant, that is, they are in equilibrium from generation to generation unless specific disturbing influences are introduced.

Helper cells: Any of the T cells that when stimulated by a specific antigen release lymphokines that promote the activation and function of B cells and killer T cells. Also called T-helper cell.

Helper plasmid: In the context of genetic transformation of plants, a helper plasmid is a plasmid present in Agrobacterium that provides functions required by the bacteria for transferring foreign DNA to a plant cell. They have been extremely important in plant genetic engineering. Generally, helper plasmids are derivatives of the Ti plasmid that contain an active virulence region, but from which the T-DNA has been removed.

Hemicellulose: A hemicellulose is any of several heteropolymers (matrix polysaccharides), such as arabinoxylans, present along with cellulose in almost all plant cell walls. While cellulose is crystalline, strong, and resistant to hydrolysis, hemicellulose has a random, amorphous structure with little strength.

Hemoglobin: Hemoglobin is the iron-containing oxygen-transport metalloprotein in the red blood cells of all vertebrates with the exception of the fish family Channichthyidae as well as the tissues of some invertebrates.

Hemolymph: Hemolymph is a fluid in the circulatory system of some arthropods (including spiders, crustaceans such as crabs and shrimp, and even some insects such as stoneflies) and is analogous to the fluids and cells making up both blood and interstitial fluid (including water, proteins, fats, sugars, and hormones) in vertebrates such as birds and mammals.

Herbicide: An herbicide, commonly known as a weed killer, is a type of pesticide used to kill unwanted plants.

Herbicide resistance: Herbicide resistance in weeds occurs as a result of changes that prevent the herbicide from effectively inhibiting the target.

Heredity: Heredity is the passing of traits to offspring (from its parent or ancestors). This is the process by which an offspring cell or organism acquires or becomes predisposed to the characteristics of its parent cell or organism.

Heterochromatin: Heterochromatin is a tightly packed form of DNA, which comes in different varieties. These varieties lie on a continuum between the two extremes of constitutive and facultative heterochromatin.

Heterogeneous nuclear RNA: Heterogeneous nuclear RNA (hnRNA) a diverse group of long primary transcripts formed in the eukaryotic nucleus, many of which will be processed to mRNA molecules by splicing.

High efficiency particulate air: A high efficiency particulate air (HEPA) filter is a type of air filter that satisfies certain standards of efficiency such as those set by the United States Department of Energy (DOE).

Histocompatibility: Histocompatibility is the property of having the same, or mostly the same, alleles of a set of genes called the major histocompatibility complex. These genes are expressed in most tissues as antigens to which the immune system makes antibodies.

Histology: Histology is the study of the microscopic anatomy of cells and tissues of plants and animals.

Homeobox: A homeobox is a DNA sequence found within genes that are involved in the regulation of patterns of anatomical development (morphogenesis) in animals, fungi, and plants.

Homeotic mutation: A mutation that causes tissues to alter their normal differentiation pattern, producing integrated structures but in unusual locations. For example, a homeotic mutation in the fruit fly, Drosophila, causes legs to develop where antennae normally form.

Homodimer: A protein is composed of two identical polypeptide chains.

Homokaryon: A bi- or multinucleate cell having nuclei all of the same kind.

Homologous recombination: Homologous recombination is a type of genetic recombination in which nucleotide sequences are exchanged between two similar or identical molecules of DNA.

Homologous: Having the same alleles or genes in the same order of arrangement: homologous chromosomes.

Human immunodeficiency virus: Human immunodeficiency virus (HIV) is a lent virus (a member of the retrovirus family) that causes acquired immunodeficiency syndrome (AIDS), a condition in humans in which progressive failure of the immune system allows life-threatening opportunistic infections and cancers to thrive.

Human leukocyte antigen system: Human leukocyte antigen system (HLA) is the name of the major histocompatibility complex (MHC) in humans. The super locus contains a large number of genes related to immune system function in humans.

Humoral immune response: The humoral immune response (HIR) is the aspect of immunity that is mediated by secreted antibodies (as opposed to cell-mediated immunity, which involves T lymphocytes) produced in the cells of the B lymphocyte lineage (B cell).

Hybrid: A genetic hybrid carries two different alleles of the same gene.

Hybridization: Hybridization is the process of establishing a noncovalent, sequence-specific interaction between two or more complementary strands of nucleic acids into a single hybrid, which in the case of two strands is referred to as a duplex

Hybridoma: A cell hybrid produced *in vitro* by the fusion of a lymphocyte that produces antibodies and a myeloma tumor cell. It proliferates into clones that produce a continuous supply of a specific antibody.

Hydrolysis: Hydrolysis is a chemical reaction during which molecules of water are split into hydrogen cations and hydroxide anions in the process of a chemical mechanism.

Ideogram: An ideogram or is a graphic symbol that represents an idea or concept. Some ideograms are comprehensible only by familiarity with prior convention; others convey their meaning through pictorial resemblance to a physical object and thus may also be referred to as pictograms.

Immediate early gene: Immediate early genes (IEGs) are genes that are activated transiently and rapidly in response to a wide variety of cellular stimuli. They represent a standing response mechanism that is activated at the transcription level in the first round of response to stimuli, before any new proteins are synthesized.

Immobilized cells: The immobilized whole cell system is an alternative to enzyme immobilization. Unlike enzyme immobilization, where the enzyme is attached to a substrate (such as calcium alginate), in immobilized whole cell systems, the target cell is immobilized.

Immortalization: Biological immortality refers to a stable rate of mortality as a function of chronological age. Some individual cells and entire organisms in some species achieve this state either throughout their existence or after living long enough.

Immune response: The immune response is how your body recognizes and defends itself against bacteria, viruses, and substances that appear foreign and harmful.

Immunoassay: An immunoassay is a biochemical test that measures the presence or concentration of a substance in solutions that frequently contain a complex mixture of substances. Analytes in biological liquids such as serum or urine are frequently assayed using immunoassay methods.

Immunodiagnostics: Immunodiagnostics is a diagnostic methodology that uses an antigen-antibody reaction as their primary means of detection. The concept of using immunology as a diagnostic tool was introduced in 1960 as a test for serum insulin. A second test was developed in 1970 as a test for thyroxine in the 1970s.

Immunogenicity: Immunogenicity is the ability of a particular substance, such as an antigen or epitope, to provoke an immune response in the body of a human or animal.

Immunoglobulin: A protein produced by plasma cells and lymphocytes and characteristic of these types of cells. Immunoglobulins play an essential role in the body's immune system. They attach to foreign substances, such as bacteria, and assist in destroying them. Immunoglobulin is abbreviated as Ig. The classes of immunoglobulins are termed immunoglobulin A (IgA), immunoglobulin G (IgG), immunoglobulin M (IgM), immunoglobulin D (IgD), and immunoglobulin E (IgE).

Immunosensor: Immunosensors act on the principle that the immune response of certain biological species (usually bacteria) to contaminants will produce antibodies, which in turn can be measured.

Immunosuppression: Immunosuppression involves an act that reduces the activation or efficacy of the immune system. Some portions of the immune system itself have immunosuppressive effects on other parts of the immune system, and immunosuppression may occur as an adverse reaction to treatment of other conditions.

Immunosuppressor: Immunosuppression involves an act that reduces the activation or efficacy of the immune system. Some portions of the immune system itself have immunosuppressive effects on other parts of the immune system, and immunosuppression may occur as an adverse reaction to treatment of other conditions.

Immunotherapy: Immunotherapy is a medical term defined as the "treatment of disease by inducing, enhancing, or suppressing an immune response.

Immunotoxin: An immunotoxin is a human-made protein that consists of a targeting portion linked to a toxin. When the protein binds to that cell, it is taken in through endocytosis, and the toxin kills the cell. They are used for the treatment of some kinds of cancer and a few viral infections.

***In situ* colony:** A procedure for screening bacterial colonies or plaques growing on plates or membranes for the presence of specific DNA sequences by the hybridization of nucleic acid probes to the DNA molecules present in these colonies or plaques.

***In vitro* fertilization:** *In vitro* fertilization (IVF) is a process by which egg cells are fertilized by sperm outside the body: *in vitro*. IVF is a major treatment in infertility when other methods of assisted reproductive technology have failed. The process involves hormonally controlling the ovulatory process, removing ova (eggs) from the woman's ovaries and letting sperm fertilize them in a fluid medium.

***In vitro* maturation:** *In vitro* maturation (IVM) is the technique of letting ovarian follicles matures *in vitro*.

***In vitro* mutagenesis:** The production of either random or specific mutations in a piece of cloned DNA. Typically, the DNA will then be reintroduced into a cell or an organism to assess the results of the mutagenesis.

***In vivo* gene therapy:** The gene therapy carried out in the living organism.

Inactivated agent: A virus, bacterium, or other organism that has been treated to prevent it from causing a disease. *See* attenuated vaccine.

Inbred line: Produced by inbreeding.

Inbreeding: Inbreeding is the reproduction from the mating of two genetically related parents, which can increase the chances of offspring being affected by recessive or deleterious traits.

Inclusion body: Inclusion bodies are nuclear or cytoplasmic aggregates of stainable substances, usually proteins. They typically represent sites of viral multiplication in a bacterium or a eukaryotic cell and usually consist of viral capsid proteins.

Incubator: A device for maintaining a bacterial culture at a particular temperature for a set length of time, in order to measure bacterial growth.

Indirect embryogenesis: Plant embryo formation on callus tissues derived from explants, including zygotic or somatic embryos and seedlings.

Indirect organogenesis: Plant organ formation on callus tissues derived from explants.

Inducer: In molecular biology, an inducer is a molecule that starts gene expression.

Inducible enzyme: An enzyme that is normally present in minute quantities within a cell, but whose concentration increases dramatically when a substrate compound is added.

Induction: A process in which a molecule (e.g., a drug) induces (i.e., initiates or enhances) or inhibits the expression of an enzyme.

Induction media: Media used to induce the formation of organs.

Infection: An infection is the colonization of a host organism by parasite species. Infecting parasites seek to use the host's resources to reproduce, often resulting in disease. Colloquially, infections are usually considered to be caused by microscopic organisms or microparasites like viruses, prions, bacteria, and viroids, though larger organisms like macroparasites and fungi can also infect.

Infectious agent: An agent capable of producing infection.

Inheritance: Inheritance is the practice of passing on property, titles, debts, and obligations upon the death of an individual. It has long played an important role in human societies. The rules of inheritance differ between societies and have changed over time.

Inhibitor: A substance that binds to an enzyme and decreases the enzyme's activity.

Initiation codon: The mRNA sequence AUG, which specifies methionine, the first amino acid used in the translation process. (Occasionally, GUG, valine, is recognized as an initiation codon).

Initiation factors: Initiation factors are proteins that bind to the small subunit of the ribosome during the initiation of translation, a part of protein biosynthesis.

Inoculate: To implant microorganisms or infectious material into a culture medium.

Inositol: Inositol is a chemical compound with formula $C_6H_{12}O_6$ or (-CHOH-)6, a sixfold alcohol (polyol) of cyclohexane.

Insecticide: An insecticide is a pesticide used against insects. They include ovicides and larvicides used against the eggs and larvae of insects respectively. Insecticides are used in agriculture, medicine, industry, and the household.

Insertion element: A section of DNA that is capable of becoming inserted into another chromosome.

Insertion mutations: A type of mutation resulting from the addition of extra nucleotides in a DNA sequence or chromosome.

Insertion sequence: An insertion sequence (also known as an IS, an insertion sequence element, or an IS element) is a short DNA sequence that acts as a simple transposable element. Insertion sequences have two major characteristics: they are small relative to other transposable elements (generally around 700–2500 bp in length) and only code for proteins implicated in the transposition activity (they are thus different from other transposons, which also carry accessory genes such as antibiotic resistance genes).

Insertion site: The point in a vein where a needle or catheter is inserted.

Insulin: Insulin is a hormone central to regulating carbohydrate and fat metabolism in the body. Insulin causes cells in the liver, muscle, and fat tissue to take up glucose from the blood, storing it as glycogen in the liver and muscle.

Intercalating agent: A chemical that can insert itself between the stacked bases at the centre of the DNA double helix, possibly causing a frame-shift mutation.

Interferon: Interferons (IFNs) are proteins made and released by host cells in response to the presence of pathogens such as viruses, bacteria, or parasites or tumor cells. They allow communication between cells to trigger the protective defenses of the immune system that eradicate pathogens or tumors.

Intergeneric: A very rare type of hybrid formed between plants of two different genera. It is indicated by the symbol × before the genus name. For example, the Leyland cypress, × *Cupressocyparis leylandii*, is a cross between *Cupressus macrocarpa* and *Chamaecyparis nootkatensis*.

Intergenic regions: An intergenic region (IGR) is a stretch of DNA sequences located between clusters of genes that contain few or no genes. Occasionally, some intergenic DNA acts to control genes nearby, but most of it has no currently known function.

Interleukin: Interleukins are a group of cytokines (secreted proteins/signaling molecules) that were first seen to be expressed by white blood cells (leukocytes).

Internal guide sequence: A polynucleotide sequence near the 5′-end of group I introns that pairs with sequences of the upstream exon in an intermediate of the self-splicing process (see also self-splicing).

Interphase: Interphase is the phase of the cell cycle in which the cell spends the majority of its time and performs the majority of its purposes including preparation for cell division. In preparation for cell division, it increases its size and makes a copy of its DNA.

Intracellular: In cell biology, molecular biology, and related fields, the word intracellular means "inside the cell."

Intracytoplasmic sperm injection: Intracytoplasmic sperm injection (ICSI) is an *in vitro* fertilization procedure in which a single sperm is injected directly into an egg.

Intraspecific: Intraspecific is a term used in biology to describe behaviors, biochemical variations and other issues within individuals of a single species, thereby contrasting with interspecific.

Introgression: Introgression, also known as introgressive hybridization, in genetics (particularly plant genetics) is the movement of a gene (gene flow) from one species into the gene pool of another by repeated backcrossing of an interspecific hybrid with one of its parent species. Purposeful introgression is a long-term process; it may take many hybrid generations before the backcrossing occurs.

Inverted repeat: An inverted repeat (or IR) is a sequence of nucleotides that is the reversed complement of another sequence further downstream.

Ion channel: Ion channels are pore-forming proteins that help establish and control the small voltage gradient across the plasma membrane of cells (see cell potential) by allowing the flow of ions down their electrochemical gradient.

Ionizing radiation: Ionizing radiation consists of particles or electromagnetic waves that are energetic enough to detach electrons from atoms or molecules, therefore ionizing them. Direct ionization from the effects of single particles or single photons produces free radicals, which are atoms or molecules containing unpaired electrons, that tend to be especially chemically reactive due to their electronic structure.

Irradiation: Irradiation is the process by which an item is exposed to radiation. The exposure can originate from any of various sources, including those occurring naturally, or as part of a mechanical process, or otherwise.

Isochromosome: An isochromosome is a chromosome that has lost one of its arms and replaced it with an exact copy of the other arm. This is sometimes seen in some females with Turner syndrome or in tumor cells.

Isoelectric focusing: Isoelectric focusing (IEF), also known as electrofocusing, is a technique for separating different molecules by their electric charge differences. It is a type of zone electrophoresis, usually performed on proteins in a gel, which takes advantage of the fact that overall charge on the molecule of interest is a function of the pH of its surroundings.

Isoenzyme: Isozymes are enzymes that differ in amino acid sequence but catalyze the same chemical reaction. These enzymes usually display different kinetic parameters or different regulatory properties.

Isoform: A protein that has the same functions as another protein but which is encoded by a different gene and may have small differences in its sequence. For example, transforming factor beta (TGF-B) exists in three versions, or isoforms (TGF-B1, TGF-B2, and TGF-B3), each of which can set off a signaling cascade that starts in the cytoplasm and terminates in the nucleus of the cell.

Isomerase: In biochemistry, an isomerase is an enzyme that catalyzes the structural rearrangement of isomers. Isomerases thus catalyze reactions of the form A → B, where B is an isomer of A.

Isotonic: Isotonic solutions have equal osmotic pressure.

Isotope: Isotopes are variants of atoms of a particular chemical element, which have differing numbers of neutrons. Atoms of a particular element by definition must contain the same number of protons but may have a distinct number of neutrons, which differs from atom to atom, without changing the designation of the atom as a particular element.

Isozyme: Isozyme is one of the multiple forms in which an enzyme may exist in an organism or in different species, the various forms differing chemically, physically, or immunologically, but catalyzing the same reaction.

Jumping genes: A fragment of nucleic acid, such as a plasmid or a transposon, that can become incorporated into the DNA of a cell.

Juvenile *in vitro* embryo technology: Juvenile *in vitro* embryo technology (JIVET) is a technology involving collection of immature eggs from young animals, *in vitro* maturation and fertilization, and the transfer of the resultant embryos into recipient females. The method is designed to achieve rapid generation turnover.

Kanamycin: Kanamycin sulfate is an aminoglycoside antibiotic, available in oral, intravenous, and intramuscular forms and used to treat a wide variety of infections. Kanamycin is isolated from *Streptomyces kanamyceticus*.

Kappa chain: A polypeptide chain of one of the two types of light chain that is found in antibodies and can be distinguished antigenically and by the sequence of amino acids in the chain.

Karyogamy: Karyogamy is the fusion of pronuclei of two cells, as part of syngamy, fertilization, or true bacterial conjugation.

Karyogram: The complete set of chromosomes of a cell or organism. Used especially for the display prepared from photographs of mitotic chromosomes arranged in homologous pairs.

Karyokinesis: During cell division, the process of partition of a cell's nucleus into the daughter cells.

Karyotype: A karyotype is the number and appearance of chromosomes in the nucleus of a eukaryotic cell. The term is also used for the complete set of chromosomes in a species or an individual organism.

Kilo base pair: Kilo base pair (kb) is a length of DNA or double-stranded RNA equal to 1000 base pairs.

Kilodalton: Kilodalton (kDa) is a unit of molecular mass equal to 1000 Da.

Kinase: In chemistry and biochemistry, a kinase, alternatively known as a phosphotransferase, is a type of enzyme that transfers phosphate groups from high-energy donor molecules, such as ATP, to specific substrates.

Kinetics: The study of biochemical reaction rates catalyzed by an enzyme.

Knockout mouse: A knockout mouse is a genetically engineered mouse in which researchers have inactivated, or "knocked out," an existing gene by replacing it or disrupting it with an artificial piece of DNA. The loss of gene activity often causes changes in a mouse's phenotype, which includes appearance, behavior, and other observable physical and biochemical characteristics.

Beta-lactamase: β-lactamases are enzymes (EC 3.5.2.6) produced by some bacteria and are responsible for their resistance to beta-lactam antibiotics like penicillins, cephamycins, and carbapenems (ertapenem) (Cephalosporins are relatively resistant to beta-lactamase.)

Lactose: Lactose is a disaccharide sugar that is found most notably in milk and is formed from galactose and glucose. Lactose makes up around 2%–8% of milk (by weight), although the amount varies among species and individuals.

Lag phase: The initial growth phase, during which cell number remains relatively constant before rapid growth.

Lagging strand: In DNA replication, the strand that is synthesized apparently in the 3′–5′ direction, but actually in the 5′–3′ direction by ligating short fragments synthesized individually. Strand of DNA being replicated discontinuously.

Lambda chain: A polypeptide chain of one of the two types of light chain that are found in antibodies and can be distinguished antigenically.

Laminar flow cabinet: A laminar flow cabinet or laminar flow closet or tissue culture hood is a carefully enclosed bench designed to prevent contamination of semiconductor wafers, biological samples, or any particle sensitive device. Air is drawn through a HEPA filter and blown in a very smooth, laminar flow towards the user. The cabinet is usually made of stainless steel with no gaps or joints where spores might collect.

Lampbrush chromosomes: Lampbrush chromosomes are a special form of chromosomes that are found in the growing oocytes (immature eggs) of most animals, except mammals.

Landrace: A landrace is a local variety of a domesticated animal or plant species, which has developed largely by natural processes, by adaptation to the natural and cultural environment in which it lives. It differs from a formal breed that has been bred deliberately to conform to a particular standard type.

Leader sequence: The sequence at the 5′ end of an mRNA that is not translated into protein. The length of untranslated mRNA from the 5′ end to the initiation codon AUG.

Leading strand: Strand of DNA being replicated continuously. In DNA replication, the strand that is made in the 5′–3′ direction by continuous polymerization at the 3′ growing tip. See also lagging strand.

Lectin: Lectins are sugar-binding proteins (not to be confused with glycoproteins, which are proteins containing sugar chains or residues) that are highly specific for their sugar moieties.

Lethal dose 50: A dose at which 50% of subjects will die.

Lethal gene: A gene whose expression results in the death of the organism, usually during embryogenesis.

Lethal mutation: A type of mutation in which the effect(s) can result in the death or reduce significantly the expected longevity of an organism carrying the mutation.

Ligand: In coordination chemistry, a ligand is an ion or molecule (see also: functional group) that binds to a central metal atom to form a coordination complex. The bonding between metal and ligand generally involves formal donation of one or more of the ligand's electron pairs.

Ligase chain reaction: The ligase chain reaction (LCR) is a method of DNA amplification. While the better-known PCR carries out the amplification by polymerizing nucleotides, LCR instead amplifies the nucleic acid used as the probe.

Ligation: In molecular biology, the covalent linking of two ends of DNA molecules using DNA ligase.

Lineage genetics: Genetic lineage is a series of mutations, which connect an ancestral genetic type (allele, haplotype, or haplogroup) to derivative type.

Linkage: Genetic linkage is the tendency of certain loci or alleles to be inherited together. Genetic loci that are physically close to one another on the same chromosome tend to stay together during meiosis and are thus genetically linked.

Linker DNA: Linker DNA is double-stranded DNA in between two nucleosome cores that, in association with histone H1, holds the cores together. Linker DNA is seen as the string in the "beads and string model," which is made by using an ionic solution on the chromatin. Linker DNA connects to histone H1 and histone H1 sits on the nucleosome core.

Lipases: A lipase is a water-soluble enzyme that catalyzes the hydrolysis of ester chemical bonds in water-insoluble lipid substrates. Lipases are a subclass of the esterases.

Lipid: Lipids are a broad group of naturally occurring molecules, which includes fats, waxes, sterols, fat-soluble vitamins (such as vitamins A, D, E, and K), monoglycerides, diglycerides, phospholipids, and others.

Lipofection: Lipofection (or liposome transfection) is a technique used to inject genetic material into a cell by means of liposomes, which are vesicles that can easily merge with the cell membrane, since they are both made of a phospholipid bilayer. Lipofection generally uses a positively charged (cationic) lipid to form an aggregate with the negatively charged (anionic) genetic material.

Liposome: Liposomes are artificially prepared vesicles made of lipid bilayer. Liposomes can be filled with drugs and used to deliver drugs for cancer and other diseases.

Liquid nitrogen: Liquid nitrogen is nitrogen in a liquid state at a very low temperature. It is produced industrially by fractional distillation of liquid air. Liquid nitrogen is a compact and readily transported source of nitrogen gas without pressurization. Furthermore, its ability to maintain temperatures far below the freezing point of water makes it extremely useful in a wide range of applications, primarily as an open-cycle refrigerant, including the cryopreservation of blood, reproductive cells (sperm and egg), and other biological samples and materials.

Litmus test: A test for chemical acidity or basicity using litmus paper.

Live vaccine: One prepared from live microorganisms that have been attenuated but that retain their immunogenic properties.

Logarithmic phase: The steepest slope of the growth curve of a culture—the phase of vigorous growth during which cell number doubles every 20–30 min.

Long terminal repeat: Long terminal repeats (LTRs) are sequences of DNA that repeat hundreds or thousands of times. They are found in retroviral DNA and in retrotransposons, flanking functional genes. They are used by viruses to insert their genetic sequences into the host genomes.

Loop bioreactors: Bioreactors in which the fermenting material is cycled between a bulk tank and a smaller tank or loop of pipes.

Luteinizing hormone: Luteinizing hormone (LH) is a hormone produced by the anterior pituitary gland.

Lux: The lux (symbol: lx) is the SI unit of illuminance and luminous emittance measuring luminous power per area. It is used in photometry as a measure of the intensity, as perceived by the human eye.

Lyase: In biochemistry, a lyase is an enzyme that catalyzes the breaking of various chemical bonds by means other than hydrolysis and oxidation, often forming a new double bond or a new ring structure.

Lymphocyte: A lymphocytes is a type of white blood cell in the vertebrate immune system.

Lymphokine: Lymphokines are a subset of cytokines that are produced by a type of immune cell known as a lymphocyte. They are protein mediators typically produced by T cells to direct the immune system response by signaling between its cells.

Lymphoma: Lymphoma is a cancer in the lymphatic cells of the immune system. Typically, lymphomas present as a solid tumor of lymphoid cells.

Lyophilize: To freeze-dry. The material is rapidly frozen and dehydrated under high vacuum. The process is termed lyophilization.

Lysis: Lysis refers to the breaking down of a cell, often by viral, enzyme, or osmotic mechanisms that compromise its integrity. A fluid containing the contents of lysed cells is called a "lysate."

Lysogenic bacteria: A bacterium that contains in its genome the DNA of a virus that is lying dormant, passively letting itself be replicated by the bacterium whenever the bacterium replicates its own genome (a lysogenic virus), but able to reactivate and destroy the bacterium at a time of the virus's choosing (becomes a lytic virus).

Lysosome: Lysosomes are cellular organelles that contain acid hydrolase enzymes to break down waste materials and cellular debris. They are found in animal cells, while in yeast and plants the same roles are performed by lytic vacuoles.

Lytic cycle: The lytic cycle is one of the two cycles of viral reproduction, the other being the lysogenic cycle.

M13: M13 is a filamentous bacteriophage composed of circular single stranded DNA (ssDNA), which is 6407 nucleotides long encapsulated in approximately 2700 copies of the major coat protein P8 and capped with 5 copies of two different minor coat proteins (P9, P6, and P3) on the ends.

Macromolecule: A macromolecule is a very large molecule commonly created by some form of polymerization. In biochemistry, the term is applied to the four conventional biopolymers (nucleic acids, proteins, carbohydrates, and lipids) as well as nonpolymeric molecules with large molecular mass such as macrocycles.

Macronutrient: Nutrients that the body uses in relatively large amounts—proteins, carbohydrates, and fats. This is as opposed to micronutrients, which the body requires in smaller amounts, such as vitamins and minerals. Macronutrients provide calories to the body as well as performing other functions.

Macrophages: Type of white blood that ingests (takes in) foreign material. Macrophages are key players in the immune response to foreign invaders such as infectious microorganisms.

Macropropagation: Production of plant clones from growing parts.

Major histocompatibility complex: The major histocompatibility complex (MHC) is a large genomic region or gene family found in most vertebrates that encodes MHC molecules. MHC molecules play an important role in the immune system and autoimmunity.

Malignant: Tending to be severe and become progressively worse, as in malignant hypertension.

Marker gene: Detectable genetic trait or segment of DNA that can be identified and tracked. A marker gene can serve as a flag for another gene, sometimes called the target gene. A marker gene must be on the same chromosome as the target gene and near enough to it, so that the two genes (the marker gene and the target gene) are genetically linked and are usually inherited together.

Marker peptide: A portion of fusion protein that facilitates its identification or purification.

Marker-assisted introgression: The use of DNA markers to increase the speed and efficiency of introgression of a new allele(s) or gene(s) into a breeding population.

Marker-assisted selection: Marker-assisted selection or marker-aided selection (MAS) is a process, whereby a marker (morphological, biochemical, or one based on DNA/RNA variation) is used for indirect selection of a genetic determinant or determinants of a trait of interest (i.e., productivity, disease resistance, abiotic stress tolerance, and/or quality). This process is used in plant and animal breeding.

Mean: The mean is the arithmetic average of a set of values or distribution; however, for skewed distributions, the mean is not necessarily the same as the middle value (median) or the most likely (mode).

Median: In probability theory and statistics, a median is described as the numerical value separating the higher half of a sample, a population, or a probability distribution, from the lower half.

Megabase cloning: The cloning of very large DNA fragments.

Megadalton (MDa): A unit of mass equal to one million atomic mass units.

Meiosis: Meiosis is a special type of cell division necessary for sexual reproduction. In animals, meiosis produces gametes (sperm and egg cells), whilst in other organisms, such as fungi, it generates spores.

Melanin: Melanin is a pigment that is ubiquitous in nature, being found in most organisms (spiders are one of the few groups in which it has not been detected). In animals, melanin pigments are derivatives of the amino acid tyrosine.

Melting temperature for DNA: Nucleic acid thermodynamics is the study of the thermodynamics of nucleic acid molecules, or how temperature affects nucleic acid structure. For multiple copies of DNA molecules, the melting temperature (T_m) is defined as the temperature at which half of the DNA strands are in the double helical state and half are in the random coil states. The melting temperature depends on both the length of the molecule and the specific nucleotide sequence composition of that molecule.

Membrane bioreactors: Membrane bioreactor (MBR) is the combination of a membrane process like microfiltration or ultrafiltration with a suspended growth bioreactor and is now widely used for municipal and industrial wastewater treatment with plant sizes up to 80,000 population equivalent.

Memory cells: Memory B cells are a B cell subtype that are formed following primary infection.

Mendel's laws: Mendel discovered that when crossing white flower and purple flower plants, the result is not a blend. Rather than being a mix of the two, the offspring was purple flowered. He then conceived the idea of heredity units, which he called "factors," one of which is a recessive characteristic and the other dominant. Mendel said that factors, later called genes, normally occur in pairs in ordinary body cells, yet segregate during the formation of sex cells.

Mendelian population: A group of interbreeding individuals; the total allelic gene content of the group is called their gene pool.

Mesoderm: In all bilaterian animals, the mesoderm is one of the three primary germ cell layers in the very early embryo.

Messenger RNA: Messenger RNA (mRNA) is a molecule of RNA encoding a chemical "blueprint" for a protein product. mRNA is transcribed from a DNA template and carries coding information to the sites of protein synthesis: the ribosomes. Here, the nucleic acid polymer is translated into a polymer of amino acids: a protein.

Metabolic cell: Undivided cell.

Metabolism: Metabolism is the set of chemical reactions that happen in living organisms to maintain life. These processes allow organisms to grow and reproduce, maintain their structures, and respond to their environments. Metabolism is usually divided into two categories. Catabolism breaks down organic matter, for example, to harvest energy in cellular respiration. Anabolism uses energy to construct components of cells such as proteins and nucleic acids.

Metabolite: Metabolites are the intermediates and products of metabolism. The term metabolite is usually restricted to small molecules.

Metaphase: Metaphase is a stage of mitosis in the eukaryotic cell cycle in which condensed and highly coiled chromosomes, carrying genetic information, align in the middle of the cell before being separated into each of the two daughter cells.

Metastasis: Metastasis or metastatic disease is the spread of a disease from one organ or part to another nonadjacent organ or part.

Methylation: In the chemical sciences, methylation denotes the addition of a methyl group to a substrate or the substitution of an atom or group by a methyl group. Methylation is a form of alkylation with, to be specific, a methyl group, rather than a larger carbon chain, replacing a hydrogen atom. These terms are commonly used in chemistry, biochemistry, soil science, and the biological sciences.

Michaelis constant: In biochemistry, Michaelis–Menten kinetics is one of the simplest and best-known models of enzyme kinetics. It is named after American biochemist Leonor Michaelis and Canadian physician Maud Menten.

Microarray: A DNA microarray (also commonly known as gene chip, DNA chip, or biochip) is a collection of microscopic DNA spots attached to a solid surface. Scientists use DNA microarrays to measure the expression levels of large numbers of genes simultaneously or to genotype multiple regions of a genome.

Microbe: A microscopic living organism, such as a bacterium, fungus, protozoan, or virus.

Microdroplet array: Microdroplet is a technique to use to simultaneously evaluate large numbers of media modifications, employing small quantities of medium into which are placed small numbers of cells or protoplasts.

Microencapsulation: Microencapsulation is a process in which tiny particles or droplets are surrounded by a coating to give small capsules many useful properties. In a relatively simplistic form, a microcapsule is a small sphere with a uniform wall around it. The material inside the microcapsule is referred to as the core, internal phase, or fill, whereas the wall is sometimes called a shell, coating, or membrane. Most microcapsules have diameters between a few micrometers and a few millimeters.

Microinjection: Microinjection refers to the process of using a glass micropipette to insert substances at a microscopic or borderline macroscopic level into a single living cell. It is a simple mechanical process in which a needle roughly 0.5–5 μm in diameter penetrates the cell membrane and/or the nuclear envelope.

Micron: A micrometer is one-millionth of a meter (1/1000 of a millimeter or 0.001 mm). Its unit symbol in the International System of Units (SI) is μm.

Micronutrient: Micronutrients are nutrients required by humans and other living things throughout life in small quantities to orchestrate a whole range of physiological functions, but which the organism itself cannot produce.

Microsatellite: Microsatellite (genetics), a repeating sequence in DNA.

Microspore: In botany, microspores develop into male gametophytes, whereas megaspores develop into female gametophytes. The combination of megaspores and microspores is found only in heterosporous organisms. In seed plants the microspores give rise to the pollen grains, and the megaspores are formed within the developing seed.

Microtubules: Microtubules are one of the active matter components of the cytoskeleton. They have a diameter of 25 nm and length varying from 200 nm to 25 μm. Microtubules serve as structural components within cells and are involved in many cellular processes including mitosis, cytokinesis, and vesicular transport.

Miniprep: Minipreparation of plasmid DNA is a rapid, small-scale isolation of plasmid DNA from bacteria. It is based on the alkaline lysis method invented by the researchers Birnboim and Doly in 1979. The extracted plasmid DNA resulting from performing a miniprep is itself often called a "miniprep." Minipreps are used in the process of molecular cloning to analyze bacterial clones. A typical plasmid DNA yield of a miniprep is 20–30 μg depending on the cell strain.

Minisatellite: A minisatellite (also referred as VNTR) is a section of DNA that consists of a short series of bases 10–60 bp. These occur at more than 1000 locations in the human genome. Some minisatellites contain a central (or "core") sequence of letters "GGGCAGGANG" (where N can be any base) or more generally a strand bias with purines (adenosine (A) and guanine (G)) on one strand and pyrimidines (cytosine (C) and thymine (T)) on the other.

Mismatch repair: DNA mismatch repair is a system for recognizing and repairing erroneous insertion, deletion and misincorporation of bases that can arise during DNA replication and recombination as well as repairing some forms of DNA damage.

Missense mutation: In genetics, a missense mutation (a type of nonsynonymous mutation) is a point mutation in which a single nucleotide is changed, resulting in a codon those codes for a different amino acid (mutations that change an amino acid to a stop codon are considered nonsense mutations, rather than missense mutations).

Mitochondrial DNA: Mitochondrial DNA (mtDNA) is the DNA located in organelles called mitochondria, structures within eukaryotic cells that convert the chemical energy from food into a form that cells can use, adenosine triphosphate (ATP). Most other DNA present in eukaryotic organisms is found in the cell nucleus.

Mitosis: Mitosis is the process by which a eukaryotic cell separates the chromosomes in its cell nucleus into two identical sets in two nuclei. It is generally followed immediately by cytokinesis, which divides the nuclei, cytoplasm, organelles, and cell membrane into two cells containing roughly equal shares of these cellular components.

Modifying gene: A gene that alters or influences the expression function of another gene, including the suppression or reduction of the usual function of the modified gene.

Molarity: The molar concentration of a solution, usually expressed as the number of moles of solute per liter of solution.

Molecular cloning: Molecular cloning refers to a set of experimental methods in molecular biology that are used to assemble recombinant DNA molecules and to direct their replication within host organisms.

Molecular genetics: Molecular genetics is the field of biology and genetics that studies the structure and function of genes at a molecular level. The field studies how the genes are transferred from generation to generation.

Monoclonal antibody: Monoclonal antibodies (mAb or moAb) are monospecific antibodies that are the same, because they are made by identical immune cells that are all clones of a unique parent cell.

Monocotyledon: Any of various flowering plants, such as grasses, orchids, and lilies, having a single cotyledon in the seed.

Monoculture: Monoculture is the agricultural practice of producing or growing one single crop over a wide area. It is also known as a way of farming practice of growing large stands of a single species.

Monosaccharide: Monosaccharides (from Greek monos: single, sacchar: sugar) are the most basic units of biologically important carbohydrates. They are the simplest form of sugar and are usually colorless, water-soluble, crystalline solids. Some monosaccharides have a sweet taste.

Morphogenesis: Morphogenesis is the biological process that causes an organism to develop its shape. It is one of three fundamental aspects of developmental biology along with the control of cell growth and cellular differentiation.

Multigene family: A set of genes descended by duplication and variation from some ancestral gene. Such genes may be clustered together on the same chromosome or dispersed on different chromosomes.

Mutagen: In genetics, a mutagen is a physical or chemical agent that changes the genetic material, usually DNA, of an organism and thus increases the frequency of mutations above the natural background level. As many mutations cause cancer, mutagens are typically also carcinogens.

Mutagenesis: Mutagenesis is a process by which the genetic information of an organism is changed in a stable manner, either in nature or experimentally by the use of chemicals or radiation. Mutagenesis as a science was developed especially by Charlotte Auerbach in the first half of the twentieth century.

Mutant: In biology and especially genetics, a mutant is an individual, organism, or new genetic character, arising or resulting from an instance of mutation, which is a base-pair sequence change within the DNA of a gene or chromosome of an organism resulting in the creation of a new character or trait not found in the wild type.

Mutation: In molecular biology and genetics, mutations are changes in a genomic sequence: the DNA sequence of a cell's genome or the DNA or RNA sequence of a virus. They can be defined as sudden and spontaneous changes in the cell.

Mycoprotein: Mycoprotein means protein from fungi. It can be used as part of any meal, particularly vegetarian.

Mycotoxin: A mycotoxin is a toxic secondary metabolite produced by organisms of the fungus kingdom, commonly known as molds. The term "mycotoxin" is usually reserved for the toxic chemical products produced by fungi that readily colonize crops. One mold species may produce many different mycotoxins and/or the same mycotoxin as another species

Myosin: Myosins comprise a family of ATP-dependent motor proteins and are best known for their role in muscle contraction and their involvement in a wide range of other eukaryotic motility processes.

Nanometer: A nanometer is a unit of length in the metric system, equal to one billionth of a meter.

Native protein: The protein inside the cell that is in its native or natural state and unaltered by denaturing agent, such as heat, chemical, enzyme action, or the exigencies of extraction.

Natural selection: Natural selection is the nonrandom process by which biologic traits become more or less common in a population as a function of differential reproduction of their bearers. It is a key mechanism of evolution.

Necrosis: Necrosis is the premature death of cells and living tissue. Necrosis is caused by factors external to the cell or tissue, such as infection, toxins, or trauma.

Negative autogenous regulation: Inhibition of the expression of a gene or set of co-ordinately regulated genes by the product of the gene or the product of one of the genes.

Nematodes: The nematodes are the most diverse phylum of pseudocoelomates, and one of the most diverse of all animals.

Neo-formation: A new and abnormal growth of tissue; tumor; neoplasm.

Neoplasm: Neoplasm is an abnormal mass of tissue as a result of neoplasia. Neoplasia is the abnormal proliferation of cells. The growth of neoplastic cells exceeds and is not coordinated with that of the normal tissues around it.

Neoteny: Neoteny is the retention, by adults in a species, of traits previously seen only in juveniles, and is a subject studied in the field of developmental biology.

Neutral mutation: In genetics, a neutral mutation is a mutation that has no effect on fitness. In other words, it is neutral with respect to natural selection. For example, some mutations in a DNA triplet or codon do not change, which amino acid is introduced: this is known as a synonymous substitution. Unless the mutation also has a regulatory effect, synonymous substitutions are usually neutral.

Neutral theory: The neutral theory of molecular evolution states that the vast majority of evolutionary changes at the molecular level are caused by random drift of selectively neutral mutants.

Nick translation: Nick translation (or Head Translation) is a tagging technique in molecular biology in which DNA polymerase I is used to replace some of the nucleotides of a DNA sequence with their labeled analogues, creating a tagged DNA sequence, which can be used as a probe in fluorescent *in situ* hybridization or blotting techniques.

Nitrification: Nitrification is the biological oxidation of ammonia with oxygen into nitrite followed by the oxidation of these nitrites into nitrates. Degradation of ammonia to nitrite is usually the rate limiting step of nitrification. Nitrification is an important step in the nitrogen cycle in soil. This process was discovered by the Russian microbiologist, Sergei Winogradsky.

Nitrocellulose: Nitrocellulose (also: cellulose nitrate, flash paper) is a highly flammable compound formed by nitrating cellulose through exposure to nitric acid or another powerful nitrating agent.

Nitrogen fixation: Nitrogen fixation is the natural process, either biological or abiotic, by which nitrogen (N_2) in the atmosphere is converted into ammonia (NH_3). This process is essential for life, because fixed nitrogen is required to biosynthesize the basic building blocks of life, for example, nucleotides for DNA and RNA and amino acids for proteins. Nitrogen fixation also refers to other biological conversions of nitrogen, such as its conversion to nitrogen dioxide.

Nitrogenous bases: The purines (adenine and guanine) and pyrimidines (thymine, cytosine, and uracil) that form DNA and RNA molecules.

Nonhistone chromosomal proteins: Chromatin consists of DNA, histones, and a very heterogeneous group of other proteins that include DNA polymerases and regulator proteins. They are often lumped together terminologically as nonhistone proteins or acidic proteins to distinguish them from the basic histones.

Northern blotting: The northern blot is a technique used in molecular biology research to study gene expression by detection of RNA (or isolated mRNA) in a sample.

Northern hybridization: A procedure in which RNA fragments are transferred from an agarose gel to a nitrocellulose filter, where the RNA is then hybridized to a radioactive probe.

Nuclear transfer: Nuclear transfer is a form of cloning. The steps involve removing the DNA from an oocyte (unfertilized egg) and injecting the nucleus which contains the DNA to be cloned.

Nuclease: A nuclease is an enzyme capable of cleaving the phosphodiester bonds between the nucleotide subunits of nucleic acids. Older publications may use terms such as "polynucleotidase" or "nucleodepolymerase.

Nucleic acid: Nucleic acids are biological molecules essential for life and include DNA (deoxyribonucleic acid) and RNA (ribonucleic acid). Together with proteins, nucleic acids make up the most important macromolecules; each is found in abundance in all living things, where they function in encoding, transmitting, and expressing genetic information.

Nuclein: Any of the substances present in the nucleus of a cell, consisting chiefly of proteins, phosphoric acids, and nucleic acids.

Nucleocytoplasmic ratio: The ratio of the volume of a nucleus of a cell to the volume of the cytoplasm. The proportion is usually constant for a specific cell type, and an increase is indicative of malignant neoplasms.

Nucleolar organizer: The chromosomal region around which the nucleolus forms, a site of tandem repeats of the rRNA gene. A region (or regions) of the chromosome set.

Nucleolus: The nucleolus is a nonmembrane bound structure composed of proteins and nucleic acids found within the nucleus. Ribosomal RNA (rRNA) is transcribed and assembled within the nucleolus.

Nucleoplasm: Similar to the cytoplasm of a cell, the nucleus contains nucleoplasm (nucleus sap) or karyoplasm. The nucleoplasm is one of the types of protoplasm, and it is enveloped by the nuclear membrane or nuclear envelope. The nucleoplasm is a highly viscous liquid that surrounds the chromosomes and nucleoli.

Nucleoprotein: A nucleoprotein is any protein that is structurally associated with nucleic acid (either DNA or RNA). Many viruses harness this protein, and they are known for being host specific, that is, they find it difficult to infect species besides the ones they normally infect.

Nucleoside: Nucleosides are glycosylamines consisting of a nucleobase (often referred to as simply base) bound to a ribose or deoxyribose sugar via a beta-glycosidic linkage. Examples of nucleosides include cytidine, uridine, adenosine, guanosine, thymidine, and inosine.

Nucleoside analogue: Nucleoside analogues are a range of antiviral products used to prevent viral replication in infected cells. The most commonly used is Aciclovir, although its inclusion in this category is uncertain, as it contains only a partial nucleoside structure, as the sugar ring is replaced by an open-chain structure.

Nucleosome: Nucleosomes are the basic unit of DNA packaging in eukaryotes, consisting of a segment of DNA wound around a histone protein core. This structure is often compared to thread wrapped around a spool.

Nucleotide: Nucleotides are molecules that, when joined together, make up the structural units of RNA and DNA. In addition, nucleotides play central roles in metabolism, in which capacity they serve as sources of chemical energy (adenosine triphosphate and guanosine triphosphate), participate in cellular signaling (cyclic guanosine monophosphate and cyclic adenosine monophosphate), and are incorporated into important cofactors of enzymatic reactions (coenzyme A, flavin adenine dinucleotide, flavin mononucleotide, and nicotinamide adenine dinucleotide phosphate).

Nucleus: In cell biology, the nucleus is a membrane-enclosed organelle found in eukaryotic cells. It contains most of the cell's genetic material, organized as multiple long linear DNA molecules in complex with a large variety of proteins, such as histones, to form chromosomes.

Null mutation: A mutation (a change) in a gene that leads to its not being transcribed into RNA and/or translated into a functional protein product. For example, a null mutation in a gene that usually encodes a specific enzyme leads to the production of a nonfunctional enzyme or no enzyme at all.

Nullisomy: A type of genome mutation in which a pair of chromosomes that are normally present in the genome is missing. Organisms that exhibit nullisomy are called nullisomes. Nullisomy, especially in higher animals, usually results in death. Viable nullisomes can be found among polyploid plants; these nullisomes are used for nullisomic analysis and for establishing new, commercially valuable strains. Nullisomic analysis is used to determine genetic linkage groups and to study the traits that are controlled by these groups. The method is also applied to tissue-culture studies of human nullisomic cells.

Nutrient film technique: Nutrient film technique or NFT is a hydroponic technique, wherein a very shallow stream of water containing all the dissolved nutrients required for plant growth is recirculated past the bare roots of plants in a watertight gully, also known as channels.

Nutrient gradient: A diffusion gradient of nutrients and gases that develops in tissues where only a portion of the tissue is in contact with the medium. Gradients are less likely to form in liquid media than in callus cultures.

Nutrient medium: A culture medium to which nutrient materials have been added.

Offspring: In biology, offspring is the product of reproduction, of a new organism produced by one or more parents.

Okazaki fragments: Okazaki fragments are short molecules of single-stranded DNA that are formed on the lagging strand during DNA replication. They are between 1000 and 2000 nucleotides long in *Escherichia coli* and are between 100 and 200 nucleotides long in eukaryotes.

Oligomer: In chemistry, an oligomer is a molecule that consists of a few monomer units in contrast to a polymer that, at least in principle, consists of an unlimited number of monomers. Dimers, trimers, and tetramers are oligomers. Many oils are oligomeric, such as liquid paraffin.

Oligonucleotide: An oligonucleotide is a short nucleic acid polymer, typically with 50 or fewer bases. Although they can be formed by bond cleavage of longer segments, they are now more commonly synthesized, in a sequence-specific manner, from individual nucleoside phosphoramidites. Automated synthesizers allow the synthesis of oligonucleotides up to about 200 bases.

Oligonucleotide ligation assay: Oligonucleotide ligation assay (OLA) is a rapid, sensitive, and specific method for the detection of known single nucleotide polymorphisms (SNPs). This method is based on the joining of two adjacent oligonucleotide probes (capture and Reporter Oligos) using a DNA ligase while they are annealed to a complementary DNA target (e.g., PCR product).

Oncogene: An oncogene is a gene that has the potential to cause cancer. In tumor cells, they are often mutated or expressed at high levels. Most normal cells undergo a programmed form of death (apoptosis). Activated oncogenes can cause those cells that ought to die to survive and proliferate instead.

Oncogenesis: The progression of cytological, genetic, and cellular changes that culminate in a malignant tumor.

Oncogenic: Oncogenic giving rise to tumors or causing tumor formation; said especially of tumor-inducing viruses.

Oncomouse: The OncoMouse or Harvard mouse is a type of laboratory mouse that has been genetically modified using modifications designed by Philip Leder and Timothy A. Stewart of Harvard University to carry a specific gene called an activated oncogene.

Ontogeny: Ontogeny describes the origin and the development of an organism for example: from the fertilized egg to mature form.

Oocyte: An oocyte, ovocyte, or rarely ocyte is a female gametocyte or germ cell involved in reproduction. In other words, it is an immature ovum or egg cell. An oocyte is produced in the ovary during female gametogenesis.

Oogenesis: Oogenesis is the creation of an ovum (egg cell). It is the female form of gametogenesis. The male equivalent is spermatogenesis. It involves the development of the various stages of the immature ovum.

Oogonium: An oogonium is an immature ovum. It is a female gametogonium. They are formed in large numbers by mitosis early in fetal life from primordial germ cells, which are present in the fetus between weeks 4 and 8.

Oospore: An oospore is a thick-walled sexual spore that develops from a fertilized oosphere in some algae and fungi.

Open continuous culture: A continuous culture system, in which inflow of fresh medium is balanced by outflow of a corresponding volume of spent medium plus cells. In the steady state, the rate of cell wash-out equals the rate of formation of new cells in the system.

Open pollination: Open pollination is pollination by insects, birds, wind, or other natural mechanisms, and contrasts with cleistogamy, closed pollination, which is one of the many types of self-pollination. Open pollination also contrasts with controlled pollination, which is controlled, so that all seeds of a crop are descended from parents with known traits and are therefore more likely to have the desired traits.

Open reading frame: In molecular genetics, an open reading frame (ORF) is a DNA sequence that does not contain a stop codon in a given reading frame.

Operon: In genetics, an operon is a functioning unit of genomic DNA containing a cluster of genes under the control of a single regulatory signal or promoter. The genes are transcribed together into an mRNA strand and either translated together in the cytoplasm, or undergo trans-splicing to create monocistronic mRNAs that are translated separately, that is, several strands of mRNA that each encode a single gene product.

Organ culture: Organ culture is a development from tissue culture methods of research, the organ culture is able to accurately model functions of an organ in various states and conditions by the use of the actual *in vitro* organ itself.

Organic chemistry: Organic chemistry is a subdiscipline within chemistry involving the scientific study of the structure, properties, composition, reactions, and preparation (by synthesis or by other means) of carbon-based compounds, hydrocarbons, and their derivatives.

Organic complex: An organic compound is any member of a large class of gaseous, liquid, or solid chemical compounds whose molecules contain carbon. For historical reasons discussed below, a few types of carbon-containing compounds such as carbides, carbonates, simple oxides of carbon, and cyanides as well as the allotropes of carbon such as diamond and graphite, are considered inorganic.

Organogenesis: In animal development, organogenesis is the process by which the ectoderm, endoderm, and mesoderm develop into the internal organs of the organism.

Organoid: A structure that resembles an organ.

Organoleptic: Organoleptic relating to the senses (taste, sight, smell, and touch) is a term also used to describe traditional USDA meat and poultry inspection techniques, because inspectors perform a variety of such procedures (involving visually examining, feeling, and smelling animal parts) to detect signs of disease or contamination. These inspection techniques alone are not adequate to detect invisible foodborne pathogens that now are the leading causes of food poisoning.

Origin of replication: The origin of replication (also called the replication origin) is a particular sequence in a genome at which replication is initiated. This can either be DNA replication in living organisms such as prokaryotes and eukaryotes, or RNA replication in RNA viruses, such as double-stranded RNA viruses. DNA replication may proceed from this point bidirectionally or unidirectionally.

Ortet: The plant from which a clone is obtained.

Osmolarity: Osmolarity is the measure of solute concentration, defined as the number of osmoles (Osm) of solute per liter (L) of solution (osmol/L or Osm/L).

Osmosis: Osmosis is the movement of solvent molecules through a selectively permeable membrane into a region of higher solute concentration, aiming to equalize the solute concentrations on the two sides. It may also be used to describe a physical process in which any solvent moves, without input of energy.

Osmotic potential: The potential of water molecules to move from a hypotonic solution (more water, less solutes) to a hypertonic solution (less water, more solutes) across a semipermeable membrane.

Outbreeding: The breeding of distantly related or unrelated individuals, often producing a hybrid of superior quality.

Ovary: The ovary is an ovum-producing reproductive organ, often found in pairs as part of the vertebrate female reproductive system. Ovaries in anatomically female individuals are analogous to testes in anatomically male individuals, in that they are both gonads and endocrine glands.

Overdominance: Overdominance is a condition in genetics where the phenotype of the heterozygote lies outside of the phenotypical range of both homozygote parents. Overdominance can also be described as heterozygote advantage, wherein heterozygous individuals have a higher fitness than homozygous individuals.

Overlapping reading frames: Start codons in different reading frames generate different polypeptides from the same DNA sequence.

Ovulation: Ovulation is the process in a female's menstrual cycle by which a mature ovarian follicle ruptures and discharges an ovum (also known as an oocyte, female gamete, or casually, an egg). Ovulation also occurs in the estrous cycle of other female mammals, which differs in many fundamental ways from the menstrual cycle. The time immediately surrounding ovulation is referred to as the ovulatory phase or the periovulatory period.

Oxygen-electrode-based sensor: Sensor in which an oxygen electrode, which measures the amount of oxygen in a solution, is coated with a biological material such as an enzyme which generates or absorbs oxygen when the appropriate substrate is present.

Packed cell volume: The ratio of the volume occupied by packed red blood cells to the volume of the whole blood as measured by a hematocrit.

Pairing gene: The two copies of a particular gene present in a diploid cell (one in each chromosome set).

Pair-rule gene: A pair-rule gene is a type of gene involved in the development of the segmented embryos of insects. Pair-rule genes are defined by the effect of a mutation in that gene, which causes the loss of the normal developmental pattern in alternating segments.

Palaeontology: Palaeontology is the study of prehistoric life, including organisms' evolution and interactions with each other and their environments (their paleoecology).

Palindrome: A palindrome is a word, phrase, number, or other sequence of units that can be read the same way in either direction (the adjustment of punctuation and spaces between words is generally permitted).

Palindromic sequence: A palindromic sequence is a nucleic acid sequence (DNA or RNA) that is the same whether read 5′ (five-prime) to 3′ (three prime) on one strand or 5′–3′ on the complementary strand with which it forms a double helix.

Panicle culture: Aseptic culture of immature panicle explants to induce microspore germination and development.

Panmictic population: A population in which mating occurs at random.

Par gene: One of a class of genes required for faithful plasmid segregation at cell division. Initially, par loci were identified on plasmids, but have also been found on bacterial chromosomes.

Paracentric inversion: An inversion not involving the centromere. A chromosomal inversion that does not include the centromere.

Paraffin wax: In chemistry, paraffin is a term that can be used synonymously with "alkane," indicating hydrocarbons with the general formula C_nH_{2n+2}. Paraffin wax refers to a mixture of alkanes that falls within the $20 \leq n \leq 40$ range; they are found in the solid state at room temperature and begin to enter the liquid phase past approximately 37°C.

Parafilm: Parafilm is a plastic paraffin film with a paper backing produced by Pechiney Plastic Packaging Company, based in Chicago, Illinois primarily used in laboratories. It is commonly used for sealing or protecting vessels (such as flasks or cuvettes). It is ductile, malleable, waterproof, odorless, thermoplastic, semitransparent, and cohesive.

Parahormone: Parahormone is a substance, not a true hormone, which has a hormone-like action in controlling the functioning of some distant organ.

Parallel evolution: Parallel evolution is the development of a similar trait in related, but distinct, species descending from the same ancestor, but from different clades.

Parasite: Traditionally parasite referred to organisms with lifestages that went beyond one host (e.g., *Taenia solium*), which are now called macroparasites (typically protozoa and helminths). Parasites can now also refer to microparasites, which are typically smaller, such as viruses and bacteria and can be directly transmitted between hosts of one species.

Parasitism: Parasitism is a type of symbiotic relationship between organisms of different species where one organism, the parasite, benefits at the expense of the other, the host.

Parasporal crystal: Tightly packaged insect pro-toxin molecules that are produced by strains of *Bacillus thuringiensis* during the formation of resting spores.

Parental generation: The first set of parents crossed in which their genotype is the basis for predicting the genotype of their offspring, which in turn, may be crossed (filial generation). In parental generation, two individuals are mated or crossed to determine or predict the genotypes of their offspring, called first filial generation (or F1 generation).

Parthenocarpy: In botany and horticulture, parthenocarpy (literally meaning virgin fruit) is the natural or artificially induced production of fruit without fertilization of ovules. The fruit is therefore seedless. It may also produce apparently seedless fruit, but the seeds are actually aborted while still small. Parthenocarpy occasionally occurs as a mutation in nature, but if it affects every flower, then the plant can no longer sexually reproduce but might be able to propagate by vegetative means.

Parthenogenesis: Parthenogenesis is a form of asexual reproduction found in females, where growth and development of embryos occurs without fertilization by a male. In plants, parthenogenesis means development of an embryo from an unfertilized egg cell and is a component process of apomixis.

Partial digest: A restriction digest that has not been allowed to go to completion and thus contains pieces of DNA with some restriction endonuclease sites that have not yet been cleaved.

Particle radiation: Particle radiation is the radiation of energy by means of fast-moving subatomic particles. Particle radiation is referred to as a particle beam if the particles are all moving in the same direction, similar to a light beam.

Parturition: Childbirth, the process of delivering the baby and placenta from the uterus to the vagina to the outside world.

Passage: Passage is the number of times cells are being multiplied under culture condition.

Passage number: The passage number simply refers to the number of times the cells in the culture have been subcultured, often without consideration of the inoculation densities or recoveries involved. The population doubling level (PDL) refers to the total number of times the cells in the population have doubled since their primary isolation *in vitro.*

Passive immunity: Passive immunity is the transfer of active humoral immunity in the form of readymade antibodies, from one individual to another. Passive immunity can occur naturally, when maternal antibodies are transferred to the fetus through the placenta and can also be induced artificially, when high levels of human (or horse) antibodies specific for a pathogen or toxin are transferred to nonimmune individuals. Passive immunization is used when there is a high risk of infection and insufficient time for the body to develop its own immune response or to reduce the symptoms of ongoing or immunosuppressive diseases.

Patent: A patent is a set of exclusive rights granted by a state (national government) to an inventor or their assignee for a limited period of time in exchange for the public disclosure of an invention.

Pathogen: A pathogen is a microbe or microorganism such as a virus, bacterium, prion, or fungus that causes disease in its animal or plant host. A pathogen introduced by deliberate human agency as in bioterrorism is termed a biological agent or bioagent. There are several substrates including pathways whereby pathogens can invade a host; the principal pathways have different episodic time frames, but soil contamination has the longest or most persistent potential for harboring a pathogen.

Pathogen-free: A term applied to animals reared for experimentation or to commence new herds or flocks of disease-free animals. Animals usually obtained as for axenic animals but are then placed into a nonsterile environment in which they become infected with a range of microorganisms, many colonizing as so-called normal flora.

Pathovar: A pathovar is a bacterial strain or set of strains with the same or similar characteristics, that is differentiated at infra-subspecific level from other strains of the same species or subspecies on the basis of distinctive pathogenicity to one or more plant hosts.

pBR322: pBR322 is a plasmid and for a time was one of the most commonly used *E. coli* cloning vectors. Created in 1977, it was named eponymously after its Mexican creators, p standing for plasmid, and BR for Bolivar and Rodriguez.

Pectin: Pectin is a structural heteropolysaccharide contained in the primary cell walls of terrestrial plants.

Pectinase: Pectinase is a general term for enzymes, such as pectolyase, pectozyme, and polygalacturonase, commonly referred to in brewing as pectic enzymes. These break down pectin, a polysaccharide substrate that is found in the cell walls of plants. One of the most studied and widely used commercial pectinases is polygalacturonase.

Pedigree: Pedigree can refer to the lineage or genealogical descent of people, whether documented or not, or of animals, whether purebred or not.

Penetrance: Penetrance the frequency with which a heritable trait is manifested by individuals carrying the principal gene or genes conditioning it.

Peptide: Peptides are short polymers of amino acids linked by peptide bonds. They have the same peptide bonds as those in proteins, but are commonly shorter in length.

Peptide bond: A peptide bond (amide bond) is a covalent chemical bond formed between two molecules when the carboxyl group of one molecule reacts with the amino group of the other molecule, causing the release of a molecule of water (H_2O), hence the process is a dehydration synthesis reaction (also known as a condensation reaction), and usually occurs between amino acids.

Peptide vaccine: A peptide vaccine is a type of subunit vaccine in which a peptide of the original pathogen is used to immunize an organism.

Peptidyl transferase: The peptidyl transferase is an aminoacyltransferase as well as the primary enzymatic function of the ribosome, which forms peptide links between adjacent amino acids using tRNAs during the translation process of protein biosynthesis.

Periplasm: The region near or immediately within a bacterial or other cell wall, outside the plasma membrane.

Permanent wilting point: Permanent wilting point (PWP) or wilting point (WP) is defined as the minimal point of soil moisture the plant requires not to wilt. If moisture decreases to this or any lower point a plant wilts and can no longer recover its turgidity when placed in a saturated atmosphere for 12 h.

Permeable: That can be permeated or penetrated, especially by liquids or gases.

Pesticide: Pesticides are substances or mixture of substances intended for preventing, destroying, repelling, or mitigating any pest. A pesticide may be a chemical substance, biological agent (such as a virus or bacterium), antimicrobial, disinfectant, or device used against any pest. Pests include insects, plant pathogens, weeds, molluscs, birds, mammals, fish, nematodes (roundworms), and microbes that destroy property, spread disease, or are a vector for disease or cause a nuisance.

Petals: Petals are modified leaves that surround the reproductive parts of flowers. They often are brightly colored or unusually shaped to attract pollinators.

Petiole: In botany, the petiole is the small stalk attaching the leaf blade to the stem. The petiole usually has the same internal structure as the stem. Outgrowths appearing on each side of the petiole are called stipules.

Petite mutant: Petite is a mutant first discovered in the yeast *Saccharomyces cerevisiae*. The 'petite" yeast has little or no mitochondrial DNA and forms small anaerobic colonies when grown on media. A neutral petite produces all wild-type progeny when crossed with wild type. Petite mutations can be induced using a variety of mutagens, including DNA intercalating agents, as well as chemicals that can interfere with DNA synthesis in growing cells. Mutagens that create petites are implicated in increased rates of degenerative diseases and in the aging process.

Petri dish: A Petri dish (or Petri plate or cell culture dish) is a shallow glass or plastic cylindrical lidded dish that biologists use to culture cells or small moss plants. It was named after German bacteriologist Julius Richard Petri.

pH: In chemistry, pH is a measure of the acidity or basicity of an aqueous solution. Pure water is said to be neutral, with a pH close to 7.0 at 25°C (77°F). Solutions with a pH less than 7 are said to be acidic and solutions with a pH greater than 7 are basic or alkaline. pH measurements are important in medicine, biology, chemistry, agriculture, forestry, food science, environmental science, oceanography, civil engineering, and many other applications.

Phage: A bacteriophage.

Phagemids: A phagemid or phasmid is a type of cloning vector developed as a hybrid of the filamentous phage M13 and plasmids to produce a vector that can grow as a plasmid and also be packaged as single stranded DNA in viral particles. Phagemids contain an origin of replication (ori) for double-stranded replication as well as an f1 ori to enable single stranded replication and packaging into phage particles.

Phagocytes: Phagocytes are the white blood cells that protect the body by ingesting (phagocytosing) harmful foreign particles, bacteria, and dead or dying cells.

Phagocytosis: Phagocytosis is the cellular process of engulfing solid particles by the cell membrane to form an internal phagosome by phagocytes and protists. Phagocytosis is a specific form of endocytosis involving the vesicular internalization of solid particles, such as bacteria, and is, therefore, distinct from other forms of endocytosis such as the vesicular internalization of various liquids.

Pharmaceutical drug: A pharmaceutical drug, also referred to as medicine, medication, or medicament, can be loosely defined as any chemical substance intended for use in the medical diagnosis, cure, treatment, or prevention of disease.

pH-electrode-based sensor: Sensor in which a standard pH electrode is coated with a biological material. Many biological processes raise or lower pH, and the changes can be detected by the pH electrode.

Phenocopy: A phenocopy is an individual whose phenotype (generally referring to a single trait), under a particular environmental condition, is identical to the one of another individual whose phenotype is determined by the genotype. In other words, the phenocopy environmental condition mimics the phenotype produced by a gene.

Phenols: In organic chemistry, phenols, sometimes called phenolics, are a class of chemical compounds consisting of a hydroxyl group (–OH) bonded directly to an aromatic hydrocarbon group. The simplest of the class is phenol (C_6H_5OH).

Phenotype: A phenotype is an organism's observable characteristics or traits: such as its morphology, development, biochemical or physiological properties, behavior, and products of behavior (such as a bird's nest). Phenotypes result from the expression of an organism's genes as well as the influence of environmental factors and the interactions between the two.

Phenylalanine: Phenylalanine is an α-amino acid with the formula $C_6H_5CH_2CH(NH_2)COOH$. This essential amino acid is classified as nonpolar because of the hydrophobic nature of the benzyl side chain. L-Phenylalanine (LPA) is an electrically neutral amino acid, one of the 20 common amino acids used to biochemically form proteins, coded for by DNA.

Pheromone: Pheromone is a secreted or excreted chemical factor that triggers a social response in members of the same species. Pheromones are chemicals capable of acting outside the body of the secreting individual to impact the behavior of the receiving individual.

Phosphatase: A phosphatase is an enzyme that removes a phosphate group from its substrate by hydrolysing phosphoric acid monoesters into a phosphate ion and a molecule with a free hydroxyl group (*see* dephosphorylation). This action is directly opposite to that of phosphorylases and kinases, which attach phosphate groups to their substrates by using energetic molecules like ATP. A common phosphatase in many organisms is alkaline phosphatase.

Phosphodiester bond: A phosphodiester bond is a group of strong covalent bonds between a phosphate group and two 5-carbon ring carbohydrates (pentoses) over two ester bonds. Phosphodiester bonds are central to most life on Earth, as they make up the backbone of each helical strand of DNA.

Phospholipase: A phospholipase is an enzyme that hydrolyzes phospholipids into fatty acids and other lipophilic substances.

Phospholipid: Phospholipids are a class of lipids and are a major component of all cell membranes as they can form lipid bilayers. Most phospholipids contain a diglyceride, a phosphate group, and a simple organic molecule such as choline; one exception to this rule is sphingomyelin, which is derived from sphingosine instead of glycerol.

Phosphorolysis: Phosphorolysis is the cleavage of a compound in which inorganic phosphate is the attacking group. It is analogous to hydrolysis. An example of this is glycogen breakdown by glycogen phosphorylase, which catalyzes attack by inorganic phosphate on the terminal glycosyl residue at the nonreducing end of a glycogen molecule. The result is glucose 1-phosphate and glycogen (or starch) (n-1) glucose units.

Phosphorylation: Phosphorylation is the addition of a phosphate (PO_4^{3-}) group to a protein or other organic molecule. Phosphorylation activates or deactivates many protein enzymes.

Photoperiod: The duration of an organism's daily exposure to light, considered especially with regard to the effect of the exposure on growth and development.

Photoperiodism: Photoperiodism can be defined as the developmental responses of plants to the relative lengths of the light and dark periods. Here, it should be emphasized that photoperiodic effects relate directly to the timing of both the light and dark periods.

Photophosphorylation: The production of ATP using the energy of sunlight is called photophosphorylation. Only two sources of energy are available to living organisms: sunlight and oxidation-reduction (redox) reactions. All organisms produce ATP, which is the universal energy currency of life.

Photoreactivation: The process whereby dimerized pyrimidines (usually thymines) in DNA are restored by an enzyme (deoxyribodipyrimidine photolyase) that requires light energy.

Photosynthate: A chemical product of photosynthesis.

Photosynthesis: Photosynthesis is a chemical process that converts carbon dioxide into organic compounds, especially sugars, using the energy from sunlight. Photosynthesis occurs in plants, algae, and many species of bacteria, but not in archaea. Photosynthetic organisms are called photoautotrophs, since they can create their own food.

Photosynthetically active radiation: Photosynthetically active radiation, often abbreviated PAR, designates the spectral range (wave band) of solar radiation from 400 to 700 nm that photosynthetic organisms are able to use in the process of photosynthesis. This spectral region corresponds more or less with the range of light visible to the human eye.

Phototropism: Phototropism is directional growth in which the direction of growth is determined by the direction of the light source. In other words, it is the growth and response to a light stimulus. Phototropism is most often observed in plants, but can also occur in other organisms such as fungi. The cells on the plant that are farthest from the light have a chemical called auxin that reacts when phototropism occurs.

Phylogeny: Phylogeny, the history of the evolution of a species or group, especially in reference to lines of descent and relationships among broad groups of organisms.

Phytochrome: Phytochrome is a photoreceptor, a pigment that plants use to detect light. It is sensitive to light in the red and far-red region of the visible spectrum. Many flowering plants use it to regulate the time of flowering based on the length of day and night (photoperiodism) and to set circadian rhythms.

Phytohormone: It is a plant hormone.

Phytoparasite: Any plant parasitic organism.

Pinocytosis: In cellular biology, pinocytosis ("cell-drinking," "bulk-phase pinocytosis," "nonspecific, nonabsorptive pinocytosis" and "fluid endocytosis") is a form of endocytosis in which small particles are brought into the cell suspended within small vesicles that subsequently fuse with lysosomes to hydrolyze, or to break down, the particles.

Pipette: A pipette (also called a pipettor or chemical dropper) is a laboratory instrument used to transport a measured volume of liquid.

Plant cell culture: Plant tissue culture is a practice used to propagate plants under sterile conditions, often to produce clones of a plant. Different techniques in plant tissue culture may offer certain advantages over traditional methods of propagation.

Plantlet: Plantlets are young or small plants used as propagules. They are usually grown from clippings of mature plants.

Plasma cells: Plasma cells, also called plasma B cells, plasmocytes, and effector B cells, are white blood cells which produce large volumes of antibodies. They are transported by the blood plasma and the lymphatic system. Like all blood cells, plasma cells ultimately originate in the bone marrow; however, these cells leave the bone marrow as B cells, before terminal differentiation into plasma cells, normally in lymph nodes.

Plasma membrane: In animals, the plasma membrane is the outermost covering of the cell, whereas in plants, fungi, and some bacteria, it is located beneath the cell wall.

Plasmid: In microbiology and genetics, a plasmid is a DNA molecule that is separate from and can replicate independently of, the chromosomal DNA. They are double-stranded and, in many cases, circular.

Plasmolysis: Plasmolysis is the process in plant cells where the plasma membrane pulls away from the cell wall due to the loss of water through osmosis. The reverse process, cytolysis, can occur if the cell is in a hypotonic solution resulting in a higher external osmotic pressure and a net flow of water into the cell. Through observation of plasmolysis and deplasmolysis, it is possible to determine the tonicity of the cell's environment as well as the rate solute molecules cross the cellular membrane.

Plastid: Plastids are major organelles found in the cells of plants and algae. Plastids are the site of manufacture and storage of important chemical compounds used by the cell. Plastids often contain pigments used in photosynthesis, and the types of pigments present can change or determine the cell's color.

Pleiotropy: Pleiotropy occurs when a single gene influences multiple phenotypic traits. Consequently, a mutation in a pleiotropic gene may have an effect on some or all traits simultaneously. This can become a problem when selection on one trait favors one specific version of the gene (allele), while the selection on the other trait favors another allele.

Ploidy: Ploidy is the number of sets of chromosomes in a biological cell.

Pluripotent: Not fixed as to developmental potentialities; especially: capable of differentiating into one of many cell types.

Point mutation: A point mutation, or single base substitution, is a type of mutation that causes the replacement of a single base nucleotide with another nucleotide of the genetic material, DNA, or RNA.

Polar auxin transport: Polar auxin transport is the regulated transport of the plant hormone, auxin, in plants. It is suggested that it involves the components of the cytoskeleton, plasma membrane, and cell wall.

Polar body: A polar body is a cell structure found inside an ovum. Both animal and plant ova possess it. It is also known as a polar cell.

Polar mutation: A polar mutation affects expression of downstream genes or operons. It can also affect the expression of the gene in which it occurs, if it occurs in a transcribed region. These mutations tend to occur early within the sequence of genes and can be nonsense, frame shift, or insertion mutations.

Polar nuclei: The two haploid nuclei found in the center of the embryo sac after division of the megaspore. They may fuse to form a diploid definitive nucleus before fusing with the male gamete to form the triploid primary endosperm nucleus.

Pole cells: In early Drosophila development, the 13 first mitosis are nuclear divisions without cell division, resulting in a multinucleate cell (a syncytium). The first mononucleate cells are created at the posterior pole, were the polar granules are tethered. These cells are called pole cells, and they will form the fly's germ line.

Pollen: Pollen is a fine to coarse powder containing the microgametophytes of seed plants, which produce the male gametes (sperm cells). Pollen grains have a hard coat that protects the sperm cells during the process of their movement between the stamens to the pistil of flowering plants or from the male cone to the female cone of coniferous plants.

Pollen grain: A structure produced by plants containing the male haploid gamete to be used in reproduction. The gamete is covered by protective layers, which perform their role until the pollen grain is capable of fertilizing when reaching the female stigma.

Pollination: Pollination is the process by which pollen is transferred in plants, thereby enabling fertilization and sexual reproduction. Pollen grains, which contain the male gametes (sperm) to where the female gamete(s), are contained within the carpel; in gymnosperms, the pollen is directly applied to the ovule itself. The receptive part of the carpel is called a stigma in the flowers of angiosperms.

Poly (A) polymerase: Poly (A) polymerase catalyzes the addition of adenine residues to the 3′ end of pre-mRNAs to form the poly(A) tail.

Polyacrylamide gel electrophoresis (PAGE): In PAGE, proteins charged negatively by the binding of the anionic detergent SDS (sodium dodecyl sulfate) separate within a matrix of polyacrylamide gel in an electric field according to their molecular weights. Polyacrylamide is formed by the polymerization of the monomer molecule-acrylamide crosslinked by *N,N′*-methylene-bis-acrylamide (abbreviated BIS). Free radicals generated by ammonium persulfate (APS) and a catalyst acting as an oxygen scavenger (-*N,N,N′,N′*-tetramethylethylene diamine [TEMED]) are required to start the polymerization, since acrylamide and BIS are nonreactive by themselves or when mixed together.

Polyadenylation: Polyadenylation is the addition of a poly(A) tail to an RNA molecule. The poly(A) tail consists of multiple adenosine monophosphates; in other words, it is a stretch of RNA that only has adenine bases. In eukaryotes, polyadenylation is part of the process that produces mature messenger RNA (mRNA) for translation. It therefore forms part of the larger process of gene expression.

Polyclonal antibodies: Polyclonal antibodies (or antisera) are antibodies that are obtained from different B cell resources. They are a combination of immunoglobulin molecules secreted against a specific antigen, each identifying a different epitope.

Polyethylene glycol: Polyethylene glycol (PEG) is a polyether compound with many applications from industrial manufacturing to medicine. It has also been known as polyethylene oxide (PEO) or polyoxyethylene (POE), depending on its molecular weight, and under the tradename Carbowax.

Polymerase chain reaction: The polymerase chain reaction (PCR) is a scientific technique in molecular biology to amplify a single or a few copies of a piece of DNA across several orders of magnitude, generating thousands to millions of copies of a particular DNA sequence.

Polymerase: A polymerase is an enzyme whose central function is associated with polymers of nucleic acids such as RNA and DNA. The primary function of a polymerase is the polymerization of new DNA or RNA against an existing DNA or RNA template in the processes of replication and transcription. In association with a cluster of other enzymes and proteins, they take nucleotides from solvent and catalyze the synthesis of a polynucleotide sequence against a nucleotide template strand using base-pairing interactions.

Polymerization: In polymer chemistry, polymerization is a process of reacting monomer molecules together in a chemical reaction to form three-dimensional networks or polymer chains.

Polymorphism: Polymorphism in biology occurs when two or more clearly different phenotypes exist in the same population of a species in other words, the occurrence of more than one form or morph.

Polynucleotide: A polynucleotide molecule is a biopolymer composed of 13 or more nucleotide monomers covalently bonded in a chain. DNA (deoxyribonucleic acid) and RNA (ribonucleic acid) are examples of polynucleotides with distinct biological function.

Polypeptide: A peptide consisting of two or more amino acids. Amino acids make up polypeptides, which, in turn, make up proteins.

Polyploidy: Polyploidy is a term used to describe cells and organisms containing more than two paired (homologous) sets of chromosomes. Most eukaryotic species are diploid, meaning that they have two sets of chromosomes one set inherited from each parent. However, polyploidy is found in some organisms and is especially common in plants.

Polysaccharide: Polysaccharides are polymeric carbohydrate structures, formed of repeating units (either mono- or di-saccharides) joined together by glycosidic bonds. These structures are often linear, but may contain various degrees of branching.

Polytene chromosomes: To increase cell volume, some specialized cells undergo repeated rounds of DNA replication without cell division (endomitosis), forming a giant polytene chromosome. Polytene chromosomes form when multiple rounds of replication produce many sister chromatids that remain synapsed together.

Polyvalent vaccine: Vaccine that is prepared from cultures or antigens of more than one strain or species.

Polyvinylpyrrolidone: Polyvinylpyrrolidone (PVP), also commonly called Polyvidone or Povidone, is a water-soluble polymer made from the monomer *N*-vinylpyrrolidone.

Population genetics: Population genetics is the study of allele frequency distribution and change under the influence of the four main evolutionary processes: natural selection, genetic drift, mutation, and gene flow. It also takes into account the factors of recombination, population subdivision, and population structure. It attempts to explain such phenomena as adaptation and speciation.

Positional cloning: Positional cloning is a technique that is used in genetic screening to identify specific areas of interest in the genome and then determine what they do. This type of genetic screening is sometimes referred to as reverse genetics, because researchers start by figuring out where a gene is and then they determine what it does, in contrast with methods which start by determining the function of a gene and then finding it in the genome. Genes related to conditions such as Huntington's disease and cystic fibrosis have been identified with this technique.

Postreplication repair: Postreplication repair is the repair of damage to the DNA that takes place after replication. DNA damage prevents the normal enzymatic synthesis of DNA by the replication fork. At damaged sites in the genome, both prokaryotic and eukaryotic cells utilize a number of postreplication repair (PRR) mechanisms to complete DNA replication.

Posttranslational modification: Posttranslational modification (PTM) is the chemical modification of a protein after its translation. It is one of the later steps in protein biosynthesis and thus gene expression for many proteins.

PPM: parts per million, 10^{-6}.

Primary antibody: Primary antibodies are antibodies raised against an antigenic target of interest (a protein, peptide, carbohydrate, or other small molecule) and are typically unconjugated (unlabeled). Primary antibodies that recognize and bind with high affinity and specificity to unique epitopes across a broad spectrum of biomolecules are available as high specificity monoclonal antibodies and/or as polyclonal antibodies. These antibodies are useful not only to detect specific biomolecules but also to measure changes in their level and specificity of modification by processes such as phosphorylation, methylation, or glycosylation. A primary antibody can be very useful for the detection of biomarkers for diseases such as cancer, diabetes, Parkinson's and Alzheimer's disease and they are used for the study of ADME and multidrug resistance (MDR) of therapeutic agents.

Primary cell wall: The primary cell wall is the part or layer of cell wall in which cell growth is permitted. Compared to secondary cell wall, this layer contains more pectin and lignin is absent until a secondary wall has formed on top of it.

Primary culture: A cell or tissue culture started from material taken directly from an organism, as opposed to that from an explant from an organism.

Primary transcript: A primary transcript is an RNA molecule that has not yet undergone any modification after its synthesis. For example, a precursor messenger RNA (pre-mRNA) is a primary transcript that becomes a messenger RNA (mRNA) after processing, and a primary microRNA (pri-miRNA) precursor becomes microRNA (miRNA) after processing.

Primer: A primer is a strand of nucleic acid that serves as a starting point for DNA synthesis. They are required for DNA replication, because the enzymes that catalyze this process, DNA polymerases, can only add new nucleotides to an existing strand of DNA. The polymerase starts replication at the 3′-end of the primer, and copies the opposite strand.

Primer DNA polymerase: A DNA polymerase is an enzyme that helps catalyze in the polymerization of deoxyribonucleotides into a DNA strand. DNA polymerases are best known for their feedback role in DNA replication, in which the polymerase "reads" an intact DNA strand as a template and uses it to synthesize the new strand.

Primer walking: Primer walking is a sequencing method of choice for sequencing DNA fragments between 1.3 and 7 kb. Such fragments are too long to be sequenced in a single sequence read using the chain termination method.

Primordium: In embryology, organ or tissue in its earliest recognizable stage of development.

Primosome: In molecular biology, a primosome is a protein complex responsible for creating RNA primers on single stranded DNA during DNA replication.

Prion: Prion is an infectious agent composed of protein in a misfolded form. This is in contrast to all other known infectious agents, which must contain nucleic acids (DNA, RNA, or both). The word prion, coined in 1982 by Stanley B. Prusiner, is a portmanteau derived from the words protein and infection.

Probability: Probability is a way of expressing knowledge or belief that an event will occur or has occurred. The concept has an exact mathematical meaning in probability theory, which is used extensively in such areas of study as mathematics, statistics, finance, gambling, science, artificial intelligence/machine learning, and philosophy to draw conclusions about the likelihood of potential events and the underlying mechanics of complex systems.

Probe DNA: A single-stranded DNA molecule used in laboratory experiments to detect the presence of a complementary sequence among a mixture of other singled-stranded DNA molecules.

Progeny testing: Progeny testing is a test of the value for selective breeding of an individual's genotype by looking at the progeny produced by different matings.

Progesterone: Progesterone also known as P4 (pregn-4-ene-3,20-dione) is a C-21 steroid hormone involved in the female menstrual cycle, pregnancy (supports gestation), and embryogenesis of humans and other species. Progesterone belongs to a class of hormones called progestogens and is the major naturally occurring human progestogen.

Prokaryote: The prokaryotes are a group of organisms that lack a cell nucleus or any other membrane-bound organelles.

Prolactin: Prolactin (PRL) also known as luteotropic hormone (LTH) is a protein that in humans is encoded by the PRL gene. Prolactin is a peptide hormone discovered by Henry Friesen, primarily associated with lactation. In breastfeeding, the act of an infant suckling the nipple stimulates the production of oxytocin, which stimulates the "milk let-down" reflex, which fills the breast with milk via a process called lactogenesis, in preparation for the next feed.

Proliferation cell: The term cell proliferation or growth is used in the contexts of cell development and cell division (reproduction). When used in the context of cell division, it refers to growth of cell populations, where one cell (the "mother cell") grows and divides to produce two "daughter cells" (M phase). When used in the context of cell development, the term refers to increase in cytoplasmic and organelle volume (G1 phase) as well as increase in genetic material before replication (G2 phase).

Promoter: In genetics, a promoter is a region of DNA that facilitates the transcription of a particular gene. Promoters are located near the genes they regulate, on the same strand and typically upstream (towards the 5′ region of the sense strand).

Pronucleus: A pronucleus is the nucleus of a sperm or an egg cell during the process of fertilization, after the sperm enters the ovum, but before they fuse.

Propagation: Plant propagation is the process of creating new plants from a variety of sources: seeds, cuttings, bulbs, and other plant parts. Plant propagation can also refer to the artificial or natural dispersal of plants.

Prophage: A prophage is a phage (viral) genome inserted and integrated into the circular bacterial DNA chromosome. A prophage, also known as a temperate phage, is any virus in the lysogenic cycle; it is integrated into the host chromosome or exists as an extra-chromosomal plasmid. Technically, a virus may be called a prophage only while the viral DNA remains incorporated in the host DNA. This is a latent form of a bacteriophage, in which the viral genes are incorporated into the bacterial chromosome without causing disruption of the bacterial cell.

Protamines: Protamines are small, arginine-rich, nuclear proteins that replace histones late in the haploid phase of spermatogenesis and are believed essential for sperm head condensation and DNA stabilization.

Protease: A protease (also termed peptidase or proteinase) is any enzyme that conducts proteolysis, that is, begins protein catabolism by hydrolysis of the peptide bonds that link amino acids together in the polypeptide chain forming the protein.

Protein: Proteins are biochemical compounds consisting of one or more polypeptides typically folded into a globular or fibrous form, facilitating a biological function. A polypeptide is a single linear polymer chain of amino acids bonded together by peptide bonds between the carboxyl and amino groups of adjacent amino acid residues. The sequence of amino acids in a protein is defined by the sequence of a gene, which is encoded in the genetic code.

Protein crystallization: Proteins, like many molecules, can be prompted to form crystals when placed in the appropriate conditions. In order to crystallize a protein, the purified protein undergoes slow precipitation from an aqueous solution. As a result, individual protein molecules align themselves in a repeating series of unit cells by adopting a consistent orientation.

Protein engineering: Protein engineering is the process of developing useful or valuable proteins. It is a young discipline, with much research taking place into the understanding of protein folding and recognition for protein design principles.

Protein kinase: A protein kinase is a kinase enzyme that modifies other proteins by chemically adding phosphate groups to them (phosphorylation). Phosphorylation usually results in a functional change of the target protein (substrate) by changing enzyme activity, cellular location, or association with other proteins.

Protein sequencing: Protein sequencing is a technique to determine the amino acid sequence of a protein, as well as which conformation the protein adopts and the extent to which it is complexed with any nonpeptide molecules. Discovering the structures and functions of proteins in living organisms is an important tool for understanding cellular processes and allows drugs that target specific metabolic pathways to be invented more easily.

Protein synthesis: Protein synthesis is the process in which cells build proteins. The term is sometimes used to refer only to protein translation but more often it refers to a multistep process, beginning with amino acid synthesis and transcription of nuclear DNA into messenger RNA, which is then used as input to translation.

Proteolysis: Proteolysis is the directed degradation (digestion) of proteins by cellular enzymes called proteases or by intramolecular digestion.

Protoclone: Regenerated plant derived from protoplast culture or a single colony derived from protoplasts in culture.

Protocol: In the natural sciences, a protocol is a predefined written procedural method in the design and implementation of experiments. Protocols are written whenever it is desirable to standardize a laboratory method to ensure successful replication of results by others in the same laboratory or by other laboratories. Detailed protocols also facilitate the assessment of results through peer review.

Proto-oncogene: A normal gene that, when altered by mutation, becomes an oncogene that can contribute to cancer. Proto-oncogenes may have many different functions in the cell. Some proto-oncogenes provide signals that lead to cell division. Other proto-oncogenes regulate programmed cell death (apoptosis).

Protoplasm: Protoplasm is the living content of a cell that is surrounded by a plasma membrane (cell membrane). Protoplasm is composed of a mixture of small molecules such as ions, amino acids, monosaccharides, and water and macromolecules such as nucleic acids, proteins, lipids, and polysaccharides.

Protoplast: A protoplast is a plant, bacterial, or fungal cell that had its cell wall completely or partially removed using either mechanical or enzymatic means.

Protoplast fusion: Protoplast fusion is a type of genetic modification in plants by which two distinct species of plants are fused together to form a new hybrid plant with the characteristics of both, a somatic hybrid.

Protozoan: Any of a large group of single-celled, usually microscopic, eukaryotic organisms, such as amoebas, ciliates, flagellates, and protozoans.

Pseudoautosomal region: The pseudoautosomal regions, PAR1 and PAR2, are homologous sequences of nucleotides on the X and Y chromosomes. The pseudoautosomal regions get their name, because any genes located within them (so far at least 29 have been found) are inherited just like any autosomal genes. PAR1 comprises 2.6 Mbp of the short-arm tips of both X and Y chromosomes in humans and other great apes (X and Y are 155 Mbp and 59 Mbp in total). PAR2 is located at the tips of the long arms, spanning 320 kbp.

Pseudogene: Pseudogenes are dysfunctional relatives of known genes that have lost their protein-coding ability or are otherwise no longer expressed in the cell. Although some do not have introns or promoters (these pseudogenes are copied from mRNA and incorporated into the chromosome and are called processed pseudogenes), most have some gene-like features (such as promoters, CpG islands, and splice sites), they are nonetheless considered nonfunctional, due to their lack of protein-coding ability resulting from various genetic disablements (stop codons, frameshifts, or a lack of transcription) or their inability to encode RNA (such as with rRNA pseudogenes).

PUC: pUC19 is a plasmid cloning vector created by Messing and coworkers in the University of California. p in the name stands for plasmid and UC represents the university in which it was created. It is a circular double stranded DNA and has 2686 base pairs. pUC19 is one of the most widely used vector molecules as the recombinants or the cells into which foreign DNA has been introduced can be easily distinguished from the nonrecombinants based on color differences of colonies on growth media. pUC18 is similar to pUC19, but the MCS region is reversed.

Pulsed-field gel electrophoresis: Pulsed field gel electrophoresis is a technique used for the separation of large deoxyribonucleic acid (DNA) molecules by applying an electric field that periodically changes direction to a gel matrix.

Pure culture: In microbiology, a laboratory culture containing a single species of organism. A pure culture is usually derived from a mixed culture (one containing many species) by transferring a small sample into new, sterile growth medium in such a manner as to disperse the individual cells across the medium surface or by thinning the sample manifold before inoculating the new medium.

Purine: A purine is a heterocyclic aromatic organic compound, consisting of a pyrimidine ring fused to an imidazole ring. Purines, including substituted purines and their tautomer's, are the most widely distributed kind of nitrogen-containing heterocycle in nature.

Pyrimidine: Pyrimidine is a heterocyclic aromatic organic compound similar to benzene and pyridine, containing two nitrogen atoms at positions 1 and 3 of the six-member ring. It is isomeric with two other forms of diazine.

Quantitative genetics: Quantitative genetics is the study of continuous traits (such as height or weight) and their underlying mechanisms. It is effectively an extension of simple Mendelian inheritance in that the combined effect of the many underlying genes results in a continuous distribution of phenotypic values.

Quantitative PCR: In molecular biology, real-time polymerase chain reaction, also called quantitative real time polymerase chain reaction (Q-PCR/qPCR/qrt-PCR) or kinetic polymerase chain reaction (KPCR), is a laboratory technique based on the PCR, which is used to amplify and simultaneously quantify a targeted DNA molecule. For one or more specific sequences in a DNA sample, real time-PCR enables both detection and quantification. The quantity can be either an absolute number of copies or a relative amount when normalized to DNA input or additional normalizing genes.

Quantitative trait: Quantitative traits refer to phenotypes (characteristics) that vary in degree and can be attributed to polygenic effects, that is, product of two or more genes, and their environment. Quantitative trait loci (QTLs) are stretches of DNA containing or linked to the genes that underlie a quantitative trait. Mapping regions of the genome that contain genes involved in specifying a quantitative trait is done using molecular tags such as AFLP or, more commonly, SNPs. This is an early step in identifying and sequencing the actual genes underlying trait variation.

Quantum: In physics, a quantum is the minimum amount of any physical entity involved in an interaction. Behind this, one finds the fundamental notion that a physical property may be "quantized," referred to as "the hypothesis of quantization.

Quarantine: Quarantine is compulsory isolation, typically to contain the spread of something considered dangerous, often but not always disease. The word comes from the Italian (seventeenth century Venetian) *quarantena*, meaning 40-day period. Quarantine can be applied to humans, but also to animals of various kinds.

Race: In biology, races are distinct genetically divergent populations within the same species with relatively small morphological and genetic differences. The populations can be described as ecological races if they arise from adaptation to different local habitats or geographic races when they are geographically isolated. If sufficiently different, two or more races can be identified as subspecies, which is an official biological taxonomy unit subordinate to species.

Radioisotope: A version of a chemical element that has an unstable nucleus and emits radiation during its decay to a stable form. Radioisotopes have important uses in medical diagnosis, treatment, and research. A radioisotope is so-named, because it is a radioactive isotope, an isotope being an alternate version of a chemical element that has a different atomic mass.

Random amplified polymorphic DNA: RAPD (pronounced "rapid") stands for random amplification of polymorphic DNA. It is a type of PCR reaction, but the segments of DNA that are amplified are random. The scientist performing RAPD creates several arbitrary, short primers (8–12 nucleotides), then proceeds with the PCR using a large template of genomic DNA, hoping that fragments will amplify. By resolving the resulting patterns, a semiunique profile can be gleaned from a RAPD reaction.

Random genetic drift: Changes in allelic frequency due to sampling error. Changes in allele frequency result because the genes appearing in offspring are not a perfectly representative sampling of the parental genes.

Random primer method: A method for labeling DNA probes, mainly for Southern hybridization experiments. A mixture of short oligonucleotides is hybridized to a single-stranded DNA probe. In the presence of DNA polymerase and deoxyribonucleotides—one of which is labeled—DNA synthesis then generates labeled copies of probe DNA.

Reca: Reca is a 38 kDa *Escherichia coli* protein essential for the repair and maintenance of DNA. Reca has a structural and functional homolog in every species in which it has been seriously sought and serves as an archetype for this class of homologous DNA repair proteins. The homologous protein in *Homo sapiens* is called RAD51.

Receptor: A receptor is a molecule found on the surface of a cell, which receives specific chemical signals from neighboring cells or the wider environment within an organism. These signals tell a cell to do something—for example, to divide or die, or to allow certain molecules to enter or exit the cell.

Recessive: In genetics, the term "recessive gene" refers to an allele that causes a phenotype (visible or detectable characteristic) that is only seen in a homozygous genotype (an organism that has two copies of the same allele) and never in a heterozygous genotype.

Recessive oncogene: A single copy of this gene is sufficient to suppress cell proliferation; the loss of both copies of the gene contributes to cancer formation.

Reciprocal crosses: In genetics, a reciprocal cross is a breeding experiment designed to test the role of parental sex on a given inheritance pattern. All parent organisms must be true breeding to properly carry out such an experiment. In one cross, a male expressing the trait of interest will be crossed with a female not expressing the trait. In the other, a female expressing the trait of interest will be crossed with a male not expressing the trait.

Recognition site: The recognition sequence, sometimes also referred to as recognition site, of any DNA-binding protein motif that exhibits binding specificity, refers to the DNA sequence (or subset thereof), to which the domain is specific. Recognition sequences are palindromes. The transcription factor Sp1 for example, binds the sequences 5′-(G/T) GGGCGG (G/A) (G/A)(C/T)-3′, where (G/T) indicates that the domain will bind a guanine or thymine at this position.

Recombinant DNA: Recombinant DNA (rDNA) molecules are DNA sequences that result from the use of laboratory methods (molecular cloning) to bring together genetic material from multiple sources, creating sequences that would not otherwise be found in biological organisms. Recombinant DNA is possible, because DNA molecules from all organisms share the same chemical structure; they differ only in the sequence of nucleotides within that identical overall structure. Consequently, when DNA from a foreign source is linked to host sequences that can drive DNA replication and then introduced into a host organism, the foreign DNA is replicated along with the host DNA.

Recombinant DNA technology: Recombinant technology begins with the isolation of a gene of interest. The gene is then inserted into a vector and cloned. A vector is a piece of DNA that is capable of independent growth; commonly used vectors are bacterial plasmids and viral phages. The gene of interest (foreign DNA) is integrated into the plasmid or phage, and this is referred to as recombinant DNA.

Recombinant protein: Recombinant protein is a protein that its code was carried by a recombinant DNA.

Recombinant toxin: A hybrid cytotoxic protein made by recombinant DNA technology, designed to selectively kill malignant cells.

Recombinant vaccine: Recombinant hepatitis B vaccine is the only recombinant vaccine licensed at present.

Recombination: Genetic recombination is a process by which a molecule of nucleic acid (usually DNA, but can also be RNA) is broken and then joined to a different one. Recombination can occur between similar molecules of DNA, as in homologous recombination, or dissimilar molecules, as in nonhomologous end joining. Recombination is a common method of DNA repair in both bacteria and eukaryotes. In eukaryotes, recombination also occurs in meiosis, where it facilitates chromosomal crossover. The crossover process leads to offspring's having different combinations of genes from those of their parents and can occasionally produce new chimeric alleles. In organisms with an adaptive immune system, a type of genetic recombination called V(D)J recombination helps immune cells rapidly diversify to recognize and adapt to new pathogens. The shuffling of genes brought about by genetic recombination is thought to have many advantages, as it is a major engine of genetic variation and also allows sexually reproducing organisms to avoid Muller's ratchet, in which the genomes of an asexual population accumulate deleterious mutations in an irreversible manner.

Regeneration: In biology, regeneration is the process of renewal, restoration, and growth that makes genomes, cells, organs, organisms, and ecosystems resilient to natural fluctuations or events that cause disturbance or damage. Every species is capable of regeneration, from bacteria to humans. At its most elementary level, regeneration is mediated by the molecular processes of DNA synthesis.

Regulatory gene: A regulator gene, regulator, or regulatory gene is a gene involved in controlling the expression of one or more other genes. A regulator gene may encode a protein, or it may work at the level of RNA, as in the case of genes encoding microRNAs.

Relaxed plasmid: A circular, double-stranded unit of DNA that replicates within a cell independently of the chromosomal DNA. Plasmids are most often found in bacteria and are used in recombinant DNA research to transfer genes between cells.

Replication: DNA replication is a biological process that occurs in all living organisms and copies their DNA; it is the basis for biological inheritance. The process starts with one double-stranded DNA molecule and produces two identical copies of the molecule. Each strand of the original double-stranded DNA molecule serves as template for the production of the complementary strand. Cellular proofreading and error toe-checking mechanisms ensure near perfect fidelity for DNA replication.

Replicative form: An intermediate stage in the replication of either DNA or RNA viral genomes that is usually double stranded, the altered, double-stranded form to which single-stranded coli phage DNA is converted after infection of a susceptible bacterium, formation of the complementary (minus) strand being mediated by enzymes that were present in the bacterium before entrance of the viral (plus) strand.

Replicon: A replicon is a DNA molecule or RNA molecule, or a region of DNA or RNA, that replicates from a single origin of replication.

Replisome: The replisome is a complex molecular machine that carries out replication of DNA. It is made up of a number of subcomponents that each provides a specific function during the process of replication.

Reporter gene: In molecular biology, a reporter gene (often simply reporter) is a gene that researchers attach to a regulatory sequence of another gene of interest in cell culture, animals or plants.

Repressible enzyme: One whose rate of production is decreased as the concentration of certain metabolites is increased.

Repressor: In molecular genetics, a repressor is a DNA-binding protein that regulates the expression of one or more genes by binding to the operator and blocking the attachment of RNA polymerase to the promoter, thus preventing transcription of the genes. This blocking of expression is called repression.

Repulsion: The tendency of some linked genetic characters to be inherited separately, because a dominant allele for each character occurs on the same chromosome as a recessive allele of the other.

Residues: In biochemistry and molecular biology, a residue refers to a specific monomer within the polymeric chain of a polysaccharide, protein or nucleic acid.

Restriction endonuclease: A restriction enzyme (or restriction endonuclease) is an enzyme that cuts double-stranded or single stranded DNA at specific recognition nucleotide sequences known as restriction sites. Such enzymes, found in bacteria and archaea, are thought to have evolved to provide a defense mechanism against invading viruses. Inside a bacterial host, the restriction enzymes selectively cut up foreign DNA in a process called restriction; host DNA is methylated by a modification enzyme (a methylase) to protect it from the restriction enzyme's activity. Collectively, these two processes form the restriction modification system. To cut the DNA, a restriction enzyme makes two incisions, once through each sugar-phosphate backbone (i.e., each strand) of the DNA double helix.

Restriction exonuclease: Exonucleases are enzymes that work by cleaving nucleotides one at a time from the end (exo) of a polynucleotide chain. A hydrolyzing reaction that breaks phosphodiester bonds at either the 3′ or the 5′ end occurs. Its close relative is the endonuclease, which cleaves phosphodiester bonds in the middle (endo) of a polynucleotide chain. Eukaryotes and prokaryotes have three types of exonucleases involved in the normal turnover of mRNA: 5′–3′ exonuclease, which is a dependent decapping protein, 3′–5′ exonuclease, an independent protein, and poly(A)-specific 3′–5′ exonuclease.

Restriction fragment: A restriction fragment is a DNA fragment resulting from the cutting of a DNA strand by a restriction enzyme (restriction endonucleases), a process called restriction. Each restriction enzyme is highly specific, recognizing a particular short DNA sequence, or restriction site, and cutting both DNA strands at specific points within this site. Most restriction sites are palindromic (the sequence of nucleotides is the same on both strands when read in the 5′–3′ direction) and are four to eight nucleotides long. Many cuts are made by one restriction enzyme because of the chance repetition of these sequences in a long DNA molecule, yielding a set of restriction fragments. A particular DNA molecule will always yield the same set of restriction fragments when exposed to the same restriction enzyme. Restriction fragments can be analyzed using techniques such as gel electrophoresis or used in recombinant DNA technology.

Restriction fragment length polymorphism: In molecular biology, restriction fragment length polymorphism, or RFLP (commonly pronounced "rif-lip"), is a technique that exploits variations in homologous DNA sequences. It refers to a difference between samples of homologous DNA molecules that come from differing locations of restriction enzyme sites and to a related laboratory technique by which these segments can be illustrated. In RFLP analysis, the DNA sample is broken into pieces (digested) by restriction enzymes, and the resulting restriction fragments are separated according to their lengths by gel electrophoresis. Although now largely obsolete due to the rise of inexpensive DNA sequencing technologies, RFLP analysis was the first DNA profiling technique inexpensive enough to see widespread application. In addition to genetic fingerprinting, RFLP was an important tool in genome mapping, localization of genes for genetic disorders, determination of risk for disease, and paternity testing.

Restriction map: A restriction map is a map of known restriction sites within a sequence of DNA. Restriction mapping requires the use of restriction enzymes. In molecular biology, restriction maps are used as a reference to engineer plasmids or other relatively short pieces of DNA and sometimes for longer genomic DNA.

Restriction nuclease: Restriction nuclease cuts nucleic acid at specific restriction sites and produce restriction fragments; obtained from bacteria (where they cripple viral invaders); used in recombinant DNA technology.

Retrovirus: A retrovirus is an RNA virus that is duplicated in a host cell using the reverse transcriptase enzyme to produce DNA from its RNA genome. The DNA is then incorporated into the host's genome by an integrase enzyme. The virus thereafter replicates as part of the host cell's DNA. Retroviruses are enveloped viruses that belong to the viral family Retroviridae.

Reverse genetics: Reverse genetics is an approach to discovering the function of a gene by analyzing the phenotypic effects of specific gene sequences obtained by DNA sequencing. This investigative process proceeds in the opposite direction of so-called forward genetic screens of classical genetics. Simply put, while forward genetics seeks to find the genetic basis of a phenotype or trait, reverse genetics seeks to find what phenotypes arise as a result of particular genes.

Reverse transcriptase: In the fields of molecular biology and biochemistry, a reverse transcriptase, also known as RNA-dependent DNA polymerase, is a DNA polymerase enzyme that transcribes single-stranded RNA into double-stranded DNA. It also helps in the formation of a double helix DNA once the RNA has been reverse transcribed into a single-strand cDNA. Normal transcription involves the synthesis of RNA from DNA; hence, reverse transcription is the reverse of this.

Rhizobacteria: Rhizobacteria are root-colonizing bacteria that form a symbiotic relationship with many plants. The name comes from the greek rhiza meaning root. Though parasitic varieties of rhizobacteria exist, the term usually refers to bacteria that form a relationship beneficial for both parties (mutualism). Such bacteria are often referred to as plant growth promoting rhizobacteria or PGPRs.

Ribonuclease: Ribonuclease (commonly abbreviated RNase) is a type of nuclease that catalyzes the degradation of RNA into smaller components. Ribonucleases can be divided into endoribonucleases and exoribonucleases.

Ribosomal-binding site: A ribosomal-binding site (RBS) is a sequence on mRNA that is bound by the ribosome when initiating protein translation. It can be either the 5′ cap of a messenger RNA in eukaryotes, a region 6–7 nucleotides upstream of the start codon AUG in prokaryotes (called the Shine–Dalgarno sequence) or an internal ribosome entry site (IRES) in viruses. The sequence is complementary to the 3′ end of the rRNA. The ribosome searches for this site and binds to it through base-pairing of nucleotides. Then, the ribosome begins the translation process and recruits initiation factors. After finding the ribosome binding site in eukaryotes, the ribosome recognizes the Kozak consensus sequence and begins translation at the +1 AUG codon.

Ribosomal RNA: Ribosomal ribonucleic acid (rRNA) is the RNA component of the ribosome, the organelle that is the site of protein synthesis in all living cells. Ribosomal RNA provides a mechanism for decoding mRNA into amino acids and interacts with tRNAs during translation by providing peptidyl transferase activity. The tRNAs bring the necessary amino acids corresponding to the appropriate mRNA codon.

Ribosome: A ribosome is an organelle (an internal component of a biological cell) the function of which is to assemble the twenty specific amino acid molecules to form the particular protein molecule determined by the nucleotide sequence of an RNA molecule.

Ribozyme: A ribozyme is an RNA molecule with a well-defined tertiary structure that enables it to catalyze a chemical reaction. Ribozyme means ribonucleic acid enzyme.

Ribulose biphosphate: Ribulose-1,5-bisphosphate (RuBP) is an organic substance that is involved in photosynthesis. The anion is a double phosphate ester of the ketose (ketone-containing sugar) called ribulose. Salts of this species can be isolated, but its crucial biological function involves this colorless anion in solution.

R-loops: A single-stranded loop section of DNA formed by the association of a section of ssRNA with the other strand of the DNA in this region whereby one DNA strand is displaced as the loop.

RNA: Ribonucleic acid or RNA is one of the three major macromolecules (along with DNA and proteins) that are essential for all known forms of life.

RNAase: Ribonuclease A (RNase A) is a pancreatic ribonuclease that cleaves single-stranded RNA. Bovine pancreatic RNase A is one of the classic model systems of protein science.

RNA editing: The term RNA editing describes those molecular processes in which the information content in an RNA molecule is altered through a chemical change in the base makeup. To date, such changes have been observed in tRNA, rRNA, mRNA, and microRNA molecules of eukaryotes but not prokaryotes. RNA editing occurs in the cell nucleus and cytosol, as well as in mitochondria and plastids, which are thought to have evolved from prokaryotic-like endosymbionts.

RNA polymerase: RNA polymerase (RNAP or RNApol) is an enzyme that produces RNA. In cells, RNAP is needed for constructing RNA chains from DNA genes as templates, a process called transcription. RNA polymerase enzymes are essential to life and are found in all organisms and many viruses. In chemical terms, RNAP is a nucleotidyl transferase that polymerizes ribonucleotides at the 3′ end of an RNA transcript.

Root apex: The tip of a tooth root, the part farthest from the incisal or occlusal side.

Root cap: The root cap is a section of tissue at the tip of a plant root. It is also called calyptra. Root caps contain statoliths that are involved in gravity perception in plants. If the cap is carefully removed, the root will grow randomly.

Ruminant animals: A ruminant is a mammal of the order Artiodactyla that digests plant-based food by initially softening it within the animal's first stomach, then regurgitating the semi-digested mass, now known as cud, and chewing it again. The process of rechewing the cud to further break down plant matter and stimulate digestion is called "ruminating."

S phase: S-phase (synthesis phase) is the part of the cell cycle in which DNA is replicated, occurring between G1 phase and G2 phase. Precise and accurate DNA replication is necessary to prevent genetic abnormalities, which often lead to cell death or disease. Due to the importance, the regulatory pathways that govern this event in eukaryotes are highly conserved. This conservation makes the study of S-phase in model organisms such as *Xenopus laevis* embryos and budding yeast relevant to higher organisms.

S_1 mapping: A method for mapping precursor or mature mRNA to particular DNA sequences using the enzyme S_1-nuclease.

S_1 nuclease: S_1 nuclease is an endonuclease that is active against single-stranded DNA and RNA molecules. It is five times more active on DNA than RNA. Its reaction products are oligonucleotides or single nucleotides with 5′ phosphoryl groups. Although its primary substrate is single-stranded, it can also occasionally introduce single-stranded breaks in double-stranded DNA or RNA, or DNA-RNA hybrids. It is used in the laboratory as a reagent in nuclease protection assays. In molecular biology, it is used in removing single stranded tails from DNA molecules to create blunt ended molecules and opening hairpin loops generated during synthesis of double-stranded cDNA.

Salmonella: Salmonella is a genus of rod-shaped, Gram-negative, non-spore-forming, predominantly motile enterobacteria with diameters around 0.7–1.5 μm, lengths from 2 to 5 μm, and flagella which grade in all directions.

Salt tolerance plant: The plant that withstands or survives at very high salt concentrations is called salt tolerance plant.

Satellite DNA: Satellite DNA consists of very large arrays of tandemly repeating, noncoding DNA. Satellite DNA is the main component of functional centromeres and forms the main structural constituent of heterochromatin.

Satellite RNA: A small, self-splicing RNA molecule that accompanies several plant viruses, including tobacco ringspot virus.

Scaffold protein: In biology, scaffold proteins are crucial regulators of many key signaling pathways. Although scaffolds are not strictly defined in function, they are known to interact and/or bind with multiple members of a signaling pathway, tethering them into complexes.

Scanning electron microscope: A scanning electron microscope (SEM) is a type of electron microscope that images a sample by scanning it with a high-energy beam of electrons in a raster scan pattern. The electrons interact with the atoms that make up the sample producing signals that contain information about the sample's surface topography, composition, and other properties such as electrical conductivity.

Screen drug candidate: Two main approaches exist for the finding of new bioactive chemical entities from natural sources: random collection and screening of material; and exploitation of ethno-pharmacological knowledge in the selection. The former approach is based on the fact that only a small part of Earth's biodiversity has ever been tested for pharmaceutical activity, and organisms living in a species-rich environment need to evolve defensive and competitive mechanisms to survive.

Secondary antibody: A secondary antibody is an antibody that binds to primary antibodies or antibody fragments. They are typically labeled with probes that make them useful for detection, purification, or cell-sorting applications.

Secondary cell wall: The secondary cell wall is a structure found in many plant cells, located between the primary cell wall and the plasma membrane. The cell starts producing the secondary cell wall after the primary cell wall is complete, and the cell has stopped expanding.

Secondary growth: In many vascular plants, secondary growth is the result of the activity of the vascular cambium. The latter is a meristem that divides to produce secondary xylem cells on the inside of the meristem (the adaxial side) and secondary phloem cells on the outside (the abaxial side).

Secondary messenger: Second messengers are molecules that relay signals from receptors on the cell surface to target molecules inside the cell, in the cytoplasm or nucleus. They relay the signals of hormones like epinephrine (adrenalin), growth factors, and others and cause some kind of change in the activity of the cell. They greatly amplify the strength of the signal. Secondary messengers are a component of signal transduction cascades.

Secondary metabolite: Secondary metabolites are organic compounds that are not directly involved in the normal growth, development, or reproduction of an organism. Unlike primary metabolites, absence of secondary metabolites does not result in immediate death, but rather in long-term impairment of the organism's survivability, fecundity, or aesthetics, or perhaps in no significant change at all. Secondary metabolites are often restricted to a narrow set of species within a phylogenetic group. Secondary metabolites often play an important role in plant defense against herbivory and other interspecies defenses. Humans use secondary metabolites as medicines, flavorings, and recreational drugs.

Secondary plant products: Plant secondary metabolites are a generic term used for more than 30,000 different substances, which are exclusively produced by plants. The plants form secondary metabolites, for example, for protection against pests, as coloring, scent, or attractants and as the plant's own hormones. It is used to be believed that secondary metabolites were irrelevant for the human diet.

Secondary root: A branch or lateral root.

Secretion: Secretion is the process of elaborating, releasing, and oozing chemicals or a secreted chemical substance from a cell or gland. In contrast to excretion, the substance may have a certain function, rather than being a waste product. Many cells contain this such as glaucoma cells.

Segment-polarity gene: A segmentation gene is a generic term for a gene whose function is to specify tissue pattern in each repeated unit of a segmented organism. In the fruit fly *Drosophila melanogaster*, segment polarity genes help to define the anterior and posterior polarities within each embryonic para-segment by regulating the transmission of signals via the Wnt signaling pathway and Hedgehog signaling pathway. Segment polarity genes are expressed in the embryo following expression of the gap genes and pair-rule genes. The most commonly cited examples of these genes are engrailed and gooseberry in Drosophila.

Selectable marker: A selectable marker is a gene introduced into a cell, especially a bacterium or to cells in culture that confers a trait suitable for artificial selection. They are a type of reporter gene used in laboratory microbiology, molecular biology, and genetic engineering to indicate the success of a transfection or other procedure meant to introduce foreign DNA into a cell. Selectable markers are often antibiotic resistance genes; bacteria that have been subjected to a procedure to introduce foreign DNA are grown on a medium containing an antibiotic, and those bacterial colonies that can grow have successfully taken up and expressed the introduced genetic material.

Selection coefficient: In population genetics, the selection coefficient is a measure of the relative fitness of a phenotype. Usually denoted by the letter "s", it compares the fitness of a phenotype to another favored phenotype and is the proportional amount that the considered phenotype is less fit as measured by fertile progeny.

Self-fertilization: Self-fertilization, fusion of male and female gametes (sex cells) produced by the same individual. Self-fertilization occurs in bisexual organisms, including most flowering plants, numerous protozoans, and many invertebrates. Autogamy, the production of gametes by the division of a single parent cell, is frequently found in unicellular organisms such as the protozoan Paramecium.

Self-pollination: The pollen grains can be carried from an anther to the stigma of the same flower is known as self-pollination.

Semiconservative replication: Semiconservative replication describes the mechanism by which DNA is replicated in all known cells.

Sense RNA: In virology, the genome of an RNA virus can be said to be either positive-sense, also known as a "plus-strand," or negative-sense, also known as a "minus-strand." In most cases, the terms sense and strand are used interchangeably, making such terms as positive-strand equivalent to positive-sense and plus-strand equivalent to plus-sense. Whether a virus genome is positive-sense or negative-sense can be used as a basis for classifying viruses.

Sepsis: Sepsis is a potentially deadly medical condition that is characterized by a whole-body inflammatory state (called a systemic inflammatory response syndrome or SIRS) and the presence of a known or suspected infection.

Sequence hypothesis: The sequence hypothesis was first formally proposed in a review "On Protein Synthesis" by Francis Crick in 1958. It states that the sequence of bases in the genetic material (DNA or RNA) determines the sequence of amino acids for which that segment of nucleic acid codes, and this amino acid sequence determines the three dimensional structure into which the protein folds.

Sequence-tagged site: A sequence-tagged site (or STS) is a short (200–500 bp) DNA sequence that has a single occurrence in the genome and whose location and base sequence are known.

Sequencing DNA: DNA sequencing includes several methods and technologies that are used for determining the order of the nucleotide bases adenine, guanine, cytosine, and thymine in a molecule of DNA.

Serology: Serology is the scientific study of blood serum and other bodily fluids. In practice, the term usually refers to the diagnostic identification of antibodies in the serum. Such antibodies are typically formed in response to an infection (against a given microorganism), against other foreign proteins (in response, for example, to a mismatched blood transfusion), or to one's own proteins (in instances of autoimmune disease).

Serum albumin: Serum albumin, often referred to simply as albumin, is a protein that in humans is encoded by the ALB gene.

Sewage treatment: Sewage treatment, or domestic wastewater treatment, is the process of removing contaminants from wastewater and household sewage, both runoff (effluents) and domestic. It includes physical, chemical, and biological processes to remove physical, chemical, and biological contaminants.

Sex chromosomes: Sex chromosome, either of a pair of chromosomes that determine whether an individual is male or female. The sex chromosomes of human beings and other mammals are designated by scientists as X and Y. In humans, the sex chromosomes comprise one pair of the total of 23 pairs of chromosomes. The other 22 pairs of chromosomes are called autosomes.

Sex determination: A sex-determination system is a biological system that determines the development of sexual characteristics in an organism. Most sexual organisms have two sexes. In many cases, sex determination is genetic: males and females have different alleles or even different genes that specify their sexual morphology. In animals, this is often accompanied by chromosomal differences. In other cases, sex is determined by environmental variables (such as temperature) or social variables (the size of an organism relative to other members of its population). The details of some sex-determination systems are not yet fully understood.

Sex hormones: Any of various hormones, such as estrogen and androgen, affecting the growth or function of the reproductive organs, the development of secondary sex characteristics, and the behavioral patterns of animals.

Sex linkage: Sex linkage is the phenotypic expression of an allele related to the chromosomal sex of the individual. This mode of inheritance is in contrast to the inheritance of traits on autosomal chromosomes, where both sexes have the same probability of inheritance. Since humans have many more genes on the X than the Y, there are many more X-linked traits than Y-linked traits.

Sex-influenced dominance: Sex-influenced dominance is the phenomenon in which the manifestation of a phenotype of a gene in heterozygosity depends on the sex of the individual.

Sex-limited genes: Sex-limited genes are genes that are present in both sexes of sexually reproducing species but turned on in only one sex. In other words, sex-limited genes cause the two sexes to show different traits or phenotypes. An example of sex-limited genes are genes that instructs male elephant seal to grow big and fight, at the same time, instructing female seals to grow small and avoid fights. These genes are responsible for sexual dimorphism.

Sexual reproduction: Sexual reproduction is the creation of a new organism by combining the genetic material of two organisms. The two main processes are meiosis, involving the halving of the number of chromosomes, and fertilization, involving the fusion of two gametes and the restoration of the original number of chromosomes. During meiosis, the chromosomes of each pair usually cross over to achieve homologous recombination.

Shake culture: A method for isolating anaerobic bacteria by shaking a deep liquid culture of an agar or gelatin to distribute the inoculum before solidification of the medium. A liquid medium in a flask that has been inoculated with an aerobic microorganism and placed on a shaking machine; action of the machine continually aerates the culture.

Shine–Dalgarno sequence: The Shine–Dalgarno sequence (or Shine–Dalgarno box), proposed by Australian scientists John Shine (1946-Present) and Lynn Dalgarno (1935–Present), is a ribosomal binding site in the mRNA, generally located 8 base pairs upstream of the start codon AUG. The Shine–Dalgarno sequence exists only in prokaryotes.

Short interspersed nuclear elements: Short interspersed elements (SINE) are short DNA sequences (<500 bases) that represent reverse-transcribed RNA molecules originally transcribed by RNA polymerase III into tRNA, rRNA, and other small nuclear RNAs. SINEs do not encode a functional reverse transcriptase protein and rely on other mobile elements for transposition.

Short-day plant: A plant requiring less than 12 h of daylight in order for flowering to occur.

Shuttle vector: A shuttle vector is a vector (usually a plasmid) constructed, so that it can propagate in two different host species. Therefore, DNA inserted into a shuttle vector can be tested or manipulated in two different cell types. The main advantage of these vectors is they can be manipulated in *E. coli* then used in a system which is more difficult or slower to use (e.g., yeast and other bacteria).

Sieve cell: An elongated cell whose walls contain perforations (sieve pores) that are arranged in circumscribed areas (sieve plates) and that afford communication with similar adjacent cells.

Sigma factor: A sigma factor (σ factor) is a prokaryotic transcription initiation factor that enables specific binding of RNA polymerase to gene promoters. Different sigma factors are activated in response to different environmental conditions. Every molecule of RNA polymerase contains exactly one sigma factor subunit, which in the model bacterium *Escherichia coli* is one of those listed below. *E. coli* has seven sigma factors; the number of sigma factors varies between bacterial species. Sigma factors are distinguished by their characteristic molecular weights. For example, $\sigma70$ refers to the sigma factor with a molecular weight of 70 kDa.

Signal sequence: A peptide present on proteins that are destined either to be secreted or to be membrane components. It is usually at the N-terminus and normally absent from the mature protein.

Signal transduction: Signal transduction is the process by which an extracellular signaling molecule activates a membrane receptor that in turn alters intracellular molecules creating a response. There are two stages in this process: a signaling molecule activates a certain receptor on the cell membrane, causing a second messenger to continue the signal into the cell and elicit a physiological response. In either step, the signal can be amplified, meaning that one signaling molecule can cause many responses.

Silencer DNA: In genetics, a silencer is a DNA sequence capable of binding transcription regulation factors termed repressors. Upon binding, RNA polymerase is prevented from initiating transcription, thus decreasing or fully suppressing RNA synthesis.

Simple sequence repeats: Simple sequence repeats (SSR), also called microsatellites, are becoming the most important molecular markers in both animals and plants.

Single-cell protein: Single-cell protein (SCP) typically refers to sources of mixed protein extracted from pure or mixed cultures of algae, yeasts, fungi, or bacteria (grown on agricultural wastes) used as a substitute for protein-rich foods, in human and animal feeds.

Single-nucleotide polymorphism: single-nucleotide polymorphism (SNP, pronounced snip) is a DNA sequence variation occurring when a single nucleotide A, T, C, or G—in the genome (or other shared sequence) differs between members of a biological species or paired chromosomes in an individual. For example, two sequenced DNA fragments from different individuals, AAGCCTA to AAGCTTA, contain a difference in a single nucleotide. In this case, we say that there are two alleles: C and T. Almost, all common SNPs have only two alleles.

Single-strand-DNA-binding protein: Single-strand binding protein, also known as SSB or SSBP, binds to single-stranded regions of DNA to prevent premature annealing. The strands have a natural tendency to revert to the duplex form, but SSB binds to the single strands, keeping them separate and allowing the DNA replication machinery to perform its function.

Site-directed mutagenesis: Site-directed mutagenesis, also called site-specific mutagenesis or oligonucleotide-directed mutagenesis, is a molecular biology technique in which a mutation is created at a defined site in a DNA molecule. In general, this form of mutagenesis requires that the wild-type gene sequence be known.

Small nuclear ribonucleoprotein: Small nuclear ribonucleoprotein (snRNPs) (pronounced "snurps") or small nuclear ribonucleoproteins are RNA-protein complexes that combine with unmodified pre-mRNA and various other proteins to form a spliceosome, a large RNA-protein molecular complex upon which splicing of pre-mRNA occurs. The action of snRNPs is essential to the removal of introns from pre-mRNA, a critical aspect of post-transcriptional modification of RNA, occurring only in the nucleus of eukaryotic cells.

Small nuclear RNA: Small nuclear ribonucleic acid (snRNA) is a class of small RNA molecules that are found within the nucleus of eukaryotic cells. They are transcribed by RNA polymerase II or RNA polymerase III and are involved in a variety of important processes such as RNA splicing (removal of introns from hnRNA), regulation of transcription factors (7SK RNA) or RNA polymerase II (B2 RNA), and maintaining the telomeres.

Sodium dodecyl sulfate: Sodium dodecyl sulfate (SDS or NaDS), sodium laurilsulfate, or sodium lauryl sulfate (SLS) ($C_{12}H_{25}SO_4Na$) is an anionic surfactant used in many cleaning and hygiene products. The salt consists of an anionic organosulfate consisting of a 12-carbon tail attached to a sulfate group, giving the material the amphiphilic properties required of a detergent.

Somaclonal variation: Somaclonal variation is the term used to describe the variation seen in plants that have been produced by plant tissue culture. Chromosomal rearrangements are an important source of this variation.

Somatic cell embryogenesis: Plant embryogenesis is the process that produces a plant embryo from a fertilized ovule by asymmetric cell division and the differentiation of undifferentiated cells into tissues and organs. It occurs during seed development, when the single-celled zygote undergoes a programmed pattern of cell division resulting in a mature embryo. A similar process continues during the plant's life within the meristems of the stems and roots.

Somatic hybridization: The production of cells, tissues, or organisms by fusion of nongametic nuclei. The phenomenon may be induced under laboratory conditions in cells that never normally fuse together and used as a plant breeding or genetic tool. It may also occur naturally, especially in fungi.

Somatostatin: Somatostatin (also known as growth hormone-inhibiting hormone [GHIH] or somatotropin release-inhibiting factor [SRIF]) is a peptide hormone that regulates the endocrine system and affects neurotransmission and cell proliferation via interaction with G-protein-coupled somatostatin receptors and inhibition of the release of numerous secondary hormones.

Somatotrophin: Growth hormone, a polypeptide containing 191 amino acids, produced by the anterior pituitary, the front section of the pituitary gland. It acts by stimulating the release of another hormone called somatomedin by the liver, thereby causing growth. Somatotropin is also known as somatropin.

Sonication: Sonication is the act of applying sound (usually ultrasound) energy to agitate particles in a sample, for various purposes. In the laboratory, it is usually applied using an ultrasonic bath or an ultrasonic probe, colloquially known as a sonicator. In a paper machine, an ultrasonic foil can distribute cellulose fibers more uniformly and strengthen the paper.

SOS response: The SOS response is a global response to DNA damage in which the cell cycle is arrested and DNA repair and mutagenesis are induced. The SOS uses the RecA protein (Rad51 in eukaryotes). The RecA protein, stimulated by single-stranded DNA, is involved in the inactivation of the LexA repressor thereby inducing the response. It is an error-prone repair system.

Southern blotting: A Southern blot is a method routinely used in molecular biology for the detection of a specific DNA sequence in DNA samples. Southern blotting combines transfer of electrophoresis-separated DNA fragments to a filter membrane and subsequent fragment detection by probe hybridization.

Spermatid: The spermatid is the haploid male gamete that results from division of secondary spermatocytes. As a result of meiosis, each spermatid contains only half of the genetic material present in the original primary spermatocyte.

Spermatocyte: A spermatocyte is a male gametocyte, derived from a spermatogonium, which is in the developmental stage of spermatogenesis during which meiosis occurs. It is located in the seminiferous tubules of the testis.

Spermatogenesis: Spermatogenesis is the process by which male primary germ cells undergo division and produce a number of cells termed spermatogonia, from which the primary spermatocytes are derived.

Spliceosomes: A spliceosome is a complex of snRNA and protein subunits that removes introns from a transcribed pre-mRNA (hnRNA) segment. This process is generally referred to as splicing.

Split genes: Genes where the genomic sequences are interrupted by intervening sequences (introns) that are spliced out of the mRNA prior to translation.

Staggered cuts: The cleavage of two opposite strands of duplex DNA at points near one another.

Standard deviation: Standard deviation is a widely used measurement of variability or diversity used in statistics and probability theory. It shows how much variation or "dispersion" there is from the average (mean, or expected value).

Standard error: The standard error is a method of measurement or estimation of the standard deviation of the sampling distribution associated with the estimation method.

Starch: Starch or amylum is a carbohydrate consisting of a large number of glucose units joined together by glycosidic bonds. This polysaccharide is produced by all green plants as an energy store.

Start codon: The start codon is generally defined as the point, sequence, at which a ribosome begins to translate a sequence of RNA into amino acids.

Stationary culture: *In vitro* cell culture without agitation.

Stem cell: Stem cells are biological cells found in all multicellular organisms that can divide through mitosis and differentiate into diverse specialized cell types and can self-renew to produce more stem cells.

Sterile room: Hygienic or bacteria free room.

Sterilization: The process of making bacteria free.

Sticky ends: DNA end or sticky end refers to the properties of the end of a molecule of DNA or a recombinant DNA molecule. The concept is important in molecular biology, especially in cloning or when sub cloning inserts DNA into vector DNA. All the terms can also be used in reference to RNA.

Stigma: The stigma is the receptive tip of a carpel, or of several fused carpels, in the gynoecium of a flower. The stigma receives pollen at pollination and it is on the stigma that the pollen grain germinates.

Stock solutions: In chemistry, a stock solution is a large volume of a common reagent, such as hydrochloric acid or sodium hydroxide, at a standardized concentration. This term is commonly used in analytical chemistry for procedures such as titrations, where it is important that exact concentrations of solutions are used.

Stop codon: In the genetic code, a stop codon (or termination codon) is a nucleotide triplet within messenger RNA that signals a termination of translation. Proteins are based on polypeptides, which are unique sequences of amino acids. Most codons in messenger RNA correspond to the addition of an amino acid to a growing polypeptide chain, which may ultimately become a protein. Stop codons signal the termination of this process by binding release factors, which cause the ribosomal subunits to disassociate, releasing the amino acid chain.

Stringent plasmid: A plasmid that only replicates along with the main bacterial chromosome and is present as a single copy, or at most several copies, per cell.

Stroma: In animal tissue, stroma refers to the connective, supportive framework of a biological cell, tissue, or organ.

Structural gene: A structural gene is a gene that codes for any RNA or protein product other than a regulatory factor (i.e., regulatory protein). It may code for a structural protein, an enzyme, or an RNA molecule not involved in regulation. Structural genes represent an enormous variety of protein structures and functions, including structural proteins, enzymes with catalytic activities, and so on.

Subcloning: In molecular biology, subcloning is a technique used to move a particular gene of interest from a parent vector to a destination vector in order to further study its functionality.

Subculture: In sociology, anthropology, and cultural studies, a subculture is a group of people with a culture (whether distinct or hidden), which differentiates them from the larger culture to which they belong.

Subspecies: Subspecies (commonly abbreviated subsp. or ssp.) in biological classification is either a taxonomic rank subordinate to species or a taxonomic unit in that rank (plural: subspecies). A subspecies cannot be recognized in isolation: a species will either be recognized as having no subspecies at all or two or more, never just one.

Subunit vaccine: A vaccine produced from specific protein subunits of a virus and thus having less risk of adverse reactions than whole virus vaccines.

Superbug: Antibiotic resistance is a type of drug resistance where a microorganism is able to survive exposure to an antibiotic. Genes can be transferred between bacteria in a horizontal fashion by conjugation, transduction, or transformation. Thus, a gene for antibiotic resistance which had evolved via natural selection may be shared. Evolutionary stress such as exposure to antibiotics then selects for the antibiotic resistant trait. Many antibiotic resistance genes reside on plasmids, facilitating their transfer. If a bacterium carries several resistance genes, it is called multiresistant or, informally, a superbug or super bacterium.

Supercoiled DNA: DNA supercoiling refers to the over- or underwinding of a DNA strand and is an expression of the strain on the polymer. Supercoiling is important in a number of biological processes, such as compacting DNA. Additionally, certain enzymes such as topoisomerases are able to change DNA topology to facilitate functions such as DNA replication or transcription. Mathematical expressions are used to describe supercoiling by comparing different coiled states to relaxed B-form DNA.

Supergene: A supergene is a group of neighboring genes on a chromosome that are inherited together because of close genetic linkage and are functionally related in an evolutionary sense, although they are rarely co-regulated genetically.

Supernatant: The usually clear liquid overlying material deposited by settling, precipitation, or centrifugation.

Suppressor: A suppressor, sound suppressor, sound moderator, or silencer is a device attached to or part of the barrel of a firearm, which reduces the amount of noise and flash generated by firing the weapon.

Suppressor mutation: A suppressor mutation is a mutation that counteracts the phenotypic effects of another mutation.

Susceptible: In epidemiology, a susceptible individual (sometimes known simply as a susceptible) is a member of a population who is at risk of becoming infected by a disease or cannot take a certain medicine, antibiotic, etc. if he or she is exposed to the infectious agent. if in one susceptibility is high, then he will respond this change in form of disease to make adaptation.

Suspension culture: The cultivation of cells suspended in the medium rather than adhering to a surface. Suspension culture is common for microorganisms but less so for the culture of the cells of most multicellular organisms. When referring to mammalian cells, suspension culture is used for the maintenance of cell types, which do not adhere, including some types of blood cells, or in order to have cells express characteristics, which are not seen in the adherent form.

Symbiotic association: Symbiotic relationships include those associations in which one organism lives on another or where one partner lives inside the other endosymbiosis, such as lactobacilli and other bacteria in humans.

Synaptonemal complex: The synaptonemal complex is a protein structure that forms between homologous chromosomes (two pairs of sister chromatids) during meiosis and that is thought to mediate chromosome pairing, synapsis, and recombination (crossing-over). It is now evident that the synaptonemal complex is not required for genetic recombination.

Synchronous culture: A synchronous or synchronized culture is a microbiological culture or a cell culture that contains cells that are all in the same growth stage.

Syndrome: In medicine and psychology, a syndrome is the association of several clinically recognizable features, signs (observed by a physician), symptoms (reported by the patient), phenomena, or characteristics that often occur together, so that the presence of one or more features alerts the physician to the possible presence of the others.

Syngamy: The process of union of two gametes to form a zygote. It involves both plasmogamy and karyogamy.

Synkaryon: The nucleus of a fertilized egg immediately after the male and female nuclei has fused.

T cells: T cells or T lymphocytes belong to a group of white blood cells known as lymphocytes and play a central role in cell-mediated immunity. They can be distinguished from other lymphocyte types, such as B cells and natural killer cells (NK cells) by the presence of a special receptor on their cell surface called T cell receptors (TCR).

T4 DNA ligase: In molecular biology, DNA ligase is a specific type of enzyme, a ligase that repairs single-stranded discontinuities in double-stranded DNA molecules; in simple words, strands that have double-strand break (a break in both complementary strands of DNA).

Tandem array: Repetitive DNA, where the repeating units are contiguous. Some genes of high copy number occur in tandem arrays, for example, ribosomal DNA.

***Taq* polymerase:** Taq polymerase is a thermostable DNA polymerase named after the thermophilic bacterium *Thermus aquaticus* from which it was originally isolated by Thomas D. Brock in 1965. It is often abbreviated as "Taq Pol" (or simply "Taq") and is frequently used in polymerase chain reaction (PCR), a method for greatly amplifying short segments of DNA.

Targeted drug delivery: Targeted drug delivery, sometimes called smart drug delivery, is a method of delivering medication to a patient in a manner that increases the concentration of the medication in some parts of the body relative to others. The goal of a targeted drug delivery system is to prolong, localize, target, and have a protected drug interaction with the diseased tissue. The conventional drug delivery system is the absorption of the drug across a biological membrane, whereas the targeted release system is when the drug is released in a dosage form. The advantages to the targeted release system are the reduction in the frequency of the dosages taken by the patient, having a more uniform effect of the drug, reduction of drug side effects, and reduced fluctuation in circulating drug levels. The disadvantage of the system is high cost that makes productivity more difficult and the reduced ability to adjust the dosages.

TATA box: The TATA box (also called Goldberg–Hogness box) is a DNA sequence (cis-regulatory element) found in the promoter region of genes in archaea and eukaryotes; approximately 24% of human genes contain a TATA box within the core promoter.

Tautomerism: Tautomerism, the existence of two or more chemical compounds that are capable of facile interconversion, in many cases merely exchanging a hydrogen atom between two other atoms, to either of which it forms a covalent bond. Unlike other classes of isomers, tautomeric compounds exist in mobile equilibrium with each other, so that attempts to prepare the separate substances usually result in the formation of a mixture that shows all the chemical and physical properties to be expected on the basis of the structures of the components.

T-cell-mediated (cellular) immune response: Cell-mediated immunity is an immune response that does not involve antibodies or complement but rather involves the activation of macrophages, natural killer cells (NK), antigen-specific cytotoxic T-lymphocytes, and the release of various cytokines in response to an antigen. Historically, the immune system was separated into two branches: humoral immunity, for which the protective function of immunization could be found in the humor (cell-free bodily fluid or serum) and cellular immunity, for which the protective function of immunization was associated with cells. CD4 cells or helper T cells provide protection against different pathogens.

T-DNA: The transfer DNA (abbreviated as T-DNA) is the transferred DNA of the tumor-inducing (Ti) plasmid of some species of bacteria such as *Agrobacterium tumefaciens* and *Agrobacterium rhizogenes*. It derives its name from the fact that the bacterium transfers this DNA fragment into the host plant's nuclear DNA genome.

Telomerase: Telomerase is an enzyme that adds DNA sequence repeats ("TTAGGG" in all vertebrates) to the 3′ end of DNA strands in the telomere regions, which are found at the ends of eukaryotic chromosomes.

Telophase: Telophase is a stage in both meiosis and mitosis in a eukaryotic cell. During telophase, the effects of prophase and prometaphase events are reversed. Two daughter nuclei form in the cell. The nuclear envelopes of the daughter cells are formed from the fragments of the nuclear envelope of the parent cell.

Template strand: When referring to DNA transcription, the coding strand is the DNA strand, which has the same base sequence as the RNA transcript produced (although with thymine replaced by uracil). It is this strand that contains codons, while the noncoding strand contains anticodons.

Terminal deoxynucleotidyl transferase: Terminal deoxynucleotidyl transferase (TDT), also known as DNA nucleotidylexotransferase (DNTT) or terminal transferase, is a specialized DNA polymerase expressed in immature, pre-B, pre-T lymphoid cells, and acute lymphoblastic leukemia/lymphoma cells.

Tetracycline: Tetracycline is a broad-spectrum polyketide antibiotic produced by the Streptomyces genus of Actinobacteria, indicated for use against many bacterial infections. It is a protein synthesis inhibitor.

Tetraploid: An individual or cell having four sets of chromosomes.

Thermophile: A thermophile is an organism a type of extremophile that thrives at relatively high temperatures, between 45°C and 80°C (113°F and 176°F). Many thermophiles are archaea. It has been suggested that thermophilic eubacteria are among the earliest bacteria.

Thermosensitivity: The central perception of temperature is located in the anterior hypothalamus (preoptic region) and in the spinal cord. The preoptic sensitivity is the more important in mammals, the spinal cord center in birds.

Thermostability: Thermostability is the quality of a substance to resist irreversible change in its chemical or physical structure at a high relative temperature.

Thymidine kinase: Thymidine kinase is an enzyme, a phosphotransferase (a kinase): 2′-deoxythymidine kinase, ATP-thymidine 5′-phosphotransferase. It can be found in most living cells. It is present in two forms in mammalian cells, TK1 and TK2. Certain viruses also have genetic information for expression of viral thymidine kinases.

Thymidine: Thymidine (more precisely called deoxythymidine; can also be labeled deoxyribosylthymine, and thymine deoxyriboside) is a chemical compound, more precisely a pyrimidine deoxynucleoside. Deoxythymidine is the DNA nucleoside T, which pairs with deoxyadenosine (A) in double-stranded DNA. In cell biology, it is used to synchronize the cells in S phase.

Thymine: Thymine (T, Thy) is one of the four nucleobases in the nucleic acid of DNA that are represented by the letters G–C–A–T. The others are adenine, guanine, and cytosine. Thymine is also known as 5-methyluracil, a pyrimidine nucleobase.

Topoisomerase: Topoisomerases are enzymes that unwind and wind DNA, in order for DNA to control the synthesis of proteins, and to facilitate DNA replication. The enzyme is necessary due to inherent problems caused by the DNA's double helix.

Totipotency: The ability of a cell, such as an egg, to give rise to unlike cells and thus to develop into or generate a new organism or part.

Toxicity: Toxicity is the degree to which a substance can damage an organism. Toxicity can refer to the effect on a whole organism, such as an animal, bacterium, or plant, as well as the effect on a substructure of the organism, such as a cell (cytotoxicity) or an organ (organotoxicity), such as the liver (hepatotoxicity). By extension, the word may be metaphorically used to describe toxic effects on larger and more complex groups, such as the family unit or society at large.

Transcription: Transcription is the process of creating a complementary RNA copy of a sequence of DNA. Both RNA and DNA are nucleic acids, which use base pairs of nucleotides as a complementary language that can be converted back and forth from DNA to RNA by the action of the correct enzymes. During transcription, a DNA sequence is read by RNA polymerase, which produces a complementary, antiparallel RNA strand.

Transcription factor: In molecular biology and genetics, a transcription factor (sometimes called a sequence-specific DNA-binding factor) is a protein that binds to specific DNA sequences, thereby controlling the flow (or transcription) of genetic information from DNA to mRNA. Transcription factors perform this function alone or with other proteins in a complex, by promoting (as an activator), or blocking (as a repressor) the recruitment of RNA polymerase (the enzyme that performs the transcription of genetic information from DNA to RNA) to specific genes.

Transcription unit: A stretch of DNA being transcribed into an RNA molecule.

Transduction: Transduction is the process by which DNA is transferred from one bacterium to another by a virus. It also refers to the process whereby foreign DNA is introduced into another cell via a viral vector. This is a common tool used by molecular biologists to stably introduce a foreign gene into a host cell's genome.

Transfection: Transfection is the process of deliberately introducing nucleic acids into cells. The term is used notably for nonviral methods in eukaryotic cells. It may also refer to other methods and cell types, although other terms are preferred: "transformation" is more often used to describe nonviral DNA transfer in bacteria, nonanimal eukaryotic cells and plant cells—a distinctive sense of transformation refers to spontaneous genetic modifications (mutations to cancerous cells (carcinogenesis), or under stress (UV irradiation)). "Transduction" is often used to describe virus-mediated DNA transfer.

Transfer RNA: Transfer RNA (tRNA) is an adaptor molecule composed of RNA, typically 73–93 nucleotides in length that is used in biology to bridge the three-letter genetic code in messenger RNA (mRNA) with the twenty-letter code of amino acids in proteins.

Transferase: In biochemistry, a transferase is an enzyme that catalyzes the transfer of a functional group (e.g., a methyl or phosphate group) from one molecule (called the donor) to another (called the acceptor).

Transformation: In molecular biology, transformation is the genetic alteration of a cell resulting from the direct uptake, incorporation, and expression of exogenous genetic material (exogenous DNA) from its surrounding and taken up through the cell membrane(s). Transformation occurs most commonly in bacteria and in some species occurs naturally. Transformation can also be effected by artificial means. Bacteria that are capable of being transformed, whether naturally or artificially, are called competent. Transformation is one of three processes by which exogenous genetic material may be introduced into a bacterial cell, the other two being conjugation (transfer of genetic material between two bacterial cells in direct contact), and transduction (injection of foreign DNA by a bacteriophage virus into the host bacterium).

Transforming oncogene: A gene that upon transfection, converts an immortalized cell, into malignant phenotype.

Transgene: A transgene is a gene or genetic material that has been transferred naturally or by any of a number of genetic engineering techniques from one organism to another.

Transgenesis: Transgenesis is the process of introducing an exogenous gene—called a transgene—into a living organism, so that the organism will exhibit a new property and transmist that property to its offspring. Transgenesis can be facilitated by liposomes, plasmid vectors, viral vectors, pronuclear injection, protoplast fusion, and ballistic DNA injection.

Transgenic animal: The term transgenic animal refers to an animal in which there has been a deliberate modification of the genome. Foreign DNA is introduced into the animal, using recombinant DNA technology, and then must be transmitted through the germ line so that every cell, including germ cells, of the animal contains the same modified genetic material.

Translation DNA: In molecular biology and genetics, translation is the third stage of protein biosynthesis (part of the overall process of gene expression). In translation, messenger RNA (mRNA) produced by transcription is decoded by the ribosome to produce a specific amino acid chain, or polypeptide, that will later fold into an active protein. In bacteria, translation occurs in the cell's cytoplasm, where the large and small subunits of the ribosome are located and bind to the mRNA.

Transmission electron microscope: Transmission electron microscopy (TEM) is a microscopy technique, whereby a beam of electrons is transmitted through an ultra-thin specimen, interacting with the specimen as it passes through. An image is formed from the interaction of the electrons transmitted through the specimen; the image is magnified and focused onto an imaging device, such as a fluorescent screen, on a layer of photographic film, or to be detected by a sensor such as a CCD camera.

Transposable genetic element: Transposons are only one of several types of mobile genetic elements. Transposons themselves are assigned to one of two classes according to their mechanism of transposition, which can be described as either "copy or paste" (class I) or "cut and paste" (class II).

Transposase: Transposase is an enzyme that binds to the ends of a transposon and catalyzes the movement of the transposon to another part of the genome by a cut and paste mechanism or a replicative transposition mechanism.

Transposon: Transposons are sequences of DNA that can move or transpose themselves to new positions within the genome of a single cell.

Transposon tagging: The term "transposon tagging" refers to a process in genetic engineering where transposons (transposable elements) are amplified inside a biological cell by a tagging technique. Transposon tagging has been used with several species to isolate genes. Even without knowing the nature of the specific genes, the process can still be used.

Trinucleotide repeat disorder: Trinucleotide repeat disorders (also known as trinucleotide repeat expansion disorders, triplet repeat expansion disorders, or codon reiteration disorders) are a set of genetic disorders caused by trinucleotide repeat expansion, a kind of mutation where trinucleotide repeats in certain genes exceeding the normal, stable, threshold, which differs per gene. The mutation is a subset of unstable microsatellite repeats that occur throughout all genomic sequences. If the repeat is present in a healthy gene, a dynamic mutation may increase the repeat count and result in a defective gene.

Trisomy: A trisomy is a genetic abnormality in which there are three copies, instead of the normal two, of a particular chromosome. A trisomy is a type of aneuploidy (an abnormal number of chromosomes).

Trypsin: Trypsin is a serine protease found in the digestive system of many vertebrates, where it hydrolyses proteins. Trypsin is produced in the pancreas as the inactive proenzyme trypsinogen.

Trypsin inhibitor: Trypsin inhibitors are chemicals that reduce the availability of trypsin, an enzyme essential to the nutrition of many animals, including humans.

Tubulin: Tubulin is one of several members of a small family of globular proteins. The most common members of the tubulin family are α-tubulin and β-tubulin, the proteins that make up microtubules. Each has a molecular weight of approximately 55 kDa. Microtubules are assembled from dimers of α- and β-tubulin.

Tumor-inducing plasmid: A giant plasmid of *Agrobacterium tumefaciens* that is responsible for tumor formation in infected plants. Ti plasmids are used as vectors to introduce foreign DNA into plant cells.

Twins: A twin is one of two offspring produced in the same pregnancy. Twins can either be identical (in scientific usage, "monozygotic"), meaning that they develop from one zygote that splits and forms two embryos, or fraternal ("dizygotic"), because they develop from two separate eggs that are fertilized by two separate sperm.

Ultrasonication: In biological applications, ultrasonication is used to disrupt or deactivate a biological material. For example, ultrasonication is often used to disrupt cell membranes and release cellular contents.

Ultraviolet radiation: Ultraviolet (UV) light is electromagnetic radiation with a wavelength shorter than that of visible light, but longer than x-rays, in the range 10–400 nm. It is named because the spectrum consists of electromagnetic waves with frequencies higher than those that humans identify as the color violet.

Upstream processing: The manufacture of human proteins by the methods of modern biotechnology is separated into two stages: upstream processing during which proteins are produced by cells genetically engineered to contain the human gene which will express the protein of interest and downstream processing during which the produced proteins are isolated and purified. Following purification of the protein of interest, the final product is formulated (meaning excipients are added to the protein), filter sterilized, filled aseptically, lyophilized, sealed, inspected, and labeled. Upstream processing, downstream processing, final drug production, and the general environment of the facility are monitored by the quality control division of the manufacturing facility.

Uracil: Uracil is one of the four nucleo-bases in the nucleic acid of RNA that are represented by the letters A, C, G, and U. The others are adenine, cytosine, and guanine. In RNA, uracil (U) binds to adenine (A) via two hydrogen bonds. In DNA, the uracil nucleobase is replaced by thymine.

Variable number tandem repeat: A variable number tandem repeat (or VNTR) is a location in a genome, where a short nucleotide sequence is organized as a tandem repeat. These can be found on many chromosomes and often show variations in length between individuals. Each variant acts as an inherited allele, allowing them to be used for personal or parental identification. Their analysis is useful in genetics and biology research, forensics, and DNA fingerprinting.

Variance: In probability theory and statistics, the variance is used as a measure of how far a set of numbers are spread out from each other. It is one of several descriptors of a probability distribution, describing how far the numbers lie from the mean (expected value). In particular, the variance is one of the moments of a distribution. In that context, it forms part of a systematic approach to distinguishing between probability distributions. While other such approaches have been developed, those based on moments are advantageous in terms of mathematical and computational simplicity.

Vascular: Vascular in zoology and medicine means "related to blood vessels," which are part of the circulatory system. An organ or tissue that is vascularized is heavily endowed with blood vessels and thus richly supplied with blood.

Vegetative propagation: Vegetative reproduction (vegetative propagation, vegetative multiplication, and vegetative cloning) is a form of asexual reproduction in plants. It is a process by which new individuals arise without production of seeds or spores. It can occur naturally or be induced by horticulturists.

Vernalization: Vernalization is the acquisition of a plant's ability to flower or germinate in the spring by exposure to the prolonged cold of winter. After vernalization, plants have acquired the ability to flower, but they may require additional seasonal cues or weeks of growth before they will actually flower.

Viability cell: Cell viability is a determination of living or dead cells, based on a total cell sample. Cell viability measurements may be used to evaluate the death or life of cancerous cells and the rejection of implanted organs. In other applications, cell viability tests might calculate the effectiveness of a pesticide or insecticide or evaluate environmental damage due to toxins.

Viability test: Test to determine the proportion of living individuals, cells or organisms, in a sample. Viability tests are most commonly performed on cultured cells and usually depend on the ability of living cells to exclude a dye.

Viral oncogene: Any virus that promotes cancer.

Viral vaccines: A vaccine is a biological preparation that improves immunity to a particular disease. A vaccine typically contains an agent that resembles a disease-causing microorganism and is often made from weakened or killed forms of the microbe or its toxins. The agent stimulates the body's immune system to recognize the agent as foreign, destroy it, and "remember" it, so that the immune system can more easily recognize and destroy any of these microorganisms that it later encounters.

Virion: An entire virus particle, consisting of an outer protein shell called a capsid and an inner core of nucleic acid (either ribonucleic or deoxyribonucleic acid RNA or DNA). The core confers infectivity, and the capsid provides specificity to the virus. In some virions, the capsid is further enveloped by a fatty membrane, in which case the virion can be inactivated by exposure to fat solvents such as ether and chloroform.

Viroid: Viroids are plant pathogens that consist of a short stretch (a few hundred nucleobases) of highly complementary, circular, single-stranded RNA without the protein coat that is typical for viruses.

Western blot: The western blot (sometimes called the protein immunoblot) is a widely used analytical technique used to detect specific proteins in the given sample of tissue homogenate or extract. It uses gel electrophoresis to separate native or denatured proteins by the length of the polypeptide (denaturing conditions) or by the 3D structure of the protein (native/nondenaturing conditions). The proteins are then transferred to a membrane (typically nitrocellulose or PVDF), where they are probed (detected) using antibodies specific to the target protein.

Wild type: Wild type refers to the phenotype of the typical form of a species as it occurs in nature. Originally, the wild type was conceptualized as a product of the standard, "normal" allele at a locus, in contrast to that produced by a nonstandard, "mutant" allele.

X-chromosome: The X chromosome is one of the two sex-determining chromosomes in many animal species, including mammals (the other is the Y chromosome). It is a part of the XY sex-determination system and X0 sex-determination system. The X chromosome was named for its unique properties by early researchers, which resulted in the naming of its counterpart Y chromosome, for the next letter in the alphabet, after it was discovered later.

Xenobiotic: A xenobiotic is a chemical that is found in an organism but which is not normally produced or expected to be present in it. It can also cover substances that are present in much higher concentrations than are usual. Specifically, drugs such as antibiotics are xenobiotics in humans, because the human body does not produce them itself, nor are they part of a normal diet.

Xenotransplantation: Xenotransplantation is the transplantation of living cells, tissues, or organs from one species to another, such as from pigs to humans. Such cells, tissues, or organs are called xenografts or xenotransplants. In contrast, the term allotransplantation refers to a same-species transplant. Human xenotransplantation offers a potential treatment for end-stage organ failure, a significant health problem in parts of the industrialized world. It also raises many novel medical, legal, and ethical issues. There are few published cases of successful xenotransplantation.

X-linked disease: Genetic disease associated with X chromosome.

X-ray crystallography: X-ray crystallography is a method of determining the arrangement of atoms within a crystal, in which a beam of X-rays strikes a crystal and diffracts into many specific directions. From the angles and intensities of these diffracted beams, a crystallographer can produce a three-dimensional picture of the density of electrons within the crystal.

Y-chromosome: The Y chromosome is one of the two sex-determining chromosomes in most mammals, including humans. In mammals, it contains the gene SRY, which triggers testis development if present. The human Y chromosome is composed of about 60 million base pairs. DNA in the Y chromosome is passed from father to son, and Y-DNA analysis may thus be used in genealogy research.

Yeast: Yeasts are eukaryotic micro-organisms classified in the kingdom Fungi, with 1500 species currently described estimated to be only 1% of all fungal species. Most reproduce asexually by mitosis, and many do so via an asymmetric division process called budding.

Yeast artificial chromosome: A yeast artificial chromosome (YAC) is a vector used to clone DNA fragments larger than 100 kb and up to 3000 kb. YACs are useful for the physical mapping of complex genomes and for the cloning of large genes. First, described in 1983 by Murray and Szostak, a YAC is an artificially constructed chromosome and contains the telomeric, centromeric, and replication origin sequences needed for replication and preservation in yeast cells.

Z-DNA: Z-DNA is one of the many possible double helical structures of DNA. It is a left-handed double helical structure in which the double helix winds to the left in a zig–zag pattern (instead of to the right, like the more common B-DNA form). Z-DNA is thought to be one of three biologically active double helical structures along with A- and B-DNA.

Zoo blot: A zoo blot or garden blot is a type of Southern blot that demonstrates the similarity between specific, usually protein-coding DNA sequences of different species. A zoo blot compares animal species while a garden blot compares plant species.

Zygonema: The synaptic chromosome formation that occurs in the zygotene stage of the first meiotic prophase of gametogenesis.

Zygote: A zygote is the initial cell formed when two gamete cells are joined by means of sexual reproduction. It is the earliest developmental stage of the embryo. A zygote is always synthesized from the union of two gametes and constitutes the first stage in a unique organism's development.

Zymogen: A zymogen (or proenzyme) is an inactive enzyme precursor. A zymogen requires a biochemical change (such as a hydrolysis reaction revealing the active site, or changing the configuration to reveal the active site) for it to become an active enzyme. The biochemical change usually occurs in a lysosome where a specific part of the precursor enzyme is cleaved in order to activate it. The amino acid chain that is released upon activation is called the activation peptide.

Index

A

F

G

H

I

L

M

Q

R

W

X

Y

AlphaBasiCs

Santa Claus

from A to Z

Bobbie Kalman

with illustrations by Barbara Bedell

Crabtree Publishing Company

AlphaBasiCs

Created by Bobbie Kalman

To Billy and his Grandpa

Author and Editor-in-Chief
Bobbie Kalman

Managing editor
Lynda Hale

Editor
Hannelore Sotzek

Computer design
Lynda Hale

Production coordinator
Hannelore Sotzek

Separations and film
Dot 'n Line Image Inc.

Printer
Worzalla Publishing Co.

Special thanks to
Macy's Department Store, Kim McQuhae and her goat

Photographs and reproductions
Bobbie Kalman: page 22 (top)
Macy's Thanksgiving Day Parade™: page 22 (bottom)
Other images by Digital Stock & Eyewire, Inc.

Illustrations
Barbara Bedell: front cover (elves), back cover, pages 3 (all except top right & middle), 4 (top), 5-11, 12 (bottom), 13 (top left, bottom), 14, 15 (top), 16 (bottom), 17 (top, bottom left) 18, 19 (bottom), 20-21, 23 (top), 24-25, 25 (top), 28, 29 (top, inset), 30 (right), 31
© Crabtree Publishing Company: page 19 (top)
Trevor Morgan: pages 24 (top), 27 (right), 30 (top)

 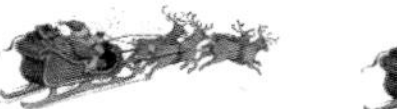

Crabtree Publishing Company

PMB 16A
350 Fifth Avenue,
Suite 3308
New York, NY
10118

360 York Road
RR 4
Niagara-on-the-Lake
Ontario, Canada
L0S 1J0

73 Lime Walk
Headington,
Oxford
OX3 7AD
United Kingdom

Cataloging in Publication Data

Kalman, Bobbie
Santa Claus from A to Z

(AlphaBasiCs series)
Includes index.

ISBN 0-86505-389-8 (library bound) ISBN 0-86505-419-3 (pbk.)
This book is an alphabetical introduction to various aspects of the legend of Santa Claus, such as "Elves," "Gifts," "Letters and List," and "Uniform."

1. Santa Claus—Juvenile literature. 2. English language—Alphabet—Juvenile literature. [1. Santa Claus. 2. Alphabet.] I. Bedell, Barbara, ill. II. Title. III. Series: Kalman, Bobbie. AlphaBasiCs.

GT4992.K35 1999 j394.62663 [E] LC 99-38618
CIP

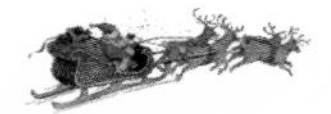       

Contents

A is for **about Santa**. This book is all about Santa Claus. Santa brings gifts to children on Christmas Eve. Santa is also called St. Nicholas. What do you know about St. Nicholas? You can read his story on the next page. Do you think St. Nicholas and Santa are the same person? Why do you think they look so different?

B is for **bishop** and **boots**. St. Nicholas was a bishop who lived a long time ago in a country that is now called Turkey. He helped poor children by filling their boots with gifts while they slept. On Christmas Eve, many children around the world put out boots for St. Nicholas to fill. Where does St. Nick leave your presents?

is for **Christmas cookies**. Do you leave cookies and milk for Santa on Christmas Eve? What kind do you leave? Are they chocolate chip, peanut butter, gingerbread, or sugar cookies? Santa loves all kinds of cookies, but if you really want to make him happy, decorate the cookies you leave with icing, cherries, and candies.

You can decorate your Christmas cookies by adding food coloring to white icing. Use the icing to draw pictures on the cookies. Put green and red cherries and candy pearls on some to make them even more festive. Hang a few cookies on your Christmas tree. They make great decorations!

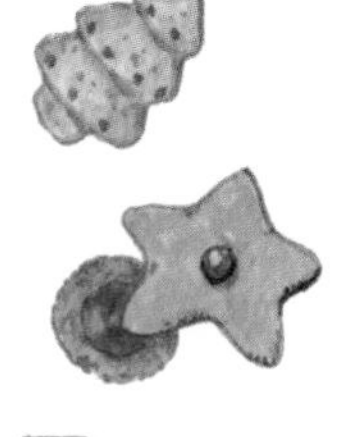

is for **decorate**. Do you decorate your tree with Santa ornaments? Here are some easy ornaments you can make of Santa, his reindeer, and his bag of toys. Follow the directions below to make these fun Christmas decorations. Santa will love them!

(right) This cork and toothpick reindeer has tiny twigs for antlers and a small cone glued on as a tail. Its nose is a red candy! Use a shiny red ribbon to hang it on the tree.

(right) This Santa is easy to make. His face is made of molding clay, and his beard is a pine cone painted white. His hat is red felt trimmed with cotton batting. His mustache is also made of cotton.

(left) Santa's bag is made from a piece of paper-twist ribbon folded in half. Tie the bag with some string and put a tiny candy cane and some pine sprigs inside.

(right) Make this Santa face from dough. Mix 1 cup of flour with 1/4 cup of salt and 6 tablespoons of hot water. Separate the dough into balls and use red food coloring to tint some of the dough. Use a toothpick to draw ridges in the beard. Ask an adult to bake it on low heat for about two hours.

is for **elves**. Santa needs a lot of help to make gifts for all the boys and girls in the world! In Santa's workshop at the North Pole, hundreds of elves work year-round to help him make toys. They work very hard sawing, hammering, painting, and wrapping. On Christmas Eve, they load the toys into Santa's sleigh.

The best part of the elves' job is trying out the new toys. Elves love to play!

A few presents are very heavy! What do you think is in this box?

The elves wrap some of the presents and put bows on others.

The elves make sure the reindeer are well rested for their long trip with Santa. They feed them and put on their harnesses and bells. After finishing all their work, the elves are very tired. Mrs. Claus brings them tea and tucks them into bed. They dream of Christmas trees, toys, and of happy children finding their presents on Christmas morning.

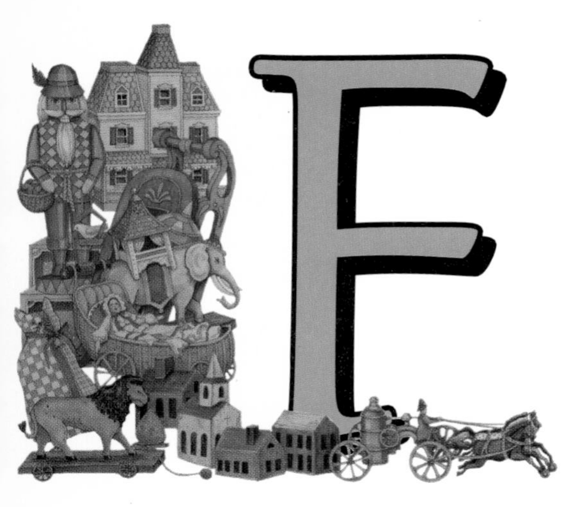

is for **fireplace**. When Santa arrives at your home, his reindeer land quietly on the roof. Santa climbs out of his sleigh, slides down the chimney, and lands in your fireplace with a big bag of toys and a THUMP. He does this without even getting dirty! How does he get back up the fireplace? No one knows! How do you think he does it?

is for **gifts**. Santa's bag is full of gifts. Some are wrapped in shiny paper and tied with ribbons. Some are unwrapped. Santa puts smaller gifts into the children's stockings. He leaves toys on the floor, too. When Santa is finished unloading his bag, he goes back up the chimney. As quick as a wink he is gone. Good-bye, Santa!

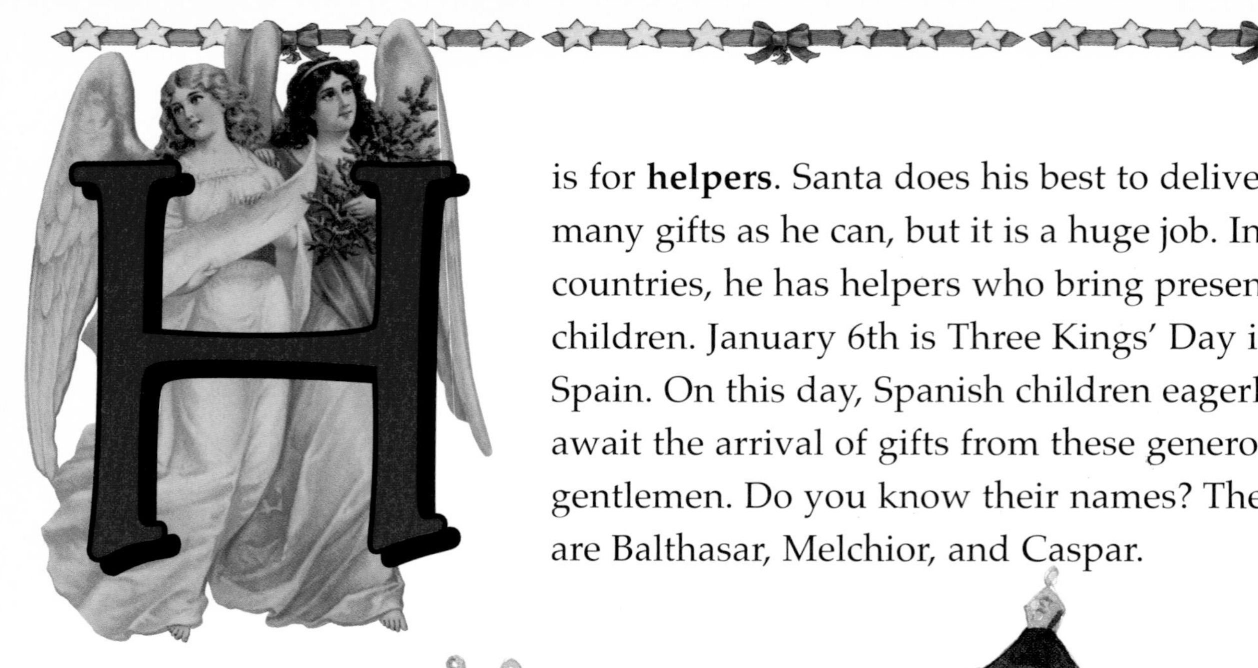

H is for **helpers**. Santa does his best to deliver as many gifts as he can, but it is a huge job. In some countries, he has helpers who bring presents to children. January 6th is Three Kings' Day in Spain. On this day, Spanish children eagerly await the arrival of gifts from these generous gentlemen. Do you know their names? They are Balthasar, Melchior, and Caspar.

In some European countries, angels bring the Christmas tree as well as gifts. An old lady named Babushka is the Christmas gift-giver in Russia. In Italy, they call her Befana. People in Russia and Italy believe that Babushka, or Befana, goes to each house to leave gifts for children.

In an old legend, Babushka is said to be searching each house for the baby Jesus so she can bring him gifts. She is still searching, so she leaves gifts behind for children instead.

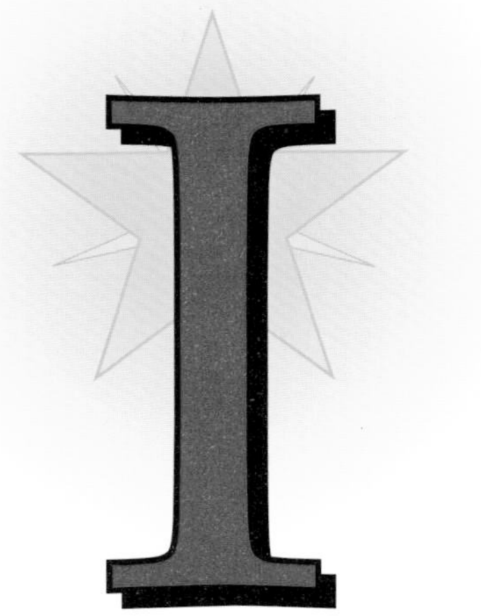

is for **illuminate**. To illuminate means to make bright. Angels help Santa illuminate Christmas trees. These angels have dressed an outdoor tree with glowing candles and stars. Have you ever had angels put lights on your tree? They don't use real candles anymore. What do they use to illuminate your tree?

J is for **jolly**. Santa is a jolly man. When he laughs, he lets out a loud "HO, HO, HO!" No one knows why he says "HO, HO, HO," but one thing is certain—Santa loves to laugh! On Christmas Eve, leave Santa a joke or two to enjoy with his cookies. When you hear his loud "HO, HO, HO," you will know that Santa liked your joke.

is for **Kriss Kringle**. Some people call Santa Kriss Kringle. The name Kriss Kringle came from a German Christmas angel called Christkindl. Over time, the name Christkindl changed to Kriss Kringle. Eventually people began to think of Kriss Kringle as St. Nicholas. Maybe St. Nicholas changed into Santa Claus in the same way. Who knows?

Christkindl *Kriss Kringle*

is for **letters** and **list**. Do you write letters to Santa? He gets millions of letters from children every year. Letters help Santa make up his list of gifts. If you don't know how to write, you can visit Santa at a shopping center, sit on his lap, and tell him what you would like for Christmas. Then Santa's elves will do their best to make your gift wishes come true.

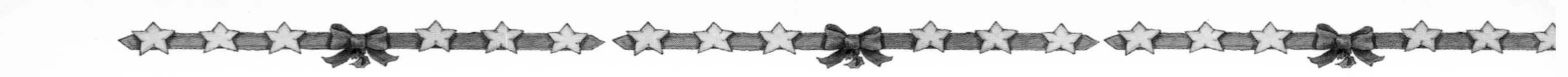

M is for **Mrs. Claus**, the wife of Santa Claus. She lives at the North Pole with Santa. Mrs. Claus has the most important job—she makes sure that Santa can deliver the gifts to every girl and boy each year. She does all the behind-the-scenes work such as ordering the materials for the clothing and gifts, caring for the elves, and mapping out Santa's route. She also plans a vacation so Santa will be well rested for the following Christmas.

N is for **North Pole**. Santa's workshop is at the North Pole in an area called the Arctic. The North Pole is a very cold place, full of ice and snow. Why would such a jolly man want to live in such a chilly place? Santa lives at the North Pole because this part of the world does not belong to any country. Instead, Santa belongs to the children of the world! Can you name any animals that live at the North Pole with Santa?

is for **one Santa**. Is there just one Santa? If there is, why does he look different in every picture, and how can he be in so many places at one time? No one has ever seen the real Santa, so people have different ideas about how he looks. Children in some countries think he wears a long coat and rides a horse. Children in other countries think he has reindeer that fly. In some places, children believe that Santa rides a goat!

Write a story about the Santas on these two pages. Draw a picture showing how you think each one delivers presents to children.

People all over the world believe in Santa Claus, but not everyone agrees on his name, how he looks, or where he lives. Are all these Santas part of one Santa? Do these Santas all live at the North Pole?

is for **parade**. Each year there are Santa Claus parades in many cities and towns. Is there one where you live? Santa Claus parades are held at the end of November or beginning of December. They remind boys and girls that Christmas is coming very soon. Children start writing letters and looking forward to Santa's visit. Waiting for Santa to come is very exciting! Draw a picture of your favorite float at the Santa parade.

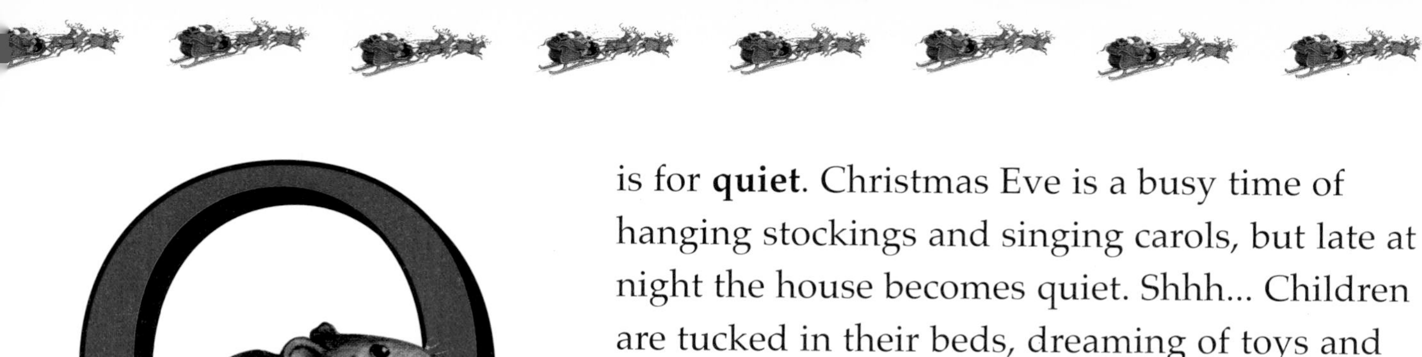

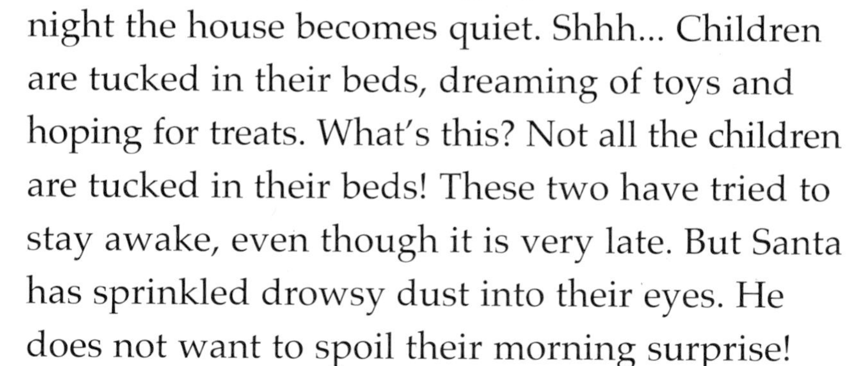

Q is for **quiet**. Christmas Eve is a busy time of hanging stockings and singing carols, but late at night the house becomes quiet. Shhh... Children are tucked in their beds, dreaming of toys and hoping for treats. What's this? Not all the children are tucked in their beds! These two have tried to stay awake, even though it is very late. But Santa has sprinkled drowsy dust into their eyes. He does not want to spoil their morning surprise!

R is for **reindeer**. Santa has just 24 hours to deliver toys to girls and boys around the world. A team of flying reindeer pull Santa and his sleigh full of goodies swiftly across the sky. Do you know the names of Santa's reindeer? They are Comet, Cupid, Dasher, Dancer, Donder, Blitzen, Prancer, and Vixen—but the most famous one of all is Rudolph, the red-nosed reindeer. Why is he especially helpful to Santa Claus?

is for **sleigh**. Santa's sleigh sails across the starry sky. In the blink of an eye he can travel all the way from Arkansas to Zanzibar and from Timbuktu to Kalamazoo. Somehow, all the tremendous toys made by his excellent elves fit into this splendid sleigh. There is a bear for Bartholomew, a horse for Harriet, and a doll for Drew. Santa's sleigh holds many wonderful surprises!

is for **tree**. Many children around the world believe that Santa Claus brings their Christmas tree and decorates it on Christmas Eve. Does he cut down a tree in the forest and then push it down the chimney? Does he decorate each tree and put presents underneath it? Do angels help illuminate the trees? Who decorates your Christmas tree?

is for **uniform**. There are pictures of Santa wearing coats of many different colors, but the Santa we see most often wears a red suit trimmed with white fur. This outfit—a coat, pants, belt, boots, hat, gloves, and a big bag in which Santa carries all his presents—has become the Santa uniform. What does Santa wear under his uniform? He wears long underwear, of course!

V is for **Virginia**. About a hundred years ago, a girl named Virginia wrote a letter to a newspaper, asking if there really was a Santa. An editor named Francis Church printed her letter and replied, "Yes, Virginia, there is a Santa Claus."

Church told Virginia that there was a Santa and that he lives forever. Santa will continue to make children happy for thousands of years to come!

W is for **white Christmas**. When people think of Christmas, they often think of snow. Many people who live in northern places hope there will be snow at Christmas, but not everyone lives in a place where it snows! People in southern parts of the world dream of spending the day at the beach or having a party with balloons and *piñatas*. Do you dream of a white Christmas or a sunny one?

is for **Xmas** and y is for **your own picture of Santa**. "Xmas" is another way of writing "Christmas." Do you send Xmas cards with pictures of Santa on them? Maybe you can make your own Santa Xmas cards. Draw some pictures of different Santas and send them to your family and friends. Include some of St. Nicholas as well as a Santa wearing a red suit.

Z is for **zigzag**, **zoom**, **zip**, and **zap**. Santa zigzags from place to place around the world, zooms down chimneys, zips up fireplaces, and zaps out of sight as quick as a wink! How does he do all that zigzagging, zooming, zipping, and zapping? Do cookies and milk help him fly? Perhaps Santa can do all these things because children believe he can. Do you believe in Santa?

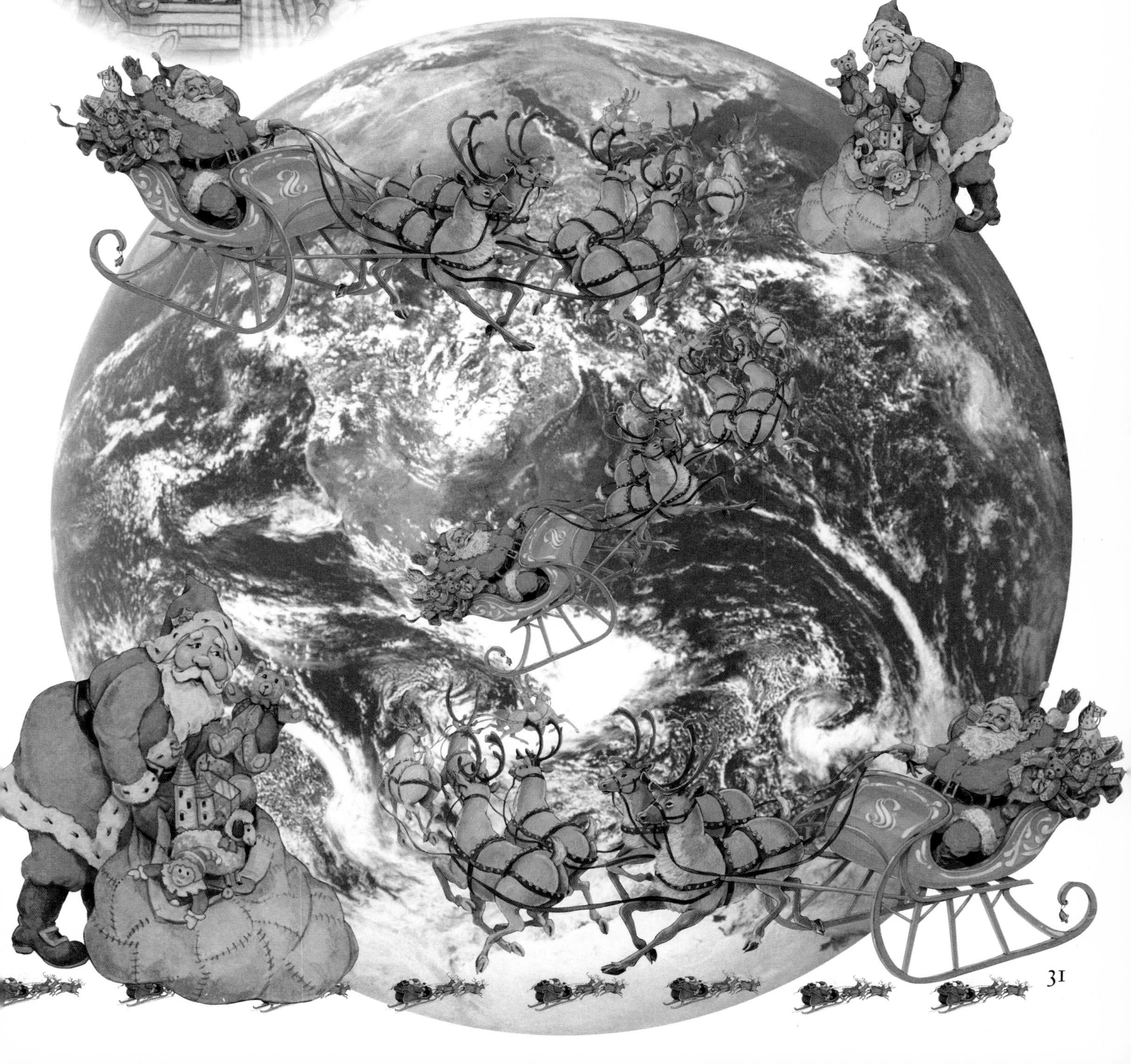

Words to know

angel A heavenly spirit who watches over people
Arkansas A state in the south-central region of the United States
behind-the-scenes Describing tasks that are unknown to the general public
bishop An important religious person who works within the Christian church
Christkindl A legendary figure who brought gifts to children in Germany
cotton batting Sheets of cotton used for stuffing furniture or for making crafts
float An exhibit on a flat platform that is pulled in a parade
harness A set of leather straps used to attach an animal to a vehicle
illuminate To light up
Jesus Christians believe that Jesus is the Son of God; His birthday is celebrated on Christmas Day.
Kalamazoo A city in the state of Michigan
North Pole The northernmost point on Earth
paper-twist ribbon Paper that is tightly twisted and rolled in a coil
piñata A colorful container filled with treats that children try to break open with a stick
Rudolph The only one of Santa's reindeer who has a glowing, red nose
Timbuktu A place in Africa near the Sahara Desert
Turkey A country in southeast Europe
Zanzibar A region of eastern Africa

Index

1 2 3 4 5 6 7 8 9 0 Printed in the U.S.A. 8 7 6 5 4 3 2 1 0 9

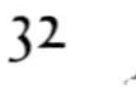